SCANNING PROBE MICROSCOPY

SCANNING PROBE MICROSCOPY

ELECTRICAL AND ELECTROMECHANICAL PHENOMENA AT THE NANOSCALE

Volume I

Sergei Kalinin
Alexei Gruverman
Editors

Springer

Sergei Kalinin
Oak Ridge National Laboratory
Oak Ridge, TN, 37831
USA
sergei2.kalininweb.com

Alexei Gruverman
Dept. Materials Science and Engineering
North Carolina State University
Raleigh, NC, 27695
USA
alexei_gruverman@ncsu.edu

Library of Congress Control Number: 2006926451

ISBN-10: 0-387-28667-5 e-ISBN-10: 0-387-28668-3
ISBN-13: 978-0387-28667-9 e-ISBN-13: 978-0387-28668-6

Printed on acid-free paper.

9 8 7 6 5 4 3 2 1

springer.com

Contents

List of Contributors

Peter M. Albrecht
Beckman Institute
405 N. Mathews
Urbana, IL 61801
Phone: 217-244-1058
Fax: 217-244-1995
E-mail: palbrech@uiuc.edu

David Alvarez
Infineon Technologies
05452 Essex Junction, VT, USA
Phone: 1-802-769-4904
E-mail: david.alvarez@infineon.com

Meredith L. Anderson
Sandia National Laboratories
P.O. Box 5800, MS1077
Albuquerque, NM 87185-1077
E-mail: mlande@sandia.gov

Steven M. Anlage
Center for Superconductivity Research
Department of Physics
University of Maryland
College Park, MD 20742-4111, USA
Phone: +001 301 405 7321
Fax: +001 301 405 3779
E-mail: anlage@umd.edu
http://www.csr.umd.edu

Arthur P. Baddorf
Oak Ridge National Laboratory
1 Bethel Valley Road, Bldg. 3025
Oak Ridge, TN 37831-6030
Phone: (865) 574-5241
Fax: (865) 574-4143
E-mail: baddorfap@ornl.gov
http://nanotransport.ornl.gov

R. Bachelot
Laboratoire de Nanotechnologie et d'Instrumentation Optique
Institut Charles Delaunay CNRS-FRE 2848
Université de Technologie de Troyes
F 10010 Troyes, France
E-mail: renaud.bachelot@utt.fr

Dayan Ban
University of Waterloo
Department of Electrical and Computer Engineering
200 University Ave. West
Waterloo, Ontario

A. Boag
Faculty of Engineering
TEL-AVIV UNIVERSITY
Tel-Aviv 69978, ISRAEL

Dawn A. Bonnell
Department of Materials Science and Engineering
The University of Pennsylvania
3231 Walnut Street, Philadelphia PA 19104
Phone: 215 898 6231
Fax: 215 573 2128
E-mail: Bonnell@lrsm.upenn.edu

Alexandre Bouhelier
Laboratoire de Physique de l'Université de Bourgogne
UMR 5027
Faculté des Sciences Mirande—9 Avenue Alain Savary
B.P. 47 870—F-21078 DIJON Cedex—FRANCE
E-mail: alexandre.bouhelier@u-bourgogne.fr

Sidney R. Cohen
Director, Surface Analysis Laboratory
Chemical Research Support
Weizmann Institute of Science
Rehovot 76100 ISRAEL
Phone: 972 8 934 2703
Fax: 972 8 934 4137
E-mail: sidney.cohen@weizmann.ac.il
http://www.weizmann.ac.il/Chemical_Research_Support/surflab/

John A. Dagata
Precision Engineering Division
National Institute of Standards & Technology
100 Bureau Drive MS 8212
Gaithersburg MD 20899-8212 USA
Phone: 301-975-3597
Fax: 301-869-0822
E-mail: john.dagata@nist.gov

St. J. Dixon-Warren
Chipworks
3685 Richmond Road, Suite 500
Ottawa, Ontario, Canada

David Dunlap
Dept. of Physics and Astronomy
Univ. of New Mexico
Albuquerque, NM 87131
Phone: (505) 277-2120
Fax: (505) 277-1520
E-mail: dunlap@unm.edu

P. Eyben
Kapeldreef 75, B-3001
Leuven, Belgium
Phone: 32 16 281 305
E-mail: eyben@imec.be

Marc Fouchier
Veeco instruments S.A.S.
Z .I. de la Gaudrée—11 rue Marie Poussepin
BP 43—91412 Dourdan—France
Phone: +33 (0)6 24 60 19 23
E-mail: marcfouchier@yahoo.fr

Thilo Glatzel
Institute of Physics, University of Basel
Klingelbergstr. 82
CH-4056 Basel
Phone: +41(0)61 267-3730
Fax: +41(0)61 267-3795
E-mail: thilo.glatzel@unibas.ch
http://www.physik.unibas.ch

Christopher Gorman
Department of Chemistry
North Carolina State University
314 Dabney Hall
Raleigh, NC 27695-8204
Phone: 919-515-4252
Fax: 919-515-8920
E-mail: Chris_Gorman@ncsu.edu

Elias Greenbaum
Chemical Sciences Division
Oak Ridge National Laboratory
Oak Ridge, TN 37831 USA

Phone: 865-574-6835
Fax: 865-574-1275
E-Mail: greenbaum@ornl.ov

Alexei Gruverman
Department of Materials Science and Engineering
North Carolina State University
Raleigh, NC 27695, USA
Phone: +1 (919) 5133319
Fax: +1 (919) 5157724
E-mail: Alexei_Gruverman@ncsu.edu

S. Hasegawa
Department of Physics
School of Science University of Tokyo
7-3-1 Hongo, Bunkyo-ku, Tokyo 113-0033, JAPAN
Phone/Fax +81-3-5841-4167
E-mail:
shuji@surface.phys.s.u-tokyo.ac.jp

K. Hinzer
Bookham
10 Brewer Hunt Way
Ottawa, Ontario

Hendrik Hölscher
Center for NanoTechnology (CeNTech)
University of Muenster
Gievenbecker Weg 11
48149 Muenster, Germany
Phone: +49 (0) 251 83-63832
Fax: +49 (0) 251 83 -63873
E-mail:
Hendrik.Hoelscher@uni-muenster.de
http://www.uni-muenster.de/nanoforce

Seungbum Hong
Semiconductor Device and Material Lab
Samsung Advanced Institute of Technology
P.O. Box 111, Suwon 440-600, Korea
E-mail: bum4won@samsung.com

Takao Ishida
Nanotechnology Research Institute (NRI),
National Institute of Advanced Industrial Science and Technology (AIST)
1-1-1 Higashi, Tsukuba, Ibaraki 305-8562, Japan
Phone: +81-29-861-7203,
Fax: +81-29-861-2786
E-mail: t-ishida@aist.go.jp

A.T. Charlie Johnson, Jr
Department of Physics and Astronomy
University of Pennsylvania,
209 S. 33rd St., Philadelphia PA 19104-6396
Phone: (215) 898-9325
Fax: (215) 898-2010
Email: cjohnson@physics.upenn.edu

Sergei V. Kalinin
Materials Sciences and Technology Division and Center for Nanophase Materials Sciences
Oak Ridge National Laboratory
1 Bethel Valley Rd
Bldg. 3025, MS6030
Oak Ridge, TN 37831
Phone: (865) 241-0236
Fax: (865) 574-4143
E-mail: sergei2@ornl.gov

Kevin F. Kelly
ECE Dept., MS-366
Rice University
PO Box 1892
Houston, TX 77005-1892
Phone: 713-348-3565
Fax: 713-348-5686
E-mail: kkelly@rice.edu

Andrei Kholkin
Department of Ceramics and Glass Engineering &

Center for Research in Ceramic and Composite
Materials (CICECO), University of Aveiro
3810-193 Aveiro, Portugal
Phone: +351 234370235
Fax: +351 234425300
E-mail: kholkin@cv.ua.pt

Jürgen Kirschner
Max-Planck Institut für Mikrostrukturphysik
Weinberg 2
06120 Halle
Germany
Phone: +49-345-5582656
Fax: +49-345-5511223
E-mail: sekrki@mpi-halle.mpg.de

J.J. Kopanski
National Institute of Standards and Technology
Semiconductor Electronics Division
100 Bureau Dr., stop 8120
Gaithersburg, MD 20899-8120
E-mail: joseph.kopanski@nist.gov

Philippe Lambin
FUNDP
Rue de Bruxelles 61
B-5000 Namur, Belgium

S. B. Kuntze
University of Toronto
Department of Electrical and Computer Engineering
10 King's College Rd.
Toronto, Ontario

Ida Lee
Oak Ridge National Laboratory
1 Bethel Valley Road
PO Box 2008
Building 4500N, MS 6194, Room A8
Oak Ridge, TN 37831
Phone: 865-241-6695 or 865-574-6183
Fax: 865-574-1275
E-Mail: ilee1@utk.edu

Minhwan Lee
Stanford University
Department of Mechanical Engineering
440 Escondido Mall
Building 530, Room 226
Stanford, CA. 94305
Phone: 650-723-6488
Fax: 650-723-5034

M. Lewis
Department of Chemistry
North Carolina State University
314 Dabney Hall
Raleigh, NC 27695-8204

Dongbo Li
Department of Materials Science and Engineering
The University of Pennsylvania
3231 Walnut Street, Philadelphia, PA 19104
Phone: 215 898 3446
Fax: 215 573 2128
E-mail: dongboli@seas.upenn.edu

Martha Ch. Lux-Steiner
Department of Heterogeneous Material Systems (SE2)
Hahn-Meitner-Institut Berlin
Glienicker Str. 100 D-14109 Berlin
Fon: +49 30 8062-2462
Fax: +49 30 8062-3199
Email: lux-steiner@hmi.de

Joseph W. Lyding
Beckman Institute
405 N. Mathews
Urbana, IL 61801
Phone: 217-333-8370
Fax: 217-244-1995
E-mail: lyding@uiuc.edu

John A. Marohn
Dept. of Chemistry and Chemical Biology
150 Baker Laboratory
Cornell University
Ithaca, NY 14853-1301
Phone: 607-255-2004
Fax: 607-255-4137
E-mail: jam99@cornell.edu

Patrick Mesquida
Division of Engineering
King's College London
Strand, London WC2R 2LS
Phone: +44 (0)20 7848 2241
Fax: +44 (0)20 7848 2932
E-mail: patrick.mesquida@kcl.ac.uk

Vincent Meunier
Computer Science and Mathematics Division
& Center of Nanophase Materials Sciences
Oak Ridge National Laboratory
1 Bethel Valley Rd.
Oak Ridge, TN 37831
E-mail: meunierv@ornl.gov

Wataru Mizutani
Nanotechnology Research Institute (NRI)
National Institute of Advanced Industrial Science and Technology (AIST)
1-1-1 Higashi, Tsukuba, Ibaraki 305-8562, Japan
Phone: +81-29-861-2434
Fax: +81-29-861-2786
E-mail: w.mizutani@aist.go.jp

Markus Morgenstern
II. Inst. of Physics B
RWTH Aachen University
D-52056 Aachen, Germany
Phone: (++)49-241-80-27075/6
Fax: (++)49-241-80-22306
E-mail: mmorgens@physik.rwth-aachen.de

E.M. Muller
Dept. of Chemistry and Chemical Biology
150 Baker Laboratory
Cornell University
Ithaca, NY 14853-1301

Yasuhisa Naitoh
Nanotechnology Research Institute (NRI)
National Institute of Advanced Industrial Science and Technology (AIST)
1-1-1 Higashi, Tsukuba, Ibaraki 305-8562, Japan
Phone: +81-29-861-7892
Fax: +81-29-861-2786
E-mail: ys-naitou@aist.go.jp

Craig Y. Nakakura
Sandia National Laboratories
P.O. Box 5800, MS1077
Albuquerque, NM 87185-1077
E-mail: cynakak@sandia.gov

Nicola Naujoks
Nanotechnology Group
ETH Zurich
Tannenstrasse 3
CH-8092 Zurich, Switzerland
Phone: +41 44 632 69 89
Fax: +41 44 632 12 78
E-mail: naujoks@nano.mavt.ethz.ch
http://www.nanotechnology.ethz.ch

Robert J. Nemanich
Department of Physics
North Carolina State University
Raleigh, NC 27695-8202
Phone: (919) 515-3225
Fax: (919) 515-7331

E-mail: Robert_Nemanich@ncsu.edu
www.ssl.physics.ncsu.edu

T.N. Ng
Dept. of Chemistry and Chemical Biology
150 Baker Laboratory
Cornell University
Ithaca, NY 14853-1301

Ryan O'Hayre
Colorado School of Mines
Department of Metallurgical and Materials Engineering
1500 Illinois St.
Golden, CO. 80401
Phone: 303-273-3770
Fax: 303-273-3795

Charles Paulson
Department of Electrical Engineering
1415 Engineering Dr.
Madison, WI 53706
E-mail: paulsonc@cae.wisc.edu

Noyeol Park
Vice President
Storage System Division
Semiconductor Business
Samsung Electronics
Suwon, Korea

Frank Peter
Research Center Juelich
52425 Juelich, Germany
Phone: +492461616479
E-mail: f.peter@fz-juelich.de

Fritz Prinz
Stanford University
Department of Mechanical Engineering
440 Escondido Mall
Building 530, Room 226
Stanford, CA. 94305
Phone: 650-723-6488
Fax: 650-723-5034

Marko Radosavljevic
Components Research
Intel Corporation, RA3-252
5200 NE Elam Young Parkway
Hillsboro, OR 97124
Phone: 503-613-4732
E-mail: marko.radosavljevic@intel.com

Bernd Reichenberg
aixACCT Systems GmbH
Dennewartstrasse 25-27
52068 Aachen, Germany
E-mail: reichenberg@aixacct.com

Christoph Renner
London Centre for Nanotechnology and Department of Physics and Astronomy
University College London
Gordon Street 17-19
London WC1E 6BT, UK
Phone: +44 (0)20 7679 3496
Fax: +44 (0)20 7679 1360
E-mail: c.renner@ucl.ac.uk
http://www.cmmp.ucl.ac.uk/~cr/

Andreas Roelofs
Seagate Technology
1251 Waterfront Place
Pittsburgh, PA 15222, USA
Phone: +1 (412) 918 7028
Fax (412) 918 7222
E-mail: Andreas.Roelofs@seagate.com

Brian J. Rodriguez
Materials Science and Technology Division
Oak Ridge National Laboratory
1 Bethel Valley Rd
Bldg. 3137, MS6057

Oak Ridge, TN 37831
Phone: (865) 574-0791
Fax: (865) 576-8135
E-mail: rodriguezbj@ornl.gov

Henrik M. Rønnow
Laboratory for Neutron Scattering
ETH-Zürich & Paul Scherrer Institute
5232 Villigen
Switzerland
Phone: +41 56 310 4668
Fax: +41 56 310 2939
E-mail: henrik.ronnow@psi.ch

Yossi Rosenwaks
Faculty of Engineering
Elec. Lab. Building, room 231
TEL-AVIV UNIVERSITY
Tel-Aviv 69978, ISRAEL
Phone: +972-3-6406248
Fax: +972-3-6423508
E-mail: yossir@eng.tau.ac.il
http://www.eng.tau.ac.il/~yossir/rosenwaks.html

Laura B. Ruppalt
Beckman Institute
405 N. Mathews
Urbana, IL 61801
Phone: 217-244-5665
Fax: 217-244-1995
E-mail: ruppalt@uiuc.edu

S. Saraf
Faculty of Engineering
TEL-AVIV UNIVERSITY
Chemical Department
Tel-Aviv 69978, ISRAEL
E-mail: mailto:saraf@eng.tau.ac.il

E.H. Sargent
University of Toronto
Department of Electrical and Computer Engineering
10 King's College Rd.
Toronto, Ontario

Uta Schlickum
Institut de Physique des Nanostructures
EPFL/SB/IPN, PHB-Ecublens, Station 3
CH-1015 Lausanne
Switzerland
Phone: +41 (0)21 69 34 742
Fax: +41 (0)21 69 33 604
E-mail: uta.schlickum@epfl.ch

Andrew R. Schwartz
Neocera Inc.
10000 Virginia Manor Road
Beltsville, MD 20705, USA
Phonr: +001 301 210 1010
Fax: +001 301 210 1042
E-mail: schwartz@neocera.com
http://www.neocera.com

Udo D. Schwarz
Department of Mechanical Engineering
Yale University
P.O. Box 208284
New Haven, CT 06520-8284
USA
Phone: +1-203-432-7525
Fax: +1-203-432-6775
E-mail: udo.schwarz@yale.edu
http://www.eng.yale.edu/nanomechanics

A. Schwarzman
Faculty of Engineering
TEL-AVIV UNIVERSITY
Tel-Aviv 69978, ISRAEL

W.R. Silveira
Dept. of Chemistry and Chemical Biology
150 Baker Laboratory
Cornell University
Ithaca, NY 14853-1301

Timothy J. Smith
Department of Chemistry and Biochemistry
University of Texas at Austin
1 University Station/MC A5300
Austin, TX 78712
E-mail: tjsmith@mail.utexas.edu

Michael Stadermann
L-234
Chemistry & Material Science
Lawrence Livermore National Laboratory
7000 East Ave
Livermore, Ca 94550

Cristian Staii
Department of Chemistry
Princeton University
Princeton, NJ 08544
Phone: 609 258-3962
E-mail: cstaii@Princeton.edu

Andreas Stemmer
Nanotechnology Group
ETH Zurich
Tannenstrasse 3
CH-8092 Zurich
SWITZERLAND
Phone: +41 44 632 4572
Fax: +41 44 632 1278
E-mail: stemmer@nano.mavt.ethz.ch

Keith Stevenson
Department of Chemistry and Biochemistry
University of Texas at Austin
1 University Station/MC A5300
Austin, TX 78712
Phone: (512)232-9160
E-mail: stevenson@mail.cm.utexas.edu

E. Strassburg
Faculty of Engineering
TEL-AVIV UNIVERSITY
Tel-Aviv 69978, ISRAEL

Kristof Szot
Institute of Physics, University of Silesia
40-007 Katowice, Poland
and Research Center Juelich
52425 Juelich, Germany
Phone: +492461616479
email: k.szot@fz-juelich.de

O. Tal
Faculty of Engineering
TEL-AVIV UNIVERSITY
Tel-Aviv 69978, ISRAEL

Vladimir V. Talanov
Neocera Inc.
10000 Virginia Manor Road
Beltsville, MD 20705, USA
Phone: +001 301 210 1010
Fax: +001 301 201 1042
E-mail: talanov@neocera.com
http://www.neocera.com

Paiboon Tangyunyong
Sandia National Laboratories
P.O. Box 5800, MS1081
Albuquerque, NM 87185-1081
E-mail: ptangyu@sandia.gov

Stephan Tiedke
aixACCT Systems GmbH
Dennewartstrasse 25-27
52068 Aachen, Germany
Phone: +492419631410
E-mail: tiedke@aixacct.com

Hiroshi Tokumoto
Nanotechnology Research Center
Research Institute for Electronic Science
Hokkaido University

Kita-ku N21 W10, Sapporo 001-002, Japan
Phone:+81-11-706 9354
Fax:+81-11-706-9355
E-mail: tokumoto@es.hokudai.ac.jp

W. Vandervorst
Kapeldreef 75, B-3001
Leuven, Belgium
tel 32 16 281 286
E-mail: vdvorst@imec.be

Rainer Waser
Research Center Juelich
52425 Juelich, Germany
Phone: +492461615811
E-mail: r.waser@fz-juelich.de

Sean Washburn
Dept Physics and Astronomy
Univ North Carolina at Chapel Hill
Chapel Hill, NC 27599-3255

Daniel van der Weide
Department of Electrical & Computer Engineering
University of Wisconsin
1415 Engineering Dr
Madison WI 53711, USA
Phone: 608-265-6561
Fax: 815-371-3407
E-mail: danvdw@engr.wisc.edu
http://vdw.ece.wisc.edu

Paul S. Weiss
Department of Chemistry and Physics
104 Davey Laboratory
The Pennsylvania State University
University Park, PA 16802-6300, USA
Phone: +1 (814) 865-3693
Fax: +1 (814) 863-5516
E-mail: stm@psu.edu
http://www.nano.psu.edu/

J.K. White
Bookham
10 Brewer Hunt Way
Ottawa, Ontario

J. Williams
Department of Chemistry
North Carolina State University
314 Dabney Hall
Raleigh, NC 27695-8204

Wulf Wulfhekel
Physikalisches Institut
Universität Karlsruhe
Wolfgang Gaede Strasse 1
76131 Karlsruhe
Germany
Phone: +49-721-6083440
Fax: +49-721-6086103
E-mail: wulf.wulfhekel@pi.uni-karlsruhe.de

M. Xu
Kapeldreef 75, B-3001
Leuven, Belgium
Phone: 32 16 288 024
E-mail: xum@imec.be

About the Editors

Sergei Kalinin is currently a research staff member at the Oak Ridge National Laboratory, Materials Sciences and Technology Division and Center for Nanophase Materials Science. He completed his Ph.D. in Materials Science at the University of Pennsylvania in 2002 (with Prof. Dawn Bonnell). His previous undergraduate and graduate work in Materials Science was conducted at Moscow State University, Moscow, Russia. The focus of his current research interests is interplay between electromechanical, transport, and mechanical phenomena in inorganic and biological systems using local probes. He serves as a member of American Vacuum Society NSTD board and member of editorial board for Journal of Nanoelectronics and Optoelectronics. As a student, he earned multiple research awards including American Vacuum Society Graduate Student Award and several Materials Research Society Graduate Student Awards. He was recognized with the Ross Coffin Purdy Award of American Ceramics Society (2003) for the development of Scanning Impedance Microscopy for characterization of frequency-dependent transport on the nanoscale. He is also a recipient of Wigner Fellowship of Oak Ridge National Laboratory (2002) and an Oak Ridge National Laboratory Early Career Accomplishment Award for Science and Technology (2005). He has authored more than 70 scientific papers, six book chapters and five patents on various aspects of SPM.

Alexei Gruverman is currently a Research Associate Professor in the Department of Material Science and Engineering, North Carolina State University. He received his PhD degree in Solid State Physics from the Ural State University in Russia. After graduation, he worked as a post-doctoral researcher at the Joint Research Center for Atom Technology (JRCAT), Japan, where he initiated a project on scanning probe microscopy studies of ferroelectrics. He then joined Sony Corporation to work on the problems related to characterization and reliability of ferroelectric memory devices by SPM. In 2000, he joined the faculty at North Carolina State University. He has published over 80 articles in refereed journals, several book chapters and edited one book. His current research interests are in the field of ferroelectric nanostructures and thin films, nanoscale phenomena in ferroic materials, mechanical properties of biomaterials and non-volatile information storage technologies. He is a recipient of the Ikeda Award (2004) for contribution to SPM studies of ferroelectrics.

Introduction

Scanning Probe Microscopy Techniques for Electrical and Electromechanical Characterization

S.V. KALININ AND A. GRUVERMAN

Progress in modern science is impossible without reliable tools for characterization of structural, physical, and chemical properties of materials and devices at the micro-, nano-, and atomic scale levels. While structural information can be obtained by such established techniques as scanning and transmission electron microscopy, high-resolution examination of local electronic structure, electric potential and chemical functionality is a much more daunting problem. Local electronic properties became accessible after the development of Scanning Tunneling Microscopy by G. Binnig and H. Rohrer in 1981 at IBM Zurich 25 years ago—an invention that earned its authors the Nobel Prize in Physics five years later [1]. Based on the quantum mechanical tunneling between an atomically sharp metallic tip and conductive surface, scanning tunneling microscopy (STM) has become the first instrument to generate real-space images of surfaces with atomic resolution and has triggered development of new classes of STM-related techniques. In 1986, Binnig, Quate, and Gerber demonstrated atomic force microscope (AFM) based on the mechanical detection of the Van der Waals forces between the tip and the surface using a pliable cantilever [2]. It was almost immediately realized that AFM could be extended to map forces of various types, such as magnetic and electrostatic forces, as well as for probing chemical interactions. This dual capability of probing currents and forces at the nanometer and atomic level has led to a rapid growth of a variety of scanning probe microscopy (SPM) techniques over the last two decades. Techniques such as AFM, magnetic force microscopy (MFM), electrostatic force microscopy (EFM), scanning capacitance microscopy (SCM), near-field scanning optical microscopy (NSOM), and others have emerged to provide capability to access local electrical, magnetic, chemical, mechanical, optical, and thermal properties of materials on the nanometer scale. It has been demonstrated that the SPM approach allows not only imaging, but also control and modification of the local structure and material functionality at the nano- and atomic level. As a consequence, the last two decades have witnessed an explosive growth in application of SPM techniques in a wide spectrum of fields of science, ranging from condensed matter physics, chemistry and materials science to medicine and biology. It will not be an exaggeration to say that the rapid development of nanoscience and nanotechnology in the last two decades was strongly stimulated by the availability of SPM techniques, and in turn constantly stimulates development of novel SPM probes.

In parallel with the development of ambient SPM techniques, a significant progress was achieved in the development of high-resolution ultrahigh vacuum (UHV) SPM systems. Compared to ambient systems, UHV operation imposes major difficulties in the sample preparation and microscope operation. However, it provides a potential for true atomic resolution imaging both in STM and AFM modes. STM sensitivity to local electronic properties provided physicists with the tool to probe electronic phenomena such as edge states in quantum Hall systems, phase separation and charge ordering in strongly correlated oxides, and transport in mesoscopic conductors. In the context of surface science and catalysis, the capability to probe chemical and photochemical processes on the atomic level have emerged. Many ambient electrostatic and electromechanical probes can be implemented in UHV conditions as well. As such, UHV SPM paves the way to the investigation of fundamental electrical and mechanical properties of materials at the atomic level.

One of the primary challenges in the SPM applications to local electrostatic and electromechanical phenomena is quantitative analysis of the acquired signal. Indeed, a typical SPM image, were it a surface potential map, electromechnical activity map, capacitance distribution or a gate potential image, provides a quantitative measure of local surface properties. However, image formation mechanisms in most SPM techniques are extremely complex due to the geometry of tip-surface system and multitude of contributing interactions. For example, noticeable progress in understanding such well-established topographic imaging techniques as non-contact AFM or intermittent contact AFM was achieved only in the last few years. The problem of quantitative interpretation of the local property map is even more difficult; further progress requires thorough understanding of the image formation mechanisms through combination of first-principles methods to describe atomistic processes at tip-surface junction, continuum mechanics, and electrostatics to describe long-range interactions and cantilever dynamics. However, once achieved, this transforms SPM from a mere imaging tool into a quantitative probe of local physical phenomena on the nanoscale, realizing an age-old dream of precise measurements of physical and chemical properties on nanometer, molecular, and ultimately atomic levels.

Tremendous growth in SPM instrumentation, theory, and applications have resulted in a large number of monographs, books, and reviews addressing general aspects of SPM [3–12], including STM studies of atomic structure and electronic phenomena [13–16], electrochemical SPM [17], theoretical aspects of AFM [18] and STM [19], nanofabrication [20], application of SPM in biology and bioengineering [21–24]. However, there is a long-due need to bring together a permanently growing knowledge base on the practical and theoretical aspects of advanced electrical and electromechanical SPM techniques. The goal of this book is to provide a comprehensive reference on the nanoscale characterization of *electrical and mechanical properties of functional materials* by SPM techniques and to make readers aware of tremendous developments in the field in the last decade. This book provides a link between well-established ambient SPM techniques and UHV SPM, materials and device applications, and theoretical basis for interpretation of SPM data. While a number of SPM techniques have been used in a variety of scientific fields ranging from semiconductors to ferroelectrics to optics, in this book various aspects of these techniques will be presented on an interdisciplinary basis. By bringing together critical reviews written by the leading researchers from the different scientific disciplines relevant information will be conveyed that will allow readers to learn more about the fundamental and technological advances and future trends in the different fields of nanoscience.

The book comprises 35 chapters divided into four parts. Part I introduces the reader to the technical and instrumental features of SPM techniques for electrical, electromechanical, transport, near-field optical and microwave characterization. Part II deals with the SPM imaging at the resolution limit in molecular systems, complex oxides, and low-dimensional structures, as well as theoretical underpinnings of SPM. Part III illustrates application of SPM to electrical and electromechanical characterization of a broad spectrum of materials ranging from semiconductors to polymers to biosystems. Finally, Part IV discusses SPM-based devices and nanofabrication methods.

Chapter I.01 by A. P. Baddorf reviews technical aspects of Scanning Tunneling Potentiometry (STP), an extension of STM to transport measurements. The chapter discusses applications of STP to probe mesoscopic transport in low-dimensional systems, defect-induced scattering, and transport tin phase-separated strongly correlated oxides. Further prospects of transport imaging with nanometer and atomic resolution are discussed. Chapter I.02 by P. Eyben *et al.*, discusses implementation, probe choice, calibration, and theoretical aspects of Scanning Spreading Resistance Microscopy, the technique that has emerged as one of the primary tools for dopant profiling and device characterization and failure analysis on the nanoscale. Chapter I.03 by J. Kopanski discusses principles of SCM and its application to electrical characterization of semiconductors and dielectric films, integrated circuit failure analysis, quantitative dopant profiling, and optical pumping for carrier mobility measurements. Chapter I.04 by T. Glatzel *et al.*, discusses Kelvin Probe Force Microscopy (KPFM), introduces principles of amplitude and frequency detection in KPFM and relevant theory. The chapter describes in detail resolution theory and probe function determination in KPFM, paving the way for quantitative measurements of electrostatic surface potentials, photovoltaic phenomena, and work function distributions. Chapter I.05 by O'Hayre *et al* introduces two SPM techniques for probing frequency-dependent transport at the nanoscale. Nanoimpedance microscopy is a current-based technique combining conventional impedance spectroscopy with AFM. Scanning Impedance Microscopy (SIM) is a force-based technique in which AFM probe detects amplitude and phase of local voltage oscillations induced by lateral periodic bias applied through macroscopic electrodes. These methods are similar to conventional 4-probe resistance measurements, in which AFM tip is used as a moving electrode, providing advantage of high spatial resolution. Piezoresponse Force Microscopy (PFM) and its application for imaging, spectroscopy, and manipulation of ferroelectric domains are reviewed in Chapter I.06 by Kholkin *et al.* The chapter briefly discusses principles and image formation mechanism in PFM, as well as its application to imaging of polar nanodomains in ferroelectric relaxors, high-density data storage, and ferroelectric capacitor characterization. Near-Field Microwave Microscopy is described in Chapter I.07 by Anlage *et al.* The authors review the basic concepts of near-field interactions between a source and sample, present a historical overview and discuss quantitative approaches to interpretation of near-field microwave images. Chapter I.08 by A. Bouchelier and R. Bachelot discusses implementation and application of apertureless Near Field Optical Microscopy and related techniques. The tip enhancement of electromagnetic field allows imaging and spectroscopy of optical phenomena on the length scales well below optical wavelength. Application of SPM techniques to the study of electrochemical interfaces is discussed in Chapter I.09 by T. Smith and K. Stevenson. These techniques include Electrochemical Scanning Tunneling Microscopy (EC-STM), Electrochemical Atomic Force Microscopy (EC-AFM), and hybrid techniques such as Scanning Electrochemical Microscopy - Atomic Force Microscopy (SECM-AFM) and Local Electrochemical Impedance Spectroscopy - Atomic Force Microscopy (LEIS-AFM). Application of electrochemical SPM in the emerging areas of energy storage and conversion, corrosion, catalysis, and electrochemical deposition processes are described. Recent theoretical and instrumental developments in high-frequency near field probes are reviewed in Chapter I.10 by C. Paulsen and D. Van der Weide, providing a unified description of a broad range of optical, infrared, and microwave probes. The chapter also provides a theoretical framework for the interpretation of near-field experiments, as well as extensive literature survey and discussion of future potential of these techniques.

Part II of the book combines a number of reviews on experimental and theoretical aspects of fundamental electrical and electromechanical phenomena on the nano- and atomic scales and describes the recent instrumental and theoretical advances in high-resolution imaging of electrical, transport, and electromechanical properties on surfaces, low-dimensional systems, nanotubes and nanowires, and molecules. In Chapter II.01, M. Morgenstern describes applications of STM and AFM for probing transport phenomena in 2D electron systems in high-mobility III-V semiconductors. Imaging nanomagnetic and spin structure of materials with nanometer and atomic resolution by Spin–polarized STM (SpSTM) is described in chapter II.02 by W. Wulfhekel *et al.* Probe fabrication, one of the key components for successful SpSTM experiment, is discussed in detail. The principles and advantages of constant-current, spectroscopic, and differential magnetic modes of SpSTM are discussed. The applications for imaging ferromagnetic and antiferromagnetic surfaces of bulk materials and thin-film systems are illustrated. Approaches for fabrication and probing of molecular electronic devices are described in Chapter II.03 by K. Kelly and P. S. Weiss. Self-assembled monolayers are utilized to isolate molecules with various electronic properties to determine the fundamental transport mechanisms, and the relationships between molecular structure, environment, connection, coupling, and electrical functionality. STM of individual conjugated molecules inserted in alkanethiol SAM illustrates a rich gamut of electronic behavior, including reversible conductance switching. SPM imaging and control of charge transport in individual carbon nanotubes and carbon nanotube networks are discussed in Chapters II.04 and II.05. In Chapter II.04, C. Staii *et al.*, discuss the application of Scanning Gate Microscopy (SGM), EFM, SIM, and thermal SPM to probe electronic structure of individual defects in carbon nanotube circuits. Applications of high-frequency SGM and SIM and memory effects in nanotube circuits are presented. In Chapter II.05, M. Stadermann and S. Washburn describe the application of conductive SPM to networks of carbon nanotubes to map current paths and differentiate semiconducting and metallic nanotubes, as well as to probe conductance decay at nanotube junctions. In Chapter II.06, V. Meunier and P. Lambin apply a density functional theory to model STM images of carbon nanotubes and grain boundaries in graphene. The transport properties of single-wall nanotubes and tip-tube interactions in SGM and EFM are modeled using non-equilibrium Green's function theory. In Chapter II.07, S. Hasegawa discusses instrumentation and application of multiple-probe STM for surface transport studies. Precise positioning of conductive tips allows probing conductance through step edges and orientation dependence of conductance. Ultimately, this technique is being developed as a probe of a surface transport Green's function. Principles and application of Dynamic Force Microscopy (DFM), also referred to as Non-Contact Atomic Force Microscopy, are described in Chapter II.08 by U. Schwarz and H. Hölscher. Application of DFM for atomic-resolution imaging of conductors, semiconductors, and insulators are described. DFM probing of surface electrostatic properties including charge distributions around charged monoatomic vacancies and individual doping atoms in semiconductors are illustrated. Application of high-resolution STM to probe atomic and electronic structure and electronic inhomogeneities in manganites are described in Chapter II.09 by C. Renner and H. Ronnow. These studies illustrate the impact high-resolution probes can have on understanding the fundamental physical phenomena in strongly-correlated complex oxides including manganites, cuprates, and relaxors.

Part III includes a series of reviews on application of ambient SPM techniques for characterization of transport, electromechanical, optical, and electrical phenomena in materials, heterostructures, and devices. A number of topics, such as transport in semiconductor optoelectronic and electronic structures, imaging and quantification of electroactive grain boundaries, dislocations and interfaces, electromechanical imaging of biological systems, and photovoltaic phenomena in photosynthetic molecules are covered. In Chapter III.01, S. Kuntze *et al.*, describe application of Scanning Voltage Microscopy (SVM) and Scanning Differential Resistance Microscopy (SDRM) for *in-situ* imaging of operational electronic devices. Examples include direct observation of the current blocking breakdown in a buried heterostructure laser, effect of current spreading inside actively biased ridge waveguide lasers, origin of anomalously high series resistance encountered in ridge lasers and electron over-barrier leakage. Instrumentation and imaging mechanisms in SVM and SDRM are discussed and sources of artifacts (such as circuit time constants and photocurrent) are analyzed. Chapter III.02 by I. Lee and E. Greenbaum describes application of KPFM and EFM to probe photovoltaic activity in

Photosystem I (PSI) reaction centers, DNAs, protein microarrays on surfaces, and PSI at the air–liquid interfaces, providing insights into fundamental mechanisms of photosynthesis. In Chapter III.03, A. Gruverman *et al.*, describe application of Piezoresponse Force Microscopy (PFM) to probe the structure of biomaterials with nanometer scale spatial resolution, utilizing ubiquitous presence of piezoelectricity in biopolymers. Examples of bioelectromechanical imaging include human tooth, femur canine cartilage, deer antler, and butterfly *Vanessa Virginiensis* wing scales. A potential of PFM to study the internal structure and orientation of the protein microfibrils is illustrated. Application of SCM for failure analysis of semiconductor devices, carrier dynamics in FET channel during device operation, and visualization of radiation effects are described in Chapter III.04 by C. Nakakura *et al.* Extensive practical experimental details and a broad range of examples render this chapter extremely useful for SCM practitioners. Application of KPFM for local studies of surface band bending, defects, and grain boundaries in semiconductors are presented in Chapter III.05 by Rosenwaks *et al.* This chapter also reviews recent KPFM studies of local electronic phenomena in quantum wells and organic light-emitting devices, and develops KPFM framework for spectroscopic studies of surface states. Chapter III.06 by B. Rodriguez *et al.*, summarizes recent applications of SPM to nanoscale studies of the electric properties of III-nitride thin films, bulk crystals and heterostructures. The chapter illustrates the complementary application of techniques such as KPFM, conductive AFM, and PFM to probe charge defects and inversion boundaries in these materials. Measurement of surface polarity and the screening mechanism of III-nitrides using SPM and Photoelectron emission Microscopy are discussed. In Chapter III.07, S. Cohen gives a broad perspective on electron flow in molecular systems. This research is motivated by advent of molecular electronics, drive to understand a charge transfer in biological systems, and functionality and performance bottlenecks in devices such as organic light emitting diodes and dye-sensitized solar cells. SPM measurements of electron flow through DNA and STM measurements of isolated molecules on a semiconductor surface are discussed in detail. The principles and instrumentation for local conductance imaging of perovskite thin films are discussed in Chapter III.08 by C. Szot *et al.* Imaging of conductive paths in insulating oxides provides real-space information on electrical activity of defects and dislocations. The role of surface adsorbates on electrical and PFM imaging is discussed. In Chapter III.09, T. Ishida *et al.*, describe methods for evaluating the electric properties of conjugated molecules embedded in alkanethiol SAMs, including electrical conduction and barrier height measurements of SAMs and single molecules using STM and conductive AFM. Finally, in Chapter III.10, W. Silveira *et al.*, review application of high-sensitivity EFM to probe charge transport and local electronic properties in organic electronic devices. Instrumentation and contrast formation in EFM are also described in detail.

Part IV of the book is devoted to the SPM-based devices and nanofabrication methods. In Chapter IV.01, N. Naujoks *et al.*, describe application of charge lithography for fabrication of nanostructures. In this method, charged patterns deposited by an AFM tip are used to attract oppositely charged nanoparticles, resulting in a stable deposited pattern. Nano-oxidation of semiconductor and metal surfaces is described in Chapter IV.02 by J. Dagata. The author introduces the method, describes relevant technical and theoretical details, and discusses the potential and limitations of this nanofabrication technique. Nanofabrication on the atomic level is introduced in Chapter IV.03 by P. Albrecht *et al.* Development of the atomic-resolution hydrogen resist technique and its application to the templated self-assembly of molecular systems on silicon are described. A mechanism of tip-induced desorption is determined through isotope studies. An integration of carbon nanotubes with silicon and the III-V compound semiconductors is explored. Ferroelectric lithography is described in Chapter IV.04 by D. Li and D. Bonnell. Several methods, including contact electrode, SPM, and e-beam are used to pattern domains on ferroelectric surfaces in the absence of a top electrode. The domain specific reactivity in metal photodeposition process and domain patterning are combined into a fabrication process that is demonstrated for several classes of magnetic and optoelectronic nanostructures. Chapter IV.05 by M. Lewis *et al.* introduces several methods of nanopatterning in SAM systems. Additive, subtractive, and exchange approaches for SAM lithography are described. Finally, in Chapter IV.06, S. Hong and N. Park describe application of resistive SPM probes for data storage in ferroelectric medium, potentially opening the pathway for Tbit density storage.

Overall, the book is intended to present a unified outlook on all aspects of modern electrical and electromechanical probes and combine practical and theoretical aspects of these techniques and applications ranging from fundamental physical studies to device characterization and failure analysis to nanofabrication. We hope that this book will develop new educational advances by helping students and postdoctoral scientists significantly improve their knowledge on the new applications of SPM and on the nanoscale properties of a number of functional materials, such as electroactive polymers, biomolecules, piezoelectrics, and so on. It is our expectation that with SPM becoming a must-know technique in many scientific disciplines, this book will become a valuable source of information for interdisciplinary research that can be used as a reference handbook.

References

1. G. Binnig, H. Rohrer, Ch. Gerber, E. Weibel, Phys. Rev. Lett. **49**, 57 (1982); Phys. Rev. Lett. **50**, 120 (1983).
2. G. Binnig, C. F. Quate, Ch. Gerber, Phys. Rev. Lett. **56**, 930 (1986).
3. D. A. Bonnell (Ed.), *Scanning Probe Microscopy and Spectroscopy: Theory, Techniques and Applications*. 2nd Edn, Wiley VCH, New York, 2000.
4. R. Wiesendanger (Ed.), *Scanning Probe Microscopy and Spectroscopy-Methods and Applications*, Cambridge University Press, Cambridge UK, 1994.
5. G. Friedbacher, H. Fuchs, Classification of Scanning Probe Microscopies—(Technical Report), Pure and Appl. Chem. **71**, 1337 (1999).
6. L. Bottomley, "Scanning Probe Microscopy," Anal. Chem., **70**, 425R-475R (1998).
7. E. Meyer, *Atomic Force Microscopy: Fundamentals to Most Advanced Applications*, Springer-Verlag New York, November 2003.

8. A. De Stefanis, A. A. G. Tomlinson, *Scanning Probe Microscopies: From Surfaces Structure to Nano-Scale Engineering*, Trans Tech Publications, April 2001.
9. E. Meyer, H. J. Hug, R. Bennewitz, *Scanning Probe Microscopy: The Lab on a Tip*, Springer 2003.
10. B. Bhushan, H. Fuchs (Eds.), *Applied Scanning Probe Methods II: Scanning Probe Microscopy Techniques, Applied Scanning Probe Methods III: Characterization, Applied Scanning Probe Methods IV: Industrial Applications*, Springer, 2006.
11. R. J. Colton, A. Engel, J. E. Frommer et al. (Eds.), *Procedures in Scanning Probe Microscopies*, John Wiley, 1998.
12. M. T. Bray, S. H. Cohen, M. L. Lightbody (Eds.), *Atomic Force Microscopy/Scanning Tunneling Microscopy*, Springer, 1995.
13. R. Wiesendanger, H. J. Guntherodt (Eds.), *Scanning Tunneling Microscopy III: Theory of STM and Related Scanning Probe Methods* (Springer Series in Surface Sciences), Springer, 1997.
14. C. Bai, *Scanning Tunneling Microscopy and Its Application* (Springer Series in Surface Sciences), Springer, 2000.
15. W. Hofer, A. Foster, A. Shluger, "Theories of Scanning Probe Microscopes at the Atomic Scale," Rev. Mod. Phys. **75**, 1287 (2003).
16. S. Morita, R. Wiesendanger, E. Meyer (Eds.), *Noncontact Atomic Force Microscopy*, Springer Verlag, 2002.
17. A. J. Bard, M. V. Mirkin, *Scanning Electrochemical Microscopy* (Monographs in Electroanalytical Chemistry and Electrochemistry Series), CRC Press, 2001.
18. R. Garcia, R. Perez, "Dynamic Atomic Force Microscopy Methods," Surf. Sci. Rep., **47**, 197–301 (2002).
19. A. Foster, W. Hofer, *Scanning Probe Microscopy: Atomic Scale Engineering by Forces and Currents* (NanoScience and Technology), Springer, 2006.
20. H. T. Soh, K. W. Guarini, C. F. Quate, *Scanning Probe Lithography* (Microsystems, Volume 7), Kluwer Academic Publishers, 2001.
21. B. P. Jena, J. K. H. Horber (Eds), *Atomic Force Microscopy in Cell Biology* (Methods in Cell Biology, Volume 68), Academic Press, 2002.
22. V. J. Morris, A. P. Gunning, A. R. Kirby, *Atomic Force Microscopy for Biologists*, World Scientific publishing, 1999.
23. P. C. Braga, D. Ricci (Eds.), *Atomic Force Microscopy: Biomedical Methods and Applications* (Methods in Molecular Biology), Humana Press, 2003.
24. Vladimir Tsukruk (Ed.), *Advances in Scanning Probe Microscopy* (Macromolecular Symposia 167), John Wiley & Sons, 2001.

I
SPM Techniques for Electrical Characterization

I.1
Scanning Tunneling Potentiometry: The Power of STM applied to Electrical Transport

A. P. BADDORF

The introduction of scanning tunneling microscopy (STM) followed by scanning tunneling spectroscopy (STS) opened experimental access to the geometric and electronic structure of materials on an atomic scale and essentially ushered in the modern field of nanoscience. The goal of scanning tunneling potentiometry (STP) is to adapt the scanning tunneling probe to measure electrical transport on the same length scale. The approach is to establish a current laterally in the sample, then to map the voltage locally by determining the tip bias that produces no tunneling at each point. The technique is similar to the macroscopic four-probe method commonly adopted for measuring electrical transport, with the inner two probes replaced by a scanning tunneling tip. This chapter describes principles, reviews approaches, and illustrates capabilities of STP.

1 Introduction

The link between the nanoscale and the human scale, be it quantum computing, nanomechanics, or nanotube devices, is almost always by electrical current. In today's world, where electronic devices proliferate, the fundamental and practical importance of understanding electronic transport in nanoscale devices is immense. Milestones in condensed matter physics which have the greatest potential to impact the world are often associated with discovery of novel electronic transport phenomena, frequently encountered in materials of nanometer length scales and reduced dimensionality. Examples abound, not the least of which date back to the birth of the transistor, or more recently high Tc superconductivity, giant magnetoresistance, and superconducting quantum interference devices (SQUID). Tools that can measure electrical conductivity at these small length scales are needed to understand the mechanisms of transport and to take full advantage of our modern electronics oriented world.

The invention of scanning tunneling microscopy (STM) by Gerd Binnig and Heinrich Rohrer [1] in 1981 at IBM Zurich (1986 Nobel Prize) opened the way to exploration of electronic states and transport with atomic resolution. In STM, a voltage is applied between a conducting tip and surface, and the current between the two is measured. (Conducting here simply means sufficient electron mobility that the current limitation comes from the tunneling gap. With very low currents, STM imaging can be performed on quite high resistance materials.) As the tip is brought closer to the surface, tunneling of the electrons through the gap produces a current before the two physically touch. Because this is a quantum mechanical process through a barrier where the electrons cannot stay, the current is exponentially dependent on the gap distance. This strong dependence on tip distance makes the current quite suitable for feedback, i.e., to maintain a constant tip-surface separation electronically. By scanning the tip, usually by deflection of piezoelectric materials, and monitoring the tunneling current, a topographical map of the surface is obtained [2].

In reality a STM image is much more complex and rich with information. Scanning with a constant current is not actually a map of the atomic topography, but more correctly a map of the electronic overlap between tip and substrate, which determines the tunneling current. For an ideal tip this provides a map of the electronic states of the substrate, often, but not always, related to the geometric structure. Exceptions led to early mistakes in structural models, for example, the "definitive" identification of p(2×1)O/Cu(110) as a buckled row structure [3]—later shown to be an added row [4]—or the interpretation of STM images of (2×3)N/Cu(110) as a missing row [5]—later shown to be a compressed, buckled row [6]. An imperfect tip can further complicate interpretation by introducing multiple tunneling sources (a double tip) or directional tunneling from *p, d,* or *f* electronic orbitals.

Imaging of electronic states is not ideal for those interested predominantly in atomic arrangement, but it is ideal if you want to know where the electrons are and what they are doing, for example, in order to study chemical bonding or conductivity. Even conventional STM can provide extraordinary information on the local electronic levels at a surface, particularly through measurements of current versus tip-surface voltage (*I-V*) at individual tunneling sites. *I-V* measurements are closely related to local transport properties when the dependence is determined at voltages near the Fermi energy, E_f. Electrical conductivity requires electron density at E_f and available states just above E_f to allow movement of charge. Insulators or poor conductors will show a gap, equivalent to an *I-V* curve with near zero slope at E_f.

It is not surprising that soon after the invention of the STM, other IBM Zurich researchers modified the technique to measure spatial distributions of electric potential. Paul Muralt, then a postdoctoral scientist, and Dieter W. Pohl published the first description of this modification in 1986 and named the new technique "scanning tunneling potentiometry," or STP [7]. The addition to conventional STM that is central to STP is a voltage bias along the surface of the sample. A schematic of the STP setup is shown in Figure 1. As the tip is scanned along the surface the tunneling current becomes highly sensitive to the local tip-surface voltage difference; since the tip voltage is known, a map to the surface voltage can be obtained. Of course, variations in tunneling current due to surface topography must

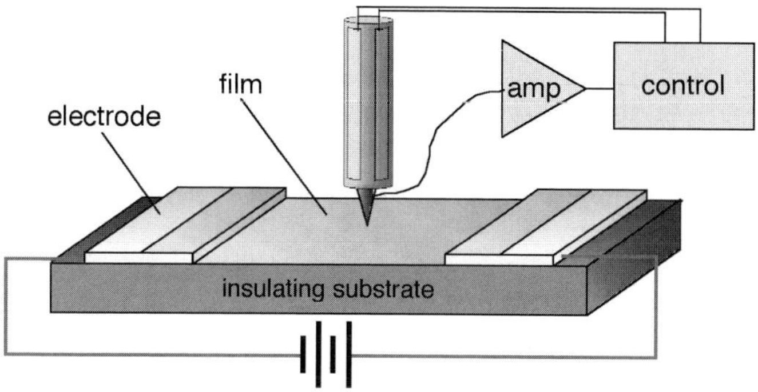

FIGURE 1. Schematic illustration of the STP technique, which combines a lateral voltage across the sample with an STM for imaging of the local surface topography and potential.

be distinguished from those due to the lateral voltage drop. Approaches to make this distinction will be discussed below.

2 STP Technique

2.1 Sample Conditions

For conventional detection electronics and tunneling equipment, voltages must be limited to the order of tens of volts to avoid arcing and to protect sensitive amplifiers. Across a typical 1-cm sample, 10 V corresponds to 1 mV/μm or 1 μV/nm. With a typical GΩ tunneling resistance, a uniform conductor would require equipment with a sensitivity of 1 pA/μm or 1 fA/nm. A femtoamp precision is too demanding for common STP circuitry, but a pA is certainly achievable. One solution that can provide a greater lateral resolution is to study smaller samples, for example, to lithographically create electrodes to apply the lateral voltage across a few tens of microns. Another solution is to study materials that are not uniform, so that much larger voltage drops and corresponding tunneling current contrasts are observed at boundaries or defects of interest. However, sensitivity to small current changes can still be a limiting factor in STP spatial resolution.

For good conductors, even 10 V/cm is too large to maintain. A 1-cm-long copper sample with 1-mm-diameter cross section will draw a little over a 1,000 A with application of 10 V, and sample heating could be a problem! This illustration reveals the need for relatively high resistance samples in STP to maintain a sufficient voltage drop across the sample. To minimize current and sample heating, sample resistances should be well over 100Ω. At the same time, STP's foundation on tunneling current requires a non-insulating sample, with resistivity well beneath a typical tunneling resistance of a GΩ. Appropriate samples include semiconductors, poor metals, thin films, and nanometer structures. By reducing the dimensions of the conductor, quite high current densities become possible. For example in

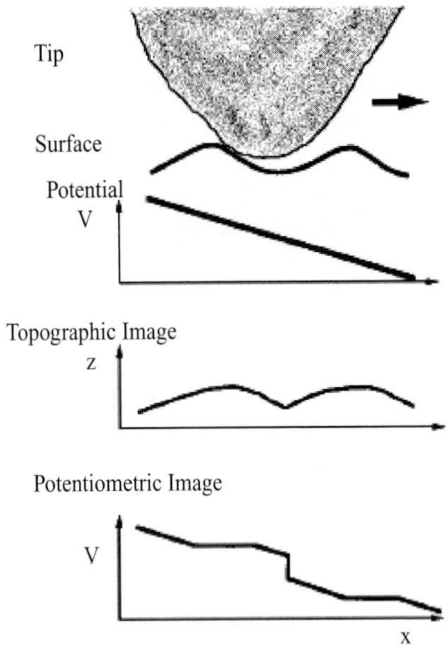

FIGURE 2. Illustration of a possible source of artifacts in STP due to the finite size of the tip. Reprinted with permission from A. D. Kent et al. [9]. Copyright 1990, AVS The Science & Technology Society.

studying Bi films with thicknesses between 20 and 30 Å on InP-based multilayers, Briner et al. [8] reported current densities up to 8×10^6 A/cm^2, avoiding sample heating since the total current was under 0.5 A.

One additional sample requirement should be noted. As with topographical STM measurements, the sample should be flat in comparison to the tip radius. Otherwise, the tip tunneling will abruptly jump from one area on the surface to another during scanning. This effect is well known in topographical studies and leads to a loss of detail or resolution in the image. In potentiometric measurements, the opposite occurs. Instead of loss of detail, new, unwanted features are added to the image, specifically abrupt discontinuities in the electrical potential. The mechanism for this is shown in Figure 2, with the top panel illustrating a geometry where the tunneling location will jump and the middle and bottom panels showing the results in topographic and potentiometric scans. Such discontinuous voltage jumps were observed in early STP studies of evaporated Au films [9] and were the focus of a publication by Pelz [10] in the same year.

2.2 Distinguishing Topography and Voltage

A key to scanning tunneling potentiometry is discrimination between the surface topography and the lateral voltage applied along the surface of the sample. The only parameter measured by STM is the current through the tip, which is a function of both the sample-tip separation and the voltage difference. A simple measurement of tip current cannot distinguish between changes in the sample distance (topography) and voltage (potentiometry). An abrupt change in the tunneling current in a scan could imply either the presence of a physical step or a voltage step, for example, at a grain boundary. In principle, topography and voltage can be distinguished by analyzing the dependence of the tunneling current on the tip-surface separation; in a simple model, tunneling current has a contribution exponential in the tip-surface separation but only proportional to the voltage difference. In practice this approach is too time consuming and complicated, and the accuracy is dependent on the electronic density of states in the tip and surface.

Three approaches have succeeded in separating signals from topography and potentiometry. The first, used in the original experiments by Muralt et al. [7,11], observes tunneling current at two frequencies which can be distinguished using lock-in technology. For simplicity, one of these tunneling currents is taken at zero frequency, i.e., dc. With this approach the topography and potentiometry signals are measured simultaneously. The second approach, introduced by Feenstra and coworkers [12], uses an interrupted feedback, first measuring the surface topography, and then retracing that topography with the feedback off to determine the potentiometry. The third approach has found limited application, but proof-of-principle has been demonstrated by Möller et al. that tunneling noise can be used as feedback for surface topography, leaving tip voltage to determine potentiometry. The following sections will discuss these three approaches in more detail.

2.2.1 Simultaneous ac and dc Feedback

A rather elegant solution to distinguish topography and potentiometry while measuring only a single parameter, tunneling current, was introduced by Muralt and Pohl in 1986 and improved later by several groups. The idea is to separate tunneling current due to topography from potentiometry by using ac current to control the tip-surface separation and dc current to identify the voltage drop across the sample. The two can be separated electronically in the feedback loop using a lock-in amplifier. A schematic of the apparatus used at the Indian Institute of Science [13–15] for simultaneous ac and dc feedback is shown in Figure 3. An ac bias, V_{ac}, is applied to the entire sample (actually both contacts) while a dc bias, V_{dc}, is applied across the sample surface. Both voltages produce tunneling current through the tip; however, the lock-in amplifier on the right isolates the ac signal to provide height feedback to the STM Z piezo and topographical information. The tunneling signal is simultaneously fed to an integrator, where the ac component is averaged away, leaving the dc signal, which represents the tunneling due to the local potential due to the lateral dc bias. The integrator serves two other functions. The first is to feed back a dc voltage to the sample that nulls the dc tunneling current. The voltage is applied to both ends of the sample (through both adding operational amplifiers in Figure 3) so the dc voltage across the sample remains V_{dc}. The second function of the integrator is to report the voltage required to null the dc tunneling current; this is the local potentiometry voltage below the tip. This approach, where the tunneling current is nulled by adjusting the potential, is reminiscent of an impedance bridge and provides high sensitivity to small variations.

Simultaneous ac and dc feedback has clear advantages of simplicity and of a constant tunneling resistance. In practice, however, the two signals cannot be

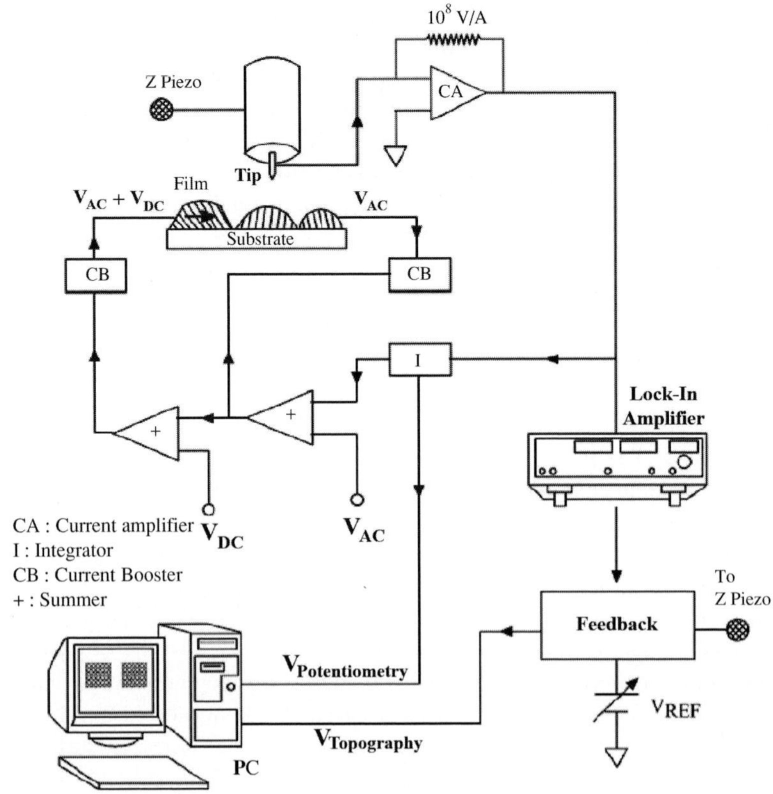

FIGURE 3. Schematic of STP using simultaneous ac and dc feedback as implemented at the Indian Institute of Science. Reprinted with permission from G. Ramaswamy et al. [15]. Copyright 1998 Springer-Verlag.

entirely separated, which may limit the voltage sensitivity. Kirtley and Washburn [16] and also Pelz and Koch [17] have pointed out that when scanning, the tunneling resistance undergoes large, broadband fluctuations because of finite gain and instability of the feedback electronics. These fluctuations will appear as broadband current noise in the tunneling and, although generated by the ac topography signal, will mix into the low-frequency signal interpreted as surface potential. Ramaswamy et al. [15] found that current noise was a practical limitation in their voltage measurements. However, with "somewhat critical adjustments of different time constants" in the lock-in and integrator circuits, which had to be different by a factor of 5–7, they were able to resolve chemical potential to 100 μeV, finally achieving the limit of their 16-bit analog-to-digital converter.

2.2.2 Interrupted Feedback

The interrupted-feedback approach to separating topographic and potential signals in STP has its origins in the "sample and hold" technique of Feenstra and coworkers [12]. The "sample and hold" method was developed to measure *I-V* curves accurately in scanning tunneling spectroscopy. First topology, or tip height for a single point, is determined in the usual way, with the feedback loop to the z piezo active. The feedback loop is then interrupted and the tip (or sample) bias used for topographic scanning switched off. With the tip held at a fixed position, *I-V* characteristics are quickly measured before drift becomes significant.

Adaptation of the "sample and hold" method to potentiometry is straightforward and was adopted by Kirtley et al. in 1988 [16]. Again the main distinction in STP is a voltage applied laterally along the sample surface. To cycle the voltage between the sample and tip, the tunneling bias is applied as a square-wave. In the portion of the cycle when the sample-tip voltage is high, the feedback to the z piezo is active and the surface topography determined. When the square-wave voltage is zero, the feedback loop is interrupted and the tip position fixed. The remaining tunneling current is dependent on the local potential of the surface due to the laterally applied voltage and can be used to create a potential map. As described, the local potential of the surface should be small compared to the square-wave voltage so it contributes little to the topographic image. This is readily accomplished by using a bridge, so that the potential across the surface remains arbitrarily large, but the local potential under the tip is close to zero, but still dependent on lateral position. The inset in Figure 4 illustrates this bridge, with V_s the lateral voltage for potentiometry and the tip bias voltage V_b referenced to a fraction of V_s so that the local voltage under the tip is close to zero.

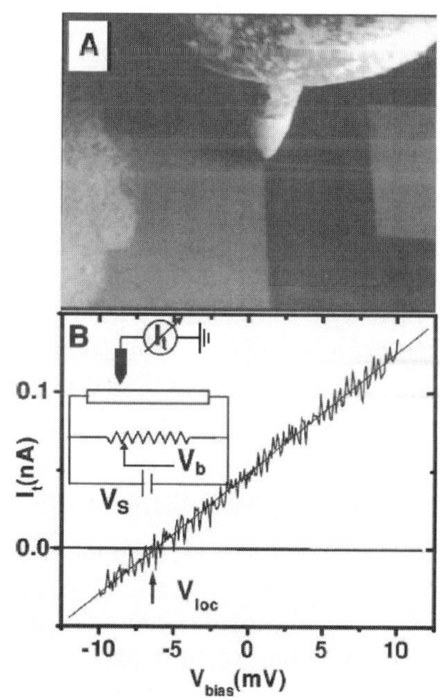

FIGURE 4. (A) Scanning electron microscope image showing the STM tip on a 20-μm patterned line. (B) Typical *I-V* curve showing the interpolation of the local potential V_{loc} value. The inset shows the circuitry used to reduce the voltage under the tip to near zero. Reprinted with permission from B. Grevin et al. [19]. Copyright 2000 by the American Physical Society.

The interrupted feedback technique was further refined in the middle of the 1990's by the IBM group of Feenstra [8,18]. This advance was made practical by faster data acquisition and computers. These groups realized that more precise measurements of the local voltage could be obtained by measuring complete *I-V* curves rather than simply measuring the residual tunneling current when the feedback is interrupted. Panel B of Figure 4 shows a typical *I-V* curve for potentiometry. The curve does not pass through the origin, since the local voltage under the tip is not zero. The x-intercept, where the measured tunneling current is zero, corresponds to the local voltage. In Figure 4 this point is labeled V_{loc}. Noise in the data requires a fit to the data to extract the voltage value and is a major source of error. Grevin et al. [19] averaged 50 spectra to find the zero current value. The width of the *I-V* scan and nature of the fit varies widely in different applications, ranging from ±5 mV with a linear fit in the work of Kent on $YBa_2Cu_3O_{7-x}$ superconductors [9] to ±650 mV and a third-order polynomial fit in a study of Bi films by Briner et al. [8]. In 2002, Feenstra's group extended this approach to include semiconductor surfaces, which have a gap near zero voltage, by searching for a constant, but nonzero current value [20].

The interrupted feedback technique has two distinct advantages over the earlier simultaneous ac and dc feedback. First, there are no requirements for a lock-in amplifier or integrator and no need to balance frequencies and filters carefully. Second, voltage measurements can be more precise due to a better signal-to-noise ratio. The dominant noise-limiting measurement is in the tunneling resistance, which increases with the tunneling voltage. The interrupted-feedback approach allows topography to be determined at a relatively high voltage, with greater noise, while the potential is measured at low voltage and low noise. Because of this, Kirtley et al. [16] claim they achieved signal-to-noise ratios that were ten times better than the earlier work with simultaneous dc and ac feedback.

Interrupted feedback is not without its faults. One notable problem inherent to the procedure is that the topography is not measured at a constant tunneling resistance and error will be introduced by inconsistencies in the tip height. Although the tunneling current and sample bias are held constant to determine the topography, local variations in the surface potential change the tunneling resistance and therefore the tip-sample separation. For low-resolution experiments distortions will not be significant, since the tunneling resistance depends exponentially on the tip-sample separation. However, the goal of atomic height resolution over an area with a significant voltage drop may not be achievable due to a distortion of several angstroms caused by variations in the tunneling resistance [16].

2.2.3 Scanning Noise Potentiometry

In 1991, Möller et al. [21] introduced a variation of scanning tunneling microscopy that they titled scanning noise potentiometry. The basis of this technique is that a voltage is not needed between the tip and sample to produce a tunneling current. Even at zero bias, thermal fluctuations, known as Johnson noise, produce measurable current between the tip and sample, although there is no net current. The key is that the tunneling resistance for these fluctuations is exponentially related to the tip-surface distance and can be used to provide feedback for topographical mapping. As the tip is scanned across the surface, any dc current observed represents a local voltage on the surface. A feedback loop consisting of an integrator and a voltage divider can be introduced to maintain the dc tunneling current at zero; the tip voltage required to null the current provides a value for the local surface voltage. Möller et al. [21] estimated the accuracy of the local potential measurements on polycrystalline silver to be about 1 μV.

The foundation of scanning noise microscopy was explored more thoroughly by Koslowski and Baur in 1995 [22]. On a practical side, their publication lists several critical adjustments to the feedback circuit (lock-in amplifier settings) to avoid artifacts in the potential map. Their research also discovered that the thermal noise current does not follow the Nyquist formula, which predicts that the square of the noise current is inversely proportional to the tunneling junction resistance [23]. This leaves unresolved the exact nature of the noise in the technique.

It may be worth noting that neither of these publications actually applied this technique to potentiometry of a laterally applied voltage and instead measured only variations in chemical potential of unbiased samples. However, the approach has been shown to map local potentials and its extension to STP is simply a matter of applying a lateral voltage. In a real sense, scanning noise microscopy is not very different from simultaneous ac and dc feedback. With noise microscopy, the ac signal is simply measured without a tip-sample bias and is consequently not periodic. The very low voltages involved do make it more susceptible to stray capacitances, for example, between the microscope leads and the sample, leading to cross-talk between topography and potential. As with the simultaneous ac and dc feedback approach, a lock-in technique can help suppress the interference caused by stray capacitance [22].

3 Applications of Scanning Tunneling Potentiometry

Scanning tunneling potentiometry has been applied to a surprisingly diverse set of phenomena. The technique is continuously evolving, so that many publications begin with descriptions of improvements in the approach. Yet the authors have also been innovative in the application of STP to a wide variety of materials and physical responses. This section will review and categorize some of these applications.

3.1 Grain Boundaries

A natural and most common application of STP has been to measure electrical transport across grain boundaries. The original work introducing STP by Muralt and Pohl was a study of granular Au on a SiO_2 substrate and falls into this category [7]. Early work on $Au_{60}Pd_{40}$ alloy films by Kirtley et al. [16] is shown in Figure 5. The potentiometric images on the right share many, but not all, of the features of the topography shown on the left. This is interpreted as a fairly constant voltage

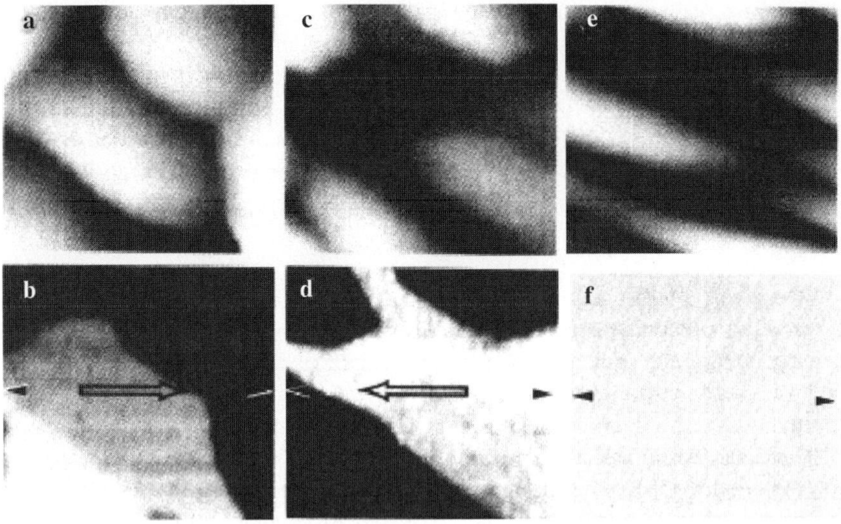

FIGURE 5. Topographic **(top)** and potentiometric **(bottom)** images of grain boundaries in a $Au_{60}Pd_{40}$ film under fields of 85, −85, and 2 V/cm. Potentiometric images have a full vertical scale of 450μV. Images are 24 × 25 nm^2. Reprinted figure with permission from J.R. Kirtley et al. [16]. Copyright 1988 by the American Physical Society.

within grains and abrupt voltage drops at grain boundaries. This pattern, with small voltage contrast within grains and large contrast at boundaries, has been observed in a wide variety of materials. A typical profile of the voltage drop at a grain boundary is shown in Figure 6, reproduced from a study of Au grains by Schneider et al. [24]. The dashed-dotted line represents the average voltage drop generated in the region by the macroscopic field. A much more abrupt voltage change of 25 μV is observed over a 2-nm distance at a grain boundary. As a cautionary reminder of complications mentioned above in the discussion of sample conditions, voltage drops observed in any region where the topography changes may be exaggerated due to tip effects.

One great feature of STP is that it provides a complete two-dimensional map of the local voltages. From the slope at each point a map of the electric field in the surface can be computed. Ramaswamy et al. have made such a map for the region around a grain boundary in a polycrystalline Pt film that is reproduced in Figure 7 [15]. Corresponding topography images are interpreted to show that the region where the electric field is large is a grain boundary and that the electric field lines are perpendicular to the boundary. From this map and knowledge of the mean free path of the electrons, the authors have determined the reflection coefficient of electrons at a surface grain boundary to range between 0.5 and 0.7. This value is lower than an average from bulk resistivity data, which the authors attribute either to significant scattering apart from grain boundaries or the reduced dimensionality of the surface.

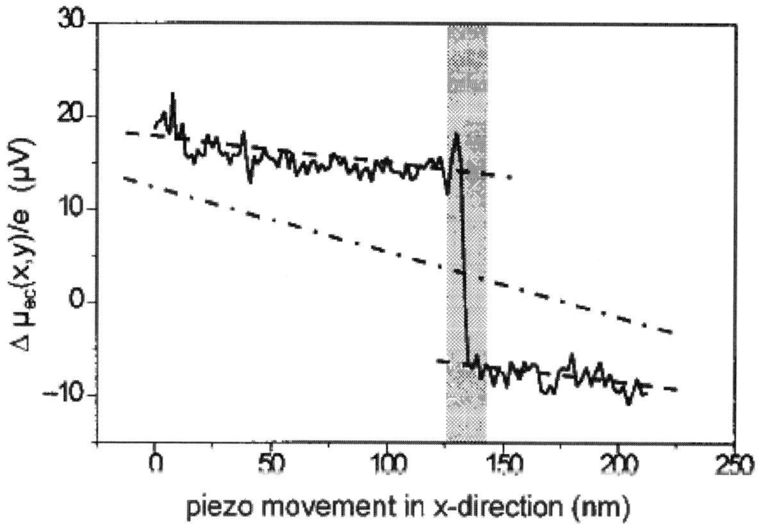

FIGURE 6. Data cross section through a grain boundary of a polycrystalline Au film. The dashed-dotted line represents the average voltage drop from the macroscopic field. At the boundary there is a 250-mV drop within the 2-nm experimental resolution. Reprinted with permission from M. A. Schneider et al. [24]. Copyright 1996, American Institute of Physics.

STP has been applied to grain boundaries in carbon resistors [25] and films of carbon [26] and AuPd [17]. Scanning noise potentiometry has been used to examine grain boundaries in silver [21]. More recently, renewed interest in the colossal magnetoresistance (CMR) in manganites has led to a several STP studies of perovskite $La_xCa_{1-x}MnO_3$ [14,27,28] and $La_xSr_{1-x}MnO_3$ [19,29] films. An

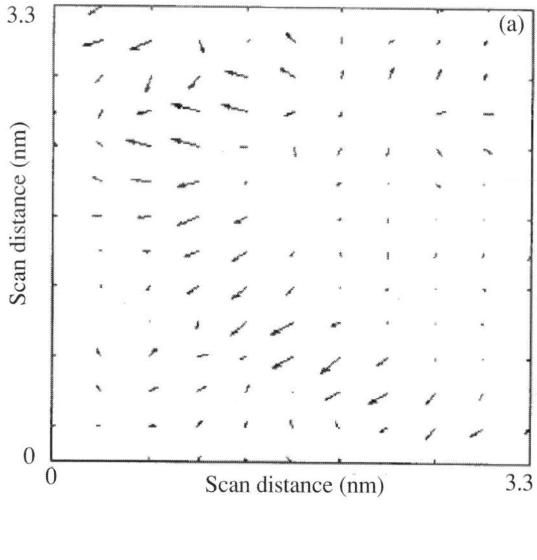

FIGURE 7. A map of the electric field evaluated in a region around a grain boundary. Direction and size of the arrows represents that of the field. Reprinted with permission from G. Ramaswamy et al. [15]. Copyright 1998 Springer-Verlag.

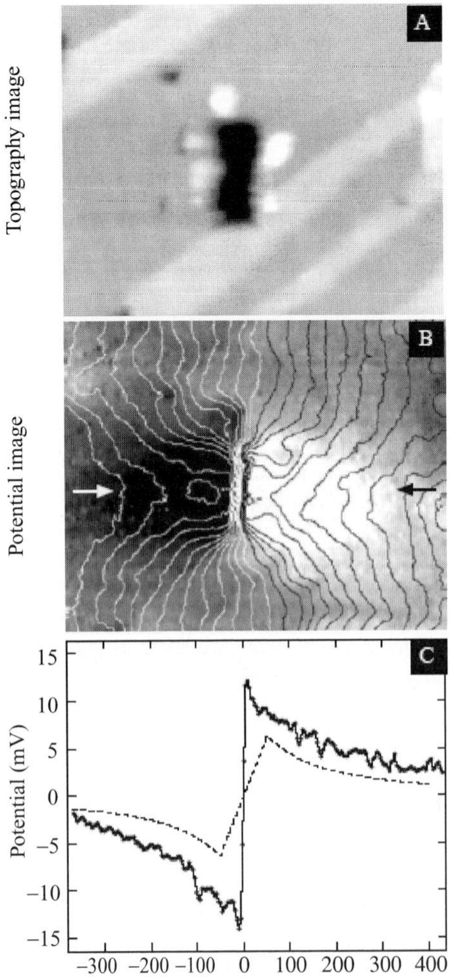

FIGURE 8. (A) The 64 × 80-nm^2 STM topograph of a 3-nm-thick Bi film with a 2.4-nm-deep void. (B) The reduced STP potential after subtraction of a linear back-ground. Contours indicate STP potential before background subtraction. (C) A cross-sectional cut along the arrows in B (solid line) and computed curve for purely diffusive transport (dashed line). Reprinted with permission from Feenstra et al. [37]. Copyright 1998, Elsevier.

example of STP images from a CMR film, $La_{0.76}Ca_{0.33}MnO_3$, is shown in Figure 8. By examining this material Paranjape et al. [27] found that annealing increased grain sizes, but also boosted grain boundary resistance, resulting in larger CMR values.

3.2 Phase Separation

Much of the controversy surrounding the nature of CMR in manganites has concerned the size and role of phase separation. From a theoretical standpoint, Dagotto et al. [30] have argued that phase separation is fundamental to CMR behavior. This viewpoint was given strong experimental support from STM experiments on $La_{0.7}Ca_{0.3}MnO_3$ by Fäth et al. [31]. Phase separation is a natural question to address with STP, which combines both spatial resolution and electronic transport. In 2002, the University of Geneva group examined epitaxial films of a similar CMR material, $La_{0.7}Sr_{0.3}MnO_3$ grown in situ on $SrTiO_3$ [32]. Unlike previous workers, they found no evidence of phase separation on the same mesoscopic scale of 100 nm. This does not rule out separation on a much smaller length scale, but from STP data the authors were able to argue that mesoscopic phase separation was not shared by all CMR materials.

3.3 Superconductivity

In addition to the manganites, complex oxides include the high-temperature superconducting cuprates; and, surprisingly, these too have been the subject of STP studies. High-temperature superconductivity holds such high scientific and technological interest that almost every technique has been applied. The surprise is that a technique for measuring local potential drops has been successfully applied to a material that by definition should have none.

The first STP studies on cuprites, published by Kent et al. [33], were on $YBa_2Cu_3O_{7-x}$ and $Y_{1-x}Pr_xBa_2Cu_3O_7$ compounds above the critical temperature where the materials remained resistive. Results showed a connection between a distribution of insulating inhomogeneities that were responsible for the high normal-state resistivities observed in these films.

To explore the nature of the superconducting state, in a later paper Kent et al. [9] added a 9.5-nm Au film on top of a 400-nm-thick $YBa_2Cu_3O_{7-x}$ film grown epitaxially on a $SrTiO_3$ substrate. The Au film remained resistive below the superconducting transition, allowing application of STP. At 77 K, below the superconducting transition, a potentiometric plot without structure was observed. However, in the center of the transition, at a temperature of 85 K, a step in the potential was observed in one region of the film. This was correlated with a topographical feature and interpreted as a normal region in the $YBa_2Cu_3O_{7-x}$ film. Their approach demonstrated the ability of STP to study normal-superconducting phase transitions.

3.4 Electromigration

Resistivity to an electrical current slows the electron flow, but also creates an electrostatic field and a drag on the scatterers that can lead to atomic migration. Electromigration is of great interest to the semiconductor industry, which would prefer their devices to remain as they were made. The electromigration kinetics of gold on a carbon film were explored using STM and STP by Besold et al. in 1994 [34]. The geometry of STP experiments allows quite high current densities; in this experiment a current density of 10^4 A/cm^2 was applied to a 1–2-nm Au film for over 200 hours. The Au atoms migrated differently in two time regimes. After the first 60 hours, the Au coalesced to form more stable (111) facets. The key result, however, was that the initial movement of the Au was toward the cathode. This meant the electrostatic force was greater than that of the momentum drag of the electrons. In this sense the kinetics of Au migration was opposite that of Al, another common electrode material for semiconductor connections.

3.5 Defects

The nature of electrical resistivity on an atomic scale, where the physical interactions are controlled by quantum mechanics, can be studied in real space by STP images at defects. In a classic paper in 1957, Landauer described the influence of scattering centers in metallic films in a novel way which still has great relevance to our views of transport today [35]. At a scattering center for ballistic transport, Landauer predicted a pile-up of charge that leads to the formation of a dipole in the electrochemical potential. Dipoles from scattering centers then contribute to the electrical field driving the current. The size of the dipoles depends on the macroscopic field. However, the magnitude of the dipoles will differ from a conventional application of Ohm's law due to the quantum mechanical nature of the interactions. A quantum mechanical analysis of defect scattering indicates that the charge density in the pile-up contains an oscillatory perturbation due to interference. This perturbation has the same origin as the Friedel oscillations screening a charged impurity in a free-electron gas [36].

This residual resistivity dipole at a defect was the focus of research by Briner, Feenstra, Chin, and Woodall [8,18,37]. Their central data was taken with STP and is shown in Figure 8. The top panel in the figure shows a STM topograph of a void in a Bi film grown on an InP-based multilayer. The central image shows the local potential measured with STP for a current applied from left to right, with an overlay of contours with equal potential. Clearly, the black-and-white contrast represents a residual resistivity dipole structure due to pile-up of electrons. Figure 8(c) shows a cross-sectional cut of the potential through the center of the dipole and a curve computed for diffusive transport. The observed dipole is significantly larger than that expected for diffusive transport, confirming the existence of a residual resistivity dipole associated with Landauer's theory. No oscillations were observed in the electron-density profile of the dipole; however, the experiments were done at room temperature, where thermal noise is probably large compared to the size of the interference effects.

A nice comparison of the electric field maps in the vicinity of defects and grain boundaries was published by Ramaswamy and Raychaudhuri in 1999 [13]. They mapped local transport field variations in a polycrystalline Pt film evaporated on glass using STP. Figure 9 shows the field maps determined for (a) a grain boundary, (b) a triple point, and (c) a void in the film. They discovered very poor electronic screening near these defects, leading to large local fields. The fields were smallest at the grain boundary, then increased at the triple point and were greatest at the void.

3.6 Thermoelectric Voltage

Dissimilar metals connected in a temperature gradient generate a thermoelectric voltage. A common application is measurement of the temperature relative to

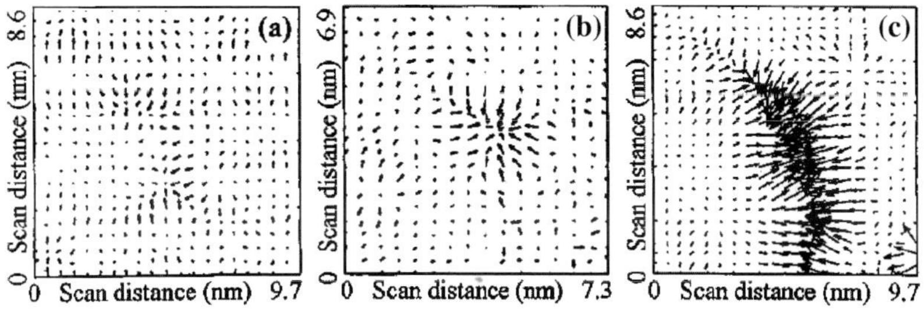

FIGURE 9. Local transport fields for (a) a grain boundary, (b) a triple point, and (c) a void in a Pt film. Reprinted with permission from Geetha Ramaswamy and A. K. Raychaudhuri [13]. Copyright 1999, American Institute of Physics.

ambient with a thermocouple. In a related manner, it has been known since 1940 that a voltage will be generated across a tunneling junction even with similar metals at different temperatures [38]. This voltage can be several μV/K and is sufficient to measure with potentiometry.

The idea of mapping the thermoelectric voltage with a scanning probe was first presented by Xu et al. in 1994 [39]. Experimental realization of the approach took six more years. Hoffman et al. [40] then used a 40-mW laser diode to heat a STM tip about 5 K and measured the thermoelectric voltage generated across the surface separating the signal from topography with simultaneous feedback. For Au films on mica, they were able to resolve individual Au atoms and steps by thermoelectric voltage. With the laser off they saw no signal. Adjacent to steps, the influence of Friedel oscillations was observed as a standing wave. From a detailed analysis of the oscillations the authors concluded that the thermoelectric signal was complementary to conventional dI/dV measurements in its sensitivity to the local density of states. They next added Ag to the Au and were able to distinguish the two elements by thermoelectric signal. Finally, a thermoelectric signal was demonstrated on a Si(111)-(7x7) surface.

3.7 Photoexcitation

Light shining on a surface can excite charge carriers through the photoelectric effect. For a semiconducting material, the charge is not immediately dissipated since there are no states available at the Fermi energy. Instead, the induced charge produces a local photovoltage. Cahill and Hamers used STP to map this photovoltage on Si(001)-(2x1) illuminated with a 10-mW He-Ne laser [41]. By modulating the laser with an electro-optic crystal and crossed polarizer, they were able to detect photovoltages under the tunneling tip using a lock-in amplifier. Results revealed that the observed voltage depended both on the intensity of the laser light and on the tunneling current in the STP. Both combine to produce band bending at the semiconductor surface. Cahill and Hamers used this information to map the local band bending across the surface, noting increases of nearly 50% near atomic scale defects.

3.8 Schotkky Barriers

Band bending also plays a prominent role in the nature of metal-semiconductor interfaces. A Schotkky barrier forms as the Fermi energy of the metal aligns with the energy of the majority carrier band edge of the semiconductor. This rectifying barrier for electrical conduction is an essential component in the successful design and operation of any semiconductor device. Interest in the height and shape of the local Schotkky barrier motivated Tanimoto and Arai to apply STP to the junction between an InGaAs film epitaxially grown on InP(100) and non-alloyed Pt/Au contacts [42]. The *I-V* characteristics of the interfaces showed linear ohmic behavior; however STP showed quite different voltage drops at the cathode and anode. With 2 V across the device, the voltage drop at the cathode was 0.34 V over 30 nm, while at the anode the voltage drop was only 0.01 V over the same region. Furthermore, the voltage drop at the anode was proportional to the total device voltage, but that at the cathode increased rapidly when the device voltage was above 4 V. These characteristics were attributed to the Schotkky barrier characteristics of the metal-semiconductor interface. Near the cathode and anode, potential drops reflected the carrier accumulation and/or depletion due to the carrier velocity dependence on the electric field.

3.9 p-n *Junctions*

One of the most basic interfaces in semiconductor electronics is the *p-n* junction. Several groups have examined these *p-n* junctions using STP. Some of the first STP experiments were performed on the interfaces between AlGaAs sandwiched between *n*- and *p*-doped GaAs in a laser diode, by the originators of the technique, Muralt et al. [11]. The edge of the interface was exposed by cleaving in ultrahigh vacuum, which formed a very flat surface. The authors identified six different regions near the active junction based on their potentiometric and *I-V* characteristics. They observed (1) *n*-type AlGaAs material, (2) a 15-nm depletion region just before the active interface, (3) a potential "waterfall" and (4) a 100-nm "plateau" in the active GaAs, (5) a drop to the *p*-doped GaAs, and (6) a region of strong *I-V* asymmetry in the GaAs away from the interface. The position of the active region was monitored as a function of lateral voltage, with changes around 10 nm for each 0.1 V.

Figure 10 reproduces line scans across a *p-n* junction formed between boron and phosphorus-doped Si from Yu et al. [43]. The figure shows the threshold voltage, $V_{turn\text{-}on}$, required for the onset of tunneling for three values of the *p-n* junction voltage, V_{pn}. $V_{turn\text{-}on}$ increases from the left to the right as the tip moves from *n*-type to *p*-type Si. The dashed line represents calculations based on a simple model of conduction and valence band positions solved by application of Poisson's equation in one-dimension and then scaled to fit the data. The model accurately reproduces the STP profiles, suggesting the band positions can be measured in this way.

Perhaps the most straightforward application of STP to *p-n* junctions was performed by Hersam et al. [44]. This group proposed exploiting STP's spatial

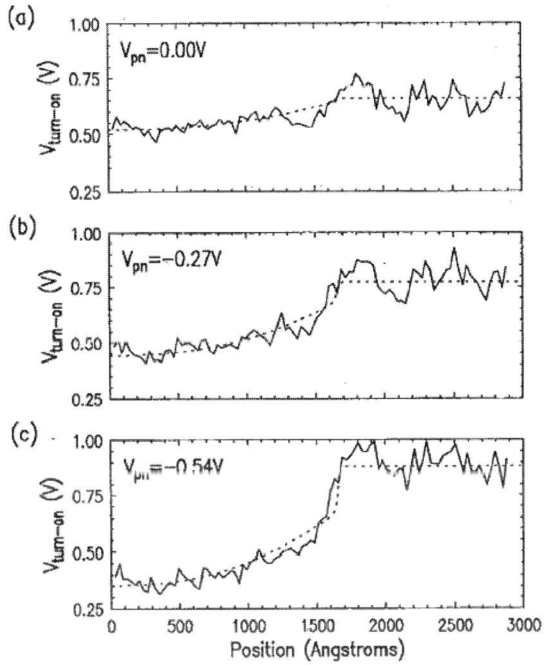

FIGURE 10. Voltage profiles from STP across a Si *p-n* junction for three biases (solid) and calculated profiles (dashed). Reprinted with permission from E. T. Yu et al. [43]. Copyright 1992, American Institute of Physics.

resolution simply to locate the position of a *p-n* junction in Si as an alignment marker for nanoscale fabrication. The measured potential across the junction differed by two orders of magnitude and revealed a depletion region of about 20 nm. This provided a clear, well-defined location marker.

4 Outlook

The 20th anniversary of the invention of scanning tunneling potentiometry is rapidly approaching. The technique has seen a number of variations and has been proven itself on a wide variety of materials and physical responses. And yet it is fair to say that STP is not a mainstream technique. One barrier is the modifications to conventional STM equipment needed to provide a properly balanced lateral voltage across the sample surface, with associated problems of contacts, and the additional apparatus and/or circuitry to differentiate topographical and potentiometric data. Second, in a review in 1996, Neddermeyer wrote that "the difficulty of surface preparation is probably one of the reasons for a comparative lack of relevant studies." He also reported that "one complication in the data analysis was that the lateral position of characteristic details in the potential distribution was dependent on the voltage" [2].

With critiques of apparatus, samples, and analysis, one might conclude that the outlook for STP is bleak. However, commensurate with our abilities to fabricate and control materials the demand for characterization of electronic transport on a nanometer scale is now greater than ever. One alternative to STP is scanning surface potential microscopy, also known as Kelvin probe microscopy, which is closely related, but uses force feedback to determine the topography instead of electronic tunneling. For measurements at an atomic scale, STP still appears to offer better promise; both force- and tunneling-based approaches share the obstacles of additional apparatus, sample quality, and data interpretation, but tunneling generally provides much better lateral resolution.

Clearly, improvements in STP need to continue to meet the demands of science and technology for an atomic scale understanding of current flow. Electronics, such as lock-in quality, and sample preparation are already far superior to those available at the inception of the technique. Little has been written about the theory of STP needed for a careful analysis of results. This is in part due to the complexity of combining electrical transport with the geometry of a tunneling microscope but also because the large length scale of many STP experiments is outside the range of first principles calculations. At smaller length scales, intriguing effects from quantum mechanics become important. In one of the few publications on the subject, Chu and Sorbello [45] have calculated interference effects between the tip and a surface defect. They found that the phase coherence between the electron wave reflected from a defect and the incident electron wave from the tip leads to Friedel-like oscillations in both the local transport field driving current and the observed tunneling voltage.

In summary, 20 years have shown continuous improvements in scanning tunneling potentiometry. Several distinct approaches have been taken to distinguish topography from potentiometry, most notably simultaneous ac and dc, interrupted, and noise feedbacks. The technique has been successfully applied to a surprisingly diverse set of physical responses on metals, complex oxides, and semiconductors. The goal of mapping true atomic-scale conductivity has been elusive, but STP appears to offer the greatest promise.

Acknowledgments. Many thanks are due to the authors who have allowed their work to be reproduced in this review. I am also grateful for the support of my colleagues in the Low Dimensional Materials By Design Group, a part of the Condensed Matter Sciences Division at the Oak Ridge National Laboratory. Oak Ridge National Laboratory is managed by UT-Battelle, LLC, for the U.S. Department of Energy under contract number DE-AC05-00OR22725.

References

1. G. Binnig and H. Rohrer, Helvetica Physica Acta **55** (6), 726 (1982); G. Binnig, H. Rohrer, Ch. Gerber, and W. Weibel, Phys. Rev. Lett. (USA) **49** (1), 57 (1982).
2. H. Neddermeyer, Reports on Progress in Physics **59** (6), 701 (1996).
3. F. M. Chua, Y. Kuk, and P. J. Silverman, Phys. Rev. Lett. (USA) **63** (4), 386 (1989).

4. Y. Kuk, F. M. Chua, P. J. Silverman, and J. A. Meyer, Phys. Rev. B, Condens. Matter (USA) **41** (18), 12393 (1990).
5. H. Niehus, R. Spitzl, K. Besocke, and G. Comsa, Phys. Rev. B, Condens. Matter (USA) **43** (15), 12619 (1991).
6. A. P. Baddorf, D. M. Zehner, G. Helgesen, D. Gibbs, A. R. Sandy, and S. G. J. Mochrie, Phys. Rev. B, Condens. Matter (USA) **48** (12), 9013 (1993).
7. P. Muralt and D.W. Pohl, Applied Physics Letters **48** (8), 514 (1986).
8. B. G. Briner, R. M. Feenstra, T. P. Chin, and J. M. Woodall, Phys. Rev. B, Condens. Matter (USA) **54** (8), 5283 (1996).
9. A. D. Kent, I. Maggio-Aprile, Ph. Niedermann, Ch. Renner, and O. Fischer, J. Vac. Sci. Technol. A, Vac. Surf. Films (USA) **8** (1), 459 (1990).
10. J. P. Pelz and R. H. Koch, Phys. Rev. B, Condens. Matter (USA) **41** (2), 1212 (1990).
11. P. Muralt, H. Meier, D. W. Pohl, and H. Salemink, Superlattices and Microstructures **2** (6), 519 (1986); P. Muralt, H. Meier, D.W. Pohl, and H. W. M. Salemink, Applied Physics Letters **50** (19), 1352 (1987).
12. A. P. Fein, J. R. Kirtley, and R. M. Feenstra, Review of Scientific Instruments **58** (10), 1806 (1987); R. M. Feenstra, W. A. Thompson, and A. P. Fein, Phys. Rev. Lett. (USA) **56** (6), 608 (1986).
13. G. Ramaswamy and A. K. Raychaudhuri, Applied Physics Letters **75** (13), 1982 (1999).
14. Mandar A. Paranjape, Doctor of Philosophy, Indian Institute of Science, 2004.
15. G. Ramaswamy, A. K. Raychaudhuri, K. Das Gupta, and G. Sambandamurthy, Applied Physics A (Materials Science Processing) **66** (suppl., pt. 1–2), 435 (1998).
16. J. R. Kirtley, S. Washburn, and M. J. Brady, Phys. Rev. Lett. (USA) **60** (15), 1546 (1988).
17. J. P. Pelz and R. H. Koch, Review of Scientific Instruments **60** (3), 301 (1989).
18. B. G. Briner, R. M. Feenstra, T. P. Chin, and J. M. Woodall, Semiconductor Science and Technology **11** (11S), 1575 (1996).
19. B. Grevin, I. Maggio-Aprile, A. Bentzen, L. Ranno, A. Llobet, and O. Fischer, Phys. Rev. B, Condens. Matter (USA) **62** (13), 8596 (2000).
20. Yang Dong, R. M. Feenstra, R. Hey, and K. H. Ploog, Journal of Vacuum Science & Technology B (Microelectronics and Nanometer Structures) **20** (4), 1677 (2002).
21. R. Moller, C. Baur, A. Esslinger, and P. Kurz, Journal of Vacuum Science & Technology B (Microelectronics Processing and Phenomena) **9** (2, pt.2), 609 (1991).
22. B. Koslowski and C. Baur, Journal of Applied Physics **77** (1), 28 (1995).
23. R. Landauer, Physica D **38** (1–3), 226 (1989).
24. M. A. Schneider, M. Wenderoth, A. J. Heinrich, M. A. Rosentreter, and R. G. Ulbrich, Applied Physics Letters **69** (9), 1327 (1996).
25. S. Morita, Y. Maita, and Y. Takahashi, Japanese Journal of Applied Physics, Part 2 (Letters) **28** (11), 2034 (1989).
26. J. Besold, R. Thielsch, N. Matz, C. Frenzel, R. Born, and A. Moebius, Thin Solid Films (Switzerland) **293** (1–2), 96 (1997).
27. M. Paranjape and A. K. Raychaudhuri, Solid State Communications **123** (12), 521 (2002).
28. M. Paranjape, A. K. Raychaudhuri, N. D. Mathur, and M. G. Blamire, Phys. Rev. B, Condens, Matter Mater. Phys. (USA) **67** (21), 214415 (2003).
29. J. J. Versluijs, F. Ott, and J. M. D. Coey, Applied Physics Letters **75** (8), 1152 (1999).
30. E. Dagotto, T. Hotta, and A. Moreo, Physics Reports **344** (1–3), 1 (2001).

31. M. Fath, S. Freisem, A. A. Menovsky, Y. Tomioka, J. Aarts, and J. A. Mydosh, Science **285** (5433), 1540 (1999).
32. B. Grevin, I. Maggio-Aprilc, A. Bentzen, O. Kuffer, I. Joumard, and O. Fischer, Applied Physics Letters **80** (21), 3979 (2002).
33. A. D. Kent, I. Maggio-Aprile, Ph Niedermann, and O. Fischer, Phys. Rev. B, Condens. Matter Mater. Phys. (USA) **39** (16), 12363 (1989); A. D. Kent, I. Maggio-Aprile, Ph Niedermann, Ch Renner, J. M. Triscone, M. G. Karkut, O. Brunner, L. Antognazza, and O. Fischer, Physica C: Superconductivity 162–64 (pt2), 1035 (1989); A. D. Kent, J. M. Triscone, L. Antognazza, O. Brunner, Ch Renner, Ph Niedermann, M. G. Karkut, and O. Fischer, Physica B: Condensed Matter 165–66 (2), 1503 (1990).
34. J. Besold, R. Kunze, and N. Matz, Journal of Vacuum Science & Technology B (Microelectronics and Nanometer Structures) **12** (3), 1764 (1994).
35. R. Landauer, IBM J. Res. Develop. **1**, 223 (1957).
36. W. Zwerger, L. Bonig, and K. Schonhammer, Phys. Rev. B, Condens. Matter (USA) **43** (8), 6434 (1991).
37. R. M. Feenstra and B. G. Briner, Superlattices and Microstructures **23** (3–4), 699 (1998).
38. M. Kohler, Ann. Physik **38**, 542 (1940).
39. J. Xu, B. Koslowski, R. Moller, K. Lauger, K. Dransfeld, and I. H. Wilson, Journal of Vacuum Science & Technology B (Microelectronics and Nanometer Structures) **12** (3), 2156 (1994).
40. D. Hoffmann, J. Seifritz, B. Weyers, and R. Moller, J. Electron Spectrosc. Relat. Phenom. **109** (1–2), 117 (2000).
41. D. G. Cahill and R. J. Hamers, Journal of Vacuum Science & Technology B (Microelectronics Processing and Phenomena) **9** (2), 564 (1991).
42. M. Tanimoto and K. Arai, Journal of Vacuum Science & Technology B (Microelectronics and Nanometer Structures) **12** (3), 2125 (1994).
43. E. T. Yu, M. B. Johnson, and J.-M. Halbout, Applied Physics Letters **61** (2), 201 (1992).
44. M. C. Hersam, N. P. Guisinger, and J. W. Lyding, J. Vac. Sci. Technol. A, Vac. Surf. Films (USA) **18** (4, pt. 1–2), 1349 (2000).
45. C. S. Chu and R. S. Sorbello, Phys. Rev. B, Condens. Matter (USA) **42** (8), 4928 (15).

I.2
Probing Semiconductor Technology and Devices with Scanning Spreading Resistance Microscopy

P. EYBEN, W. VANDERVORST, D. ALVAREZ, M. XU, AND M. FOUCHIER

1 Introduction to Profiling Needs for Nanoscale Semiconductor Technology

As the downscaling in semiconductor industry continues, the correct operation of devices becomes critically dependent on the precise location and activation of the dopants in two dimensions. For Si technology, Duane had already determined in 1996 that, in the 0.25-μm CMOS technology, a 10-nm decrease in the channel length was responsible for a more than 10% increase in the gate-to-drain overlap capacitance [1]. In the 65-nm devices processed today, effects of shifts in lateral position of a few nanometers or of variations in concentration of a few percent in the channel are drastically more pronounced.

This necessitates the development of quantitative tools to measure the carrier distribution with a sufficient spatial resolution and accuracy. One dimensional tools such as secondary ion mass spectroscopy (SIMS) [2] and spreading resistance probe (SRP) [3–5], or more recently carrier illumination [6], have been used successfully in the last few years to understand and calibrate the increasingly complex implantation, diffusion, and activation mechanisms involved in semiconductor processing and in particular CMOS technology processing. They have also been used to calibrate TCAD (technology computer-aided design) process simulators, which are used to generate accurate predictions of the spatial dopant distribution before the initiation of costly silicon processing. However, most recent devices are very sensitive to lateral diffusion mechanisms' influencing the effective channel length and to implantations realized within the channel (as halo and threshold voltage adjustment implants), which cannot be studied with one-dimensional techniques. They are also sensitive to the stress and stress-induced lateral diffusion. Due to the complexity of the underlying physics, TCAD software typically involves a lot of process parameters, which need to be calibrated through direct two-dimensional (2D) dopant measurements before reliable TCAD predictions can be made. Therefore new methods are required to measure 2D carrier profiling. In addition, there is a growing need for techniques able to characterize specific devices on wafers in case of failure.

Hence, the Semiconductor Industry Association (SIA) roadmap has pointed out the need for analysis techniques capable of measuring accurately the spatial (2D) electrical carrier (and dopant) profile with a high spatial resolution (2–3 nm), a high sensitivity (5%), and a large dynamic range (10^{15} to 10^{21} at/cm^3).

2 Dopant Profiling in Si with SSRM

2.1 Basic Concepts of SSRM

2.1.1 Principle

The scanning spreading resistance microscopy (SSRM) technique was conceived by Vandervorst et al. [7, 8] and implemented by De Wolf [9, 10] at IMEC starting in 1994.

Different techniques have been proposed to measure the current flowing from a conducting AFM probe to a semiconductor sample [11]. The originality of the SSRM technique is the use of a high force. As a consequence, the measured resistance is dominated by the spreading resistance and not by the contact resistance.

The SSRM concept is in fact born from the motivation to extend the capabilities of SRP by replacing the large SRP probes (1 μm) with a small probe (radius about 10 nm) mounted on an AFM system [9–12]. The motivation was to allow direct measurements on the sample cross sections (in order to avoid the beveling-related carrier-spilling effects) as well as two-dimensional carrier profiling without special test structures.

In SSRM, a hard conductive probe is scanned in contact mode across the sample, while a dc bias is applied between a back-contact on the back side of the sample and the tip (see Figure 1). The resulting current is measured using a logarithmic current amplifier providing a typical range of 10 pA to 0.1 mA.

The advent of SSRM is intimately linked with the use of silicon probes coated with doped diamond. These probes, mounted on stiff cantilevers (10 to 100 N.m), are the only ones that can survive the forces required for SSRM. In the last few years, the advent of solid diamond probes [13, 14] has allowed significant progress in SSRM.

The main characteristic of SSRM is, as previously explained, that the technique uses a high force (≥μN) to realize an intimate contact between the probe and the silicon sample such that spreading resistance dominates the contact resistance

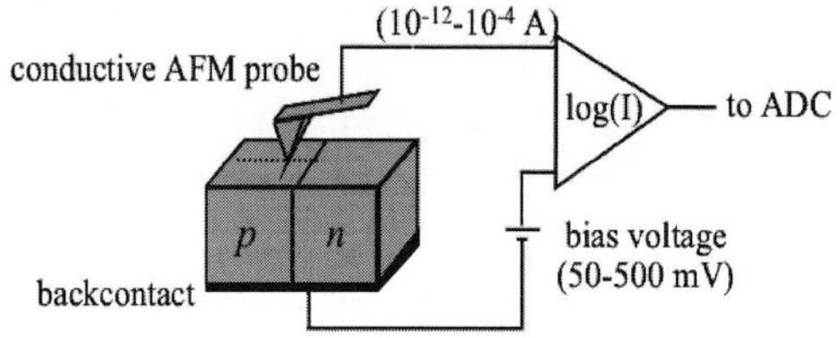

FIGURE 1. Illustration of the SSRM basic scheme.

FIGURE 2. Resistance measured as a function of load.

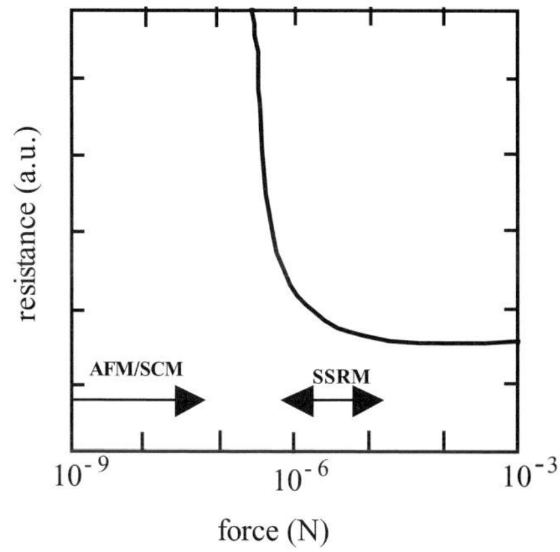

(Figure 2). Clarysse et al. have established that for such pressures, the probe punches through the silicon oxide and the underlying silicon undergoes a plastic deformation and phase transformation [15]. In analogy with SRP, the theoretical equation linking the measured resistance to the resistivity of the sample under the probe is then

$$R = \frac{\rho}{4a} \tag{1}$$

where ρ is the resistivity of the silicon sample under the probe and a is the electrical radius of the probe. This simplistic model is, however, not fully satisfactory and contains various inaccuracies and deficiencies, as the electrical properties of the probe–semiconductor contact depend on many different parameters, such as the surface states' density and their energy distribution, the exact probe shape, or the pressure. It is almost impossible to express mathematically all those non-idealities. Comparisons with calibration samples should therefore be performed, and the deviation from the ideal curve is classically expressed with a barrier resistance component $R_{barrier}$ as proposed by De Wolf et al. [16]:

$$R = \frac{\rho}{4a} + R_{barrier}(\rho) \tag{2}$$

Moreover, for non-uniformly doped samples, the resistance value at every position is no longer exclusively determined by the carrier concentration at this same place, as the spreading of the current might be influenced by neighboring conducting or insulating regions (see Figure 3). Hence, a correction factor (CF) is introduced, as proposed by De Wolf et al. [16]:

$$R = CF(a, \rho)\frac{\rho}{4a} + R_{barrier}(\rho) \tag{3}$$

Eyben et al. [17] have studied these phenomena in more detail, as discussed section 2.2.4.

However, despite a few restrictions and corrections, the SSRM technique allows a direct link to the carrier distribution information and offers the possibility of

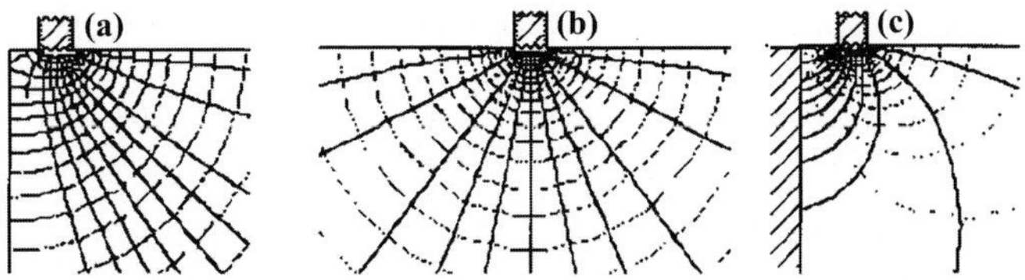

FIGURE 3. Current distribution (solid lines) and potential contour lines (dashed lines) under a SSRM tip for three extreme situations: (a) nearby isolating boundary, (b) semi-infinite uniform layer, (c) nearby conducting boundary [16].

realizing an easy, first-order quantification of the measurements [18], as will be shown in section 2.3.

Somewhat counterintuitively, but as demonstrated by Eyben et al. [19], one of the main advantages of the high-pressure SSRM technique is its unique spatial resolution (1–3 nm) and dopant gradient resolution (1–2 nm/dec) combined with an excellent signal-to-noise ratio (around 20) and adequate reproducibility (5–10%) as compared to all other low-force SPM-based techniques. The SSRM technique is also relatively robust and less sensitive to sample preparation conditions than SCM. The importance of a precise and specific SSRM sample preparation for high-resolution measurements has nevertheless been evidenced [20]. SSRM measurements may be performed on both polished and cleaved structures.

Like SRP, SSRM benefits in theory from an extremely large dynamic range (10^{14} to 10^{21} at/cm^3). In practice, this range is, however, limited in the highly doped range ($>10^{20}$ at/cm^3) by limited conductivity of the diamond probe, which is measured in series with the spreading resistance. Unlike the metallic SRP probe, the resistance of the AFM diamond probe is not negligible (about 1 kΩ). In low-doped areas ($<10^{17}$ at/cm^3), the SSRM measurements are disturbed by the presence of surface charges, as will be discussed in section 2.2.2.1. Note nevertheless that this surface charges influence is not specific to the SSRM technique but is related to the sample sectioning process also present with KPFM or SCM [21].

An important drawback of the SSRM technique is the damage to the sample and to the probe due to the high forces employed while scanning (in particular due to the presence of an important shear force). A solution named pulsed force SSRM has been proposed and implemented. More details may be found in [22].

2.1.2 Sample Preparation

Application of two-dimensional electrical characterization techniques based on AFM usually requires measurements on the cross section of the sample, making the sample preparation (cleavage and/or polishing) the first step. Although SSRM is less sensitive to the sample surface than, for instance, scanning capacitance microscopy (SCM) [23], it has been found that better reproducibility and reduced noise level can be obtained with improved sample preparation. For instance, a very flat surface enables more stable scanning and a drastic reduction of the force

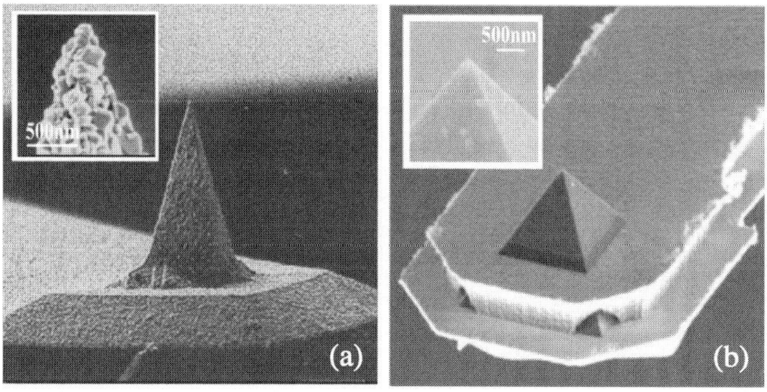

FIGURE 4. SEM pictures of diamond-coated (a) and full diamond (b) probes. Respective encapsulated zoom pictures clearly show the difference in terms of diamond grains, with the presence of large grains on diamond-coated tips that may affect the probe radius.

needed for a stable electrical contact. The latter is very important because high force is responsible for probe and sample damage and has a strong impact on the spatial resolution, probably because lower force leads to a reduced "effective" tip radius.

In our optimized procedure, chemical-mechanical polishing is not used in order to avoid rounding effects and doping-level-sensitive polishing that may affect the results. All last polishing steps are realized with Al_2O_3 lapping films with decreasing grain size, ending with a final polish using a 50-nm grain size. Aluminium oxide is preferred compared to diamond due to the fact that it is softer and reduces the presence of scratches. The typical RMS value provided with this sample preparation procedure (in the bulk and near the surface) is 0.2–0.3 nm. The use of this procedure allows stable current measurements for relatively low forces (expressed in terms of deflection set-point voltage values). Classical diamond-coated probes (stiffness about 50 N/m) typically give a stable electrical contact at about 1 V or 2 V versus 5 V with the previous procedure.

A short HF exposure (a few seconds) just before the SSRM measurement also allows a small decrease in the force used (about two times) but may introduce some topography artifacts and thus has to be used carefully.

2.1.3 Tips for SSRM

At present most probes for SSRM are based on diamond-coated silicon tips, as the use of diamond has been shown to be essential to sustain the high forces used in SSRM [24].

Basically, a thin layer (100 to 200 nm) of boron-doped diamond (3 to 3.3 ppm) is grown by CVD on top of n-doped silicon probes[1]. Those probes are commercially produced [25] (see Figure 4(a)). However, they have a number of limitations. Their

[1] P-type Si probes were produced in the past but they are no longer available.

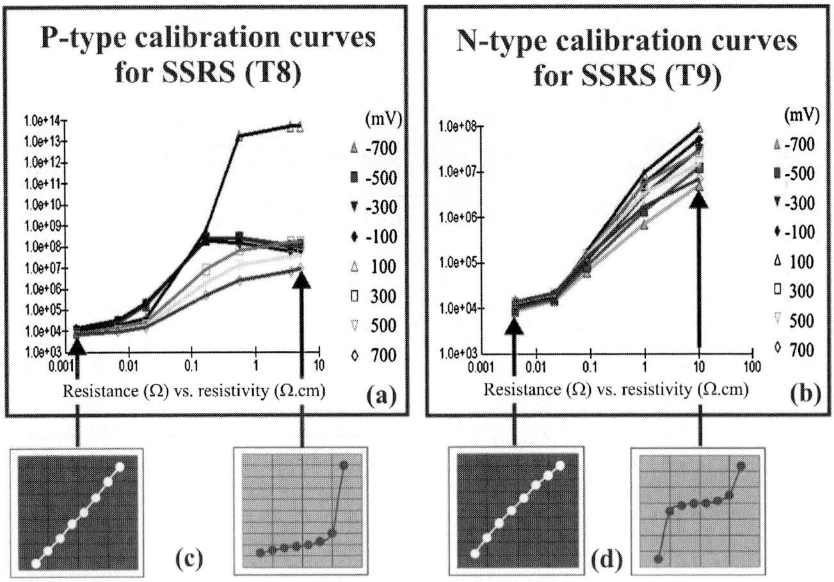

FIGURE 5. Calibration curves on the T8 (a) and T9 (b) staircase structures with a bias range between −700 and 700 mV obtained using MicroQuanti. I-V curves corresponding to extreme layers of T8 (c) and T9 (d).

resistivity appears to be too high to allow study of highly doped areas ($>10^{20}$ at/cm^3) and saturation is visible in the calibration curves (see Figure 5) as the probe resistance typically lies above 1 kΩ on Pt for a radius above 5 nm.

Due to the final diamond deposition procedure, the outer surface of the diamond-coated tips is not flat but covered with sharp diamond crystals (see Figure 4(a)). The electrical contact is realized through one of those grains. Note that, since the contact is defined by the faceted diamond crystal, the effective electrical tip radius is better than the overall radius visible in the image (i.e., 3-nm-thick buried oxide layers have already been detected with a factor 2 resistance increase), but it is usually difficult to predict. Moreover, the probability of having a multiple contact is not negligible. As the silicon tip itself has a rather high aspect ratio, cleavage of the tip due to the high shear forces exerted is frequently observed. Increasing the diamond film thickness can reduce this effect at the expense of the tip radius as the thicker coating tends to become smoother [26].

In order to overcome these limitations, low-aspect tips have been developed [24]. They are molded pyramidal tips, and their outer diamond surface (deposited first within the mold) is smooth (see Figure 4(b)). Recent improvements of this process [14], have demonstrated a significantly enhanced spatial resolution (around 1 nm) at moderate forces (1–5 μN) [27] (see section 2.4 for details).

Although they are superior in terms of spatial resolution, tip lifetime seems to be less as compared to the commercial diamond probes. Tip failures primarily occur due to small diamond grains' being removed from the solid pyramid. Clearly improved diamond deposition is required, leading to a stronger cohesion of the diamond film. Faced with these limitations in tip lifetime, it is imperative to limit the shear force, i.e., providing a very well polished surface and working at very low force. The measurements also have to be done on small scan sizes (a few square microns maximum) and at low speed to reduce the damage to the probe.

2.2 Physics of SSRM

2.2.1 Introduction

A major objective in scanning spreading resistance microscopy is to understand and (if possible) to describe quantitativaly the nature of the nanocontact between a probe and a semiconductor sample. The quantification procedure linking the macroscopic resistance measured with SSRM to the resistivity of the semiconductor under investigation is determined by the nature of the nanocontact. We will first present the electrical and mechanical properties before proceeding to a nanocontact modeling.

2.2.2 Experimental Observations

2.2.2.1 Electrical Properties of the SSRM Nanocontact

- Calibration curves
  Scanning spreading resistance spectroscopy (SSRS) measurements (see [18]) performed on p-type and n-type staircase structures (in this case samples T8 and T9 [28]) are used for probe assessment and quantification and referred to further on as "calibration curves." In Figure 5(a) and Figure 5(b), we may observe that the calibration curves are not linear and that they vary a lot with the bias (in this case between -700mV and $+700$mV). Moreover, non-monotonic calibration curves are observed on n-type material for all negative biases (applied to the sample).
- Current-voltage curves
  From local I–V curves extracted from SSRS measurements—Figure 5(c) and Figure 5(d))—one notices that the contact varies from an ohmic-like shape (in highly doped areas) to a rectifying shape (in low-doped areas). Furthermore, surface states (SS) induced by the sample preparation affect the I–V curves [29]. This effect is particularly pronounced in low-doped p-type areas (see Figure 6).
- Junction position
  Junction positions measured with SSRM on highly doped materials (source/well or drain/well implants) are in perfect agreement with the expected theoretical zero-field position, and a very good conformity with SIMS and SRP results can be obtained (see section 2.3.1.4)[2]. When studying p^+n junctions with a

[2] It should be stressed that SSRM measures the electrically active profile and thus the electrical junction (EJ) rather than the metallurgical junction (MJ) measured by SIMS.

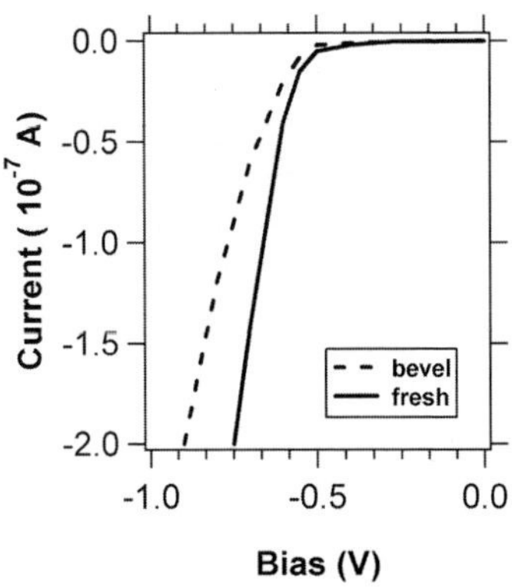

FIGURE 6. SSRM I–V curves on low-doped (10 Ω.cm) p-type material measured on fresh and beveled surfaces.

low-doped side (source/substrate and well/substrate) a large shift of the SSRM electrical junction position away from the zero-field position towards the surface is, however, observed. For n^+p junctions, a complete disappearance of the junction peak can sometimes be observed (see Figure 15 and Figure 16).

2.2.2.2 *Mechanical Properties of the SSRM Nanocontact*

- Radius of curvature
  The radius of curvature of the AFM probes (see *R* in Figure 7) can be determined using direct SEM observations or through measurements on $SrTiO_3$ samples [30]. Such measurements were performed on different generations of diamond-coated and full diamond probes dedicated for SSRM, and typical radii varying between 20 and 50 nm were measured. For most recent generations of full diamond probes, radii of approximatively 10 nm were observed for the sharpest tips.
- Spatial resolution
  The SSRM spatial resolution is difficult to determine quantitatively; however, it clearly appears to be drastically better than the radius of curvature of the probe.

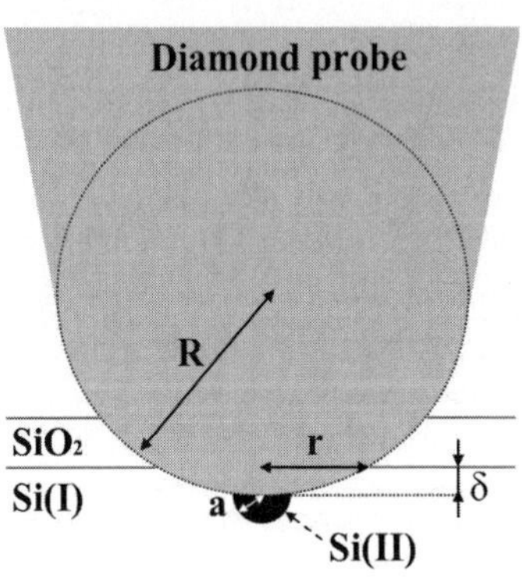

FIGURE 7. Schematic diagram of a spherical SSRM indentation in a flat silicon surface with elasto-plastic deformation and beta-tin Si(II) formation.

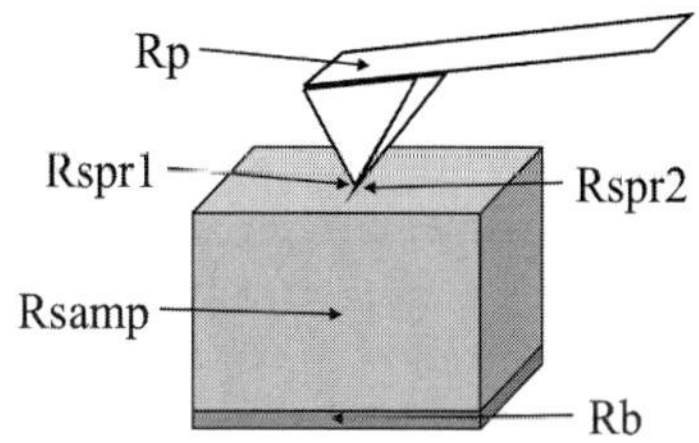

FIGURE 8. Schematic representation of the various resistance components involved in SSRM measurements.

Experiments that have been realized on dedicated buried oxide structures (see section 2.4), as well as on device structures, demonstrate an SSRM spatial resolution clearly lying in the sub-5-nm range [20, 27]. Evidence for sub-nanometer spatial resolution has even been reported (see section 2.4)

2.2.3 Classical Ohmic Nanocontact Modeling

Observations presented in the previous section show that the SSRM nanocontact is not ohmic. We, however, present in this section a classical ohmic nanocontact model as it is a simple analytical model and a reasonable first approximation in relatively highly doped materials (above 10^{18} at/cm^3).

2.2.3.1 Maxwell Ohmic Nanocontact Modeling

The total measured resistance R_{tot} in scanning spreading resistance microscopy experiments (see Figure 8) includes the resistance of the probe R_p, the probe nanocontact resistance (also called *spreading resistance*) R_{spr1}, the sample nanocontact resistance R_{spr2}, the sample resistance R_{sam},, and the back-contact resistance R_b :

$$R_{tot} = R_p + R_{spr1} + R_{spr2} + R_{samp} + R_b \tag{4}$$

The resistance of the probe (R_p) and of the sample (R_{samp}) are given by the classical Pouillet equation,

$$R = \rho.\frac{L}{A} \tag{5}$$

where ρ is the resistivity of the material (typically diamond for the tip and silicon for the sample), L its length, and A its area.

The resistance of a macroscopic circular constriction (or nanocontact) separating two homogeneous conductors is described by the Maxwell formula [31],

$$R_{spr} = R_{spr1} + R_{spr2} = \frac{\rho_1}{4a} + \frac{\rho_2}{4a} \tag{6}$$

where ρ_1, ρ_2 are the resistivities of the materials in contact and a is the radius of the contact area (also called *electrical radius*).

In appropriate conditions (see requirements on probe and back-contact in section 2.1), the dominant resistance in scanning spreading resistance microscopy measurements on silicon is nevertheless the sample nanocontact resistance (or

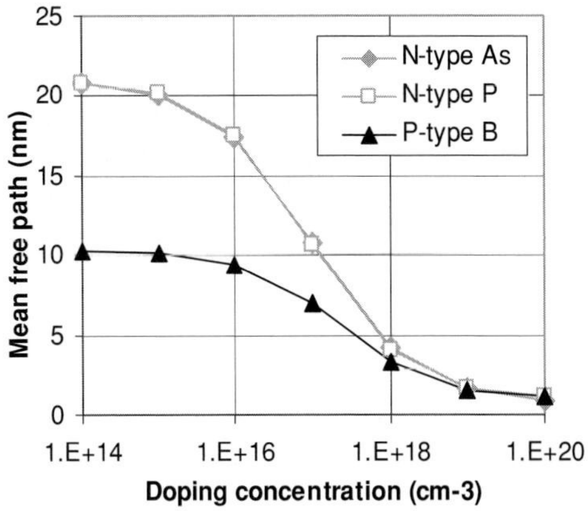

FIGURE 9. Simple calculation of the mean free path as a function of doping concentration for As-, P-, and B-doped silicon. Classical ohmic nanocontact modeling.

spreading resistance), and we obtain the classical Maxwell spreading resistance formula:

$$R = \frac{\rho_{samp}}{4a} \tag{7}$$

2.2.3.2 *Sharvin Ohmic Nanocontact Modeling*

The Maxwell formula is, however, only valid when the radius (a) is large compared to the mean free path of electrons and holes (λ). In Figure 9, the mean free path of electron and holes in silicon has been calculated for different concentrations of arsenic, phosphorus, and boron using simple mobility equations without electric field saturation [32]. The dependence of the mean free path on carrier type (electron or holes) and doping concentration (10^{14} to 10^{20} at/cm^3) is illustrated (see also [33]).

In the extreme situation of a contact area reduced to a few atoms, electron transport is ballistic, and the conductivity is described by the Landauer–Büttiker formula [34],

$$G_c = \frac{2e^2}{h} \sum_{i=1}^{N_c} T_i \tag{8}$$

where N_c is the number of conductance channels through the contact and T_i is the transmission coefficient of the i-th channel. The number of conductance channels is proportional to the ratio of contact area to Fermi wavelength λ_F. Transmission coefficients are close to unity when no scattering sites such as impurities or grain boundaries are involved at the constriction. Quantized conductance can be observed for small contacts when only a few channels are involved.

When the contact radius a is large compared to the Fermi wavelength λ_F, the resistance converges to the Sharvin equation [35,36]:

$$R_{Sharvin} = \frac{h}{2e^2} \frac{\lambda_F^2}{\pi^2 a^2} \tag{9}$$

The Sharvin resistance can be expressed in fully classical terms by substituting λ_F with ρ and using the electron mean free path λ:

$$R_{Sharvin} = \frac{4\rho\lambda}{3\pi a^2} \tag{10}$$

In Figure 10, the dependence of the Sharvin and Maxwell formulas on the contact radius is presented. We clearly observe that the Sharvin formula is dominant for small contacts (~1 nm), whereas it becomes negligible compared to the Maxwell formula for large contacts (~100 nm). In the transition region, both contributions are comparable.

The increase in the quality of the measurement conditions (better probes and lower forces used) as well as latest experiments on electrical radius (see section 2.4.1) prove that the SSRM nanocontact size lies more in the Sharvin regime (at least in low-doped areas). Note that the Sharvin resistance equation has the same property as the Maxwell spreading resistance equation: the resistance increases monotonically with the sample resistivity, providing a highly sensitive method for carrier profiling (see Figure 11). Note also that the curvature present in the ballistic calibration curves is not sufficient to give explanation to the curvature in the experimental calibration curves. This curvature is mainly linked to influence of the surface states, as will be demonstrated in section 2.2.4.

2.2.4 Schottky Contact And Surface Modeling

As discussed in section 2.2.2.1, experimental observations have revealed that the ohmic nanocontact modeling is not fully satisfactory. To understand the discrepancies between the classical ohmic theory and experimental observations, a more extended physical model of the contact and of the surface involving a Schottky-like contact with tunneling and surface states is introduced. Using this model (implemented in a device simulator [37]), the I-V curves, calibration curves, and even full SSRM profiles can be calculated.

2.2.4.1 Basic Description of the Model

The introduction of the Schottky contact is based on the observation of rectified I-V curves (see Figure 5). The necessity to introduce this Schottky contact is not surprising given the nature of the materials involved in the SSRM nanocontact (diamond-coated probe, beta-tin phase of silicon, and silicon sample), none of which are real metals. The barrier height $E_b = e\phi_m - e\chi$ has been chosen to obtain a rectifying contact ($e\phi_m < e\phi_s$) for all p-type impurity concentrations and a weakly rectifying contact ($e\phi_m > e\phi_s$) for most n-type impurity concentrations (see Figure 12).

In highly doped areas, the ohmic-like I-V curves observed experimentally (with large currents) may only be explained by the presence of a dominant tunneling current through the thin potential barrier of the contact.

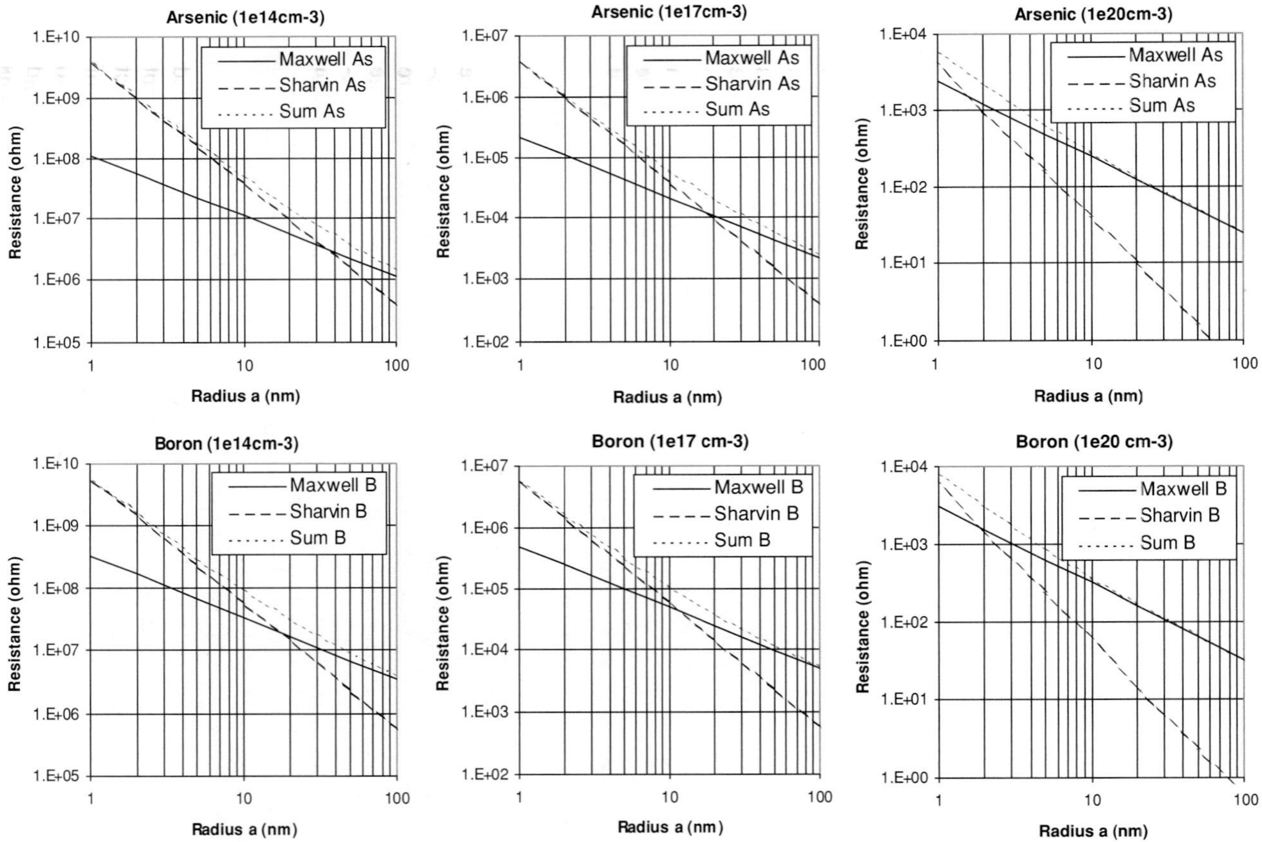

FIGURE 10. Diffusive ($R_{Maxwell}$) and ballistic ($R_{Sharvin}$) resistance as a function of the contact radius (a) for two different dopants (As and B) and three different doping levels (10^{14}, 10^{17} and 10^{20} at/cm^3). The effective resistance is bounded by the sum (R_{sum}) and the single values. The calculations assumed bulk resistivities for a contact between a perfect probe and silicon.

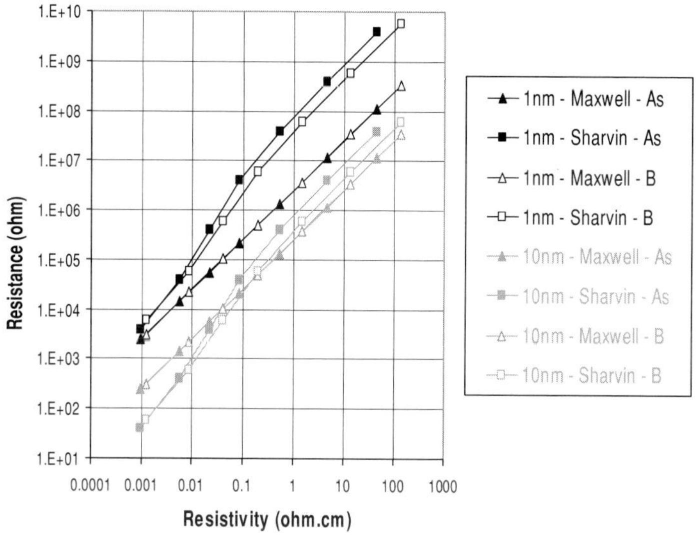

FIGURE 11. Diffusive ($R_{Maxwell}$) and ballistic ($R_{Sharvin}$) resistance as a function of the resistivity (ρ) for two different dopants (As and B) and two different tip radii (1 nm and 10 nm).

FIGURE 12. Schematic representation of the contact model for a n-type silicon, where e ϕ_m is the metal work function, e ϕ_s the semiconductor work function, and eχ the semiconductor electron affinity. The SS zero-level is situated above the midband gap.

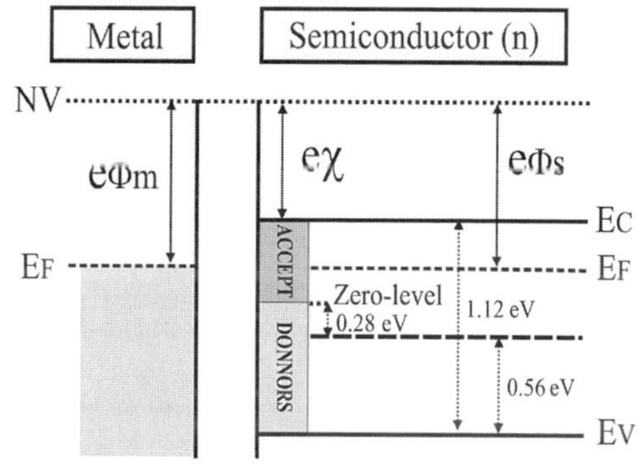

The sample preparation (polishing) has an important impact on the total SS concentration and distribution.[3] It is common knowledge that the surface charges

[3] At the surface of the silicon sample, the number of neighbor atoms is reduced (three instead of four) and each atom shows a non-saturated orbital called a *dangling bond*. Those dangling bonds generate two-dimensional energy bands in the band gap that form the so-called *surface states* (SS). SS are also created through the reconstruction of the surface (that can be compared to the Peierls' transition). Besides these "intrinsic" states, "extrinsic" states are also present and play an important role in SSRM. They are typically linked to the presence of foreign atoms (like carbon or oxygen) and irregularities and defects at the surface.

TABLE 1. Summary of the different parameters involved in the contact model

Parameters		Values
Surface states	Zero-level	0.28 eV above mid-band gap energy
	SS concentration	10^{13} eV/cm^2
Schottky contact	Schottky barrier $E_b = e\phi_m - e\chi$	0.36 eV
	Tunneling distance	10 nm
Contact size	Tip radius	20 nm

create variations of doping concentration at the surface changing the I-V curves, shifting the junction and even creating inversion layers [38]. For the SS in our model we have chosen a zero-level above the mid-band gap in order to fit with the experimental observations showing a lower impact of SS on n-type silicon (see Figure 12) as compared to classical SS modeling [39].

2.2.4.2 *Calibration of the New Contact Modeling*

Based on the experimental observations and ISE/DESSIS device simulations [37], the different parameters have been extracted (see Table 1).

2.2.4.2.1 I-V Curves. The model parameters have been calibrated by comparing both experimental I-V and calibration curves. In highly doped areas, the effect of surface states (SS) is negligible, and the barrier is transparent, implying that the ohmic contact model is still applicable. The saturation (Figure 5) which can be observed for very high concentrations results from the fact that the probe tip resistance (which is not dependent on the silicon resistivity) and the spreading resistance are becoming the same order of magnitude. A good qualitative correspondence between measured and simulated curves (see Figure 13) can be obtained. Note that in order to simplify the comparison with the simulations, the bias in the x-coordinate

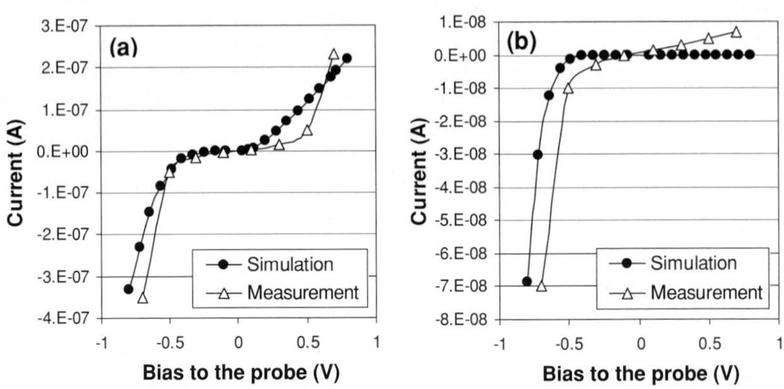

FIGURE 13. Comparison between simulated and measured I-V curves in low-doped n-type (10.5 Ω.cm) (a) and p-type (5.2 Ω.cm) (b) homogeneous Si samples. A qualitative agreement may be observed.

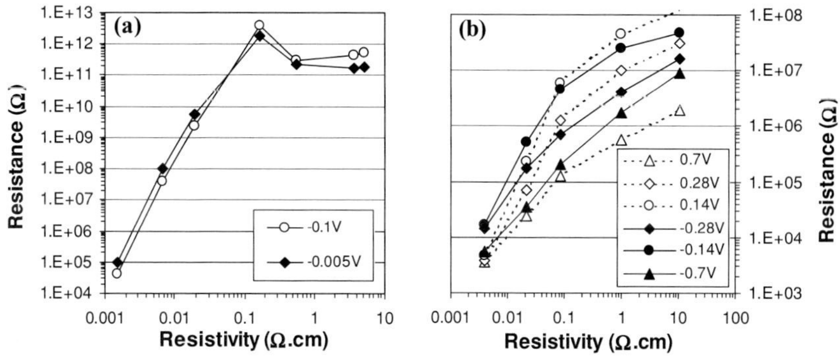

FIGURE 14. Simulated calibration curve. (a) simulation of the calibration curves on p-type Si for small negative biases to the back-contact. The non-linearity observed in experiments (see Figure 7) is reproduced. (b) simulation of the calibration curves on n-type Si for biases ranging from −700 mV to 700 mV. The qualitative agreement with the measured curves is once again clearly visible.

in Figure 13 is not the one applied to the sample Figure 5). The presence of large currents for negative biases to the probe in low-doped n-type samples is linked with the Schottky contact.

2.2.4.2.2 Calibration Curves. Applying the same model (parameters), to simulate calibration curves, the non-monotonic behavior for p-type samples can be observed (Figure 14(a)). The increase of the resistance (n-type) in lowdoped areas with decreasing bias is correctly reproduced when SS are included (Figure 14(b)).

2.2.4.2.3 Asymmetric Junction Structures. Using this model, we can elucidate the role of SS on the actual SSRM profiles. For this purpose, we present four typical p/n, n/p junction structures (Table 2).

An SSRM measurement on the cross section of those structures was performed first. Then an SSRM measurement was simulated. For that purpose, we used the SIMS measurements (for B and As) as input dopant profiles for the ISE/DESSIS device simulator whereby full activation of the dopants was verified with SRP profiling. Then, the spreading resistance (in fact the current flowing from the

TABLE 2. Measured and simulated SSRM electrical junction positions for the test structures: comparison with SIMS measurements and zero-field EJ

	Metall. junction	Electrical junction on cross section		
Struct.	SIMS meas. (nm)	ZF simul. (nm)	SSRM meas. (nm)	SSRM simul. (nm)
$p^{++}.n$	660	1040	340 ± 30	370
$n^{++}.p$	650	780	No peak	No peak
$p^{++}.n^{+}$	180	215	210 ± 10	200
$n^{++}.p^{+}$	210	210	230 ± 10	240

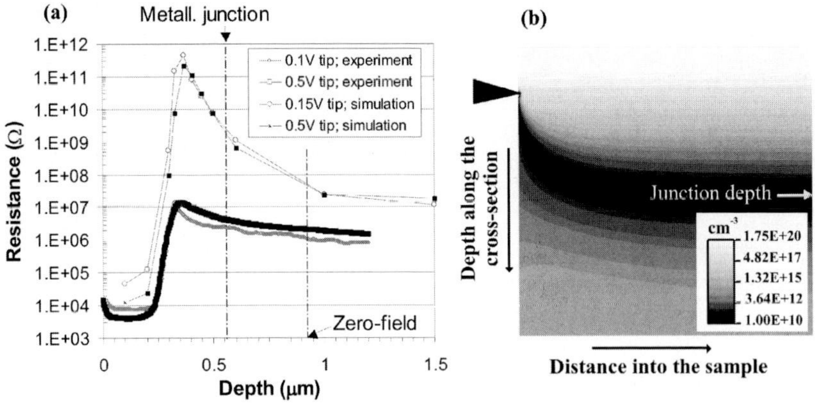

FIGURE 15. (a) Resistance profiles for $p^{++}n$. Comparison between experiment (SSRM on cross section) and simulations (with the new contact model). (b) Carrier density 2D map (500 × 500 nm) for sample $p^{++}n$ as simulated with ISE/DESSIS for the new contact model.

probe to back-contacts on the back side of the sample for a given bias voltage) was simulated for different tip positions (on the cross section) using the new contact and surface model. The curve with all the simulated spreading resistances for the different lateral positions gives the simulated SSRM profile. To reduce the calculation time, we performed two-dimensional simulations. This has no influence on the junction position (our main concern here) but only on the resistance levels. Hence a quantitative comparison of the calculated resistance values is not possible.

- *Case* 1 : $p^{++}n$
  Sample $p^{++}n$ is a p-type (B) implant ($>10^{20}$at/cm^3 at the surface) on an n-type (As) substrate (4×10^{14} at/cm^3). Figure 15(a) presents the measured and simulated SSRM profiles. Both show the same drastic shift of the junction peak toward the surface compared to the zero-field position (see Table 2). The latter can be understood by looking to Figure 15(b), which shows the calculated carrier density (holes/electrons) as a function of distance perpendicular to the cross-sectional plane. In this case, the large bending of the electrical junction position results in a much shallower junction position close to the cross-sectioned surface. Note that far from the surface, the surface states have no influence and the zero-field position for the electrical junction position is found. The displacement of the junction position due to surface states results from the accumulation induced in low-doped n-type material since in this case the SS zero-level.
- *Case* 2 : $n^{++}p$
  Sample $n^{++}p$ is an n-type (As) implant ($>10^{20}$at/cm^3 at the surface) on a p-type (B) substrate (2×10^{15} at/cm^3). The experiment and simulation indicate the absence of junction peak. Only a slow gradual increase of the resistance with depth is observed (Figure 16(a)). This is caused by the fact that the surface

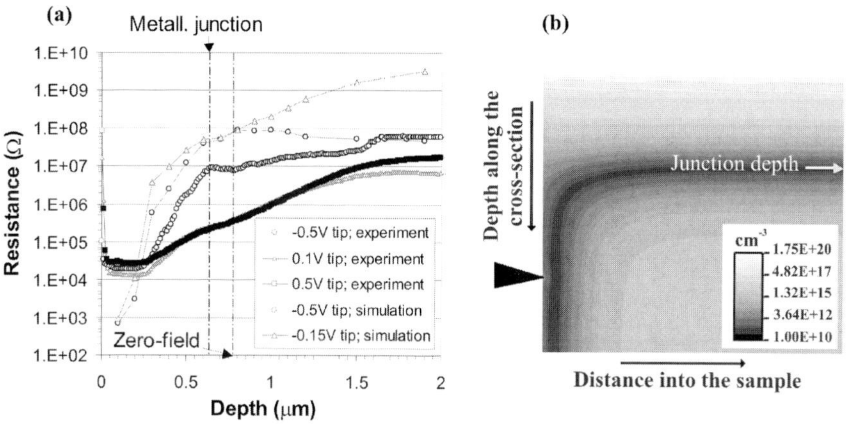

FIGURE 16. (a) Resistance profiles for $n^{++}p$. Comparison between experiment (SSRM on cross section) and simulations (with the new contact model). (b) Carrier density 2D map (500 × 500 nm) for sample $n^{++}p$ as simulated with ISE/DESSIS for the new contact model.

states induce an inversion layer along the cross section in the p-type region (see Figure 16(b)). The electrical junction does not intersect the cross-sectional surface of the sample and thus cannot be detected. Far from the surface, the surface states and the Schottky contact effect are minimal and the zero-field position for the electrical junction position is found. The disappearance of the junction due to surface states results from the inversion they induce in the low-doped p-type region. For p-type material, the Fermi level is below the SS zero-level, causing a transfer of electrons from the surface states to the bulk. Because of the low doping level in the bulk and because of a SS zero-level at 0.28 eV above the midband energy, a large depletion layer forms. Bands bend enough so that conduction electrons become the majority carriers near the surface and form the inversion layer.

- *Case* 3 : $p^{++}n^{+}$
  Sample $p^{++}n^{+}$ is a p-type (B) implant ($>10^{20}$ at/cm^3 at the surface) on an n-type (As) implant (2×10^{17} at/cm^3) similar to source/drain implant in an n-well. Due to the high doping levels, the surface states impact is drastically reduced and the difference between the zero-field and the simulated positions is negligible and lies within the error bar (Figure 22(b)).
- *Case* 4 : $n^{++}p^{+}$
  Sample SRP1 is an n-type (As) implant ($>10^{20}$ at/cm^3 at the surface) on a p-type (B) highly doped substrate (5×10^{18} at/cm^3) such as source/drain implant in p-well. As in the previous case, the difference between the zero-field and the simulated positions is almost negligible due to the high doping levels (Figure 22(a)). However, a small shift toward the depth remains visible.

The conclusion of these simulations is that for junctions formed in low-doped substrate, the SS may influence the profile significantly (in particular in the junction region). At around 10^{17} at/cm^{3,} this phenomenon becomes drastically reduced and the role of the SS disappears. It needs to be stressed that this phenomenon is true for any technique using cross sectioning as sample preparation.

2.2.4.3 Role of Pressure in Contact Modeling

The SSRM contact modeling is a particularly challenging task: the contact size is extremely small (noticeably below 10 nm), the materials in contact are complex, and the current density in the nanocontact vicinity is extremely high. The good agreement between the new model and the measurements tends to prove that none of those parameters have a drastic influence. We will, however, in this section study the most striking missing parameter, the large pressure (in the GPa range) between the probe and the sample.

All models within this section have (apparently) considered a zero-pressure situation. However, as it has been shown in section 2.1, one of the major characteristics of the SSRM technique is the use of high pressure (GPa) between the probe and the sample to penetrate the native silicon oxide and to obtain a stable electrical contact dominated by the sample spreading resistance.

Experiments such as load–unload force versus penetration depth curves for spherical indenter [40] or electrical resistance measurements on rectifying gold-chromium contacts on silicon [41] have demonstrated that as in the SRP situation, the large pressure involved in SSRM measurement was responsible for elasto-plastic or even fully plastic deformations of the silicon sample indented and local phase transformation of silicon to so-called the β-tin phase [15]. The β-tin structure (also named Si(II)) is a denser structure than the cubic structure Si(I) (22% volume reduction) formed at moderately high pressures [42].

Under pure hydrostatic conditions, the transformation occurs at about 12 GPa, but this is reduced to values as low as 8 GPa when shear stresses are present (as is the case in the scanning contact mode SSRM) The Si(II) properties are substantially different from those of Si(I). In particular, this phase is metallic [43].

The electrical behavior of the point contact under pressure has been studied by measuring the current-voltage (I-V) characteristics at different loads for metal and silicon [44]. The presence of a larger threshold pressure necessary to obtain a stable resistance on silicon has been observed and interpreted as formation of Si(II) under the probe [45].

Contrary to an earlier suggestion by De Wolf [46], we believe that the presence of the "metallic" β-tin pocket under the probe does not result in crossover from Schottky to ohmic contact behavior. Indeed, if the contact between the probe and the Si-II is ohmic, the dominant SSRM nanocontact (linked to the silicon resistivity) is the contact between the Si(II) pocket and the Si(I) (see Figure 7). Realizing high force measurements (inducing a full plastic deformation under the indenter), Marchand et al. have demonstrated that such an internal contact was Schottky-like [47].

As discussed in section 2.4.1, the resolution of the SSRM technique is governed by the size of the β-tin pocket created under the probe, rather than by the

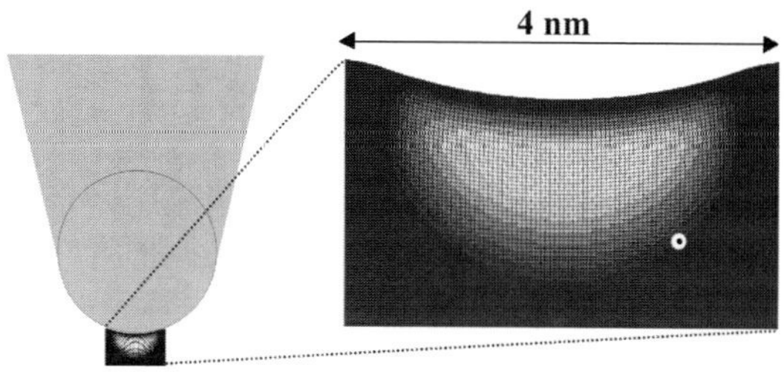

FIGURE 17. Distribution of the plastic strain in the silicon as simulated with MARC. The radius of curvature is 5 nm and the force on the indenter is 5 μg. The grayscale pocket corresponds to the plastically deformed region and the black area to the elastically deformed region.

probe itself, giving an explanation to the nanometer spatial resolution of the SSRM technique. The β-tin pocket can thus be considered as a *virtual SSRM probe*. Nevertheless, the shape of the probe–sample contact is of crucial importance, as it will determine the stress distribution in the silicon sample. Two-dimensional (axisymetric) finite element simulations [48] provide the von Mises stress distribution under the probe for a typical probe configuration (see Figure 17) and three different radii of curvature (5, 10, and 50 nm). The simulations show three important steps: appearance of the plastic deformation, establishing of the contact between the probe and the plastically deformed pocket, and appearance of the fully plastic deformation. An important parameter obtained from the simulations is the ratio between the contact radius and the radius of the plastically deformed area, the latter determining the final resolution of the technique.

The simulations show that the initial size of this pocket is almost independent from the radius of curvature of the probe and that it does form below the probe, rather than directly at the contact, in agreement with earlier observations by Laursen [49]. Note, however, that the pockets do not appear at the same depth when the radius of curvature is varying. A depth reduction is indeed observed when the radius of curvature is decreased implying that for a smaller radius, the contact tip-pocket will be established sooner (= lower force). Since the size of the pocket increases with force, a smaller tip will lead to a smaller electrical contact. The most important result in Figure 18 is that the pocket is substantially smaller the tip radius ($\approx 3\times$ for tips varying between 5 and 50 nm).

2.3 *Quantification*

2.3.1 Calibration Curves

2.3.1.1 The Concept

As pointed out in section 2.2, a straight convertion of spreading resistance (R) to local resistivity (ρ) can be made using the ideal (flat, pure-ohmic contact) formulas $R = \rho/4a$ (Maxwell) or $R = 4\rho\lambda/3\pi a^2$ (Sharvin), depending on the tip radius (a). The latter already provides a reasonable accuracy. In practice a true straight-line dependence is never observed, among other reasons due to the influence of the nanocontact nature (Schottky), the SS at the surface, and the

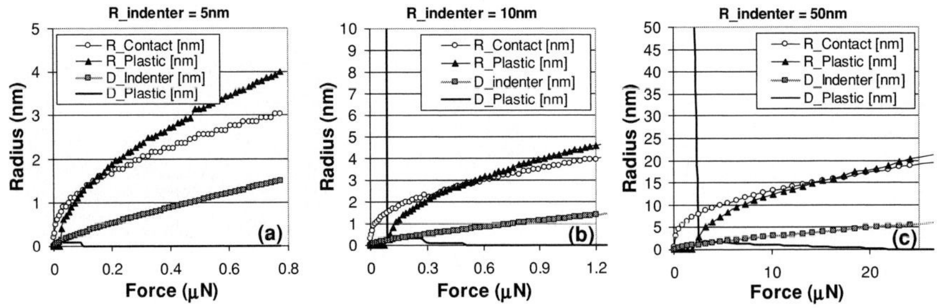

FIGURE 18. Analysis of the evolution with the force of important parameters (from the MARC stress simulation) as the contact radius (R_Contact), the electrical radius (R_Plastic), the indenter or probe penetration (D_Indenter), and the pocket depth (D_Plastic) for radii of curvature (R_Indenter) of 5 nm (a), 10 nm (b), and 50 nm (c), respectively.

probe resistance. Although these effects can be modeled (cf. above), taking them into account would make data conversions very slow. A more pragmatic approach is to perform quantification using of a look-up procedure based on calibration curves. Provided a high reproducibility is maintained (see section 2.4.2) this assures quantification accuracy as good as 20–30%.

2.3.1.2 The Implementation

2.3.1.2.1 Automatic Generation of Calibration Curves. As the probes used to measure on the unknown structures have varying characteristics, the first step of the quantification procedure is the acquisition of two calibration curves (for both impurity types) for each probe. Those calibration curves are constructed by collecting the resistances measured for different homogeneous calibration samples with a well-known resistivity.

Whereas originally separate samples of different resistivity levels were used to establish the calibration curves (5–6 for each impurity type), dedicated epitaxial test-structures (one for each type) with a 1D staircase-structure are now used to obtain a complete calibration curve from a single scan. [28, 50]. This drastically reduces the time required to establish a calibration curve. In these structures, the resistivity levels range from 1×10^{-3} to 10 Ω.cm (within concentration range of interest) as determined by the spreading resistance probe (SRP) and secondary ion mass spectrometry (SIMS) techniques.

To eliminate the manual extraction of the resistances corresponding to each staircase step, a dedicated program, called *MicroQuanti*, was developed to build the calibration curves automatically based on the raw data collected for the staircase calibration structures. In Figure 19, an illustration of this procedure is presented for p-type (T8) and n-type (T9) staircase structures.

The *MicroQuanti* routine has also been extended in order to perform automatic calibration procedure on other dedicated epitaxial test structures such as the MAYA structure [50].

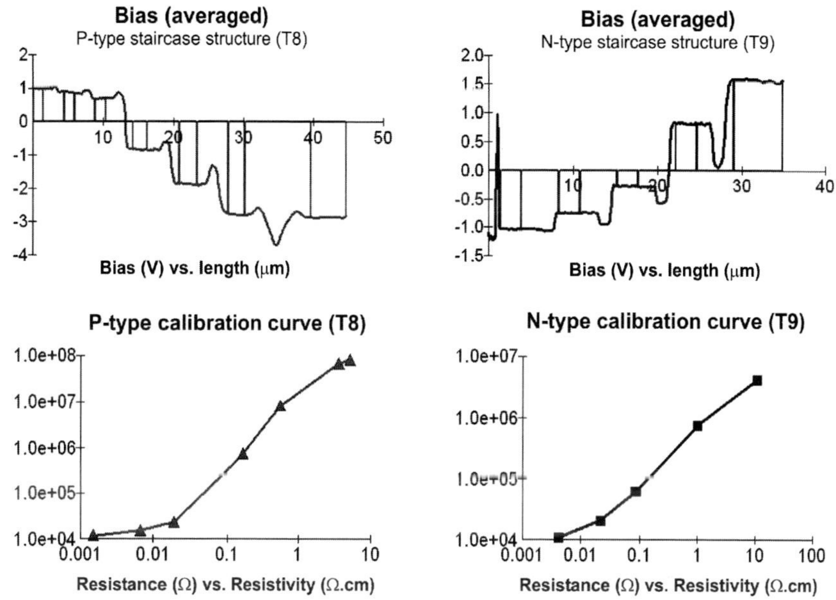

FIGURE 19. Screen shot from *MicroQuanti* illustrating the automatic generation of calibration curves for the T8 and T9 staircase structures.

2.3.1.2.2 Automatic Quantification for 1D And 2D Carrier Profiles. A fast and very easy way to obtain resistivity profiles from the raw resistance data is to use a calibration procedure. In this procedure, the resistance profile of the unknown sample is measured with the same probe, and using linear interpolation on the appropriate calibration curve (n- or p-type), the resistivity profile of the unknown sample can be reconstructed. The use of the experimental calibration curve obviously eliminates non-linearities due to the nature of the contact, the probe, and the back-contact resistance.

In the *MicroQuanti* program a measured 1D or 2D spreading resistance profile is converted using the calibration curve, thereby ignoring all 2D-interactions. From the calculated resistivity profile, a carrier concentration profile is constructed using known values for the carrier mobility (Thurber mobility) [51].

This procedure is very fast and provides interesting quantitative information. In Figure 20, the one-dimensional quantification procedure for a lateral section through the channel of a 0.18-μm n-MOS transistor (with 0.25-μm gate size) is illustrated. In Figure 21, the two-dimensional quantification procedure for a 0.18-μm n-MOS transistor (with 0.25-μm gate size) is illustrated. The quantification software produces a two-dimensional active dopant (carrier) map using p-type calibration and conversion curves whereby the extension of the p-type halo implants within the channel are quantified.

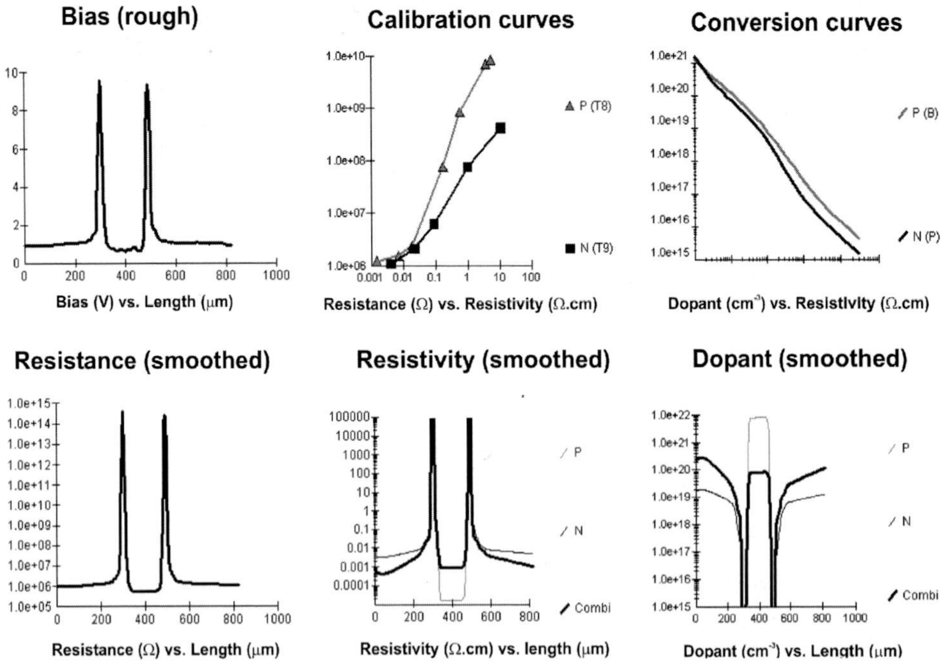

FIGURE 20. Screen shot from *MicroQuanti* illustrating the calibration quantification procedure for a one-dimensional profile.

2.3.1.3 Limitations

For further quantitative refinements, a more complex (iterative) procedure is required to account for the current spreading effects induced by the dopant gradients. Improved contact and surface modeling also has to be taken into account (specially in low-doped areas), as explained in the next section, to explain carrier redistribution near the sample surface.

2.3.1.4 Application and Comparison with 1D Techniques

Secondary ion mass spectroscopy (SIMS) is nowadays the standard reference technique for one-dimensional dopant profiling. A major difference between SSRM and SIMS is the fact that SIMS measures the position and the concentration of the dopant ions, rather than charge carriers. Even in the case of a fully activated profile, the dopant and the carrier distribution are not identical, due to the outdiffusion of the mobile carriers (difference between the EJ and the MJ). The classical procedure to compare SIMS and SSRM results is thus to calculate the carrier profile from the SIMS dopant profile solving the Poisson equation and then to compare this calculated profile with the SSRM quantified profile. In Figure 22(a) and Figure 22(b) corresponding, respectively, to n^{++} p^{+} and p^{++} n^{+} structures, very good agreement

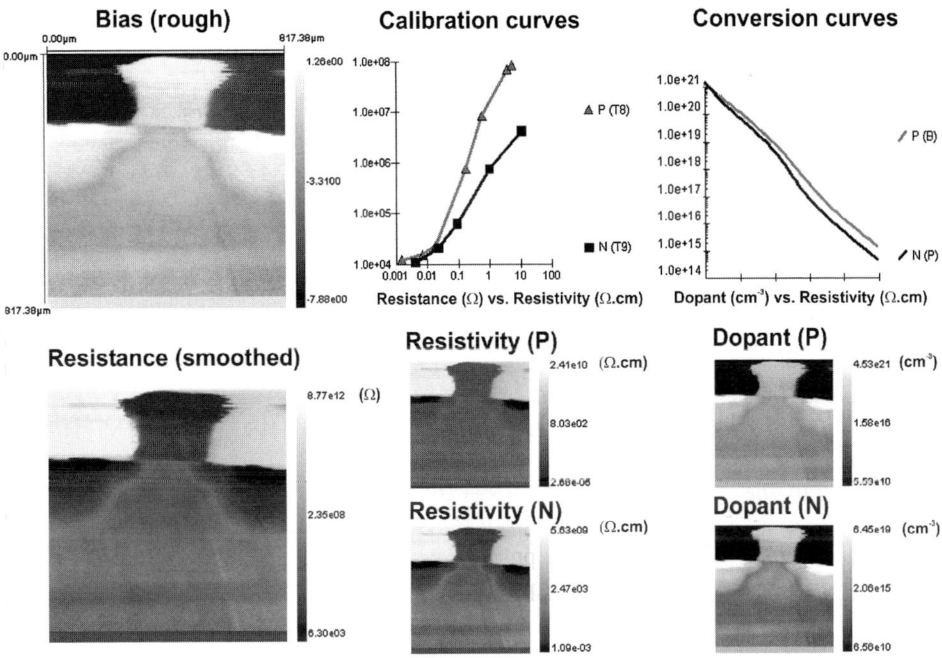

FIGURE 21. Screen shot from *MicroQuanti* illustrating the calibration quantification procedure for a two-dimensional profile.

in terms of concentration level and electrical junction position is observed between SIMS and quantified SSRM.

The SRP technique is a carrier profiling tool similar to SSRM. However, SRP measurements are realized on beveled samples. As has been explained by Hu, the beveling procedure results in a shift of the junction (carrier spilling) [52].

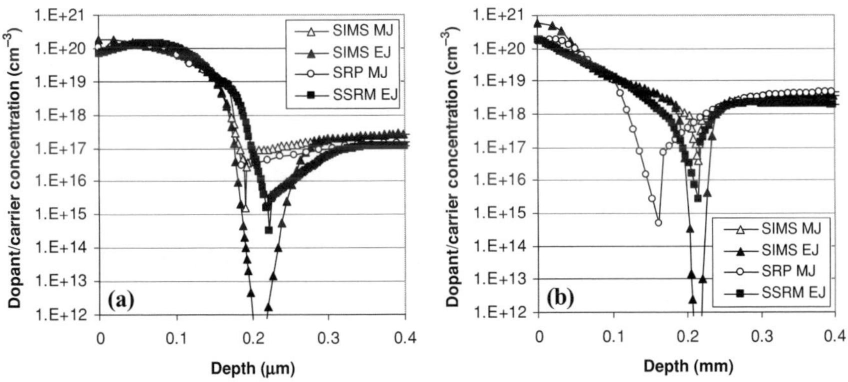

FIGURE 22. Comparison between SIMS, SRP, and SSRM dopant and carrier profiles on the $n^{++}p^{+}$ SRP16 structure (a) and on the $p^{++}n^{+}$ SRP1 structure (b).

Comparison between cross-sectional SSRM or SIMS and on-bevel SRP profiles is thus not always straightforward. For a detailed discussion see [53]. In Figure 22(a) and Figure 22(b) good agreement between SRP and quantified SSRM (using MicroQuanti) in terms of carrier concentration is observed. The shallower junctions of SRP are representative of the carrier spilling effect that could not be completely corrected.

Small differences between SIMS and SSRM exist in the most heavily doped part of the profiles (measured carrier concentration slightly lower with SSRM than with SIMS). SRP results evidence that this difference is linked to the presence of a not fully activated implant and not to a saturation problem of the technique (due to the predominance of the probe resistance). The profiles in the depletion zone also differ. The peak is less pronounced with SSRM than with SIMS. This is most probably linked to the perturbation of this zone by the voltage used in the SSRM measurements, resulting in non-zero carrier concentration. When the SSRM bias is increased the junction peak is further reduced.

Overall, the agreement between SSRM and SIMS EJ positions (the latter being calculated from the SIMS MJ position) is very good, and one can conclude that SSRM is a technique for carrier profile determination with a very high spatial resolution (higher than SRP) and quantification accuracy.

2.3.2 2D Corrections

2.3.2.1 Influence of Current Flow

For steep 2D profiles, additional corrections are necessary to account for the two-dimensional current flow induced by nearby conductive or insulating layers (see Figure 3). Introduction of current flow correction factors (CF) had initially been developed for SRP [54]. Schumann and Gardner have proposed a deconvolution algorithm for on-bevel measurements [55].

In SSRM, the situation is different, since the measurements are performed on cross sections. Moreover, the gradient is drastically reduced, as the contact radius is two to three orders of magnitude smaller. De Wolf et al. have nevertheless developed and patented a special quantification procedure using a database containing correction factors for SSRM [56,57]. It is important to note that this quantification algorithm is based on the ohmic contact model thus limiting the applicability of this approach.

Note that gradient correction factors are only important for distances of a few times the electrical probe radius. Moreover, preliminary calculations by De Wolf have shown that even then these are fairly limited (<50%) even for very close layers (<1–2 tip radii) (see Figure 23) [58]. This is also in line with the observed spatial resolution (see section 2.4.1.1) in the 1–3-nm range.

2.3.2.2 Influence of Surface States

Physical modelling of the probe–semiconductor contact, including surface effects, has provided information on the potential profile distortions induced by surface

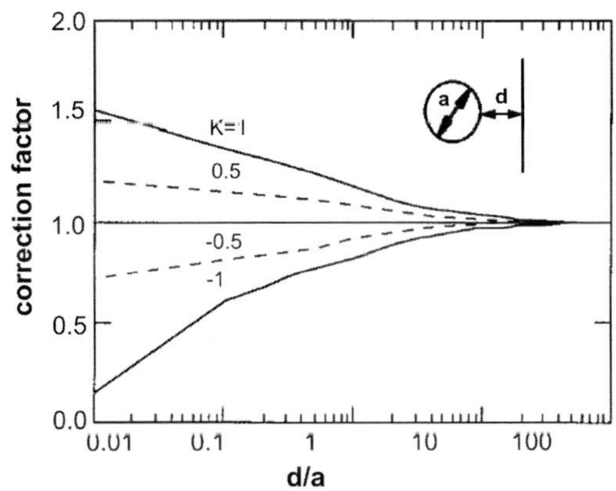

FIGURE 23. Evolution of the SSRM correction factor with the distance (normalized to the tip radius) to a perfectly conducting (bottom curve) or perfectly isolating (top curve) boundary. The dashed lines show the correction factor for different resistivity ratios $K = (\rho_2 - \rho_1)/\rho_2 + \rho_1)$ [58].

states (in particular in low-doped areas [17]). Note that the latter is a universal problem not limited to SSRM but equally present in all technologies using sectioned devices such as SCM or electron holography [59].

For asymmetrical junctions correction procedures should thus also be introduced in order to account for the SS distortions on the low-doped side. Further modeling is required to establish appropriate correction procedures.

2.4 Performance

2.4.1 Resolution

In order to assess the performance of the SSRM technique, it is essential to establish unique and verifiable definition of resolution. The definition of resolution is, however, not straightforward for carrier profiling techniques and has to be divided into three main characteristics: the spatial resolution, the dopant gradient resolution, and the concentration sensitivity.

In defining the spatial resolution, simple length scale can be used when discussing the ability to resolve the precise location of a p-n junction, the thickness of an oxide etc. For dopant profiling applications an equally (even more) relevant number is the resolvable dopant gradient (nm/dec), as a length scale is relatively meaningless unless one specifies the concentration difference between the points one is trying to resolve. To complicate the situation even further, the dopant gradient needs to be specified around a particular doping level as the "natural" outdiffusion of the mobile carriers ($\sim$ Debye length) will limit the detected dopant gradient at low concentrations, regardless of the intrinsic performance of the analysis technique. Finally, the third parameter, which enters into the assessment, is the concentration sensitivity, which can be viewed as the ability to detect marginally different concentration levels.

For AFM-based techniques, the spatial resolution will, among others, be set by the tip radius, whereas the dopant gradient will also depend on the sampling volume and the concentration sensitivity. Concentration sensitivity is set by the intrinsic properties of the technique and the signal-to-noise ratio occurring in its actual implementation.

2.4.1.1 Spatial Resolution

For spatial resolution analysis, a special set of test structures with thin buried oxide layers (ranging from 10 nm to 0.8 nm) embedded between a highly doped silicon layer and a highly doped poly-silicon layer has been realized. The thickness of the oxides has been verified using cross-sectional TEM measurements. The latter represents an ideal case for the assessment of the length scale resolution, as there is no carrier outdiffusion from the oxide region.

2.4.1.1.1 Assessment of the SSRM Spatial Resolution. From Figure 24(b), which represents single-line scans, SSRM is clearly capable of detecting a 3.3–3.5-nm buried oxide. The presence of the (non-conducting) oxide leads to a drastic increase in the resistance (almost a factor 10). If the tip diameter is less than the oxide thickness, one would expect to measure (at least in one data point) the intrinsic resistance of an oxide (in our setup usually more than 100 GΩ). However, this logic ignores any surface conduction as well as any tunneling through the (extremely thin) oxide surrounding the tip. Hence even when the tip diameter is less than the oxide thickness one does not expect to measure the full oxide resistance. Detailed calculations (ignoring any tunneling and high field saturation) on the expected response as a function of tip radius by De Wolf show that one does expect a factor two of increase when the tip radius equals the oxide thickness [60]. Based on this criterion, the radius of the tip used in Figure 24(b) should be less than 3 nm. The impact of the tunneling and/or surface conduction can be assessed by comparing the detected resistance increase versus the thickness of the oxide (Figure 24(c)) Without tunneling, the theoretical calculations predict an extremely rapid increase

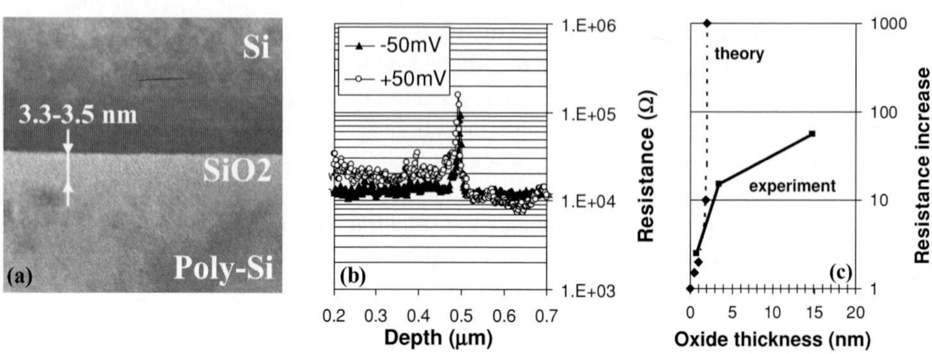

FIGURE 24. (a) TEM measurement of a 3.3–3.5-nm buried oxide structure. (b) Single-line scan SSRM measurement across this buried oxide (realized with a full diamond probe and 50 mV/–5 0 mV as bias potential on the probe). (c) Calculated (assuming no tunneling, tip radius 1 nm) and experimentally observed increase in resistance measured in the oxide layers, as a function of the thickness of the oxide.

when the oxide thickness is larger than the tip diameter. The experimental data show a much slower increase.

Interestingly, the shape of the resistance profile on the left and right side of the oxide layer is very different. In principle, this increase is solely due to the impact of the insulating layer on 2D-current distribution and should be identical for both sides, as the doping levels are nearly identical. Potential explanations for the asymmetry could be the differences in interfacial roughness, presence of interface charge causing some depletion, (asymmetrical) tip artifacts, or variations in the dopant profile in the near-interface region. As multiple tips and scan directions were used, tip artifacts can be excluded. Moreover, the pyramidal tips rarely show such an asymmetrical shape. Since both sides are highly and almost equally doped, an asymmetrical carrier profile resulting from interfacial charges is highly unlikely. It would also require an extremely high interface charge, which is not expected as the sample was grown in a standard, high quality CMOS-line. Finally, it needs to be mentioned that the dopant profiles in the poly-silicon are completely flat with a small interfacial peak. Hence a reduced dopant level in the near surface region is not an explanation either.

There are two possible explanations: i.e., either a gradient in the activated dopant profile in the poly-silicon (regardless of the flat dopant profile) or a larger interfacial oxide roughness on the poly-silicon side. The TEM-results in Figure 24(c) show a difference in interfacial roughness between the poly-silicon and oxide versus the silicon and the oxide, suggesting that the steeper increase of the SSRM signal on the silicon side (right) versus the more gradual increase on the poly-silicon side (left) is a measure of the interfacial roughness. The ability to probe these small differences in interfacial roughness then illustrates the excellent spatial resolution of the SSRM technique.

An even more challenging test is presented by the case of an extremely thin oxide (0.7–0.9 nm, see Figure 25). Even for this very small oxide thickness, the SSRM

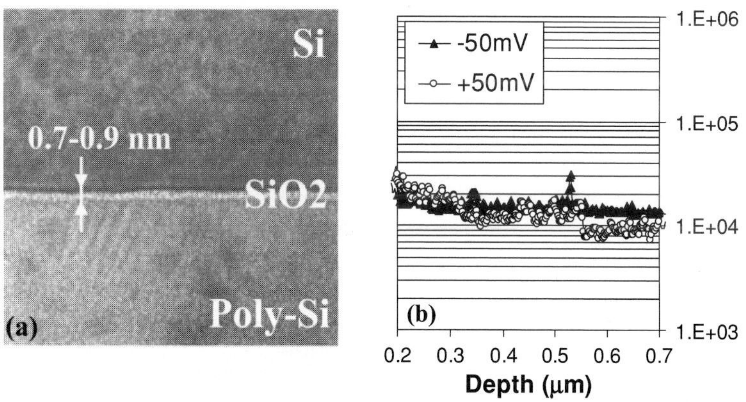

FIGURE 25. (a) TEM measurement of a 0.7–0.9 nm buried oxide structure. (b) Single-line scan SSRM measurement across this buried oxide (realized with a full diamond probe and 50 mV/–50 mV as bias potential on the probe).

line scan does show a pronounced increase in resistance peak, still exceeding a factor two. Based on this result and Figure 24, one would estimate the tip radius to be as small as 0.7–0.9 nm. In this particular measurement, the choice of an optimum bias voltage between the probe and the sample (+50 mV) aids in reducing the noise and thus peak identification.

Basically, the SSRM response function can be characterized by the height and the width of the SSRM peak. A larger tip gives a lower peak (see Figure 24), whereby, for a factor two increase, the oxide thickness equals the tip radius [60]. The height criterion is a straightforward approach which univocally sets the electrical tip radius, in this case to less than 1 nm. Note that even higher spatial resolutions have been observed with the latest generation of full diamond probes.

2.4.1.1.2 Comparison with the SCM Spatial Resolution. Applying SCM to this structure, the SCM signal, in the highly doped region as well as in the oxide, is fairly low (zero for the oxide), thus limiting the available contrast. When only single-line scans are used, the electrical noise is too high and even the 8-nm oxide sample is hardly discernible (Figure 26). Averaging over 100 lines leads to a better

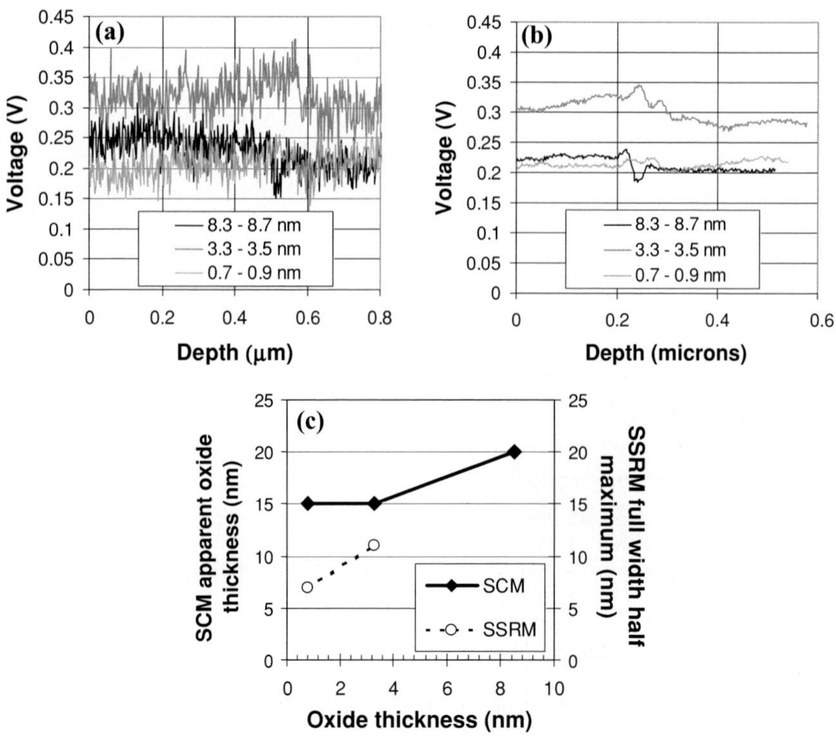

FIGURE 26. (a) Single-line SCM scan above 8.3–8.8-nm, 3.3–3.5-nm, and 0.7–0.9-nm buried oxide layers. (b) Average line SCM scan above 8.3–8.8 nm, 3.3–3.5 nm, and 0.7–0.9 nm buried oxide layers. (c) Real oxide thickness compared to apparent oxide thickness as measured with SCM (left) and to full width half-maximum with SSRM (right).

signal-to-noise ratio, allowing the scan to reveal distinct features which could be interpreted as characteristic for the oxide. In principle, the signal should be zero when the tip is above the oxide, which in Figure 26(c) is noticeable as a small dip in the signal. For thicker oxides (up to 50 nm), a reduction of the signal to 0 V is indeed observed [61]. Taking the width of the region with low SCM-signal as representative for the apparent thickness of the oxide, one finds a value of 15 nm for the 0.8- and 3.4-nm samples. For the 8.5-nm case, the measured thickness increases to about 20 nm. This saturation at 15-nm is a clear indication of the relatively large tip radius and/or sampling volume in SCM, limiting its spatial resolution. Note also that the amplitude of the signal reduction is very small and depends only weakly on the oxide thickness. This indicates that the measurements are severely limited by the sampling volume of SCM and that very fine features will be convoluted with the response function of SCM. The achieved spatial resolution ($\sim$ 15 nm) is in line with previous observations on the spatial resolution of SCM being inferior to SSRM.

For comparing the SSRM and the SCM results, we use the full width half-maximum (FWHM). The latter measure underestimated the resolution, as it includes the known resistance increase outside the oxide due to the insulating boundary, which typically extends over a few tip radii. Looking at the difference between the FWHM for various oxide thicknesses, the difference between the 3.4 and 0.8 nm structures is clearly visible (4 $\pm$ 1 nm) and close to the expected 2.8 nm value (see Figure 26(c) right).

The observed limitation in SCM spatial resolution is not linked to the sample structure but has been equally observed by Douheret et al. [62]. They compared the ability of SCM and SSRM to resolve thin (5 and 10 nm) InGaAs/InP quantum wells, separated by variable distances (10 to 200 nm). In agreement with our results, it was found that the SSRM resolution (sub-5 nm) is much better than the SCM resolution (sub-30 nm).

2.4.1.2 Concentration Sensitivity

In most recent technologies, the electrical characteristics of the components are extremely sensitive to small variations of concentration. For profiling techniques, high sensitivity to small variations is thus a crucial requirement.

Sensitivity is by definition the ratio of the change in the instrument response (i.e., the measured resistance) to a corresponding change in carrier concentration. The concentration sensitivity of the technique is thus its ability to detect marginally different concentration levels and is thus determined by the intrinsic response function of the technique, as limited by its signal-to-noise ratio. For assessing the concentration sensitivity use can be made of the staircase structures (T8 for p-type and T9 for n-type) introduced originally for the calibration of SRP-profiles and now also used for SSRM and SCM [18,28].

2.4.1.2.1 Assessment of the SSRM Concentration Sensitivity. In Figure 27 (a), single-line scan SCM and SSRM measurements on the n-type T9 staircase structure are shown. In Figure 27 (b), standard deviation on each step for the different

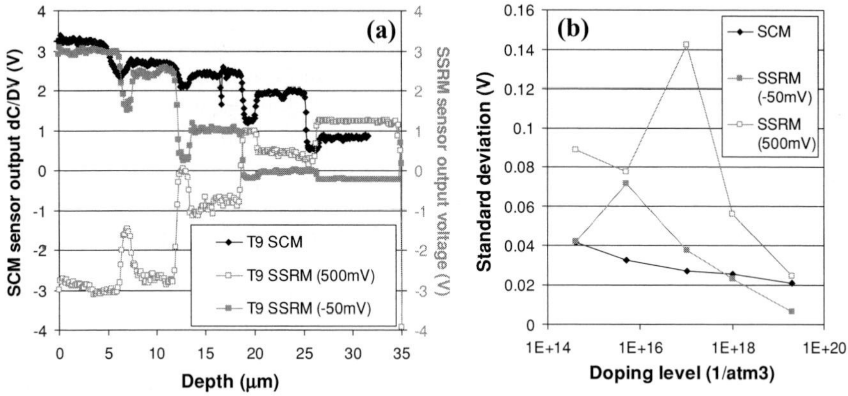

FIGURE 27. (a) Single-line scan across the T9 n-type staircase structure for SCM and SSRM (for two different biases). (b) Standard deviation on each step for those scans.

single-line scans is presented, showing that for SSRM, the signal-to-noise ratio is strongly bias dependent and somewhat better for highly doped layers versus low-doped ones. This may be explained by the nature of the tip–semiconductor contact, which can be described as Schottky contact combined with a tunneling barrier. In highly doped areas, the contact is ohmic-like (the barrier is transparent), while it is Schottky-like and thus more affected by electrical noise in low-doped areas. This model suggests that for each impurity type in principle an optimum bias voltage exists for reducing the electrical noise.

2.4.1.2.2 Comparison with the SCM Concentration Sensitivity. Figure 27 illustrates that the SCM signal-to-noise ratio is significantly better than in Figure 26 and independent of the dopant concentration. This is due to the larger sampling volume of SCM. As the structure here is composed of large constant dopant level steps (thick layer), the additional stray signal from the same doping levels improves the signal-to-noise ratio. Conversely, a low signal-to-noise ratio is observed in SCM on small structures with different concentration levels, whereas in SSRM the sensitivity is independent of the layer thickness.

A more detailed discussion is presented in [61], from which it is clear that SSRM is by far superior in terms of ability to probe detailed halo/voltage adjust implants due to the clear delineation of the junction position (versus the broad bands in SCM) and its intrinsic sensitivity to small dopant concentration variations.

2.4.1.3 Dopant Gradient Resolution

The dopant gradient resolution of the technique is by definition the steepest variation in dopant concentration that can be measured around a doping level. This value can be dependent on the doping level. It is expressed in nanometer of profile length per decade of concentration variation.

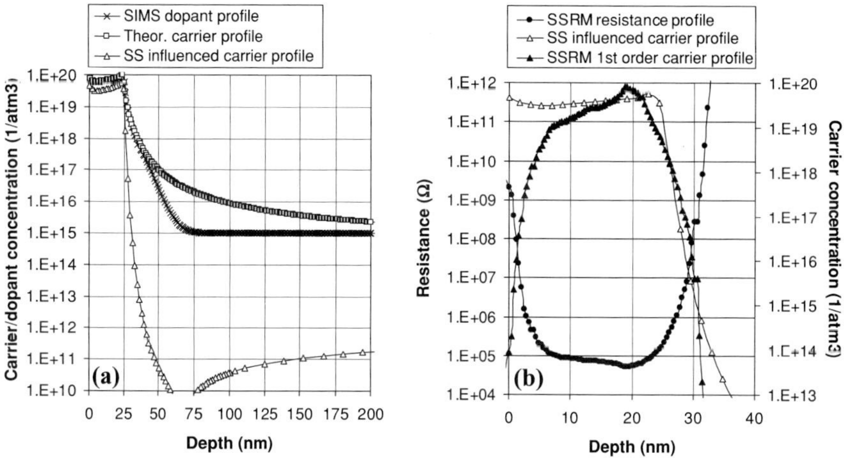

FIGURE 28. (a) Comparison between SIMS dopant profile, theoretical carrier profile, and SS perturbated carrier profile for the CVD1.12 structure. (b) Comparison on the same structure between SSRM resistance profile and the SS perturbated carrier profile.

2.4.1.3.1 Assessment of the SSRM Dopant Gradient Resolution. The ideal test would of course be a perfect rectangular profile. For that purpose, a sub-30-nm p^+p (5×10^{19} at/cm^3 of boron on 1×10^{15} at/cm^3 p-type substrate) box-like profile (labeled CVD 1.12) has been studied with SSRM (see Figure 28(b)). Before discussing the SSRM results it is important to realize that one first has to estimate the carrier distribution correlated with this dopant profile, as the latter are not identical and SSRM is probing the carrier profile. The difference originates from the outdiffusion of the mobile carriers, which can be calculated directly by solving the Poisson equation (square markers in Figure 28(a)) using the SIMS dopant profile (cross markers in Figure 28(a)) as input. The differences between carriers and dopant are quite large in those cases where a steep dopant profile is present on top of a lowly doped substrate. The thus calculated carrier profile represents the situation whereby one would be able to look inside the sample without causing any distortion. Unfortunately in order to perform a SSRM (and SCM) measurement, one needs to make a (polished) cross section. This polished surface is far from ideal and contains a lot of surface states. Since SSRM and SCM probe the carrier distribution on this polished cross section, any impact of the surface on the near-surface carrier profile will be reflected in the SSRM and SCM results. In order to assess the impact of the surface states on this profile, we have calculated the final carrier distribution as it will be found on the polished surface (triangle markers in Figure 28(a)). The details of this surface state model and its impact on the carrier profile have been discussed in detail [17]. The calculations show that in this case the surface states enhance the profile steepness in the (very) low doping regions with virtually no impact on the width of the highly doped plateau (~ 25 nm).

The SSRM measurement shows a slightly smaller plateau (about 22 nm) (probably linked to top surface edge rounding effects and/or an improper calibration of the piezo) and an increase of the resistance near the top surface (probably linked to the vicinity of the non-conducting glue layer). The slope of this first region amounts to a gradient of about 1–2 nm/dec. In terms of sensitivity (and spatial resolution) it is important to point out that within the thin 20-nm plateau region, SSRM is still able to detect the light gradient in the dopant (carrier) profile (see the slope on the SSRM-profile between 5 and 20 nm versus the slope on the SIMS profile). At the interface (20–22 nm) the small dopant peak at the end of the plateau is also visible in the SSRM, indicating once again a very nice spatial resolution and carrier concentration sensitivity.

When comparing the quantified SSRM-profile with the simulated one based on the surface state model, a reasonable agreement is observed. This emphasizes again that the sectioning and polishing procedure used in SSRM and SCM does cause some distortions in lowly doped regions due to the presence of surface states on the polished surface. Remaining discrepancies are related to the improper tuning of the surface states model, non-linearities in the piezo-system causing and near-surface depletion due to the glue. The gradient in the near-surface region is around 1 nm/dec, whereas at the end of the plateau a gradient of 2.2 nm/dec is found.

2.4.1.3.2 Comparison with the SCM Dopant Gradient Resolution. Ultra-shallow (sub-50-nm) box profiles, like the one used to study the SSRM dopant gradient resolution, cannot be measured using SCM (due to the large integration volume of the technique). From careful analysis of SCM measurements on steep profiles presented in the literature, one can, however, conclude that the SCM dopant gradient resolution clearly remains above 5–10 nm/dec.

2.4.2 Repeatability and Reproducibility

2.4.2.1 Repeatability

Repeatability is the ability to reproduce the measurement results using exactly the same experimental conditions (sample, probe, scan speed, force, etc.). It has been studied by measuring different staircase calibration structures and the repeatability (for each point of every single-line scan) has been defined as the normalized deviation from the average value:

$$\text{Repeatability} = (R_{\text{single line}} - R_{\text{average}})/R_{\text{average}}(\%) \tag{11}$$

First, the repeatability between different single-line sections within one SSRM scan is analyzed in Figure 29(a) and Figure 29(b). Three different SSRM single scan lines (within one SSRM two-dimensional scan) are presented for the n-type (T9) and the p-type (T8) staircase structures, respectively (black curves). The repeatability is also presented (gray curves) and typically remains better than 20% in the different layers. Note that the bias used (50 mV and −500 mV to the tip, respectively) is typical for SSRM measurements and has not been chosen specifically to optimize the repeatability. The lower bias used for the T9 structure is

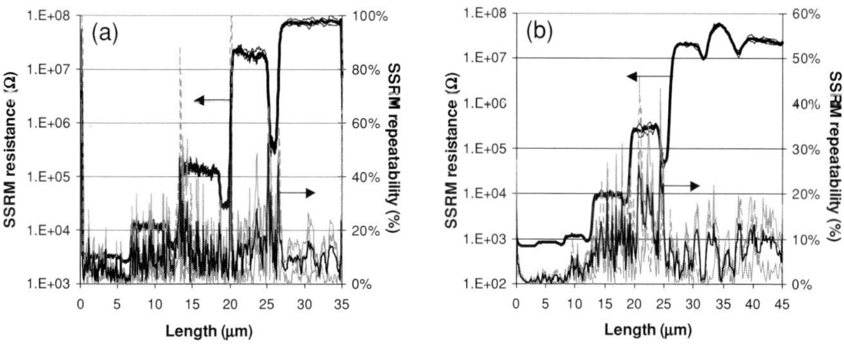

FIGURE 29. (a) Left axis: Three single SSRM line scans across the T9 n-type staircase structure within one SSRM scan and average SSRM line scan. Right axis: Corresponding repeatability for each of the three line scans. (b) Left axis: Three single SSRM line scans across the T8 p-type staircase structure within one SSRM scan and average SSRM line scan. Right axis: Corresponding repeatability for each of the three line scans.

responsible for the slightly higher noise level (and thus poorer repeatability) in this measurement. Note also that in the highly doped areas, the repeatability is typically better than in the low-doped areas (mainly visible with the p-type structure here due to the larger bias used). The latter effect may be enhanced by the saturation.

The previous structures are extremely large for SSRM. The question can be raised whether this has a negative influence on the repeatability. Another p-type staircase structure composed of a smaller double staircase (Maya 4 or M4 [50]) has been analyzed with the same bias voltage (−500 mV) but at a much lower force (typically 5–10 times lower). Lower force allows better resolution but also of larger noise level. As can be seen in Figure 30, the repeatability remains very good and even appears to be a little bit better (between 10% and 15%). This is a clear indication that SSRM repeatability on very small structures (typically below the square micron) is potentially better than 10%.

For some measurements, an average can be realized and it is interesting to study the repeatability of average sections corresponding to different scans. In this study, five successive scans have been realized at two different positions. In Figure 30(b), different average sections corresponding to two different positions and to different scans are compared. Repeatability is here typically better than 5%.

2.4.2.2 Reproducibility

Reproducibility is defined as the ability to reproduce the measurement results when successive measurements are made on different samples originating from the same wafer. An additional parameter that can be varied in reproducibility experiments is the use of different probes.

As long as all other parameters are maintained constant (force, probe, scan speed, scan dimensions, etc.) and the wafer is homogeneously processed, absolutely no

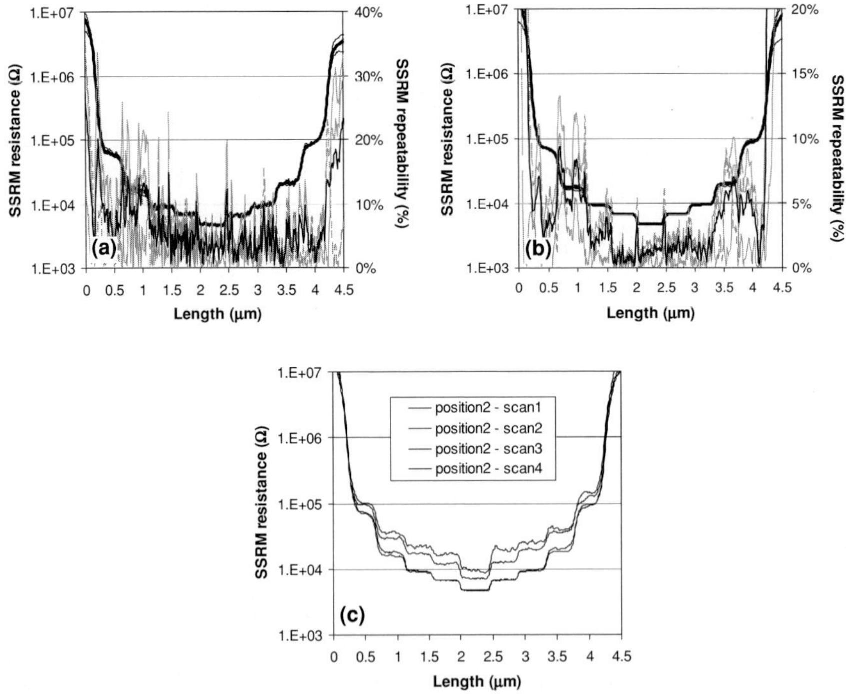

FIGURE 30. (a) Left axis: Three single SSRM line scans across the M4 p-type staircase structure within one SSRM scan and average SSRM line scan. Right axis: Corresponding repeatability for each of the three line scans. (b) Left axis: Three average scans across the M4 p-type staircase structure (corresponding to two different scan positions and to different scans for each position) and average SSRM line scan. Left axis: Corresponding repeatability for each of the three average scans. (c) Average SSRM line scans across the M4 p-type staircase structure for two different positions evidencing possible reproducibility problems.

additional deviation to repeatability is observed while measuring various samples originating from the same wafer and polished with the same procedure. This is an indication of the very good reproducibility of the sample preparation procedure and relative robustness of the technique to sample preparation.

However, by performing successive engagement procedures, one increases the possibility of probe damage, and sudden changes in the contact characteristics due to probe degradation may occur. This problem is likely to be solved with maturation of the diamond probe technology (see Figure 30(c)).

When different probes are used, the reproducibility of the raw data is drastically reduced. This limitation is representative of the variations in probe-tip properties and is not an intrinsic limitation of the SSRM technique itself. It does imply, however, that for each tip, separate calibration curves need to be established.

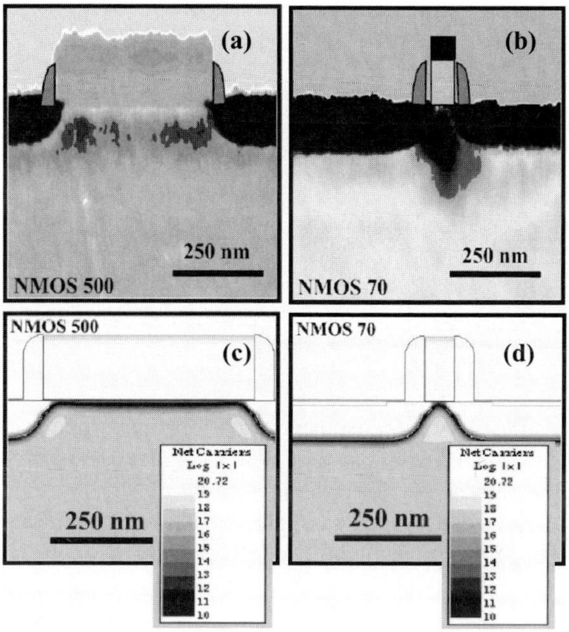

FIGURE 31. SSRM measurement (a,b) (black is more conductive) and simulation (c,d) (white is more conductive) for 500-nm and 70-nm n-MOS structures. Separation of the halo pockets for 500-nm gate size is visible for simulation and measurement.

2.5 Applications

2.5.1 Calibration of Process Simulator for 90-nm n-MOS Devices

Transistors from a 90-nm node technology with gate sizes ranging from 500 nm down to 70 nm have been measured, and the SSRM results are compared with the results from a process simulator, which was initially run using a default set of parameters (in Figure 31). The results emphasize the ability of SSRM to probe fine details of the 2D-carrier distribution within the channel area, which result from the halo and threshold voltage (V_T) adjust implants. Small concentrations variations (<25%) within the channel as well as in-between structures with different gate sizes can be observed. The change in 2D-carrier profile with gate size can clearly be observed illustrating the increasing overlap of the pocket implants with decreasing gate size.

Comparisons of the lateral extent of the halo profiles between SSRM and simulations indicate that the simulations underestimate the lateral width of the halo profiles (while the vertical depth is similar to the SSRM results). Quantitative comparisons of channel profiles taken on large structures also indicate a difference in absolute concentration of around 25% between SSRM and simulations (see Figure 32). The deficiency of the simulations becomes also apparent when comparing experimental and simulation device characteristics (in this case the

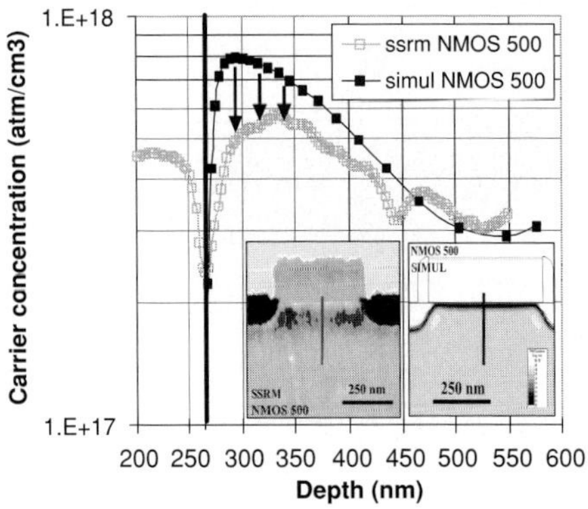

FIGURE 32. Comparison between simulation and quantified profiles of the threshold voltage adjustment implant for a 500-nm n-MOS transistor.

Vt-roll off Figure 33(a)). An improvement can be obtained when using the SSRM results, i.e., adapting the simulations to produce a higher lateral diffusion and decreased channel concentration. As shown in Figure 33(b) (representing the relative Vt variation between 0.5-μm and 1-μm gate size before and after taking into account the different SSRM indications), the difference between the calculated and experimental Vt is then strongly reduced.

2.5.2 Lateral Diffusion in 40-nm n-MOS Devices

Comparisons between SSRM and scanning tunneling microscopy (STM) have been established on a 40-nm technology n-MOSFETs transistors [63]. In particular,

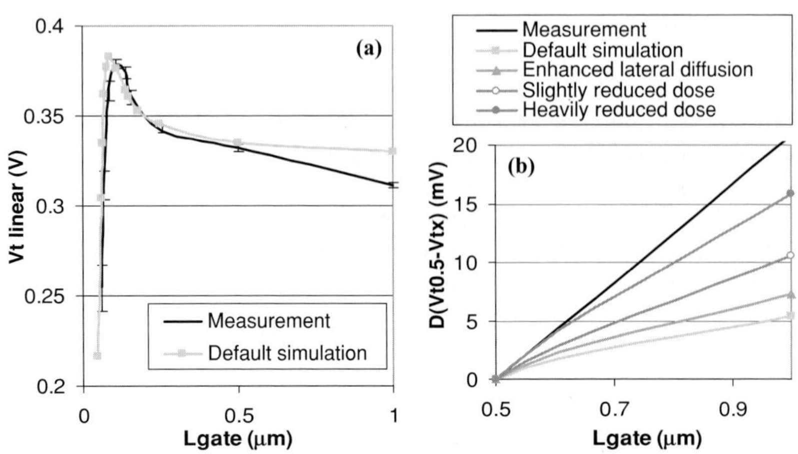

FIGURE 33. Comparison between measured and simulated (default settings) V_t vs. L_{gate} curves (a). The discrepancies observed between 0.5- and 1-μm gate sizes are highlighted (b) by plotting the Vt variation between these two gate sizes (taking the value at 0.5 μm as a reference). When introducing an enhanced lateral diffusion and a reduced Vt dose (leading to a lower channel concentration) an enhanced variation, representative for a reduced discrepancy between simulation and measurement, is obtained.

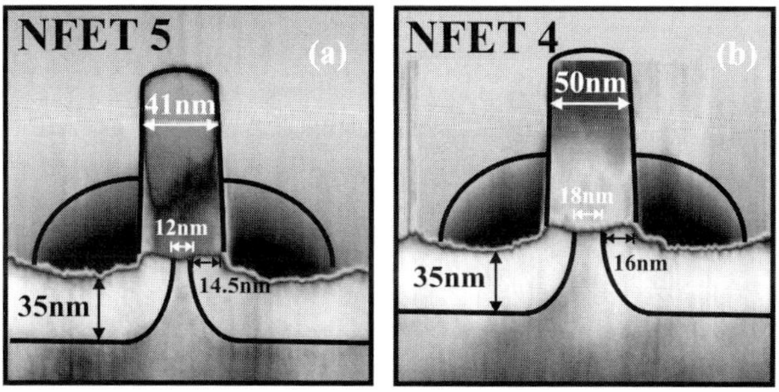

FIGURE 34. SSRM resistance map of 40-nm n-MOSFETs transistors with (a) and without (b) nitridation step.

the influence on lateral diffusion in a nitridation-processing step has been studied. In Figure 34, typical SSRM two-dimensional resistance maps are presented for a 40-nm nominal gate size for transistors with (NFET 5) and without (NFET 4) the nitridation step. The results extracted from these maps (i.e., the source/drain implant electrical junction depth and the lateral underdiffusion for the different gate sizes) are summarized in Table 3. For completeness and assessment, these results are also presented with the one obtained for transistors with a larger (150 nm) nominal gate size [64]. Please note the excellent reproducibility of the measurements over the different structures. The variation of the vertical junction depth is below 1 nm.

In Table 4, a comparison between SSRM and STM results is presented which shows an excellent qualitative agreement (1–2-nm suppression of lateral diffusion induced by the nitridation processing step) and a reasonable quantitative agreement (3–4-nm difference between STM and SSRM). The STM results relate to the averaged values from several transistors to eliminate the effects due to fluctuation of the fabrication process. The slightly increased lateral diffusion measured with SSRM is probably due to the relatively large bias used in this study (500 mV) as well as to the different procedures in STM and SSRM to determine the (apparent) junction position. In STM the junction position has been defined as the edge of the

TABLE 3. SSRM vertical junction depths and lateral diffusion lengths on 40-nm n-MOSFETs transistors

	Vertical depth [err = 4 nm]	Lateral diffusion [err = 2 nm]
NFET4 - 150nm	35 nm	15 nm
NFET4 - 40nm	35 nm	16 nm
NFET5 - 150nm	35 nm	14 nm
NFET5 - 40nm	35 nm	14.5 nm

TABLE 4. Comparison between STM and SSRM measured lateral diffusion lengths on 40-nm n-MOSFETs transistors

	STM	SSRM [err = 2nm]
NFET4	12 nm	15 nm
NFET5	10 nm	14 nm

depletion layer, whereas in SSRM the junction position is derived from the maximum in resistance peak, which is typically located in the middle of the depletion layer. For junctions with a low concentration on one side, the difference can be significant in view of the relatively large dimensions of the depletion layer. This explains the larger difference between STM and SSRM for the vertical junction measurement.

The measured suppression of lateral diffusion with the nitridation step is also in agreement with the electrical characteristics of the devices and is most likely to be linked with a nitrogen incorporation during the nitridation step (see [63]).

2.5.3 Activation Mechanisms in SPER MOS Devices

For deep sub-100-nm CMOS technology, it is of major importance to produce ideal box-shaped, ultra-shallow, and highly doped profiles, in order to maintain reasonable source/drain series resistance and short channel effects (SCE). As it is very difficult to achieve this with classical implantation and annealing technologies (in particular for boron due to solid solubility and transient enhanced diffusion problems [65]), in the last few years, use of *pre-amorphization implantation* (PAI) followed by *solid phase epitaxiale regrowth* (SPER) has been proposed as an alternative.

Junctions formed with PAI followed by SPER result, however, in poly-silicon doping deactivation and suffer from drive current degradation problems that have been discussed by Severi [66]. SSRM and TEM have revealed an additional problem of SPER junctions as an obvious relationship between the presence of defects in the *end of range* (EOR) region and halo implant de-activation has been observed. The SSRM image in Figure 35(a) clearly illustrates a deactivated area (dark spot inside circle) in a region corresponding to defects (circle in the TEM picture), which is a possible explanation for the poor electrical characteristics of this device. Note that no back-end or SiO_2-protecting layer was present on this sample, thus explaining the poor delineation with gate and source/drain implants.

We have also measured p-MOS structures with a SPER junction formation (with a thermal budget below 650°C) for both lowly doped diffusion (LDD) also named extension and highly doped diffusion (HDD), and with a Ni-FUSI (fully silicided) gate electrode. Figure 35(b) shows a TEM image (left) of a FUSI gate transistor with SPER junctions. A comparison with SSRM (right) reveals the sharp profile of the

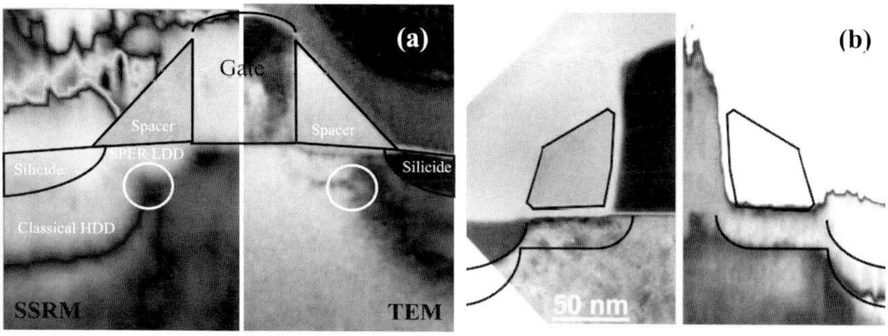

FIGURE 35. (a) Comparison of SSRM (**left**, white is more conductive) and TEM (**right**) results on a SPER p-MOS device (HDD implant is realized classically but LDD implant is a SPER implant). Deactivated area is indicated with a white circle and corresponds to the EOR region. (b) XTEM image of a FUSI gate p-MOS transistor (**left**). SSRM view of SPER LDD and HDD junctions profile (**right**). A important detail is the absence of diffusion of the HDD under the spacer. Note also the presence of a super-halo implant (white circled spot under the extension).

junctions and also the good junction overlap under the gate. Deep SPER junctions completely suppress the boron diffusion under the spacer. Super halo profiles are obtained with SPER, which opens up the halo design space for accurate short channel effect (SCE) control. Note that the junction leakage is greatly reduced by engineering the damage region away from the junction depletion region. Details on this successful integration of truly diffusionless junction formation by SPER in p-MOSFETs in combination with Ni-FUSI gates are to be found in [67].

2.5.4 SOI-Transistors

Together with gate-length decrease and the source/drain implant depth reduction in standard bulk MOSFETs, device concepts like ultrathin silicon-on-isolator (SOI) transistors are introduced to improve the device performance. SOI devices fabricated on SOI wafers with an Si layer of 45 nm and a 100-nm buried oxide were studied by SSRM. The gate lengths ranged from 200 to 50 nm and the gate oxide was 2.5 nm thick. Doped channels with boron (B) 9×10^{17} cm^{-3} and phosphorus (P) 1×10^{19} cm^{-3} were implemented for nMOS and pMOS devices, respectively [68].

The SSRM measurements for the p-MOSFETs (Figure 36(a,b)) show a lateral diffusion for source/drain under the gate region of ~25 nm, leading to a shortage for devices with gate lengths of 50 nm. Correlation with the electrical measurements of the devices was found (Figure 36(c)), with non-working characteristics for devices with gate lengths of 50 nm. For the n-MOSFETs, good performance was found up to gate length of 50 nm, in agreement with electrical measurements (Figure 37(b)); and SSRM could successfully resolve the gate length for such small transistors (Figure 37). Different lateral diffusion/abruptness was found at different depths

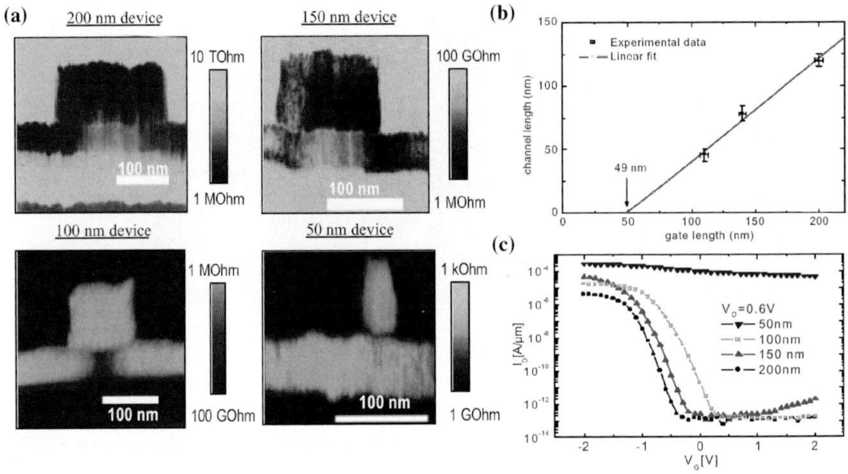

FIGURE 36. (a) SSRM of pMOS SOI devices with different gate lengths. (b) Gate length as measured by TEM versus channel length as measured by SSRM. Lateral underdiffusion (~25 nm) of this particular process leads to a complete shortage for 50-nm devices. (c) Electrical characteristics for different gate lengths (pMOS).

of the SOI layer for this process. The resistance profile shows a variation of 7 nm between the underdiffusion at 10 nm and 25nm from the top edge of the Si layer (Figure 37(c)) at half the variation of the spreading resistance with respect to the channel.

Fine details in the dopant profile in the SOI layer are predicted by process simulations (Athena) and resolved by SSRM. The bump-like profile originating from the As implant conditions is clearly resolved in the resistance profile (Figure 37(c)) in agreement with the predicted shape. The relative variation in underdiffusion for the two mentioned processes (p-MOS and n-MOS) was quantified for 100-nm transistors and was found to be 18 nm (Figure 38), providing the necessary feedback for correction/modification of the processing conditions.

Another particularly exciting illustration of the resolution and capabilities of SSRM is shown in Figure 39. This (35nm Si) SOI-device was showing anomalous conduction in its normally OFF-state. The SSRM image clearly shows the presence of an inversion layer (without gate bias) as the cause of this high leakage.

2.5.5 Multiple Gate Transistors (MuGFETs)

Another device concept that has attracted much attention recently is the multiple gate transistor (MuGFET), which hhas been proposed to reduce the short channel effects in narrow gate length ($L_g < 40$ nm) transistors. Different architectures of MuGFETs have been proposed (double gate, triple gate, surrounding gate, FinFET,etc). As these structures are no longer planar devices, simple

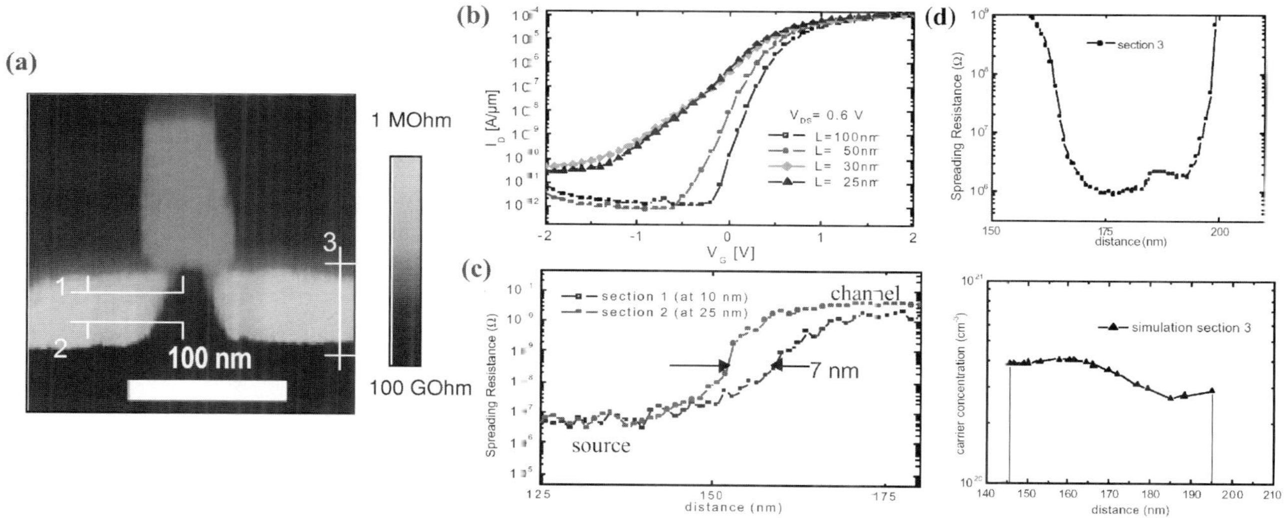

FIGURE 37. (a) SSRM measurement of a 50-nm nMOS SOI device. (b) Electrical characteristics for different gate lengths (nMOS) (c) Section lines 1 and 2, comparing the underdiffusion at different depths of the SOI layers (variation = 7 nm). (d) Resistance depth profile for the SOI layer (section 3) and simulated profile.

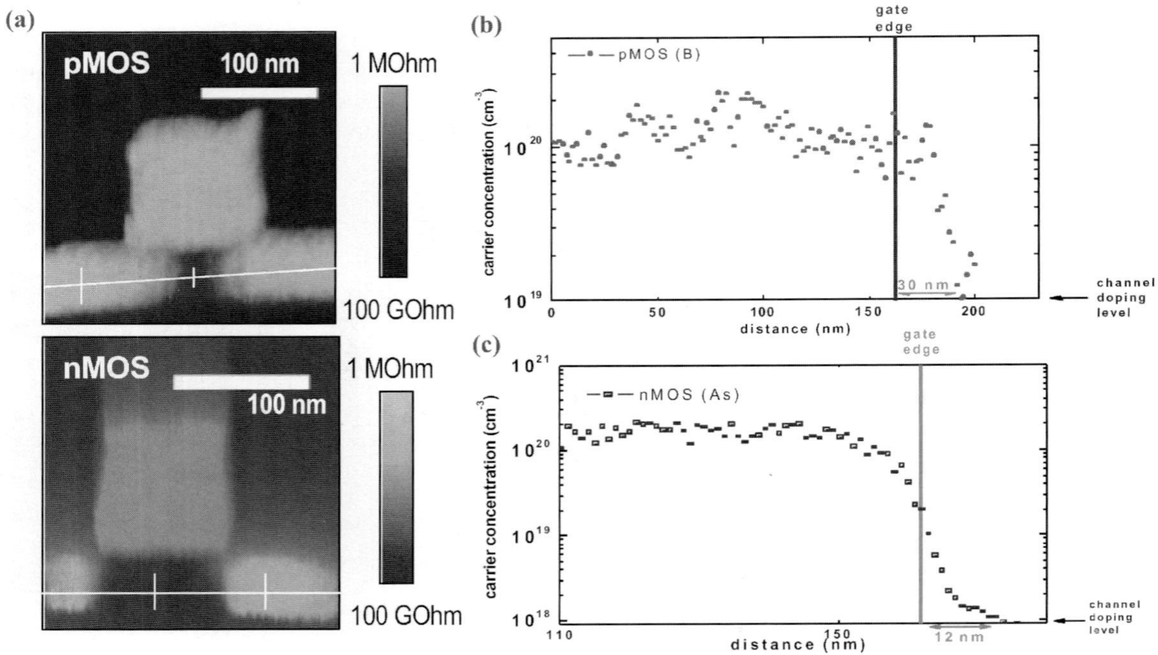

FIGURE 38. (a) SSRM image of two 100-nm SOI devices (pMOS and nMOS). (b) Section lines comparing underdiffusion for both processes, the difference was found to be: $30 - 12 = 18$ nm.

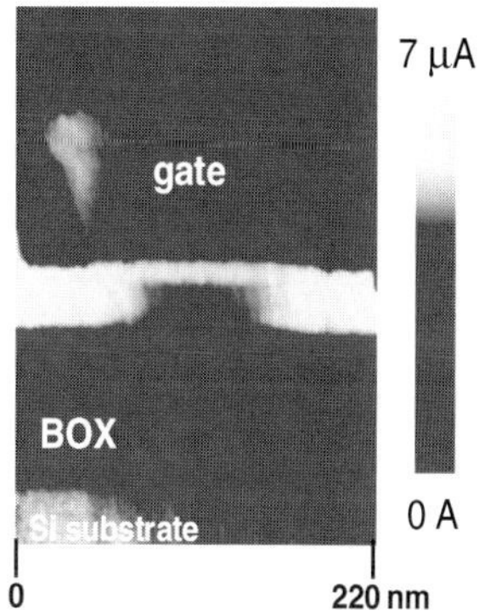

FIGURE 39. Visualization of an inversion layer in an SOI device.

one-dimensional characterization becomes obsolete and two-dimensional profiling tools as SSRM become necessary.

In Figure 40, two-dimensional profiling is illustrated through the analysis of a B implant in a Si nanoridge (or fin). A 2D effect is observed (reduction of junction depth along the Si sides), and associated to a reduction in TED due to an increased annihilation of interstitials at the lateral Si surfaces [69].

3 Dopant Profiling in InP with SSRM

Although originally developed for 2D profiling in Si, SSRM can equally well be used in other areas such as III-V materials.

Here, SSRM applications to InP-based lasers are discussed in more detail. Due to its intrinsic material properties, InP is an indispensable material for optoelectronics for fiber telecommunications. In state-of-the-art InP-based optoelectronics, e.g., InP-based laser devices for 1.3 and 1.5-μm wavelengths, complex device structures are crucial for performance and device reliability. Because these 2D features usually have a relatively large dimension with respect to contemporary CMOS technology, SSRM on InP does not demand nanometer spatial resolution, as it does with the silicon CMOS technology. However, the different nature of the point

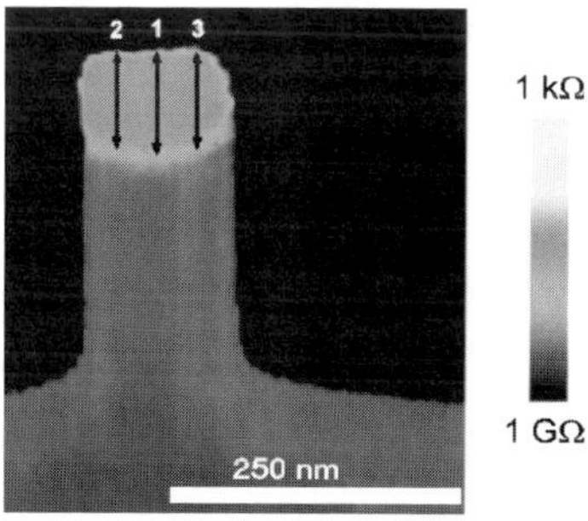

FIGURE 40. SSRM image of the Si ridge after implantation of 1×10^{15} at cm^{-2} B and annealing with a 0 s anneal at 1050°C. There is a clear difference in the depth of the dopant profile in the middle of the structure (1) (97 nm), and at 30 nm from the sides (2) and (3) (91 and 89 nm, respectively).

contact between the probe and the InP sample necessitates new experimental and theoretical explorations. In this section, a brief introduction is given to this topic, with a few application examples.

3.1 Calibration Curves and Resistance Response Relationship

An experimental optimization of SSRM on InP requires forming a stable electrical contact between the probe and the sample, avoiding the unwanted tip-induced oxidation during the SSRM scan [70]. A diamond tip proved to be essential as it can easily penetrate the oxide layer grown on InP during the measurement and thus maintain a stable contact. The capability to penetrate (and to remove) the oxide layer in a controllable way is shown in Figure 41. The topographical result shows a 2×2-μm^2 box-like pit in the center of the region, which corresponds to a previous SSRM scan. Note that the removed InP material is attached to the probe or is found as a pile-up at the edge of the scanning region.

Using this approach, Figure 41(b) shows the response curve of the measured resistance to the load force on a homogenously doped sample under a typical bias condition. The measured resistance decreases dramatically when the load force increases until it reaches a rather stable region. Such a transient curve can be divided into two parts. A fast drop region is from 0 to 0.35 (the load force is of the order of μN), with the measured resistance changing from 10^{10} to 10^6 Ω. A stable region starts from 0.35, with only small variations in the measured resistance.

As the load force increases, the oxide thickness between the probe and InP decreases due to the mechanical deformation, resulting in increased tunneling of carriers through this contact. Due to the exponential nature of the tunneling transmission coefficient through an interfacial oxide [71], the decrease of the measured resistance is rather rapid. This is observed as a fast drop in Figure 41(b). As the probe penetrates through the oxide layer forming a metal-semiconductor contact with negligible oxide thickness, the change in the measured resistance slows. Instead, the further increase of the load force results in an increase of the contact area, which shows a slow drop of the measured resistance in the stable region.

This two-phase behavior in the response curve can be modeled when considering penetrating the interfacial oxide layer. A simple linear relation between the (remaining) interfacial oxide thickness and the load force is assumed for a first region, while a linear relation between the contact size and the load force is assumed for the second phase. Taking measured data points in Figure 41(b) to fit the above relationships, the dashed line in Figure 41(b) can approximate the response curve.

Similar to the Si case, quantification is based on calibration curves. Depending on the operational conditions (metal versus diamond probe), the calibration curves on n-type and p-type samples can be very different (Figure 42) [70, 72]. The calibration curves before the optimization show −0.3 decade/decade for the n-type sample and almost flat for the p-type sample, which means a very limited concentration sensitivity. In contrast, the slope of calibration curves by using the

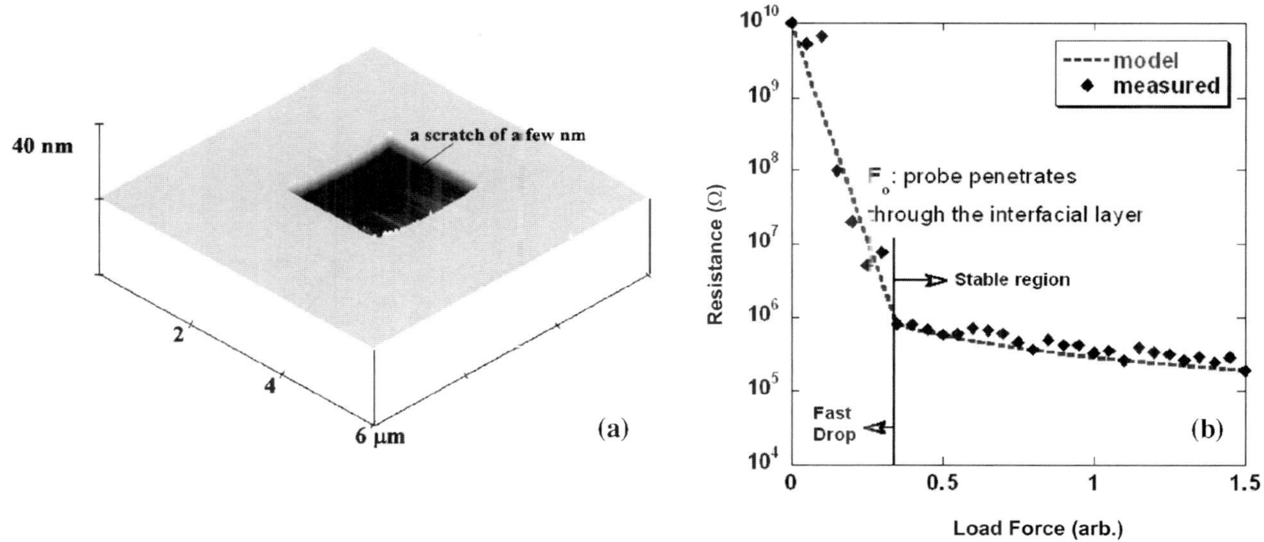

FIGURE 41. (a) An AFM topographical measurement shows a 2 × 2-μm² scratch (a few nm in depth) formed after an SSRM scan in this optimized procedure. (b) The response curve between the load force and the measured resistance can be divided into two distinguished regions.

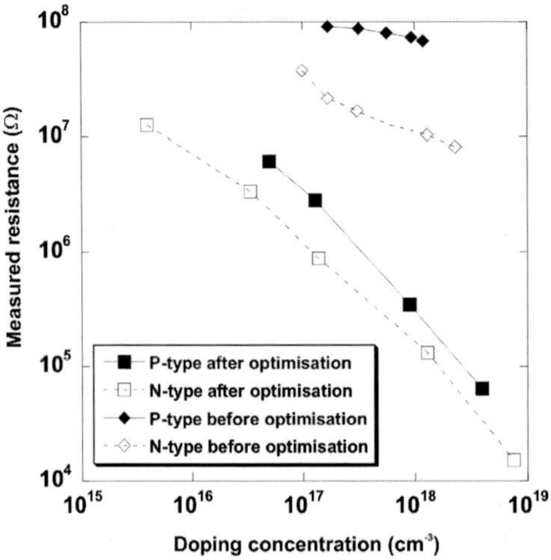

FIGURE 42. The calibration curves after optimizations show a close-to-unity sensitivity for both n-type and p-type InP, which are compared with the calibration curves before optimizations.

optimized SSRM is around −1.0 decade/decade for each carrier, which seems to coincide with the prediction of spreading resistance dependence on bulk carrier concentration in the theory, although this is not necessarily universal, as will be explained later. Another obvious difference between these two situations is that the measured resistance value before the optimization is more than one decade larger than it after optimization at the same carrier concentration. This is explained by a large contact barrier resistance due to the finite interfacial oxide on InP.

3.2 SSRM Point Contact Modeling on InP

Classical metal-semiconductor studies were based on the contact formed on a large scale, which could be approximated by a one-dimensional (1D) simplification. The metal-semiconductor contact modeling was also extended to include the tunneling effect through the barrier [71], interfacial oxide layer effect [73], etc, providing the background for this work. However, due to the extreme small size of the point contact formed in SSRM, the SSRM point contact modeling on InP is based on 2D/3D device simulations that combine the advanced device models with multi-dimensional boundary conditions, providing a powerful tool to tackle this problem.

As the conductive probe penetrates the oxide layer in SSRM, the point contact can be modeled as an intrinsic metal-semiconductor contact on InP. Unlike its original application on Si, SSRM on InP is dominated by a Schottky barrier resistance, instead of a spreading resistance, at most carrier concentrations. Hence, a dedicated point contact model is developed to describe the electrical properties of SSRM on InP using as boundary conditions a contact radius of 10 nm and a Schottky barrier of 0.4 eV.

Also, the polished (or cleaved) surface is far from ideal, and the presence of SS needs to be accounted for. Hence a SSRM model must include the Schottky barrier tunneling effect and the InP surface pinning effect.

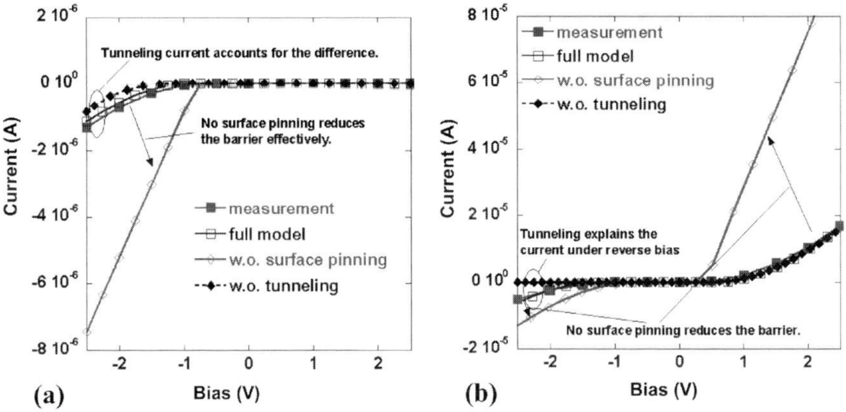

FIGURE 43. The measured I-V curve on p-type (a) and n-type (b) InP is compared with different point contact models.

In Figure 43(a) and Figure 43(b), I-V curves on a p-type sample doped at 1.3×10^{17} cm^{-3} and an n-type sample doped at 1.4×10^{17} cm^{-3} are simulated and compared with measurements. The dedicated point contact model includes both the tunneling effect through the Schottky barrier and surface/interface pinning. It is also compared with the other two cases neglecting either the tunneling effect or surface pinning around the metal-semiconductor point contact illustrating the importance of these effects.

The n-type I-V (Figure 43(b)) shows a considerable conduction in the reversed bias range, which is attributed to the tunneling current through the barrier. The simulation confirms this by the result without tunneling which shows little current in reverse bias condition. If no surface pinning is included around the point contact, a large amount of current is observed in the simulated result; i.e., the effective barrier height of the metal-semiconductor contact at the SSRM (nm) scale is reduced significantly, resulting in an excessive amount of current in both forward and reverse bias conditions.

The p-type I-V (Figure 43(a)) shows a typical Schottky-like behavior in the experimental data as well as in the three simulated results. It is clear from the simulation that tunneling through the Schottky barrier is not sufficient to cause high conduction for p-type because of a high Schottky barrier height and a large hole mass. Again, the surface pinning conditions are still important at this contact size, as reduced barrier height provides more current in the forward condition.

The calibration curves scanning a large dynamic range have been simulated using the same parameters. These results are compared with experimental data in Figure 44(a). These data are extracted from simulations of I-V at different doping concentrations at the bias of -2 V. A reasonably good agreement is observed in the full dynamical range, while the n-type simulated curve apparently shows a slightly lower sensitivity than the experimental data.

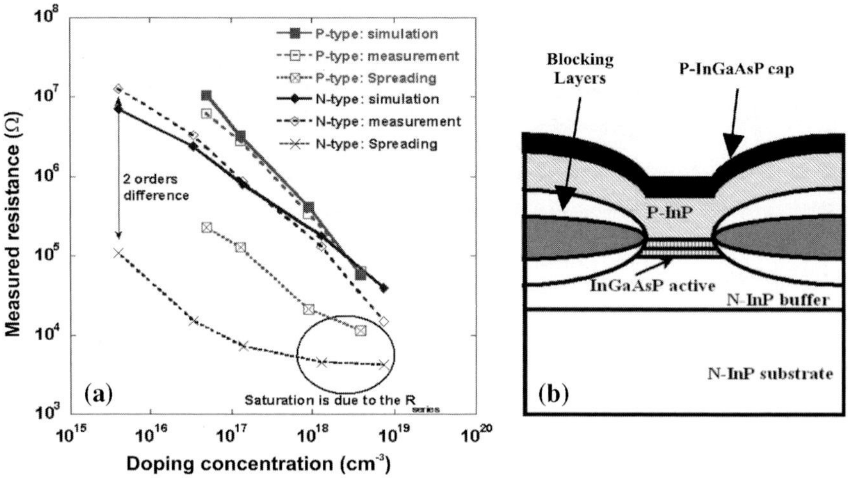

FIGURE 44. (a) The measured calibration curves are compared with simulated results of both the dedicated point contact model and the spreading resistance dominated model. (b) Schematics of InP-laser made with a selective EPI growth.

When SSRM is applied to Si, the resistance is dominated by the spreading resistance due to the contact formed by a metallic phase transition in silicon underneath the probe. In InP, this metallic transition is absent and the resistance is strongly influenced by the complexity of the contact barrier. A simple ohmic approach would lead to the "virtual" calibration curves in Figure 44(a) (10-nm radius). The p- versus n-type difference originates from the differences in mobility of the two carrier types in InP. Their absolute value is, however, about 100 times lower than the experimental results.

When using the model explored in Figure 43(a) and Figure 43(b), a good agreement with experimental data can be obtained. Hence SSRM on InP is not strictly a spreading resistance dominated technique but is based on the Schottky barrier and tunneling effect at a small scale. The non-linear contact resistance originating from the Schottky barrier explains the measured carrier concentration dependence of SSRM on InP. Again, most of these non-linearities are removed in the quantification procedure by relying on a conversion using the calibration curves.

3.3 Application to Laser Structures

In order to achieve the high-speed data communication, InP-based lasers are widely used as the transmitter and the optical amplifier. Their fabrication leads to a complex 2D structure (Figure 44(b)).

Dedicated InP layers are used to passivate the surface, forming a high-quality interface. Moreover, dedicated structures (i.e., doping profiles) formed in the selective EPI growth (SEG) phase can provide both current and optical confinement.

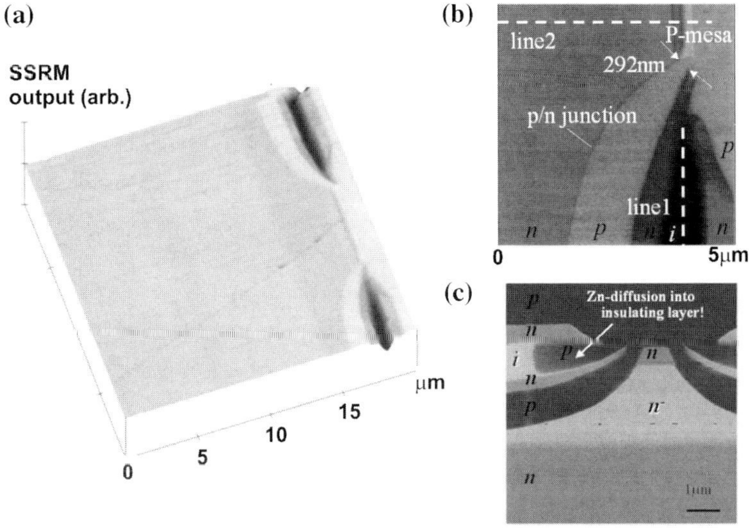

FIGURE 45. The first type of InP-based lasers measured by SSRM: (a) a SSRM measurement at a large scale, (b) a close-up image close to the mesa of interests, and (c) a similar InP laser structure from an old process for a comparison.

Different types of structures can be used but should block the current by using either semi-Insulating (SI) layers or reversed p/n junctions. Due to the different dielectric permittivities between the active region and the blocking regions from SEG, it can also provide optical confinement for the vertical lasering mode, provided a suitable structure is carefully designed. Additionally, the high resistive blocking layers can also reduce the parasitic capacitance of the total structure, which is critical for high-speed applications [74–76].

However, due to the complexity of the growth rate and the dopants' incorporation in the SEG on a non-planar InP substrate, it can have a complicated and unpredictable 2D dopant/carrier distribution. Hence, SSRM offers here a unique capability, providing a detailed 2D analysis of the structure, overcoming the limitations of conventional technique [77, 78].

Figure 45(a) shows a typical example of InP-based laser structures measured by SSRM. The SSRM result shows that it has an etched mesa ($\sim$6 µm in width) on top of the active region, with a multilayered blocking region on each side of the mesa. The cross-sectional line analysis along the multilayered blocking region shows a p-n-SI-n-p-n structure, which agrees with the SIMS result.

A close-up look at Figure 45(b) near the mesa resolves more details such as a small opening between the p-type mesa and the p-type layer. It is desirable to get a precise width, as this determines the behavior of the device. A clear p/n junction is also resolved by SSRM between the n-type substrate and p-type layer (shown as thin dark line). This dark curve is due to the depletion region at the junction. As a comparison, another similar InP-based laser structure from a old process,

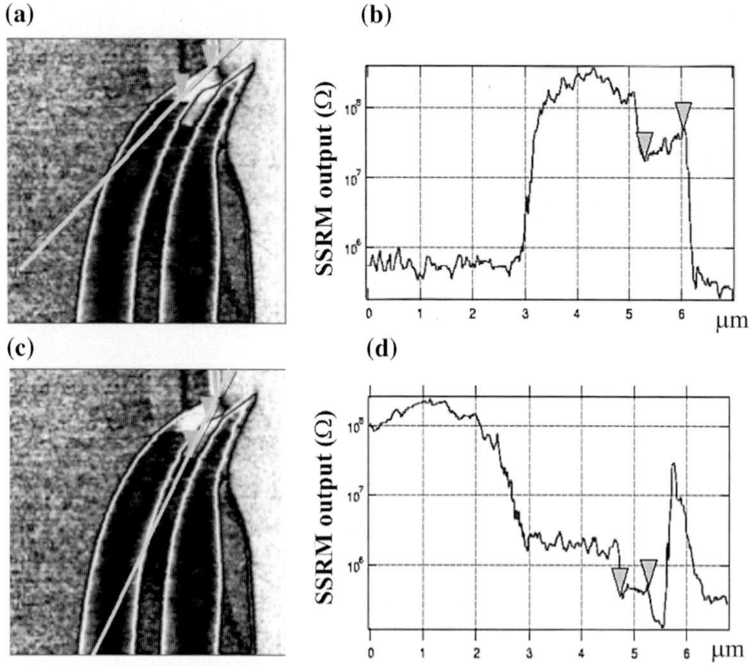

FIGURE 46. The second type of InP-based lasers measured by SSRM: (a) SSRM measurement with a line analysis along the SI layer, (b) the corresponding SSRM output in (a) showing a strong increase of the resistance close to the mesa as marked by two triangles, (c) SSRM measurement with another line analysis along the n-type region, and (d) the corresponding SSRM output in (c) showing an unknown decrease of the resistance close to mesa as marked by two triangles.

which is suffering from the inter-diffusion between Zn in the mesa and Fe in the SI layer, is shown in Figure 45(c) [72]. An inter-diffusion tail up to a few microns was measured by SSRM. It is clear that the laser device in Figure 45(a) does not have the strong inter-diffusion in the new process.

Figure 46 shows another type of laser device measured in our work. Of interest here is the different structure close to the mesa which consists of a stack of n-SI-n-SI-n-p junctions. Basically, two SI layers act to block the current rather than one as in the first type of laser device. However, SSRM shows a lower measured resistance ($\sim 10^8$ Ω) in these two SI layers. The cross-sectional analysis in the SI region in Figure 46(b) shows a strong increase of resistance close to the mesa (~1 μm) marked by two triangles. Meanwhile, in Figure 46(d), the line along the n-type region between the two SI layers (see Figure 46(c)) has an unknown increase in carrier concentrations when it is close to the mesa. These two observations indicate a very complex dopant interaction in the lateral direction in this process and suggest an increased carrier concentration in SI due to some Zn out-diffusion from the mesa. However, the increased carrier concentration in n-type

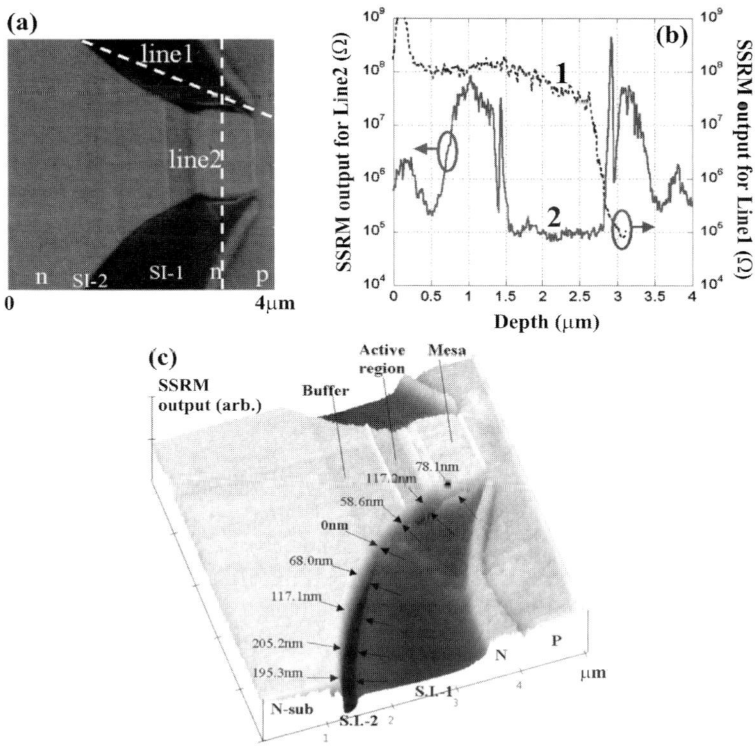

FIGURE 47. The type 3 of InP-based lasers measured by SSRM: (a) the SSRM result close to the mesa, (b) the corresponding line analyses in (b), and (c) the 3D vision revealing a critical region in SI-2 in this device.

region close to mesa seems due to n-type impurity segregation, which is hard to explain with a simple 2D diffusion model.

Figure 47(a) shows the third type of InP lasers measured by SSRM. Two SI layers are grown by SEG in this structure. The insulating layer SI-2 is non-homogeneous and becomes ultrathin at some critical points. The measured thickness is illustrated in Figure 47(c). Note that a critical zero thickness is found in the middle of SI-2 inside the device. Further details are revealed by line analyses. Line1 in Figure 47(b) shows that the measured resistance by SSRM is not homogeneous in the SI-1 region, which is much smaller than the measured resistance (more than 10^9 Ω) in SI-2. When line2 crosses the mesa, two highly conductive slices next to the mesa, are seen linked with the p-type cap layer. The gap between the slice and the mesa is the extension of SI-2, which has a high SSRM resistance up to 10^9 Ω. Moreover, the extension slice is also symmetrically visible in the other side of this mesa. From this analysis, it reveals the extremely complicated SEG mechanisms of two SI layers around the mesa. Due to the 2D nature in the growth rate and the dopant/carrier profiles, neither conventional 1D profiling techniques (e.g. SIMS,

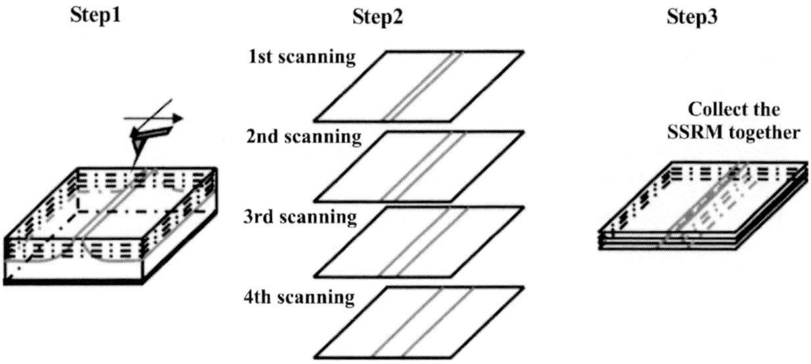

FIGURE 48. Schematic illustrating how to extend SSRM to the third dimension in a 3D mode.

etc.) nor 2D metrology techniques (e.g., SEM, etc.) are capable of this type of analysis.

As illustrated in the analyses of a few typical InP-based laser structures, SSRM shows its unique capability in resolving problems with extremely 2D fine details, which is particularly useful for the InP-based laser devices using complex SEG fabrication technique. These semi-quantitative analyses not only help us to characterize and understand the physical properties of these devices, but also can give direct feedback to device and processing engineers.

3.4 3D Profiling Capability on InP

So far SSRM has proven to be a promising 2D carrier profiling tool with high spatial resolution and a large dynamical range for its application to InP-based devices. A key point in the successful application of SSRM to InP has been the introduction of diamond probes: hardness and the sharpness of coated diamond grains result in the removal of some layers of InP material, which counteracts the effect of oxide growth during each scan. The latter effect can also be used to expand the application of SSRM from a conventional 2D mode toward a 3D carrier profiling capability. In this section, we present 3D SSRM on InP-based devices.

The basic idea of 3D SSRM is illustrated in Figure 48. As the first step, a series of SSRM measurements are taken from the top surface (instead of the cross section of the sample). In view of the hardness of the diamond probe and the removal of InP material, the scanning probe actually progressively digs into the sample in SSRM measurements. Thus at each scan the probe goes one step deeper, providing the sampling capability in the third dimension (i.e., relative to the normal 2D scanning plane). The full 3D SSRM profile is than constructed by collecting SSRM data during each scan and by combining them with the depth of each scanning phase from the topographical measurements at the same time. In the third step a 3D profile is recovered from a set of separated SSRM measurements as the schematic.

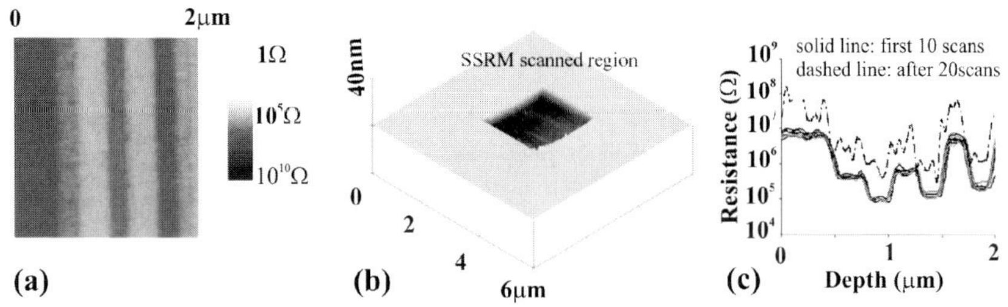

FIGURE 49. (a) A typical SSRM measurement is obtained on an InP sample. (b) The topographical measurement by AFM shows the InP removal from the previous SSRM scan. (c) Multiple SSRM scans illustrate good repeatability for at least first 10 scans.

Figure 49(a) shows a typical SSRM result taken from an epitaxial multiplayer far away from the mesa in InP laser sample. The topographical result measured afterward, over the previously scanned region, shows a well-defined square pit coincident with previous SSRM measurements in Figure 49(b). Repeated scanning on the same area will thus provide the measured resistance as a function of depth. Note that good control of progressive scans over the InP sample and maintaining reliable SSRM measurements simultaneously are the keys for successful 3D profiling. As exemplified by the line scans taken across these results (Figure 49(c)), these successive measurements show good repeatability at least for the first 10 measurements. Further repeated SSRM measurements (especially after 20 scans) begin to show a larger discrepancy and more noise. We attribute this degradation to the wearing out of the tip and/or the pile-up of removed material around the pit. This is confirmed by measuring a complex staircase multiple times (thus eroding the cross-sectional plane).

The 3D mode is not only useful for real 3D structures but can also resolve some of the depth resolution limitations when using conventional 2D SSRM profiling on the cross section of a device. Since the resolution of SSRM is limited by the tip radius (around 20 nm for the diamond-coated tips), extremely fine layer structures cannot be resolved in this mode. In addition, making a cross section of structures with large built-in stress proves to be difficult. A solution to this problem is given by 3D SSRM. In this mode, the probe scans on the top surface (instead of the cross section) of the sample with decreasing scan sizes, so that a well-defined staircase pit (Figure 50(a)) is formed. A cross-sectional analysis in Figure 50(b) shows clear each step is around 20 nm, which could be controlled by adjusting the load force on the probe. Note that the pile-up of material disappears in this image, because contact mode AFM scanning removes them to the outside of imaged region. Subsequently SSRM measurements are performed along the layered surface of the etched pit and a topographical result and a SSRM result are collected simultaneously. In this way we can construct a resistance profile as a function of the depth by combining data from both results. Since the staircase-like scratched pit basically acts as the bevel magnifying the vertical scale, limitations in resolving very thin layers are no longer related to the tip radius.

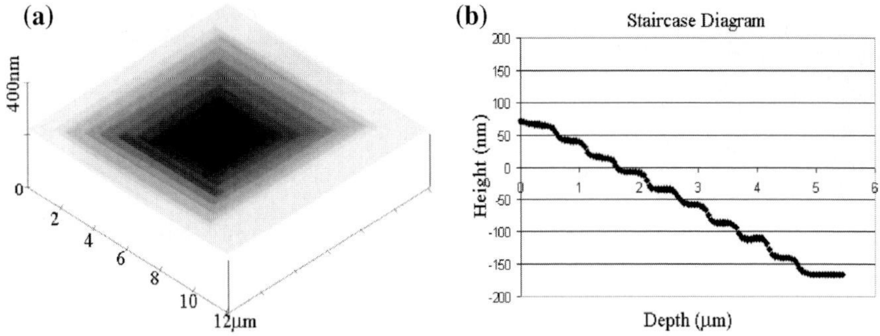

FIGURE 50. (a) Topographical image after a series SSRM scans by diamond probe. (b) Corresponding cross-sectional analysis of (a).

This is illustrated in the case of an EPI sample composed of a stack of alternating p-type and n-type doped layers where the layer thickness is decreasing toward the surface (for dopant profile as measured by SIMS in Figure 51(a)). The conventional SSRM taken on the cross section is able to delineate the thicker layer but cannot distinguish the fine layers close to the edge and to the top surface, whose size is less than 20 nm. Obviously the fact that the probe starts contacting more than one fine layer at the same time limits the resolution for the thinnest layer, which leads to a loss in layer definition and the amplitude of the oscillation in measured resistance profile (Figure 51(a)). The results taken in the depth-profiling mode on the other hand show quite clearly a superior resolution and demonstrate that layers as thin as 10 nm can be adequately resolved (Figure 51(b)). Also layers very close to the top surface are visible. A substantial improvement in the spatial resolution is obvious from the comparison between results of traditional mode and new mode and with the corresponding SIMS profile.

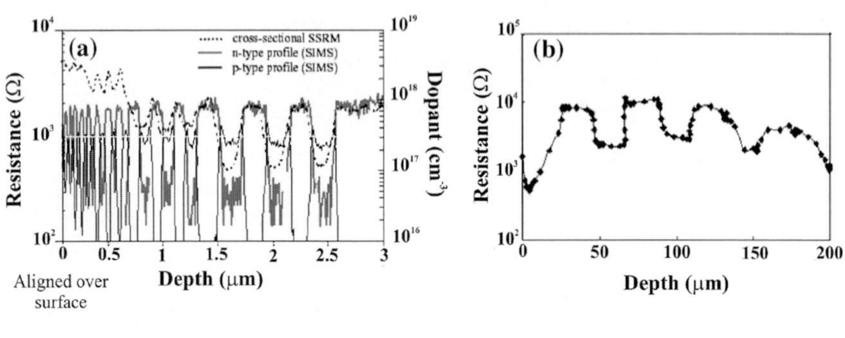

FIGURE 51. (a) InP profiles with varying layer sizes measured by SIMS and normal SSRM illustrating the limitation in spatial resolution in the latter. (b) 3D SSRM measurement on the same sample proving its superior spatial resolution in the third dimension.

In conclusion, it is shown that SSRM is able to perform 3D measurements on InP-based samples with an optimized experimental condition. The key to this operation is the use of a highly conductive diamond-coated probe, which provides a good electrical contact and at the same time leads to a controlled removal of the InP material. The application can range from a full 3D profiling to a "depth profiling" mode where the limitations of the cross-sectional profiling (resolution limited by tip radius, proximity of the surface, difficulty in cleaving sample with high built-in stress) can be overcome. We believe that SSRM, with its capability of 3D profiling, is a promising tool to meet the requirement of characterizing multidimensional microelectronics devices in the future. However, quantitative characterization based on a well-defined calibration sample and 3D device modeling is required for further evaluation.

4 Conclusion

SSRM has emerged as one of the most important 2D-profiling methods in the semiconductor industry, as it provides a unique spatial resolution (1–3 nm) combined with high concentration sensitivity and ease of quantification. Essential in its application to Si is the formation of a metallic phase due to the high pressures used, which provides a electrical contact with a radius smaller than the actual tip radius. Diamond as a tip material (coated or molded) proved to be the key to its success. Numerous applications range from process optimization and TCAD calibration to high-resolution imaging of inversion layers in devices. Its application is not restricted to Si: it has also been proven to be highly beneficial for the analysis of complex InP-lasers SSRM.

References

1. M. Duane, P. Nunan, M. ter Beek, and R. Subrahmanyan, JVST **B14**, 218 (1996).
2. R. G. Wilson, F. A. Stevie and C. W. Magee, Wiley, New York (1989).
3. R. G. Mazur and D. H. Dickey, J. Electrochem. Soc. **113**, 255 (1966).
4. W. Vandervorst and T. Clarysse, JVST B10(1), **302** (1992).
5. M. Pawlik, JVST B10(1), **388** (1992).
6. P. Borden, L. Bechtler, K. Lingel and R. Nijmeijer in *Handbook of Silicon Semiconductor Metrology,* edited by A. C. Diebold, Dekker Inc., NY (2001).
7. W. Vandervorst and M. Meuris, EP-466274, Jan. 15 (1992).
8. W. Vandervorst and M. Meuris, US-5369372, Nov. 29 (1994).
9. P. De Wolf, J. Snauwaert, T. Clarysse, W. Vandervorst, and L. Hellemans, Appl. Phys. Lett. **66**, 1530 (1995).
10. P. De Wolf , Ph.D. thesis , University of Leuven, Belgium (1998).
11. J. N. Nxumalo, D. T. Shimizu, and D. J. Thomson, JVST B **14**, 386 (1996).
12. C. Shafai, D. J. Thomson, M. Simard-Nomardin, G. Mattiussi, and P. J. Scanlon, Appl. Phys. Lett. **64**, 342 (1994).
13. T. Hantschel, S. Slesazeck, P. Niedermann, P. Eyben, W. Vandervorst, Microelec. Eng. **57–58**, 749 (2001).

14. M. Fouchier, P. Eyben, D. Alvarez, N. Duhayon, M. Xu, J. Lisoni, S. Brongersma and W. Vandervorst, Proc. of SPIE **5116**, 607 (2003).
15. T. Clarysse, P. De Wolf, H. Bender and W. Vandervorst, JVST B **14**(1), 358 (1996).
16. P. De Wolf , T. Clarysse and W. Vandervorst, JVST B **16**(1), 320 (1997).
17. P. Eyben, S. Denis, T. Clarysse and W. Vandervorst, Mat. Sci. & Eng. B, **102**(1–3), 132 (2003).
18. P. Eyben, M. Xu, N. Duhayon, T. Clarysse, S. Callewaert and W. Vandervorst, JVST B **20**(1), 471 (2002).
19. P. Eyben, N. Duhayon, D. Alvarez and W. Vandervorst, AIP Conf. Proc. **683**, 678 (2003).
20. P. Eyben, D. Alvarez, T. Clarysse and W. Vandervorst, AIP Conf. Proc. **683**, 685 (2003).
21. J. Yang and F. C. J. Kong, Appl. Phys. Lett. **81**, 4973 (2002).
22. P. Eyben, M. Fouchier, P. Albart, J. Charon-Verstaffen and W. Vandervorst Mat. Res. Soc. Proc. **717**, C7.7 (2003).
23. D. Goghero, F. Giannozzo, V. Raineri, E-MRS Proc. 2002, (2002).
24. T. Hantschel, P. Niedermann, T. Trenkler, W. Vandervorst, Appl. Phys. Lett. **76**, 1603 (2000).
25. See: http://www.nanosensors.com/ diamondcoatedsensors.html
26. T. Trenkler, T. Hantschel, R. Stephenson, P. De Wolf, W. Vandervorst, L. Hellimans, A. Malavi, D. Büchel, E. Oesters-chulze, W. Kulisch, P. Niedermann, T. Sulzbach and O. Ohlsson. JVST B **18**, 418 (2000).
27. D. Alvarez, J. Hartwich, M. Fouchier, P. Eyben and W. Vandervorst, Appl. Phys. Lett. **82**(11), 1724 (2003).
28. T. Clarysse, M. Caymax, P. De Wolf, T. Trenkler, W. Vandervorst, J. S. McMurray, J. Kim, C. C. Williams, J. G. Clark and G. Neubauer, JVST B **16**, 394 (1998).
29. T. Clarysse, P. Eyben, N. Duhayon, and W. Vandervorst, JVST **B21**(2), 729 (2003).
30. S. S. Sheiko, M. Moeller, E. M. C. M. Reuvekamp and H. W. Zandbergen, Phys. Rev. B **48**(8), 5675 (1993).
31. R. Holm, "Electrical Contacts Handbook", Springer, Berlin, 17 (1958).
32. See: http://ece-ww.colorado.edu/~bart/book/book/chapter2/ch3_7.htm
33. S. M. Sze, John Wiley & Sons (1981).
34. M. Büttiker, Y. Imry, R. Landauer, S. Pinhas, Phys. Rev. B **31**, 6207 (1985).
35. Y. V. Sharvin, JETP (21), 655 (1965).
36. M. Brandbyge, J. Schiøtz, M. R. Sørensen, P. Stoltze, K. W. Jacobsen, J. K. Nørskov, L. Olesen, E. Laegsgaard, I. Stensgaard and F. Besenbacher, Phys. Rev. **B52**, 8499 (1995).
37. See: http://www.ise.ch/products/dessis/
38. S. Denis, Master's Thesis, Université de Liège, 15 (2002).
39. E. H. Pointdexter, G. J. Gerardi, M.-E. Rueckel, P. J. Caflan, N. M. Johnson and D. K. Biegelsen, JAP **56**(10), 2844 (1984).
40. E. R. Weppelmann, J. S. Field, and M. V. Swain, J. Mater. Res. **8**, 830 (1993).
41. G. M. Pharr, W. C. Oliver, R. F. Cook, P. D. Kirchner, M. C. Kroll, T. R. Dinger and D. R. Clarke. J. Mat. Res. **7**, 961 (1992).
42. J. Z. Hu, L. D. Merkle, C. S. Menoni, and I. L. Spain, Phys. Rev. B **34**, 4679 (1986).
43. M. C. Gupta and A. L. Ruo, JAP **51**, 1072 (1980).
44. P. De Wolf, Ph.D. thesis , University of Leuven, Belgium, 97–101 (1998).
45. P. De Wolf, Ph.D. thesis , University of Leuven, Belgium, 103–106 (1998).
46. P. De Wolf, Ph.D. thesis , University of Leuven, Belgium, **91** (1998).
47. J. J. Marchand and Van Khiem Truong, JAP **54** (12), 7034 (1983).
48. See: http://www.marc.com/

49. T. A. Laursen and J. C. Simo, Mat. Res. **7**, 618 (1992).
50. C. Raskon internal report, LETI (2003).
51. Thurber W. R., Mattis R., Liu Y. M. and Filliben J. J, Semic. Meas. Techn., NBS Special Publi. 400–64 (1981).
52. S.M. Hu, JAP **53**, 1499 (1982).
53. T. Clarysse, D. Vanhaeren, I. Hoflijk and W. Vandervorst. Mat. Sci. & Eng. Reports **47**(5-6), 123 (2004).
54. P. De Wolf, T. Clarysse and W. Vandervorst, JVST B **16**, **335** (1998).
55. P. A. Schumann and E. E. Gardner, Solid St. Electron. **12,** 371 (1969).
56. M. Meuris, W. Vandervorst, and P. De Wolf, US-5585734 (Dec. 17, 1996) and EP-466174 (Oct. 18, 1996) (1996).
57. P. De Wolf, T. Clarysse, and W. Vandervorst, US-5995912 (Apr. 29, 1997), (1997).
58. P. De Wolf, Ph.D. thesis , University of Leuven, Belgium, **142** (1998).
59. J. Yang and F. C. J. Kong, Appl. Phys. Lett. **81**(26), 4973 (2002).
60. P. De Wolf , Ph.D. thesis , University of Leuven, Belgium, 133–134 (1998).
61. N. Duhayon, T. Clarysse, D. Alvarez, P. Eyben, M. Fouchier, W. Vandervorst and L. Hellemans, Proc. ALTECH 2003, J.**1**, 743 (2003).
62. O. Douheret, S. Bonsels and S. Amand, JVSTB **23**(1), 61 (2005).
63. H. Fukutome,T. Saiki, M. Hori, T. Tanaka, R. Nakamura and H. Arimoto, Jpn. J. Appl. Phys. **43**(4B) (2004).
64. P. Eyben, H. Fukutome, D. Alvarez and W. Vandervorst, Proc. ESSDREC 2004, 101–104 (2004).
65. S. C. Jain, W. Schoenmaker, R. Lindsay, P. A. Stolk, S. Decoutere, M. Willander and H. E. Maes, JAP **91**(11), 8919 (2002).
66. S. Severi, K. Henson, R. Lindsay and K. De Meyer, Proc. ULIS 2004, Leuven, **69** (2004).
67. S. Severi, K. G. Anil, J. B. Pawlak, R. Duffy, K. Henson, R. Lindsay, A. Lauwers, A. Veloso, J.-F. deMarneffe, J. Ramos, B. Sijmus, K. Devriendt, R. A. Camillo-Castillo, P. Eyben, W. Vandervorst, M. Jurezak, S. Biesemans and K. DeMeyer, Proc. IEDM 2004 (2004).
68. W. Vandervorst, D. Alvarez, P. Eyben, M. Fouchier, J. Hartwich, S. Slesazeck, P. Verheyen and S. Hogg, Proc. ULIS 2004, **63** (2004).
69. S. Kluth, D. Alvarez, St. Trellenkamp, J. Moers, S. Mantl, J. Kretz, W, Vandervorst, JVST B. **23**(1), 76 (2005).
70. M. W. Xu, T. Hantschel and W. Vandervorst, Proc. 2001 Int. Conf. on InP and Rel. Mat., Nara, Japan, 2001.
71. C. Y. Chang and S. M. Sze. **13**, 727 (1970).
72. P. De Wolf, M. Geva, C. L. Reynolds, T. Hantschel, W. Vandervorst and R. B. Bylsma, JVST A.**17**, 1285 (1999).
73. Ching-Yuan Wu. JAP **51**, 3786 (1980).
74. P. Speier, K. Wunstel, and F. Tegude. Electron. Lett. **23**(25), 1363 (1987).
75. D. Bertone and G. Fornuto, Proc. 7th Intern. Conf. on InP and Rel. Mat., 279 (1995).
76. T. Tanbun-ek, R. A. Logan, and J. P. van der Ziel. Electron. Lett. **24**(24), 1483 (1988).
77. T. Kallstenius, U. Smith, and B. Stoltz. Journal of the Electrochem. Soc., **146**(2), 749 (1999).
78. P. Harde, R. Gibis, R. Kaiser, H. Kizuki, and H. Kunzel, J. of Crystal Growth **201–202**, 719 (1999).

I.3
Scanning Capacitance Microscopy for Electrical Characterization of Semiconductors and Dielectrics

J. J. KOPANSKI

A scanning capacitance microscope (SCM) combines an atomic force microscope (AFM) with a 1-GHz tuned inductance-capacitance-resistance (LCR) circuit to measure the capacitance between a conducting tip and sample. When applied to a semiconductor sample, an ac voltage at around 10 kHz is used to induce a depletion region within the semiconductor. The resulting differential capacitance is measured with a lock-in amplifier. The SCM contrast is proportional to this differential capacitance, which in turn is proportional to the inverse square root of the dopant concentration in the semiconductor beneath the tip. In this way dopant gradients in semiconductors, either natural or process induced, can be imaged. The differential capacitance measured by the SCM is also dependent on the properties of any native oxide or deposited dielectric film on the semiconductor surface, and the carrier mobility, which may be degraded near defects. This chapter will first review the history, principles, and modes of operation of the SCM. The remainder of the chapter will discuss major applications of scanning capacitance microscopy (*SCM*) for the electrical characterization of semiconductors and dielectric films, including qualitative characterization for integrated circuit failure analysis, quantitative dopant profiling and models for interpreting *SCM* images as dopant profiles, applications to semiconductors other than silicon, characterization of dielectric films, and optical pumping for carrier mobility measurements.

1 Introduction

The most common implementation of a scanning capacitance microscope (SCM [1]), combines an atomic force microscope (AFM) with a high-frequency capacitance sensor. (The instrument, the scanning capacitance microscope, will be abbreviated as SCM in this paper, while the technique, scanning capacitance microscopy, will be abbreviated as *SCM*.) A capacitance sensor that can detect very small ($\approx 10^{-21}$ F) changes in capacitance is used. A conducting AFM tip is electrically connected to the capacitance sensor. The AFM is used to acquire an image of topography and to control tip position, while the sensor is used to measure the capacitance between the tip and the sample under test simultaneously. Commonly, an ac voltage (around 1 $V_{peak\text{-}to\text{-}peak}$ at 10 kHz) is used to induce and modulate a depletion region in a semiconductor sample. The depletion region changes in response to the ac voltage and produces a varying, or differential, *SCM* tip-to-sample capacitance. The capacitance sensor detects the varying capacitance and produces a proportional output voltage. A lock-in amplifier that is referenced to the modulation frequency is used to measure the magnitude and phase of the sensor output voltage. The signal measured by the SCM is the output voltage of the lock-in amplifier, which is proportional to the induced tip-to-sample differential capacitance. As the SCM/AFM tip is scanned over the sample surface, simultaneous images of topography and differential capacitance are obtained.

Differential capacitance imaging of this sort is entirely dependent on the presence of a semiconductor sample. If the surface of the semiconductor sample is electrically passivated with a thin insulating layer, the charge on the surface of the semiconductor can be modulated with an externally applied electric field. In this case, the metallized SCM tip forms a metal-oxide-semiconductor (MOS) capacitor with the semiconductor sample under test. The differential capacitance signal arises from the depletion region capacitance generated in the semiconductor by the AC voltage applied to the tip. If insulators or conductors are imaged in this mode, no differential capacitance signal will be produced. The MOS capacitor is a critical part of a MOS field effect transistor (MOSFET) and the physics of such a device are well understood [2,3]. As will be discussed later, the three-dimensional (3D) nature of the SCM tip necessitates a 3D model to predict the capacitance between the tip and the sample adequately.

Much of the early development of the SCM was driven by the quest for a quantitative two-dimensional (2D) carrier/dopant profiling tool for MOSFET source/drain regions for use by the silicon integrated circuit (IC) industry. For this application, the ICs are cross-sectioned to expose the structure of individual transistors and other electronic devices. The differential capacitance measured by the SCM is related to the carrier concentration in the silicon. Regions of high carrier concentration, such as the source and drain of a transistor, produce a low differential capacitance signal level, while regions of low carrier concentration, such as the transistor channel, produce a high signal level. Thus, *SCM* images of cross-sectioned transistors contain information on the details of the structure of the transistor that is not usually visible in an optical or topographic image. The qualitative images can be converted into quantitative dopant profiles via modeling software. The C-V response of a MOS capacitor depends critically on the charges and properties of the insulating layer, as much as the properties of the underlying semiconductor. Recent work has exploited this fact to use *SCM* as a tool to characterize insulating layers on semiconductors.

The need and specifications of 2D dopant profiling are described in the International Technology Roadmap for Semiconductors [4]. If accurate 2D carrier profiles could be measured, they could be used to enhance the predictive capabilities of technology computer-aided design (TCAD). However, for this application, the performance goals for 2D carrier profiling tools are very challenging. Carrier profiles of the source and drain regions of the transistors need to be known with 5-nm spatial resolution and ±5% accuracy in 2005—the requirements rise to 0.6-nm spatial resolution and ±2% accuracy by 2014. The status of quantitative carrier profiling with *SCM* will be reviewed later in this chapter.

This chapter will first review the history, principles, and modes of operation of the SCM. The remainder of the chapter will discuss major applications of *SCM* for the electrical characterization of semiconductors and dielectric films, including qualitative characterization for integrated circuit failure analysis, quantitative dopant profiling and models for interpreting *SCM* images as dopant profiles, applications to semiconductors other than silicon, characterization of dielectric films, and optical pumping for carrier mobility measurements.

2 Development of Scanning Capacitance Microscopy

The first SCM predates the AFM. Matey and Blanc described an early version of the SCM in 1985 [5,6]. This instrument measured capacitance using a sensor removed from an RCA[1] SelectaVision VideoDisc System [7,8] and used a 2.5 × 5-μm probe that was guided in a groove. The VideoDisc was a competitor to the video cassette recorder (VCR) for home video playback. The VideoDisc retrieved for playback an analog video signal that had been mechanically imprinted onto a capacitance electronic disc, or CED. The VideoDisc player featured a capacitance detector that operated at 915 MHz and detected small changes in capacitance by the shift induced in the resonant peak of the response of an inductance-capacitance-resistance (LCR) resonator. The principle of operation of the capacitance detector will be discussed in detail later. Capacitance sensors similar in design to the VideoDisc sensor are used in modern commercial SCMs.

There followed a generation of SCMs that controlled tip height by maintaining a constant capacitance between the tip and the sample [9–14]. These instruments featured a sharpened metal tip and a piezoelectric scanning system like that used for the scanning tunneling microscope (STM). By using the capacitance as the feedback signal, these SCMs were able to measure the topography of insulating surfaces. In contrast, STMs can only measure surfaces that are relatively good conductors.

Slinkman, Wickramasinghe, and Williams of IBM patented a SCM based on AFM and a RCA VideoDisc sensor in 1991 [15]. This instrument was similar in configuration to current SCMs. Rather than using the capacitance signal to control the tip-sample separation, this instrument used the AFM to control tip position, while the capacitance measurement was made independent of the AFM. Combinations of commercial AFMs and RCA capacitance sensors to form SCMs

[1] Certain commercial equipment, instruments, or materials are identified in this paper in order to specify the experimental procedure adequately. Such identification does not imply recommendation or endorsement by NIST, nor does it imply that the materials or equipment used are necessarily the best available for the purpose.

were demonstrated shortly thereafter [16–18]. The first turnkey commercial SCMs were introduced in 1995 by Digital Instruments, Inc.

Since 1992, the National Institute of Standards and Technology (NIST) has had a *SCM* program with the goal of making *SCM* a quantitative 2D carrier profiling technique [16–30]. The NIST program has developed models of the *SCM* measurement based on 3D finite element solutions of Poisson's equation [26, 27] and Windows-based software for rapid carrier profile extraction from *SCM* images [30]. Researchers at the University of Utah [31–41] and at the ETH Zurich [42–44] have also developed quantitative *SCM* image to dopant profile conversion software.

Today, numerous groups pursue research into further development of *SCM*. Much, but not all, of this work is focused on developing *SCM* as a tool to measure the 2D carrier profiles of the source/drain regions of silicon transistors. Since this author's last review of scanning capacitance microscopy [1], recent research concerning scanning capacitance microscopy has sought to gain insight into the non-idealities seen in the *SCM* signal. The phenomena of contrast reversal (that is, *SCM* signal that is non-monotonic with dopant concentration) has been modeled [45] and correlated with interface traps and residual surface polish damage [46]. The effects of interface traps [47] and external illumination [48–50] have also been simulated and experimentally investigated.

3 Principles of Operation of the Scanning Capacitance Microscope

Figure 1 shows a block diagram of a typical scanning capacitance microscope operated in the constant delta voltage (ΔV) mode. The SCM consists of four major components: (1) AFM, (2) conducting tip, (3) capacitance sensor, and (4) signal detection electronics. The AFM controls and reports the x-, y-, and z-positions of

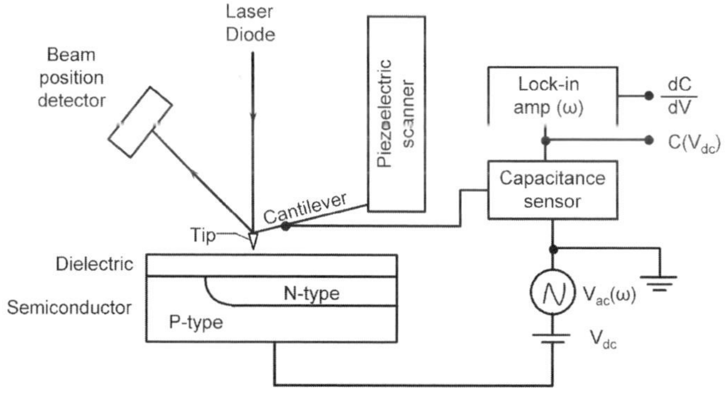

FIGURE 1. Block diagram of a scanning capacitance microscope configured for constant ΔV mode operation [16]. (Copyright American Institute of Physics 1996, used with permission.)

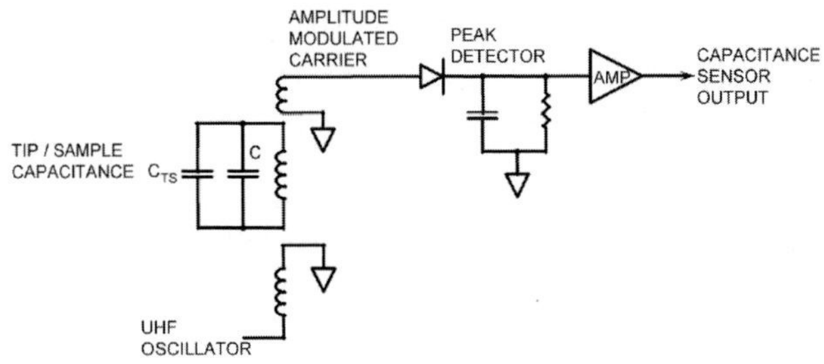

FIGURE 2. Capacitance detection circuit used in the RCA VideoDisc sensor and later modified for use in scanning capacitance microscopy (adapted from reference [7]).

the SCM tip. The conducting tip is similar to ordinary AFM tips, except that it needs to be of high electrical conductivity and in electrical contact with both the capacitance sensor and the sample. The radius of the tip determines the ultimate spatial resolution of the *SCM* image. Commercially available metal-coated silicon tips, tips made completely of very highly doped silicon or diamond, or solid metal tips have been used.

The capacitance sensor enables the scanning capacitance microscope operation. The archetypical SCM uses a capacitance sensor similar to that used in the RCA VideoDisc player [7,8]. Commercial SCMs use sensors that are similar in concept, though unique and proprietary in design. The sensor measures capacitance using a driven LCR resonant circuit. The capacitance to be measured is incorporated in series as part of the LCR circuit. The amplitude of the oscillations in this resonant circuit provides a measure of the capacitance. Figure 2 shows a simplified schematic of the capacitance sensor circuitry. The circuit consists of three inductively coupled circuits: an ultra-high frequency (UHF) oscillator, the central LCR resonant circuit, and a peak detection circuit.

The UHF ($\omega_{\mathrm{hf}} \approx 915$ MHz for the VideoDisc) oscillator is used to drive the LCR resonator at a constant frequency. The oscillator circuit contains a mechanically tunable potentiometer and capacitor to allow some tuning of the UHF signal frequency and amplitude. The magnitude of the UHF signal voltage, V_{hf}, which is applied between the tip and sample, is an important variable for *SCM* image interpretation.

The capacitance between the SCM tip and the sample, C_{TS}, is incorporated into the central LCR circuit. As shown in Figure 3, the LCR circuit displays a characteristic bell-shaped response versus frequency, with the peak response at the resonant frequency. Changes in C_{TS} cause a shift in the resonant frequency of the response curve, in turn causing the amplitude of the oscillations in the resonant circuit to change. The total capacitance in the sensor LCR circuit includes the tip-sample capacitance, the stray capacitance between the sample and sensor, and a variable tuning capacitance. In an RCA style sensor, a varactor diode is used to provide a voltage-variable capacitance. The varactor capacitance can be used to adjust the total capacitance in the LCR circuit, which in turn changes the circuit's

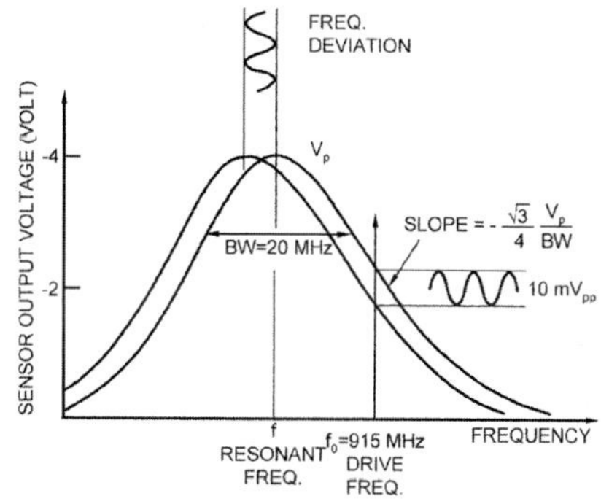

FIGURE 3. Capacitance sensor LCR circuit response illustrating how the shift in resonant frequency is converted into an output voltage (adapted from reference [8]).

resonant frequency. In this way, the sensor can be "tuned" to produce a similar response for a range of tip-sample capacitances.

SCM image contrast is due to the voltage-dependent capacitance between the conducting SCM tip and a semiconductor sample under test. When imaging silicon with the SCM, this capacitor is usually considered a metal-oxide-semiconductor (MOS) capacitor, though a metal-semiconductor Schottky contact may also be formed [51]. The quality of the *SCM* measurement of silicon depends on the formation of a thin oxide and establishment of an unpinned semiconductor surface. A MOS capacitor displays a characteristic voltage dependent capacitance due to the "field effect." The field effect is also the basis of the technologically important metal-oxide-semiconductor field effect transistor (MOSFET). The physics of the field effect and the MOS capacitor are discussed in extensive detail in many textbooks, for example [2, 3].

The SCM detection electronics are used to measure the difference capacitance, ΔC, between the tip and sample. SCMs are commonly operated in one of two different modes, the constant ΔV mode or the constant ΔC mode.

Constant ΔV *(or open loop) mode*—In the constant delta voltage (ΔV) mode of *SCM* [11–13], an ac sinusoidal voltage with a magnitude of V_{ac} at a frequency of ω_{ac} is applied between the sample and the SCM tip, which is a "virtual" ground; i.e., the tip is inductively coupled to ground. The inductance is such that it passes the ω_{ac} signal to ground, but does not ground the ω_{hf} signal, which is used to detect the capacitance. V_{ac} is typically around 1 V at a frequency of 10 kHz to 100 kHz.

Figure 4 illustrates the voltage dependence of capacitance for an MOS structure and the mechanism of *SCM* contrast generation. The V_{ac} signal induces a change in tip-to-sample capacitance, $\Delta C(\omega_{ac})$, that depends on the slope of the C-V curve between the tip and the semiconductor. When the tip is over a semiconductor with high dopant concentration, the C-V curve changes slowly with voltage and V_{ac} produces a relatively small ΔC. When the tip is over a semiconductor with low doping concentration, the C-V curve changes rapidly with voltage and V_{ac} produces a relatively large ΔC. When V_{ac} is applied between the tip and the sample, the capacitance sensor responds with a voltage output signal that varies at ω_{ac} in proportion to ΔC.

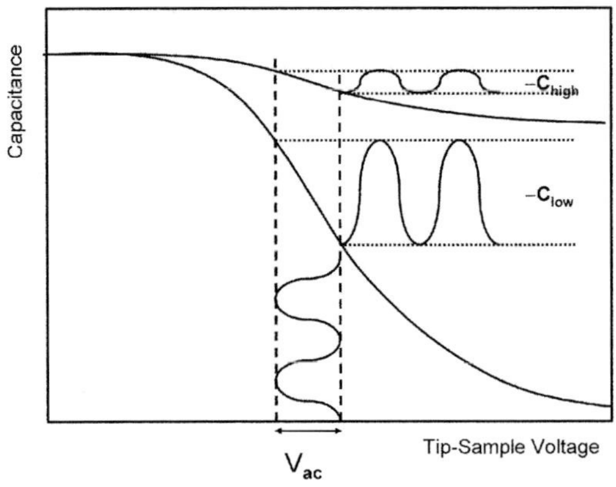

FIGURE 4. Schematic illustration of how SCM image contrast is generated for two different dopant densities of silicon.

As shown in Figure 1, the magnitude of the sensor output is detected using a lock-in amplifier that is referenced to ω_{ac}. Since n-type and p-type semiconductors produce C-V characteristics that are mirror images of each other, they produce distinctly different responses to the SCM [20, 21]. In general, n-type silicon produces a response that is in-phase with the drive signal and p-type silicon produces a response that is 180° out-of-phase with the drive signal. When using a lock-in amplifier in x/y mode, n-type silicon produces a positive signal and p-type silicon produces a negative signal.

The capacitance sensor output provides a relative measure of differential capacitance—larger values of ΔC produce larger sensor output voltages. Because the tip shape and the tuning of the capacitance sensor are variable, the absolute value of capacitance (in farads) measured by the SCM is difficult to determine. When measuring an unknown, the absolute capacitance is determined by comparison to measurements of a reference whose capacitance can be calculated based on its known properties. Several operational factors strongly influence the *SCM* signal:

1. *AC bias voltage, V_{ac}*—Larger values of V_{ac} result in larger values of ΔC. The larger change in capacitance is due to larger number of carriers responding to V_{ac}. Larger values of V_{ac} also result in larger depletion volumes and lower spatial resolution when the SCM is used to measure quantitative carrier profiles.
2. *DC bias voltage, V_{dc}*—The dc voltage, V_{dc}, applied between the sample and the tip determines where on the C-V curve that ΔC is measured. The effect of V_{dc} is relative to the flatband voltage of the MOS capacitor formed by the tip and the sample. V_{fb} will vary from sample to sample. The peak *SCM* response occurs when $V_{dc} \approx V_{\mathrm{fb}}$ [27]. In a sample containing a dopant gradient, V_{fb} varies with dopant concentration. In practice, V_{dc} is usually set equal to the voltage that produces the maximum *SCM* response in the region with the lowest dopant concentration.
3. *Sensor high-frequency voltage, V_{hf}*—The magnitude of the high-frequency voltage, V_{hf}, in the LCR loop of the capacitance sensor is proportional to the magnitude of the output signal. Increasing V_{hf} increases the capacitance sensor

FIGURE 5. SCM sample geometry that reduces the effect of both stray light and stray capacitance [55]. (Copyright American Institute of Physics 2004, used with permission.)

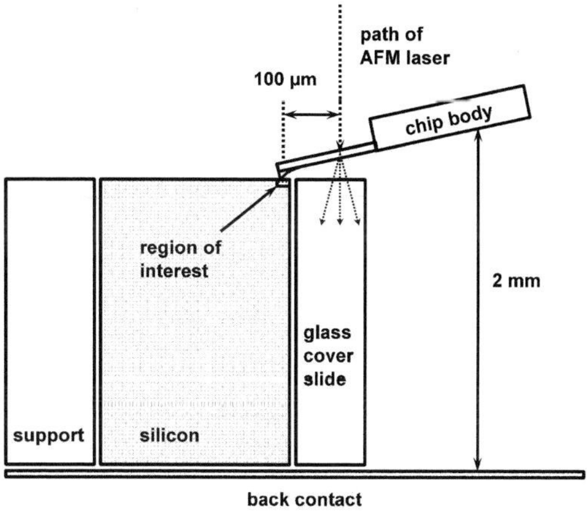

output and the signal-to-noise of the capacitance measurement. However, similar to the case for V_{ac}, the measured *SCM* signal is averaged over its values at the voltages spanned by V_{hf}. Large values of V_{hf} generate large signal at the cost of spatial resolution.

4. *Sensor-to-sample coupling*—The capacitance sensor forms a grounded LCR loop that includes the tip-sample capacitance. The capacitance between the sample and the grounded outer shield of the capacitance sensor completes the loop. Subtle variations in the sample geometry can influence this coupling capacitance. Figure 5 shows a sample geometry where the sensor-to-sample coupling is minimized.

Constant ΔC *(or closed loop) mode*—For carrier profiling applications, the SCM can also be operated in a constant ΔC mode to control the volume depleted of carriers better. A feedback loop (hence, closed loop mode) is added to the SCM to adjust the magnitude of V_{ac} automatically to keep ΔC constant in response to changes in dopant concentration as the SCM tip is scanned over a dopant gradient [21,31,33]. The lower the doping concentration beneath the tip, the smaller the V_{ac} needed to induce the same ΔC. Since ΔC remains constant, the volume depleted of carriers, and therefore the spatial resolution of the image, is less dependent on the dopant concentration in the semiconductor. In the constant ΔC mode of *SCM*, the *SCM* signal is the value of V_{ac} that the feedback loop establishes to maintain a constant ΔC. The feedback loop of the constant ΔC mode produces a signal that changes monotonically with carrier concentration only for dopant gradients in like-type semiconductors. When a p-n junction is present, the changes in sign of the *SCM* signal when transitioning between n-type to p-type material cause the feedback loop to saturate near the junction.

Intermittent-contact SCM—The SCM can also be operated in intermittent-contact (or tapping) mode [23,24]. In this case, a differential capacitance is generated by the change in tip-sample distance due to the vibrating tip. Figure 6 shows

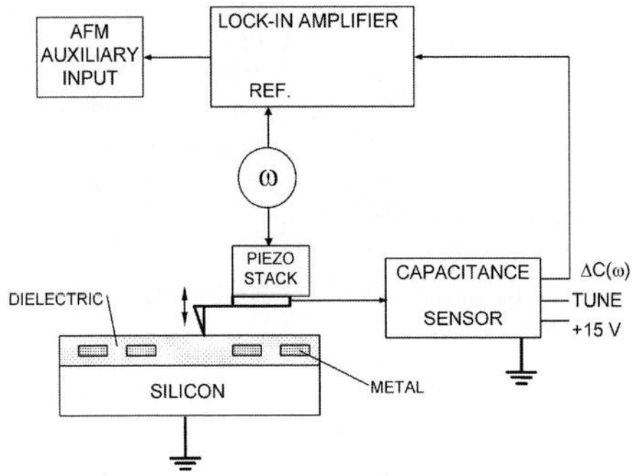

FIGURE 6. Block diagram of the intermittent-contact mode scanning capacitance microscope [23]. (Copyright American Institute of Physics 1998, used with permission.)

a block diagram of the IC-SCM. The capacitance sensor output is measured with a lock-in amplifier using the tip vibration frequency as the reference.

In the IC-*SCM* mode, no modulation of a depletion region in a semiconductor is required to generate a differential capacitance signal. Since all materials display a tip-sample spacing dependent capacitance, the IC-SCM extends the technique to measurements on non-semiconductors. IC-*SCM* can detect metal lines beneath an insulating layer, even if the insulating layer has been planarized to remove any indication of the metal lines in the surface topography [23]. This ability may have applications in lithographic overlay metrology. IC-*SCM* is also able to detect variations in the dielectric constant of thin films, with the difference in SiO_2 ($\varepsilon = 3.9$) and Si_3N_4 ($\varepsilon = 7.5$) clearly resolvable. This capacity may have applications for evaluation of deposited dielectric films used in the semiconductor industry, such as candidate alternative gate dielectrics and for capacitors in memory circuits.

Capacitance-voltage curve measurement—The measurement of capacitance versus dc voltage (i.e., the C-V curve) is the usual method of characterizing any MOS capacitor. Measurement of the SCM tip-sample C-V characteristics reveals the suitability of the sample preparation for quantitative carrier profiling. SCM tip-sample C-V curves can be measured with a boxcar averager [31] or a digital oscilloscope operated in x-y mode. An AC voltage at around 1 kHz and spanning the range of the desired voltage of the C-V curve is applied between the tip and the sample. The AC voltage is displayed on the x-channel, while the capacitance sensor output is displayed on the y-channel. The averaging feature of the digital oscilloscope is used to improve the signal-to-noise. To be suitable for quantitative carrier profiling, the C-V curve must display clear accumulation and depletion regions, with little hysteresis between the forward and reverse sweep, and with no increase in capacitance in the depletion region.

The ΔC versus V curve can also be measured directly with the SCM [28]. In this case, a *SCM* image is acquired while slowly changing the value of V_{dc} so that the desired range for the ΔC-V curve is swept out once in the time it takes to acquire the image. Sections of such an image taken parallel to the slow scan direction reveal the ΔC-V curve. A high-quality surface should yield a single peak (a positive peak for n-type and negative peak for p-type) at the flatband voltage.

4 Sample Preparation

SCM cross-sectional sample preparation is derived from techniques developed for scanning electron microscope (SEM) and transmission electron microscopy (TEM). A detailed common procedure is described in [52]. Sample preparation usually involves four steps: (1) cross-section exposure, (2) backside metallization, (3) mechanical polishing, and (4) insulating layer formation. Prior to cross-sectioning, the region of interest is usually glued face-to-face with a glass cover slide or silicon dummy. This cover prevents rounding and chipping of the surface of interest during mechanical polishing. Thin (a few mm) cross-sections are cut from the sample-cover sandwich with a mechanical dicing saw. To improve the electrical contact between the back of the sample and the microscope sample holder, the backside is usually coated with a Cr layer (for adhesion), followed by an Au layer (for low resistance and corrosion resistance). The sample is attached to a metal sample holder using a conducting silver-filled epoxy. Front-to-back sample resistance of no more than a few ohms indicates a good backside contact.

An additional aspect of cross-section preparation is the creation of a sample geometry that minimizes the capacitive coupling between the active parts of the sensor and ground, and that also reduces the stray light from the AFM laser that may impinge on the semiconductor under test. Figure 5 shows an idealized sample geometry. Here the sample is wide enough so that the active parts of the SCM tip, cantilever, and body are physically separated from ground by several millimeters. This reduces the coupling capacitance and prevents excessive loading of the sensor. Because the sample is terminated by an insulating glass cover slide, most stray light from the AFM laser does not impinge on the semiconductor sample, which would generate excess carriers and distort the *SCM* image.

The mechanical polishing step must achieve both a highly polished surface and a surface with low residual crystalline damage. Sample preparation optimized for SEM image may produce a smooth surface, but leave behind substantial crystalline damage. The final step of silicon sample preparation is the formation of a thin, passivating, insulation layer on the surface to be imaged. This layer is necessary to reduce the number of surface states and to serve as the oxide of the MOS capacitor that the SCM tip forms with the sample. Several processes have been used to form this layer including a short polish with colloidal silicon [52], a 300°C anneal for 60 minutes with UV light [38] or without [53], oxidation in an UV-ozone photoreactor [53], and a 200°C anneal in 5% ozone/ 95% O_2 [54]. Regardless of the process to form the surface-insulating layer, the quality of the insulating layer for *SCM* is determined by measuring the C-V with the SCM.

5 Qualitative Imaging for Integrated Circuit Failure

SCM is used as a *qualitative* analysis tool with multiple applications for failure analysis. *SCM* can measure the shape of dopant profiles, ascertain if certain

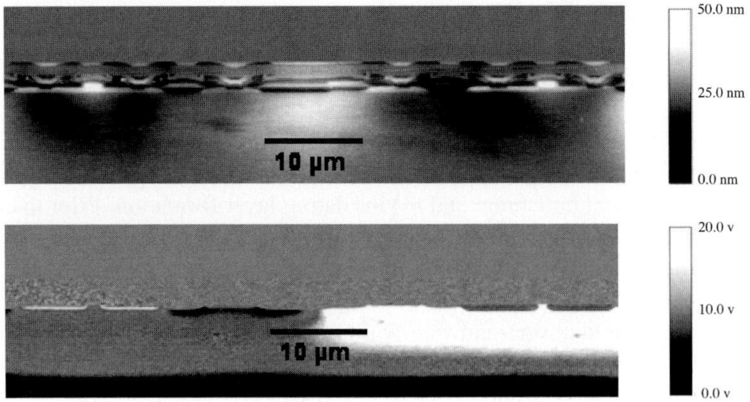

FIGURE 7. (a) Atomic force microscopy image of a polished cross section of a CMOS pair showing polysilicon gates and metal contact layers. No contrast is seen from the doped structures. Region imaged is 50 μm across. (b) SCM image of the CMOS pair, acquired simultaneously with (a), showing contrast from source/drain, channel, and n-well dopants.

implants have been completed, verify the conduction type of implanted regions, and look for unexpected shorts or opens in doped regions of a device. To obtain *quantitative* 2D dopant profiles from *SCM* images of dopant gradients requires careful attention to the data acquisition conditions and processing of the measured *SCM* image using quantification software. Qualitative imaging is less demanding since a larger AC voltage can be used to enhance image contrast.

Figure 7(a) shows an example of an AFM image and Figure 7(b) the corresponding *SCM* image of a silicon CMOS pair of transistors. All the components of the CMOS cell are discernible. In the AFM image, only the metal contact plugs and polysilicon gates are clearly visible because sample preparation has etched the insulating layers surrounding these structures leaving them slightly elevated above the silicon relative to the rest of the layers. No features are visible in the underlying silicon in the AFM image. In the *SCM* image, features due to dopant differences in the silicon are visible. The contrast in the *SCM* image is proportional to dopant concentration in the following way: p-type material produces negative (dark) *SCM* signal, the most lightly doped p-type material is darkest; and n-type material produces positive (bright) signal, the most lightly doped n-type material is brightest. The deepest layer in Figure 7 (b), with the darkest shading, is the lightly doped p-type silicon substrate. Above this layer, is a more highly doped and corresponding less *SCM* contrast layer due to a p-type epitaxial layer. The n^+ source drain regions, p-channel, and doped p^+ contacts to the epitaxial layer are visible on the left side of the image. The region to the right with the brightest contrast is the n-well. Within the n-well, the p^+ source drain regions, the n-channel, and doped n^+ contacts to the well are visible.

6 Quantitative Semiconductor Dopant Profiling

6.1 Imaging Conditions

Three voltages are applied between the SCM tip and the sample under test. All three voltages must be known, relative to the flatband voltage of the sample, for accurate dopant profile extraction. The AC voltage, V_{ac}, determines the width of the tip-sample C-V curve that is used to generate the *SCM* differential capacitance signal, ΔC. The larger the value of V_{ac}, the larger the volume depleted of carriers and the poorer the spatial resolution of the *SCM* image. The DC voltage, V_{dc}, determines the center point on the C-V curve, relative to the flatband point, where ΔC is acquired. The optimal value of V_{dc} is the flatband voltage of the surface at the dopant density of most interest. The high-frequency voltage, V_{hf}, is used by the SCM capacitance sensor to generate the capacitance signal. The effect of V_{hf} is to average over voltage the portion of the tip C-V curve used to generate the *SCM* ΔC signal. While models of the SCM do account for V_{hf}, large values of V_{hf} also reduce the spatial resolution of the *SCM* image. These three voltages all must be optimized in order to extract meaningful quantitative dopant profiles from SCM images [55].

Several software packages have been developed to extract 2D dopant profiles from *SCM* images, including *DPAK* from the University of Utah [56], *SCaMsim* from ETH in Zurich [44], and *FASTC2D* from the National Institute of Standards and Technology [30]. These packages all use different models for calculating the C-V curve between the SCM tip and sample. However, they all use some variation of the calibration curve method to extract dopant profiles from *SCM* images.

6.2 Calibration Curve Method

The calibration curve method (CCM) uses a calculated calibration curve of *SCM* signal as a function of dopant concentration to convert the *SCM* signal to dopant concentration. This is illustrated schematically in Figure 8 by a screen shot from the *FASTC2D* program. The calibration curve is calculated from the known model and image acquisition parameters [57]. Model parameters include the tip shape, the sample oxide thickness and dielectric constant, and the flatband voltage. Data acquisition parameters are the applied SCM operating biases. Several assumptions are implicit in the calibration curve approach:

1. *Each data point is treated independently.* Every data point in the *SCM* image is assigned a dopant concentration from the calibration curve without regard to the values of adjacent points. The local 2D dopant gradient is not taken into account. This approach is valid only if the dopant gradient is changing slowly with respect to the size of the SCM tip diameter.
2. *Normalization to a known dopant concentration required.* The *SCM* signal is measured in arbitrary units with an unknown proportionality constant between the measured signal level and the actual differential capacitance in F/V. To align

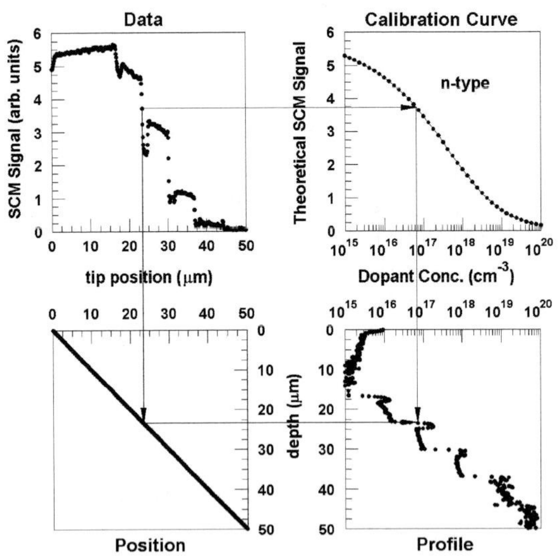

FIGURE 8. Graphical illustration of the calibration curve method for dopant profile extraction from SCM images as implemented by the *FASTC2D* program [57]. SCM data **(upper left)** are aligned with the calibration curve **(upper right)** by means of a known ΔC at a known N normalization point. Dopant concentration **(lower right)** is then obtained for each SCM data point by interpolation from the calibration curve.

the calibration curve with the measured *SCM* signal profile, the *SCM* signal must be known at one or more dopant concentrations. This is often the peak dopant concentration from a 1D secondary ion mass spectroscopy (SIMS) dopant profile, but the normalization point also can be obtained from a reference sample included with the sample under test. If dopant concentration is known at more than one point within the profile (for example, the peak dopant concentration and the substrate concentration), then a regression can be performed on one model parameter to find the value of the parameter that generates a calibration curve that intersects both known points.

3. *Calibration curves are calculated from a database of C-V curve solutions.* The time-consuming part of the calibration curve method is the calculation of the tip C-V curves over a wide range of dopant densities for any given set of model parameters. The CCM is sped considerably by using a database of C-V curves calculated offline for a set of representative model parameters. The *SCM* signal for a set of specific model parameters can then be quickly and accurately found from interpolation versus the database. The CCM has the advantage of being very fast, less than a minute on a current-technology personal computer, to convert a 512 × 512 pixel *SCM* image to a 2D dopant profile.)

6.3 Regression Method

Finding the dopant profile from the *SCM* image by means of a regression procedure represents the second generation of *SCM* interpretation and provides a significant improvement over the calibration curve method. A regression method uses repeated solutions of the forward problem (calculation of the *SCM* signal from candidate dopant profiles) to solve the reverse problem (calculation of the dopant profile from the measured *SCM* signal). In this approach, a candidate dopant profile is used to calculate the expected *SCM* signal. The calculated *SCM* signal is then compared to the measured *SCM* signal. Based on this comparison, the candidate dopant profile is adjusted. This procedure is repeated until the calculated *SCM* signal agrees with the measured *SCM* signal to the desired degree of accuracy. McMurray et al. implemented an early regression type solution using a quasi-3D model [38].

The NIST group has developed an optimized regression procedure to find the *SCM* reverse solution based on numerical solution of the Poisson's equation [58]. The technique uniquely uses two meshes to achieve conversion in only a few iterations (five steps of the coarse mesh and two steps of the fine mesh for a model problem). Reverse solutions determined by this technique fully include the effects of the local dopant gradient and gradient curvature on the *SCM* measurement of the dopant profile. The method can handle dopant profiles in similar substrates or in opposite type substrates (i.e., across p-n junctions). The procedures developed for the 2D model are expected to be straightforward to extend to 3D models.

Figure 9 shows a model dopant profile and the results of recovering the dopant profile using the calibration curve method (Figure 9(a)) or reverse regression (Figure 9(b)). In this experiment, our Poisson solver is used to calculate the expected *SCM* signal for the model dopant profile. Spatial resolution of sharp features is limited with the CCM; the step at 10^{17} cm^{-3} is only faintly resolved. When we use reverse regression, the recovered dopant profile clearly resolves all three steps in dopant concentration with spatial resolution near the mesh size.

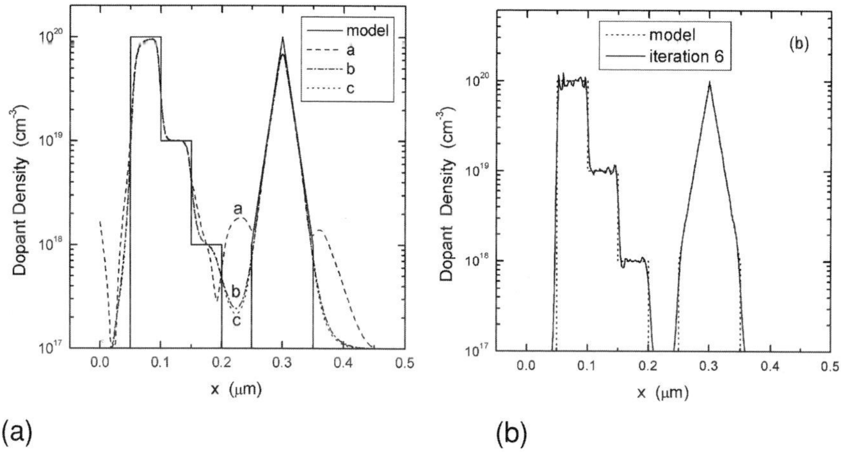

FIGURE 9. (a) Dopant concentration profile recovered from theoretical simulation of a SCM signal using the calibration curve method at three different bias offset values: *a)* no flatband offset, *b)* flatband at $N_d = 10^{17}$ cm^{-3}, and *c)* flatband at $N_d = 10^{18}$ cm^{-3} [13]. (Copyright American Institute of Physics 2002, used with permission). (b) Dopant concentration profile recovered from a SCM signal using a regression procedure to solve the reverse problem [13]. (Copyright American Institute of Physics 2002, used with permission).

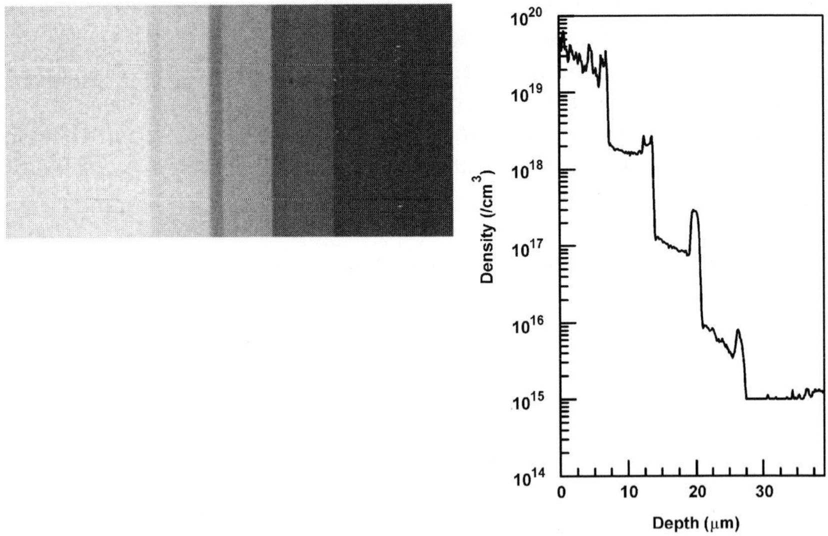

FIGURE 10. (a) SCM image of the IMEC n-step staircase structure. Lowest dopant concentration is on the right side of the image [57]. (b) One-dimensional dopant profile extracted from the center of the SCM image of the staircase structure [57].

6.4 Example: IMEC Staircase Structure

As an example, consider the dopant staircase structures available from IMEC [59]. The n-type version of this structure consists of five distinctly doped levels from 4.2×10^{14} cm^{-3} up to 1.7×10^{19} cm^{-3}, a total change in dopant concentration of about a factor of 40,000. The square root of N dependency predicts a change in *SCM* signal measured over this structure of $\times 200$. Actual *SCM* measurements recorded a signal change of $\times 40$ to $\times 45$. Figure 10(a) shows a *SCM* image of the five dopant concentration levels. Figure 10(b) shows a 1D dopant concentration profile of this structure determined with the *FASTC2D* software. The initial fit parameters were $r_{tip} = 30$ nm, $t_{ox} = 4$ nm, $\varepsilon_{ox} = 3.9$, $V_{ac} = 0.4$ V, $V_{hf} = 2000$ mV, and $V_{dc} = V_{fb}$ at $N_d = 10^{15}$ cm^{-3}. The fit was improved by adding 2% stray signal to the *SCM* ΔC signal and increasing V_{hf} to 2,200 mV; no other parameters were changed. The *SCM* determined dopant concentration levels always matched the reported values to within 50 % across the total dopant concentration range.

6.5 Example: NIST SCM Tutorial Sample

NIST has constructed a basic *SCM* test structure to serve as an semi-quantitative research material to share among users of *SCM* [57]. These have been designated as "*SCM* Tutorial Samples" (STS). These structures are available with the *FASTC2D* code along with a detailed set of tutorial instructions describing the procedure to measure a *SCM* image suitable for quantitative extraction of the 2D dopant profiles. The structures were fabricated by the MOSIS service using the AMI-Semiconductor 1.5 μm CMOS process. This is a relaxed geometry with respect to current state-of-the-art devices. Even beginning users of *SCM* should be able to acquire images of these structures that are suitable for quantitative interpretation. The dopant profiles present in these devices (approximately a four order of magnitude change in dopant concentration over 200 nm) are sufficiently broad that the calibration curve method will be largely valid throughout the entire profile range.

The STS are 2,000 μm long for easy cross sectioning and angle lapping, if desired. The STS contain the following structures: (1) A CMOS pair of transistors constructed using the complete process. The n-channel device consists of arsenic-doped source/drain regions in a boron-doped substrate. The p-channel device consists of boron-doped source/drain regions in a phosphorus-doped n-well. Figure 7(a) is an atomic force microscopy image of these structures; Figure 7(b) is the corresponding *SCM* image. (2) The same CMOS transistor structures, but with no metallization. Since metallization can sometimes interfere with *SCM* measurements, these devices allow any effects due to metallization to be removed from the image. (3) Transistor-like structures utilizing the same implants and geometry as the CMOS pair but with the type of the source/drains reversed relative to the substrate. That is, the boron implants are made into the p-type substrate, and the arsenic implants are made into the n-well. Since the pn-junction can have a large influence on the appearance of the *SCM* image, these devices allow any effects due to the pn-junctions to be removed from the image. Given that all implants are made at the same time and processed under the same conditions, they can be expected to produce nearly identical dopant profiles, regardless of the metallization or the type of the substrate. (4) 500 × 500-μm SIMS test pads for the p-type substrate, the n-well, the p-type source/drain ion implanted regions, and the n-type source/drain ion implanted regions.

As an example of the type of data that can be obtained with *SCM*, consider the following analysis of the boron implants from the STS. Figure 11 shows the SIMS

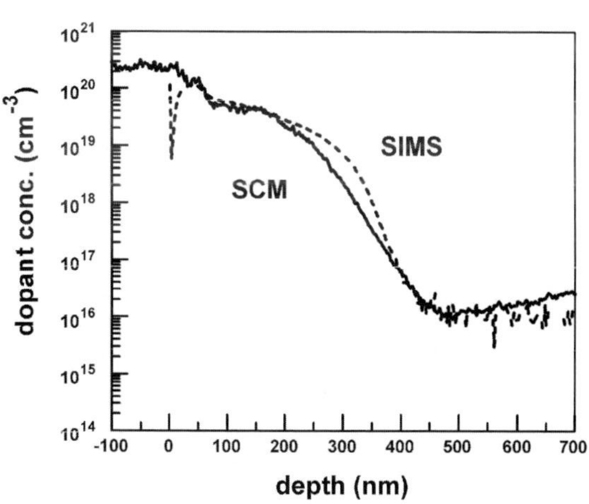

FIGURE 11. Dopant profiles measured with SIMS (dotted line) and SCM (solid line) for the boron implant of the STS [55]. (Copyright American Institute of Physics 2004, used with permission.)

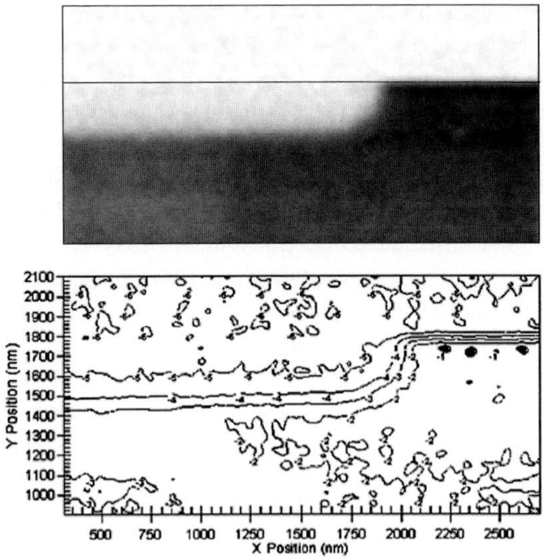

FIGURE 12. (a) SCM image of boron implant into the p-type epitaxial layer [55]. (Copyright American Institute of Physics 2004, used with permission.) (b) 2D dopant contours extracted from a SCM image using the *FASTC2D* software. Contours are spaced an order of magnitude in dopant concentration apart, starting from a base of $Na = 10^{16}$ cm^{-3} [55]. (Copyright American Institute of Physics 2004, used with permission.)

and *SCM* measured dopant profiles for the B implants into the n-type region. *SCM* data were acquired with a scan length of 1.02 μm, resulting in a point-to-point measurement spacing of 2 nm. The *SCM* signal tracks the SIMS B profile well near the peak doping ($\sim 6 \times 10^{20}$ cm^{-3}). Under these imaging conditions, the back end of the implant can be seen with *SCM* down to a dopant concentration of 3×10^{16} cm^{-3}, even with the presence of a pn-junction. The *SCM* signal will change sign in the vicinity of the pn-junctions; the beginnings of this are evident in the increase in *SCM* determined dopant concentration at the end of the profile. Figure 12 shows a *SCM* image of the corner of one of the B implants into the p-type epitaxial layer. Figure 12(b)

7 Applications to Semiconductors Other than Silicon

A variety of other semiconductors have also been imaged with *SCM* with the same general goals as for silicon. Silicon carbide (SiC), a wide bandgap semiconductor, has proven difficult to characterize by traditional electrical dopant concentration profiling techniques, which often result in large differences between SIMS chemical dopant profiles and measured electrically active dopant profiles. Quantitative carrier profiles of SiC using cross-sectional *SCM* has allowed these differences to be more fully explained [60]. Other materials examined with *SCM* include various light emitters such as AlGaN/GaN heterostructures [61, 62], InGaN/GaN quantum wells [63], InGaAS/InP quantum wells [64], and images of InP-based laser structures [65–67]. *SCM* imaging of InP laser structures has been used to evaluate regrown material quality, interface locations, and the uniformity of doped structures formed by liquid phase epitaxy (LPE) and metal-organic vapor phase epitaxy (MOVPE).

8 Dielectric Film Characterization

SCMs have been used to study the charge trapping and detrapping dynamics of silicon dioxide or silicon nitride insulating layers on silicon [68–75]. A voltage pulse through a SCM tip can be used to inject charge into a nitride-oxide-silicon (NOS) structure, which becomes trapped. The trapped charge induces a depletion region in the silicon whose capacitance can be detected by the SCM. A reverse bias pulse can remove the stored charge. Much of this work has been conducted with the goal of determining the potential for using charge trapping and detection with capacitive probes as a non-volatile semiconductor memory. Because the small tip size enables high recording densities, it has been hypothesized that such a technique may provide ultrahigh density data storage.

The potential of scanning capacitance microscopy for the nondestructive evaluation of silicon dioxide on silicon has been addressed by several groups [46,47,69,73,76]. Some qualitative information and relative comparisons between samples can be obtained from the shape of the SCM measured ΔC-V curves. The ΔC-V curves typically display a peak response as a function of V_{dc}; the peak being near, but not necessarily exactly at, the flatband voltage [27]. The relative fixed charge density, N_f, between MOS samples can be evaluated from the relative positions of the peaks in ΔC signal. The width of the peak in ΔC is related to D_{it} [76]. D_{it} can also result in side peaks in the ΔC-V characteristics [78]. As with MOS capacitors with deposited contacts, hysteresis between forward and reverse ΔC-V voltage sweeps and shifts in flatband (ΔC peak) position with a stress voltage are related to the susceptibility of the insulating layer to charge trapping.

The degree to which interface trapped charge interacts with the SCM measurement has been investigated [47, 77]. SCM measured ΔC-V curves often show features that can be interpreted as being due to interface traps. Most traps would not be able to respond to the $\approx$ 0.9-GHz high-frequency voltage utilized by the SCM sensor to measure capacitance. However, some traps would be able to follow the $\approx$10 kHz signal used to generate the ΔC signal. In the calculations described below, no effort is made to simulate the frequency response of the traps; rather all traps in the input trap distribution are assumed to be active and contribute to C-V curve stretch out.

To illustrate the kind of MOS information that may be obtained from *SCM* measurements, C-V and ΔC-V curves with various N_f and D_{it} were simulated using the NIST quasi-3D model of the SCM [30,78]. The quasi-3D model simulates the 3D nature of the SCM tip by breaking the tip into a series of concentric rings distributed along the tip surface. The C-V characteristic of the SCM tip is obtained by summing the contributions of each individual ring calculated with a 1D model. In this case, we used a 1D analytical model of the MOS capacitor that includes Fermi-Dirac statistics and quantum mechanical effects, as well as an arbitrary distribution of interface traps [79].

Figure 13 shows calculated ΔC-V characteristics with high densities of interface traps. Shown are curves with interface trapped charge distributions of (a) no

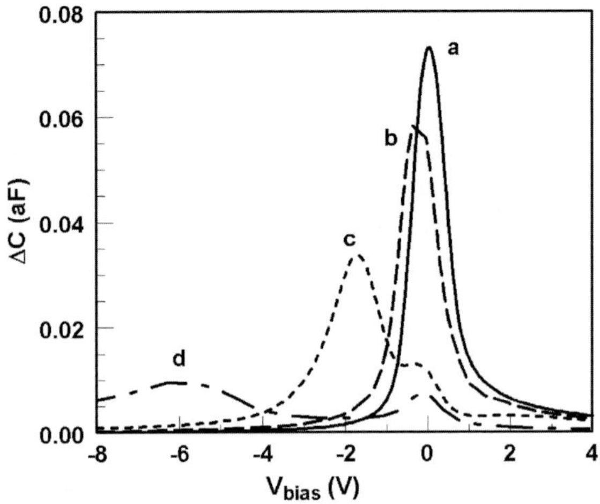

FIGURE 13. Theoretical ΔC-V curves calculated for various interface trap distributions. Curves are calculated with: (a) no interface traps; (b) a trap distribution with a single peak of 2×10^{12} cm^{-2}eV^{-1} acceptors at 0.27 eV above the valence band edge (and a corresponding peak in donors below the conduction band); (c) peak trap concentrations of 1×10^{13} cm^{-2}eV^{-1}; and (d) peak trap concentrations of 6×10^{13} cm^{-2}eV^{-1}. (Reproduced with permission from Interfaces in Electronic Materials, Electrochemical Society Symposium, Orlando, FL. Copyright 2003, The Electrochemical Society.)

interface traps; (b) a trap distribution with a single peak of 2×10^{12} cm^{-2}eV^{-1} acceptors at 0.27 eV above the valence band edge (and a corresponding peak in donors below the conduction band); (c) peak trap concentrations of 1×10^{13} cm^{-2}eV^{-1}; and (d) peak trap concentrations of 6×10^{13} cm^{-2}eV^{-1}. The traps broaden the ΔC peak full width at half-maximum (FWHM) by 0.05 V for D_{it} distribution b, 2.5 V for distribution (*c*), and ≈ 11 V for distribution d. Larger effects are seen for higher dopant concentrations and thicker oxides; for example, increasing the oxide thickness to 50 nm at $N_a = 10^{16}$ cm^{-3} broadens the peak FWHM to 0.5 V for distribution (*b*). Again, the equivalent number of traps involved in these calculations amounts to just a few charges. While the simplicity of this calculation is inadequate to model the experimental situation completely, it does give some indication of the size of the expected effects.

9 Optical Pumping of the SCM

The silicon MOS capacitor is an extremely light-sensitive device. The SCM depends on use of an AFM with a laser optical lever detection system. The light from this laser floods the vicinity of the tip contact point where the *SCM* signal is being acquired and strongly affects the measured *SCM* signal [49, 50]. There are two

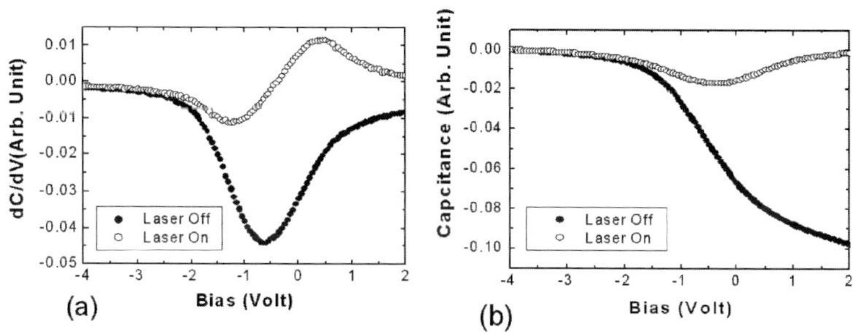

FIGURE 14. (a) dC/dV curves with the AFM laser on and with the laser off (after achieving equilibrium). (b) C-V curves obtained by integrating the dC/dV curves [50]. (Copyright American Institute of Physics 2003, used with permission).

basic effects of light on the measured C-V characteristics. The first is a decrease in the time-constant of the minority-carrier generation in the inversion layer, often resulting in a low-frequency characteristic, even under high-frequency *SCM* C-V measurement conditions. The second is a decrease in the surface potential, resulting in a reduction of the width of the depletion layer, with a corresponding increase in the capacitance at strong inversion.

A conventional AFM laser can easily generate excess carriers underneath the probing tip even in a well-aligned, normal condition, resulting in distortion of some electrical measurements. However, at least for the SCM, obtaining C-V curves or *SCM* images under true dark condition can be achieved by switching off the AFM laser, since the capacitance measurement is totally separate from the AFM. Therefore, the AFM laser itself can be used as an active component for light-pumping of the SCM to measure some optical properties of a semiconductor. Figure 14(a) shows the dC/dV versus voltage curves measured just prior to (open circles) and just after (filled circles) turning off the AFM laser [50]. A high-quality metal-oxide-semiconductor (MOS) sample with long carrier lifetime was used for the experiment, and the measurement was performed directly after thermal oxidation (900°C for 14 min) of a virgin wafer (boron-doped, nominally 15 Ω-cm to 20 Ω-cm). Illuminated conditions were obtained with the laser in the normal AFM position, centered on the cantilever and directly above the tip. Dark conditions were obtained with all applied light sources off (laser and the incandescent light for the optical alignment microscope) and the AFM enclosed in a light tight box. Figure 14(b) shows the C-V curves obtained by numerically integrating the C-V curves in Figure 14(a).

Figure 15(a) shows single sweep dC/dV curves of the same sample as above [49]. The vertical dotted line in Figure 15(a) indicates the flat-band bias condition of the MOS structure formed between the tip and sample. Figure 15(b) shows the capacitance transient behavior at flatband condition when illumination is changed from on to off. The sum of detected AFM signal (so-called A + B) is simultaneously

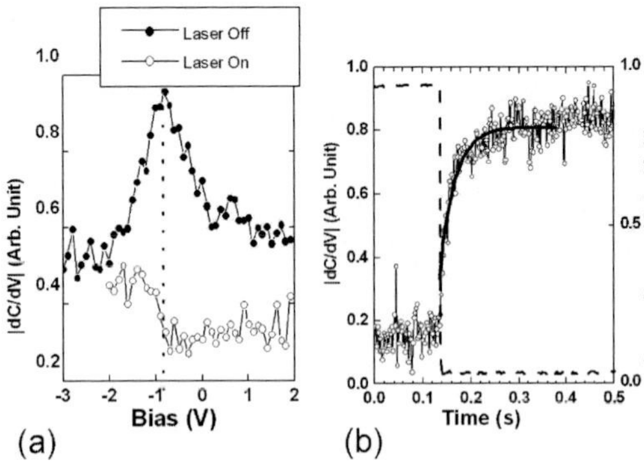

FIGURE 15. (a) dC/dV curves measured with illumination and in the dark. (b) Transient capacitance signal measured as a function of time. The laser power intensity (dashed line) is monitored to synchronize precisely the measured transient with the change in illumination [49]. (Copyright American Institute of Physics 2003, used with permission).

measured (dashed line) and gives the precise time (when the illumination condition is changed), which is essential for properly fitting the time constant of the measured capacitance transient. From these two curves, the transient curve is fitted to a first order exponential decay curve (thick solid line), giving the effective decay time directly (i.e., the effective carrier recombination lifetime) to be 24 ms. The AFM tip is very localized (10 nm), but will gather carriers from within the silicon diffusion length.

Because the tip is small, this technique can be used to acquire carrier lifetime images that are limited in spatial resolution only by the diffusion length (a physical property of the semiconductor) and not at all by the instrumentation.

10 Conclusion

The SCM is a versatile instrument for the electrical characterization of semiconductors and dielectric films with spatial resolution on the order of 10 nm. *SCM* has already proven useful for qualitative characterization for integrated circuit failure analysis, quantitative dopant profiling of the source drain regions in silicon MOSFETs, and the characterization of silicon dioxide and high-k dielectrics on silicon. *SCM*'s sensitivity to capacitance changes in the attoFarad range means that it can detect charges due to just a few electrons. Since *SCM* uses a low contact force AFM to make an electrical contact, it is a most promising tool for the electrical characterization of emerging nanostructures, such as carbon nanotubes, semiconductor nanowires, and nanoparticles.

Acknowledgments. This work is supported by the National Semiconductor Metrology Program, through the Office of Microelectronics Programs at the National Institute of Standards and Technology (NIST). Contribution of NIST is not subject to U.S. copyright.

References

1. J. J. Kopanski, "Capacitive Probe Microscopy," *Wiley Encyclopedia of Imaging Science and Technology*, (John Wiley & Sons, New York, 2005), pp. 16–31.
2. E. H. Nicollian and J. R. Brews, *MOS (Metal Oxide Semiconductor) Physics and Technology*, (John Wiley & Sons, New York, 1982).
3. S. M. Sze, *Physics of Semiconductor Devices*, (John Wiley & Sons, New York, 1981).
4. The International Technology Roadmap for Semiconductors is published by International SEMATECH, 3101 Industrial Terrace Suite 106, Austin, TX 78758. The 2004 update is available at http://www.itrs.net/Common/2004Update/2004Update.htm.
5. J. R. Matey and J. Blanc, J. Appl. Phys. **57**, 1437–1444 (1985).
6. J. R. Matey, SPIE Scanning Microscopy Technologies and Applications **897**, 110–117 (1988).
7. J. K. Clemens, RCA Review **39**, 33–59 (1978).
8. R. C. Palmer, E. J. Denlinger, and H. Kawamoto, RCA Review **43**, 194–211 (1982).
9. C. D. Bugg and P. J. King, J. Phys. E, Sci. Instrum. **21**, 147–151 (1988).
10. H. P. Kleinknecht, J. R. Sandercock, and H. Meier, Scanning Microscopy **2**, 1839–1844 (1988).
11. C. C. Williams, W. P. Hough, and S. A. Rishton, Appl. Phys. Lett. **55**, 203–205 (1989).
12. C. C. Williams, J. Slinkman, W. P. Hough, and H. K. Wickramasinghe, J. Vac. Sci. Technol. A **8**, 895–898 (1990).
13. C. C. Williams, J. Slinkman, W. P. Hough, and H. K. Wickramasinghe, Appl. Phys. Lett. **55**, 1662–1664 (1989).
14. M. D. Kirk, S. I. Park, and I. R. Smith, "3-Dimensional dopant profiling using scanning capacitance microscopy," Final Technical Report on DARPA Contract #DAAH01-91-C-R118, September 9, 1991.
15. J. A. Slinkman, H. K. Wickramasinghe, and C. C. Williams, "Scanning capacitance-voltage microscopy," U. S. Patent No. 5,065,103, Nov. 12, 1991.
16. J. J. Kopanski, J. F. Marchiando, and J. R. Lowney in *Semiconductor Characterization: Present Status and Future Needs*, W. M. Bullis, D. G. Seiler, and A. C. Diebold, Eds. (AIP, New York, 1996), pp. 308–312.
17. J. J. Kopanski, J. F. Marchiando, and J. R. Lowney, J. Vac. Sci. Technol. B **12**, 242–247 (1996).
18. G. Neubauer, A. Erickson, C. C. Williams, J. J. Kopanski, M. Rodgers, and D. Adderton, J. of Vac. Sci. Technol. B **14**, 426–432 (1996).
19. J. J. Kopanski, J. F. Marchiando, and J. R. Lowney, Mat. Sci. Eng. B **44**, 46–51 (1997).
20. J. J. Kopanski, J. F. Marchiando, and R. Alvis in, *Diagnostic Techniques for Semiconductor Materials and Devices*, Vol 97-12, P. Rai-Choudhury, J. L. Benton, D. K. Schroder, and T. J. Shaffner, Eds., (The Electrochemical Society, Inc, Pennington, New Jersey, 1997), pp. 183–193.
21. J. F. Marchiando, J. R. Lowney, and J. J. Kopanski, Scanning Microscopy **11(1)**, 205–224, (1997).
22. J. J. Kopanski, J. F. Marchiando, D. W. Berning, R. Alvis, and H. E. Smith, J. Vac. Sci. Technol. B **16**, 339–343 (1998).
23. J. J. Kopanski and S. Mayo, Appl. Phys. Lett. **72**, 2469–2471 (1998).
24. S. Mayo, J. J. Kopanski, and W. F. Guthrie in *Characterization and Metrology for ULSI Technology*, D. G. Seiler, A. C. Diebold, W. M. Bullis, T. J. Shaffner, R. McDonald, and E. J. Walters, Eds., AIP Conf. Proceedings 449 (AIP, New York, 1998), pp. 567–572.

25. J. J. Kopanski, J. F. Marchiando, and B. G. Rennex, J. Vac. Sci. Technol. B **18**, 409–413 (2000).
26. J. F. Marchiando, Intl. J. Num. Meth. Engineer. **39,** 1029–1040 (1996).
27. J. F. Marchiando, J. J. Kopanski, and J. R. Lowney, J. Vac. Sci. Technol. B **16**, 463–470 (1998).
28. J. J. Kopanski, J. F. Marchiando, J. Albers, and B. G. Rennex in *Characterization and Metrology for ULSI Technology*, D. G. Seiler, A. C. Diebold, W. M. Bullis, T. J. Shaffner, R. McDonald, and E. J. Walters, Eds. AIP Conf. Proceedings 449, (AIP, New York, 1998), pp. 725–729.
29. J. F. Marchiando, J. J. Kopanski, and J. Albers, J. Vac. Sci. Technol. B **18**, 414–417 (2000).
30. B. G. Rennex, J. J. Kopanski, and J. F. Marchiando in *Characterization and Metrology for ULSI Technology: 2000*, D. G. Seiler, A. C. Diebold, T. J. Shaffner, R. McDonald, W. M Bullis, P. J. Smith, and E. M. Secula, Eds., AIP Conference Proceedings 550 (AIP, New York, 2001), pp. 635–640.
31. Y. Huang and C. C. Williams, J. Vac. Sci. Technol. B **12**, 369–372 (1994).
32. Y. Huang, C. C. Williams, and J. Slinkman, Appl. Phys. Lett. **66**, 344–346 (1995).
33. Y. Huang, C. C. Williams, and H. Smith, J. Vac. Sci. Technol. B **14**, 433–436 (1996).
34. J. S. McMurray, J. Kim, and C. C. Williams, J. Vac. Sci. Technol. B **15**, 1011–1014 (1997).
35. J. S. McMurray, J. Kim, C. C. Williams, and J. Slinkman, J. Vac. Sci. Technol. B **16**, 344–348 (1998).
36. J. Kim, J. S. McMurray, C. C. Williams, and J. Slinkman, J. Appl. Phys. **84**, 1305–1309 (1998).
37. V. V. Zavyalov, J. S. McMurray, and C. C. Williams, J. Appl. Phys. **85**, 7774–7783 (1999).
38. J. S. McMurray and C. C. Williams in *Characterization and Metrology for ULSI Technology*, D. G. Seiler, A. C. Diebold, W. M. Bullis, T. J. Shaffner, R. McDonald, and E. J. Walters, Eds., AIP Conf. Proceedings 449, (AIP, New York, 1998), pp. 731–735.
39. V. V. Zavyalov, J. S. McMurray, and C. C. Williams in *Characterization and Metrology for ULSI Technology*, D. G. Seiler, A. C. Diebold, W. M. Bullis, T. J. Shaffner, R. McDonald, and E. J. Walters, Eds. AIP Conference Proceedings 449 (AIP, New York, 1998), pp. 753–756.
40. V. V. Zavyalov, J. S. McMurray, and C. C. Williams, Review of Scientific Instruments **70**, 158–164 (1999).
41. V. V. Zavyalov, J. S. McMurray, S. D. Stirling, C. C. Williams, and H. Smith, J. Vac. Sci. Technol. B **18**, 549–554 (2000).
42. L. Ciampolini, M. Ciappa, P. Malberti, and W. Fichtner in *Proceedings of the 1st European Workshop on Ultimate Integration of Silicon*, Grenoble, France (January 20–21, 2000).
43. L. Ciampolini, M. Ciappa, P. Malberti, and W. Fichtner in *Proceedings of the 3rd International Conference on Modeling and Simulation of Microsystems*, San Diego, CA (March 27–29, 2000).
44. L. Ciampolini, Scanning Capacitance Microscope Imaging and Modeling, *Series in Microelectronics* **130**, (Hartung-Gorre, Verlag, Konstanz, 2002).
45. J. Smoliner, B. Basnar, S. Golka, E. Gornik, B. Loffler, M. Schatzmayr, and H, Enichlmair, Appl. Phys. Lett. **79**, 3162–3164 (2001).
46. D. Goghero, V. Raineri, and F. Giannazzo, Appl. Phys. Lett. **81**, 1824–1826 (2002).
47. J. Yang and F. C. J. Kong, Appl. Phys. Lett. **81**, 4973–4975 (2002).

48. M. N. Chang, C. Y. Chen, F. M. Pan, J. H. Lai, W. W. Wan, and J. H. Liang, Appl. Phys. Lett. **82**, 3955–3957 (2003).
49. G. H. Buh and J. J. Kopanski, Appl. Phys. Lett. **83**, 2486–2488 (2003).
50. G. H. Buh, J. J. Kopanski, J. F. Marchiando, A. G. Birdwell, and Y. Kuk, J. Appl. Phys. **94**, 2680–2685 (2003).
51. Y. Li, J. N. Nxumalo, and D. J. Thomson, J. Vac. Sci. Technol. B **16**, 457–462 (1998).
52. Scanning Capacitance Microscopy, Support Note No. 224, Rev. B, Digital Instruments Inc., 520 E. Montecito St., Santa Barbara, CA (1996).
53. D. E. McBride, J. J. Kopanski, and B. J. Belzer in *Characterization and Metrology for ULSI Technology 2000*, D. G. Seiler, W. M. Bullis, A. C. Diebold, R. McDonald, T. J. Shaffner, P. Smith, and E. M. Secula, Eds., (AIP, New York, 2001), pp. 657–661.
54. G. D. Wilk and B. Brar, IEEE Electron Dev. Lett. **20**, 132–134 (1999).
55. J. J. Kopanski, J. F. Marchiando, B. G. Rennex, D. Simons, and Q. Chau, J. Vac. Sci. Technol. B. **22**, 399–405 (2004).
56. C. C. Williams and J. S. McMurray, Conversion Algorithm and Software for converting Scanning Capacitance Microscopy Data to 2D Dopant Density—DPAK—Version 3.01 (September 2000) is available through the University of Utah Technology Transfer Office.
57. J. J. Kopanski and J. F. Marchiando in *Proceedings of the 29th International Symposium for Testing and Failure Analysis,* 2-6 November 2003, Santa Clara, California, pp. 398–405.
58. J. F. Marchiando and J. J. Kopanski, J. Appl. Phys. **92**, 5798–5809 (2003).
59. R. Stephenson, A. Verhulst, P. DeWolf, M. Caymax, and W. Vandervorst, J. Vac. Sci. Technol. B. **18**, 405 (2000).
60. F. Giannazzo, L. Calcagnao, V. Raineri, L. Ciampolini, M. Ciappa, and E. Napolitani, Appl. Phys. Lett. **79**, 1211–1213 (2001).
61. D. M. Schaadt, E. J. Miller, E. T. Yu, and J. M. Redwing, Appl. Phys. Lett. **78**, 88–90 (2001).
62. K. V. Smith, E. T. Yu, C. R. Elsass, B. Heying, and J. S. Speck, Appl. Phys. Lett. **79**, 2749–2751 (2001).
63. X. Zhou, E. T. Yu, D. Florescu, J. C. Raner, D. S. Lee, and E. A. Armour, Appl. Phys. Lett. **85**, 407–410 (2001).
64. K. Maknys, O. Douheret, and S. Anand, Appl. Phys. Lett. **83**, 4205–4207 (2001).
65. S. Anand, C. F. Carlström, E. R. Messmer, S. Lourdudoss, and G. Landgren, Appl. Surf. Sci. **144-145**, 525–529 (1999).
66. O. Bowallius, S. Anand, M. Hammar, S. Nilsson, and G. Landgren, Appl. Surf. Sci. **144-145**, 137–140 (1999).
67. S. Anand, IEEE Circuits and Devices **16**, 12–18 (2000).
68. R. C. Barrett and C. F. Quate, J. Appl. Phys. **70**, 2725–2733 (1991).
69. J. W. Hong, S. M. Shin, C. J. Kang, Y. Kuk, Z. G. Khim, and S.-I. Park, Appl. Phys. Lett. **75**, 1760–1762 (1999).
70. C. J. Kang, G. H. Buh, S. Lee, C. K. Kim, K. M. Mang, C. Im, and Y. Kuk, Appl. Phys. Lett. **74**, 1815–1817 (1999).
71. C. J. Kang, C. K. Kim, Y. Kuk, and C. C. Williams, Appl. Phys. A **66**, S415–S419 (1998).
72. R. Yamamoto, K. Sanada, and S. Umemura, Jpn. J. Appl. Phys. **35**, 5284–5287 (1996).
73. H. Tomiye, T. Yao, H. Kawami, and T. Hayashi, Appl. Phys. Lett. **69**, 4050–4052 (1996).
74. H. Tomiye and T. Yao, Appl. Phys. *A* **66**, S431–S434 (1998).

75. M. Dreyer and R. Wiesendanger, Appl. Phys. A **61**, 357–362 (1995).
76. W. K. Chim, K. M. Wong, Y. T. Yeow, Y. D. Hong, Y. Lei, L. W. Teo, and W. K. Choi, IEEE Electron Dev. Lett. **24**, 667 (2003).
77. Y. D. Hong, Y. T. Yeow, W.-K. Chim, K.-M. Wong, and J. J. Kopanski, IEEE Trans. Electron Dev. **51**, 1496–1503 (2004).
78. J. J. Kopanski, W. R. Thurber, and M. L. Chun, in *Interfaces in Electronic Materials, Electrochemical Society Symposium*, Orlando, Fla., October 13–16, 2003.
79. E. M. Vogel, C. A. Richter, and B. G. Rennex, Solid-State Electron. **47**, 1589 (2003).

I.4
Principles of Kelvin Probe Force Microscopy

TH. GLATZEL, M.CH. LUX-STEINER, E. STRASSBURG, A. BOAG,
AND Y. ROSENWAKS

In this chapter we describe and discuss Kelvin probe force microscopy (KPFM), a scanning probe microscopy technique designed to obtain laterally resolved work function images by measuring the electrostatic forces between probe and sample surface. By operating the microscope in ultrahigh vacuum, even absolute work function measurements with very high lateral and energy resolution can be realized.

The method of Kelvin probe force microscopy is introduced in the first section and the two main measuring modes are discussed in detail. The amplitude modulation technique has the advantage of requiring only small modulation voltages; however, due to long-range electrostatic forces the lateral resolution is limited. The frequency-modulation technique, on the other hand, is sensitive to the force gradient, which provides better lateral resolution, but requires a much larger sampling voltage.

The second section examines the factors affecting the sensitivity and resolution of KPFM in general, and in semiconductors measurements in particular. An efficient numerical analysis of the electrostatic interaction between the measuring AFM tip and the semiconductor surface has allowed us to derive a point spread function of the measuring tip and to reconstruct the actual surface potential from KPFM images in almost real time. Several examples of recent applications will be reviewed in our second chapter.

1 Introduction

The Kelvin probe force microscopy (KPFM) is a method for measuring the surface potential distribution with nanometer resolution. Since its introduction in 1991 by Nonenmacher et al. [1] KPFM has found many diverse applications in several fields due to the direct and quantitative measurement of the surface potential distribution. In materials research it has been used for work function mapping [1], ordering measurements in III-V compound semiconductors [2], local surface photovoltage and surface photovoltage spectroscopy [3], surface states and defects under different ambient conditions [4], domain characterization in ferroelectric materials

[5], measurement of organics and self-assembled monolayers [6], and many more applications. KPFM has also proved to be a very important tool for potential mapping of passive and active semiconductor devices such as *pn* junctions [7], resistors and *n-i-p-i* heterostructures [8], high-electron-mobility transistors (HEMT) [9,10], light-emitting diodes [11,12], solar cells [13,14], and organic and polymer-based transistors [15]. Recently several groups have reported on high-resolution measurements of low-dimensional structures such as 5-nm CdSe nanocrystals [16], and single and multi-quantum well structures [17]. Although atomic scale features in KPFM were reported five years ago [18,19], their origin is poorly understood mainly due to the long-range nature of the electrostatic forces discussed in detail in the next section. Another important aspect of local potential measurements is related to quantitative two-dimensional dopant mapping. Several groups have used KPFM for two-dimensional dopant profiling [20]; however, it was found that it is not yet the ideal tool for this purpose. The resolution of KPFM compared with scanning capacitance microscopy (SCM) [21] and scanning spreading resistance microscopy (SSRM) [22] is smaller; and with a sensitivity of a few millivolts, it is challenging to quantify doping levels in the range required by the microelectronic industry.

The KPFM method as well as the basic detection method are introduced in sections 1.1 and 1.2. The two main measuring modes used commonly in KPFM are described and discussed in sections 1.3 and 1.4. Section 2 examines the factors affecting the sensitivity and resolution of KPFM in general, and in semiconductors measurements in particular. An efficient numerical analysis of the electrostatic interaction between the measuring AFM tip and the semiconductor surface has allowed us to derive a point spread function of the measuring tip and to reconstruct the actual surface potential from Kelvin probe microscopy images in almost real time.

1.1 Kelvin Probe Force Microscopy

The contact potential difference (CPD) between two materials, for example, between an AFM tip and a sample, is defined as

$$V_{CPD} = \frac{\Phi_{sample}}{q} - \frac{\Phi_{tip}}{q} \equiv \Delta\Phi/q, \tag{1}$$

where Φ_{tip} and Φ_{sample} are the work functions of the tip and the sample, respectively, and q is the elementary charge. Therefore, if an AFM tip and a semiconductor with different work functions are held in close proximity, an electrostatic force will developed between them, due to the potential difference V_{CPD}; this is schematically described in Figure 1. When the two materials are not connected their local vacuum levels are aligned but there is a difference in their Fermi levels. Upon electrical connection, the Fermi levels will align through electron current flow, as shown in Figure 1(a). The two materials (electrodes) are now charged and there is a

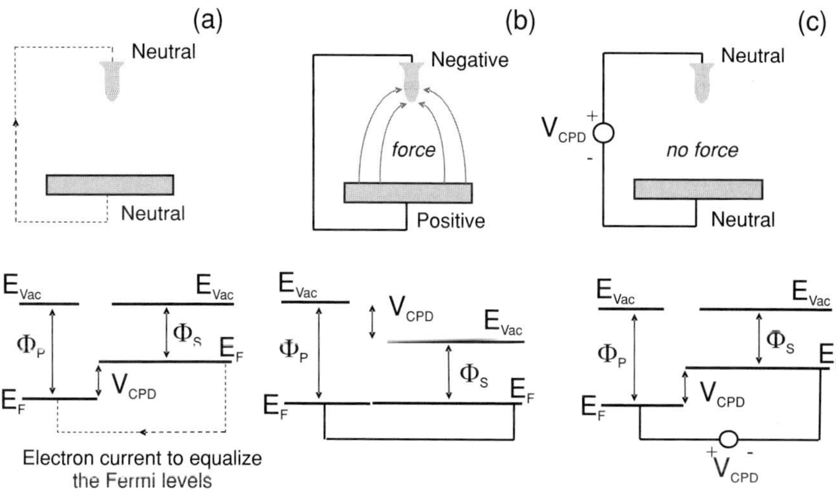

FIGURE 1. Definition and basic measurement setup of contact potential difference (CPD).

difference in their local vacuum levels. Due to the charging of the tip and the sample, an electrostatic force develops as shown in Figure 1(b).

This force can be nullified by applying an external bias between the tip and the sample. The magnitude of this bias is the contact potential difference and its sign depends whether it is applied to the sample or to the tip.[1] A typical KPFM measurement is conducted in the following way. An ac bias at a frequency ω is applied between the tip and the sample. It is shown below [23] that the force component at this frequency is proportional to the CPD and therefore can be nullified using a feedback loop whose input is the component of the electrostatic force at a frequency of ω. The simplest approximation of this force is to treat the tip-sample system as a parallel plate capacitor with one plate being the tip apex, and the other the sample underneath it [24].

1.2 Detection in Kelvin Probe Force Microscopy

In Figure 2 the schematic setup of an ultrahigh vacuum-atomic force microscope (UHV-AFM) with the main electronic components for KPFM is shown. In most noncontact AFM setups for UHV the cantilever is oscillated at its resonance frequency f_1 with a constant amplitude (automatic gain controll, AGC) and scanned across the sample at a constant distance by maintaining a constant frequency shift Δf_1 to the free resonance frequency (frequency modulation mode, FM-mode).

[1] The sign of the measured V_{CPD} will be positive (negative) if the nullifying voltage is applied to the sample (tip) respectively, see Ref. [11, 12]

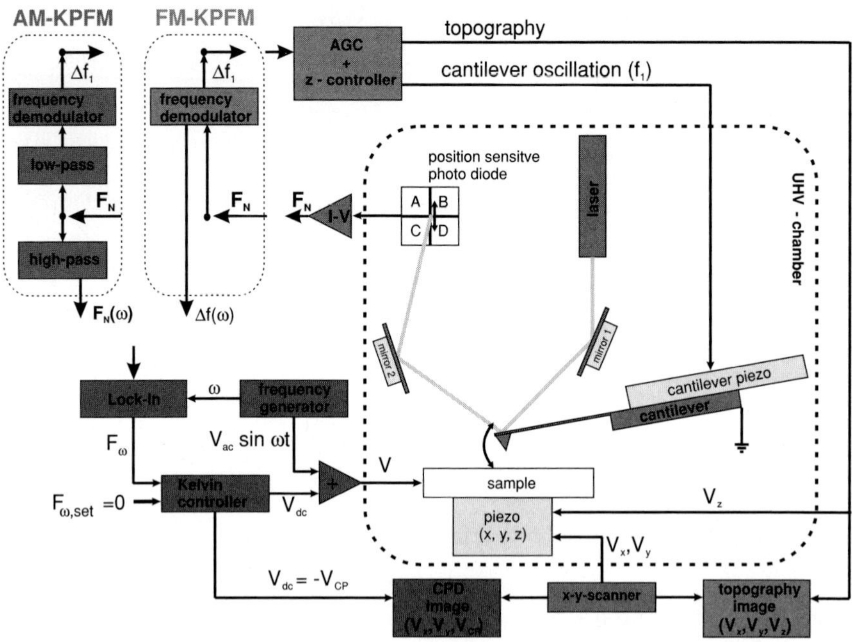

FIGURE 2. Schematic setup of an UHV-AFM with the main electronic components for the AM-mode and FM-mode detection in KPFM.

The measured frequency variation Δf_1 is approximately proportional to the force gradient, given by [25–27]

$$\Delta f_1 \approx -f_1 \frac{\partial F}{\partial z}, \tag{2}$$

where F is the force acting on the cantilever and z the tip-sample distance.

To measure the electrostatic force F_{es} separately from other forces (mainly van der Waals and chemical forces), an ac voltage V_{ac} at a frequency ω is superimposed on the tip-sample voltage V_{dc}. The application of this voltage $V = (V_{dc} - \Delta\Phi/q) + V_{ac}\sin(\omega t)$ results in an oscillating electrostatic force, inducing an additional oscillation of the cantilever at ω as mentioned before. The force which is the derivative of the electrostatic energy with respect to the tip-sample separation, z, is given by

$$F_{es} = -\frac{1}{2}\frac{\partial C}{\partial z}\left[(V_{dc} - \Delta\Phi/q) + V_{ac}\sin(\omega t)\right]^2, \tag{3}$$

where $\partial C/\partial z$ the gradient of the capacitance. Equation (3) can be written as

$$F_{es} = F_{dc} + F_{\omega} + F_{2\omega} \tag{4}$$

with the spectral components

$$F_{dc} = -\frac{\partial C}{\partial z}\left(\frac{1}{2}(V_{dc} - \Delta\Phi/q)^2 + \frac{1}{4}V_{ac}^2\right) \quad (5)$$

$$F_{\omega} = -\frac{\partial C}{\partial z}(V_{dc} - \Delta\Phi/q)V_{ac}\sin(\omega t) \quad (6)$$

$$F_{2\omega} = +\frac{\partial C}{\partial z}\frac{1}{4}V_{ac}^2\cos(2\omega t). \quad (7)$$

The dc part F_{dc} (Eq. (5)) contributes to the topographical signal, the term F_{ω} at the frequency ω (Eq. (6)) is used to measure the CPD, and the contribution $F_{2\omega}$ (Eq. (7)) can be used for capacitance microscopy [28].

Equation (6), although used by many authors, is strictly correct only for a metallic tip-sample system. Hudlet et al. [23] have presented a detailed one-dimensional analysis of the electrostatic force between a tip and a semiconductor and have shown that the force at a frequency of ω can be expressed as

$$F_{\omega} = \frac{Q_s}{\epsilon_0}C_{eff}V_{ac}\sin\omega t, \quad (8)$$

where Q_s is the semiconductor surface charge, ϵ_0 is the dielectric constant, and C_{eff} is the effective capacitance of the electrode-air/vacuum-semiconductor system. For a metallic sample where Q_s is replaced by $-C(V_{DC} - \Delta\Phi/q)$ and C_{eff} by C (where C is the tip-metallic sample capacitance), Eq. (8) reduces to Eq. (6). It must be emphasized here that, based on Eq. (6), the sign of V_{CPD} will be different if the nullifying voltage is applied to the tip or to the sample. The posteriori dc voltage difference V_{CPD}^P is thus given for the two cases as

$$V_{CPD}^P = -\frac{\Phi_{tip}}{q} - \left(\frac{\Phi_{sample}}{-q} + V_{dc}\right) = V_{CPD} - V_{dc} \quad (9)$$

$$V_{CPD}^P = -\frac{\Phi_{tip}}{q} - \frac{\Phi_{sample}}{-q} + V_{dc} = V_{CPD} + V_{dc}, \quad (10)$$

where Eqs. (9) and (10) are valid for the cases of voltage applied to the sample and to the tip, respectively. Following the nullifying procedure, i.e., when $V_{CPD}^P = 0$, we obtain $V_{dc} = \pm V_{CPD}$, where the "+" and "−" refer to the external bias applied to the sample and the tip, respectively.

The force component at the frequency F_{ω} (Eq. (6)) is proportional to the CPD and therefore can be nullified by a compensation voltage V_{dc} applied by the so called Kelvin controller using a feedback loop (Figure 2). The input of this feedback loop depends on the detection mode. In the amplitude-modulation-detection mode (AM-mode) [7,29–31], the component of the electrostatic force at a frequency of ω provided by the signal $F_N(\omega)$ is used, whereas the frequency-modulation detection mode (FM-mode) as first presented by Kitamura et al. [32,33] uses the variation of the force gradient $\Delta f_1(\omega)$ (Eq. 2). Both techniques can be used to detect the samples' topography and the CPD image simultaneously.

In the following sections both techniques are presented. The main difference between the two methods is the sensitivity to the electrostatic force; in FM-mode

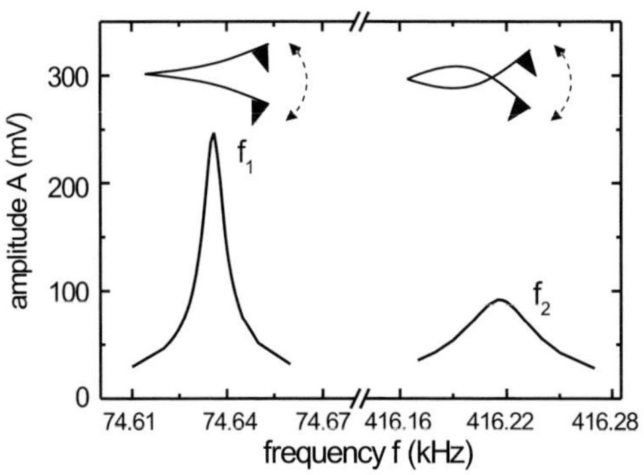

FIGURE 3. Typical resonance spectra of a silicon cantilever with the first f_1 and the second f_2 resonance. The quality factor is in the range of $Q = 10{,}000$. The drawings visualize the type of oscillation.

the force gradient and in AM-mode the electrostatic force are detected. A different contribution of the cantilever and the tip to the signal is expected and therefore absolute and relative work function resolutions are different. To quantify this effect, measurements with both techniques on the same structure using different types of cantilevers are presented.

1.3 Amplitude-Modulation Detection

In the AM-mode the amplitude of the cantilever oscillation at ω induced by the electrostatic force is measured directly by lock-in technique. This method is used in most KPFM setups under ambient conditions using frequencies on the order of 10–30 kHz [11,12]. However, to achieve sufficient sensitivity, ac voltages in the range of 1–5 V have to be used.

In the UHV environment, the cantilevers Q value is increased by at least a factor of 10 [34]. A vibration amplitude-frequency characteristic of a cantilever including the second resonance frequency f_2 is shown in Figure 3. It is obvious that the sensitivity of the potential measurement is improved when the ac-frequency ω is tuned to the second resonance frequency of the cantilever. Another reason for using the second resonance peak is to facilitate separation of the two signals.

The resonance frequency shift due to the forces acting on the cantilever also occurs at the second resonance. However, its sensitivity is far lower than that of the first one. Thus the first resonance is used for the tip height control and the second resonance for the potential measurement [29]. The limiting factor in this mode is the bandwidth of the photodiode used for the detection of the cantilever oscillation [35]. Normally photodiodes with a bandwidth of around 500 kHz are used; therefore, the stiffest cantilevers for this detection mode have the first resonance frequency in the range of 70 to 80 kHz, and the second resonance around 400 to 470 kHz. The main benefit of this resonant detection is the very high sensitivity achieved using ac voltages as small as 100 mV [30,31].

In Figure 4 the topography (a) and the work function images (b) of Au islands on highly oriented pyrolytic graphite (HOPG) measured in AM-mode at $\omega = f_2$ and $V_{ac} = 100$ mV are shown. The HOPG sample was freshly cleaved, and

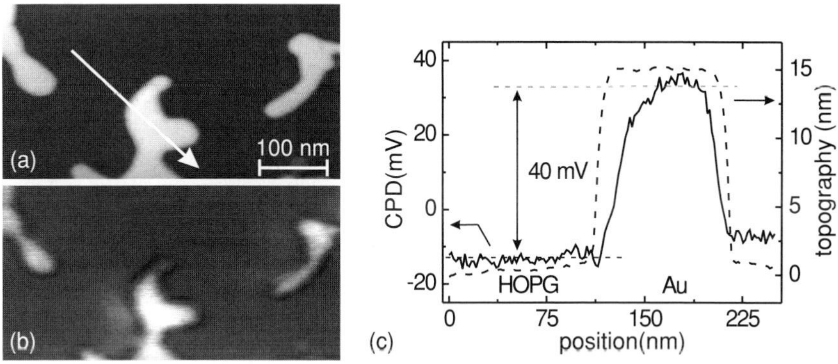

FIGURE 4. KPFM measurement on a HOPG sample with Au islands. (a) Topography ($\Delta z =$ 17 nm) and (b) contact potential ($\Delta CPD = 50$ meV) obtained in AM-mode ($\omega = f_2$, $V_{ac} =$ 100 mV) with a silicon cantilever. Line scans through the images are presented in (c).

approximately three monolayers of gold were thermally evaporated at a sample temperature of 150°C. After annealing in UHV at 110°C for 1 h, dendritic Au islands are formed.

The measurement shows height variations in the range of $\Delta z = 16$ nm in the topography and elevated work function values ($\Delta\Phi = 50$ mV) for the Au islands. The work function increases toward the center of the island and only reaches a constant value on the big island in the center of the image. Furthermore, a decrease in the work function along the Au steps in the measurements is visible. This effect can be explained by a reduction of the work function due to dipoles located at step edges of metallic surfaces [36]. In Figure 4(c) a line scan of the topography (dotted) and the work function (straight) through a gold island indicated by arrows in the images are presented. The noise level within this measurement was determined to be $\approx$5 mV.

In AM-mode the lateral energy resolution is reduced by an averaging effect of the tip and the cantilever due to the detection of long-range electrostatic force. A further improvement can be achieved by using stiffer cantilevers, and as a consequence, using a photodiode with a higher bandwidth. Another possibility is the heterodyne laser beam deflection method, which enables conventional AFM setups to use frequencies of several MHz, as recently suggested by Fukuma et al. [37]. Stiffer cantilevers allow smaller tip-sample distances, thereby increasing tip apex contributions to the signal. In addition, to reduce the contribution of the cantilever, the tip should be as long as possible and the opening angle should be as small as possible. The importance of the tip geometry will be shown in the next section.

1.4 Frequency-Modulation Detection

In the FM-mode, the applied ac voltage induces a modulation of the electrostatic force which is detected by the oscillation at ω of the frequency variation of f_1

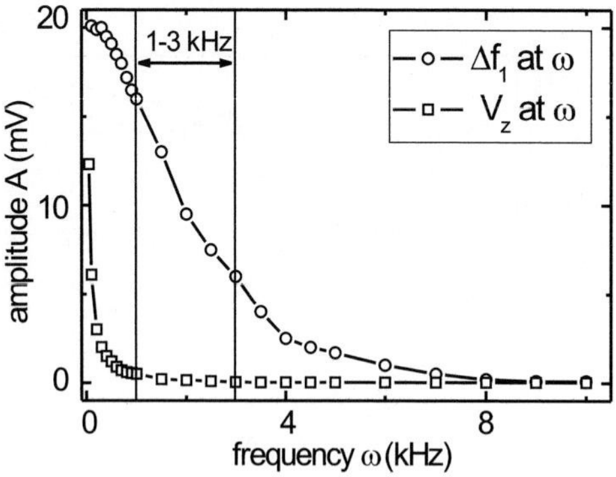

FIGURE 5. Restrictions in the FM-mode: Dependence of the frequency shift Δf_1 amplitude and the hight control signal of the topography V_z at the frequency ω of the ac voltage. The measurements were obtained with a silicon cantilever on a HOPG sample.

(Figure 2). The measured signal is approximately proportional to the force gradient, as can be deduced from Eqs. (2,6):

$$\Delta f_1(\omega) \propto \frac{\partial F_\omega}{\partial z} = -\frac{\partial^2 C}{\partial z^2}(V_{dc} - \Delta\Phi/q)V_{ac}\sin(\omega t). \tag{11}$$

The frequency ω has to be chosen in an appropriate range. The lower limit is dictated by increasing cross-talk to the topography signal, which will cause the tip-sample distance to start to oscillate at ω. The higher the frequency, the lower the coupling between the topography and CPD; however, the bandwidth of the frequency demodulator determines the upper limit of the frequency range. In Figure 5 the amplitude at ω of the oscillation of Δf_1 and of the oscillation of the piezo-voltage V_z, which controls the tip-sample distance, is plotted. With increasing frequency the cross-talk to the topography signal decreases, but the signal intensity of the electrostatic force also decreases due to the restricted bandwidth of the frequency demodulator. Furthermore, ac voltages larger than $V_{ac} = 2$ V are necessary to obtain sufficient sensitivity. However, the ac voltage also contributes quadratically to the topography signal (Eq. (5)). Moreover, on semiconducting samples, a voltage-induced band bending is possible [31]; thus, ac voltages as low as possible should be used.

Figure 6 shows a KPFM measurement of HOPG with Au islands comparable to the one in the previous section (Figure 4), now acquired in the FM-mode. A high-frequency cantilever ($f_1 \approx 250$–350 kHz) with a tip sample bias of $V_{ac} = 2$ V at a frequency of $\omega = 2$ kHz was used. The topography shows height variations in the range of $\Delta z = 16$ nm, comparable to the AM-mode measurement. However, the work function values for the Au islands are up to $\Delta\Phi = 290$ mV higher than the HOPG background. Furthermore, the potential image scanned in FM-mode shows a homogeneous work function on the islands in contrast to the AM-mode.

In the FM-mode the electrostatic interaction mainly takes place between the tip apex and the sample due to the detection of the short ranged gradient of the electrostatic force (Eq. (11)). This results in a lateral resolution in the range of the size of the tip apex.

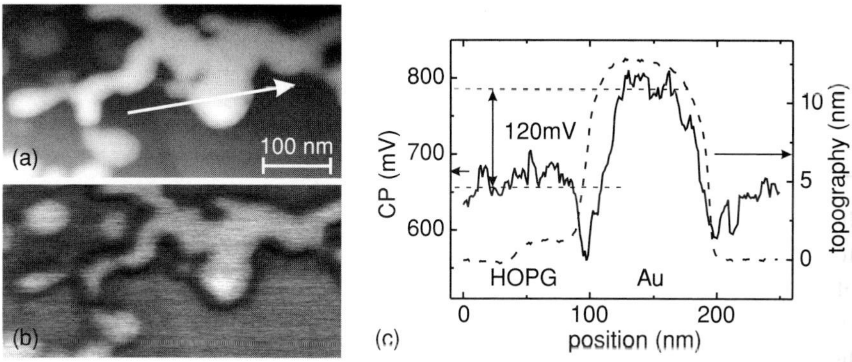

FIGURE 6. KPFM measurement on a HOPG sample with Au islands. (a) Topography (Δz = 15 nm) and (b) contact potential (ΔCPD = 290 meV) obtained in FM-mode ($\omega = f_2$, V_{ac} = 2 V) with a high frequency cantilever. Line scans through the images are presented in (c).

In both detection modes point spectroscopy measurements above the center of a Au island and above HOPG were performed. Different cantilevers were used, and the work function difference of both materials as a function of the tip-sample distance was determined (Figure 7). The expected difference in work function between Au and HOPG is $\Delta\Phi = 400$ meV, which was not resolved in either AM- or FM-mode. In Figure 7(a) the distance dependence of the work function difference measured by a cantilever with a short (3–6 μm) silicon tip is plotted. In AM-mode only a slight increase in the difference with decreasing tip-sample distance was detected. This can be explained by the averaging effect in AM-mode, also the surrounding area of the measured point contributes to the signal. In contrast, in FM-mode, where the force gradient is detected, a decreasing tip-sample distance reduces the area which contributes to the detected signal.

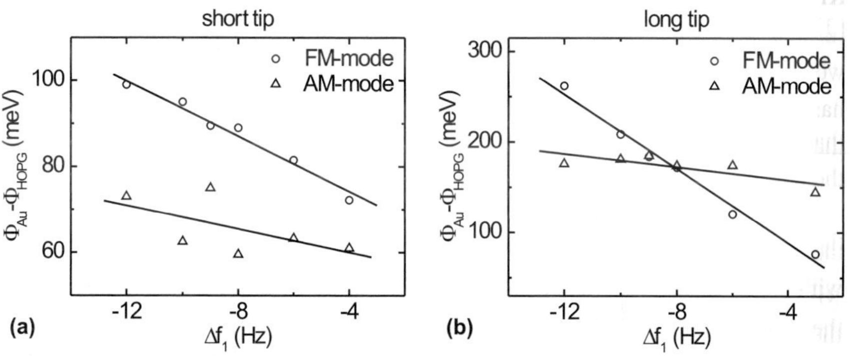

FIGURE 7. Point spectroscopy measurements in AM- and FM-mode on a HOPG sample with Au islands. The work function difference of Au and HOPG are plotted obtained with different silicon cantilevers with a tip length of (a) 3–6 μm and (b) 15–20 μm.

In Figure 7(b) the results of similar measurements conducted with a longer (15–20 μm) tip is presented. The behavior in AM- and FM-mode remain similar; however, due to a smaller contribution of the cantilever and the tip shape, a higher work function difference is obtained. This shows that with the choice of an appropriate cantilever-tip geometry the resolution can be significantly improved: long tips are better for KPFM experiments, which was predicted also by simulations [38].

2 Sensitivity and Spatial Resolution in Kelvin Probe Force Microscopy

2.1 Tip-Sample Electrostatic Interaction

It is well known that the finite tip size in scanning probe-microscopes has a profound effect on the measured image. In electrostatic-force-based microscopies, the effect of the measuring tip is much larger because the measured forces have an infinite range. Tip effects in electrostatic force and Kelvin probe force microscopy have been discussed and analyzed by several authors. One of the simplest models was suggested by Hochwitz et al. [39], who modeled the tip by a series of (staircase) parallel plate capacitors. Hudlet et al. [40] have presented an analytical evaluation of the electrostatic force between a conductive tip and a metallic surface, while Belaidi et al. [41] have calculated the forces and estimated the resolution in a similar system. Jacobs et al. [38] have extended the calculations for the case of a semiconductor sample, by replacing its surface by a set of ideal conductors with mutual capacitances between them. In addition they have shown that the KPFM images of infinitely large conducting surfaces are two-dimensional (2D) convolutions of the actual surface potential distribution with a transfer function, defined by the tip geometry [42].

To the best of our knowledge almost all the papers that analyze semiconductor KPFM measurements replace the semiconductor sample by a surface with a fixed or variable potential. This is only valid for the case of a weakly interacting tip-sample system, i.e., when there is no tip-induced band-bending phenomenon. As described above, a typical KPFM measurement is conducted by nullifying the electrostatic force at a frequency of ω, using an external bias V_{DC}. Thus a calculation of the KPFM signal amounts to finding the voltage applied to the tip or to the sample [23], V_{DC}, that minimizes the total electrostatic force at the frequency ω. Below we calculate the static (dc) electrostatic force; i.e., it is assumed that the ac voltage has a negligible effect on the tip-sample forces. This is justified based on the fact that the frequency ω used in the measurements (>300 kHz) is high and in addition the ac modulation amplitude applied to the tip is very low (100 mV).

The electrostatic force was calculated as follows. First the electric field, E, for the semiconductor-air/vacuum-tip system is calculated using a Poisson equation with appropriate boundary conditions. Assuming that the tip is a perfect conductor, the electrostatic force is then calculated by integrating the Maxwell stress over the entire tip surface:

$$\mathbf{F} = \frac{1}{2\epsilon_0} \int_{S'} \rho^2(\mathbf{r})\hat{\mathbf{n}}\, ds, \tag{12}$$

where σ is the tip surface charge density, ds is a tip surface element, and $\hat{\mathbf{n}}$ is the outward normal unit vector. A similar approach for electrostatic force calculations was applied in the past for a metallic sample where the sample surface potential was assumed to be constant [41]. However, here we calculate whether the presence of the tip has any significant effect on the semiconductor surface potential: i.e., is there any tip-induced band bending at the semiconductor surface? Thus, our calculation takes into account the electrostatic energy present when the semiconductor energy bands are not flat due to the presence of surface states and/or due to tip-induced band bending. Hudlet et al. [23] have shown that in the one-dimensional case, the tip-sample system can be modeled as two capacitors in series. Thus, when the distance between the sample and the tip is reduced, the sample capacitance cannot be neglected, and it changes the force acting on the tip and the measured CPD.

As an example, we have calculated the surface-induced band bending for a tip with a potential of $0.1V$ higher than the sample surface ($n = 5 \times 10^{17}\text{cm}^{-3}$ GaP with no charged surface states, or zero band bending) and located 5 nm above it; the result is shown in Figure 8. The figure shows that the surface band bending, expressed as $(E_F - E_{is})$, where E_{is} is the surface Fermi level position, due to the presence of the biased tip, is zero everywhere except for a small region in the middle, where the band bending is less than 6 mV. A similar calculation conducted

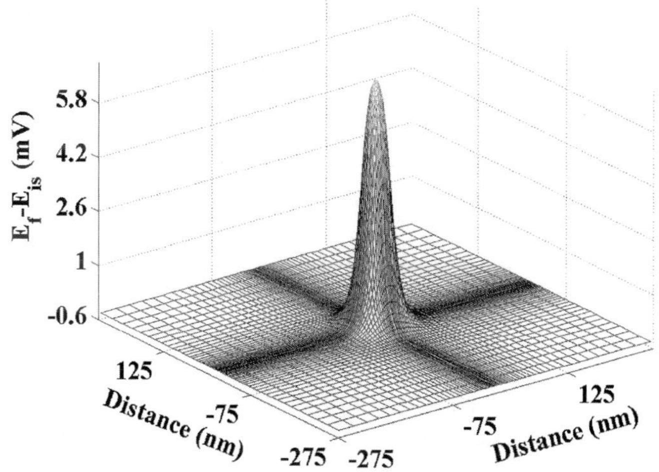

FIGURE 8. Calculated local band bending, expressed as $(E_F - E_{is})$ for a GaP with no surface states, tip-sample distance of 5 nm, and an applied bias of $V_{tip} = 0.1$ V between tip and sample. The protrusion in the center is the tip-induced band bending at the GaP surface.

for a tip-sample surface potential difference of 0.6 V resulted in an surface-induced band bending of around 38 mV [43].

The small tip-induced band bending effect can be explained as follows. The tip-vacuum-semiconductor system can be modeled as two capacitors—tip-vacuum-semiconductor surface and the semiconductor space charge region (SCR) connected in series. Thus, an external voltage (in the present case it is the semiconductor-tip CPD) will drop mainly on the smaller of the two capacitors. The SCR capacitance is typically much larger (except for very low doped semiconductors) thus causing the voltage to drop mainly between the tip and the sample surface, and hence inducing a negligible band bending in the SCR. It must be re-emphasized that in KPFM measurements the CPD between the tip and the sample is nullified; so typically the potential difference between any point on the tip, and on the sample surface in a close distance to it, will be even lower than the 0.1 V used in calculating the potential in Figure 8. This conclusion simplifies tremendously the simulation of semiconductors KPFM measurements. This is because it allows us to represent the whole semiconductor sample by a surface with a varying potential; the three-dimensional potential of the tip-sample system can then be calculated using an integral-equation-based boundary element method (BEM), which is a much faster numerical process, as shown below.

2.2 A Fast Algorithm for Calculating the Electrostatic Force

Figure 9 shows schematics of the tip-sample system used for the BEM calculations. A perfectly conducting tip with a surface S^t is at a height d above a perfectly planar conducting sample surface S located at the $z = 0$ plane. The tip is scanning the sample, and the vector $\rho_t = (x_t, y_t)$ represents the horizontal shift of the tip in the global coordinate system.

We use a notation where for any point, $\mathbf{r} = (\rho, z)$, the two-dimensional (2D) vector $\rho = (x, y)$ represents the horizontal coordinates, whereas z denotes the vertical one. In addition, we define a local coordinate system attached to the tip. In the global

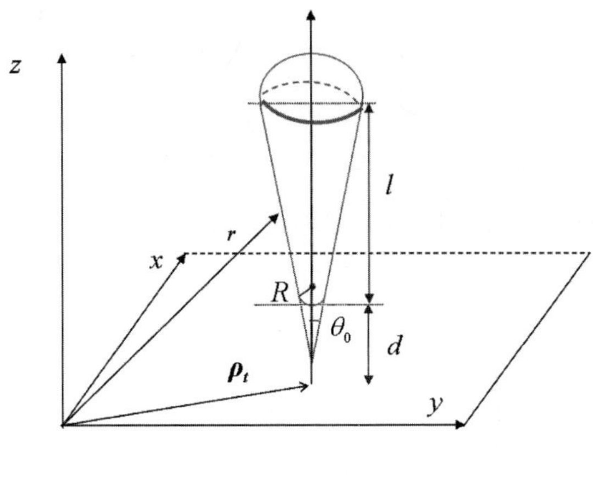

FIGURE 9. Schematics of the tip-sample system used for the electrostatic calculations. A perfectly conducting tip with a length l, half aperture angle θ_0, and spherical apex radius R, is scanning the sample at a height d. The vectors ρ_t and r represent the position of the tip axis in the (x, y) plane and a point on the tip surface in the in the (x, y, z) space, respectively.

coordinate system, a point on the tip surface $\mathbf{r} \in S^t$ is represented as $\mathbf{r} = \tilde{\mathbf{r}} + \rho t$, where $\tilde{\mathbf{r}}$ are the coordinates of the point in the tip fixed coordinate system. In KPFM measurements, the CPD at each tip position, ρ_t, is the potential applied to the tip $V(\rho_t)$ to give minimum vertical (z-directed) force; this is equivalent to nullifying the first harmonic modulated force [40]. Let $\Phi(\mathbf{r}|\rho_t)$ denote the electrostatic potential at point $\mathbf{r}$ when the tip is located at ρ_t. Assume that the presence of the tip is not modifying the surface potential, i.e., $\forall \rho_t \Phi(\rho, z = 0^+|\rho_t) = \Phi^S(\rho)$. The surface of the tip is assumed equipotential, namely, $\Phi(\mathbf{r}|\rho_t)|_{\mathbf{r}\in S'} = V(\rho_t)$. Below we calculate the actual surface potential, $\Phi^S(\rho)$, based on the tip potential measurements, $V(\rho_t)$.

The surface potential is represented by a dipole layer of density $\eta(\rho) = \Phi^S(\rho)/\epsilon_0$ located on the top of a grounded plane (the sample) at $z = 0$. The total potential at a point $\mathbf{r}$ on the tip surface, is given by the superposition

$$\Phi^t(\mathbf{r}|\rho_t) = \Phi^t(\mathbf{r}|\rho_t) + \Phi^d(\mathbf{r}), \tag{13}$$

where $\Phi^t(\mathbf{r}|\rho_t)$ is the potential produced by the charges residing on the tip, and $\Phi^d(\mathbf{r})$ is the potential of the dipole layer. The potential of the tip charges and their images can be represented as

$$\Phi^t(\mathbf{r}|\rho_t) = \int_{S^t} [G(\mathbf{r} - \mathbf{r}') - G(\mathbf{r} - \mathbf{r}'')]\, \sigma(\mathbf{r}'|\rho_\mathbf{t})\, ds' \equiv \mathcal{A}\sigma, \tag{14}$$

where $\mathbf{r}'' = (\rho', -z')$ is the image point of $\mathbf{r}' = (\rho', z')$, $G(\mathbf{r}) = 1/(4\pi\epsilon_0|\mathbf{r}|)$, is the three-dimensional electrostatic Green's function, and $\mathcal{A}$ is an operator yielding the contribution to the tip potential due to the charge density σ on the tip surface. Let $\Phi^d(\tilde{\mathbf{r}}|\rho_t) = \Phi^d(\mathbf{r} = \tilde{\mathbf{r}} + \rho_t)$ represent the dipole layer contribution at a point $\tilde{\mathbf{r}}$ on the tip located at ρ_t. We can express the dipole contribution to the potential as a convolution

$$\Phi^d(\tilde{\mathbf{r}}|\rho_t) = D(\tilde{\mathbf{r}}|\rho_t) * \Phi^S(\rho_t), \tag{15}$$

where

$$D(\tilde{\mathbf{r}}|\mathbf{x}') = 2\epsilon_0 \left.\frac{\partial G(\mathbf{r} - \mathbf{r}')}{\partial z'}\right|_{z'=0} \tag{16}$$

and the asterisk represents a 2D convolution with respect to ρ_t.

Imposing the boundary condition $\Phi(\mathbf{r}|\rho_t)|_{\mathbf{r}\in S'} = V(\rho_t)$ and substituting Eq. (14) into Eq. (13), we obtain an integral equation for the tip-surface charge density:

$$\mathcal{A}\sigma = V(\rho_t) - \Phi^d(\tilde{\mathbf{r}}|\rho_t) \quad \tilde{\mathbf{r}} \in S^t. \tag{17}$$

Equation (17) can be approximated via the boundary element method (BEM) and expressed in a matrix notation as

$$[A]\,\vec{\sigma} = V\,\vec{1} - \vec{\Phi}^d, \tag{18}$$

where $[A]$ is an N by N matrix representing operator $\mathcal{A}$, and $\vec{\sigma}$, $\vec{1}$, and $\vec{\Phi}^d$ are N-vectors representing the tip charge density, a vector whose elements are all 1, and the dipole layer potential, respectively. The system of linear equations represented by Eq. (18) can be formally solved as

$$\vec{\sigma} = V\,\vec{a} - [A]^{-1}\,\vec{\Phi}^{d}, \tag{19}$$

where $[A]^{-1}$ is the inverse matrix of $[A]$ and $\vec{\sigma} \equiv [A]^{-1}\,\vec{1}$ is a sum of all columns of $[A]^{-1}$.

The tip potential, V, is determined based on the minimum force condition employed in the KPFM measurements described above. The z-component of the electrostatic force acting on the tip is given by

$$F_z = \frac{1}{2\epsilon_0}\int_{S^t} \sigma^2(\mathbf{r})(\hat{\mathbf{n}}\cdot\hat{\mathbf{z}})\,ds \equiv \mathcal{B}\sigma^2, \tag{20}$$

where $\hat{\mathbf{n}}$ is the outward normal unit vector and we define $\mathcal{B}$ as a linear functional relating the vertical force to the charge density. In discrete form, the vertical force can be approximated as

$$F_z = \vec{\sigma}'\,[B]\,\vec{\sigma}, \tag{21}$$

where $[B]$ is an N by N diagonal matrix stemming from the BEM discretization of $\mathcal{B}$. The minimum force condition $\partial F_z/\partial V = 0$ can be expressed via Eq. (21) as

$$\vec{a}^{\,t}\,[B]\,\vec{\sigma} = 0, \tag{22}$$

where, based on Eq. (19), we made a substitution $\partial\,\vec{\sigma}\,/\partial V = \vec{a}$. Substituting Eq. (19) in Eq. (22), we obtain an explicit expression for the measured tip potential in terms of the potential produced by the dipole layer:

$$V(\bar{\mathbf{x}}) = \frac{\vec{a}^{\,t}\,[B][A]^{-1}\,\vec{\Phi}^{d}\,(\bar{\mathbf{x}})}{\vec{a}^{\,t}\,[B]\,\vec{a}}. \tag{23}$$

Finally, using Eq. (15), we can write the measured tip potential $\Phi^S(\rho_t)$ as a convolution of the actual sample surface potential with a continuous point spread function (PSF) as

$$V(\rho_t) = h_c(\rho_t) * \Phi^S(\rho_t), \tag{24}$$

where

$$h_c(\rho_t) \equiv \frac{\vec{a}^{\,t}\,[B][A]^{-1}\vec{D}(\rho_t)}{\vec{a}^{\,t}\,[B]\,\vec{a}} \tag{25}$$

and $\vec{D}(\rho_t)$ is an N-vector obtained via BEM discretization of $\vec{D}(\tilde{\mathbf{r}}|\rho_t)$.

In practice, scanning of the sample is conventionally performed with a discrete step Δ, such that $\rho_t = (m_x\Delta, m_y\Delta)$, where m_x and m_y are integers. By the same token, we can use a discrete approximation of the surface potential,

$$\Phi^S(\rho) = \sum_{m_x,m_y} \Phi_d^S(\mathbf{m})b(\rho - m_x\Delta\hat{\mathbf{x}} + m_y\Delta\hat{\mathbf{y}}) \tag{26}$$

where $\Phi_d^S(\mathbf{m})$ with $\mathbf{m} = (m_x, m_y)$ are the discrete samples of the surface potential and $b(\rho)$ is a pixel interpolation function. Substituting $\rho_t = m_x \Delta\hat{\mathbf{x}} + m_y \Delta\hat{\mathbf{y}}$ and Eq. (26) in (24) and exchanging the order of summation and integration, we obtain the discrete counterpart of Eq. (24),

$$V(\mathbf{m}) = h_d(\mathbf{m}) * \Phi_d^S(\mathbf{m}), \tag{27}$$

where the asterisk denotes a 2D discrete convolution with respect to $\mathbf{m}$, and $h_d(\mathbf{m})$ is the discrete point spread function computed by convolving the continuous PSF given by Eq. (25) with the pixel interpolation function. Equation (27) and its one-dimensional forms constitute the basis for the reconstruction of the actual surface potentials presented in the following section. The convolutional nature of Equation 27 is of major importance in terms of the numerical feasibility of the proposed approach.

2.3 Deconvolution of KPFM Images

The electrostatic analysis presented in the previous sections did not include the noise inevitably present in any KPFM measurement. Therefore, Eq. (27) has to be modified in the presence of noise to

$$V(\mathbf{m}) = h_d(\mathbf{m}) * \Phi_d^S(\mathbf{m}) + n(\mathbf{m}), \tag{28}$$

where n is an additive white Gaussian noise (AWGN) which is assumed to be uniform across the sample. Note that the 2D Fourier transform of the point spread function h_d is concentrated at low spatial frequencies, whereas the noise spectral density is uniform. Thus, using a straightforward deconvolution is not suitable here, since the noise would be amplified above the restored potential particularly at high frequencies. In the case of additive noise, the Wiener filter [44] provides an optimal estimator in terms of minimum square error. The filter is defined in the spatial frequency domain as

$$W_{opt}(\mathrm{f}) = \frac{S_{VV}(\mathbf{f}) - S_{nn}(\mathbf{f})}{S_{VV}(\mathbf{f})} \frac{1}{H(\mathbf{f})}, \tag{29}$$

where S_{VV} and S_{nn} are the spectral power densities of V and n, respectively, and $\mathbf{f}$ is the 2D spatial frequency. Also in Eq. (29), H is the 2D Fourier transform of the point spread function h_d. The actual surface potential is then restored as

$$\hat{\Phi}_d^S(\mathbf{m}) = \mathbf{F}^{-1}(W_{opt}\mathbf{F}(V)), \tag{30}$$

where $\mathbf{F}$ and $\mathbf{F}^{-1}$ stand for the fast Fourier transform (FFT) and its inverse, respectively.

Figure 10 shows a typical two-dimensional point spread function (in arbitrary units) calculated using Eq. (26) and the following tip parameters: $l = 4\ \mu\text{m}, d = 10$ nm, $\theta_0 = 10$ deg, and $R = 10$ nm.

One-dimensional point spread functions calculated for various tip parameters are presented in Figure 11. The results show that the tip-sample distance, d, and the half-aperture angle, θ_0, have a very large effect on the PSF magnitude, and

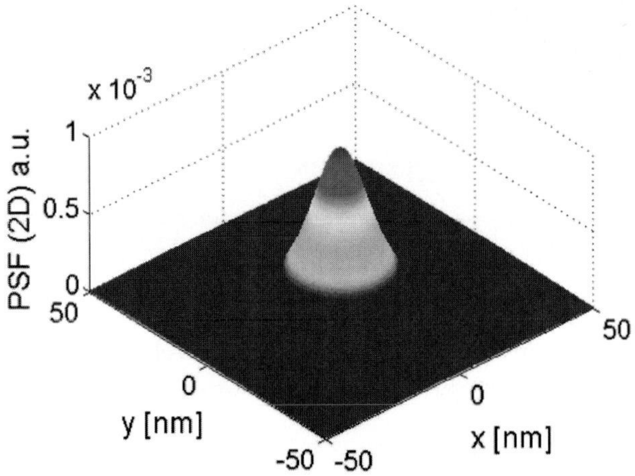

FIGURE 10. A typical two-dimensional point spread function (PSF) calculated for a tip with $l = 4$ μm, $\theta_0 = 10$ deg, and $R = 10$ nm located at a height $d = 10$ nm above a sample.

thus on the sensitivity in KPFM measurements. For example, Figure 11 (a) shows that if we increase the tip-sample distance from 5 to 20–30 nm, the KPFM signal represented by the PSF decreases by about a factor of 3.

This implies that the sensitivity of UHV-KPFM measurements, typically carried out at a distance $d = 5$ nm, is about three times larger then air KPFM measurements, as we reported recently [4]. Similar results were reported by Kalinin et al., who have used carbon nanotubes as a tip calibration standard for electrostatic force microscopy [45]. The effect of the tip length on the PSF shown in Figure 11 (b) is weak; this is due to the fact that at such a small tip-sample separation (10 nm), the electrostatic force is mostly due to the tip apex, as was also reported by others

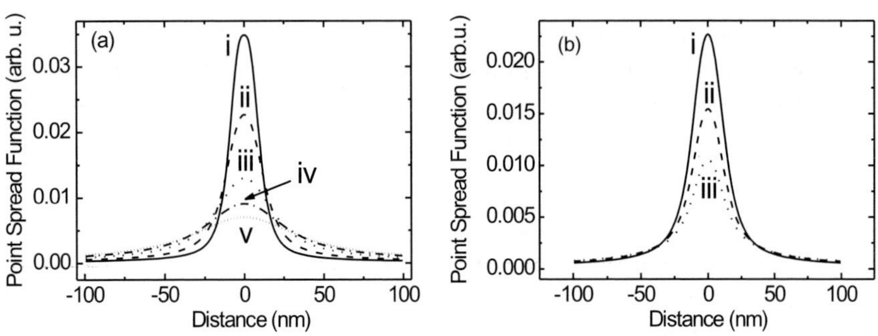

FIGURE 11. One-dimensional point spread functions calculated as a function of (a) tip-sample distances, d of (i) 5, (ii) 10, (iii) 20, (iv) 30, and (v) 40 nm; and (b) tip aperture angles, θ_0, of (i) 10, (ii) 20, and (iii) 30 degrees.

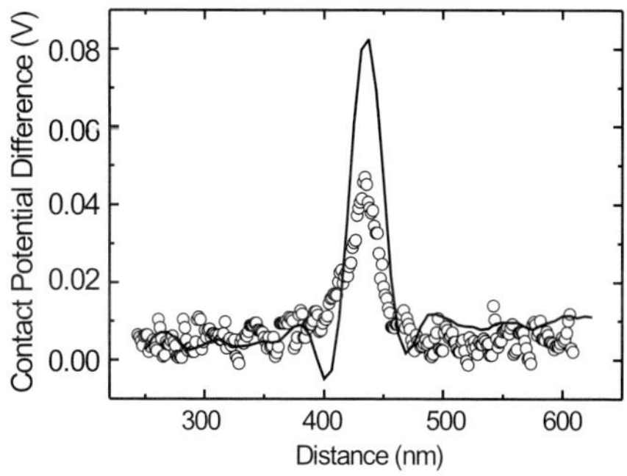

FIGURE 12. Measured (circles) and restored (solid) surface potential of an atomic step on a GaP(110) surface. The KPFM measurement, described in detail in [4], was conducted under UHV conditions where the tip-sample distance, d, was estimated as 5 nm. The sharp work function change at the step is due to localized surface charge.

[40]. The tip apex radius, R, affects mainly the width of the PSF, in agreement with previous works [38].

Figure 12 shows a measured (circles) and noise-filtered and restored (solid line) surface potential line scan of an atomic step on a GaP(110) surface. The KPFM measurement, described in detail in [4], was conducted under UHV conditions where the tip-sample distance, d, was estimated as 5 nm. The sharp work function change at the step is due to localized surface charge, as discussed in detail elsewere [8]. The potential restoration, based on the method developed here, shows that the reconstructed CPD increases by more then a factor of 2 (relative to the measured one) even at the small tip-sample separation used in UHV measurements. This result demonstrates the importance of the restoration in KPFM measurements of small features; in ambient KPFM measurements the effect is much larger, as discussed above.

3 Summary and Conclusions

In this chapter, we have described and discussed in detail Kelvin probe force microscopy (KPFM), which is a method capable of obtaining high laterally resolved work function images. It was shown that by operating the microscope in ultrahigh vacuum, even absolute work function measurements can be realized. The amplitude-modulation technique has the advantage of requiring only small modulation voltages; however, due to long-range electrostatic forces the lateral resolution is limited. The frequency-modulation technique, on the other hand, is sensitive to the force gradient, which provides better lateral resolution, but requires a much larger sampling voltage. The factors affecting the sensitivity and resolution of KPFM in general, and in semiconductors measurements in particular, are analyzed. We have introduced an efficient numerical algorithm for calculating the electrostatic interaction between the measuring AFM tip and the semiconductor surface. This has allowed us to derive a point spread function of the measuring tip and to reconstruct the actual surface potential from Kelvin probe microscopy images in almost real time.

Y.R. and A.B. acknowledge discussions with the group of L. Yaroslavsky, and financial support from Israel Science Foundation (ISF grants 1118/04 and 224/03) and the German Israeli Science Foundation (GIF grant 801/03).

References

1. M. Nonnenmacher, M. P. O'Boyle, and H. K. Wickramasinghe, Appl. Phys. Lett. **58**, 2921 (1991).
2. Y. Leng, C. C. Willams, L. C. Su, and G. B. Stringfellow, Appl. Phys. Lett. **66**, 1264 (1995).
3. S. Saraf, R. Shikler, J. Yang, and Y. Rosenwaks, Appl. Phys. Lett. **80**, 2586 (2002).
4. Y. Rosenwaks, R. Shikler, T. Glatzel, and S. Sadewasser, Phys. Rev. B **70**, 085320 (2004).
5. M. Shvebelman, P. Urenski, R. Shikler, G. Rosenman, M. Molotskii and Y. Rosenwaks, Appl. Phys. Lett. **80**, 1806 (2002).
6. T. Glatzel, H. Hoppe, N. S. Sariciftci, M. C. Lux-Steiner, and M. Komiyama, Jap. J. Appl. Phys. **44**, 5370 (2005).
7. A. Kikukawa, S. Hosaka, and R. Imura, Appl. Phys. Lett. **66**, 3510 (1995).
8. A. Chavez-Pirson, O. Vatel, M. Tanimoto, H. Ando, H. Iwamura and H. Kanbe, Appl. Phys. Lett. **67**, 3069 (1995).
9. T. Mizutani, M. Arakawa, and S. Kishimoto, IEEE Elec. Dcv. Lett **18**, 423 (1997).
10. M. Arakawa, S. Kishimoto, and T. Mizutani, Jap. J. Appl. Phys. **36**, 1826 (1997).
11. R. Shikler, T. Meoded, N. Fried, B. Mishori, and Y. Rosenwaks, J. Appl. Phys. **86**, 107 (1999).
12. R. Shikler, T. Meoded, N. Fried, and Y. Rosenwaks, Appl. Phys. Lett. **74**, 2972 (1999).
13. T. Glatzel, D. F. Marrón, T. Schedel-Niedrig, S. Sadewasser, and M. C. Lux-Steiner, Appl. Phys. Lett. **81**, 2017 (2002).
14. S. Sadewasser, T. Glatzel, M. Rusu, A. Jäger-Waldau, and M. C. Lux-Steiner, Appl. Phys. Lett. **80**, 2979 (2002).
15. L. Bürgi, H. Sirringhaus, and R. Friend, Appl. Phys. Lett. **80**, 2913 (2002).
16. T. D. Krauss, S. O'Brien, and L. E. Brus, J. Phys. Chem. B **105**, 1725 (2001).
17. O. Douhéret, S. Anand, T. Glatzel, K. Maknys, and S. Sadewasser, Appl. Phys. Lett. **85**, 5245 (2004).
18. S. Kitamura, K. Suzuki, M. Iwatsuki, and C. B. Mooney, Appl. Surf. Sci. **157**, 222 (2000).
19. K. Okamoto, Y. Sugawara, and S. Morita, Appl. Sur. Sci. **188**, 381 (2002).
20. N. Duhayon, P. Eyben, M. Fouchier, T. Clarysse, W. Vandervorst, D. Alvarez, S. Schoemann, M. Ciappa, M. Stangoni, P. Formanek, V. Raineri, F. Gian- nazzo, D. Goghero, Y. Rosenwaks, R. Shikler, S. Saraf, S. Sadewasser, N. Barreau, Th. Glatzel, M. Verheijen, S.A.M. Mentink, M. von Sprekselen, T. Maltezopoulos, R. Wiesendanger and L. Hellemans, J. Vac. Sci. Tech. B **22**, 385 (2004).
21. C. C. Williams, Annual Review of Materials Science **29**, 471 (1999).
22. P. De Wolf, M. Geva, T. Hantschel, W. Vandervorst, and R. B. Bylsma, Appl. Phys. Lett. **73**, 2155 (1998).
23. S. Hudlet, M. S. Jean, B. Roulet, J. Berger, and C. Guthmann, J. Appl. Phys. **77**, 3308 (1995).
24. A. K. Henning, T. Hochwitz, J. Slinkman, J. Never, S. Hoffmann, P. Kaszuba and C. Daghlian, J. Appl. Phys. **77**, 1888 (1995).

25. F. Giessibl, and M. Tortonese, Appl. Phys. Lett. **70**, 2529 (1997).
26. F. Giessibl, Phys. Rev. B **56**, 16010 (1997).
27. F. Giessibl, H. Bielefeldt, S. Hembacher, and J. Mannhart, Appl. Surf. Sci. **140**, 352 (1999).
28. D. Abraham, C. Williams, J. Slinkman, and H. Wickramasinghe, J. Vac. Sci. Technol. B **9**, 703 (1991).
29. A. Kikukawa, S. Hosaka, and R. Imura, Rev. Sci. Instrum. **67**, 1463 (1996).
30. C. Sommerhalter, T. W. Matthes, T. Glatzel, A. Jäger-Waldau, and M. C. Lux-Steiner, Appl. Phys. Lett. **75**, 286 (1999).
31. C. Sommerhalter, T. Glatzel, T. Matthes, A. Jäger-Waldau, and M. C. Lux-Steiner, Appl. Surf. Sci. **157**, 263 (2000).
32. S. Kitamura, K. Suzuki, and M. Iwatsuki, Jpn. J. Appl. Phys. **37**, 3765 (1998).
33. S. Kitamura, K. Suzuki, and M. Iwatsuki, Appl. Surf. Sci. **140**, 265 (1999).
34. T. Albrecht, P. Grütter, D. Horne, and D. Rugar, J. Appl. Phys. **69**, 668 (1991).
35. T. Glatzel, S. Sadewasser, and M. C. Lux-Steiner, Appl. Surf. Sci. **210**, 84 (2003).
36. T. Glatzel, S. Sadewasser, R. Shikler, Y. Rosenwaks, and M. C. Lux-Steiner, Mat. Sci. Engineer. B **102**, 138 (2003).
37. T. Fukuma, K. Kimura, K. Kobayashi, K. Matsushige, and H. Yamada, Appl. Phys. Lett. **85**, 6287 (2004).
38. H. Jacobs, P. Leuchtmann, O. J. Homan, and A. Stemmer, J. Appl. Phys. **84**, 1168 (1998).
39. T. Hochwitz, A. Henning, C. Levey, C. Daghlian, and J. Slinkman, J. Vac. Sci. Technol. B **14**, 457 (1996).
40. S. Hudlet, M. S. Jean, C. Guthmann, and J. Berger, Eur. Phys. J. B **2**, 5 (1998).
41. S. Belaidi, F. Lebon, P. Girard, G. Leveque, and S. Pagano, Appl. Phys. A **66**, 239 (1998).
42. H. O. Jacobs and A. Stemmer, Surf. Interface Anal. **27**, 361 (1999).
43. R. Shikler, PhD thesis, Tel-Aviv University, 2003.
44. A. K. Jain, *Fundamentals of Digital Image Processing* (Prentice Hall, 1989).
45. S. V. Kalinin, D. A. Bonnell, M. Freitag, and A. T. Johnson, Appl. Phys. Lett. **81**, 754 (2002).

I.5
Frequency-Dependent Transport Imaging by Scanning Probe Microscopy

RYAN O'HAYRE, MINHWAN LEE, FRITZ B. PRINZ,
AND SERGEI V. KALININ

Frequency-dependent transport measurement, including impedance spectroscopy and capacitance-voltage measurement, is a key tool for materials and device characterization and failure analysis. The rapid development of nanomaterials and nanoscale devices necessitates extending our fundamental understanding of transport phenomena to the nanoscale as well. The ability to conduct nanoscale impedance measurements provides a promising route to extend our understanding of nanoscale transport physics and device operation. In this chapter, we describe two paradigms for scanning probe microscopy (SPM)-based impedance measurement [1–5]. In a first configuration, the SPM tip is used as a current probe in a manner similar to conventional impedance spectroscopy (nanoimpedance microscopy). In a second configuration, the SPM tip is used as a moving voltage electrode in a manner similar to four probe impedance measurements (scanning impedance microscopy). Applying these two configurations, nanometer scale measurement and visualization of impedance is demonstrated for a wide variety of materials systems, including solid electrolytes, semiconductors, electroceramics, corrosion research, and fuel cell systems. Future prospects for SPM-based impedance measurement are discussed.

1 Introduction

1.1 Impedance Basics

Impedance (Z) is a measure of a system's ability to impede the flow of electrical current, providing an extension of resistance to time or frequency-dependent phenomena. Impedance is defined as the ratio between an applied sinusoidal voltage perturbation, $V(t) = V_o e^{i\omega t}$, and a system's resultant current response, $I(t) = I_0 e^{i(\omega t - \phi)}$. Laplace transformations of V and I from the time domain to the frequency domain, $V(t) \rightarrow \tilde{V}(\omega)$ and $I(t) \rightarrow \tilde{I}(\omega)$, yield Z as a function of the frequency ω:

$$Z = \frac{\tilde{V}(\omega)}{\tilde{I}(\omega)} \tag{1}$$

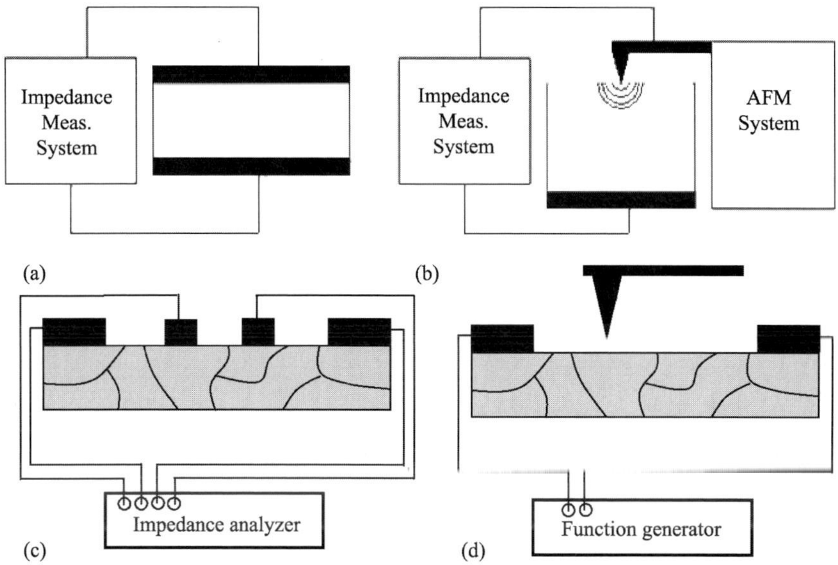

FIGURE 1. Bulk vs. SPM impedance measurement. (a) General concept of a two-electrode impedance measurement. An impedance measurement system acquires an ac impedance spectrum from a sample of interest sandwiched between two bulk electrodes. (b) In SPM impedance measurement, impedance is measured between a local probe (the SPM tip) and a bulk electrode. (c) Four probe impedance measurements allowing minimization of contact contribution to measured impedance. (d) In Scanning impedance microscopy, tip acts as a force-based moving voltage electrode, enabling spatial resolution.

By measuring a system's impedance over a range of frequencies, an impedance spectrum can be constructed. The frequency-resolved modulus (Z_0, the amplitude of impedance) and phase (ϕ, the angular lag between applied voltage and resulting current) information obtained from an impedance spectroscopy (IS) measurement allows characteristic relaxation times of a system to be identified. These relaxation times can be associated with various physicochemical processes including ionic transport, charge transfer, and diffusion [6]. Typical measurement frequencies range from 1 mHz to 1 MHz, thus electronic relaxation processes (which have much faster relaxation times) are usually not resolved.

Impedance measurements are generally performed in either a two-electrode or a four-electrode configuration. In the two-electrode configuration, a sinusoidal voltage perturbation is applied across a sample of interest through two contacting electrodes and the resulting sinusoidal current response is measured (Figure 1(a)). In this set-up, the measured impedance includes the response from the system and the electrodes. An alternative is the four-probe impedance measurement, in which electrode contribution is separated from the system response (Figure 1(c)). Because standard IS measurements utilize bulk electrodes, the technique produces bulk, or system-averaged, results. In other words, standard IS measurements do not provide spatial resolution. As this chapter will illustrate, due to the rapid development of electrical scanned-probe techniques, it is now possible to combine the capabilities of SPM with impedance measurement to acquire *spatially resolved* impedance measurements (Figs. 1(b,d)). In the last 10 years, a number of other approaches for local impedance measurements have been also developed, as summarized below.

1.2 Localized Impedance Measurements

In the last decade, several local electrochemical impedance techniques have been developed based on the use of miniaturized impedance probes. Localized impedance measurement has progressed especially quickly in the field of electrochemical investigations. As an electrochemical tool, SPM impedance measurement may be classified as one of a series of micro-electrochemical techniques that have largely emerged in the last 20 years. Issacs et al. and others [7–9] have demonstrated localized electrochemical impedance spectroscopy methods (now commercialized) that function in an aqueous electrolyte and are capable of acquiring impedance data with a resolution of about 30 μm. Most recently, Pilaski et al. [10] have developed a scanned technique based on a capillary liquid-electrolyte droplet cell with an apparent resolution of around 100 μm. Other major micro-electrochemical techniques include ultramicroelectrodes (UMEs) [11–13], microcontact impedance [14–18], and scanning electrochemical microscopy/electrochemical AFM (SECM/EC-AFM) [11,19,20]. Layson et al. [21] have obtained ac impedance data on poly(ethylene) oxide films directly through a conductive AFM tip at individual points on the film surface, but do not report impedance imaging.

In the solid-state arena, Fleig et al. [22,23] have extended two probe impedance measurements by using patterned arrays of microelectrodes to acquire spatially resolved impedance data from polycrystalline ceramics with a lateral resolution of 15–20 μm. In this case, the impedance of individual microstructural elements such as grain boundaries can be determined. However, the necessity to fabricate microelectrodes and low spatial resolution limits the applicability of this technique. Because of the nascent status of these various techniques, the general field of microelectrochemistry is still relatively disjointed and dispersed. The intent of Figure 2 is therefore to place the diverse set of micro-electrochemical techniques onto a common basis by illustrating the general scaling trends that exist for all electrochemical systems at the micro- and nanoscale. This figure summarizes the spatial and temporal domains that are accessible by the various micro-electrochemical techniques. The relevant time and length scales for these techniques are mapped out along with typical cell constant scaling, Faradaic relaxation times, diffusion scaling, and other physical parameters of importance. In general, this diagram visually illustrates how smaller probe sizes provide access to smaller spatial and temporal regimes. Because of the nanoscale dimensions of the probes used in SPM systems, SPM impedance measurement can provide access to frequency-dependent phenomena on a nanometer scale. The general feature of this approach is the use of a spatially localized impedance probe that ultimately provides information on probe-surface impedance, enabling the extraction of basic materials properties on the sub-micron length scale.

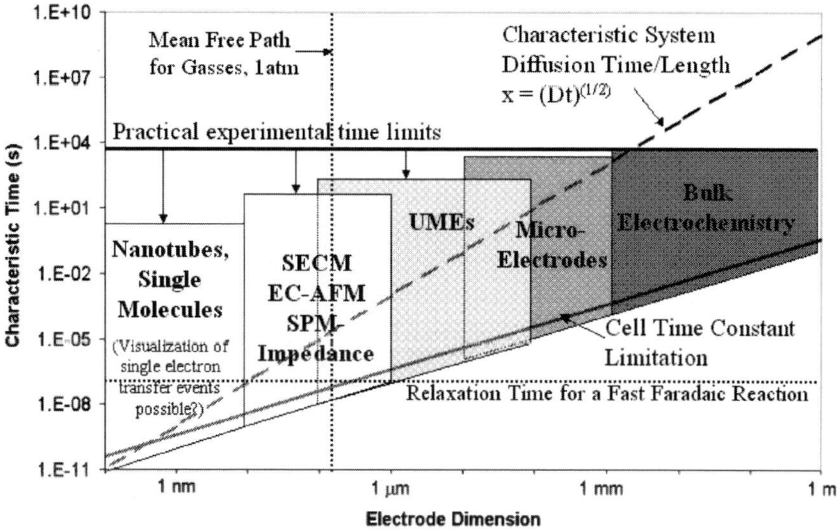

FIGURE 2. Summary of micro-electrochemical techniques. Small electrodes allow experimental access to faster relaxation processes. Often, the kinetics of the fastest faradaic processes can only be studied with microelectrodes. Typical double layer capacitance, electrolyte conductivity, and diffusion values were used in this simulation. ($C^0 = 50\ \mu F/\text{cm}^2$, $\sigma = 0.1(\Omega\text{cm})^{-1}$, $D_0 = 10^{-5}\ \text{cm}^2/\text{s}$.)

In parallel with the development of localized electrochemical impedance methods, a broad range of SPM electrical characterization techniques have been developed. Scanning spreading resistance microscopy (SSRM) [24–26], conductive or current sensing AFM [27–29] and tunneling-AFM [30,31] allow dc characterization of materials. Other techniques make use of the long-range electrostatic forces between a sample and a conductive noncontact SPM tip to extract surface potential images [32,33] (scanning surface potential microscopy, SSPM), or capacitive information on semiconductor oxide surfaces [34–36] (scanning capacitance microscopy, SCM).

1.3 SPM-Based Impedance Measurement

In a manner entirely analogous to macroscopic impedance measurements, SPM-based impedance measurements can be performed in two ways. The SPM tip can be used as a microscopic analog of a macroscopic current probe. This approach is referred to as nanoimpedance microscopy or contact SPM impedance measurement [4]. In this case, a conventional force detection mechanism is used to trace surface topography, and current detection is used for the impedance measurements. Alternatively, the SPM tip can be employed as a moving voltage probe using electrostatic force detection (scanning impedance microscopy) [1–3]. In this case, the contribution of the tip-surface contact resistance to impedance measurement is minimized. Parts I and II of this chapter discuss these two complementary methods in further detail.

2 Nanoimpedance Microscopy

In the nanoimpedance microscopy mode, impedance measurements are acquired directly through a conductive AFM tip which is in hard contact with a sample surface. In contrast to the non-contact impedance modes, contact impedance provides direct, local impedance results. Nanoimpedance microscopy is capable of mapping impedance distributions as a 2D image with sub-100-nm resolution. The AFM tip, which must be conductive, serves as a moveable, localized, nano-scale electrode for the IS measurements. This local probe electrode replaces one of the bulk electrodes used in a traditional two-probe impedance measurement; the nanoscale dimension of the tip ensures local impedance characterization. Successive measurements across a sample surface are obtained at regularly and densely spaced points on the surface. For this, custom developed software automates the communication and synchronization between impedance measurement hardware and an AFM system. Due to the time requirements for impedance acquisition, the probe must move in a point-by-point fashion rather than in a continuous scanning manner during the measurement. Single-frequency or complete impedance spectra can be obtained at each point during the scans. In certain circumstances, these results can be interpreted to yield meaningful, quantitative information. In spite of this exciting possibility, however, the technique also introduces a host of distinct experimental issues and requirements that require careful consideration. These experimental issues will now be discussed in detail.

2.1 Experimental Details

2.1.1 Local Characterization

Because AFM nanoimpedance measurements are conducted in contact mode, the resolution of the method is governed by the size of the conductive AFM tip. Typical AFM tip dimensions are much smaller than typical sample dimensions, thus the spreading resistance associated with a small volume of material near the tip/sample contact point often dominates the total impedance response of the system (Figure 1(b)). This localized response may be approximated by the spreading resistance formula for a circular point contact of radius r between the tip and sample: $R_{SR} = \rho/4r$. In this formula, ρ is the resistivity of the sample. In inhomogeneous samples, additional non-local contributions to the impedance response can arise from high-impedance elements (such as blocking grain boundaries). In certain cases, these non-local contributions can overshadow the localized impedance response.

The localized impedance response originates from a small volume of material, roughly equivalent in size to the radius of the tip/sample point contact. Typical conductive AFM tip contact radii are on the order of 10–100 nm (at the high forces required for the technique), thus setting the absolute lower limit for AFM impedance imaging resolution. Small contact points permit high resolution, but

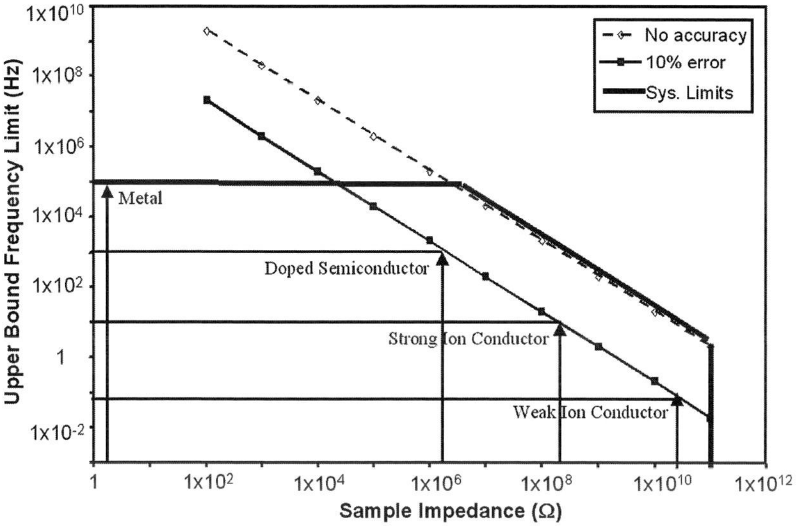

FIGURE 3. Frequency-bandwidth limitations of the current SPM impedance system. Chart shows the envelope of available impedance/frequency space over which accurate measurements can be made. The characteristic impedance moduli of several sample materials are shown on the chart for reference. (R. O'Hayre, M. Lee, F. B. Prinz, J. Appl. Phys. **95**, 8382 (2004))

also lead to high impedances. The ability to measure extremely high impedance therefore becomes a chief limitation of the nanoimpedance technique. For metals, this is not a constraint, assuming an AFM tip contact radius of 10 nm and a typical metal resistivity value $\rho_{metal} = 10^{-4}\ \Omega\cdot\text{cm}$, $R_{SR} = 25\ \Omega$. However, the resistivity of typical ionic conductors is many orders of magnitude larger. For $\rho_{\text{ic}} = 10^4\ \Omega\cdot\text{cm}$, $R_{SR} = 2.5\ \text{G}\Omega$. These large resistances, when combined with unavoidable stray capacitances present in a measuring system, can lead to RC time constants that may obscure meaningful processes. This can restrict the available frequency bandwidth over which reliable impedance data can be extracted.

Figure 3 gives an example of the approximate frequency measurement range for various sample materials ranging from metals to ceramic ion conductors. This figure is based on the typical stray capacitance value in our AFM impedance imaging system (measured to be around 10–100 pF and largely originating from cantilever/surface capacitive coupling).

2.1.2 Tip-Sample Model

Quantitative AFM impedance measurement depends critically on the nature of the tip/sample contact and the presence of cantilever-surface capacitance. Figure 4 proposes a simple equivalent circuit model of the tip/sample contact. The contact is modeled as a parallel *RC* element with additional series resistors accounting for the tip and sample spreading resistances. The resistance of the tip (R_{tip}) is often significant given the small dimensions of AFM probes. A conductive diamond-coated tip,

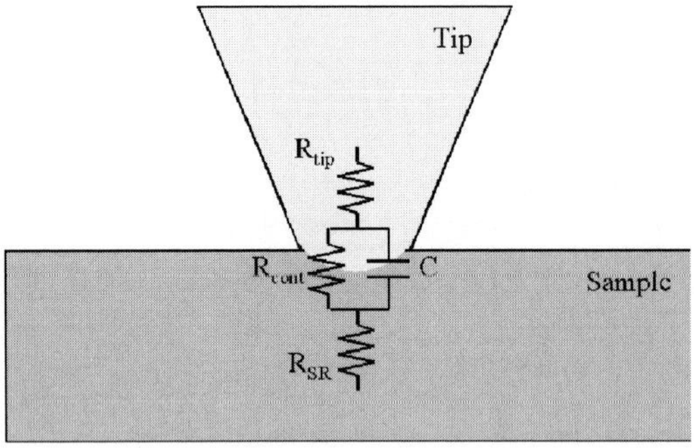

FIGURE 4. Simple equivalent circuit model of the tip/sample contact. R_{tip} represents the resistance contribution from the tip, R_{cont} accounts for the tip/sample interfacial resistance, which may include Schottky or Faradaic effects if the contact is non-ohmic. C represents the tip/sample capacitance. R_{SR} accounts for the spreading resistance associated with a small volume of material just beyond the tip/sample contact. Reprinted with permission from [5].

for example, has $R_{tip} \approx 3{,}000\ \Omega$. The tip/sample contact resistance (R_{cont}) models the interfacial resistance between the tip and the sample. Obviously this contact resistance is a strong function of the tip/sample contact area. For a purely ohmic contact, R_{cont} is a constant. For a nonohmic contact, R_{cont} might also be a function of any dc bias between the tip and the sample (i.e., the tip/sample contact may exhibit Schottky behavior). In the case of an electrochemically active interface, R_{cont} represents the resistance to Faradaic charge transfer at the tip/sample interface. Faradaic resistance scales with interfacial area. For electrochemically active interfaces, the tip/sample capacitance (C) contains a contribution from the double layer capacitance C_d. Like Faradaic resistance, double-layer capacitance scales with contact area. However, C also contains a nonlinear contribution from stray capacitive coupling to the extended tip area. Simple estimations of the tip/sample capacitance in non-electrochemical systems ($C \sim 1$ pF) are considerably less than the typical measured system capacitance (10 – 100 pF), indicating that the extended tip area contributes significantly to the capacitive response of the system. This extraneous capacitive response can impose significant limitations on the quantitative utility of the nanoimpedance technique. The final resistor in the model (R_{SR}) represents the spreading resistance associated with a small volume of sample just beyond the tip/sample contact.

2.1.3 System Configuration

The AFM impedance-imaging system developed at Stanford consists of a Molecular Imaging PicoPlus AFM in combination with a computer-controlled Gamry PC4/750 potentiostat/impedance measurement system. The available impedance measurement frequency range of the system is 1 mHz to 100 kHz. Coordination between the impedance measurement system and the AFM system is accomplished by using the flexible scripting capabilities of both systems. The TCP/IP protocol (internet protocol) and a file interchange scheme are used for communication between the two systems. Impedance data and the corresponding coordinates of tip position are stored together in a data file, followed by post-processing for 2D display. Optical access for alignment and feature location is provided by a large working distance lens system coupled to a CCD camera output. The AFM hardware is enclosed in a vibration isolation chamber to limit acoustic noise, and is placed inside a Faraday cage to limit electrical noise. Measurement cables are shielded and cable lengths are balanced and minimized to limit stray capacitance. Impedance measurements are conducted with ac excitation signals ranging from 10 to 100 mV, and dc bias ranges from 0 to 8 V.

2.1.4 Point Measurements Versus Images

Both full-spectrum impedance at a single position and single-frequency impedance maps made up of multiple position measurements are attainable through nanoimpedance measurement. As with bulk impedance measurements, quantitative values such as contact resistance, Faradaic resistance, and stray capacitance can be deduced from full-spectrum impedance measurements. Single-frequency impedance maps, on the other hand, show spatial contrasts of impedance across sample surfaces but sacrifice the detailed quantitative transport information provided by a full spectrum. Two-dimensional impedance maps are usually acquired at a single frequency to limit experiment time and sample drift. Normally at least 30×30 points are necessary to obtain a 2D map. Assuming a two-minute acquisition time for a full impedance spectrum, 30 hours ($= 30 \times 30 \times 2$ min) would be necessary to obtain a full spectrum map from a 30×30 pixel grid as opposed to a half-hour ($= 30 \times 30 \times 2$ sec) for a single-frequency map. Before obtaining a 2D impedance map, full impedance spectra are generally taken at several points of interest to choose a frequency that will give clear contrast in the single-frequency impedance mode. In this way, we benefit from the virtues of both full-spectrum information and single-frequency mapping while reducing the measurement time. Examples of single-frequency maps at various frequencies and full impedance spectra at discrete points can be found later in this chapter.

2.1.5 Tip Selection

To ensure quality nanoimpedance results, appropriate AFM tip selection is crucial. As will be shown later, reproducible impedance measurements require large tip/sample contact forces (in the μN range). Therefore, cantilevers with high spring constants ($k > 10$ N/m) must be used. Many researchers have noted that large tip/sample contact forces can lead to rapid wear of both the conductive coating and the silicon probe tip [37,38]. This problem is mitigated by the use of boron-doped diamond-coated probes. The diamond coating is extremely wear resistant, and the high-dose boron doping gives degenerate conductivity. Typical probe lifetime is extended from minutes/hours for a metal-coated probe to days/months for a doped diamond-coated probe. Employing diamond-coated probes confers several additional advantages. Diamond probes exhibit extremely low voltammetric background currents and double-layer capacitances [39]. Diamond probes are also highly stable, exhibiting a remarkably wide electrochemical potential window [40]. These are all highly desirable traits for an impedance probe, especially in electrochemical systems. For all the reasons above, the best nanoimpedance results to date have been acquired with conductive diamond-coated non-contact AFM cantilevers (Nanosensors GmbH CDT-NCHR); with typical force constants of around 50 N/m and typical tip resistances of around 3,000 Ω. Furthermore, as is discussed in the next section, the best impedance results are obtained using high tip/sample force values. Typically, a constant tip/sample force of at least 1.00 μN should be maintained during impedance measurement; during topographic scans this force should be reduced by a factor of at least ten to limit wear and minimize surface damage.

2.1.6 Force Selection and Measurement Repeatability

Because the contact SPM impedance technique is performed in contact mode, the measured impedance is sensitive to the tip/sample contact force. Figures 5 and 6 summarize the results of impedance repeatability versus force experiments conducted on a simple gold-coated silicon sample (single-point impedance measurements at 100 Hz). The sharp bend of the curve in Figure 5 suggests that impedance measurements made at sufficiently high contact forces (e.g., >1 μN) should be relatively insensitive to small errors or inaccuracies in tip/sample contact. In other words, at high contact forces, the measured impedance values change little with small force variations. This hypothesis is further reinforced by the results in Figure 6. In Figure 6(a), 150 impedance measurements were collected while the AFM tip was held with a constant contact force on the gold film for approximately 0.5 h. The experiment was conducted at two different force set points: 0.50 μN and 1.25 μN. Use of a high (>1 μN) force set point dramatically improves data consistency.

During nanoimpedance imaging, the tip is continually retracted and then extended point by point across a sample of interest. Under these conditions, although a constant force set point is maintained at each measurement point, the repeated establishment of new surface contacts combined with slight variations in the SPM z-scanner control and tip/surface interaction could affect the tip/sample contact and thus the measured impedance. The results of Figure 6(b) show that at least for a smooth gold surface, this is not the case. These data were acquired for a stepwise scan (10 nm steps) on the same gold film as in Figure 6(a).

Both the absolute values and statistical variability of the scanned-tip data (Figure 6(b)) are comparable to those of the fixed-tip data (Figure 6(a)), indicating that variations due to tip/sample contact fall within the experimental noise of the impedance measurement. Of course, a smooth gold film provides the almost ideal case for repeatable tip/surface contact. Repeatability on rough samples or sharp topographical features will certainly be worse. In addition, dramatically softer or harder samples will likely exhibit very different force/impedance behavior, obligating larger or smaller force set points to acquire optimal contact nanoimpedance data.

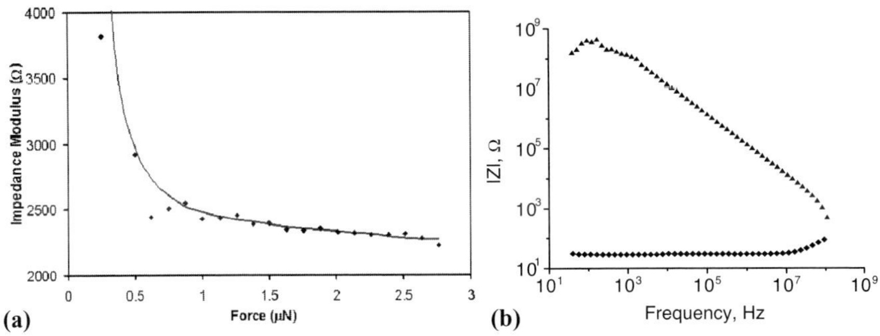

FIGURE 5. (a) Impedance modulus vs tip/sample contact force for a conductive diamond AFM tip in contact with a gold-coated silicon sample. The measured impedance quickly drops with increasing tip/sample contact force, and then stabilizes, suggesting that measurements at high force values are more repeatable. Reprinted with permission from [4]. (b) Impedance spectra with the frequency range from 40 Hz to 110 MHz of a tip/Au surface contact for a good tip-coating (•) and a damaged coating (▲). Reprinted with permission from [3].

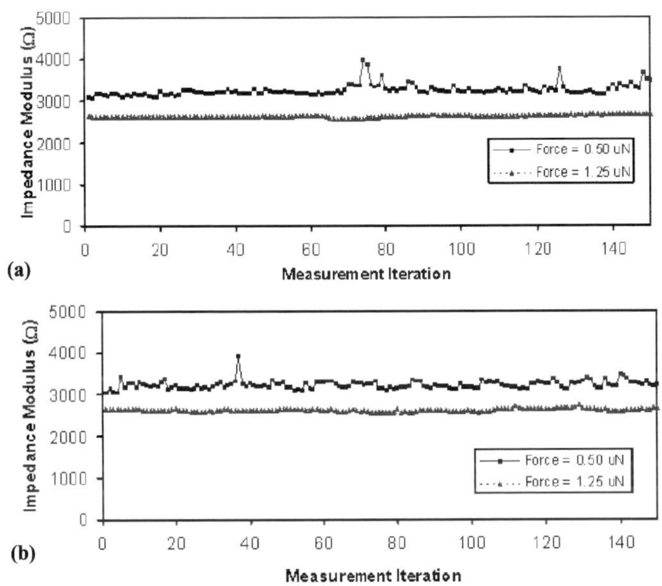

FIGURE 6. Variability of single point AFM impedance measurements for two different force set points between a conductive diamond AFM tip and a gold coated silicon sample. (a) Impedance measurements are acquired while the tip is held at a fixed position in contact with the gold film. (b) Impedance measurements are acquired while the tip is stepped in 10 nm increments across the gold surface. Reprinted with permission from [4].

2.1.7 Measuring Speed and Resolution

The absolute resolution limit of the nanoimpedance technique is set by the magnitude of the tip/sample contact spreading resistance. In practice, however, two other factors often conspire to limit the achievable detail of impedance images. These are (1) capacitive coupling between the sample and the extended tip and (2) the experimental measurement time. The constraints on measurement time are intimately linked to the problem of drift. Sample drift leads to an accumulating, time-related distortion in the acquired image data. To reduce measurement time and limit drift, it is therefore currently necessary to acquire two-dimensional (2D) impedance maps at a single measurement frequency (for example 100 Hz). As mentioned previously, full impedance spectra are first taken at several points across the surface to determine the primary frequencies of interest, i.e., characteristic frequencies that correspond to spectral peaks or valleys in the impedance spectra. Then, single-frequency impedance images are acquired at these frequencies. It is assumed that the critical frequencies do not shift significantly across the sample or in time. Measurement speed is currently 1–3 sec per point for frequencies greater than 10 Hz. At lower frequencies, the measurement time increases commensurately. This measurement time includes a settling period to allow the tip (which is stepping from pixel to pixel across the surface) to establish good contact with the sample.

In our experience, the 50×50 pixel array size provides a good compromise between image detail and measurement time. Most of the 2D impedance images presented in the application section of this chapter are constructed from single-frequency impedance scans taken in 50×50 pixel arrays. Typical measurement times for 50×50 pixel arrays are 1–4 h. The raw impedance modulus (Z_0) versus position and phase angle (ϕ) versus position data are post-processed in MATLAB (color range, contrast, pixel smoothing, interpolation) to produce 2D images.

2.2 Nanoimpedance Imaging of Polycrystalline ZnO

A practical example of nanoimpedance imaging is provided by a study of grain/grain boundary transport in commercial polycrystalline ZnO varistors. Following Figure 7(a), the impedance of a cross-sectioned commercial ZnO varistor was probed laterally between the AFM tip and a bulk top electrode. Thus, in addition to the local impedance response at the AFM tip, non-local impedance contributions from any intervening grain boundaries between the tip and the bulk electrode were also probed.

Coupled SEM, AFM topography deflection, and AFM Z_0 images from a 50-μm region of the ZnO varistor are shown in Figure 8. The Z_0 image was acquired with a 100-mV excitation signal under +5V dc bias at 1 kHz. Several distinct ZnO grains are visible in the images. The ZnO grains at the upper left of the image show purely ohmic behavior at a +5V dc bias. These grains are closest to the bulk top electrode, which was positioned approximately 30 μm above and to the left of the image field of view.

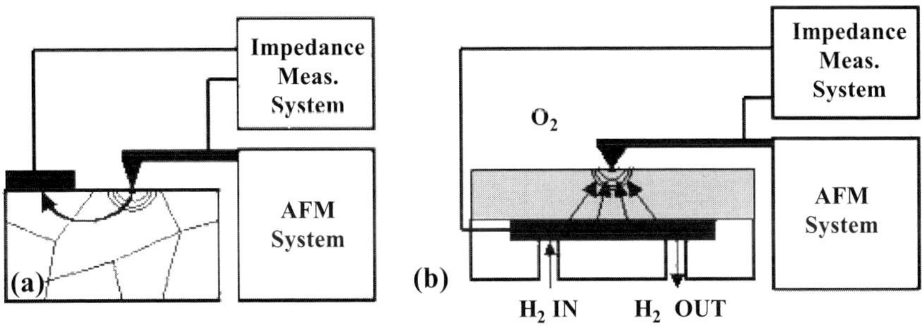

FIGURE 7. (a) Lateral SPM impedance measurement probes intervening microstructural features between the probe tip and the bulk electrode. This configuration may be used to probe, for example, individual grain boundaries in electro-ceramic materials such as ZnO. (b) Through-sample configuration used for the Pt/Nafion impedance experiment. Reprinted with permission from [4].

The highly nonlinear I-V properties of ZnO varistors arise from double-Schottky like barriers formed at the grain boundaries of the material. Below a critical grain-boundary breakdown voltage (typically 3–4 V), transport across the boundary is almost purely capacitive and the boundary is highly insulating. Above the critical voltage, however, transport across the grain boundary becomes ohmic [41,42].

Figure 9 shows a set of Z_0 and ϕ images for the same 50-μm area at five different bias voltages ranging from 0V dc to +8 V DC. (Measurements acquired at 1 kHz with a 100-mV excitation.) Note how individual grain boundary barriers break down sequentially during the dc bias voltage ramp, starting from the upper left with the grains closest to the electrode. This grain-by-grain cascade visibly demonstrates the highly nonlinear I-V characteristics of the polycrystalline varistor. The first grains to exhibit ohmic transport characteristics do so at 3–4 V DC, indicating

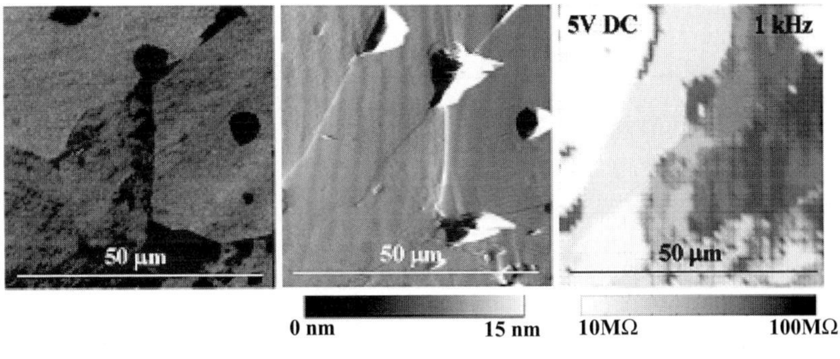

FIGURE 8. (a) SEM image of a 50-μm region of a polished, cross-sectioned commercial ZnO varistor. (b) AFM topography deflection image of the same 50 μm region. (c) Impedance modulus image of the same region. The bulk electrode for this measurement is located approximately 30 μm above and to the left of the image field of view. Reprinted with permission from [4].

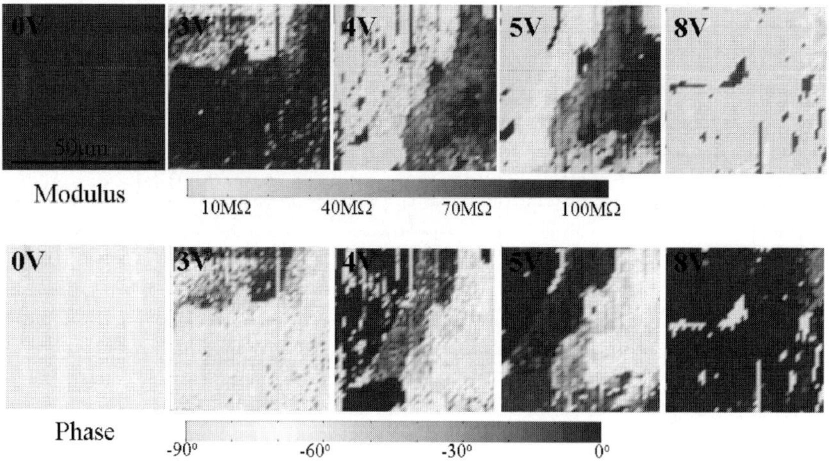

FIGURE 9. Impedance modulus **(top row)** and impedance phase **(bottom row)** images measured as a function of dc bias for the same 50 μm ZnO region documented in figure 8. From left to right, the images are acquired with increasing dc bias; 0 V, 3 V, 4 V, 5 V, 8 V. All images acquired at 1 kHz with a 100-mV excitation signal. Note the grain-by-grain "cascade" with increasing dc bias as the grains become conductive, starting with the grains on the upper left closest to the bulk electrode. Reprinted with permission from [4].

that they are removed by a single grain boundary from the top electrode. The other grains become ohmic between 5–8 V DC, indicating that they are probably separated by two grain boundaries from the top electrode. This result is reasonable given the varistor's 40-μm average grain size and the location of the top electrode 30 μm above and to the left of the images.

Figure 10 demonstrates the sub-micron resolution capabilities of the nanoimpedance imaging technique with a series of "zoom-in" magnifications on a ZnO triple junction. The small triangular shaped region at the junction between the three ZnO grains (clearly visible in the 6-μm image) is a Bi_2O_3 second-phase inclusion, as confirmed by energy dispersive x-ray analysis (EDX). Bi_2O_3 is added to ZnO varistors to control the grain-boundary properties of the material. Excess Bi_2O_3 typically phase-segregates to the ZnO triple junctions, which is nonconductive.

NIM point spectroscopy on a similar system is illustrated in Figure 11(a), showing single-point Cole–Cole (Nyquist) spectra from a ZnO ceramic. The experimental data can be well approximated by two RC elements in series, due to grain boundary and metal-semiconductor contact [3]. On increasing the dc bias, the curve collapses, indicative of the grain boundary breakdown. In comparison, shown in Figure 11(b) is the bias dependence of the local impedance measured at different separations from the macroscopic electrode contact corresponding to varying number of the grain boundaries between the two. Note that the breakdown voltage increases with increasing separation, as expected for an increasing number of grain boundaries between the two.

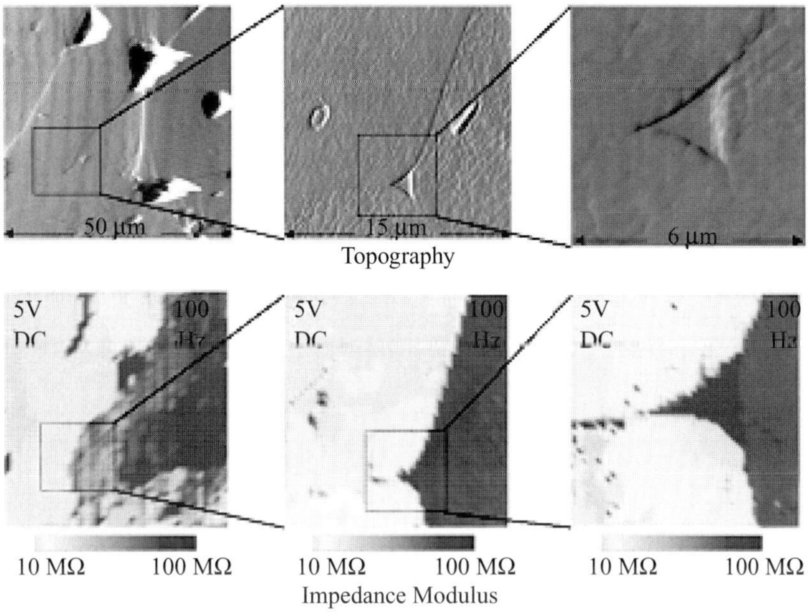

FIGURE 10. AFM topography **(top)** and impedance modulus **(bottom)** images, increasing in magnification from left to right. (50 µm, 15 µm, and 6 µm scan regions, respectively.) Impedance-modulus images are acquired at 1 kHz with a 100 mV excitation signal and 5 V dc bias. The images "zoom" into a triple-junction region between 3 ZnO grains. The V-like intrusion between the 3 grains is a highly insulating Bi_2O_3 second phase inclusion. (by EDX analysis.) Reprinted with permission from [4].

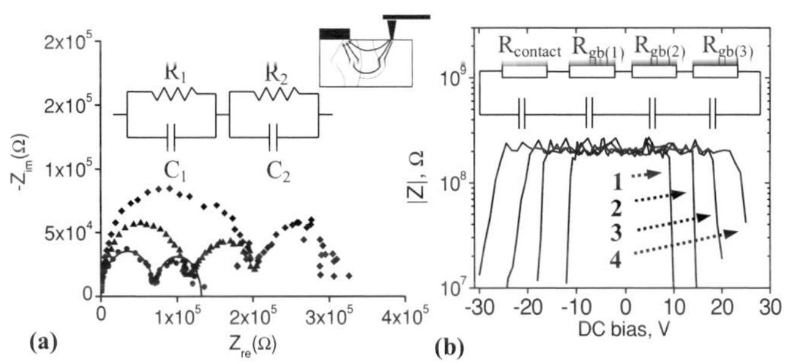

FIGURE 11. (a) Cole–Cole plot of local impedance spectra between the SPM tip and the top electrode (nanoimpedance spectroscopy) at tip/sample biases of +5 V (•), +3 V (Δ), +2 V (♦) for fixed tip location. Solid line is the fit to the equivalent circuit of two RC elements in series. Reprinted with permission from [3]. (b) Bias dependence of tip-electrode impedance for different tip locations on the surface. Curves 1–4 correspond to transport through 1, 2, 3, and 4 grain boundaries. Courtesy of R. Shao.

2.3 Quantitative Nanoimpedance Measurement

The nanoimpedance microscopy detailed in the previous section provided qualitative 2D maps or images of impedance variations across sample surfaces. In this section, a methodology for the extraction of quantitative data from the nanoimpedance technique is developed. Furthermore, this quantitative approach can be widely extended to a variety of other contact-based local-probe techniques, including SSRM, SSPM, and SCM. The methodology depends critically on the ability to characterize the AFM-tip/sample contact quantitatively. Using contact mechanics models, tip/sample contact forces measured in the AFM can be converted into tip/sample contact area estimates. Then, using a reasonable model for the tip/sample contact, quantitative impedance results may be extracted [5].

2.3.1 Contact Area Estimation

As mentioned above, the key to extracting quantitative impedance values is to determine tip/sample contact area accurately. Unfortunately, tip/sample contact area is not directly measurable in a standard AFM experiment. However, it can be estimated from known tip/sample contact forces. This methodology is now briefly described.

Assuming that the hardness (H) of a material is constant, contact area estimates can be obtained from contact force data using the standard hardness relation, $H = P/A$, where H = hardness (N/m^2); P = applied load (applied normal force) (N); and A = projected or surface area of contact (m^2) [43–45]. H is a constant for noncrystalline materials and geometrically similar probes (such as Vickers and Berkovich indenters). H is also constant for crystalline materials with small indentation size effect (ISE) [46]. However, hardness is tip-geometry dependent. Because the tips used in an SPM differ from the standard indenter geometries used in hardness experiments, a simple hardness relation like the one defined above is not immediately applicable. Instead, we require a hardness equation that explicitly accounts for the geometry of the tip. Thus, we define a tip-geometry dependant hardness function, $H(\theta)$, where: $(H\theta) = P/A_{proj}$. In this equation, the hardness $H(\theta)$ depends on the sharpness (θ = tip semi-angle) of the tip. For a given load and tip semi-angle, $H(\theta)$ pgay be used to determine the projected area of contact between the tip and the sample. A reasonable $H(\theta)$ function can be developed for most SPM tips by applying a conical approximation. Conical tips possess the property of geometric similarity; therefore a simplified version of Johnson's expanding cavity model can be used to relate hardness to the materials properties of the sample and the sharpness of the tip [46]:

$$H(\theta) = \frac{4}{3}Y + \frac{2}{3}Y \ln\left[\frac{\dfrac{E_r}{Y\tan\theta} + 4(1-2\nu)}{6(1-\nu)}\right] = \frac{2Y}{3}\,[2 + \ln(\eta)] \qquad (2)$$

where $H(\theta)$ = material hardness measured by a conical tip of sharpness θ (N/m^2); Y = yield strength of the sample (N/m^2); E_r = reduced modulus or indentation modulus (N/m^2); θ = tip semi-angle from the vertical (°); and ν = Poisson's ratio of the sample. E_r, the reduced modulus, is defined as

$$\frac{1}{E_r} = \frac{1-\nu_t^2}{E_t} + \frac{1-\nu^2}{E} \tag{3}$$

where ν_t = Poisson's ratio of the tip material; E_t = Young's modulus of the tip material (N/m^2); ν = Poisson's ratio of the sample; and E = Young's modulus of the sample (N/m^2). In general (especially for a diamond-coated SPM tip), $E_t >> E$, so $Er \approx E/(1-\nu^2)$, and

$$\eta \approx \frac{1}{6(1-\nu)}\left[\frac{E}{(1-\nu^2)Y\tan\theta} + 4(1-2\nu)\right] \tag{4}$$

For most metallic materials, ν is ~1/3; for most polymers, ν is about 0.5, thus for metals we have

$$\eta_{metal} \approx \frac{9E}{32Y\tan\theta} + \frac{1}{3} \tag{5}$$

and for polymers

$$\eta_{polymer} \approx \frac{4E}{9Y\tan\theta} \tag{6}$$

Figure 12(a) shows *H/Y* versus tip semi-angle for different *E/Y*, based on Eq. (2). As the figure indicates, *H* decreases with increasing tip semi-angle. The figure also demonstrates that the results are insensitive to ν. For a specific tip semi-angle (70.3°), Figure 12(b) shows curves of (*H* for ν)/(*H* for $\nu = 0.5$) versus ν for various values of *E/Y*, indicating again that *H* is relatively insensitive to ν.

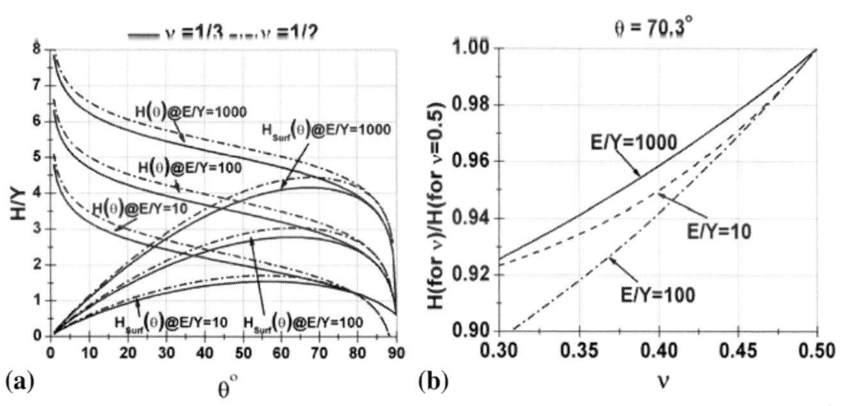

FIGURE 12. (a) H/Y and H_{surf}/Y vs. the tip semi-angle for different E/Y; $E/Y = 10$ (typical of polymers, ceramics), $E/Y = 100$, $E/Y = 1000$ (typical of ductile metals). Based on Eqs. (3) and (7). (b) (H for ν)/(H for ν 0.5) vs. ν for different materials tested with a Berkovich indenter. Reprinted with permission from [5].

Since quantification of SPM measurements often requires an estimate of the true contact surface area between the tip and the sample rather than the projected contact area, an additional modification is required. We define $H_{surf}(\theta)$ as a surface-area related hardness value, which provides an estimate of the tip/sample contact surface area (rather than the projected area) for a given load: $H_{surf}(\theta) = P/A_{surf}$. In this definition, A_{surf} (m^2) represents the true contact surface area between the tip and the sample. For the case of perfect conical tip geometry, we have

$$H_{surf}(\theta) = H(\theta)\sin\theta \tag{6}$$

The second set of curves in Figure 12(a), then, are curves of $H_{surf}(\theta)$ versus tip semi-angle for different E/Y. As the indicates, a broad maximum occurs in the $H_{surf}(\theta)$ curves. Around this maximum, $H_{surf}(\theta)$ is fairly insensitive to the tip geometry. At $E/Y = 10$ and $30^\circ < \theta < 70^\circ$, for example, $H_{surf}(\theta)$ is nearly constant. This fortuitous tip-geometry insensitivity is due to two opposing geometric correction factors that partially cancel one another out. This result implies that the contact area conversion methodology should be applicable for most SPM probe geometries if the SPM tip angle is not far from the curve maximum.

Equations (2–6) provide the necessary adjustments to produce contact surface area estimates for conical SPM tips of arbitrary sharpness based on nanoindentation hardness measurements or materials properties. If the materials properties of the sample are well known, they can be inserted into Eqs. (2–5) to generate an approximate hardness function. More accurately, nanoindentation experiments on the sample of interest can be used to obtain E_r and $H(\theta)$ values at known tip angles θ. For example, Berkovich nanoindentation experiments give $H = 70.3^\circ$. Estimating the Poisson's ratio of the material, Eq. (2) may be solved to produce the yielding stress Y. Combining Eqs. (2) and (6), $H_{surf}(\theta)$ estimates can then be obtained at any tip angle.

Several significant assumptions have been made to derive these relations. Among the most important are the following:

1. AFM tip geometry is perfectly conical.
2. Hardness is constant (No ISE).
3. Material is elastic—perfectly plastic.

Most materials reasonably satisfy assumption three. However, in the case of significant strain-hardening rates, modifications to the presented model can be applied [5]. For large contact depths (large contact areas), the first two assumptions are also satisfied. However, due to sample surface effects and the finite sharpness of real SPM tips, these assumptions become problematic for extremely small contacts. Most SPM tips have tip radii on the order of 10 nm. Therefore, for contacts below 10 nm (radius) in size, spherical rather than conical contact mechanics theory should likely be applied. Equation (2) may be replaced by [46]:

$$H(r) = \frac{2Y}{3}\left[2 + \ln\left[\left(\frac{E_r r}{YR} + 4(1-2\nu)\right)\Big/ 6(1-\nu)\right]\right] \tag{7}$$

where r = contact radius (m) and R = radius of the spherical tip (m).

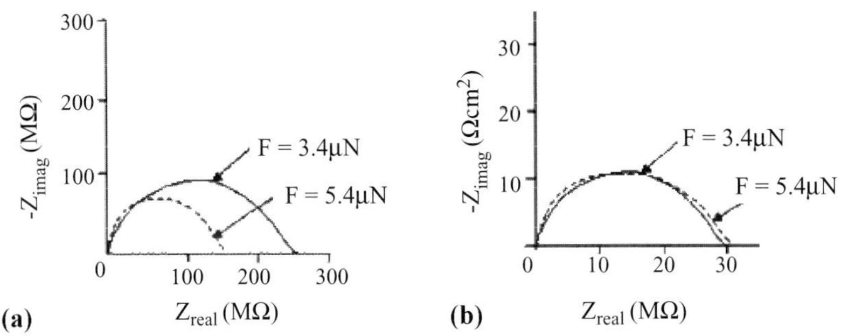

FIGURE 13. (a) Example Nyquist impedance spectra results obtained at two different force settings from the Pt/Nafion experiment configuration described in Figure 7(b). The size of the impedance loop provides information on the kinetics of the oxygen reduction reaction occurring at the tip/Nafion interface. (b) Area-specific Nyquist impedance spectra converted from (a) based on the calculated area estimates for the tip/Nafion contacts.

At the 10-nm length scale, surface effects and sample roughness likely become significant. Thus, the methodology developed in this section is recommended only for contact sizes larger than about 10 nm (contact area >100 nm^2).

2.3.2 Quantitative Nanoimpedance Measurement of ORR Kinetics

An example application of the quantitative nanoimpedance measurement is provided by an investigation of the oxygen reduction reaction (ORR) kinetics at a nanoscale Pt/Nafion interface.

Nafion is an important ionic conductor for fuel cell and electrolysis systems. The development of commercial fuel cell systems requires a nanoscale understanding of fuel cell reaction kinetics. A nanoscale impedance experiment, such as the one diagrammed in Figure 7(b), can help provide this kinetic understanding. In this experiment, a platinum-coated AFM tip is used as a local probe of the oxygen reduction kinetics on a Nafion 117 membrane. The impedance response measured in this experiment is related to the kinetics of the ORR. Figure 13(a) gives typical impedance measurement results at two different tip/sample contact force settings. Both Nyquist impedance spectra in this figure show a clear semicircular loop. The diameter of this semicircular loop provides information about the rate of the Faradaic charge transfer reaction at the tip/Nafion interface. However, because the two spectra were acquired at different tip/sample contact forces, they show drastically different loop diameters! In order to provide quantitative rate information we must normalize the impedance spectra in Figure 13(a) by tip/sample contact area.

To arrive at an area-normalized version of Figure 13(a), tip/sample contact areas must be estimated for both measurements. Following the methodology outlined in the previous section, these tip/sample contact estimates can easily be produced. We briefly detail the procedure for the two impedance measurements above. First, from SEM inspection, the semi-angle of the AFM tip used in the above Nafion experiments is determined to be approximately 40°. Then, from a set of standard nanoindentation experiments on Nafion samples, an $H(\theta)$ curve for Nafion is established. (Alternatively, $H(\theta)$ for Nafion can be estimated from the E, Y, and ν for Nafion based on Eq. (2).) Using Eq. (6), the $H(\theta)$ curve is then converted into a $H_{surf}(\theta)$ curve, and $H_{surf}(40°)$ is determined. For Nafion, $H_{surf}(40°)$ is determined to be approximately 2.8×10^7 Pa [5]. Finally, knowing the applied force used in the measurement, the tip/sample contact area may be estimated as

$$A_{surf} = \frac{P}{H_{surf}(\theta)} \tag{8}$$

The impedance spectra in Figure 13(a) were acquired with tip/sample contact forces $F_1 = 3.4$ μN and $F_2 = 5.4$ μN. Therefore $A_{surf,1} \approx 0.12$ μm^2 and $A_{surf,2} \approx 0.192$ μm^2. Using these contact area estimates, the impedance spectra in Figure 13(a) (Z in units of Ω) can be converted to area-specific spectra (Z in units of Ω·cm^2), as shown in Figure 13(b). As Figure 13(b) illustrates, the kinetic response of the system is virtually identical in the two experiments after contact area normalization. From either of these area-specific impedance spectra, the exchange current density, j_o (an important kinetic parameter in fuel cell systems) can be extracted. The spectra in Figure 13(b) yield $j_o \approx 2.6 \times 10^{-7}$ A/cm^2 [5]. This value lies well within the range of values obtained by macroscopic studies of Pt/Nafion ORR kinetics. The match between bulk kinetic measurements and those obtained by the nanoimpedance method affirm the quantitative capability of the nanoimpedance technique.

3 Noncontact SPM Impedance Measurement

In the nanoimpedance microscopy and spectroscopy method discussed above the AFM tip acts as a moving current electrode, similar to conventional two-probe impedance measurements. However, in this case the total impedance is a sum of the system impedance and the impedance of the tip-surface contact coupled with the cantilever capacitance, which in most cases will dominate the response of the system. Alternatively, the SPM can be set-up in a four-probe impedance configuration, where the current is applied across the system using macroscopic electrodes and the SPM tip is used to detect the local amplitude and phase of voltage oscillations. This measurement configuration eliminates the tip-surface junction impedance. These measurements can be performed in the current detection mode as discussed in detail in chapter 3.1. However, this approach requires tip-surface impedance to be comparable to the input impedance of the voltmeter, limiting its applicability to highly conductive materials. An alternative approach, scanning impedance microscopy (SIM), is based on AFM force detection, where the AFM cantilever is used as a force sensor to detect periodic electrostatic forces between the tip and the surface. Principles, data interpretation, and resolution limits in SIM as well as applications to several materials systems are discussed below.

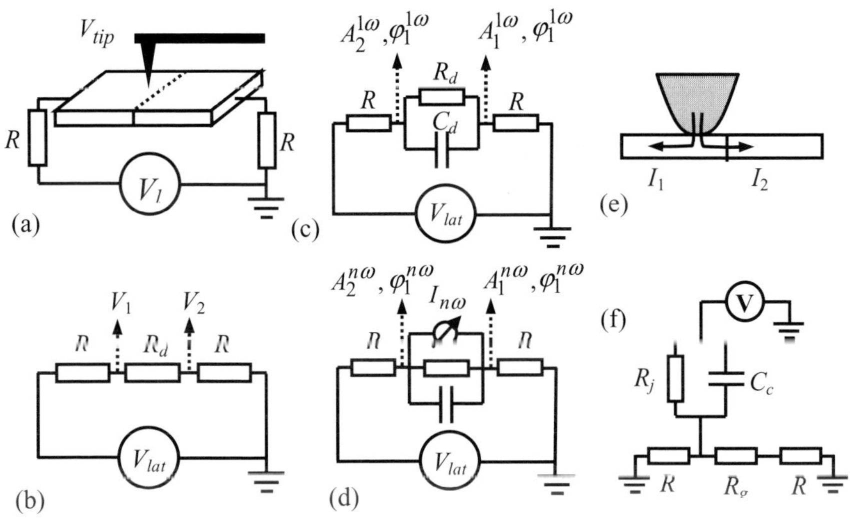

FIGURE 14. (a) Experimental set-up for dc and ac force-based SPM transport measurements. (b) Equivalent circuit for SSPM-based dc transport measurements. (c) Equivalent circuit for linear SIM measurements. (d) Equivalent circuit for non-linear SIM. (e,f) Experimental set-up and equivalent circuit for current based SPM transport measurements. For clarity, only resistive components of the sample equivalent circuit are shown. Reprinted with permission from [55].

3.1 Principles

In an SPM transport experiment, the experimental setup is configured similarly to standard four-probe resistance or impedance measurements, as illustrated in Figure 14(a), where the tip acts as a moving voltage electrode. In dc transport measurements by scanning surface potential microscopy (SSPM, also referred to as Kelvin probe force microscopy), the tip measures the dc potential distribution induced by a lateral bias applied across the sample, thus imaging *resistive* elements of the equivalent circuit (Figure 14(b)) [47–49]. In SIM, the tip measures the distribution of the phase and amplitude of the ac voltage, thus imaging both the *resistive* and *capacitive* elements of the equivalent circuit (Figure 14(c)) [1,2]. SIM can also be extended to the nonlinear domain, measuring the higher harmonics [50] or mixed-frequency signals [51] of potential oscillations in the sample generated due to frequency mixing on nonlinear interfaces. A corresponding equivalent circuit is shown in Figure 14(d), where a nonlinear interface acts as a current source at mixed-frequency harmonics of the applied bias. Note that c-AFM can also be configured for potential measurement techniques by nulling the tip-surface current (scanning potentiometry, or SP). However, the information obtained in SSPM and SP is different—the former measures electrochemical potential; the latter measures electrostatic potential.

In the force-based techniques, acquisition of electrostatic force data requires contributions from elastic and Van der Waals interactions to be minimized. Correspondingly, SSPM and SIM are typically implemented in a dual-pass imaging mode. The tip scans the surface in the contact or intermittent contact mode, determining the position of the surface. Electrostatic data are then collected in the second scan, in which the tip retraces the topographic profile while maintaining a constant tip-surface separation. In SIM the tip is held at constant bias V_{dc} and a lateral bias $V_{lat} = V_{dc} + V_{ac}(x)cos(\omega t)$ is applied across the sample. This lateral bias induces an oscillation in the sample surface potential

$$V_{surf} = V_s(x) + V_{ac}(x)\cos(\omega t + \varphi(x)), \tag{9}$$

where $\varphi(x)$ and $V_{ac}(x)$are the position-dependent phase shift and voltage oscillation amplitude and $V_s(x)$ is the dc surface potential. The variation in surface potential results in a capacitive force acting on the tip

$$F_{1\omega}^{cap}(z) = C_z'(V_{dc} - V_s(x))V_{ac}(x), \tag{10}$$

that results in a cantilever deflection detected by the optical beam-deflection system. The magnitude, $A(\omega)$, and phase, φ_c, of the cantilever response to the periodic force induced by the voltage are [52]

$$A(\omega) = \frac{F_{1\omega}}{m}\frac{1}{\sqrt{(\omega^2 - \omega_{rc}^2)^2 + \omega^2\gamma^2}} \quad \text{and} \quad \tan(\varphi_c) = \frac{\omega\gamma}{\omega^2 - \omega_{rc}^2} \tag{11a,b}$$

where m is the effective mass, γ the damping coefficient, and ω_{rc} is the resonant frequency of the cantilever. Equations (10) and (11) imply that the local phase shift between the applied voltage and the cantilever oscillation is $\varphi(x) + \varphi_c$ and the oscillation amplitude $A(\omega)$ is proportional to the local voltage oscillation amplitude $V_{ac}(x)$. Therefore, variation in the phase shift (phase image) along the surface is equal to the variation of the true voltage phase shift with a constant offset due to the inertia between the sample and tip. The spatially resolved phase shift signal thus constitutes the SIM phase image of the device. The tip oscillation amplitude is proportional to the local voltage oscillation amplitude and constitutes the SIM amplitude image.

SIM is complemented by scanning surface potential microscopy, which can access the dc potential distribution $V_s(x)$ along the surface. In SSPM the tip is biased directly by $V_{tip} = V_{dc}^t + V_{ac}^t(x)\cos(\omega t)$, where V_{ac}^t is referred to as the driving voltage. The capacitive force between the tip and the surface is given by Eq. (10), where $V_{ac}(x) = V_{ac}^t$ is now position independent. Feedback is used to nullify $F_{1\omega}^{cap}$ by adjusting the dc component of tip bias and mapping the nulling potential V_{dc}^{t0} thus yields a surface potential image. Note that the force-nulling, rather than current-nulling mechanism in SSPM means that the measured potential is $V_{dc}^{t0} = V_s + \Delta\ CPD$, where $\Delta\ CPD$ is the work function difference between the tip and the surface.

In nonlinear SIM, higher harmonics of the tip deflection signal or a mixed frequency signal are detected [52,53]. In this case, the detection frequency is chosen so as to coincide with the resonance of the cantilever in order to amplify the otherwise weak mixed-frequency signal.

To determine the absolute value of local amplitude, $V_{ac}(x)$, from SIM and NL-SIM data, the microscope is reconfigured to the open-loop SSPM mode, in which the feedback is disengaged, and the tip oscillation response to the ac bias applied to the tip is determined. The local voltage oscillation amplitude is then

$$V_{ac}(x) = \frac{V_{ac} A_{sim}(x)\left(V_{surf}(x) - V_{tip}^{sspm}\right)}{A_{sspm}(x)\left(V_{surf}(x) - V_{tip}^{sim}\right)} \tag{12}$$

where A is oscillation amplitude, V_{tip} is the tip dc bias, V_{ac} is the tip ac bias, and sim and sspm refer to the SIM and open-loop SSPM modes, respectively. $V_{surf}(x)$ is the surface potential, which varies with x in the presence of a lateral bias and can be determined by SSPM.

The driving frequency in SIM must be selected far from the resonant frequency of the cantilever to minimize the variations of the phase lag between tip and surface due to electrostatic force gradients related to the non-uniform surface potential. In practice, however, SIM is used to obtain frequency-dependent phase shift and amplitude; therefore, the data is acquired in a broad range of frequencies and the vicinity of the resonance can be excluded. The frequency range of SIM is limited by the bandwidth of the optical detector to 2–5 MHz. In addition, at high frequencies dynamic stiffening effects become important, minimizing response to electrostatic forces.

3.2 *Quantification*

Here, we present the formalism to quantify the dc and ac transport properties from SSPM and SIM data. Unlike conventional two or four probe resistivity measurements, SSPM is sensitive to variations in local potential, while (local) current is generally unknown. However, for single interfaces such as in bicrystals or metal-semiconductor junctions, the system can be represented by a 1D equivalent circuit, which defines the current. In an SSPM transport experiment, a biased interface is connected to a voltage source in series with current limiting resistors to prevent accidental current flow to the tip. For a system with a single electroactive interface, the total resistivity of the sample R_{Σ}, is $R_{\Sigma} = 2R + R_{gb}(V_{gb})$, where V_{gb} is the potential across the interface, $R_{gb}(V_{gb})$ is the voltage-dependent resistivity of the interface and R is the resistivity of the current limiting resistors.

The applied bias dependence of the potential drop at the interface is directly assessable by SSPM and is referred to as the voltage characteristic of the interface. To compensate for potential variations due to differences in local work function, images under applied lateral bias should be corrected by the grounded surface potential values. In general case, interface current-voltage characteristics, $I_{gb}(V_{gb})$, can then be obtained as

$$I_{gb}\left(V_{gb}\right) = \left(V - V_{gb}\right)/2R, \tag{13}$$

provided the values of current limiting resistors are known. The current limiting resistors, R, can be varied to determine the presence of stray resistances in the circuit (e.g. contact and bulk resistances). Alternatively, the current, I_{gb}, can be measured directly using a current-voltage converter. Similar analyses can be performed for the transport in multiple interface systems [53]. Given the unavoidable limitations for the resolution and precision of potential measurements by SPM, this approach for I-V curve reconstruction is applicable if the interface resistance and circuit termination resistance are well matched (i.e., $R_{gb}/R \sim 0.01 \div 1$).

For strongly nonlinear interfaces such as metal-semiconductor junctions, the I-V curve is highly asymmetric and effective resistance varies by many orders of magnitude. In this case, the analysis can be performed using the (known) analytical form for I-V curve. For transport across a Schottky barrier, the current is

$$I_d = I_0 \left(\exp\left(\frac{qV_d}{nkT} \right) - 1 \right) + \sigma V_d \tag{14}$$

where V_d is potential across the junction, $q = 1.6 \cdot 10^{-19}$ C is electron charge, n is ideality factor, $k = 1.38 \cdot 10^{-23}$ J/K is the Boltzmann constant, T is temperature, and σ is the leakage conductivity. In the limit of a large forward bias Eq. (13) simplifies to $V_d = (kT/q)\ln(V_{dc}/2RI_0)$. Therefore, for a positively biased diode the potential drop at the interface is expected to be small and hardly detectable by SSPM. For a large negative bias, however, the potential drop occurs primarily at the interface. The crossover between the two regimes is expected at a lateral bias $V = -2RI_0$ and in this limit Eq. (13) becomes

$$V_d = (V + 2RI_0)/(1 + 2R\sigma) \tag{15}$$

Equation (15) implies that for finite conductivity in a reverse-biased diode, the potential drop occurs both at the diode and at current limiting resistors. Therefore, experimental voltage characteristics of the interface can be used to obtain both the saturation and leakage current components of diode resistivity.

3.2.1 ac Transport Properties by SIM

For a single electroactive interface the analysis of the SIM-imaging mechanism is similar to that of SSPM. For the equivalent circuit in Figure 14(c) the total impedance of the circuit, Z_Σ, is $Z_\Sigma = 2R + Z_{gb}$, where Z_{gb} is the grain boundary impedance. The grain boundary equivalent circuit is represented by a parallel R-C element and the impedance is $Z_{gb}^{-1} = R_{gb}^{-1} + i\omega C_{gb}$, where R_{gb} and C_{gb} are the voltage-dependent interface resistance and capacitance. Experimentally accessible and independent of the tip properties are the interface phase shift, $\varphi_{gb} = \varphi_2 - \varphi_1$, and the amplitude ratio, A_1/A_2, across the interface. Interface phase shift is calculated from the ratio of impedances from each side of the interface, $\beta = R/(Z_{gb} + R)$, as $\tan(\varphi_{gb}) = \mathrm{Im}(\beta)/\mathrm{Re}(\beta)$ (impedance divider effect). For the equivalent circuit in Figure 14(d)

$$\tan(\varphi_d) = \frac{\omega C_d R_d^2}{(R + R_d) + R\omega^2 C_d^2 R_d^2} \tag{16}$$

TABLE 1. Frequency dependence of interface phase shift and amplitude ratio

Frequency, ω	Phase shift, $\tan(\varphi_{gb})$	Amplitude ratio, A_1/A_2
Low-frequency limit, $\omega \ll \omega_r$	$\omega C_{gb} \frac{R_{gb}^2}{(R+R_{gb})}$	$\frac{R+R_{gb}}{R}$
High-frequency limit, $\omega \ll \omega_r$	$\frac{1}{\omega C_{gb} R}$	1
Crossover frequency, $\omega_r = R_{gb}^{-1} C_{gb}^{-1} \sqrt{1 + R_{gb}/R}$	$\frac{R_{gb}}{2\sqrt{R(R+R_{gb})}}$	$\sqrt{\frac{R+R_{gb}}{R}}$

The voltage oscillation amplitude ratio, $A_1/A_2 = |\beta|^{-1}$, is

$$\beta^{-2} = \frac{\left\{(R + R_d) + R\omega^2 C_d^2 R_d^2\right\}^2 + \omega^2 C_d^2 R_d^4}{R^2\left(1 + \omega^2 C_d^2 R_d^2\right)^2} \tag{17}$$

The high- and low-frequency limiting behaviors for Eqs. (16,17) are summarized in Table 1. In the high frequency limit phase shift at the interface is determined by the interface capacitance and circuit termination only. Thus, SIM phase imaging at frequencies *above* the interface relaxation frequency, $\omega \gg \omega_r = R_{gb}^{-1} C_{gb}^{-1}$, provides a quantitative measure of interface capacitance.

Similarly to conventional impedance spectroscopy [54], the interface phase shift and the amplitude ratio can be used simultaneously to determine the interface transport properties. For frequency independent R_{gb}, C_{gb} (Model 1), the frequency dependence of the interface phase shift and the amplitude ratio can be fitted to Eqs. (16,17), where C_{gb} and R_{gb} are now fitting parameters. Alternatively, frequency-dependent interface resistance and capacitance $R_{gb}(\omega)$, $C_{gb}(\omega)$ (Model 2) can be calculated at each frequency from the experimental phase shift and amplitude ratio. Application of this analysis to frequency-dependent transport at electroactive interfaces in $SrTiO_3$ is reported elsewhere [55]. Such data are expected to be particularly important for the characterization of interfaces possessing significant frequency dispersion of the interface transport properties, e.g., due to interface states or deep traps at semiconductor grain boundaries [56] or due to several relaxation processes in ionic conductors, for which interpretation of conventional impedance spectroscopy results is not straightforward.

3.3 Resolution

One of the key characteristics of any SPM technique is spatial resolution. For current-based techniques, the resolution is ultimately determined by the contact area between the tip and the surface, which is on the order of several nanometers as discussed in section I. However, in force-based techniques such as SIM and SSPM, the probing volume is determined by relatively long range capacitive interactions between the tip and the surface. The algorithms for the determination of spatial resolution in SSPM and magnetic force microscopy have been extensively studied using both direct calculations from known probe geometry [57,58], mathematical modeling [59], and reconstruction methods [60].

The analysis of thc image formation mechanism in SPM is greatly simplified if the imaging is *linear*, i.e., a measured image, $I(\mathbf{x})$, where $\mathbf{x}$ is a set of spatial coordinates, can be represented as a convolution of the ideal image $I_0(\mathbf{x}-\mathbf{y})$ and the microscope resolution function, $F(\mathbf{y})$, as

$$I(\mathbf{x}) = \int I_0(\mathbf{x}-\mathbf{y})F(\mathbf{y})d\mathbf{y} + N(\mathbf{x}) \tag{18}$$

where $N(\mathbf{x})$ is the noise function. The Fourier transform of Eq. (18) is

$$I(\mathbf{q}) = I_0(\mathbf{q})F(\mathbf{q}) + N(\mathbf{q}) \tag{19}$$

where $I(\mathbf{q}) = \int I(\mathbf{x})e^{i\mathbf{kx}}d\mathbf{x}$, $I_0(\mathbf{q})$, $F(\mathbf{q})$ and $N(\mathbf{q})$ are the Fourier transforms of the measured image, ideal image, resolution function, and noise, respectively. The instrument resolution function can be determined directly provided that the ideal image $I_0(\mathbf{q})$ is known. Once the resolution function is determined for a known calibration standard, it can be used to evaluate the ideal image, $I_0(\mathbf{x})$, from the measured image, $I(\mathrm{x})$ for an arbitrary sample. The width of $F(\mathbf{y})$ provides a quantitative measure of lateral resolution.

From Eq. (19), the resolution function can be determined from suitable standard for which the ideal image is known. However, measurements in ambient are prone to the formation of conductive surface water layers and local contamination, resulting in a significant smearing of potential contrast even in well-defined systems such as interdigitated electrode systems, two-phase materials, or ferroelectric domains. An alternative approach for the calibration of resolution can be based on point-source type standards, such as carbon nanotubes [61]. In this case, the lateral size of the nanotubes is well below both the geometric and electrostatic radius of the tip, providing an ideal calibration standard of the tip properties.

For a system consisting of carbon nanotubes on a substrate, the force between the tip and the surface can be written as [55]

$$2F_z = C_{ts}^{'}(V_t - V_s)^2 + C_{ns}^{'}(V_n - V_s)^2 + C_{tn}^{'}(V_t - V_n)^2 \tag{20}$$

where V_t is the tip potential, V_n is the nanotube potential and V_s is the surface potential, C_{ts} is the tip-surface capacitance, C_{ns} is the nanotube-surface capacitance and C_{tn} is the tip-nanotube capacitance, and the derivative is taken with respect to the z direction. When an ac bias is applied to the nanotubes (SIM), $V_n = V_0 + V_{ac}\cos(\omega t)$ and $V_s = V_0$. Therefore, the first harmonic of tip-surface force is

$$F_{1\omega} = C_{tn}^{'}V_{ac}(V_t - V_0) \tag{21}$$

In comparison, application of an ac bias to the tip, $V_t = V_{dc} + V_{ac}\cos(\omega t)$, yields

$$F_{1\omega} = C_{tn}^{'}V_{ac}(V_{dc} - V_0) + C_{ts}^{'}V_{ac}(V_{dc} - V_s) \tag{22}$$

Therefore, applying an ac bias directly to the carbon nanotube allows the tip-surface capacitance to be excluded from the overall force.

Equation (21) can be generalized in terms of the tip-surface transfer function, $F(x, y) = C_z'(x, y)$, defined for SIM and SSPM as the capacitance gradient between the tip and a region $dxdy$ on the surface as $F_{1\omega} = (V_t - V_0) \int F(x, y) V_{ac}(x, y)\, dx\, dy$, which has the form of Eq. (18) for linear imaging. For a nanotube oriented in the y-direction and taking into account the small width, w_0, of the nanotube compared to the tip radius of curvature, Eq. (18) can be integrated as

$$F_{1\omega}(a) = w_0 V_{ac}(V_t - V_0) \int F(a, y)\, dy \tag{23}$$

where a is the distance between the projection of the tip on the surface and the nanotube. Assuming a rotationally invariant tip, the resolution function is $F(x, y) = F(r)$, where $r = \sqrt{x^2 + y^2}$ and Eq. (23) can be rewritten as a function of a single variable, a. Therefore, the resolution function can be found by numerically solving Eq. (23) using experimentally available force profiles across the nanotube, $F_{1\omega}(a)$.

The validity of the proposed standardization technique is illustrated in Figure 15. If the measurements are made sufficiently far (1–2 μm) from the biasing contact, the image background and potential distribution along the nanotube are uniform indicating the absence of contact-probe interactions. The height of the nanotube is $a = 2.7$ nm, while apparent width is $H = 40$ nm due to the convolution with the tip shape. Simple geometric considerations, $H = 2\sqrt{Ra}$, yield a tip radius of curvature of $R \approx 75$ nm. The SIM amplitude profile is significantly broader, with a full width at half-maximum (FWHM) of at least ~100 nm, which increases with tip-surface separation.

Experimentally measured SIM amplitude profiles perpendicular to the nanotube, $F_{1\omega}$, were found to have Lorentzian shape,

$$F_{1\omega}(x) = F_0 + \frac{2A}{\pi} \frac{w}{4(x - x_c)^2 + w^2}, \tag{24}$$

where F_0 is an offset, A is the area below the peak, w is the peak width and x_c is the position of the peak. The offset F_0 provides a direct measure of the non-local contribution to the SPM signal due to the cantilever and conical part of the tip [51,62–64]. The distance dependence of peak height $h = 2A/\pi w$ for large tip-surface separations is $h \sim 1/d$. The distance dependence of width, w, is shown in Figure 15(f) and is almost linear in distance for $d > 100$ nm. The integral Eq. (23) can be solved analytically and the radially symmetric resolution function in SIM/SSPM is

$$F(r) = \frac{2A}{\pi} \frac{w}{(4r^2 + w^2)^{3/2}} \tag{25}$$

where A and w are z-dependent parameters determined in Eq. (25) and r is radial distance.

Equation (25) can be used to determine the tip-shape contribution to electrostatic SPM measurements in systems with arbitrary surface potential distributions. For example, for a stepwise surface potential distribution such as grain boundary or metal-semiconductor interface, $V_{surf} = V_1 + (V_2 - V_1)\theta(x)$, where $\theta(x)$ is a Heaviside step function, provided that the intrinsic interface width is significantly smaller than the microscope resolution. In this case, the measured potential profile will be $V_{eff} = V_1 + V_2 \arctan(2x/w)/\pi$, provided that the cantilever contribution to the measured potential is small. Figure 15(f) shows the width of the SIM phase profile across a grain boundary in a Nb-doped $SrTiO_3$ bicrystal. From independent measurements the double Schottky barrier width is <20 nm, i.e., well below the SPM resolution; thus the observed profile width is largely due to the instrumental resolution. The distance dependence of profile width for the nanotube standard and SIM phase image of the $SrTiO_3$ grain boundary are compared in Figure 15(f), demonstrating excellent agreement. The profile width determined from SSPM measurements is significantly larger, indicative of the mobile surface charge contribution to the profile width [65].

Thus, a carbon-nanotube-based standard provides a simultaneous measure of topographic and electrostatic resolution, as well as the quantitative resolution function for electrostatic SPM. In contrast to traditional SPM measurements (tip is ac biased) in which the tip interacts both with the DC-biased nanotube and the substrate, the latter interaction is effectively excluded. Moreover, surface and oxide trapped charges contribute to the signal for AC tip biasing, as discussed in detail below.

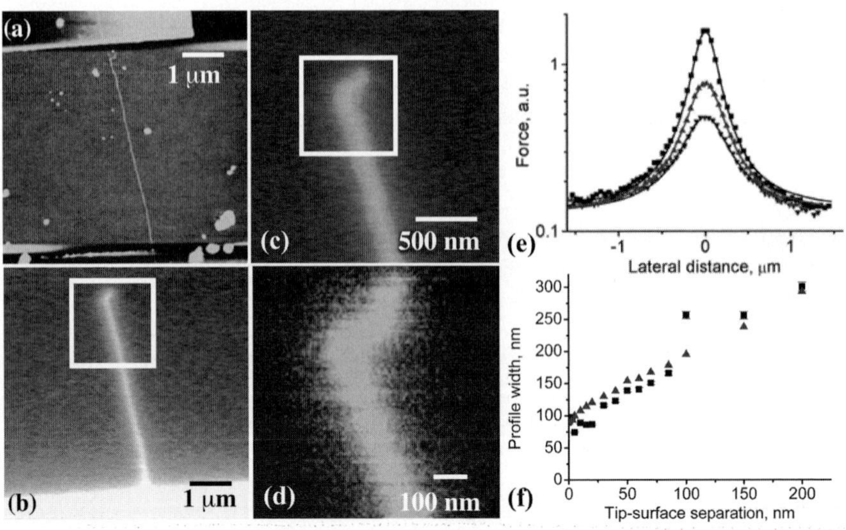

FIGURE 15. Surface topography (a) and SIM amplitude images (b-d) for a carbon nanotube circuit. The contrast is uniform along the tube. Scale is 10 nm. (e) Force profiles at lift heights of 10 nm (■), 30 nm (▲), and 100 nm (▼) and corresponding Lorentzian fits. (f) Profile width for a carbon nanotube standard (▲) and profile width for an SIM phase image of a SrTiO3 grain boundary (■) as a function of lift height. Reprinted with permission from [61].

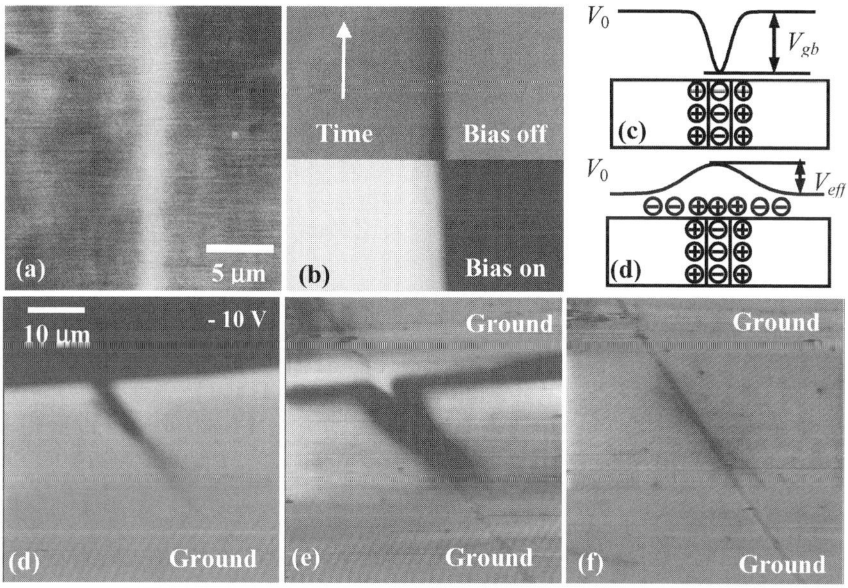

FIGURE 16. (a) Equilibrium surface potential at $SrTiO_3$ bi-crystal grain boundary. (b) Grain boundary potential evolution in the turn-off experiment. The arrow indicates time direction. Inset shows magnified potential profile after turn-off exhibiting positive feature to the right of the negatively charged interface. Schematic representation of charge and potential distributions at a pristine (c) and screened (d) grain boundary-surface junction. (e) Surface potential at 10 V negative bias and (f) immediately after the bias is off illustrates the formation of charged negative halo around SnO_2 nanowire. (g) After a one-week time period, the halo disappears. The vertical scale is 10 V (d) and 1 V (e,f). Reprinted with permission from [55] (a–c) and [71] (d–f).

3.4 Mobile Charge Effect on SIM and SSPM

The presence of a water layer on sample surfaces in ambient is a well-known phenomenon that affects, e.g., tip-surface adhesion forces [66], electrostatic tip-surface interactions [67,68], tip-induced electrochemical reactions [69], etc. Correspondingly, the water layer can be expected to affect force-based electrical SPM techniques, and specifically SPM-based transport measurements. Evidence for such behavior can be obtained from lateral biasing experiments, as illustrated in this section.

Shown in Figure 16(a) is the surface potential distribution measured above the grain boundary in a Nb-doped $SrTiO_3$ bi-crystal [57]. Macroscopic I-V and C-V of a transport measurements have shown that the grain boundary is depletion type, with a characteristic built-in potential of ~ -500 mV and a depletion width of $\sim$20 nm. Strikingly, the sign of the grain boundary potential feature as observed by SSPM is positive and the magnitude of grain boundary potential is $\sim +20$ mV, while the width is $\sim$500 nm. The broadening and decrease of the observed potential can be explained due to the convolution with the probe resolution function. However, this explanation alone cannot account for the observed positive sign of potential feature, indicative of an accumulation type grain boundary.

Insight into this behavior can be obtained from dynamic measurements, i.e. under variable temperature, electric field, or atmosphere. Shown in Figure 16 (b) are surface potential maps acquired during the application of electric bias and immediately after the bias was turned off. Note that the grain boundary potential feature becomes negative and the magnitude decreases with time along the scan. Similar behavior was observed in variable temperature experiments, during which the potential of the grain boundary in semiconducting $BaTiO_3$ increased on heating and inverted on cooling [70].

This behavior can be explained by introducing a screening model for the surface-interface junction, as shown in Figure 16(c,d). In this model, accumulation of charged adsorbates at the surface-interface junction results in the widening of the grain boundary potential feature and, most notably, in the sign inversion. In the biasing experiment, the application of $\sim$1 V lateral bias across the interface results in a high electric field across the interface region ($\sim 10^7$ V/m) and removal of the screening charges from the surface-interface junction area. After the bias is switched off, the true sign of the grain boundary is therefore observed. Note the formation of a weak positive feature on the right side of the interface in Figure 16(b), suggesting the mobile species are positively charged and move preferentially in the direction of lower potentials. This potential distribution is metastable; over time, the accumulation of screening charges reduces the magnitude of the negative feature and eventually leads to sign reversal. Corresponding relaxation times are large (30 minutes to several hours) and strongly depend on the surface treatment prior to the experiment. Recently, this model was further validated by observations of grain boundary potential under ultra high vacuum (UHV) conditions [3,49].

Similar behavior is observed in other ambient SPM-based transport experiments. Figure 16(e,f) shows the surface potential evolution in a turn-off experiment on a voltage-biased SnO_2 nanowire. Here, the negative potential feature can be observed when one of the electrodes is negatively biased and potential drops at the defect and grounded electrode can be clearly seen. From the bias dependence of these potential drops, the I-V curves of the individual defects can be reconstructed [71]. After the negative bias is turned off, a dark halo can be seen along the nanowire and the edge of biased electrode in Figure 16(e), indicating the presence of mobile charges. After a relaxation for a period of one week, the halo disappears, as shown in Figure 16(f). Notice that in this case the mobile charges are injected from the electrode and the sign of the charge is that of the bias; i.e., the surface and the electrode are coupled resistively.

This surface charge effect can be also observed in macroscopic transport measurements as a response to the back gate bias. The relaxation time of these charges is similar to SPM observations; however, in this case the sign of the screening charge is opposite to back gate bias, as can be expected given that the back gate and surface are coupled capacitively.

Thus, the presence of mobile charges strongly affects force-based electrical SPM measurements under ambient conditions. In particular, mobile charges can result in *surface potential inversion*, where the observed surface potentials are significantly reduced and are of the opposite sign to the corresponding bulk potentials. Similar behavior was recently observed for ferroelectric surfaces, on which the equilibrium surface potential as observed by SSPM has the sign of the screening charge rather than the polarization charge [65,72]. These observations suggest that surface screening is a universal feature of oxide surfaces in air and great care should be taken in the direct interpretation of the results of ambient electrostatic SPMs.

Mobile surface charges redistribute under the dc bias, resulting in "smearing" of the dc potential or electrostatic profiles. Charge redistribution results in long-term memory effects in the electrical characteristics of devices and the formation of charged haloes can be visualized in SPM experiments. Depending on whether the charges are injected directly on the surface or attracted by the action of a back gate, the sign of the charges can be the same or opposite to the bias. However, given that the characteristic relaxation times for surface charges in air are relatively high [73–75] (on the order of minutes), the surface charge dynamics do not contribute to SIM measurements performed at relatively high ($\sim$10–100 kHz) frequencies. This results in a significant increase in the resolution and signal-to-noise ratio in SIM compared to SSPM, even though the distance dependence of electrostatic forces in SIM and SSPM is similar.

3.5 Transport Across a Model Metal-Semiconductor Interface

The use of SIM and SSPM for active device characterization is illustrated with a conventional Schottky diode example [5]. The device was cross-sectioned and polished; the I-V curves before and after the preparation were compared to ensure the veracity of observed transport behavior. Surface topography is shown in Figure 17(a), with the metal and semiconductor regions marked. Figure 17(b) shows the surface potential distribution when a slow (0.002 mHz) triangular wave is applied across the device. For a positively biased interface no potential drop is observed (bright regions), while for the negatively biased device potential drops at the Schottky barrier are observed.

To quantify the dc transport behavior, the potential drop across the interface was measured as a function of lateral bias for several circuit termination resistances, as shown in Figure 18(a). Both the linear segment under reverse bias and the almost zero potential drop under forward bias predicted by Eqs. (14,15) are clearly seen. The linear part of the plot can be approximated as $y = a + bx$. The saturation current can be determined from the intercept, $I_0 = a/2Rb$, as $\sim$7 μA, very close to the value from macroscopic I-V measurements, $I_0 = 7.83$ μA. The deviation of slope from unity in the linear part of the voltage characteristics results from a leakage current contribution to the diode conductivity. The associated leakage resistance $R_l = 1/\sigma$ can be calculated from the voltage characteristics of the interface as$R_l = 2Rb/(1 - b)$. Calculated values of leakage resistivity ($\sim$700 kΩ for

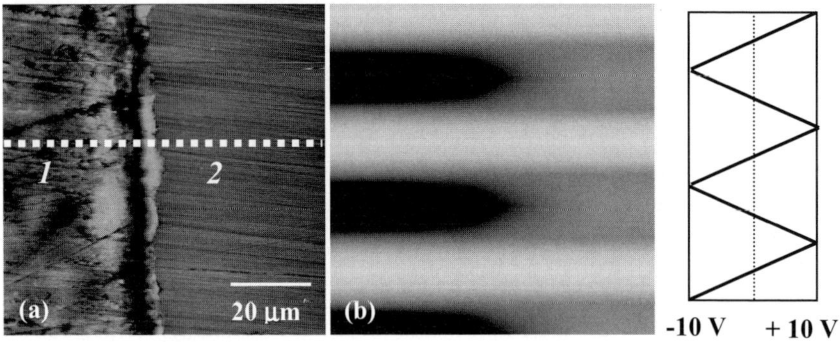

FIGURE 17. Surface topography(a) of the cross-sectioned diode. The potential and phase profiles were acquired along the dotted line. The changes of potential, phase and amplitude were determined from positions 1 and 2. Surface potential (b) during a 0.002 Hz triangular voltage ramp to the sample for $R = 500\ \Omega$. Note there is no potential drop at the interface under the forward bias conditions, while negative bias results in a significant potential drop at the interface. The scale is 300 nm (a) and 10 V (b). Reprinted with permission from [2].

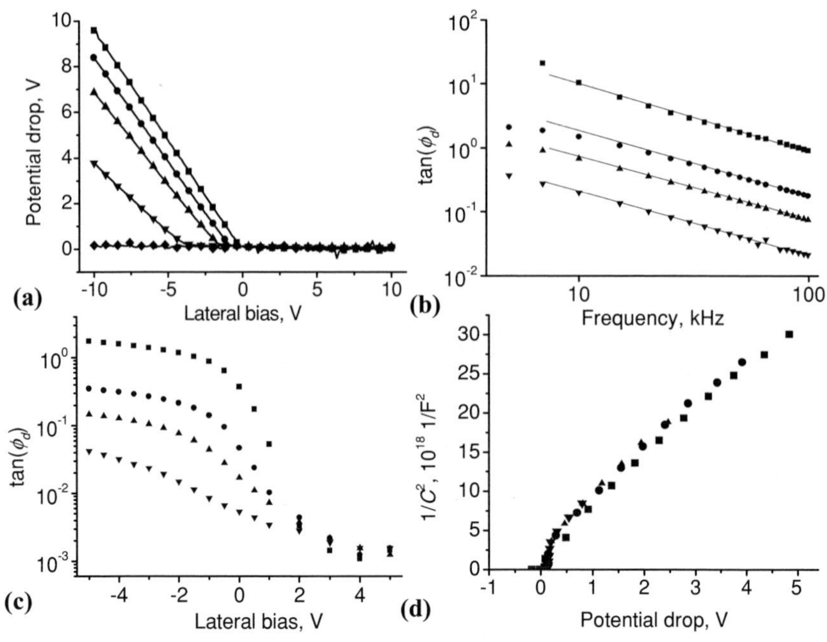

FIGURE 18. (a) Potential drop at the interface as a function of lateral bias for circuit terminations of (■)10 kΩ, (•) 47 kΩ, (▲) 100 kΩ, (▼) 220 kΩ, and (♦) 1 MΩ. (b) Frequency dependence of the phase shift for same terminations. Solid lines are linear fits. (c) Voltage phase angle as a function of lateral bias. (d) Calculated $1/C^2$ vs. V_d for different circuit terminations. Though $\tan(\varphi_d)$ varies by 2 orders of magnitude, $1/C^2$ exhibits universal behavior. Reprinted with permission from [2].

10 kΩ termination) are very close to the resistivity calculated from the macroscopic I-V curve under reverse bias conditions (~650 kΩ).

To determine the frequency-dependent transport characteristics of a device, the SIM amplitude and phase shifts across the interface were measured under similar conditions. The frequency dependence of the phase shift for different circuit terminations is shown in Figure 18(b). From macroscopic impedance spectroscopy the relaxation frequency of the junction is estimated as 1.5 kHz at −5V reverse bias. Therefore, SIM measurements are performed in the high-frequency region, in which $\tan(\varphi_d)$ is inversely proportional to frequency with a proportionality coefficient determined by the product of the interface capacitance and the circuit termination resistance (Table 1). In agreement with theoretical expectations, the slopes in Figure 18(b) are within experimental error of the theoretical value −1. The intercepts provide the value of the interface capacitance.

In order to quantify the microscopic C-V behavior, the interface phase shift was measured as a function of lateral dc bias for different circuit terminations. Under reverse bias, $\tan(\varphi_d)$ changes by almost two orders of magnitude from $tan(\varphi_d) = 1.8$ for $R = 10$ kΩ to $\tan(\varphi_d) = 0.042$ for $R = 220$ kΩ. The interface capacitance can be calculated from the data in Figure 18(c), while the potential drop at the interface is directly accessible from SSPM measurements. Combination of the two determines the C-V characteristics of the interface, which is shown in Figure 18(d). The resulting curve exhibits universal behavior independent of the current limiting resistance, $1/C^2 = (4.0 \pm 0.6)10^{18} + (6.1 \pm 0.3)10^{18} V_d$. From the intercept of the $1/C^2 - V$ dependence [76], the Schottky barrier height can be estimated as $\phi_B = 0.6 \pm 0.1$ V. This value agrees with the barrier height obtained from macroscopic I-V measurements as $\phi_B = 0.55$ V. From the slope, the dopant concentration for the material is estimated as $N_b = 1.06 \times 10^{24}$ m^{-3}.

These results demonstrate that local interface imaging of metal-semiconductor interfaces yields junction properties consistent with that determined by macroscopic techniques, thus verifying the quantitative nature of local probe approach employed in SIM. Also note that to obtain quantitative data, SSPM and SIM are used sequentially, when the phase and amplitude shifts across the interface determined by SIM are quantified as a function of local potential drop (rather then lateral bias applied across the system) determined by SSPM. Similarly, in NL-SIM the data on nonlinear interface behavior can be analyzed only if corresponding SSPM and SIM data are available.

3.6 Transport in Polycrystalline Materials

Transport and dielectric properties of polycrystalline oxides and semiconductors are often determined by the presence of electroactive grain boundaries and interfaces. While the averaged properties can be studied using macroscopic techniques, the transport characteristics of individual interfaces and, in some cases, even the origin of observed frequency-dependent transport behavior, are unknown. Here, we demonstrate applicability of SIM and SSPM for characterization of two polycrystalline oxide ceramic systems, $BiFeO_3$ and $CaCu_3Ti_4O_{12}$ (CCTO). In both cases, real space imaging of ac and dc potential distributions combined with macroscopic impedance spectroscopy measurements allows the electroactive microstructural elements to be visualized directly, providing insight into the origins of observed transport behavior.

3.6.1 Transport in $BiFeO_3$

Bismuth ferrite, $BiFeO_3$, has attracted broad attention as multiferroic material simultaneously exhibiting both ferroelectric ($T_C = 830°C$) and long-range antiferromagnetic G-type ordering ($T_N = 370°C$) [77]. Because of this magnetoelectric coupling, it has been proposed that $BiFeO_3$ ceramics systems could be used to develop novel memory-device applications. Extensive structural, magnetic, and electric studies of various $BiFeO_3$ solid solutions systems have been reported [78–80]. Particularly of interest for these applications are electric and dielectric properties of $BiFeO_3$, which could be strongly affected by small amounts of impurities. In particular, impedance spectroscopy studies in $BiFeO_3$ have identified low-frequency relaxation process that can be attributed either to the presence of grain boundary potential barriers or ferroelectric domain walls.

The combination of SSPM, SIM, and piezoresponse force microscopy (PFM) is used to establish the origins of dielectric properties in $BiFeO_3$ ceramics. The surface topography and surface potential at a $BiFeO_3$ surface under different bias conditions are shown in Figure 19. The topographic image exhibits a number of spots due to contaminants and depressions due to inter- and intragranular pores. Grain boundaries can be seen due to selective polishing of grains with different orientations. The surface potential of the grounded $BiFeO_3$ surface exhibits large-scale potential variations due to ferroelectric domains and surface contaminants (Figure 19(b)). On application of a 10-V lateral bias, the potential drops at the grain boundaries become evident (Figure 19(c)). The contrast inverts on application of a bias of opposite polarity, as shown in Figure 19(d). Note that the potential features related to ferroelectric polarization are independent of the applied bias. Ramping the dc bias across the sample has shown that the potential drop at the interface is linear in external bias and the grain boundaries exhibit ohmic behavior for small interface biases ($\Delta V_{gb} < 50$ mV)).

The surface topography, SIM phase and amplitude images at 70 kHz for the same region are shown in Figure 19(e,f). Note that the phase image exhibits well-defined phase shifts at the grain boundaries, while the amplitude image shows a uniform decrease of amplitude across the surface. Positive phase shifts at the grain boundary and a negative phase shift in the bulk are clearly observed in agreement with theoretical arguments. For higher frequencies phase shifts in the grain interior are not observed due to the resistive component in the experimental circuit. At the same time, the amplitude decreases linearly in the direction of current flow indicating that the experimental frequency range (10–100kHz) is above the RC relaxation frequency of the interface.

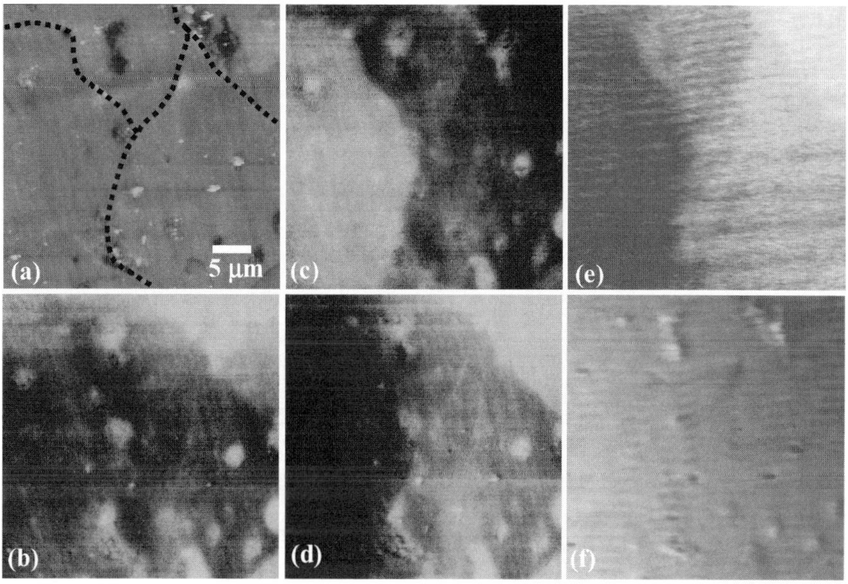

FIGURE 19. Surface topography (a), surface potential of the grounded surface (b), and surface under lateral bias of 10 V (c) and –10 V (d). Scale is 200 nm (a), 50 mV (b–d). SIM phase (e) and amplitude (f) images of the same region at 70 kHz. Scale is 0.2° (e). Reprinted with permission from [53].

This transport behavior can be correlated with the ferroelectric domain pattern as determined by PFM shown in Figure 20. Note that no potential drop at the 180° domain walls is observed, indicative of the dominant contribution of grain boundaries to low-frequency transport behavior.

To establish the relationship between macroscopic properties and SIM, the sample was studied by impedance spectroscopy and the corresponding spectra

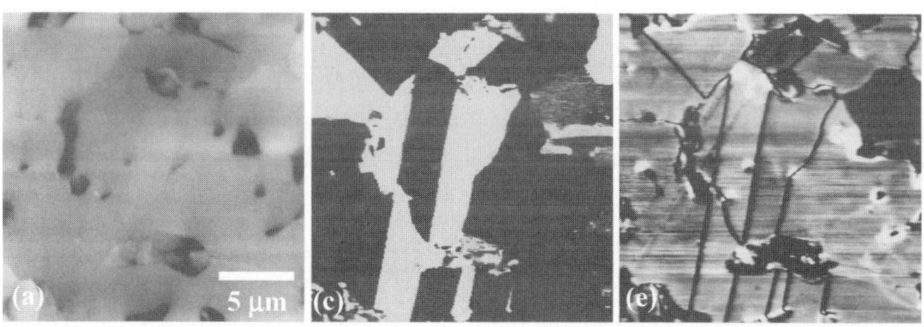

FIGURE 20. Surface topography (a), piezoresponse phase (b), and amplitude (c) images of polycrystalline $BiFeO_3$ illustrating the presence of ferroelectric domains. Note that extremely clear PFM contrast is observable despite relatively high (~100 kΩ) conductivity of the sample.

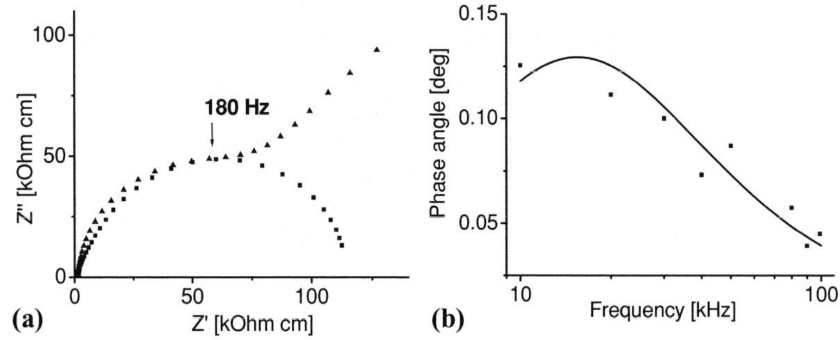

FIGURE 21. (a) Cole–Cole plots of as prepared $BiFeO_3$ pellets (■) and the rectangular sample (▲) used for scanning probe microscopy studies. (b) Experimental SIM phase shift across the interface and theoretical curve calculated from the impedance data. Reprinted with permission from [53].

are shown in Figure 21. From the impedance spectroscopy data, the average grain boundary resistivity and capacitance are estimated as $R_{gb} = 116\ \text{k}\Omega\cdot\text{cm}$ and $C_{gb} = 7.6\ \text{nF/cm}$, while the grain interior resistivity and capacitance are $R_{gi} = 812$ $\Omega\cdot$cm and $C_{gi} = 7$ pF/cm. Figure 21(b) shows the calculated grain boundary phase shift vs. frequency dependence as compared to experimental SIM data [55]. The only free parameter in the calculations is the effective grain number. The best fit is obtained for $n = 210$ grains, which is comparable with grain number N ~ 70 estimated from the grain size ($\sim$20–30 μm) and the distance between measurement point and left contact ($\sim$1–2 mm). The discrepancy between the two is due to the uncertainty in the bulk resistance and variation in grain boundary properties and orientation. Note the excellent agreement between phase-angle frequency dependences obtained from the local measurements and the bulk impedance spectroscopy measurements.

3.6.2 Transport in $CaCu_3Ti_4O_{12}$

Recently, $CaCu_3Ti_4O_{12}$ (CCTO) has attracted broad attention due to the discovery of ultrahigh dielectric constant behavior [81–83]. A number of attempts to interpret this behavior in terms of intrinsic material properties [84,85] as well as impurity and interface mediated phenomena [86] have been reported. One of the primary explanations for ultrahigh ($\sim 10^5$) dielectric constants is the formation of nonconductive grain boundaries between conductive grains, i.e., grain boundary layer capacitor behavior, common in doped perovskite titanates [87]. This model was recently supported by impedance spectroscopy [88–90]. Recently, low-frequency transport properties in CCTO and the resistive barriers at grain boundaries were imaged using scanning probe microscopy (SPM) [91,92].

The surface topography of a CCTO sample and the corresponding surface potential distribution are shown in Figure 22(a,b). The surface potential of the same region under a lateral 10-V bias is shown in Figure 22(c). The surface potential

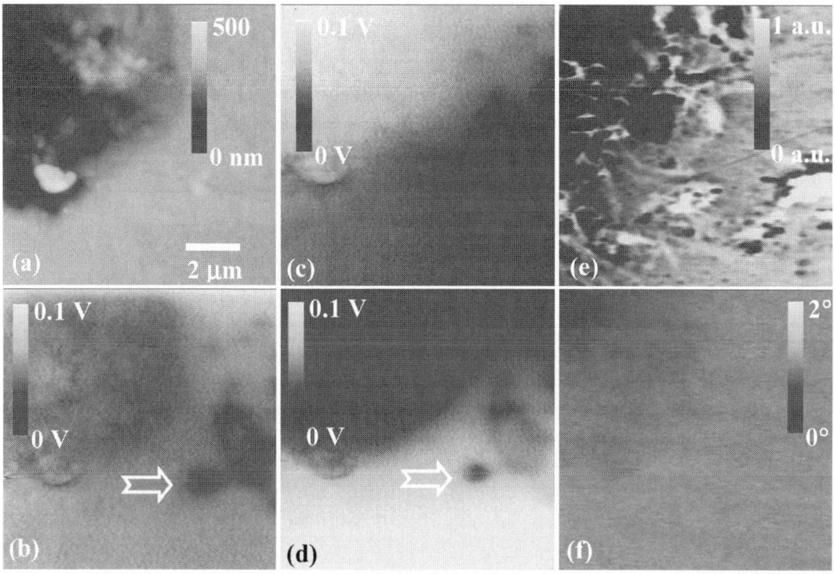

FIGURE 22. (a) Topography and (b) potential of the grounded CCTO surface measured with scanning probe microscopy. Surface potential of the sample laterally biased by +10 V (c) and −10 V (d). SIM amplitude (e) and phase (f) images of the same region. Reprinted with permission from [92].

exhibits well-defined potential drops, represented by sharp intensity contrast, within the image. The direction of the contrast, and therefore the voltage drops, inverts when the bias is switched, as illustrated in Figure 22(d). The magnitude of the potential drops is ~30-50 mV, comparable with that expected for a 10-V bias applied across ~2 mm for an average grain size of ~10 μm. This behavior indicates a significant contribution from grain boundary resistance to the dc transport in the material.

SIM amplitude and phase images of the same region obtained at 99 kHz are shown in Figure 22(e,f). The amplitude image shows significant influence of the surface topography, since the signal is directly proportional to the tip-surface capacitance gradient. Unlike dc transport behavior, both amplitude and phase data show only minor variations at interfaces.

Frequency-dependent macroscopic transport properties of the material were studied with impedance spectroscopy. The frequency dependence of the impedance magnitude and phase are shown in Figure 23. The equivalent circuit is dominated by a single RC element with a resistance of 30.7 MΩ (R_1) and capacitance 0.4 nF (C_1), corresponding to a dielectric constant $\kappa = 1,830$. The RC relaxation frequency is $\omega_1 = 1/R_1C_1 = 16_{\text{Hz}}$. Thus, at high (~99 kHz) frequencies the sample impedance is more than three orders of magnitude lower than the dc value due to capacitive coupling across the interfaces. Hence, the ac voltage decreases at the electrodes and in the bulk are comparable to the decreases at the grain boundaries and the

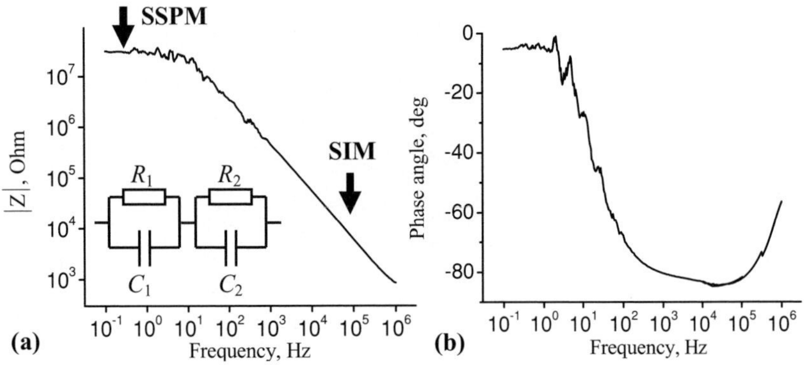

FIGURE 23. (a) Impedance magnitude and (b) phase for bulk CCTO sample as a function of frequency. The inset shows a typical equivalent circuit with grain boundary and bulk contributions. Note that without additional information the elements of the equivalent circuit cannot be unambiguously associated with individual microstructural elements such as grain boundaries, bulk, or contacts. Reprinted with permission from [92].

SIM images exhibit weak amplitude and phase contrast. This illustrates that grain boundaries in CCTO act as resistive barriers at low frequencies and are conductive due to capacitive coupling at high frequencies. Combined with the impedance data, these results show unambiguously that the low frequency RC element in the equivalent circuit is due to grain boundary behavior.

4 Conclusion

The rapid development of nanophase materials and nanoelectronic devices necessitates spatially resolved studies on dc and frequency-dependent transport behavior on the nanoscale. NIM, SSPM, SIM and NL-SIM provide a set of complementary techniques for the quantification of the nanoscale transport properties, as microscopic analogs of conventional C-V and two- and four-probe impedance spectroscopy measurements. While conventional current-based transport measurements allow significantly higher sensitivity and precision and are capable of measurements over a larger (1 mHz–100 MHz) frequency range, the frequency range in SIM and NIM is limited by the scan rate, bandwidth of the optical detector, and stray cantilever-surface capacitances to $\sim$1–10 MHz. With calibration and correction procedures described above, SIM and SSPM can provide transport information within $\sim$10% error, this value being primarily determined by the cantilever contribution to measured signal. The primary advantage of SIM and SSPM is that these techniques allow spatial localization of microstructural elements with resistive and capacitive behavior, which can then be compared to AFM, optical, or electron microscopy images. SIM/SSPM provides best results in conjunction with current based transport measurements, in which the global frequency-dependent impedance of the system and the local behavior of the individual structural element are determined simultaneously. In addition to electronic device characterization and fundamental physical studies, localized ac impedance and transient potential techniques may be especially critical in the domain of electrochemical systems. Faradaic reactions, ionic transport, and diffusion are best studied with transient techniques, making SPM-impedance measurement extremely attractive for these applications.

Continued improvements to the speed and upper-bound impedance measurement limits of the technique are needed. Long measurement times currently introduce a host of associated problems, including temperature stability, tip and sample drift issues, image distortion, and non-stationary impedance concerns (especially when imaging electrochemical systems). These issues can be partially alleviated through the use of closed-loop scanner technology. Measurement bandwidth may be improved by novel tip and tip-holder designs, including shielded probes and cantilevers [93], which can reduce stray capacitance and other sources of spurious ac response.

References

1. S. Kalinin and D. A. Bonnell, Appl. Phys. Lett. **78**, 1306 (2001).
2. S. Kalinin and D. A. Bonnell, J. Appl. Phys. **91**, 832 (2002).
3. R. Shao, S. V. Kalinin, and D. A. Bonnell, Appl. Phys. Lett. **82**, 1869 (2003).
4. R. O'Hayre, M. Lee, F. B. Prinz, J. Appl. Phys. **95**, 8382 (2004).
5. R. O'Hayre, G. Feng, W. D. Nix, F. B. Prinz, J. Appl. Phys. **96**, 3540 (2004).
6. J.R. Macdonald, W.R. Kenan, *Impedance Spectroscopy: Emphasizing Solid Materials and Systems*, Wiley, NY (1987).
7. R. S. Lillard, P. J. Moran, and H. S. Issacs, J. Electrochem. Soc. **139**, 1007 (1992).
8. F. Zou, D. Thierry, and H. S. Issacs, J. Electrochem. Soc. **144**, 1957 (1997).
9. E. Bayet, F. Huet, M. Keddam, K. Ogle, and H. Takenouti, J. Electrochem. Soc. **144**, L87 (1997).
10. M. Pilaski, T. Hammelmann, A. Moehring, and M. M. Lohrengel, Electrochim. Acta **47**, 2127 (2002).
11. A.J. Bard, *Electrochemical Methods*, 2nd ed., Wiley, NY (2001).
12. R. M. Wightman, Anal. Chem., **53**, 1125A (1981).
13. M. Fleischmann, S. Pons, D. R. Rolison, and P. P. Schmidt. Ultramicroelectrodes. Datatech Systems, Morgantown, NC, USA (1987).
14. J. Fleig, S. Rodewald, and J. Maier, J. of Appl. Phys, **87**, 2372 (2000).
15. J. Fleig and J. Maier. Solid State Ionics, **85**, 9 (1996).
16. S. Rodewald, J. Fleig, and J. Maier. J. of the European Ceramic Soc., **19**, 797 (1999).
17. J. Fleig and J. Maier. Phys. Chem., Chem. Phys., **1**, 3315 (1999).
18. V. Brichzin, J. Fleig, H. U. Habermeier, and J. Maier. Electrochem. SolidState Lett., **3**, 403 (2000).
19. C. E. Gardner and J. V. Macpherson. Anal. Chem., **74**, 576A (2002).
20. J. V. Macpherson, P. R. Unwin, A. C. Hillier, and A. J. Bard. J. Amer. Chem. Soc. , **118**, 6445 (1996).
21. A. Layson, S. Gadad, and D. Teeters, Electrochim. Acta **48**, 2207 (2003).
22. J. Fleig and J. Maier, Solid State Ionics **85**, 9 (1996).

23. J. Fleig, S. Rodewald, and J. Maier, Solid State Ionics **136**, 905 (2000).
24. P. De Wolf, J. Snauwaert, L. Hellemans, T. Clarysse, W. Vandervorst, M. D'Olieslaeger, and D. Quacyhaegens, J. Vac. Sci. Technol. A **13**, 1699 (1995).
25. A. C. Diebold, M. R. Kump, J. J. Kopanski, and D. G. Seiler, Proc. Electrochem. Soc. **94**, 78 (1994).
26. P. De Wolf, M. Geva, T. Hantschel, W. Vandervorst, and R. B. Bylsma, Appl. Phys. Lett. **73**, 2155 (1999).
27. T. W. Kelley, E. L. Granstrom, and C. D. Frisbie, Adv. Mater. (Weinheim, Ger.) **3**, 261 (1999).
28. A. Bietsch, M. A. Schneider, M. E. Welland, and B. Michel, J. Vac. Sci. Technol. B **18**, 1160 (2000).
29. M. Freitag, M. Radosavljevic, W. Clauss, and A. T. Johnson, Phys. Rev. B **62**, R2307 (2000).
30. S. J. Oshea, R. M. Atta, M. P. Murrell, and M. E. Welland, J. Vac. Sci. Technol. B **13**, 1945 (1995).
31. A. Olbrich, B. Ebersbergerm, and C. Boit, Appl. Phys. Lett. **73**, 3114 (1998).
32. J. M. R. Weaver and D. W. Abraham, J. Vac. Sci. Technol. B **9**, 1559 (1991).
33. M. Nonnenmacher, M. P. O'Boyle, and H. K. Wickramasinghe, Appl. Phys. Lett. **58**, 2921 (1991).
34. C. C. Williams, W. P. Hough, and S. A. Rishton, Appl. Phys. Lett. **55**, 203 (1989).
35. C. C. Williams, J. Slinkman, W. P. Hough, and H. K. Wickramasinghe, Appl. Phys. Lett. **55**, 1662 (1989).
36. A. C. Diebold, M. R. Kump, J. J. Kopanski, and D. G. Seiler, J. Vac. Sci. Technol. B **14**, 196 (1996).
37. T. Trenkler et al., J. Vac. Sci. Technol. B **18**, 418 (2000).
38. T. Hantschel, P. Niedermann, T. Trenkler, and W. Vandervorst, Appl. Phys. Lett. **76**, 1603 (2000).
39. H. B. Martin, A. Argoitia, U. Landau, A. B. Anderson, and J. C. Angus, J. Electrochem. Soc. **143**, L133 (1996).
40. N. Vinokur, B. Miller, Y. Avyigal, and R. Kalish, J. Electrochem. Soc. **143**, L238 (1996).
41. D. R. Clarke, J. Am. Ceram. Soc., **82**, 485 (1999).
42. R. Einzinger, Appl. of Surf. Sci., **3**, 390 (1979).
43. D. Tabor, *The Hardness of Metal,* Clarendon, Oxford (1951).
44. B. Bhushan, *The Handbook of Micro/Nano Tribology* 2nd Ed., CRC Press, Boca Raton, FL. (1999).
45. W. D. Nix and H. J. Gao, J. Mech. Phys. Solids **46**, 4577 (1998).
46. K. L. Johnson, *Contact Mechanics*, Cambridge University Press, Cambridge (1987).
47. O. Vatel and M. Tanimoto J. Appl. Phys. **77**, 2358 (1995).
48. T. Meoded, R. Shikler, N. Fried, and Y. Rosenwaks, Appl. Phys. Lett. **75**, 2435 (1999).
49. B. D. Huey and D. A. Bonnell, Appl. Phys. Lett. **76**, 1012 (2000).
50. J. Shin, V. Meunier, A. P. Baddorf, and S. V. Kalinin, Appl. Phys. Lett. **85**, 4240 (2004).
51. B. Rodriguez, S. Jesse, A. P. Baddorf, and S. V. Kalinin, to be submitted.
52. P. M. Chaikin and T. C. Lubensky, *Principles of Condensed Matter Physics* (Cambridge University Press, New York, 1997).
53. S. V. Kalinin, M. R. Suchomel, P. K. Davies, and D. A. Bonnell, J. Am. Ceram. Soc. **85**, 3011 (2002).
54. *Impedance Microscopy: Emphasizing Solid Materials and Systems*, ed. J. R. Macdonald (John Wiley, New York, 1987).
55. S. V. Kalinin and D. A. Bonnell, Phys. Rev. B **70**, 235304 (2004).

56. G. Blatter and F. Greuter, Phys. Rev. B **33**, 3952 (1986).
57. H. O. Jacobs, P. Leuchtmann, O. J. Homan, and A. Stemmer, J. Appl. Phys. **84**, 1168 (1998).
58. E. Strassburg, A. Boag, and Y. Rosenwaks, Rev. Sci. Instrum. **76**, 083705 (2005).
59. J. Colchero, A. Gil, and A. M. Baro, Phys. Rev. B **64**, 245403 (2001).
60. P. J. A. van Schendel, H. J. Hug, B. Stiefel, S. Martin, and H. -J. Guntherodt, J. Appl. Phys. **88**, 435 (2000).
61. S. V. Kalinin, D. A. Bonnell, M. Freitag, and A. T. Johnson, Appl. Phys. Lett. **81**, 754 (2002).
62. A. K. Henning, T. Hochwitz, J. Slinkman, J. Never, S. Hoffmann, P. Kaszuba, C. Daghlian, J. Appl. Phys. **77**, 1888 (1995).
63. S. V. Kalinin and D. A. Bonnell, Phys. Rev. B **63**, 125411 (2001).
64. G. Koley, M. G. Spencer, and H. R. Bhangale, Appl. Phys. Lett. **79**, 545 (2001).
65. S. V. Kalinin and D. A. Bonnell, Phys. Rev. B **62**, 10419 (2000).
66. D. Bonnell (Ed.), *Scanning Probe Microscopy and Spectroscopy*, Wiley, 2000.
67. J. Hu, X. D. Xiao, and M. Salmeron, Appl. Phys. Lett. **67**, 476 (1995).
68. H. Sugimura, Y. Ishida, K. Hayashi, O. Takai, and N. Nakagiri, Appl. Phys. Lett. **80**, 459 (2002).
69. J. A. Dagata, T. Inoue, J. Itoh, K. Matsumoto, and H. Yokoyama, J. Appl. Phys. **84**, 6891 (1998).
70. S. V. Kalinin and D. A. Bonnell, Nano Letters **4**, 555 (2004).
71. S. V. Kalinin, J. Shin, S. Jesse, D. Geohegan, A. P. Baddorf, Y. Lilach, M. Moskovits, and A. Kolmakov, J. Appl. Phys. **98**, 044503 (2005).
72. S. V. Kalinin, C. Y. Johnson, and D. A. Bonnell, J. Appl. Phys. **91**, 3816 (2002).
73. S. Cunningham, I. A. Larkin, and J. H. Davis, Appl. Phys. Lett. **73**, 123 (1998).
74. G. H. Buh, H. J. Chung, and Y. Kuk, Appl. Phys. Lett. **79**, 2010 (2001).
75. T. Tybell, C. H. Ahn, and J. -M. Triscone, Appl. Phys. Lett. **75**, 856 (1999).
76. S. M. Sze, *Physics of Semiconductor Devices* (Wiley, New York, 1981).
77. P. Fischer, M. Polomska, I. Sosnowska, and M. Szymanski, J. Phys. C: Solid St. Phys., **13** 1931 (1980).
78. M. Mahesh Kumar, A. Srinivas, S. V. Suryanarayana, and T. Bhimasankaram, Phys. Stat. Sol. (a), **165**, 317 (1998)
79. M. Polomska, W. Kaczmarek, Z. Pajak, Phys. Stat. Sol. (a) **23**, 567 (1974).
80. J. B. MacChesney, J. J. Jetzt, J. F. Potter, H. J. Williams, and R. C. Sherwood, J. Am. Ceram. Soc. **49**, 644 (1966).
81. M. A. Subramanian, D. Li, N. Duan, B. A. Reisner, and A. W. Sleight, J. Solid State Chem. **151**, 323 (2000).
82. A. P. Ramirez, M. A. Subramanian, M. Gardel, G. Blumberg, D. Li, T. Vogt, and S. M. Shapiro, Solid State Commun. **115**, 217 (2000).
83. C. C. Homes, T. Vogt, S. M. Shapiro, S. Wakimoto, and A. P. Ramirez, Science **293**, 673 (2001).
84. L. He, J. B. Neaton, D. Vanderbilt, and M. H. Cohen, Phys. Rev. B **67**, 012103 (2003).
85. L. He, J. B. Neaton, M. H. Cohen, D. Vanderbilt, and C. C. Homes, Phys. Rev. B **65**, 214112 (2002).
86. M. H. Cohen, J. B. Neaton, L. He, and D. Vanderbilt, J. Appl. Phys. **94**, 3299 (2003).
87. L. L. Hench and J. K. West, *Principles of Electronic Ceramics*, Wiley, (1990).
88. D. C. Sinclair, T. B. Adams, F. D. Morrison, and A. R. West, Appl. Phys. Lett. **80**, 2153 (2002).
89. T. T. Fang and H. K. Shiau, J. Am. Ceram. Soc. **87**, 2072 (2004).

90. T. B. Adams, D. C. Sinclair, and A. R. West, Adv. Mater. **14**, 1321 (2002).
91. S. Y. Chung, I. D. Kim, and S. L. Kang, Nature Mat. **3**, 774 (2004).
92. S. V. Kalinin, J. Shin, G. M. Veith, A. P. Baddorf, M. V. Lobanov, H. Runge, and M. Greenblatt, Appl. Phys. Lett. **86**, 102902 (2005).
93. B. T. Rosner and D. W. van der Weide, Rev. Sci. Instrum. **73**, 2505 (2002).

I.6
Review of Ferroelectric Domain Imaging by Piezoresponse Force Microscopy

A. L. KHOLKIN, S. V. KALININ, A. ROELOFS, AND A. GRUVERMAN

This chapter describes the principles, theoretical background, recent developments, and applications of a local probe-based technique for nondestructive high-resolution ferroelectric domain imaging and manipulation—piezoresponse force microscopy (PFM). This technique has proven to be a powerful tool for the characterization of ferroelectric thin films, ceramics, and single crystals. Recent advances in application of PFM for studying a mechanism of polarization reversal at the nanoscale, domain dynamics, degradation effects, and size-dependent phenomena in ferroelectrics are reviewed in detail. Examples of using PFM for the characterization of various polar materials such as ferroelectric films, piezoelectric semiconductors, and ferroelectric relaxors are given.

1 Introduction

Ferroelectric materials attract significant attention because they have a number of applications in electronic devices. One of the main features of ferroelectrics is the presence of two or more thermodynamically equivalent switchable polarization states, which can be employed in nonvolatile random access memory (FeRAM) devices and for high-density data storage. FeRAM is one of the most promising and advanced nonvolatile memory technologies in use today. It offers such benefits as high reading and writing speed, low power consumption, high endurance, and better scalability than other nonvolatile memories [1]. For FeRAM to become high-volume, mainstream devices competitive with other types of nonvolatile memories, two major goals must be accomplished. First is the ability to manufacture at a high level of device integration (i.e., memory density in the gigabit range at least), where the constituent ferroelectric capacitors should be reduced to submicron dimensions. Second, it is necessary to ensure that the FeRAM devices *of these dimensions* can achieve the necessary performance, reliability, and lifetime requirements. This necessitates that several fundamental issues be addressed, including the effects of film thickness and lateral size of the ferroelectric capacitor on ferroelectric properties such as remanent polarization and coercive field. Also, of fundamental interest is the relationship between the capacitor size and the mechanism of degradation effects such as retention, imprint and polarization loss during operation (fatigue).

To answer these questions, one has to probe ferroelectric domain configurations and local properties at micro- and nanometer scales, since the investigation of domain structure and its evolution during polarization reversal provides full and direct information on the static and dynamic properties of ferroelectrics. Scanning probe microscopy (SPM) offers several different possibilities to reveal information on the domain configurations of ferroelectric thin films and single crystals. Several reviews of SPM-based methods for characterization of ferroelectric domains are available in the literature [2,3]. A difference in mechanical, structural, electrochemical, dielectric, and piezoelectric properties of opposite ferroelectric domains may provide distinct domain contrast in the SPM contact mode. In the noncontact regime, domains can be visualized by detecting surface polarization charge and surface potential.

Currently, the most widely used SPM method for imaging domain structure of ferroelectric surfaces is piezoresponse force microscopy (PFM) [4–10], based on monitoring piezoelectric surface displacements induced by the electrically biased probing tip. This method was introduced in 1992 by Guethner and Dransfeld [11] to detect polarized regions in ferroelectric copolymer films and, soon after that, proved to be the most effective approach for the nanoscale study and control of ferroelectric domains in bulk crystals and thin films [12,13]. Figure 1 illustrates PFM's

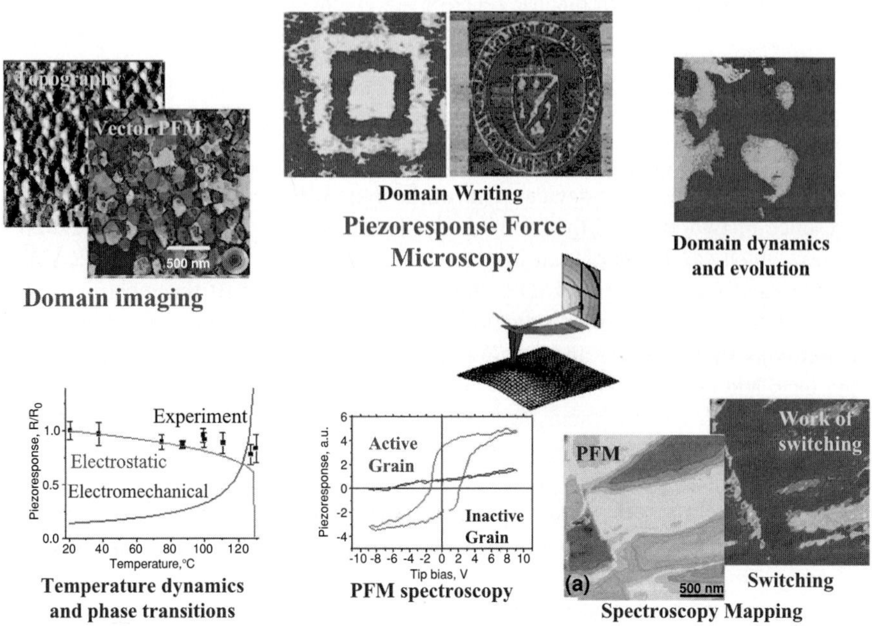

FIGURE 1. Schematic illustration of PFM capabilities in nanoscale characterization of ferroelectrics. (See also Plate 2 in the Color Plate Section.)

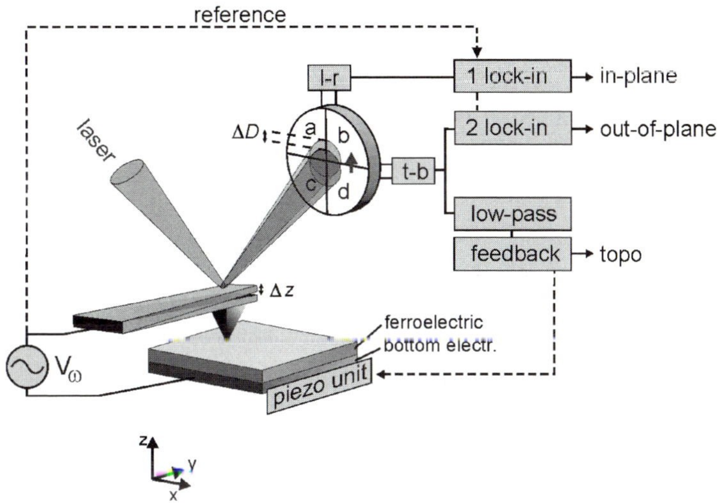

FIGURE 2. PFM experimental setup for simultaneous acquisition of topography and vertical and lateral polarization components. A function generator is used to apply an ac voltage V_ω between the tip and the bottom electrode. The voltage-induced cantilever deflection is detected by a reflected laser beam on a four quadrant photodiode. The two signals: $(a + c) - (b + d)$ and $(a + b) - (c + d)$ are demodulated with two lock-in's representing the in- and out-of-plane signals, respectively. The average force on the cantilever during the scanning process is kept constant by a feedback loop.

capabilities for obtaining information on the nanoscale properties of static domain configurations, domain switching behavior, slow relaxation processes, and local hysteresis spectroscopy. In this chapter, we summarize the basic principles of PFM, show what kind of information can be obtained from PFM experiments, delineate the limitations of PFM signal interpretation relevant to quantitative imaging, and review recent advances in PFM with respect to nanoscale ferroelectric research and applications.

2 Principle of PFM

2.1 Experimental Setup

The standard experimental PFM setup is usually based on a commercial scanning probe microscope equipped with a four-quadrant photo detector, a conductive probing tip, a function generator, and two lock-in amplifiers (as illustrated in Figure 2).

A ferroelectric sample under investigation is placed between the bottom electrode and the conductive PFM tip, serving as a movable top electrode. The most frequently used probes are either metal-coated silicon probes with conducting coatings (Pt/Ir, WC_2, Au, Rh) or highly doped silicon cantilevers. Measurements are carried out in the contact mode and allow information on the out-of-plane component of electromechanical surface response—i.e., normal to the film plane (vertical PFM, or VPFM), as well as on the in-plane component via the frictional forces (lateral PFM, or LPFM) [14], to be obtained. Application of an ac electric bias to the probing tip results in a sample surface oscillation due to the converse piezoelectric effect. Electronic feedback (that maintains the "average" force on the cantilever constant) facilitates accurate tracking of the surface motion by using the lock-in technique with simultaneous recording of the surface topography. The effect of the Maxwell stress (capacitive forces) compared to the piezoelectric response (electromechanical forces) is discussed in the following section. Typical operation frequencies for VPFM vary from 1–3 kHz to 2–5 MHz, as limited by the bandwidth of optical detector. In LPFM, the transduction of signal from surface to the tip necessitates operation in the 1–30 kHz range.

Note that, in order to prevent attenuation of the out-of-plane piezoelectric response signal, the feedback loop has to operate at a frequency lower than the frequency ω of the applied voltage V_ω. This provides the possibility to separate the PFM signal from the topographical information. To avoid the cross-talk between piezoelectric deformation and topographic image, a low pass filter is typically put in front of the feedback. At the same time, high operating frequency ensures sufficient sampling of each image point.

In terms of exciting the piezoelectric vibration of the sample, there are two major approaches in PFM. In a local excitation approach, the vibration is generated locally by applying a modulation voltage between the bottom electrode and the conductive SPM tip which scans the bare surface of the film without a deposited top electrode (Figure 3(a)). An advantage of this approach is the possibility of establishing correlations between domain configurations and film microstructure. In addition, this method can be used for nanoscale domain control via highly localized polarization reversal and direct investigation of domain wall interaction with microstructural features, such as defects and grain boundaries, for local spectroscopy measurements, and for the investigation of electrical and mechanical coupling between adjacent grains. Furthermore, this approach offers extremely high resolution, potentially allowing for the investigation of the microscopic mechanism of the domain wall motion.

Note that the electric field generated by the SPM tip in this configuration is highly inhomogeneous, which makes quantitative analysis of the field- induced signal extremely difficult. In other words, PFM measurements in the sample without the extended top electrode collect signal only from a surface layer of an unknown thickness that is a function of dielectric permittivity and contact conditions.[1] This effect has been understood by directly comparing PFM images taken on epitaxial PZT thin films containing *a*-domains with a cross-sectional TEM image

[1] Only when the diameter of the tip-sample contact area is of the same order of magnitude as the film thickness can the field be treated as approximately homogeneous. This is valid for standard cantilevers and films with the thicknesses below ~20 nm.

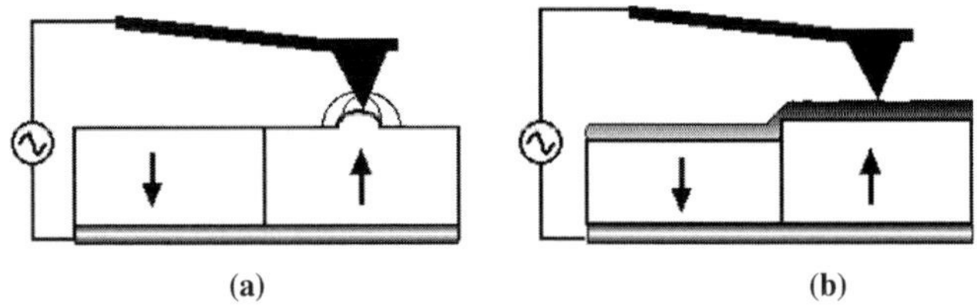

FIGURE 3. Local (a) and integral (b) methods of excitation in PFM.

[15]. It was found that the a-domain width observed in the PFM measurement complies with the a-domain width near the surface determined from the TEM image. This proves that the PFM measurement is not integrating over the whole film thickness but images the polarization only several nanometers below the surface.

In a global excitation approach, domain structure can be visualized through the top electrode of a ferroelectric capacitor (although at the expense of lower lateral resolution). In this case, the piezoelectric vibration is generated in a film region underneath the deposited top electrode which is much larger that the tip-sample contact area. The modulation voltage can be applied either by using an external wire attached to the top electrode or, in the case of a micrometer size electrode, directly through the conductive SPM tip. As in the former case, the piezoelectric displacement is probed locally by the SPM tip (Figure 3(b)). In such a configuration, a homogeneous electric field is generated throughout the ferroelectric film, which allows quantitative treatment of domain wall dynamics and investigation of polarization reversal mechanism in ferroelectric capacitors. Due to the reduced time constant, fast pulse switching and transient current measurements can be accomplished in submicron capacitors, thus making PFM suitable for ferroelectric memory device testing. Hereafter, to distinguish between samples without and with top electrodes we will refer to the latter as "capacitors."

2.2 Elementary Theory of PFM

Even though the PFM technique was introduced more than ten years ago, the quantitative analysis of PFM measurements is still under debate [16–20]. The three major issues involved in PFM contrast formation mechanism are (i) voltage-dependent contact mechanics of piezoelectric material that determines the PFM signal, (ii) dynamic properties of the cantilever and transduction mechanism between the electromechanical response of the surface and electrostatic forces and resulting vertical and lateral PFM signal, and (iii) electroelastic field distribution in the material that determines resolution, domain switching behavior, and hysteresis loop acquisition mechanism.

Luo et al. [21] have found that the piezoelectric (1st harmonics) signal of ferroelectric triglycine sulfate does not diverge at the phase transition but gradually decreases, following the same trend as the spontaneous polarization. From this fact the authors concluded that the signal was dominated by the capacitive forces, rather than by electromechanical ones, since the electromechanical signal was expected to be proportional to the relevant piezoelectric coefficient that should strongly increase in the vicinity of the phase transition. The influence of the capacitive force was further supported by the finite vibration amplitude measured on non-ferroelectric surfaces [22]. On the other hand, the ability to detect the in-plane piezoelectric signal, the absence of PFM signal attenuation due to charge screening, and the possibility of domain imaging in ferroelectric capacitors through the top electrode are all indicative of a strong contribution of the electromechanical forces to the observed PFM contrast.

As such, taking both the electromechanical response and the long-range capacitive forces into account, we can present the total response of a cantilever in a contact with ferroelectric in a form: $A = A_{piezo} + A_{cap} + A_{nl}$, where A_{piezo} represents the electromechanical response of the surface due to converse piezoelectric effect, A_{cap} is the contribution due to the electrostatic force between tip and sample, and A_{nl} is the contribution due to non-local electrostatic force between the cantilever and the sample. Typically, when the domain size is much smaller than the cantilever length, non-local interaction results only in a constant background (offset) that can be easily subtracted from PFM image data.

In order to get reliable piezoelectric response data, one has to maximize the electromechanical contribution and to minimize the electrostatic one. One of the most serious problems affecting the electromechanical signal is an interfacial layer (dielectric gap) between the PFM tip and the ferroelectric surface [23] due to either sample surface contamination or oxidation of the Si probe. To get a good electrical contact between the tip and the sample surface one has to clean the surface of the ferroelectric by thermal treatment prior the scanning process (heating in vacuum at 200°C removes water, CO, and CO_2 [24]) and to use metal-coated instead of doped silicon cantilevers. Tip-sample contact force is also an important factor that determines the efficiency of PFM imaging. To overcome the contact problem, Kalinin and Bonnell [8] suggested to use stiff cantilevers (spring constant $k > 1$ N/m) and apply high contact forces (on the order of 10 to 1,000 nN depending on the tip radius and the thickness of the dielectric gap). However, this should be used with caution since at high contact forces the direct piezoelectric effect may counteract with the converse piezoelectric effect and additional effects such as higher-order ferroic switching [25] or stress-induced suppression of piezoelectricity [26,27] related to the phase transition under the high mechanical stress exerted by the tip [28]. This makes the analysis of the PFM measurements more complex. In the following, we will briefly analyze the forces acting on the tip while applying a voltage.

2.2.1 Capacitive forces

The system comprising the conducting tip in a contact with the dielectric surface represents a capacitor. Therefore, an external voltage applied between the tip and the counter electrode results in an additional (capacitive or Maxwell) force that can be, in general, determined by the following equation:

$$F_{cap} = \frac{dW_{cap}}{dz} = \frac{1}{2}V^2\frac{dC}{dz}, \quad (1)$$

where $W_{cap} = V^2C/2$ is the energy stored in the capacitor C and z is the vertical distance. It follows that, if a combination of the dc and ac voltages ($V = V_{\rm dc} + V_\omega \cos \omega t$) is applied between the tip and the counter-electrode, the capacitive forces will consist of a static force F_0, a force oscillating at a driving frequency $\omega(F_\omega = V_{dc}V_\omega dC/dz)$ and a force $F_{2\omega}$ appearing at a frequency 2ω. These forces can be essential even in contact mode [18] and must be compared with other forces, including the electromechanical one due to electric field-induced deformation of the sample.

The total capacitance C of the sample-cantilever system can be written as the sum of the capacitance of the cantilever (C_{cant}) and the capacitance of the tip (C_{tip}). The former gives rise to so-called nonlocal contribution to the capacitive force because of the macroscopic dimensions of the cantilever. The capacitance of the cantilever can be estimated by using the analytical expression derived by Prume et al. [29]:

$$\frac{dC_{cant}}{dz} = \frac{\varepsilon_0 d_c}{\tan\alpha}\left(\frac{1}{h_c} - \frac{1}{L_c \tan\alpha + h_c}\right), \quad (2)$$

where d_c is the width and L_c is the length of the cantilever. h_c and α are the gap and the angle between the cantilever and the sample surface, respectively. Equation 2 describes the non-local electrostatic force contribution $F_{\rm nl}$ of the cantilever to the PFM signal and can result in a signal offset appearing at 1st harmonics (frequency ω)[2]. The typical value of the dc force for nonlocal capacitive contribution (standard Si cantilevers, $L_c = 450$ μm, $d_c = 50$ μm) estimated using Eq. (2) is quite small (∼0.1 nN) and can be neglected for many practical cases.

In addition to capacitive forces, Coulomb attractive forces between the tip and the charged ferroelectric surface can be observed in PFM experiments (Hong et al. [30]). This force can be estimated by using an equation:

$$F_{coul} = \frac{\sigma C V}{2\varepsilon_0}, \quad (3)$$

where σ is the surface-bound charge density and V is the voltage applied to the sample. Surface polarization charges giving rise to electrostatic interaction are screened by the absorbed charges of the opposite sign and thus are normally smaller than the electromechanical forces for many practical applications. For the typical polarization charge of 1 μC/cm^2, tip radius ≈40 nm, and bias voltage of $V \approx 1$ V, the corresponding force can be estimated in the range of few nN.

[2] Note that the PFM signal is monitored at a driving frequency ω.

We should finally note that the major attractive forces between the tip and the sample surface are adhesion forces due to Van der Waals interaction. Rabe et al. [31] reported values of 50–500 nN for different types of cantilevers in contact with dielectric and metallic surfaces. This proved that the dominant forces in the PFM measurements should be both adhesive forces and strong short-range repulsive forces F_{rep}. These forces compel the cantilever to follow the motion of the ferroelectric surface caused by the converse piezoelectric effect with Coulomb interaction playing only a minor role.

2.2.2 Electromechanical Forces

Applying an electric field between a cantilever and a dielectric material leads to capacitive forces as discussed above. A piezoelectric material will provide an additional response to an applied ac electric field due to the converse piezoelectric and electrostrictive effects. For a homogeneously polarized (in the z- direction), stress-free ferroelectric material, its vertical (Z) displacement can be expressed as follows:

$$\Delta z = d_{33}V + \frac{M_{333}}{t}V^2, \tag{4}$$

where V is the applied voltage, t is the sample thickness, and d_{33} and M_{333} are the piezoelectric and electrostrictive constants, respectively. The second term in Eq. (4) is typically much smaller than the first one in a polarized state. [32] Under the external voltage $V = V_{\text{dc}} + V_\omega \cos(\omega t)$ the surface displacement will consist of a dc component (that cannot be measured with a lock-in), and the signals at first and second harmonics:

$$\Delta z_\omega = d_{33}V_\omega + 2\frac{M_{333}}{t}V_{dc}V_\omega, \tag{5}$$

$$\Delta z_{2\omega} = \frac{1}{2}V_\omega^2\frac{M_{333}}{t}. \tag{6}$$

The sign of the converse piezoelectric signal (first term in Eq. (5)) depends on the relative orientation of the polarization and applied electric field. When a positive voltage is applied to the PFM tip and the polarization is pointing at negative z-direction (electric field is parallel to the spontaneous polarization), the sample will expand, thus pushing up the cantilever and allowing measurements of d_{33} (Figure 4(a)). In this case, a positive sign is attributed to the displacement Δz. As we apply an alternating voltage, this means that the electric field and the piezoelectric signal are in phase. These measurements are referred to as *out-of-plane* (or vertical PFM) measurements. If the polarization points at the positive z-direction, the applied electric field will be antiparallel to the spontaneous polarization, causing the ferroelectric to contract with the consequent lowering of the cantilever (Figure 4(b)). The electric field and the piezoresponse signal are shifted in phase by 180°. Additionally, ferroelectric grain may contract in the film plane, which is caused

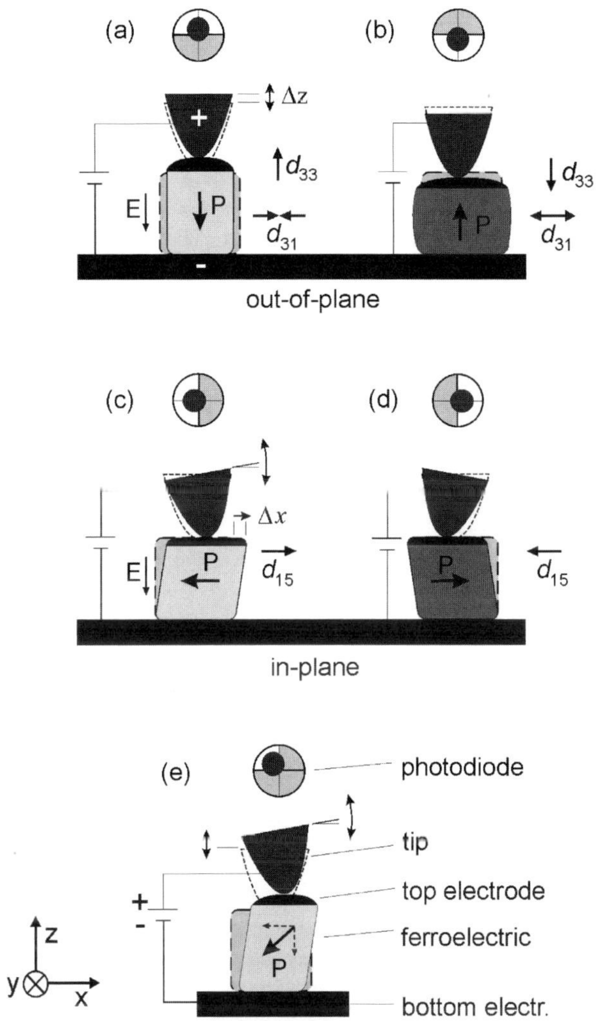

FIGURE 4. Piezoelectric effect in ferroelectric perovskite investigated by PFM. (a) Electric field aligned parallel to the spontaneous polarization leads to a lifting of the cantilever due to the d_{33} effect (out-of-plane signal). It causes additional lateral contraction of the ferroelectric via the d_{31} piezoelectric coefficient. (b) The antiparallel alignment of the electric field and the spontaneous polarization leads to a vertical contraction and a horizontal expansion of the ferroelectric. (c, d) Electric field applied orthogonal to the polarization results in a shear movement due to the d_{15} coefficient. This movement causes a torsional deformation of the cantilever forcing the laser spot to move horizontally (in-plane signal). (e) A grain polarized in the x-z-plane will contribute to the in-plane as well as to the out-of-plane signal.

by the transverse piezoelectric coefficient d_{31}.[3] The electrostrictive effect does not depend on the polarization direction: it causes only a constant background but does not influence the detection of the polarization orientation. As was already mentioned, the electrostrictive term is typically much smaller that the piezoelectric one and vanishes when no dc field is applied to the sample. Thus it can be concluded that the VPFM measurements enable one to deduce the polarization direction of the individual domain or grain with a nanometer resolution.

Similarly, the direction of the polarization for the in-plane polarized ferroelectric grain can be deduced via a relevant (shear) piezoelectric coefficient (Figure 4(c), (d)). In this case, the applied electric field causes a shear deformation of the grain, which is transferred via the friction forces to the torsional movement of the cantilever. These measurements will be further denoted as *in-plane* (or lateral PFM, or LPFM) measurements.

Figure 4(e) shows a grain with the polarization aligned in the x-z-plane. In this case, the applied field will cause both vertical as well as torsional movement of the cantilever. A strong in-plane and a weak out-of-plane signal will indicate that the polarization has only a small component of the polarization pointing into the z-direction (P_z) and is mainly aligned in the film plane. Due to the cantilever asymmetry, polarization in the y-direction P_y can only be recorded by physically rotating the sample by 90° along the z-axis and repeating the in-plane measurement. By acquiring all three components of the piezoresponse signal, it is possible to perform at least semiquantitative reconstruction of polarization orientation. However, precise orientation of polarization can be calculated only if all the components of the piezoelectric tensor are known. The first attempt to relate the amplitude of the piezoresponse signal to the orientation of the ferroelectric polarization has been undertaken by Harnagea et al. [33], and detailed formalism has been later developed by Kalinin et al. [34].

2.2.3 Calibration of the PFM

The PFM can be calibrated by using a commercially available *x-cut*[4] quartz. The textbook value of its piezoelectric coefficient $d_{11} = -2.3$ pm/V means that the application of a bias of 1 V to the major faces of an *x-cut* quarts leads to a contraction of 2.3 pm along the x-direction. To measure the converse piezoelectric effect, gold electrodes are deposited onto both sides of the crystal. Afterwards, the

---

[3] This effect could lead to a torsional movement of the cantilever in the case when extended top electrodes are used and the cantilever is not positioned in the center of the grain. This would make the analysis of the three-dimensional polarization distribution more complicated or even impossible, whereas using the tip as a movable top electrode will only cause a symmetrical contraction of the ferroelectric around the tip and consequently causing no torsional movement. The in-plane and out-of-plane measurements will thus exhibit almost no crosstalk.

[4] When a plane parallel plate is cut from a single crystal, the term *cut* is used to indicate the direction of the normal to the major faces. Thus in the case of *x-cut* the normal to its major faces is parallel to the x-axis of the crystal.

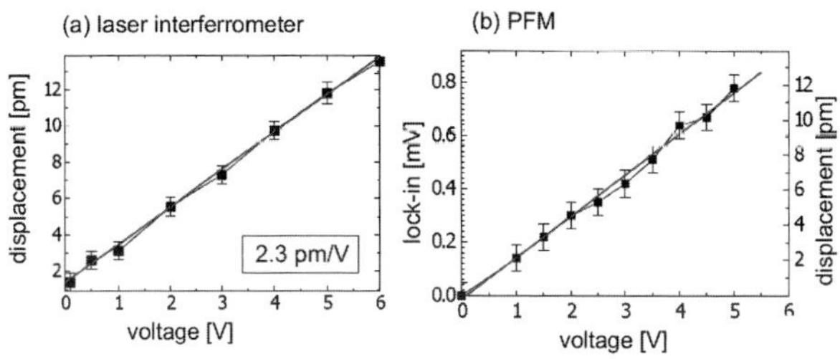

FIGURE 5. Calibration of PFM carried out with the aid of an *x-cut* quartz plate. (a) Measurement of the quartz plate in a double-beam laser interferometer to check the value of the piezoelectric coefficient d_{11}. (b) Measurement of the lock-in output in PFM. The PFM signal is interrelated to the known dilation of the quartz sample.

quality of the crystal is checked by performing converse piezoelectric measurements by using a sensitive displacement meter, for example, a double-beam laser interferometer (Figure 5(a)). Then the crystal is placed on the AFM stage and the lock-in output of the PFM is measured as a function of the applied voltage (at a fixed frequency) calibrating it to a known dilation of the quartz (Figure 5(b)). For accurate quantitative measurements, the PFM calibration is repeated for each new mounted cantilever, as the lock-in output is sensitive to the laser alignment and the exact cantilever parameters.

3 Nanoscale Domain Switching and PFM Spectroscopy

PFM provides a unique opportunity for direct studies of the domain structure evolution under an external electric field, which cannot be matched by previously available techniques. A conductive probing tip can be used not only for domain visualization but also for modification of the initial domain structure. Application of a small dc voltage between the tip and bottom electrode generates an electric field of several hundred kilovolts per centimeter, which is higher than the coercive voltage of most ferroelectrics, thus inducing local polarization reversal. An application of the positive or negative dc bias to the tip can induce 180° polarization switching, orienting polarization upward or downward. Thus, created domains can be imaged using PFM, which, therefore, provides both "storage" and "read-out" capabilities. Given that the width of domain walls in ferroelectrics is typically very small (on the order of 1–3 unit cells), this ferroelectric recording [35] potentially allows extremely high data storage density [36,37], well above that achievable by conventional magnetic recording. In fact, 40-nm bit size had been demonstrated as early as 2002 by Tybell et al. [36], while a similar level was achieved using thermomagnetic recording by Wickramasinghe group only in 2004 [38]. By the end of 2005, Cho et al. [39] had demonstrated 8-nm bit size corresponding to 10 TBit/inch^2 storage density. These considerations have resulted in significant interest to ferroelectric data storage from major industrial corporations, and the review of the current state of the field is provided by Hong and Park in Chapter IV.6.

Other important applications, such as domain engineering for electrooptical devices, have also emerged [40]. A further extension of this approach is based on the fact that surface chemical reactivity of ferroelectric materials in, e.g., acid dissolution processes is strongly polarization-dependent [41]. Recently, similar behavior was observed for metal photodeposition [42]. The combination of polarization-dependent chemical reactivity and domain patterning gives rise to ferroelectric lithography—a novel approach for nanoscale structure fabrication [43–45]. See also Chapter IV.4 by Bonnell, Lei, and Li in this book.

3.1 Thermodynamics of Domain Switching

Applications of PFM for high-density data storage and ferroelectric lithography necessitate studies of both thermodynamics and kinetics of the switching process. The first analysis of the thermodynamics of domain switching in PFM capturing the essential features of this process including finite nucleation bias was given by Abplanalp [20]. The closed-form solution (Molotskii model) for domain switching in the point charge approximation was given by Molotskii et al. [46,47]. The thermodynamics of domain switching using rigorously derived [48] electroelastic fields was given by Kalinin et al. [49]. Subsequently, the analysis of this problem was further elaborated by Morozovska and Eliseev [50] and by Emelyanov [51]. In all cases, it was realized that the spatial extent of the field produced by the tip is finite; therefore, the switching will be limited to a small volume of material. For point-charge type models with charge located on or below the ferroelectric surface, the field in the vicinity of the tip is infinite and domain nucleation is induced at arbitrarily small biases [47,52]. In the models including finite tip-surface separation or based on realistic field geometries [47,49–51,53], the finite nucleation bias is predicted (Figure 6). Below, we analyze the thermodynamics of domain switching in a more detail.

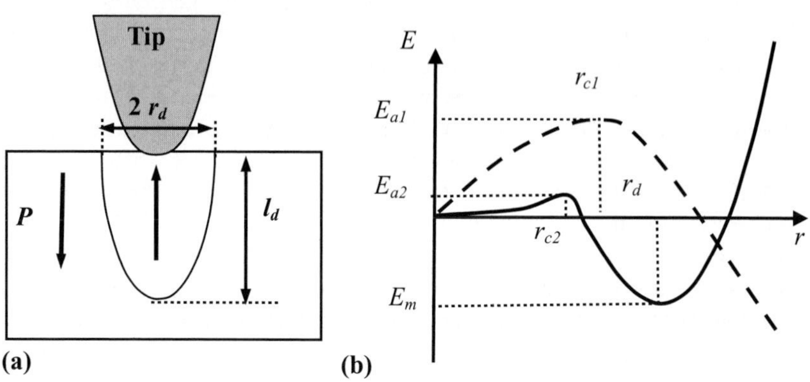

FIGURE 6. (a) Domain geometry during tip-induced switching. (b) Free energy as a function of the lateral domain size. Dashed line, in a uniform electric field; solid line, in a tip-induced electric field.

The driving force for the 180° polarization switching process in ferroelectrics is due to change in the bulk free energy density [20,25]:

$$\Delta g_{bulk} = -\Delta P_i E_i - \Delta d_{i\mu} E_i X_\mu, \tag{7}$$

where P_i, E_i, X_μ, and $d_{i\mu}$, are the components of the polarization, electric field, stress and piezoelectric constants tensor, respectively, $i = 1, 2, 3$, and $\mu = 1, \ldots, 6$. The first and second terms in Eq. (7) describe ferroelectric and ferroelectroelastic switching, respectively. For materials such as $LiNbO_3$ and lead zirconate-titanate (PZT), the signs of the corresponding free energy terms are opposite and the polarities of the domains formed by ferroelectric and ferroelectroelastic switching are opposite, thus providing an approach to distinguish these switching mechanisms.

The free energy of the nucleating domain is

$$\Delta G = \Delta G_{bulk} + \Delta G_{wall} + \Delta G_{dep}, \tag{8}$$

where the first term is the change in bulk free energy, $\Delta G_{bulk} = \int \Delta g_{bulk} dV$, the second term is the domain wall energy, and the third term is the depolarization field energy. In the Landauer model of switching, the domain shape is approximated as a half-ellipsoid with the small and large axes equal to r_d and l_d, correspondingly (Figure 6(a)). The domain wall contribution to the free energy in this geometry is $\Delta G_{wall} = b r_d l_d$, where $b = \sigma_{wall}\pi^2/2$ and σ_{wall} is the direction-independent domain wall energy. The depolarization energy contribution is $\Delta G_{dep} = c r_d^4 / l_d$, where

$$c = \frac{4\pi P_s^2}{3\varepsilon_{11}} \left[\ln \left(\frac{2l_d}{r_d} \sqrt{\frac{\varepsilon_{11}}{\varepsilon_{33}}} \right) - 1 \right] \tag{9}$$

has only a weak dependence on the domain geometry [54].

In the uniform field case, the free energy surface (as a function of l_d, r_d) has a saddle point character and the domain grows indefinitely once the critical size corresponding to activation barrier for nucleation E_a is reached [55]. The critical domain size and activation energy for nucleation can be obtained from minimization of Eq. (9) as $r_c = 0.83b/a$, $l_c = 1.86bc^{1/2}a^{-3/2}$, and $E_a = 0.518b^3c^{1/2}a^{-5/2}$, where $a = 4\pi P_s E/3$. For typical ferroelectric materials such as $BaTiO_3$ ($\sigma = 7$ mJ/m^2, $P_s = 0.26$ C/m^2, $\varepsilon_{11} = 2000$, $\varepsilon_{33} = 120$ [41]) in the uniform field $E = 10^5$ V/m the corresponding values are $E_a = 2.4 \cdot 10^5$ eV and $l_c = 16.4$ μm, $r_c = 0.264$ μm. Thus, for relatively weak fields corresponding to experimental coercive fields, homogeneous domain nucleation is impossible, which explains why in typical ferroelectric materials domain nucleation occurs on the surface or at interface defects.

The opposite is true for the tip-induced switching, when the small radius of curvature of the tip results in large (10^6–10^9 V/m) electric fields localized at the tip apex. The corresponding domain free energy can be determined from electroelastic field distribution generated by the PFM tip as

$$\Delta G_{bulk} = \int_V \Delta g_{bulk}(\vec{r})dV = 2\pi \int_0^{l_d} dz \int_0^{r(z)} \Delta g_{bulk}(r, z)rdr, \quad (10)$$

where $r(z) = r_d\sqrt{1 - z^2/l^2d}$. The analysis of the switching process can be greatly simplified in the point charge model [46,47] applicable if domain sizes $l_d, r_d \gg R, a$, where R is the tip radius and a is the contact radius and provided that the singularity at the origin is weak enough to ensure convergence of the integral in Eq. (10). For ferroelectric switching induced by a point charge q_s located on the surface, the integral in Eq. (10) can be taken analytically and $\Delta G_{bulk} = dr_d l_d/(l_d + \gamma r_d)$, where $d = 2P_s q_s/(\varepsilon_0 + \sqrt{\varepsilon_{11}\varepsilon_{33}})$ and $\gamma = \sqrt{\varepsilon_{33}/\varepsilon_{11}}$. In this case, domain size and energy are $r_e = 0.342d^{2/3}(bc)^{-1/3}$, $l_e = 0.2d/b$, and $E_m = -0.205d^{5/3}(bc)^{-1/3}$ [46, 47]. The activation energy for domain nucleation in this approximation is zero, due to the infinite field at the origin. This analysis predicts that domain shape in the switching process follows the invariant relation $r_e^3/l_e^2 = b/c$.

The applicability of the point-charge approximation to the thermodynamics of domain switching on the large length scales is limited by the contribution of the electrostatic fields produced by the conical part of the tip, which decay much slower than that produced by the point charge. At smaller length scales comparable to the tip radius of curvature, the thermodynamics of switching process requires an exact electroelastic field structure to be taken into account. The evolution of a free energy surface for macroscopic switching in a uniform field (10^5 V/m) and point charge field are shown in Figure 7(a), (b). Free energy surfaces for several point-charge magnitudes and $h = 10\ nm$ are shown in Figure 7(c–e). For $q_a = 100$ e$^-$, the free energy is positive for all l_d, r_d and a domain does not form. For $q_a = 200$ e$^-$ the free energy surface develops a kink. Finally, for $q_a = 400$ e$^-$ the free energy minimum corresponding to a stable domain and a saddle point corresponding to the activation energy for nucleation are clearly seen. Finally, the free energy surface for a tip radius $R = 50$ nm and bias $V = 5$ V in the weak indentation regime is shown in Figure 7(f). Note that domain nucleation requires certain threshold bias of the ot order of 0.1–1 V, corresponding to nonzero activation energy for nucleation (of order of $\sim kT$) [49].

In the last several years, a number of reports have become available on high-order ferroic switching in PFM [5,25]. Due to the rapid decay of corresponding electroelastic fields, the use of Eq. (13) in point-charge/force approximation results in singularity in the contact area, necessitating exact electroelastic field structure to be taken into account. Using rigorous electroelastic solutions, it was shown [48] that for higher-order ferroic switching (e.g., ferroelectroelastic), the domain size is limited by the tip-sample contact area, thus allowing precise control of domain size.

3.2 Time-Dependent Measurements in PFM

Applications of ferroelectric domain patterning for data storage, electrooptical devices, and ferroelectric lithography necessitate fundamental studies of the domain

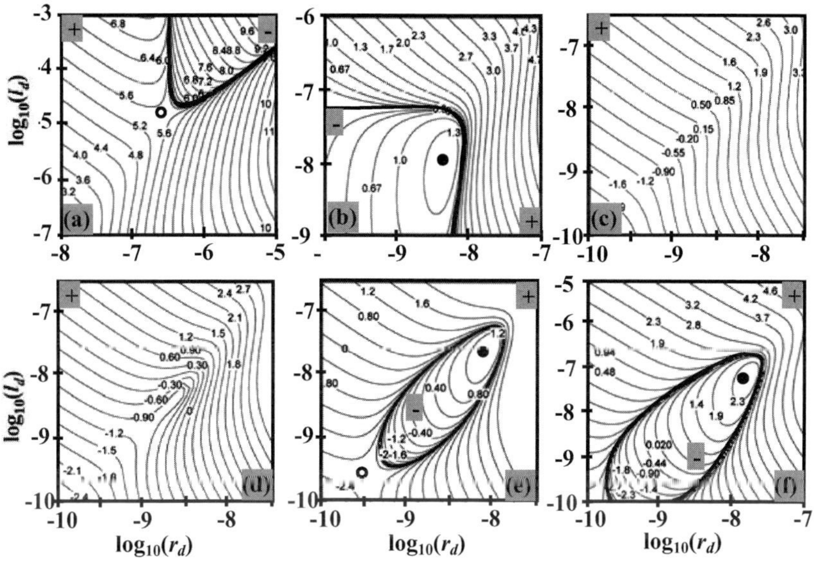

FIGURE 7. The free energy surface for domain switching in (a) uniform electric field and (b) field produced by a point charge on the surface and (c–e) above the surface. Relevant parameters are given in the text. (f) The free energy surface for a realistic tip shape. Plotted is the logarithm of the absolute value of the energy in eV. Solid lines separate regions of opposite signs, indicated by the plus and minus signs. Saddle points (o) and local minima (•) are shown. Reprinted with permission from [49]. Copyright 2005, American Institute of Physics.

switching process, including thermodynamics and kinetics of domain nucleation, growth, and relaxation. Although the high spatial resolution of PFM provides a unique opportunity for investigation of domain dynamics in thin films, its poor time resolution, which is determined by the time required for image acquisition (at the scan rate of 2 Hz and image resolution of 256 × 256 pixels, it takes about 2 min to acquire an image), makes the *in situ* measurements of domain dynamics during fast switching processes difficult. While PFM can be readily used to investigate slow polarization relaxation processes with characteristic times on the order of minutes and above, it is a great challenge to deduce the mechanism of ferroelectric domain transformation when polarization reversal occurs in a matter of microseconds and faster.

This problem can be partly circumvented by studying domain dynamics in a quasi-static regime using a so-called *step-by-step switching* approach developed and applied to PFM (see, e.g., [4]). In this approach, partial reversal of polarization is generated by applying a voltage pulse shorter than the time required for full switching but with the pulse amplitude fixed above threshold voltage. By applying a sequence of such short pulses of incrementally increasing duration with PFM domain imaging after each pulse, a consistent picture of domain dynamics can

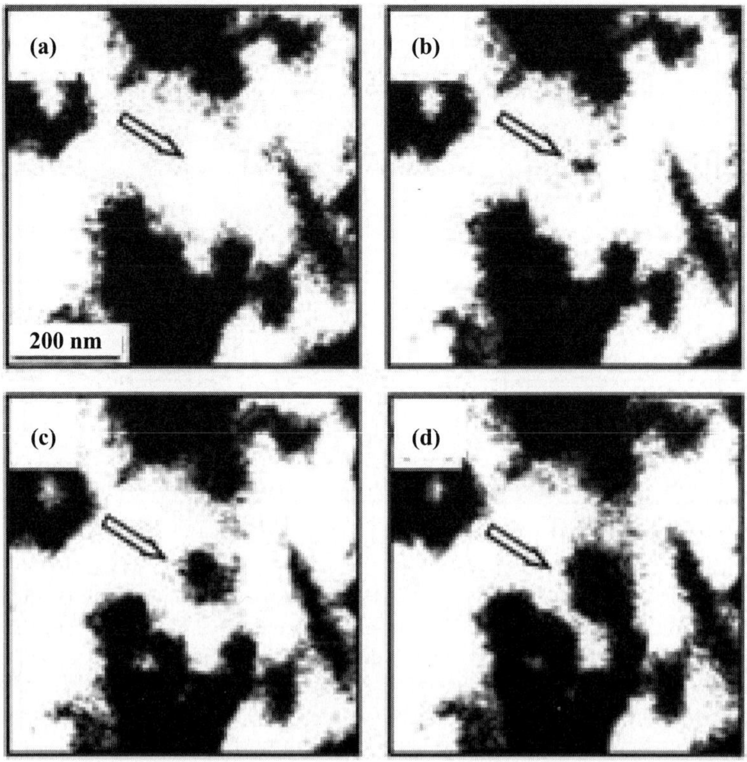

FIGURE 8. Sequence of PFM images of PZT film showing a sidewise growth of a domain in the center of the image. The growth is induced by application of 5 V voltage pulses of increasing duration to the same spot by a fixed probing tip. (a) Original domain structure; (b–d) domain images after application of voltage pulses. Pulse duration: (b) 0.25 s, (c) 0.5 s, (d) 1 s. Reprinted with permission from [2]. Copyright 2004, American Scientific Publishers.

be obtained. This approach can provide information on the domain wall velocity, its spatial anisotropy, and its field dependence. To avoid data misinterpretation due to spontaneous back-switching between the pulses, stability of the produced intermediate patterns should be checked by acquiring domain images at different time intervals after single pulse application.

An example of this approach is illustrated in Figure 8, which shows a sequence of piezoresponse images of a growing domain in $Pb(Zr,Ti)O_3$ (PZT) films. Based on the contrast of the growing domain, it is assumed that the domain fully extends through the film thickness after the first pulse and further growth occurs via the sidewise expansion only. The diameter of the smallest stable domain produced by the first voltage pulse of 5 V varied from 20 nm to 40 nm. Since the electric field generated by a probing tip quickly decreases with distance, the rate of domain growth, after fast initial expansion, slows down until the domain stops growing. To

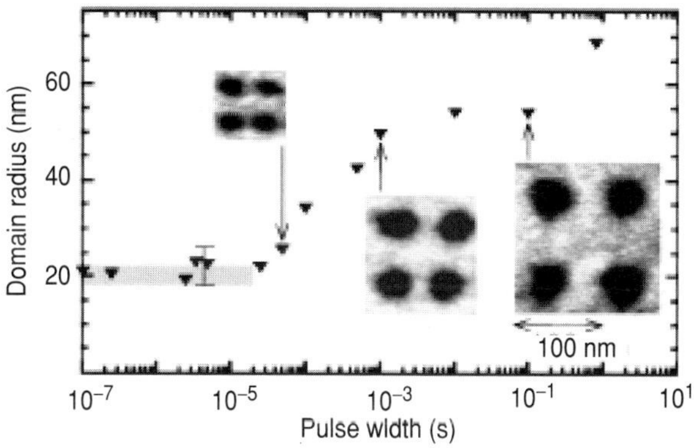

FIGURE 9. The dependence of the domain size on the switching time in $Pb(Zr_{0.2}Ti_{0.8})O_3$ films of 29 nm thick. Insets show the corresponding images of scanned domains. Reproduced with permission from [36]. Copyright 2002, American Physical Society.

describe the sidewise expansion of the domain, it is necessary to take into account the field dependence of the domain wall velocity and the spatial distribution of the electric field generated by the probing tip.

In the step-by-step approach, the time resolution is determined by the pulse length increment, which can be in the nanosecond range. This is particularly important from the viewpoint of studying domain switching dynamics in thin-film ferroelectric capacitors used in memory devices where the switching time is typically below 100 ns. In this case, the time resolution will also depend on the rise time of the input pulse generated by the voltage source and on the size of the capacitor, i.e. its *RC*-time constant [56,57].

3.2.1 Domain Growth During Tip-Induced Switching

The measurement of the size of switched domains versus the magnitude of applied voltage and time can give us insight into the mechanism of polarization switching in ferroelectric films. Although the PFM measurements are quite slow, it is possible to distinguish between different mechanisms that control the propagation of the domain wall in such a complex object as ferroelectric thin film with defects. Tybell et al. [36] by studying the time dependence of the size of switched domains by PFM tip could assign the motion of the domain wall to the creep mechanism. Figure 9 presents some of their results along with the images of switched domains in $Pb(Zr_{0.2}Ti_{0.8})O_3$ epitaxial films prepared by magnetron sputtering on $SrTiO_3$:Nb substrates that simultaneously served as a bottom electrode.

These measurements allowed calculations of the domain wall velocity as a function of the electric field. Within the limitation of the simple point-charge model, the results showed that even in a high-quality thin ferroelectric film the domain wall motion is limited by the creep mechanism with the critical exponent close to unity. In the subsequent work of the same group [58] these results were confirmed by the

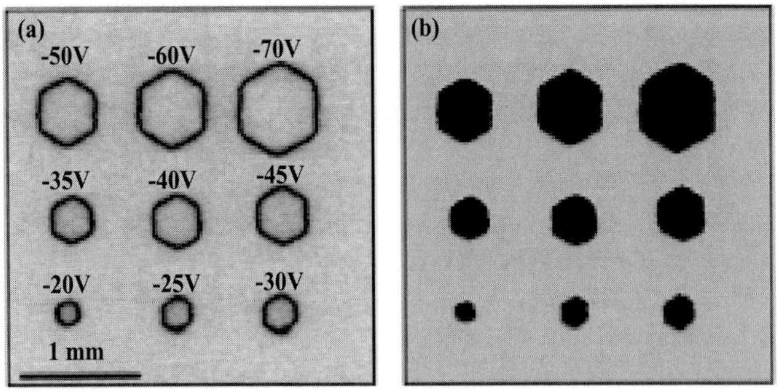

FIGURE 10. PFM (a) amplitude and (b) phase images of ferroelectric domains fabricated by 10-ms voltage pulses of different amplitudes in $LiNbO_3$ single crystals. Reprinted with permission from [59]. Copyright 2005, American Institute of Physics.

measurements of roughness of the domain walls, however, the critical exponent was found to be less than unity. Pinning defects limited domain wall propagation could not be identified in this work, however, the intrinsic pinning mechanism was ruled out. The results are useful for the development of high-density data storage devices.

In another work, Rodriguez et al. [59] studied domain switching kinetics in stoichiometric $LiNbO_3$ single crystals by applying the voltage pulses of different amplitudes and durations over 5 orders of magnitude. The minimum size of the created domains was much larger than in the case of thin PZT films, and the mechanism of domain wall motion in this case was found to be different. Figure 10 demonstrates the written domains patterns obtained with poling under different voltages.

The domain wall velocity was calculated as a function of the domain radius and local electric field using distribution function defined by a charge sphere model. It was shown that at sufficiently large domain sizes, the motion of the domain wall is the electric-field activated process, while for the small sizes it is not activated since the applied electric field is much greater than the coercive field. The domain wall velocity v was satisfactorily fitted to a simple equation $V(r) \propto \exp[-\alpha/E(r + r_0)]$, thus confirming the time-dependent switching results. Here α is the activation field and r_0 is the fitting parameter. It was interpreted as the domain radius at which an activated process begins. As expected, r_0 increases with increasing applied voltage reaching 110 nm under 100 V. Along with the width of the domain wall, high values of r_0 may limit the use of $LiNbO_3$ single crystals for dense data storage.

3.2.2 Domain Dynamics in Capacitor Configuration

The PFM technique also allows investigation of the domain dynamics in a "normal," i.e., plane capacitor situation, where the electric field distribution is homogeneous across the thickness of the ferroelectric layer. In this case, the PFM

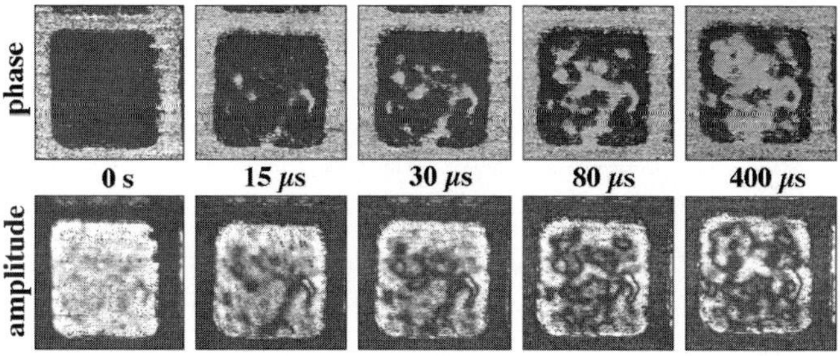

FIGURE 11. PFM phase and amplitude images of instantaneous domain configurations developing in the $3 \times 3\mu m^2$ PZT capacitor at different stages of polarization reversal under a 1.1 V bias. Reprinted with permission from [60]. Copyright 2004, American Institute of Physics.

results can be directly compared with the switching data obtained by conventional measurements. Direct studies of domain dynamics during polarization reversal in 180-nm-thick PZT capacitors have been performed by Gruverman et al. [60] (Figure 11). These results have been used to elucidate the switching mechanism in PZT capacitors. It has been shown that at low voltages a Kolmogorov–Avrami–Ishibashi mechanism [61] is applicable, while at longer times the broad distribution of the relaxation times should be taken into account. Thus, the PFM method can be used for direct experimental verification of the existing theoretical models of switching in a broad range of the switching times.

3.2.3 Domain relaxation

Along with the study of the mechanisms of the domain nucleation and growth, the PFM method is also applicable to the investigation of domain relaxation, i.e., the process always accompanying depolarization. The stability of the polarization state created by the external voltage is a prerequisite for the realization of SPM-based ferroelectric memories, since it directly controls retention time of the stored bits. In addition, polarization relaxation is important for the understanding of aging phenomena in piezoelectric ceramics. The advantage of PFM as compared to macroscopic measurements of depolarization is that it allows understanding this effect locally rather than by the averaging over many backswitching events. Unfortunately, the time resolution of PFM measurements is significantly smaller than in macroscopic measurements preventing studying initial stages of polarization loss.

Several studies have been recently performed on the relaxation phenomena by PFM. Despite these efforts, there is still a lack of reliable data on most of ferroelectric materials useful in device applications, including the time dependence of the retention loss in ferroelectric films. While some authors reported a log-linear dependence of the polarization relaxation in PZT films [62], other groups deduced a stretched exponential dependence [63] in the same material. Moreover, polarization relaxation has been found to differ in epitaxial films, where a spontaneous

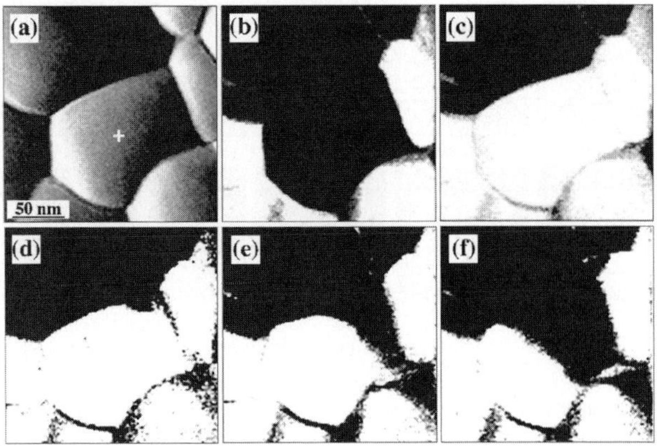

FIGURE 12. Retention loss in PZT(20/80) film on LSCO/TiN/Si measured after poling with 6 V applied for 0.2 s. (a) Topographic image with cross indicating poling point, (b) as-grown domain structure, (c) domains immediately after poling, and (d–f) evolution of domain structure after 4, 90, and 140 min after poling. Reprinted with permission from [4]. Copyright 1998, Annual Reviews.

nucleation of 180° domain walls on existing twin boundaries has been found to modify the relaxation process. [9] In this case, it was possible to use a modified Avrami–Kolmogorov kinetics [61] for the description of depolarization process in ferroelectric thin films. Figure 12 illustrates the polarization loss in PZT films studied by PFM [4]. A voltage pulse of 6 V was applied for 0.2 s in the center of the middle grain and dynamics of domain relaxation was observed during subsequent scanning. As can be seen the domain relaxation proceeds mainly through the sidewise motion of the domain walls.

It has been found that the domain walls measured in dynamics appear much thicker than in static conditions. This is a natural result caused by the intrinsic disadvantage of PFM: slow scan rate. From a quantitative analysis of domain images, it follows that the domain wall slows down with time, giving rise to a familiar logarithmic dependence of the fraction of reversed domains: $P(t) = P_0 - m \log(t/t_o)$. As usual, the log time dependence suggests a broad distribution of relaxation times in polycrystalline ferroelectric films. Note that the shape of the shrinking domain becomes irregular due to the spatial variation of domain wall velocity. PFM retention measurements in SBT films [64] were found to be highly asymmetric, i.e., the application of positive voltage pulses gave much better stability of domain structure as compared to negative ones. Such asymmetric retention (called generally an imprint) was found also in macroscopic measurements. On the other hand, the same SBT films grown on RuO_2 electrodes showed no sign of polarization retention, irrespectively of the sign of applied voltage. Based on these experiments. Gruverman and Tanaka [64] developed a qualitative model of polarization retention in SBT-type materials.

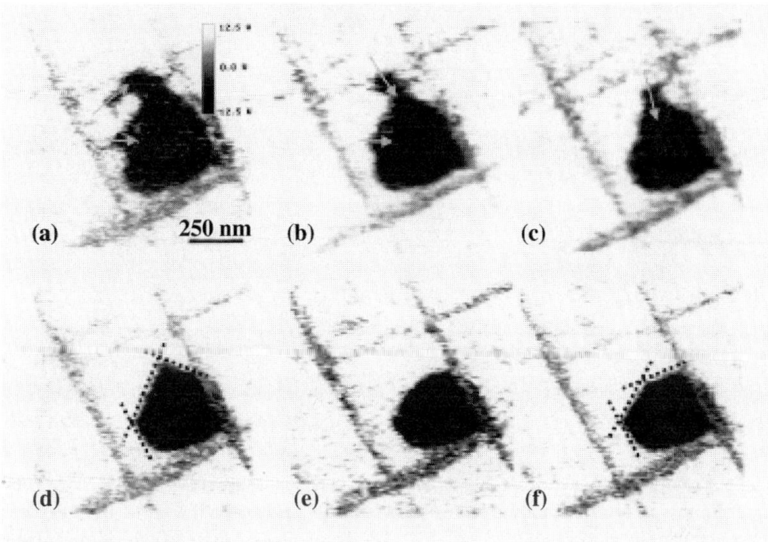

FIGURE 13. (a–f) Relaxation of the domains in epitaxial PZT films after poling during scanning. Arrows indicate the direction of the velocity of moving wall in areas with high curvature. Faceting is shown in the final stage of domain relaxation (dotted lines). Reprinted with permission from [65]. Copyright 2001, American Physical Society.

In another experiment [65], the authors studied the polarization relaxation via PFM in epitaxial $PbZr_{0.2}Ti_{0.8}O_3$ films on LSCO electrodes. These films were found to have a large number of 90° domains (twins) that strongly contributed to domain wall relaxation. Figure 13 gives an example of local depolarization process in such films.

The essential feature of such relaxation observed after poling under scanning with a dc bias applied to the tip is that the back-switching mainly starts near the existing 90° domain wall and propagates with a velocity that depends on the curvature of domain wall. In the final stage of domain relaxation, faceting of the domain wall is seen that is a result of slowing down the relaxation along certain crystallographic directions. Investigation of the relaxation of such domain walls opens up new horizons for the understanding domain-wall propagation in such complicated objects as anisotropic ferroelectrics with defects.

3.3 Local Piezoelectric Hysteresis Loops

Along with the sequental polarization switching performed with a series of voltage pulses applied to the tip followed by consequent imaging, PFM can be used in a spectroscopy mode when the measurements are done on a fixed tip position (local d_{33} acquisition) under a dc voltage swept in the cyclic manner. This dependence of the local piezoelectric vibration on applied bias is referred to as a

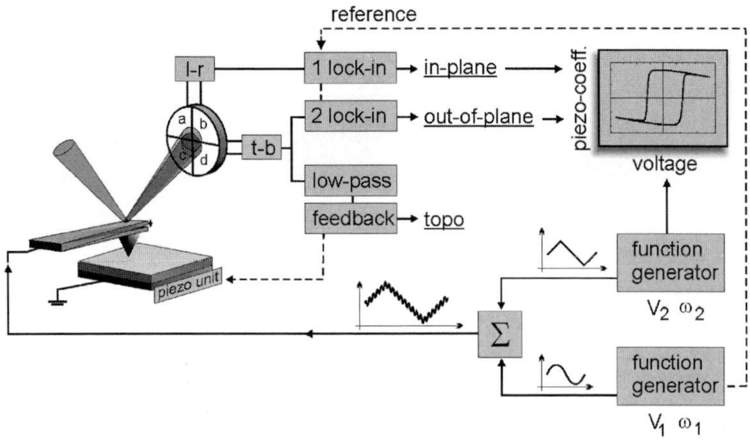

FIGURE 14. Setup to measure the piezoelectric hysteresis loop with the PFM. Two function generators and a summation amplifier are used to superimpose the PFM imaging voltage $V_1(\omega_1)$ to the slow triangle voltage $V_2(\omega_2)$. The in-plane and out-of-plane signals from the PFM are plotted in an oscilloscope versus the switching voltage V_2 to obtain the piezoelectric hysteresis loops.

piezoelectric hysteresis loop. If performed on a macroscopic scale, this loop corresponds to a weak-field piezoelectric coefficient tuned by a continuously varying bias field. According to a linearized electrostriction equation, the piezoelectric coefficient can be often expressed as $d_{33} = 2Q\varepsilon_{33}P_3$, where Q is the longitudinal electrostriction coefficient and ε_{33} and P_3 are the corresponding dielectric constant and spontaneous polarization values. Therefore, the d_{33} variation reflects polarization switching with corresponding tuning of the dielectric permittivity ε(E) and polarization P(E), which both affect piezoelectric coefficient [66]. Piezoelectric hysteresis loops are indeed very useful for the identification of the switching mechanism and ferroelectric degradation both in thin film [67] and bulk ferroelectrics [68]. It should be stressed that at the nanoscale level the shape of piezoelectric loops is different [69], reflecting different physical mechanisms involved. Measurements of local hysteresis loops are of great importance in inhomogeneous or polycrystalline ferroelectrics because they are able to quantify polarization switching on a scale significantly smaller than the grain size or inhomogeneity variation (typically few tens of nm).

Figure 14 depicts a typical setup to measure the piezoelectric hysteresis loop by PFM. In addition to the conventional PFM configuration (Figure 2), a second function generator, a summing amplifier, and an oscilloscope are needed. The first function generator provides the PFM imaging voltage $V_1(\omega_1)$, and the second function generator supplies the switching voltage $V_2(\omega_2)$[5]. Adding these two signals

[5] Here a saw voltage variation is used to initiate switching, as this is the standard waveform for macroscopic hysteresis measurements.

results in a slowly varying switching voltage ($\omega_2 \gg$ mHz) with the superimposed PFM imaging voltage ($\omega_1 >$ 10–20 kHz). Feeding the lock-in output of the PFM into an oscilloscope (y-axis) and using the switching voltage V_2 as the x-axis input, the piezoelectric hysteresis loop can be recorded. Acquiring both signals, the in-plane and the out-of-plane, enables one to obtain two complementary hysteresis loops describing the evolution of the piezoresponse signal upon switching.

Macroscopically, the switching occurs via the nucleation and growth of a large number of reverse domains in the situation where the applied electric field is uniform. Therefore, the d_{33} hysteresis reflects the switching averaged over entire sample under the electrode. In PFM experimental conditions, as discussed in section 3.1, the electric field is strongly localized and inhomogeneous, therefore, the polarization switching starts with the nucleation of a single domain just under the tip [70]. When the applied dc bias increases, this newly formed domain elongates to the bottom electrode, simultaneously expanding in lateral dimensions until reaching an equilibrium size, which depends on the value of maximim applied voltage.

The shape of the local piezohysteresis measured by PFM can be understood by scanning the area adjucent to the tip in-between the voltage increments, provided no significant relaxation occurs. In addition, useful information on the propagation of domain walls in ferroelectrics can be obtained. An example of the vizualization of the piezoelectric hysteresis in polycrystalline PZT films prepared by sol-gel is shown in Figure 15 [71]. The measurements were done by the setup shown in Figure 14, but the switching voltage was periodically interrupted for scanning. The images are shown near the corresponding points of the hysteresis loop.

It is seen that the sudden contrast change (arrow at V^*) corresponds to the appearance of the stable inverse domain. Thus nucleation voltage $V^* \approx 11$ V can be considered as an analog of the coercive voltage in local measurements rather than the voltage where $d_{33} = 0$. A rapid increase of d_{33} with increasing V_{dc} is explained by the forward and lateral growth of the nucleated domain, progressively contributing to the PFM signal. When the size of the inverse domain is much greater than the penetration depth of the weak ac field, the tip senses only a fully polarized area with aligned polarization and d_{33} is well saturated. Figure 15 also illustrates the propagation of domain wall in polycrystalline ferroelectrics where local inhomogeneities and stresses make the domain boundary strongly irregular.

It is instructive to compare local piezoelectric hysteresis with that obtained in macroscopic configuration (homogeneous electric field). Figure 16 represents macroscopic d_{33} loop (measured by laser interferometry as a deflection of metal electrode of macroscopic dimensions under the same combination of dc and ac voltages) in comparison with the local one for $PbZr_{0.45}$ $Ti_{0.55}O_3$ films prepared by sol-gel method [69]. Local d_{33} is given in arbitrary units as the calibration procedure (described in section 2.2.3) has not been performed.

This result demonstrates that both the shape of the loop and values of the coercive voltage are strongly affected by the measurement conditions. A notable decrease of the piezoresponse after saturation in the case of local measurements is caused by much greater depolarizing field in PFM conditions due to the influence of low-

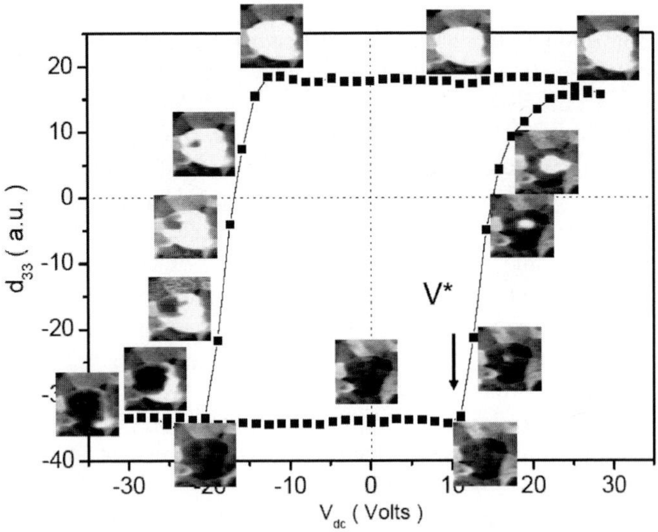

FIGURE 15. PFM hysteresis loop of thick PZT films prepared by sol-gel. The insets (scan size 1 μm) show the static domain structures obtained by scanning with the ac voltage of the amplitude of 1 V in the corresponding points of the hysteresis (adapted from [71]).

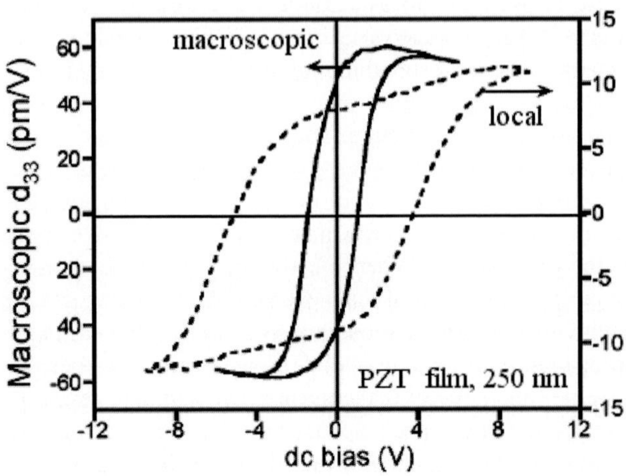

FIGURE 16. Comparison of macroscopic (interferometry) and local piezoelectric hysteresis loops measured by PFM (adapted from [69]).

dielectric-constant layer on the surface. On the other hand, the "coercive voltage" (zeroing of d_{33}) is much smaller in macroscopic conditions illustrating its different physical meaning.

The crystallographic orientation dependence of hysteresis loop was studied by Kalinin et al. [70]. Comparison of the film surface topography (Figure 17(a))

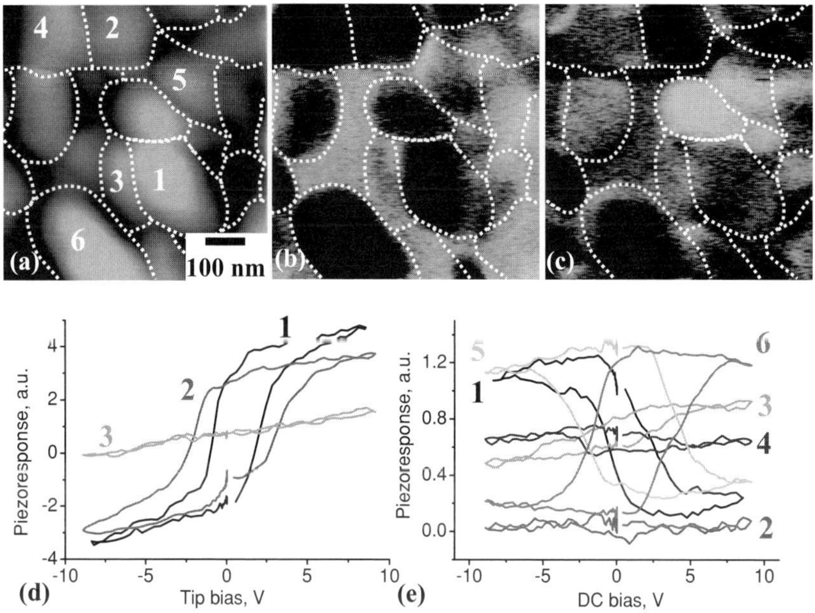

FIGURE 17. Sample surface topography (a), vertical (b), and lateral (c) PFM signal of SBT thin film. Vertical (a) and lateral (b) piezoelectric hysteresis loops for selected grains. For clarity, vertical loops are shown only for grains 1, 2, and 3. Reprinted with permission from [70]). Copyright 2005, American Institute of Physics.

with vertical (Figure 17(b)) and lateral (Figure 17(c)) PFM images (VPFM and LPFM, respectively) suggests that most grains exist in a single domain state. Grains that exhibit strong VPFM contrast also have distinctive hysteresis loops (Figure 17(d)), indicative of a large out-of-plane polarization component. For grain 3, the VPFM hysteresis loop is linear, indicative of a non-ferroelectric nature of the grain or purely in-plane polarization corresponding to (001) orientation [72]. Complementary LPFM hysteresis loops of individual grains are shown in Figure 2(e). Unlike the vertical PFM hysteresis loops, lateral loops can be oriented in both directions depending on the in-plane polarization orientation with respect to the cantilever axis. This results in higher variability of the LPFM loops. Comparison of the angular dependence of piezoresponse signal and coercive bias suggests that for the off-axis orientation of the grains the response decreases and coercive bias increases, as can be expected from simple energy arguments. In the limiting case of the (001)-oriented grain piezoresponse and coercive bias become zero and infinity, respectively.

The analysis of the hysteresis loop shape can be performed in the thermodynamic [70,73] and kinetic limits [74]. In the thermodynamic case, the domian size is equal to the thermodynamically limited size as analyzed in section 3.1. In the kinetic limit, the domain size is limited by the velocity of the forward and lateral growth of the domain walls.

Domain switching in the thermodynamic limit can be analyzed using the point-charge model developed by Molotskii [47]. The electric field of the tip is approximated by the point charge on the smaple surface, coinciding with strong indentaiton limit [75]. In this case, the length of domain l (perpendicular to the free surface of a ferroelectric) is proportional to the applied voltage:

$$l = d/5b, \tag{11}$$

where $b = \sigma_{wall}\pi^2/2$ and $d = d_0 V_{dc} = 2P_s C_c V_{dc}/(\varepsilon_0 + \sqrt{\varepsilon_{11}\varepsilon_{33}})$. σ_{wall} is the direction - independent domain wall energy, P_s is the polarization, and $\sqrt{\epsilon_{11}\epsilon_{33}}$ is the dielectric constant of the material. If we consider an intermediate situation (new domain does not reach bottom electrode), the measured piezoresponse will consist of two contributions having opposite sign: from the newly formed domain and from the remaining nonswitched part of the rest of the sample using a 1D model [76]. The shape of the hysteresis loop in PFM is given by $PR = d_{eff}\{V(0) - 2V(l)\}$, where $V(0)$ is the potential at the surface and $V(l)$ is the potential at the domain boundary below the tip, and d_{eff} is the effective electromechanical response of the material [48]. It follows that the measured local piezoelectric coefficient equals zero not under the coercive voltage as in macroscopic conditions but simply when the contribution from the top domain cancels out the contribution of unswitched area $[V(0) = 2V(l)]$.

In the strong indentation regime, potential decays inversely with distance as $V(l) = \beta V(0)\alpha/l$, in which β is a proportionality coefficient on the order of unity reflecting the field distribution in the material, l is the distance from the center of the contact area, and a is the contact radius. In the strong indentation limit, the capacitance of the tip-surface junction is $C_c \approx 4a\sqrt{\varepsilon_{11}\varepsilon_{33}}$, thus yielding $d_0 \approx 8aP_s$. It follows that the tail of a single branch of the hysteresis response can be simply described by Equation:

$$PR = d_{eff}\{1 - 2\beta a/l(V_{dc})\}\,. \tag{12}$$

In this thermodynamic limit, the shape of the hysteresis loop is determined solely by the domain wall energy and spontaneous polarization. Note that in the complete absence of pinning, the thermodynamic model suggests that the hysteresis will proceed along the solid line in Figure 18(a), i.e., the process is reversible and the domain should grow and shrink instantaneously in response to tip bias (Figure 18 (c)). However, if a weak pinning is included in the model, the actual domain wall boundary will lag behind that predicted by Eq. (12), and nucleation of a new domain will dominate the hysteresis loop, as illustrated in Figure 18(d). The saturation and remanent responses in this approximation are equal, and the onset of switching corresponds to the domain nucleation below the tip ($V_c^+ = V_c^- = 0$ in the point charge limit when the field below tip is infinite).

From Eq. (12), the magnitude of the coercive bias can be derived as $V^+ = V^- = 5\beta\sigma_{wall}\pi^2/8P_s$ and for $BaTiO_3$ ($\sigma = 7mJ/m^2$, $P_s = 0.26\ C/m^2$) [74] the estimated coercive bias is $V^+ = 0.166$ V (for $\beta = 1$). Note that while Eqs. (11) and (12) are

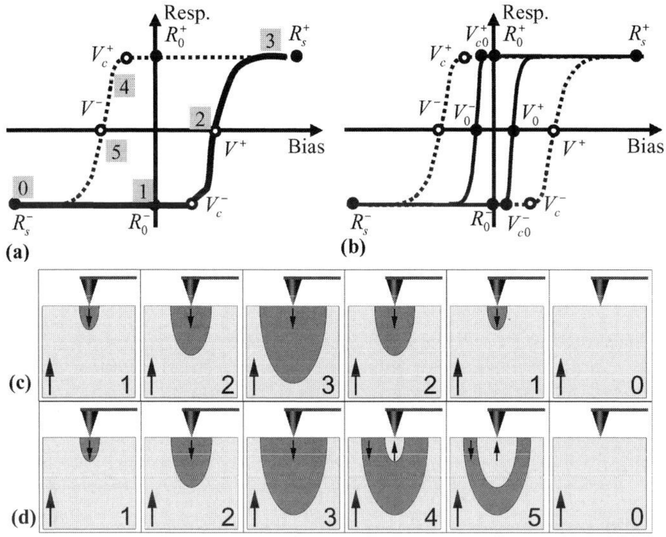

FIGURE 18. (a) Electromechanical hysteresis loop in the no pinning (solid line) and weak pinning (dotted line) limits. (b) Hysteresis loop (solid line) in the thermodynamic and kinetic (dotted line) limits. Forward and reverse coercive voltages, V^+ and V^-, nucleation voltages, V_c^+ and V_c^-, and forward and reverse saturation and remanent responses, R^+, R^-, R_s^+, and R_s^-, are shown. The effective work of switching A_s is defined as the area below the loop. The imprint bias and maximum switchable polarization are defined as $\mathrm{Im} = (V^+ + V^-)/2$ and $R_m = R_s^+ - R_s^-$, correspondingly. (c) Schematic representation of domain growth for the thermodynamic model without pinning and (d) with pinning. The numbers correspond to points on the hysteresis curve in (a). Arrows indicate domain orientation.

derived in the 1D approximation, in the general case, the effective piezoresponse is zero for some fixed domain size, l_{crit}, for which the response from the nascent domain below the tip and the surrounding material are compensating each other, giving rise to similar behavior.

However, the thermodynamic analysis ignores the effects related to the finite velocity of the domain wall required to explain time dependences in section 3.2. In this kinetically limited process, the hysteresis loop broadens compared to thermodynamic case (Figure 18(b)). Experimentally, domain switching in ferroelectric materials has been shown to follow the phenomenological dependence $r(V,t) = k\ V \ln t$, and domain wall velocity is thus $\dot{r} = k_s V/t$, where k_s is the kinetic constant related to the pinning strength in the material [59]. Note that, while the kinetic data was obtained for the lateral domain size, the time dependences for vertical and lateral sizes can be expected to be commensurate.

In the spectroscopic PFM measurements, the envelope of the tip bias is ramped linearly with time, $V = bt$. Hence, $\dot{r} = k_s b$ and $r = k_s bt = k_s V$. Therefore, the domain size is independent of the ramp velocity and is determined by the pinning strength of the material alone. Thus, we recover the functional form of Eq. (12), which provides a robust description of hysteresis loop shape. The coercive bias is now $V^+ = V^- = 2\beta a/k_s$, and is controlled by the pinning in the material.

With extension of this analysis to the experimental data, it is possible to analyze additional factors affecting the loop shape. Hysteresis loop broadening can occur due to imperfect tip-surface contact, i.e., dielectric gap, resulting in the attenuation of effective surface potential, $V(0) = \alpha V_{tip}$ [23]. This gap effect will also result in the decrease of electromechanical response, $PR = d_{eff}/\alpha$. For frequency-independent dielectric constant of the gap, $\varepsilon_{dc} = \varepsilon_\omega$, the product of coercive bias and saturated electromechanical response is independent of the gap effect, $R^+V^+ = const(\alpha)$. Thus, the *area* within the hysteresis loop (rather than loop width) provides a robust measure of the pinning strength in the material. The electrostatic and electrostrictive contributions to the PFM signal are conservative (for linear elastic material) and do not contribute to the area within the loop. Thus, the measure of energy losses during switching and pinning strength in the material are all included in the loop area. Also, from the simplified Eq. (12) we can find that the tilt of the measured hysteresis loop is roughly proportional to the applied dc voltage and spontaneous polarization P_s. The measurements of piezoelectric hysteresis in PZT films of different compositions [73] confirmed that the slope of PFM hysteresis is really a function of the P_s value, thus giving a possbility of evaluating local ferroelectric properties.

Following progress continuous in hysteresis loop measurements by PFM, a technique referred to as switching spectroscopy PFM (SS-PFM), was developed recently by the ORNL group [77]. In SS-PFM, the local hysteresis loop is acquired at each image point, yielding a 3D data array. The relevant switching characteristics such as coercive biases, imprint, switchable polarization, and effective work of switching as defined in Figure 19 are extracted and plotted as 2D maps.

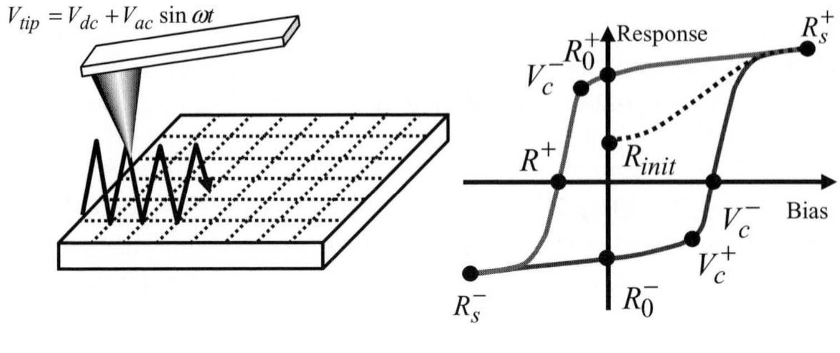

FIGURE 19. (a) In SS-PFM, the local hysteresis loop is collected at each point on NxM mesh. (b) The single-point probing waveform in SS-PFM and data acquisition sequence. Forward and reverse coercive voltages, V_0^+ and V_0^-, nucleation voltages, V_{c0}^+ and V_{c0}^-, and forward and reverse saturation and remanent responses, R_0^+, R_0^-, R_s^+, and R_s^-, are shown. The work of switching A_s is defined as the area within the loop. The imprint bias and maximum switchable polarization are defined as $\mathrm{Im} = (V_0^+ + V_0^-)/2$ and $R_m = R_s^+ - R_s^-$ correspondingly.

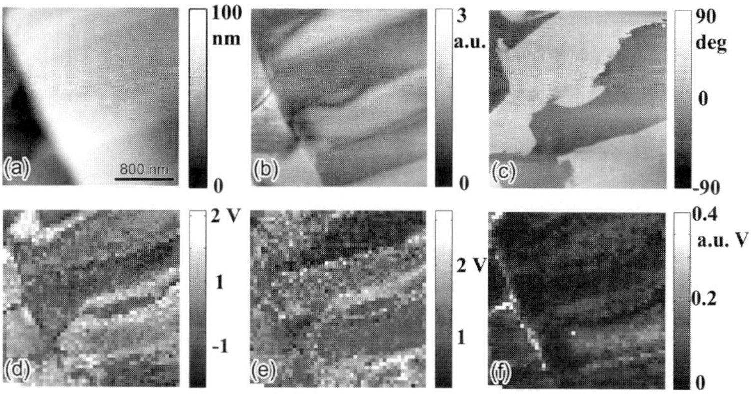

FIGURE 20. (a) Surface topography, piezoresponse (b) amplitude and (c) phase and SS-PFM (d) imprint, (e) hysteresis loop width, and (f) effective work of switching maps of polycrystalline PZT. Note the enhanced pinning in the vicinity of the grain boundary.

The operation principle of SS-PFM and several imaging examples are shown in Figure 20.

4 Applications of PFM

Since the middle of 1990s, the PFM method has been developed by several groups, giving rise to an ever-increasing number of publications (>120 publications in 2004). This method is becoming the most popular tool in the science and technology of ferroelectrics and is considered as a main instrument for getting information on ferroelectric properties at the nanoscale. Below we briefly review recent activities on nanoscale studies in various ferroelectric systems, focusing on the application of PFM for solving fundamental and technological problems.

4.1 Size Effects in Ferroelectrics

Recent advances in the science and technology of ferroelectrics have resulted in ferroelectric materials and devices with nanostructures with length scales on the order of or less than several hundred nanometers [78]. Typical examples include, but are not limited to, thin films for dynamic random access memories, microactuators and rf/microwave phase shifters, to name a few. Ferroelectrics have also become a candidate gate-dielectric material for field effect transistor applications due to their high permittivity. As device dimensions are scaled down, a size range is reached where ferroelectric materials show pronounced size effect. Size effects in ferroelectrics lead to a significant deviation from bulk properties, which is a consequence of a size-driven instability in the polar phase followed by the suppression of ferroelectricity below a critical size. Research in this field has gained great momentum in recent years because of the importance of nanophase ferroelectrics in an array of applications mentioned above. Concurrently, many phenomenological and *ab initio* calculations have been directed to this area. However, until recently, the experiments were limited to scattering techniques such as high-resolution TEM or X-ray/neutron scattering. Direct techniques, such as PFM, are ideally suited for the polarization detection in small objects such as ultrathin films and ferroelectric nanostructures because the effective tip diameter can be as small as few nanometer. Clearly, the assessment of size-governed evolution of the ferroelectric state can be done by PFM under the influence of electric filed and mechanical forces. Recent advances of using PFM for the size effect investigations are outlined below.

In general, size effects can be defined as changes of the material properties due to the change of their geometrical dimensions. Keeping all parameters of the materials constant, just shrinking its size can result in a variety of different characteristics. Following parameters were found to change in ferroelectrics as their size decreases: (i) reduction of the remanent polarization, dielectric permittivity, and, consequently piezocoefficient; (ii) increasing coercive field; (iii) change in the domain structure; (iv) lower phase transition temperature and smearing of the phase transition.

All these features can be, in principle, assessed by PFM if the measurements are properly performed. Many theoretical approaches have been used to calculate and to predict the critical size or thickness for ferroelectricity. For example, using a mean-field Ginzburg–Landau approach, the free energy of the finite system, including volume and surface terms, was calculated and minimized. [79]. These and other calculations revealed a material-dependent critical thickness and correlation volume below which ferroelectric is loosing its polar nature. Using this approach, a critical thickness of 20 nm at room temperature has been predicted for PZT films. However, recent *ab initio* calculations [80] were able to show that, with appropriate boundary conditions, ferroelectricity can occur at the monolayer level.

Using the most advanced structuring techniques like electron-beam lithography, e-beam direct writing, or focused-ion-beam etching, the smallest achievable ferroelectric structures were fabricated to study the intrinsic size effects. However, the majority of reported size effects in the devices structured from thin films can be attributed to a reduction of the in-plane strain and the influence of the structuring method has to be taken into account, e.g., damage from etching, doping with sputter ions, etc. A straightforward method to fabricate a ferroelectric capacitor was introduced by Stanishevsky et al. [81], using focused-ion-beam equipment to "cut out" the device from an electrode/ferroelectric/electrode heterostructure. Different samples were prepared by varying the milling depth. To fabricate a capacitor structure, the multilayer was milled either to the ferroelectric layer (only the top electrode was structured) or down to the bottom electrode. By optimizing the process it was possible to achieve Pt/SBT/Pt capacitors of $100 \times 100\ \text{nm}^2$ and Pt/LSCO/PNZT/LSCO/Pt capacitors of $70 \times 70\ \text{nm}^2$ [82,83]. Even for the smallest capacitors, no intrinsic size effects were detected by PFM. For capacitors milled down to the bottom electrode a higher piezoelectric response was measured by PFM compared to capacitors where only the top-electrode was structured

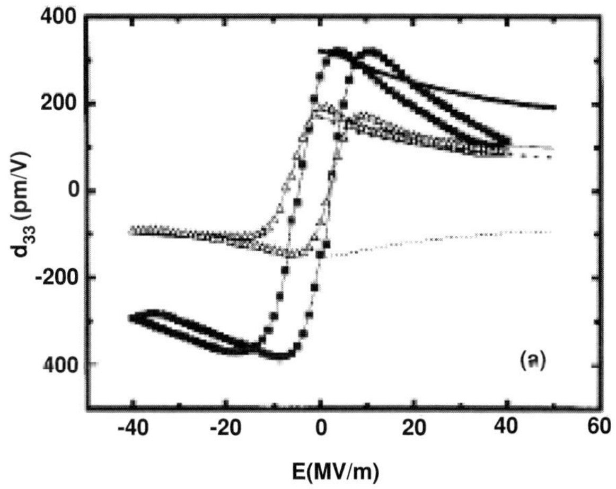

FIGURE 21. Piezoelectric hysteresis measured via PFM for PZT 50/50 film 200-nm thick. Solid symbols are used for a patterned 500 × 500-nm capacitor, while open symbols denote hysteresis for unpatterned film. Reprinted with permission from [82]. Copyright 1999, American Institute of Physics.

(Figure 21). This result was correlated to the in-plane constraints in the continuous film, rather than to intrinsic size effects. The hystereis loops obtained by PFM in PZT capacitors of different dimensions. No change of the intrinsic properties were detected except of the reduction due to clamping effect of the substrate. Similar results were obtained by Bühlmann et al. [84]. They deposited an epitaxial PZT thin film and used e-beam lithography to fabricate a Cr hard mask by means of a lift-off process. Subsequent reactive ion etching (RIE) led to ferroelectric island of sizes from 200×200 nm^2 down to 100×100 nm^2. The piezoelectric response showed a strong increase with decreasing feature size. They explained the increase of the piezoelectric response by a decreasing fraction of *a*-domains for the PZT islands, whereas the continuous film exhibited 15% of *a*-domains, which again is correlated to different mechanical strain.

A direct electron-beam writing technique (EBDW) to achieve a submicron metal pattern was introduced by Craighead and Schiavone [85]. First, an organometallic polymer thin film was locally irradiated with an electron beam, rendering the exposed area insoluble in an organic solvent. In a second step the unexposed areas were removed with a solvent. A heating process is added to remove the organic material and leave the metal pattern. Enhancing the EBDW method, Alexe et al. [86] could fabricate SBT and PZT structures as small as 100×100 nm^2. The coercive field was found to be size-independent. Contrary to the above-referenced papers, the authors of [86] reported a decrease of the piezoelectric response with decreasing size.

While studying size effects in devices structured from thin films, one of the biggest challenges is to distinguish between effects caused by the fabrication method, i.e., damage or contamination of the surface layer of the ferroelectric and the intrinsic size effects. The electron beam direct writing is an exceptional method were no obvious damage or contamination is expected to occur. Nevertheless, like the other structuring methods, no intrinsic changes in the ferroelectric properties can be observed down to the achievable device sizes.

Another technique frequently used to fabricate ferroelectric nanostructures is self-assembly. In this case, less damage is expected, and the minimum feature size can be as small as a few nanometer. Seifert et al. [87] and Fujisawa and coworkers [88] studied the growth of ferroelectric thin films in the early stage of a deposition process. The smallest PZT and $PbTiO_3$ islands investigated by Fujisawa were 80 nm in diameter and exhibited a clear piezoelectric hysteresis loop without notable size effects.

Self-assembly approaches use the microstructural instability observed by Seifert et al. for the epitaxial growth relationship between very thin $PbTiO_3$ films and $SrTiO_3$ single crystalline substrates. It led to the formation of single-crystal islands instead of continuous films [87]. The normally undesirable process was adapted to silicon substrates with platinum bottom electrodes in a tailored way to generate separated 3D nanograins [89]. In this way, highly diluted $PbTiO_3$ precursor solutions were deposited onto Pt(111)/TiO_2/SiO_2/Si substrates, subsequently pyrolized, and finally crystallized [91]. By controlling sol-gel procedure, i.e., the concentration of the stock solution, different grain size and grain size distributions, ranging from 200 nm to about 10 nm in diameter and 20 nm down to 2 nm in height were obtained. PFM measurements revealed an apparent dissappearance of the contrast indicating possible size effect in grains finer than 20 nm. An example of the PFM measurements on such structure is shown in Figure 22.

Although self-assembly methods have the advantage of being noninvasive, one major drawback remains as the registration and size distribution of the deposited nanograins is governed by thermodynamics and, therefore, is random. Based on the fact that ultra-thin TiO_2 seed layers promote the nucleation and growth of (111) textured PZT films on (111) oriented platinized substrates, Schneller et al. proposed a template technique to gain control on the arrangement of perovskite structures in the nanoscale regime prepared by CSD [90]. The idea was to prepare predefined nanosized TiO_2 patterns onto a bottom electrode to act as seeds for the subsequent deposition of $PbTiO_3$ or PZT. PFM measurements turned out to be very useful for studying size effects in self-assembled films, too.

4.2 Imprint and Fatigue in FERAM Capacitors

As was shown above, the functioning of several ferroelectric devices is determined by domain switching and reorientation. Therefore, the domain studies are of great practical importance. This concerns mainly ferroelectric random access memories (FeRAM) in which domain studies are indispensable for the understanding of mechanisms of polarization fatigue and imprint. Fatigue is defined as a loss of switchable polarization under prolonged cycling with a strong ac or pulsed electric

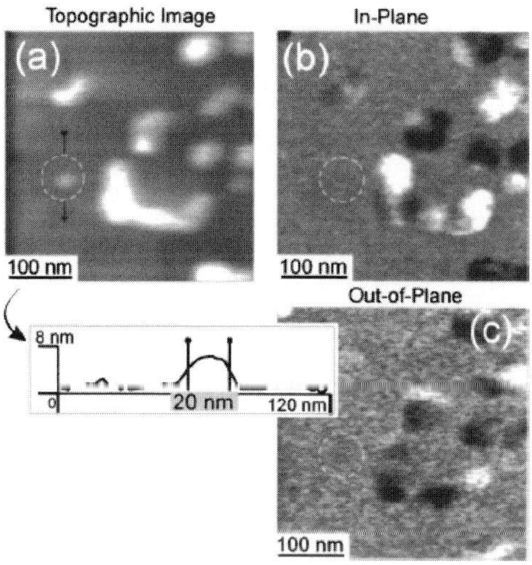

FIGURE 22. The topographical image (a) shows several $PbTiO_3$ grains of sizes between 100 nm down to 20 nm, indicated by circles. In the line scan over the grain denoted with an arrow, shown in the bottom, the size of the grain can be determined. In the PFM images (b) in-plane and (c) out-of-plane piezoresponse of 20 nm grain is not visible. the PFM image (b) the grains of 18 nm (indicated by the circles) are not visible. Reprinted with permission from [91]. Copyright 2002, American Institute of Physics.

field. The problem of fatigue is difficult to solve because it involves a complex of interrelated physical and chemical phenomena occurring mainly at the interface between the electrode and ferroelectric thin film. Until recently, the direct measurements of domain configurations before and after fatigue were impossible to do because of the lack of suitable technique for domain visualization. Only with the advent of PFM, it became feasible to address this issue. Gruverman et al. [92] first observed an effect of fatigue on the domain structure of PZT films on Pt-coated substrates. When applying a high ac voltage to the tip, continuous switching was imitated. It has been calculated that each point received about 10^3 switching cycles. After this, a dc poling was done to the same area in an attempt to switch all ferroelectric domains. As seen in Figure 23, not all domains were switched into the "negative" (bright) state.

These observations [92] have proved that fatigue is always accompanied by strong domain wall pinning, so that certain domains become frozen and do not participate in switching anymore. The results were complemented by local piezoelectric hysteresis measurement that showed a progressive shift of the d_{33} loop in both vertical and horizontal directions (observed also by macroscopic technique [67]). This shift corresponds to a significant imprint of the polarization state one polarization state is preferred over the other).

Later, Colla et al. [93] confirmed the results by Gruverman and co-authors [92] using a removable electrode. In this way, the problem of inhomogeneous electric

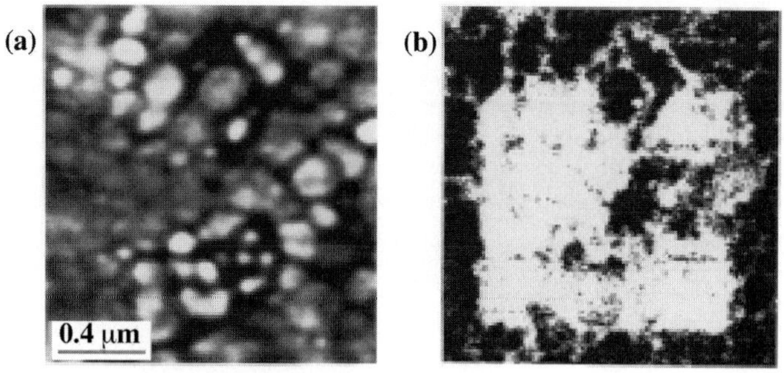

FIGURE 23. Simultaneously obtained (a) topographic and (b) piezoresponse images of the fatigued PZT/Pt heterostructure. The bright square of $1 \times 1\ \mu m^2$, produced by moving the PFM tip while applying a negative 15 V dc voltage, represents a negatively polarized region. Unswitchable crystallites appear as dark areas inside the white square region. Reproduced with permission from [92]. Copyright 1996, American Institute of Physics.

field was alleviated since PZT films were fatigued in real (FERAM) functioning conditions. The image was taken in "normal" PFM conditions (see Figure 3(a)) with the resolution limited only by the tip diameter.

4.3 Ferroelectric Relaxors by PFM

Recently, ferroelectric relaxor materials such as $PbMg_{1/3}Nb_{2/3}O_3$ (PMN), $PbZn_{1/3}Nb_{2/3}O_3$ (PZN) and its solid solutions with $PbTiO_3$ became objects of increasing interest due to their high dielectric and electromechanical properties. Exceptionally high electrostrictive and piezoelectric strain of relaxor materials is especially attractive because of the absence of hysteresis, which is detrimental to high-precision devices. It is widely accepted that unusual piezoelectric, ferroelectric, and dielectric properties of relaxors are determined by the dynamics of nanoscale polar clusters that appear far above the temperature of the dielectric maximum. These clusters may coarsen and coalesce at low temperatures, but normal ferroelectric domains may develop only after poling with a dc bias. PFM is ideally suited for studying relaxors since the effective tip diameter can be as small as the polarization cluster size and the polarization value can be assessed through d_{33}. An example of the PFM study in PMN-0.2PT single crystals [94] is given in Figure 24.

The coexistence of rhombohedral domains (elongated strips) with remnants of polar clusters (small nanodomains embedded in between) is evident in this picture. It also illustrates the fractal nature of micron-sized domains when disorder is present. Nanodomains were shown to be important for the understanding of giant piezoelectric response in PZN-based materials [95]. In this work, nanodomains were only observed in PZN-0.045PT single crystals of certain crystallographic orientations, where the giant d_{33} is found.

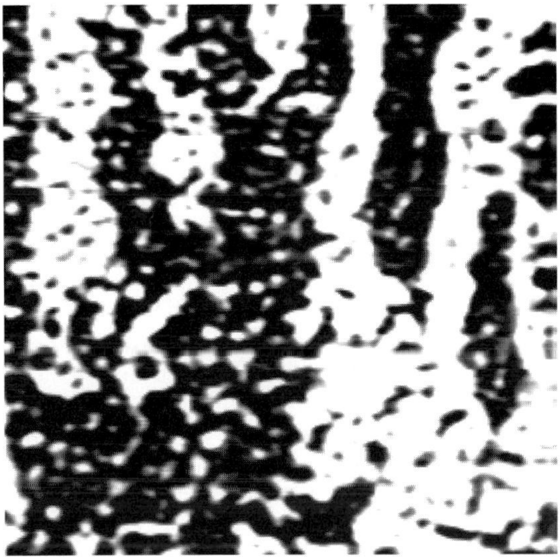

FIGURE 24. Domain structure in PMN-20%PT single crystal illustrating a coexistence of "normal" micron-sized domains with remnants of polar clusters appearing due to random fields induced by B-site nonstoichiometry. The scan size is $5 \times 5\ \mu m^2$. Reprinted with permission from [94]. Copyright 2004, American Physical Society.

In relaxor films the switching into the terroelectric state was found to be influenced by the properties of the individual grains (due to their different orientation, chemical composition, mechanical stress, etc.). Typical topography and piezoelectric images of PMN-0.1PT films are presented in Figure 25 [96].

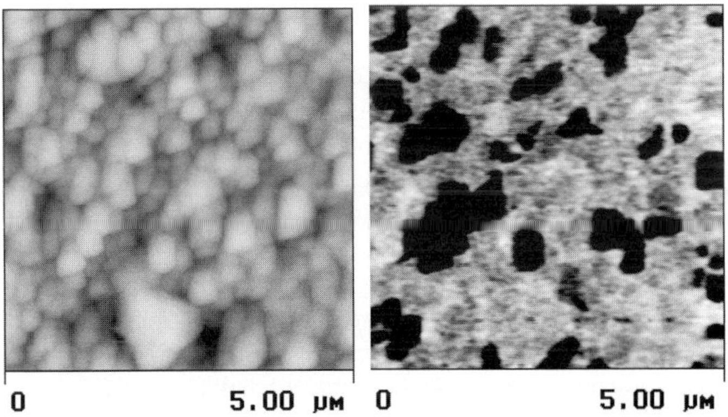

FIGURE 25. Topography **(left)** and piezoresponse image **(right)** of PMN-0.1PT film prepared by laser ablation. Reproduced with permission from [96]. Copyright 2002, American Institute of Physics.

The majority of the grains exhibit gray contrast (weak piezoelectric activity), while others have relatively high d_{33}. It means that they are strongly polarized without any bias. Only gray contrast is expected for the relaxor ferroelectric even for the temperature below the dielectric maximum. However, strong dark contrast in some grains suggests the existence of highly polarized material.

Those grains possess significant polarization without external bias, i.e. are self-polarizated. Two conclusions were made: (i) self-polarization can be found at the nanoscale, while it is hardly found in microscopic measurements; and (ii) since the piezoelectricity in relaxors is possible only under the bias field, the distribution of self-polarization represents also the distribution of internal field in these materials. Thus, relaxor thin film on the microscopic level behaves as a nanocomposite, the macroscopic response of which can be derived by the averaging of local responses of the individual grains. Self-polarization was recently found in pure PMN films [97], proving that self-polarization in relaxor films is not related to local composition inhomogneity due to fluctuations of $PbTiO_3$ content. Domain structure in single-crystal and ceramic relaxors has been consequently studied by several groups [98–100] proving that the PFM is a powerful tool for the investigation of random polarization contrast due to disorder.

4.4 Study of Piezoelectric Semiconductors by PFM

The piezoelectric character of III-nitride semiconductors yields a novel degree of freedom in designing devices for modern micro- and nanoelectronic applications [101]. As such, measurement of the piezoelectric properties in these materials with high spatial resolution is of primary importance for the design and performance of GaN-based electronic and optoelectronic devices. It follows that the investigation of the piezoelectric behavior of nitride thin films and heterostructures at the nanoscale level is a prerequisite for determining how interfaces, dislocations, stacking faults, and inversion domain boundaries can affect electrical properties. It is known that in the wurzite phase GaN and other similar materials have sufficiently high piezoelectric response depending on whether Ga or N terminates the surface and on the quality of the material. The local piezoelectric coefficients can be thus serve as a figure of merit for various applications. The lower values of piezoelectric coefficients (d_{33} is much smaller than in PZT) can be counteracted by applying relatively high ac-voltage to the tip. No switching is expected in piezoelectric semiconductors since they are not ferroelectric.

The example of the polarization imaging is shown in Figure 26 for GaN thin films deposited either directly on sapphire substrate (N termination) or on the AlN nucleation layer (Ga face) [102]. The film was deposited in such a way that both terminations are seen on the image as bright and dark areas.

By imaging GaN with PFM it was possible to judge the quality of the interface by measuring piezoelectric coefficients, since they are influenced by stress. It was shown that the quality of the Ga-terminated surface is lower than that for N-face.

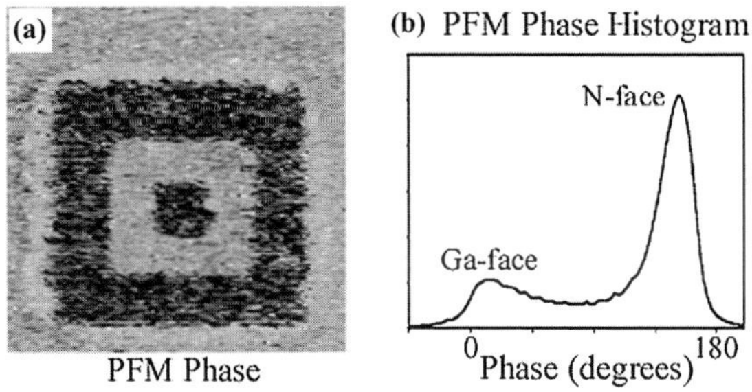

FIGURE 26. (a) PFM phase image of a GaN-based lateral polarity heterostructure, (d) histogram of the PFM phase image. The innermost $5 \times 5\ \mu m^2$ square is the Ga-face region. There is a sharp contrast difference in the piezoresponse phase image, demonstrating inversion in film polarity. The magnitudes of the Ga- and N-face peaks correspond to the relative areas of each region. Reprinted with permission from [102]. Copyright 2002, American Institute of Physics.

5 Outlook

As shown above, during the past decade piezoelectric force microscopy has grown from a mere imaging tool for domain visualization in ferroelectric materials to a powerful technique for quantitative characterization and manipulation in polar materials at the nanoscale. PFM is currently being implemented in many commercial systems (see, e.g., [103]) with a possibility of adding additional modes (imaging mode, piezoelectric spectroscopy, hysteresis, etc). Recently demonstrated usefulness of PFM for imaging of biological materials such as electromechanically active proteins in calcified and connecting issues [104] suggests a much broader range of materials (other than ferroelectric) that can be investigated by this technique. Below we will briefly mention some new challenges offered by this increasingly important measurement and manipulation technique.

The crucial part of the PFM setup is the shape and material of the probe in contact with the ferroelectric surface. In order to be used as a movable electrode, the electrical and mechanical properties of the probe as well as its shape should be carefully controlled. Conventional probes (metal-coated or made from doped Si) all suffer from metal peel-off, contamination with hydrocarbonates and other species on the surface, and uncontrolled shape. Also, the resistivity of the doped Si probes is not low enough for quantitative determination of piezoelectric coefficients at the nanoscale. Platinum (frequently used as a metal) causes electrochemical reactions on the surface of ferroelectric oxides, thus resulting in deterioration of contrast and loss of resolution. We envisage that the development of all-metal probes or Si with implanted metal coatings will drastically increase the capabilities of PFM allowing, for example, the measurements at high temperatures or in different environments. The resolution (determined by the effective diameter of the probe) can be increased by using diamond tips with controlled shape and aspect ratio. These new probes combined with the thorough control of the surface conditions should drastically increase the performance of PFM with a potential of atomic resolution.

Until now, the PFM was used mainly as a high-resolution imaging tool, but the advantages of local polarization reversal have never been utilized except of the few attempts to build SPM-based ferroelectric data storage systems [105]. With the continuing increase of the spatial resolution, PFM can be also used for a new task: i.e., intentional creation of polar structures on the surface for assembling various nanostructures via a so-called polarization-driven mechanism [106]. It has been long known that the chemical reactivity of ferroelectric surfaces depend on the polarization direction (e.g., the dissolution in acids is always depend whether positive or negative end of domain is etched). Recently, the combination of polarization-dependent chemical reactivity and domain patterning gave rise to a novel technology called ferroelectric lithography [106]. The advantage of this lithography is that, depending on mechanism of adsorption, a variety of nanostructures can be, in principle, built. Further, local polarization patterning on the scale of few nanometers can be used for the recognition of specific polar molecules and groups of many biologically important materials, such as glycolipids or certain proteins. Along with sub-nanometer resolution for probing biological materials via piezoelectric or electrostrictive effects [104], it provides a powerful tool to modern nanotechnology.

References

1. J. F. Scott, *Ferroelectric Memories* (Springer, Heidelberg, 1993).
2. A. Gruverman in *Encyclopedia of Nanoscience and Nanotechnology*, edited by H. S. Nalwa (American Scientific Publishers, Los Angeles. 2004) Vol. 3, pp. 359–375.
3. L. M. Eng, M. Bummerlin, C. Loppacher, M. Guggisberg, R. Bennewitz, R. Luthi, E. Meyer, T. Huser, H. Heinzelmann, and H.-J. Guntherodt, Ferroelectrics **222**, 153 (1999).
4. A. Gruverman, O. Auciello, and H. Tokumoto, Annu. Rev. Mater. Sci. **28**, 101 (1998).
5. M. Abplanalp, L. Eng, and H. Güntherodt, Appl. Phys. A **66**, S231 (1998).
6. E. L. Colla, S. Hong, D. V. Taylor, A. K. Tagantsev, N. Setter, and K. No, Appl. Phys. Lett. **72**, 2763 (1998).
7. A. Roelofs, U. Boettger, R. Waser, F. Schlaphof, S. Trogisch, and L. Eng, Appl. Phys. Lett. **77**, 3444 (2000).
8. S. V. Kalinin and D. A. Bonnell, Phys. Rev. B **65**, 125408 (2002).
9. C. S. Ganpule, A. L. Roytburd, V. Nagarajan, B. K. Hill, S. B. Ogale, E. D. Williams, R. Ramesh, and J. F. Scott, Phys. Rev. B **65**, 014101 (2002).
10. C. Loppacher, F. Schlaphof, S. Schneider, U. Zerweck, S. Grafstrom, L. M. Eng, A. Roelofs, and R. Waser, Surface Science **483**, 532, (2003).
11. P. Guthner and K. Dransfeld, Appl. Phys. Lett. **61**, 1137 (1992).
12. M. Alexe and A. Gruverman (eds.), *Nanoscale Characterization of Ferroelectric Materials* (Springer-Verlag, Berlin, 2004).

13. S. Hong (ed.), *Nanoscale Phenomena in Ferroelectric Thin Films* (Academic Publishers, Norwell MA, 2004).
14. L. Eng, H. Güntherod, G. Schneider, U. Köpke, and J. M. Saldana, Appl. Phys. Lett. **74**, 233 (1999).
15. C. S. Ganpule, V. Nagarajan, B. K. Hill, A. L. Roytburd, E. D. Williams, R. Ramesh, S. P. Alpay, A. Roelofs, R. Waser, and L. M. Eng, J. Appl. Phys. **91**, 1477 (2002).
16. C. Durkan, M. Welland, D. Chu, and P. Migliorato, Appl. Phys. Lett. **76**, 366 (2000).
17. C. Harnagea, Ph.D. thesis, Martin-Lüther-Universität Halle, Germany, 2001.
18. S. V. Kalinin, Ph.D. thesis, University of Pennsylvania, USA, 2002.
19. S. Hong, J. Woo, H. Shin, J. U. Jeon, Y. E. Pak, E. Colla, N. Setter, E. Kim, and K. No, J. Appl. Phys. **89**, 1377 (1998).
20. M. Abplanalp, Ph.D. thesis, Swiss Federal Institute of Technology, Switzerland, 2001.
21. E. Luo, Z. Xie, J. Xu, I. Wilson, and L. Zhao, Phys. Rev. B **61**, 203 (2000).
22. J. W. Hong, K. H. Noh, S. Park, S. I. Kwun, and Z. G. Khim, Phys. Rev. B **58**, 5078 (1998).
23. S. V. Kalinin and D. A. Bonnell, Appl. Phys. Lett. **78**, 1116 (2001).
24. F. Peter, K. Szot, R. Waser, B. Reichenberg, S. Tiedke, and J. Szade, Appl. Phys. Lett. **85**, 2896 (2004).
25. M. Abplanalp, J. Fousek, and P. Günter, Phys. Rev. Lett. **86**, 5799 (2001).
26. G. Zavala, J. H. Fendler, and S. McKinstry, J. Appl. Phys. **81**, 7480 (1997).
27. A. L. Kholkin, V. V. Shvartsman, A. Y. Emelyanov, R. Poyato, M. L. Calzada, and L. Pardo, Appl. Phys. Lett. **82**, 2127 (2003).
28. A. Y. Emelyanov, N. A. Pertsev, and A. L. Kholkin, Phys. Rev. B **66**, 214108 (2002).
29. K. Prume, A. Roelofs, T. Schmitz, B. Reichenberg, S. Tiedke, and R. Waser, Jpn. J. Appl. Phys. **41**, 7198 (2002).
30. J. W. Hong, D. S. Kahng, J. C. Shin, H. J. Kim, and Z. G. Khim, J. Vac. Sci. Technol. B **16**, 2942 (1998).
31. U. Rabe, K. Janser, and W. Arnold, Rev. Sci. Instrum. **67**, 3281 (1996).
32. B. Jaffe, W. R. Cook, and H. Jaffe, *Piezoelectric Ceramics* (Academic, London, 1971).
33. C. Harnagea, A. Pignolet, M. Alexe, D. Hesse, Integr. Ferroelectr. **38**, 667 (2001).
34. S. V. Kalinin, B. J. Rodriguez, S. Jesse, J. Shin, A. P. Baddorf, P. Gupta, H. Jain, D. B. Williams, and A. Gruverman. *Microscopy and Microanalysis* **12**, 206 (2006).
35. T. Tybell, C. H. Ahn, and J.-M. Triscone, Appl. Phys. Lett. **75**, 856 (1999).
36. T. Tybell, P. Paruch, T. Giamarchi, and J.-M. Triscone, Phys. Rev. Lett. **89**, 097601 (2002).
37. K. Terabe, M. Nakamura, S. Takekawa, K. Kitamura, S. Higuchi, Y. Gotoh, and Y. Cho, Appl. Phys. Lett. **82**, 433 (2003).
38. H. F. Hamann, Y. C. Martin, and H. K. Wickramasinghe, Appl. Phys. Lett. **84**, 810 (1994).
39. Y. Cho, S. Hashimoto, N. Odagawa, K. Tanaka, and Y. Hiranaga, Appl. Phys. Lett. **87**, 232907 (2005).
40. Y. Rosenwaks, D. Dahan, M. Molotskii, and G. Rosenman, Appl. Phys. Lett. **86**, 012909 (2005).
41. F. Jona and G. Shirane, *Ferroelectric Crystals* (Dover Publications, New York, 1993).
42. J. Giocondi and G. S. Rohrer, Chem. Mater. **13**, 241 (2001).
43. S. V. Kalinin, D. A. Bonnell, T. Alvarez, X. Lei, Z. Hu, J. H. Ferris, Q. Zhang, and S. Dunn, Nano Letters **2**, 589 (2002).

44. S. V. Kalinin, D. A. Bonnell, T. Alvarez, X. Lei, Z. Hu, R. Shao, and J. H. Ferris, Adv. Mat. **16**, 795 (2004).
45. X. Liu, K. Terabe, M. Nakamura, S. Takekawa, and K. Kitamura, J. Appl. Phys. **97**, 064308 (2005).
46. M. Molotskii, A. Agronin, P. Urenski, M. Shvebelman, G. Rosenman, and Y. Rosenwaks, Phys. Rev. Lett. **90**, 107601 (2003).
47. M. Molotskii, J. Appl. Phys. **93**, 6234 (2003).
48. S. V. Kalinin, E. Karapetian, and M. Kachanov, Phys. Rev. B **70**, 184101 (2004).
49. S. V. Kalinin, A. Gruverman, B. J. Rodriguez, J. Shin, A. P. Baddorf, E. Karapetian, and M. Kachanov, J. Appl. Phys. **97**, 074305 (2005).
50. A. N. Morozovska and E. A. Eliseev, Phys. Stat. Sol. (b) **242**, R79 (2005).
51. A. Y. Emelyanov, Phys. Rev. B **71**, 132102 (2005).
52. M. I. Molotskii and M. M. Shvebelman, Philos. Magaz. A **85**, 1637 (2005).
53. A. N. Morozovska and E. A. Eliseev, Physica B **373**, 54 (2006).
54. Eq.(3) is valid only for $l_d > r_d$. To avoid this limitation, used here was the expression for the demagnetization factor for prolate ellipsoid from J. A. Osborn, Phys. Rev. **67**, 351 (1945).
55. R. Landauer, J. Appl. Phys. **28**, 227 (1957).
56. J. Li, B. Nagaraj, H. Liang, W. Cao, C. H. Lee, and R. Ramesh, Appl. Phys. Lett. **84**, 1174 (2004).
57. C. Dehoff, B. J. Rodriguez, A. I. Kingon, R. J. Nemanich, A. Gruverman, and J. S. Cross, Rev. Sci. Instrum. **76**. 023708 (2005).
58. P. Paruch, T. Giamarchi, and J.-M. Triscone, Phys. Rev. Lett. **94**, 197601 (2005).
59. B. J. Rodriguez, R. J. Nemanich, A. Kingon, A. Gruverman, S. V. Kalinin, K. Terabe, X. Y. Liu, and K. Kitamura, Appl. Phys. Lett. **86**, 012906 (2005).
60. A. Gruverman, B. J. Rodriguez, C. Dehoff, J. D. Waldrep, A. I. Kingon, R. J. Nemanich, and J. S. Cross, Appl. Phys. Lett. **87**, 082902 (2005).
61. A. N. Kolmogorov, Izv. Akad. Nauk SSSR, Ser. Math. **3**, 355 (1937).
62. W. Jo, D. C. Kim, and J. W. Hong, Appl. Phys. Lett. **76**, 390 (2000).
63. A. Gruverman, H. Tokumoto, A. S. Prakash, S. Aggarwal, B. Yang, M. Wuttig, R. Ramesh, O. Auciello, and T. Venkatesan, Appl. Phys. Lett. **71**, 3492 (1997).
64. A. Gruverman and M. Tanaka, J. Appl. Phys. **89**, 1836 (2001).
65. C. S. Ganpule, A. L. Roytburd, V. Nagarajan, B. K. Hill, S. B. Ogale, E. D. Williams, R. Ramesh, and J. F. Scott, Phys. Rev. B **65**, 014101 (2001).
66. A. L. Kholkin, E. K. Akdogan, A. Safari, P.-F. Chauvy, and N. Setter, J. Appl. Phys. **89**, 8066 (2001).
67. A. L. Kholkin, E. L. Colla, A. K. Tagantsev, D. V. Taylor, and N. Setter, Appl. Phys. Lett. **68**, 2577 (1996).
68. Y. Zhang, I. S. Baturin, E. Aulbach, D. C. Lupascu, A. L. Kholkin, V. Y. Shur, and J. Rödel, Appl. Phys. Lett. **86**, 012910 (2005).
69. V. V. Shvartsman, N. A. Pertsev, J. M. Herrero, C. Zaldo, and A. L. Kholkin, J. Appl. Phys. **97**, 104105 (2005).
70. S. V. Kalinin, A. Gruverman, and D. A. Bonnell, Appl. Phys. Lett. **85**, 795 (2004).
71. A. L. Kholkin, I. K. Bdikin, V. V. Shvartsman, A. Orlova, D. Kiselev, and V. Bogomolov, MRS Proceedings **838E**, O7.6 (2005).
72. A. Gruverman, A. Pignolet, K. M. Satyalakshmi, M. Alexe, N. D. Zakharov, and D. Hesse, Appl. Phys. Lett. **76**, 106 (2000).
73. A. Wu, P. M. Vilarinho, G. Suchaneck, and A. L. Kholkin, Nanotechnology **16**, 2587 (2005).

74. S. V. Kalinin, Appl. Phys. Lett. (submitted).
75. S. V. Kalinin and D. A. Bonnell, Phys. Rev. B **63**, 125411 (2001).
76. C. S. Ganpule, V. Nagarjan, H. Li, A. S. Ogale, D. E. Steinhauer, S. Aggarwal, E. Williams, R. Ramesh, and P. Wolf, Appl. Phys. Lett. **77**, 292 (2000).
77. P. Gupta, H. Jain, D. B. Williams, S. V. Kalinin, J. Shin, and A. P. Baddorf, Appl. Phys. Lett. **87**, 172903 (2005).
78. R. Waser (ed), *Nanoelectronics and Information Technology: Advanced Electronic Materials and Novel Devices* (Wiley-VCH, Berlin, 2005).
79. J. F. Scott, H. M. Duiker, P. D. Beale, B. Pouligny, K. Dimmler, M. Parris. D. Butler, and S. Eaton, Physica B+C **150**, 160 (1988).
80. P. Ghosez and K. M. Rabe, Appl. Phys. Lett. **76**, 2767 (2000).
81. A. Stanishevsky, S. Aggarwal, A. Prakash, J. Melngailis, and R. Ramesh, J. Vac. Sci. Technol. B **16**, 3899 (1998).
82. C. Ganpule, A. Stanishevsky, S. Aggarwal, J. Melngailis, E. Williams, R. Ramesh, V. Joshi, and C. P. Araujo, Appl. Phys. Lett. **75**, 3874, (1999).
83. C. Ganpule, A. Stanishevsky, Q. Su, S. Aggarwal, J. Melngailis, E. Williams, and R. Ramesh, Appl. Phys. Lett. **75**, 409 (1999).
84. S. Bühlmann, B. Dwir, J. Baborowski, and P. Muralt, Appl. Phys. Lett. **80**, 3195 (2002).
85. H. Craighead and L. Schiavone, Appl. Phys. Lett. **48**, 1748 (1989).
86. M. Alexe, C. Harnagea, W. Erfurth, D. Hesse, and U. Gösele, Appl. Phys. A **70**, 247 (2000).
87. A. Seifert, A. Vojta, J. S. Speck, and F. F. Lange, J. Mat. Res. **11**, 1470 (1996).
88. H. Fujisawa, K. Morimoto, M. Shimizu, and H. Niu, Jpn. J. Appl. Phys. **39**, 5446 (2000).
89. R. Waser, T. Schneller, S. Hoffmann-Eifert, and P. Ehrhart, Integr. Ferroelectr. **36**, 3 (2001).
90. T. Schneller and R. Waser, Ferroelectrics **267**, 293 (2002).
91. A. Roelofs, T. Schneller, K. Szot, and R. Waser, Appl. Phys. Lett. **81**, 5231 (2002).
92. A. Gruverman, O. Auciello, and H. Tokumoto, Appl. Phys. Lett. **69**, 3191 (1996).
93. E. L. Colla, I. Stolichnov, P. E. Bradely, and N. Setter, Appl. Phys. Lett. **82**, 1604 (2003).
94. V. V. Shvartsman and A. L. Kholkin, Phys. Rev. B **69**, 014102 (2004).
95. I. K. Bdikin, V. V. Shvartsman, and A. L. Kholkin, Appl. Phys. Lett. **83**, 4232 (2003).
96. V. V. Shvartsman, A. Y. Emelyanov, A. L. Kholkin, and A. Safari, Appl. Phys. Lett. **81**, 117 (2002).
97. V. V. Shvartsman, A. L. Kholkin, M. Tyunina, and J. Levoska, Appl. Phys. Lett. **86**, 222907 (2005).
98. S. B. Vakhrushev, A. A. Naberezhnov, B. Dkhil, J.-M. Kiat, V. Shvartsman, A. Kholkin, B. Dorner, and A. Ivanov, AIP Conf. Proc. **677**, 74 (2003).
99. F. Bai, J. Li, and D. Viehland, Appl. Phys. Lett. **85**, 2313 (2004).
100. V. V. Shvartsman, M. Wojtas, S. Vakhrushev, and A. L. Kholkin, Mat. Res. Soc. Symp. Proc. **785**, D4.11 (2004).
101. O. Ambacher, J. Majewski, C. Miskys, A. Link, M. Hermann, M. Eickhoff, M. Stutzmann, F. Bernardini, V. Fiorentini, V. Tilak, B. Schaff, and L. F. Eastman, J. Phys.: Condens. Matt. **14**, 3399 (2002).
102. B. J. Rodriguez, A. Gruverman, A. I. Kingon, and R. J. Nemanich, Appl. Phys. Lett. **80**, 4166 (2002).

103. http://www.ntmdt.ru/SPM-Techniques
104. B. J. Rodrigues, S. V. Kalinin, J. Shin, J. Jesse, V. Grichko, T. Thundat, A. P. Baddorf, and A. Gruverman, J. Appl. Phys. (in press).
105. Y. Hiranaga and Y. Cho, Jpn. J. Appl. Phys. **44**, 6960 (2005).
106. S. V. Kalinin, D. A. Bonnell, T. Alvarez, X. Lei, Z. Hu, R. Shao, and J. H. Ferris, Adv. Mat. **16**, 795 (2004).

I.7 Principles of Near-Field Microwave Microscopy

STEVEN M. ANLAGE, VLADIMIR V. TALANOV, AND ANDREW R. SCHWARTZ

Near-field microwave microscopy is concerned with quantitative measurement of the microwave electrodynamic response of materials on length scales far shorter than the free-space wavelength of the radiation. Here we review the basic concepts of near-field interactions between a source and sample, present an historical introduction to work in the field, and discuss a novel quantitative modeling approach to interpreting near-field microwave images. We discuss the spatial resolution and a number of concrete applications of near-field microwave microscopy to materials property measurements, as well as future prospects for new types of microscopy.

1 Introduction

Much of our understanding of materials comes from studying the interaction of electromagnetic fields with matter. The optical properties of metals, semiconductors, and dielectrics have revealed many aspects of charge and lattice dynamics in condensed matter and the study of materials at lower frequencies has also been a fruitful area of investigation [1]. For example, early evidence for a gap in the spectroscopic properties of superconductors came from microwave transmission experiments. Evidence for case-II coherence effects was also seen in measurements of the complex conductivity of superconductors [2]. The study of magnetization dynamics in ferromagnets came through measurements of ferromagnetic resonance and anti-resonance at microwave frequencies [3–5].

One thing that all of these techniques have in common is that they are carried out with measurement systems that are on the scale of the free-space wavelength of the radiation employed. For example, transmission experiments are done in the far-field of a source, and they typically require a sample on the scale of the wavelength in size. Many other experiments are carried out in resonant cavities, which are on the order of the wavelength in at least one dimension. One result is that the electrodynamic properties of the sample are averaged over macroscopic length scales. The properties of the material may in fact be varying on much shorter length scales, even into the nanometer range. Hence to measure the intrinsic response, such traditional measurements require remarkably large, pure, and homogeneous samples to study. Most materials of interest today are complex multi-component compounds or nanoscale composites and can rarely be made homogeneous on the millimeter- or centimeter-length scales required for traditional electrodynamics measurements. In addition, the typical dimensions of devices into which some of these functional materials are integrated are orders of magnitude smaller than the wavelength at the frequency of operation of the device.

Additional constraints of traditional measurements are imposed by the sample geometry. Ideally, one would like to measure the properties of ellipsoidally shaped samples to eliminate de-magnetization and de-polarization effects, but such samples are rarely available, particularly for novel and interesting materials. Thin-film samples often have such large de-magnetization factors that results of measurements in perpendicular fields are extremely challenging to interpret, particularly on anisotropic materials. A related issue concerns the edges and corners of superconductors in the Meissner state, at which screening currents are known to become singular [6]. These extraordinarily large current densities result in an extreme type of weighted average of the sample properties. The properties of edges and corners, which may not be representative of the bulk, will dominate such measurements. This is an issue for single crystal, ceramic and thin-film samples.

A new paradigm of electrodynamics measurements in condensed matter physics has emerged in recent years. This is associated with the concept of so-called *near-field* interactions between a source and a sample in which near-zone fields and/or evanescent waves with high spatial frequency are created and interact with the sample. The recovery of these signals gives insight into the localized electrodynamic properties of the sample. The art of near-field microwave microscopy of materials is one aspect of this new paradigm. We are motivated in part by the deficiencies of traditional far-field measurements, as outlined above. We are also motivated by the desire to examine new physics on nanometer-length scales present in highly correlated electron systems and biological systems, as well as technologically motivated investigations of semiconductor and other functional materials on the nanometer scale.

In this review we arbitrarily confine ourselves to situations where the following four conditions are simultaneously met: (1) measurements are performed at frequencies between 100 MHz and 100 GHz (this ranges from the bottom of A-band to the top of M-band in the EU/NATO spectrum or I-band to W-band in the IEEE US electronic warfare spectrum); (2) measurements of material properties are performed; (3) there is scanning of the probe relative to the sample; and (4) the spatial resolution of the measurement is substantially ($\leq 10^{-1}$) less than the free-space wavelength. Even with such stringent constraints, there is a great deal of interesting work to discuss.

Although we have attempted to be comprehensive in our coverage of this field, it is inevitable that important work has been accidentally overlooked. We apologize to the reader in advance for these oversights. Some other review articles on microwave near-field imaging that are of possible interest to the reader include references [7–9]. We do not consider localized microwave electromagnetic field probing in this review, as the reader can find coverage of this topic in [7] and [9] and in the

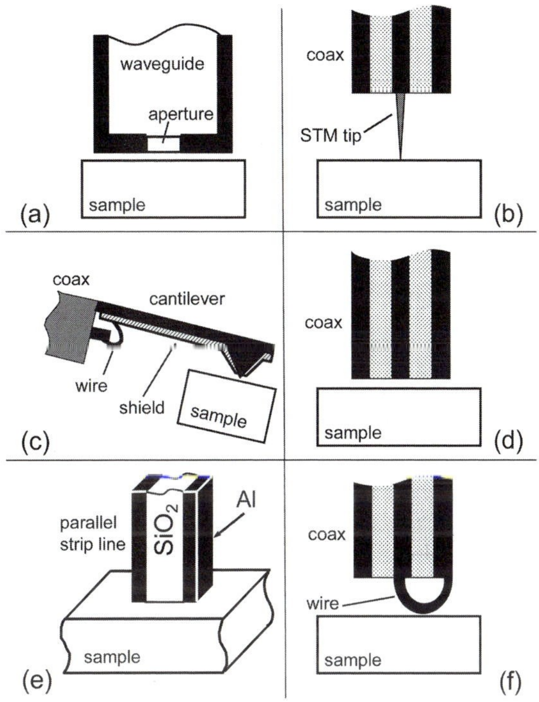

FIGURE 1. The main types of near-field microwave probes: (a) aperture in a waveguide, (b) STM tip [12], (c) AFM tip [13], (d) open end of coaxial line [14], (e) parallel strip transmission line [15], (f) magnetic loop [16].

chapter by Paulson and van der Weide in this book. We also do not consider the use of scanned near-field microwave probes for nondestructive testing of structural materials, as the reader can find coverage of this in [10] and [11].

1.1 What is the "Near-Field"?

Most near-field microwave microscopes consist of a sub-wavelength antenna-like feature, i.e., the microscope's *tip*, which is scanned in close proximity to a sample. The tip can be formed by an aperture in an opaque screen (Figure 1(a)), an electric wire formed by a sharpened rod or scanning tunneling microscopy (STM) tip (Figure 1(b)), an atomic force microscopy (AFM) tip (Figure 1(c)), an electrically open flush end of a transmission line (Figures 1d, 1e), a magnetic loop (Figure 1(f)), or a variety of other geometries. It is typically attached to some type of microwave detection apparatus such as an LC-oscillator, transmission line resonator, waveguide, etc. One monitors the electromagnetic response of the detection system as the tip is scanned over the sample either in contact or at a separation h typically much less than the tip characteristic length-scale D, in order to maintain a reasonable signal from the sample.

The study of *small* antennas, in which the size D is small compared to the wavelength of the radiation λ, naturally breaks into three categories: near-zone (static), intermediate-zone (induction), and far-zone (radiation) properties [17].

The antenna properties in the far zone (defined as $D \ll \lambda \ll r$, where r is the radius vector), are governed by the outgoing propagating waves with wavenumber $k_0 = \omega/c$, where ω is the angular frequency of the source and c is the speed of light in vacuum. The ratio of the electric and magnetic field magnitudes is $|E|/|H| = Z_v$, where Z_v is the vacuum impedance ($Z_v = 120\pi\ \Omega$), the fields are in-phase, transverse to the radius vector r, and fall off as $1/r$.

In the near zone of an antenna, the structure of electric and magnetic fields is more complicated, however. They have a static character while oscillating harmonically as $exp(\boldsymbol{i}\omega t)$ and their distribution strongly depends on the antenna geometry as well as electrodynamic properties of the surrounding region. The ratio $|E|/|H|$ could be much less or much greater than Z_v, the fields are not transverse and more nearly in quadrature, and they fall off as $1/r^2$ or faster, depending on the multipole order of the antenna. Retardation effects are minimal, simplifying the equations governing the fields. It has to be pointed out that in conventional antenna theory the near zone is typically defined as $D \ll r \ll \lambda$ [17], while we are interested in investigation of the *near-fields* occupying the range $D \leq r \ll \lambda$. Theoretically, by increasing the multipole order of the antenna system its near-zone field can be more strongly confined and the far field radiation reduced as much as desired [18,19].

Near-fields with purely exponential dependence on r (not due to dissipation) are often called *evanescent*. They are created as a result of scattering of electromagnetic radiation in geometries such as a cutoff waveguide [20], total internal reflection (which is employed in near-field optical microscopy), a sub-wavelength aperture in an opaque screen (can be used for near-field microwave microscopy) [21], or a lens with negative index of refraction [22]. These waves with imaginary wavenumbers do not carry energy away [17,23] and instead decay exponentially on a length-scale $\sim D$ [8], where D is the characteristic size of the scatterer (e.g., sharp tip or aperture).

Fundamentally, the electrodynamic response of a near-field microwave probe is due to the ability of either near-zone or evanescent fields to store *reactive* energy, electric and/or magnetic, in the vicinity of the microscope's tip (see Sec. 4.1 below). When a dielectric or permeable sample is brought in close proximity to the tip this energy changes, which in turn affects the electromagnetic response of the microscope's detection system.

In order to create a true near-zone or evanescent field *in the sample*, the tip characteristic size D must be small enough not only compared to the free-space wavelength λ, but also to provide $|k_s|D \ll 1$, where $k_s = \omega(\varepsilon_0\varepsilon_s\mu_0\mu_s)^{1/2}$ is the complex wave number of the material under test (see also [8]). In this case the tip-sample interaction can be viewed as a "cloud" of the probing electric or magnetic field penetrating the sample. The size of this cloud is on the order of the tip size D due to the static character of the near-field. Therefore the microscope spatial resolution—both lateral and in-depth—is mostly governed by the tip geometry rather than the electrodynamic properties of the sample, as will be illustrated in Sec. 4.5 through numerical simulations. However, the spatial resolution can be a function of the sample properties too as the electric fields between a sharp

AFM- or STM-like tip and a *high* permittivity sample are enhanced due to the field concentration just below the apex [8].

1.2 What Can Be Learned at Microwave Frequencies?

As mentioned above, quantitative microwave measurements have been very fruitful in the investigation of condensed matter systems. Because these measurements are so highly refined and precise, one would like to develop new experimental techniques to make them on deep sub-wavelength length scales. Optical radiation interacts with materials through quantum interactions, plasmon excitation, lattice dynamics, etc., and these interactions are strongly wavenumber dependent. As a result, it is often difficult to obtain quantitative information about optical properties of materials from near-field interactions. On the other hand, microwave frequency radiation interacts with matter in a much more straightforward classical manner. In this review we are primarily interested in quantitative imaging of materials properties, such as conductivity, dielectric constant, polarization, etc. We believe that the relative simplicity of the detected signal interpretation is one of the most important advantages of near-field microwave microscopy, and enables many new types of quantitative investigations of local physics in condensed matter systems.

2 General Principles of Microwave Microscope Design

As mentioned above, a microwave microscope consists of a scanned near-field antenna structure that is monitored by a detection system. In general, the detection system can be resonant, non-resonant or self-oscillating. The near-field antenna structure can be aperture-based or apertureless (Figure 1). In the latter case, one can prepare field-concentrating features that enhance either electric or magnetic fields. Rather than attempt to gather all of the super-resolution information simultaneously (as one would do in a traditional imaging system), near-field microscopes employ pixel-by-pixel scanning. In this way the signal is acquired from only a small part $\sim D$ of the sample at one time. This scanning method is tedious and slow, but effective. Here we discuss some general principles of microwave microscope design illustrated through an historical review of the field.

2.1 Early Microwave Microscopes

The original idea of near-field microscopy is often attributed to Synge [21], who proposed using an opaque screen with a small sub-wavelength diameter hole (10 nm in diameter), held about 10 nm above the surface of a smooth flat sample. An optically transparent sample was passed just beneath this aperture, and the transmitted light is collected in a point-by-point scan and raster fashion. This theoretical paper was decades ahead of its time in terms of the technology required to carry out the experiment. Detailed calculations of the field distribution near the aperture were later done by Bethe [24] and Bouwkamp [25]. These near fields contain a large fraction of high-spatial frequency evanescent waves that can lead to deep sub-wavelength spatial resolution [26,27].

Synge's idea lay dormant in the literature until the late 1950s, when Frait developed the first ferromagnetic resonance (FMR) microwave microscope [3]. Frait recognized that a conventional FMR resonant cavity simply averaged over the properties of a sample placed *inside* the cavity. He inverted the design by placing the sample *outside* the cavity but retaining coupling by means of a small hole (500-μm diameter) in a thin face of a metallic cavity operating at a wavelength of 3 cm (10 GHz). A $Ni_{78}Fe_{22}$ ferromagnetic thin-film sample was scanned beneath the hole and showed contrast in the uniaxial anisotropy parameter. Independent and similar results with a 500-μm diameter hole were obtained by Soohoo on a 350-nm-thick Permalloy film at 5.5 GHz [4]. Similar work with manganite ferromagnetic material has been published more recently [28,29].

Ash and Nicholls performed another Synge-style experiment at microwave frequencies using a quasi-optical hemispherical resonator as the detection system [30]. This experiment utilized the Synge geometry of a small aperture (1.5 mm diameter) scanned over a sample with a microwave signal at 10 GHz. The sample was harmonically distance modulated at a fixed frequency, and the reflected signal was phase-sensitively detected at this modulation frequency to improve sensitivity to sample contrast. Similar distance modulation techniques are still in use today. This microscope demonstrated contrast sensitivity to metal films on dielectric substrates, as well as bulk dielectrics with dielectric constant differing by only about 10%.

Non-resonant scanned aperture probes were also developed in the 1960s. In this case a transmission line (coaxial or waveguide) delivers a microwave signal to the probe aperture. The fringe electric and magnetic fields from the aperture interact with the sample. Some part of the signal is stored locally in evanescent and near-zone waves, some is absorbed by the sample, some is reflected back up the transmission line, and some is scattered away as far-field radiation. By monitoring the scattered or reflected signals as a function of probe height and position, an image of the sample response can be made. An early example of this type of scanned probe was developed by Bryant and Gunn [31]. They measured the reflected signal from a tapered open-ended coaxial probe with an inner conductor diameter of 1 mm. The probe was used for quantitative measurements of semiconductor resistivities between 0.1 and 100 Ω-cm. A similar, but resonant, method was developed by Bosisio *et al.* to measure the moisture content of paper using measurements of the phase shift of a scanned resonant coaxial probe [32].

The development of near-field optical microscopes using optical waveguides beyond cutoff [33,34] spurred the development of new microwave microscopes. The first of these microwave systems was developed by Fee, Chu and Hänsch by measuring the reflected signal from a non-resonant open-ended coaxial probe [35]. Although similar in design to the probe of Bryant and Gunn, this paper was the first to present the results in the language of near-field microscopy. They employed a 500-μm-diameter center conductor coaxial probe which was extended and sharpened to a tip with 30 μm radius. Operating at 2.5 GHz, they measured the phase of the reflected signal as a function of position over the sample. An image of copper grid lines showed a spatial resolution of about 30 μm (or λ/4,000). Other microscopes based on a scanned aperture in a waveguide structure were developed by Golosovsky *et al.* [36,37] and Bae *et al.* [38,39].

2.2 Aperture-Based Probes and Apertureless Probes

The design of near-field probes can basically be divided into two main categories. The first are aperture-based probes, essentially described by Synge in his seminal paper [21]. The idea is to create a very small interaction volume between the probe and sample by simply constraining the lateral extent of the probing field by means of a sub-wavelength aperture (see Figure 1(a,e)). An alternative approach is to allow a propagating wave to irradiate some large portion of the sample, but to employ a field-concentrating feature, e.g., tip, to enhance the probe–sample interaction locally (Figure 1(b) and Figure 2, inset). The latter apertureless probe design usually results in much stronger signals, albeit with a large background signal.

The designs of Synge, Frait, Soohoo, and Ash and Nicholls are classic aperture-based approaches to microwave microscopy. The designs of Bryant and Gunn, and Fee, Chu, and Hänsch are prototypical apertureless or field-concentrating

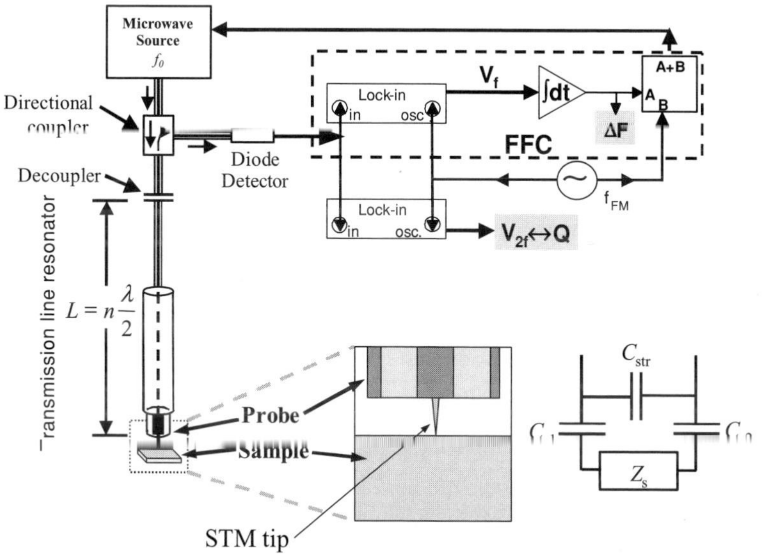

FIGURE 2. The key components of a typical near-field microwave microscope are shown schematically in this figure. The FFC feedback circuit consists of a lock-in, amplifier integrator, and adder circuit. Two output signals are generated: the frequency shift ΔF and the $2f_{FM}$ voltage, V_{2f}. The insets show a close view of the coaxial probe, STM tip and sample, as well as a lumped element circuit model for the tip-sample interaction.

designs. In general, the field-concentrating objects are sharp tips, such as those used for STM or AFM. The electric fields between a sharp tip and a flat sample with high permittivity can be greatly enhanced due to the induced image of the probe, thus leading to strongly localized measurement of sample properties. STM-tip field-concentrating features have been employed by many authors at microwave frequencies [12,40–59], and at far-infrared frequencies [60]. Although the field strengths can be large, the effects of local sample heating are a concern. In at least one case local heating has been shown to be negligible [61]. AFM tip field concentration has been accomplished by a number of groups [62–66]. There have also been efforts to combine an aperture probe with a field-concentrating feature through a tip on an aperture [67], and a waveguide aperture with a tip projection [68]. There is also interest in creating concentrated microwave magnetic fields in a small area of the sample. Seminal work on superconducting quantum interference device (SQUID) microscopy at RF, microwave, and millimeter-wave frequencies showed that this was possible [69,70]. Work with enhanced magnetic field probes was also used to study the local magnetic permeability of materials [16], and localized superconducting nonlinearities [71–75].

2.3 Operating Modes

Microwave microscopes can be operated in one of several detection modes. Nonresonant microscopes typically involve measurements of reflection and/or transmission at a fixed frequency chosen to maximize the signal level at the sample [36,52,53]. Resonant microscopes can be operated in either a fixed-frequency mode [76], with probe/sample distance modulation [77], or with a feedback system to stabilize the source frequency to the variable resonant frequency of the probe/detector system [12,14].

The fixed-frequency resonant mode involves monitoring the reflection or transmission properties of the resonator at a point of maximum slope of the response curve. As the sample properties change, the stored energy and dissipated power in the resonant detection system change. This in turn changes the resonant frequency and quality factor of the detection system. By simply monitoring the off-resonant response (through the reflection coefficient S_{11}, or the transmission coefficient S_{21}), these systems are sensitive to small changes in sample properties.

Because a resonant microwave microscope produces two independent streams of data (e.g., frequency shift and quality factor change, or magnitude and phase, or resistance and reactance), it is advantageous to develop a method to measure both quantities simultaneously. The most common method to do this uses a feedback circuit (such as a phase-locked loop) to force the microwave source to "follow" the time-dependent resonant frequency of the detection system [14], as discussed in section 3. A related technique continuously modifies the distance between the probe and sample to keep the microscope resonant frequency locked to that of a synthesized microwave source [78].

2.4 Tip-Sample Distance Control in Microwave Microscopy

One of the strengths of near-field microscopy of any kind is that measurements can be made without any physical contact between the near-field tip and the sample or device under test. However, this requires that the separation between the tip and the surface of the sample be small compared to the characteristic tip size, and that it be kept constant. The precision of any near-field measurement is directly related to the precision with which the tip-sample distance can be maintained, and a good rule of thumb is that the tip should be held reliably at a distance above the sample that is less than one-tenth of the tip size D. Although the absolute value of the mean height is not critical (provided that it meets the criteria above), the variance must be less than 1% of the tip size in order to obtain high precision measurements. For example, for a tip with a characteristic size of 0.1 μm, the tip-sample separation will have to be maintained at approximately 10 nm, with a precision of ±1 nm. There are a limited number of techniques that can achieve this precision. Possible solutions that have been employed on various microwave microscopes include shear force, as used in near-field scanning optical microscopy (NSOM), atomic force microscopy (AFM), scanning tunneling microscopy (STM), and height modulation. The challenge is integration of each technique with a suitable near-field microwave microscopy probe. Each of these distance control mechanisms is discussed briefly below, with reference to microscopes on which they have been employed.

2.4.1 Shear Force Based Distance Control

Shear force is a distance control mechanism that was developed for use in near-field scanning optical microscopy (NSOM) [79,80]. The basic idea is that an optical fiber is flexible and can therefore be mounted onto and dithered by a piezoelectric-element or a quartz tuning-fork oscillator (TFO) [81] with amplitude from a few nanometers down to a few angstroms. As the tip of such a fiber is brought into close proximity to the sample surface, the amplitude and frequency of the tip oscillations are changed by interactions between the tip and the sample surface. The motion of the tip is detected optically or electrically and a feedback loop allows for precise distance control down to 1 nm. In addition, the height at which this control can be performed is a function of the amplitude of the oscillation—the smaller the amplitude, the smaller the control distance. It has been demonstrated that this type of force feedback can be effectively applied to probe structures other than optical fibers, making it applicable for use in near field microwave microscopy [15,82,83].

2.4.2 Atomic Force Based Distance Control

Atomic force microscopy (AFM) is an established technique for measuring surface profiles with atomic resolution. In the simplest embodiment, a cantilever is fabricated with a small cone-like tip at one end. This tip is brought into close proximity to the sample surface, and the displacement of the cantilever caused by the repulsion force between the tip and the sample is measured using an optical beam deflection technique as the tip is scanned over the surface. In order to enhance

AFM sensitivity and decrease the tip pressure on the sample, a so-called tapping mode is usually employed. The AFM approach allows for atomic-scale lateral resolution and sub-nanometer precision in height control. AFM distance control has been used to image ferroelectric domains at 1.3 GHz [84]. Microwave transmission lines have been successfully integrated onto traditional AFM cantilevers [66,85].

2.4.3 Scanning Tunneling Based Distance Control

Scanning tunneling microscopy (STM) is a well-established technique for surface studies of conducting, semiconducting and superconducting samples with atomic or near atomic resolution [86]. The conducting probe is brought near the sample (with dc bias applied to either the tip or the sample) at a nominal distance of one nanometer, so that a quantum mechanical vacuum tunnel junction can be established between the tip and the sample. With a fixed voltage applied between the tip and the sample, the quantum mechanical tunnel current decays exponentially as the tip-to-sample distance is increased. It is this exponential decay of current with distance that makes STM sensitive enough for atomic scale imaging [87]. Integrating a near-field microwave microscope with constant tunnel current mode STM allows for distance control. During surface scanning the STM tip remains close to the surface (maintaining the nominal height of 1 nm) and the same tip can be used to perform near-field microwave microscopy. As previously mentioned in section 2.2, a number of researchers have employed STM-based microwave microscopes [12,40–60].

2.4.4 Height Modulation Distance Control

A number of microwave microscopes have employed some form of the height-modulated distance control [88–92]. The general principal is that the tip-sample separation is modulated with fixed amplitude (greater than the separation for tapping mode) and frequency, and the modulated signal is used to control the separation. In one embodiment [78] the tapping mode has been combined with frequency modulation to dramatically improve sensitivity and signal-to-noise ratio of images, resulting in enhanced spatial resolution.

3 Detailed Description of One Microscope

We discuss one class of microwave microscope in some detail to illustrate the basic operation. The essential elements of this microscope (shown in Figure 2) are a microwave source, a coaxial resonator coupled to the microwave generator (either via a decoupling capacitor or inductor), a detector to measure the reflected signal from the resonator, and a frequency-following circuit (FFC).

Figure 2 shows the circuit diagram of the microscope. The generator creates a time harmonic signal that is sent to the microscope resonator. The resonator used is a coaxial transmission line, which is capacitively coupled (via a decoupling capacitor) to the feedline on one side and couples to a sample on the other side.

One particular transmission line used is 2.16 mm in outer diameter with a 500-μm diameter inner conductor (standard 0.085-in. coaxial cable). The inner and outer conductors are copper and the dielectric material used is Teflon. A field-concentrating feature, such as an STM tip (Figure 1(b)), can be added to the center conductor.

The transmission line resonator response is measured in reflection using a directional coupler that allows the signal from the source to reach the resonator and the reflected signal from the resonator to be directed to the diode detector. The diode detector converts the measured power at microwave frequency into a voltage signal (this output, V_{diode}, is proportional to the input microwave power in the range of power of interest). The microwave generator is frequency modulated at a rate f_{FM} to run the FFC. The diode voltage signal is sent to two lock-in amplifiers that phase-sensitively detect the microscope resonator response at the modulation frequency f_{FM} and twice the modulation frequency $2f_{FM}$. The primary function of the FFC is to keep the generator locked onto the resonant frequency of the resonator. The $2f_{FM}$ lock-in measures the curvature of the $V_{diode}(f)$ curve, and gives an output that can be related to the quality factor of the resonator, Q [14,93,94].

The inset of Figure 2 presents a simple model of tip-to-sample interaction (in terms of a lumped-element model), and is discussed further in the next section. This resonator has been integrated onto the probe arm of an Oxford Instruments STM probe. An STM feedback loop was also incorporated into the microscope so that the probe-sample distance control could be maintained through a constant tunnel current while scanning [57,59].

4 Theory of Near-Field Microwave Microscopy

The electrodynamic properties of the sample under study, such as complex permittivity and permeability, affect the detection system parameters such as reflection/transmission coefficient, quality factor, and/or resonant frequency. The ultimate goal for theoretical analysis of a near-field microwave probe is to find a relationship between these detectable quantities and the sample properties. The theoretical study can often be broken in two parts: the sample-to-tip interaction and the tip-to-circuit interaction. The former can be treated within either a lumped element or full-wave/evanescent approach, while a theoretical model for the detection circuit depends on its geometry and can be a lumped element, transmission line, cavity perturbation, or other approach. Below we review some of these approaches, present a novel lumped element model for the sample-to-tip interaction, and obtain general analytical expressions relating the detectable probe parameters to the sample properties in the common case of a detection circuit formed by a transmission line resonator.

4.1 Lumped Element Model for Sample-to-Tip Interaction

Consider the near-field tip as a two-terminal, linear, passive system (e.g., antenna) connected to the detection apparatus such as a transmission line, LC-oscillator, etc. If the sample is lossless and only the near-zone and/or evanescent field exists beyond the plane of the terminals, then no real power is transmitted into the sample. Such fields store reactive energy, however, and therefore the tip is seen as a reactance by the probe detection circuit. If the sample and/or tip are lossy, then some of this energy dissipates and the impedance gains a resistive part. Thus the interaction between the sample and the probe can be described in terms of the lumped element impedance $Z_{\mathrm{t}} = R_{\mathrm{t}} + \boldsymbol{i} X_{\mathrm{t}}$. Generally, Z_{t} depends on the tip geometry, the sample electrodynamic properties, and the tip-sample distance. The complex Poynting theorem yields the following expression for the reactance of the near-field tip (i.e. *electrically small* antenna) [17]:

$$X_{\mathrm{t}} = \frac{4\omega}{|I_i|^2} \int\limits_V (w_{\mathrm{m}} - w_{\mathrm{e}}) \mathrm{d}^3 x \tag{1}$$

where I_i is the harmonic input current at the tip terminals (such that input voltage $V_i = Z_{\mathrm{t}} I_i$), $w_m = \boldsymbol{B} \cdot \boldsymbol{H}^*/4$ and $w_{\mathrm{e}} = \boldsymbol{E} \cdot \boldsymbol{D}^*/4$ are the magnetic and electric energy densities, respectively, and the integral is taken over the entire volume beyond the plane defined by the terminals (i.e. the probe sampling volume). Depending on the dominant type of reactive energy stored in this field, the tip is considered either *electric* or *magnetic*. The real part of the tip impedance, R_t, can be due to conduction losses (Joule heat) and electric and magnetic absorptive dissipation in the sample and/or the tip [17,20]:

$$R_{\mathrm{t}} = \frac{\omega}{|I_i|^2} \int\limits_V \left(\frac{\sigma}{\omega} |\boldsymbol{E}|^2 + \varepsilon_0 \varepsilon'' |\boldsymbol{E}|^2 + \mu_0 \mu'' |\boldsymbol{H}|^2 \right) \mathrm{d}^3 x \tag{2}$$

as well as conventional far-field radiation generated by the tip (i.e., radiation resistance) that can often be neglected for most near-field tips. Note that expressions similar to Eqs. (1) and (2) can be derived for the tip complex admittance as well [17].

4.1.1 Impedance of an Electric Tip

Generically, the impedance Z_{tE} of an electric tip formed by the open end of a two-conductor transmission line (e.g., flush open, terminated with STM tip, etc.), can be represented as a network of the tip-to-sample coupling capacitance $1/C_{\mathrm{c}} = 1/C_{\mathrm{c1}} + 1/C_{\mathrm{c2}}$, the sample "near-field" impedance Z_{s}, and the tip stray capacitance C_{str}, (Figure 2, inset):

$$(Z_{\mathrm{tE}})^{-1} = \left(\frac{1}{\boldsymbol{i}\omega C_{\mathrm{c}}} + Z_s \right)^{-1} + \boldsymbol{i}\omega C_{\mathrm{str}} \tag{3}$$

Z_s is due to the energy stored and/or dissipated in the sample under test, and can be found by integrating Eqs. (1) and (2) over the entire *sample* volume (see, for example, [95]). Both C_c and C_{str} depend on the tip-sample distance h. To obtain high enough sensitivity to the sample properties (i.e., to make $Z_{\mathrm{tE}} \sim Z_{\mathrm{s}}$), it is imperative to have both $1/\omega C_{\mathrm{c}} Z_{\mathrm{s}}$ and $\omega C_{\mathrm{str}} Z_{\mathrm{s}}$ much smaller than or at least on the order of unity, which can typically be achieved by making $h \ll D$, where D is the characteristic tip size (see section 1.1).

Under the above conditions let us estimate the near-field impedance due to a homogeneous bulk sample, Z_{sb}, while assuming the true near-field situation $|k_s|D \ll 1$, $k_s = \omega(\varepsilon_0\varepsilon_s\mu_0)^{1/2}$. Due to its static nature, the sampling near-field occupies a volume governed by D, and the integral in Eq. (1) is on the order of $-\varepsilon_0\varepsilon'|E_0|^2D^3$, where $|E_0| \sim V_i/D$ is an estimate for the average electric field in the sample. Then Eqs.(1) and (2) yield for $Z_{sb} = R_{sb} + iX_{sb}$,

$$Z_{sb} \approx \frac{1}{i\omega\varepsilon_0\varepsilon_s D} \tag{4}$$

which is basically the impedance of a capacitor with a geometrical capacitance $\varepsilon_0 D$ filled up with material of complex relative permittivity $\varepsilon_s = \varepsilon' - i\varepsilon''$. Let us illustrate this concept for variety of materials.

Dielectric. If the sample is a low-loss dielectric with $\varepsilon_s = \varepsilon'(1 - i\tan\delta)$, $\tan\delta \ll 1$, then the near-field impedance is

$$Z_{sb} = \frac{\tan\delta}{\omega\varepsilon_0\varepsilon' D} - i\frac{1}{\omega\varepsilon_0\varepsilon' D} \tag{5}$$

Its reactive part is capacitive and the probe is sensitive to the sample dielectric constant as well as the loss tangent (tanδ).

Semiconductor. At microwave frequencies semiconductors can exhibit $\varepsilon' \sim \varepsilon'' = 1/\varepsilon_0\omega\rho$, i.e. contributions from the two conduction mechanisms, displacement and physical, are comparable. Then

$$Z_{sb} = \frac{1}{D/\rho + i\omega\varepsilon_0\varepsilon' D} \tag{6}$$

so the sample resistance and reactance are of the same order. The sample reactance is again capacitive in this case.

Normal metal. A conductive metal can be characterized by the relative permittivity $\varepsilon_s = \varepsilon_{nm} \approx -i/\varepsilon_0\omega\rho$. Thus Z_{sb} represents the dc resistance of the tip sampling volume:

$$Z_{sb} = \rho/D \tag{7}$$

Unlike the conventional surface impedance of a bulk metal (see [96]), Z_{sb} has *no* reactive part because the geometrical inductance is negligible in the case of an electric probe. Eq. (7) implies that when D is much smaller than the metal skin-depth the probe spatial resolution is governed by the tip size rather than the sample skin-depth. This creates a unique opportunity for near-field microwave microscopy to study just the subsurface portion of the material, unlike conventional far-field techniques that are sensitive to the entire skin-depth layer.

Plasma. In alternating electromagnetic fields some materials can be conveniently characterized by a plasma-like permittivity [96]:

$$\varepsilon_p = 1 - \frac{\omega_p^2}{\omega(\omega - i\nu)} \tag{8}$$

where ω_p is the plasma (or Langmuir) frequency, and ν is the collision frequency. One example is a *superconductor* with a permittivity at frequencies up to ~100 GHz and temperatures below ~$0.95T_c$ given by [97]

$$\epsilon^{sc} = -\frac{1}{\epsilon_0\mu_0\omega^2\lambda_{sc}^2} - i\frac{\sigma_1}{\epsilon_0\omega} \tag{9}$$

Here λ_{sc} is the magnetic penetration depth, σ_1 is the normal conductivity, and $1/\varepsilon_0\mu_0\omega^2\lambda_{sc}^2 \gg \sigma_1/\omega\varepsilon_0$ (i.e. the response is dominated by the superconducting carriers). Then the near-field sample impedance of a superconductor is

$$Z_{sb} \approx i\omega\mu_0\frac{\lambda_{sc}^2}{D} \tag{10}$$

Unlike the cases of a dielectric and semiconductor, the near-field reactance of a superconductive sample is *inductive* because the probe is sensitive to the *kinetic* inductance of the superconducting carriers, while again there is no geometrical inductance in Eq. (10).

Another situation when the tip-sample interaction can be described in terms of the lumped element impedance arises if the tip size is such that $|k_s|D \gg 1$ but $|k_0|D \ll 1$. Then $Z_s = G \cdot Z_{sur}$ in Eq. (3), where G ~1 is the tip-dependent geometrical factor, and Z_{sur} is the sample complex *surface* impedance. The typical example is a flush open coaxial tip (see Figure 1(d)) above a bulk metal or thin non-transparent metallic film. In this case the tip impedance is

$$\begin{aligned} Z_t &= \frac{1}{i\omega C_c} + Z_{sur}\frac{1}{2\pi}\ln\frac{r_{diel}}{r_{in}} \\ &= \frac{d}{i\omega\varepsilon_0\pi r_{in}^2} + \frac{d}{i\omega\varepsilon_0\pi\left(r_{out}^2 - r_{diel}^2\right)} + Z_{sur}\frac{1}{2\pi}\ln\frac{r_{diel}}{r_{in}} \end{aligned} \tag{11}$$

where Z_{sur} is the surface impedance in the case of a bulk sample and the effective surface impedance [98,99] in the case of a thin-film sample, and r_{in}, r_{diel}, and r_{out} are the outer radii of the inner conductor, dielectric, and outer conductor, respectively. The above equation holds for small tip-sample distances $h \ll r_{in}, r_{out} - r_{diel}$ which is typically the case for $h \ll D$.

4.1.2 Impedance of a Magnetic Tip

In the case of $|k_s|D \gg 1$ the lumped element impedance of the magnetic tip can be obtained by the method of images [16,100]:

$$Z_{tM} = i\omega L_0 + \frac{\omega^2 M^2}{Z_s + i\omega L_1} \tag{12}$$

where L_0 is the loop self-inductance, L_1 is the sample effective inductance, M is the tip-sample mutual inductance (i.e. tip-to-sample coupling), and Z_s is the sample surface impedance. Away from the sample the tip impedance approaches $i\omega L_0$.

4.2 Transmission Line Model of the Probe

When a transmission line terminated with either an electric or magnetic tip forms a near-field microwave probe, the tip-sample system can be treated as a discontinuity in this line. Consider a near-field tip with impedance Z_t terminating the probe transmission line of characteristic impedance Z_0. The impedance of a substantially electric tip should satisfy the condition $|Z_{tE}| \gg Z_0$, while for a substantially magnetic tip $|Z_{tM}| \ll Z_0$. Then the complex reflection coefficient from the electric tip can be found as follows:

$$\Gamma_E = |\Gamma_E| \exp(i\theta_E) = \frac{Z_{tE} - Z_0}{Z_{tE} + Z_0} \cong \exp\left(-\frac{2Z_0 R_{tE}}{R_{tE}^2 + X_{tE}^2}\right) \exp\left(i\frac{2Z_0 X_{tE}}{R_{tE}^2 + X_{tE}^2}\right) \tag{13}$$

and the reflection coefficient from the magnetic tip is

$$\Gamma_M = |\Gamma_M| \exp(i\theta_M) = \frac{Z_{tM} - Z_0}{Z_{tM} + Z_0} \cong -\exp\left(-\frac{2R_{tM}}{Z_0}\right) \exp\left(-i\frac{2X_{tM}}{Z_0}\right) \tag{14}$$

To obtain the exponential forms in the right hand side above we used the approximation $(1 - x)/(1 + x) \approx \exp(-2x)$, which is better than 1% accurate for $x < 0.25$. Note that an equation similar to Eq. (13) can be written more conveniently in terms of the tip admittance, in which the phase and magnitude dependences on the conductance and susceptance separate like those in Eq. (14) for a magnetic tip.

In order to validate our lumped element model, we compare it to results of numerical modeling for the near-field tip geometry shown in Figure 3(a) and for two types of samples: Si wafers with and without implants. As will be discussed below, both the bulk Si resistivity and the implant sheet resistance were varied. The tip was formed by the open end of a parallel strip transmission line (see Figure 1(e)) that consists of two metallic electrodes supported by a rectangular quartz bar (see [15,114]). For the purposes of modeling the tip size was on the order of 100 μm, and the tip-sample separation was 250 nm. The modeling was performed using Ansoft's High Frequency Structure Simulator (HFSS™), a full-wave high-frequency 3D finite element modeler (FEM).

To find the reflection coefficient for the case of bulk Si we substitute Eq. (6) into (3) and the result into (13), which yields the real and imaginary parts of the complex reflection coefficient as a function of the Si bulk resistivity. We are also able to extract these two quantities from the FEM. This modeling was performed for a variety of probe sizes, and representative results are shown in Figure 3(b), where the Si resistivity ranges from 0.01 to 1000 Ω·cm. It is clear that over a wide range of Si resistivities, the analytical lumped element model agrees exceptionally well

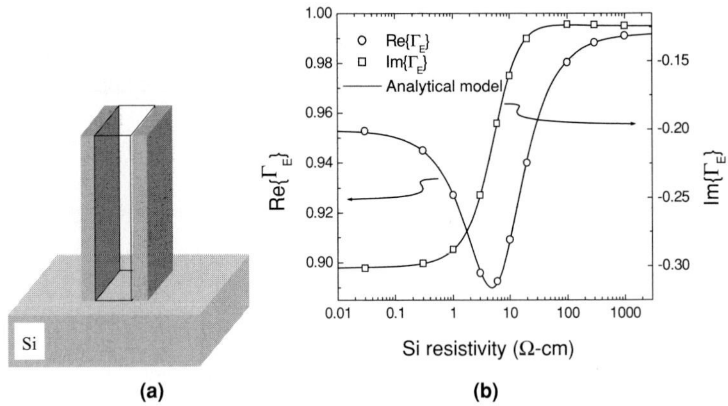

FIGURE 3. (a) Parallel strip transmission line probe. (b) Comparison of analytical lumped element model and finite element numerical data at 4 GHz for the real and imaginary parts of the reflection coefficient as a function of silicon resistivity for bulk Si wafers. The open symbols are the results of the FEM and the lines are fits to the FEM data using the lumped element analytical model, as described in the text.

with the FEM. The two free parameters used to fit the lumped element model to the FEM data are D and C_c, while the characteristic line impedance Z_0 was obtained from 2D modeling of the strip-line cross-section. By fitting data for various tip sizes we find that the characteristic tip size D is governed by the entire cross section of the tip. In addition, C_c was found to be in good agreement with the parallel plate capacitor estimate.

Next we included a 0.5-μm-thick implanted layer on top of the Si wafer to study the response of the probe as a function of the implant sheet resistance R_{sh}. The wafer resistivity ρ_{sub} was fixed at 20 Ω·cm in this case and Z_{sub} is given by Eq. (6). The lumped element model shown in Figure 4(a) now includes the implant impedance $Z_{imp} = G \cdot R_{sh}$, where $G \sim 1$ is the geometrical coefficient dependent on the probe geometry (e.g., see Eq. (11) for the coaxial probe). The results for both lumped element and FEM are shown in Figure 4(b). Again, the agreement between them is excellent over the range of implant sheet resistances from 100 to 800 Ω/sq. In this case the values of D and C_c were fixed as determined from the fitting of FEM data for bulk Si, and the geometrical coefficient G was the only fitting parameter. The fact that the FEM data can be fit so well with only a single fitting parameter is indicative of the robustness of the model and also of the fact that both D (~tip cross section) and C_c depend only on the tip geometry and tip-sample separation.

4.3 Resonant Transmission Line

To increase the measurement sensitivity many microwave microscopes employ a distributed circuit resonator which reduces the impedance mismatch, $|Z_{tE}| \gg Z_0$ or $|Z_{tM}| \ll Z_0$, between the tip and the probe feed line. Using standard transmission

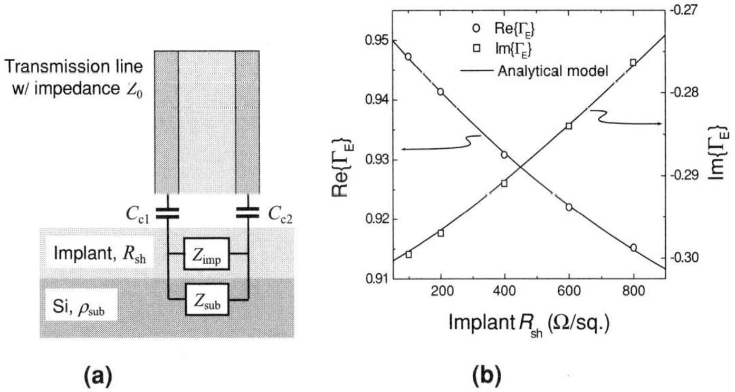

FIGURE 4. (a) Lumped element model for implanted Si in the case of implant thickness $t_{imp} \ll D$, $|k_{imp}|D \ll 1$ and $|k_{sub}|D \ll 1$. (b) Comparison of the lumped element model and high frequency finite element numerical data for the real and imaginary parts of the reflection coefficient as a function of the implant sheet resistance for implanted Si wafers with bulk resistivity of 20 Ω-cm. The open symbols are the FEM data and the lines are fits to the FEM data using the analytical model.

line theory, one can solve for the voltage at any point in the microscope, including an external detector [76,101,102]. Such models can include many details of the microscope design and are very successful at reproducing its gross and fine properties. The disadvantage of such models is the lack of clear intuitive analytical expressions for the measured quantities. Here we suggest an approach that provides general analytical expressions relating the sample electrodynamic properties to the probe resonant frequency and Q-factor.

The resonant condition of a transmission line resonator can be written as follows [103]:

$$\exp(-i2hL)\,\Gamma\Gamma_0 = \exp(-i2\pi n) \tag{15}$$

where L is the resonator length, $h = h' - ih'' = \omega(\varepsilon_0\varepsilon_{\text{eff}}\mu_0)^{1/2} - ih''$ $(h' \gg h'')$ is the transmission line complex propagation constant, ε_0 is vacuum permittivity, μ_0 is vacuum permeability, ε_{eff} is the effective dielectric constant of the transmission line, Γ is given by Eq. (13) or (14), Γ_0 is the complex reflection coefficient from the resonator opposite end (i.e., short, open, or coupling impedance), and $n = 1,2,\ldots$ is the mode number. As an example consider an electric tip terminating a half-wavelength resonator. Substitution of $\Gamma_0 = 1$ and (13) into (15), using the complex angular frequency $\omega = \omega' + i\omega''$, and separating the real and imaginary parts yields the following expressions for the probe resonant frequency $F = \omega'/2\pi$ and the unloaded Q-factor:

$$F = \frac{\pi n}{2\pi L\sqrt{\varepsilon_0\mu_0\varepsilon_{\text{eff}}}}\left[1 + \frac{Z_0 X_{\text{tE}}}{\pi n\left(R_{\text{tE}}^2 + X_{\text{tE}}^2\right)}\right] \tag{16}$$

$$Q = \frac{\omega'}{2\omega''} = \left(\pi n + \frac{Z_0 X_{\text{tE}}}{R_{\text{tE}}^2 + X_{\text{tE}}^2}\right)\Big/\left(2h''L + \frac{2Z_0 R_{\text{tE}}}{R_{\text{tE}}^2 + X_{\text{tE}}^2}\right) \tag{17}$$

From Eq. (16) one can find the relative shift in the probe resonant frequency versus change in the tip capacitance C_t for a low-loss sample as follows:

$$\frac{\Delta F}{F} = -\frac{Z_0}{L\sqrt{\varepsilon_0 \mu_0 \varepsilon_{\text{eff}}}} \Delta C_t \tag{18}$$

Equation (18) provides an estimate for the microscope's sensitivity to changes in the tip capacitance. For typical probe parameters $L \sim 1$ cm, $Z_0 \sim 50\ \Omega$, $\varepsilon_{\text{eff}} \sim 2$ and typical precision in $\Delta F/F$ measurement $\sim$0.1 ppm, ΔC_t is on the order of 10^{-19} F = 100 zF. Capacitance resolution down to a few zepto-Farads has been demonstrated in state-of-the-art experiments [104].

According to Eq. (17) there is a small change in the Q-factor due to change in the probe resonant frequency ω'. In practice however, the losses are not uniform in the transmission line, i.e., h'' depends on z. In this case there will be an extra change in Q due to redistribution of the resonator currents near the tip (see also [105]), even in the case of a lossless sample.

Using the above approach for the magnetic tip one can obtain expressions similar to Eqs. (16) and (17), which are omitted here due to limited space.

4.4 Lumped Element/Series and Parallel Models

Lumped element models are useful for understanding the properties of a single resonance of the microscope. The simple inductor-resistor-capacitor (LRC) series or parallel models have proven successful for generating simple intuitive formulas for microscope behavior [89,106–108]. The tip impedance Z_t should be added in parallel with the capacitor for electric probes [77] or in series with the inductor for magnetic probes [16,100].

4.5 Full-Wave Analysis

If the near-field tip forms a discontinuity in the probe transmission line or waveguide it can be analyzed by the modal analysis technique [20] from which the equivalent lumped element circuit for the discontinuity can be derived using one of the standard discontinuity representations [109]. Then one can proceed to Eqs. (13) and (15) to relate the sample electrodynamic properties to the probe detectable quantities.

Another way to carry out a full-wave analysis is to use numerical simulations of the problem [59,100]. One example is shown in Figure 5 where the fringe electric field near the open end of a parallel strip transmission line (see Figure 1(e)) was simulated using Ansoft's HFSS™, a full-wave 3D finite element modeler of Maxwell's equations. The parallel strip transmission line is formed by 1-μm-thick and 4-μm-wide metallic strips separated by a 4-μm-thick quartz dielectric. Figure 5(a) shows the E-field in the case when no sample is present, i.e., the sample being vacuum. In Figure 5(b) the tip is placed 100 nm above a bulk Si sample with

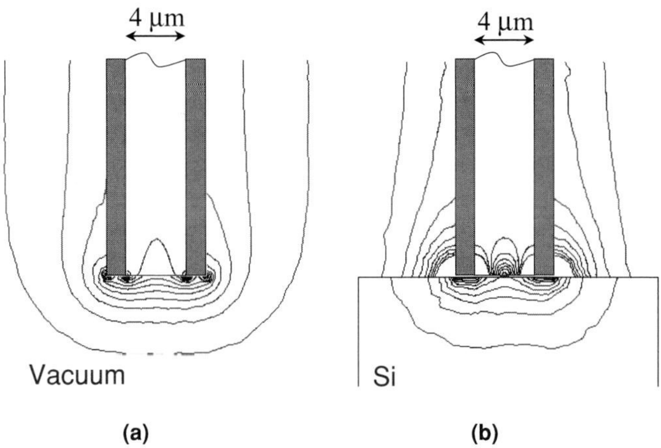

FIGURE 5. Numerical simulation of the fringe electric field at 5 GHz near the open end of the parallel-plate strip-line probe for (a) vacuum and (b) 0.01 Ω·cm resistivity Si samples. The solid lines are intensity contour plots of the E-field magnitude.

resistivity 0.01 Ω-cm. Such a resistivity at 5 GHz provides $\varepsilon' \ll \varepsilon'' = 1/\varepsilon_0\omega\rho$, while the skin-depth is still much greater than the tip size. The solid lines are the intensity contour plots for the E-field. The scale is chosen such that the outermost line indicates where the probing field is reduced by 20 times compared to its maximum value at the sample surface, which differs by almost four orders of magnitude for the two samples. One can see that the field probing depth is basically the same for both samples. It is unaffected by the sample electrodynamic properties and is governed primarily by the tip geometry.

4.6 Cavity Perturbation Approach

Typically, the volume occupied by the probing fields is much less than the overall resonator volume. Thus a cavity perturbation approach [20] can be employed to calculate changes in the resonant frequency and Q-factor of the probe due to its interaction with the sample [105,110]. For example, to calculate changes in the probe resonant frequency F and Q-factor associated with measurement of a low-loss homogeneous dielectric thin-film of unknown complex permittivity $\varepsilon_{\text{film}} = \varepsilon'_{\text{film}}(1 - \boldsymbol{i} \tan\delta)$ backed by an arbitrary substrate, one proceeds as follows. Assume that there are two resonant probes that have the same geometry but differ in the sample that is being measured. One sample is the thin-film sample under test, and the other is an air film with relative permittivity $\varepsilon_{\text{air}} = 1$ of the same thickness and backed by the same substrate as the film under test. One can then calculate the electric fields in the two cases, assuming that everything else about the measurement is identical. The two field solutions shall be called $\boldsymbol{E}_1$ and $\boldsymbol{E}_2$.

Cavity perturbation theory shows that

$$\frac{\Delta F}{F} \cong -\left(\varepsilon'_{\text{film}} - 1\right) \frac{\varepsilon_0}{4W} \int_{V_s} \mathbf{E}_1 \cdot \mathbf{E}_2 \, dV \tag{19}$$

$$\Delta\left(\frac{1}{Q}\right) = -\tan\delta \frac{2\varepsilon'_{\text{film}}}{\varepsilon'_{\text{film}} - 1} \frac{\Delta F}{F} \tag{20}$$

where W is the total energy stored in the resonator, and V_s is the volume of the thin film.

The electric fields $\boldsymbol{E}_1$ and $\boldsymbol{E}_2$ can be calculated by a number of means. One popular choice is to invoke the electrostatic approximation that holds at distances much less than the wavelength from the source, and use the solutions to Poisson's equation for the electric field. Another choice is to use a Fourier optics decomposition of the fields due to a source (such as a sphere above a plane [111], or an electrically small dipole above an interface [112,113]) and directly calculate $\Delta F/F$ through the integrated Poynting vector. A third approach is to use full-wave numerical solutions to Maxwell's equations (see 4.7) for the perturbation calculation. If only the film loss tangent is to be determined from the experiment then no field solution is required if both $\Delta(1/Q)$ and $\Delta F/F$ are measured simultaneously and $\varepsilon'_{\text{film}}$ is known.

4.7 Spatial Resolution

To discuss the spatial resolution of a near-field probe it is convenient to divide the subject into the *qualitative* (or imaging) and *quantitative* (e.g., metrological) resolutions (see [114]). Empirically, the imaging resolution can be defined as the size of the smallest feature that can be observed on a high contrast sample such as alternating strips of high and low permittivity (e.g., metal/dielectric) or permeability material for electric or magnetic probes, respectively. This concept has been widely employed in practice [57,65] and is somewhat similar to using a Au/C combination to define the ultimate resolution of the scanning electron microscope. The metrological resolution however shall be defined as the size of an area or volume such that the probe response is insensitive to the material properties outside of it. In other words, the volumes that confine, say, 50% and nearly 100% of the energy stored in the probing field determine the microscope's qualitative and quantitative resolutions, respectively. While imaging resolution is generally sample dependent, the metrological one is basically defined by the probe geometry due to the static character of a near-field (see also Figure 5).

The imaging resolution of a near-field microwave probe is typically governed by the probe's smallest feature size such as the curvature of an STM or AFM tip apex. Other authors have discussed features governing microwave microscope spatial resolution as well [59,66,106,115–117]. It has been argued elsewhere [118] that for metallic tips the tip geometry rather than the skin-depth of the tip material is the limiting length scale. Limits in the range of a few nanometers have been discussed by several authors [57,65,88,119]. Such an extreme resolution is generally due to redistribution of the currents/charges at the tip in the presence of a high permittivity sample where an image of the probe is induced. At the same time, the metrological resolution of apertureless probes could be on the order of a few hundred microns as governed by the length of the entire STM tip wire or AFM cantilever (see Figure 1(b,c)). This is basically due to the energy stored in the probe parasitic stray field or the "non-tip-end field component" [120]. In the STM-tip based resonant probe, for example, this could account for up to 50% [120] of the total frequency shift caused by the sample and therefore, according to the resonant cavity perturbation theory [20], a similar portion of the reactive *probing* energy as well.

To take full advantage of small antenna feature sizes, the tip must be kept as close as possible to the sample surface. Scanning tunneling, atomic force, and shear-force distance control mechanisms or just a mechanical contact can be used to achieve ultra-resolution, as discussed in section 2.4.

4.8 Quantitative Materials Property Imaging

One advantage of near-field microwave microscopy is the ability to image properties of materials in a quantitative manner. This is possible because of the relative simplicity of the physics of electromagnetic interactions with materials at microwave frequencies. There are two approaches to quantitative imaging. The semi-empirical approach employs standard samples of known properties to establish a calibration data set. The properties of unknown samples are then found by converting the measured quantity into the unknown value through some type of interpolation scheme. These methods are best when they are backed up with a model that justifies the calibration and interpolation routine, (see for example [15,93,102,105,121]). The other approach is to rely entirely on calculations of the probe-sample interaction and to work backward from measured results to deduce the unknown sample property (see, for example, [122]). One drawback of such an approach is that it typically requires knowledge of the exact tip geometry and tip-sample separation.

5 Experimental/Empirical Results in Microwave Microscopy

Here we present an overview of experimental results of microwave materials property investigation with scanning near-field microwave microscopes at sub-wavelength resolution between 100 MHz and 100 GHz. The overview is broken down in terms of materials properties, including dielectrics and ferroelectrics, semiconductors, metals, superconductors, and biological specimens.

5.1 Dielectrics and Ferroelectrics

Insulating materials can display interesting and unique polarization dynamics due to their electrically charged internal constituents. They can also be of technological interest for their ability to store information in the form of ferroelectric polarization, and to serve as either low- or high-dielectric constant materials in wiring networks or capacitors. It is of interest to characterize the dielectric properties of these materials on short length scales because of their use in highly integrated electronics applications. Fundamental materials concerns include the microscopic origins of dielectric response and its sensitivity to defects, the origins of dielectric loss, and the ultimate limit to the size of a stable and switchable ferroelectric domain. Materials issues include the engineering of materials with controlled dielectric response (usually with very high or very low dielectric constant), minimizing dielectric losses, and maximizing ferroelectric polarization, as well as piezoelectric, and magneto-electric effects. Near-field microwave microscopes offer opportunities to investigate all of these properties. We discuss first the localized linear response measurements, followed by the nonlinear response, and finally measurements of dielectric loss.

5.1.1 Linear Dielectric Response

The use of open-ended coaxial transmission lines and resonators to measure linear dielectric response has a very long history [123,124]. The first scanned measurements of dielectric contrast at microwave frequencies used lumped element [64], coaxial [65,125–128], and microstrip resonators [115] but did not extract quantitative measurements of the dielectric constant. Later quantitative modeling of the probe–sample interaction led to the development of quantitative local dielectric constant measurements [12,105,110,120,129–136]. Most of these techniques employ an electric probing tip in contact with the sample. These models relate the sample dielectric constant to the measured frequency shift, and rely on knowledge of the probe and sample geometry, as well as an accurate model of the electric field structure at microwave frequencies. Recently, a novel non-contact technique for dielectric constant metrology of thin films has been developed by Talanov *et al.* [15,114]. It requires no knowledge of either the tip geometry or absolute tip-sample distance, while a shear-force distance control mechanism is used to hold the tip in close proximity to the sample.

Most models employed to date assume that the electrostatic field structure is adequate for these calculations, although experimental and numerical work to investigate finite frequency effects is underway [137]. This method of imaging has proven very useful for the evaluation of sample properties in combinatorial dielectric libraries [138–144]. The imaging of combinatorial libraries and phase diagrams has been reviewed in references [145,146].

5.1.2 Nonlinear Dielectric Response

The microscopes developed for linear dielectric response also prove useful for nonlinear dielectric imaging. Because some of these microscopes use field-concentrating features, such as STM and AFM tips, in contact with the film, a low-frequency electric field can be locally applied to the sample in addition to the microwave electric field. The additional field can influence the dielectric response of the material. This method of imaging was pioneered by Cho, who showed how higher-order terms in the electric displacement versus electric field $D(E)$ relationship could be measured with a scanning coaxial microscope [64,65]. An important advantage of nonlinear imaging is the fact that the sample response is more and more localized near the tip as higher order nonlinearities are measured [147]. The higher-order dielectric constants are proportional to higher powers of the electric field. Because the near-field decays on a scale on the order of the tip size, these higher powers are more concentrated beneath the tip, both laterally and in terms of depth. However, the higher order nonlinear terms become smaller with increasing order, making the measurements more sensitive to noise. Nonlinear dielectric imaging has also been pursued by other groups [91,110,130,131,148–151], including quantitative results for the higher-order terms in the nonlinear $D(E)$ expansion [110,129,130].

It was also found that the nonlinear dielectric response is sensitive to internal de-polarizing fields in polarized materials. The internal field has the effect of shifting the nonlinear dielectric constant versus applied field curve. The amount of shift is proportional to the local value of polarization, making it possible to image a quantity proportional to polarization. This technique was employed in reference [131] to image ferroelectric domains in periodically poled $LiNbO_3$, to image locally switched domains of deuterated triglycine sulfate (DTGS), and to image domain formation and coarsening at the paraelectric to ferroelectric phase transition in DTGS at the Curie temperature $T_C \approx 340$ K, as shown in Figure 6 [102,131].

5.1.3 Dielectric Losses

It was shown above that microwave microscopes have the ability to determine losses in a sample through measurement of the quality factor, Q. The minimum value of $\tan\delta$ that can be measured is determined by the ratio of the energies stored in the tip sampling volume and the probe resonator as well as the experimental capability to measure small differences in Q. Eq. (20) shows that for a typical frequency shift $\Delta F/F \sim 10^{-3}$ and state-of-the-art sensitivity in Q-factor measurement $\Delta(1/Q) \sim 10^{-6}$, loss tangents down to 10^{-3} can be detected.

Many techniques have been developed to measure dielectric loss in stationary conditions on bulk samples [123]. A scanned probe measurement of dielectric loss was first presented by reference [105], where they measured dielectric loss values between 1.5×10^{-3} and 5×10^{-1} in oxide single crystals, and between 0.01 and 0.2 in $SrTiO_3$ and Ba-Sr-Ti-O thin films. Other researchers have measured loss contrast [128,132,152] and values of $\tan\delta$ as small as 1.4×10^{-3} in yttria-stabilized zirconia [133].

5.2 Semiconductors

Semiconducting materials are intermediate between insulators and metals in that they offer two parallel conduction channels: displacement and physical. As the doping density increases, the two conduction mechanisms can become comparable

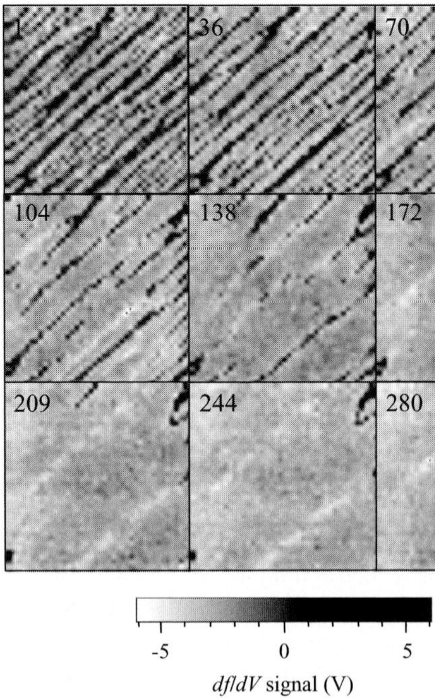

FIGURE 6. Domain formation in a deuterated triglycine sulfate crystal at 320 K. The nonlinear dielectric response images are of the same 100 × 100-μm region of the sample, at different times following cooling from 380 K (above T_C) to 320 K (below T_C). The signal is proportional to the local vertical component of polarization (down is dark, up is bright). The elapsed time in minutes is shown in the upper left corner of each image. For further details, see references [102,131].

at microwave frequencies (see Eq. (4)). Hence microwave microscopes have been pursued as a sensitive way to measure localized doping profiles in semiconductors.

5.2.1 Linear Response

Some of the earliest work on near-field microwave microscopy was devoted to sheet resistance measurements in semiconductors [31,153] and employed scanned coaxial probes [154,155]. Further work explored high-resolution sheet resistance contrast and imaging [57,77,92,156]. Dopant profiling was done through capacitance imaging [157] and spatially resolved measurements of recombination lifetimes have also been done [107,158,159]. Semiconductor thermography [160] and carrier concentration [107] can be measured through near-field techniques as well.

5.2.2 Nonlinear Response

Semiconductor samples also show a nonlinear response. The applied RF and microwave electric fields can create carrier concentration enhancement or depletion near surfaces and buried p/n junctions. This has the effect of changing the capacitance between the probe and sample in a way that depends on the instantaneous value and peak amplitude of the applied oscillating electric field. Such variations can give rise to harmonic generation, mixing, or intermodulation of signals in a localized area around the microscope probe. Nonlinear spectroscopy measurements were carried out at high resolution by [46] and [161]. Later [162] demonstrated that the nonlinear response could be used to significantly enhance the spatial resolution of the microwave microscope. Nonlinear semiconductor spectroscopy was also used to image variation of local carrier density in indium tin oxide [117].

5.3 Metals

The imaging of metals by near-field microwave microscopes has been a very fruitful area of investigation. The microscopes are sensitive to probe-sample capacitance and sample topography, sheet resistance of thin films, and spin properties of ferromagnetic metals and impurities.

5.3.1 Capacitance/Topography

Scanning capacitance microscopy (SCM) has benefited greatly from the use of resonant RF and microwave techniques. The earliest use of a "scanning microwave microscope" for topographic and capacitive imaging was the RCA VideoDisk player [163,164]. This instrument used a 1-GHz lumped LC oscillator that employed a metal-coated tip as one part of the capacitor. As the tip moved over the surface of a corrugated disc, the capacitance variations would be detected through variation of the resonant frequency of the circuit, and information could be retrieved from the topography of the disc. The microscope could detect topography on the order of 0.3 nm with a spatial resolution of 100 nm. Other microwave frequency scanning capacitance microscopes have followed from this invention [165–168]. In most cases, the probe-sample capacitance forms part of a capacitor in a resonant lumped LC circuit. One measures the resonant frequency shift as the probe moves over the sample. This can yield sensitivities to changes in capacitance on the order of zepto-Farads, as demonstrated above. Other near-field microwave microscopes have been employed for quantitative topographic imaging [56,57,76,82,101,104,126,169,170]. Generally speaking, scanned microwave dielectric measurements are basically just capacitance measurements combined with detailed knowledge of the probe and sample geometry. The reader can find more on scanning capacitance microscopy in the chapter by J. Kopanski in this book.

5.3.2 Thin Films/Sheet Resistance/Hall Effect

The microwave microscope has found application as a quantitative measurement of metallic thin-film sheet resistance. After early experiments demonstrating contrast from metallic thin films [76,152,153], Steinhauer *et al.* showed that systematic and quantitative results could be obtained for sheet resistance and topography

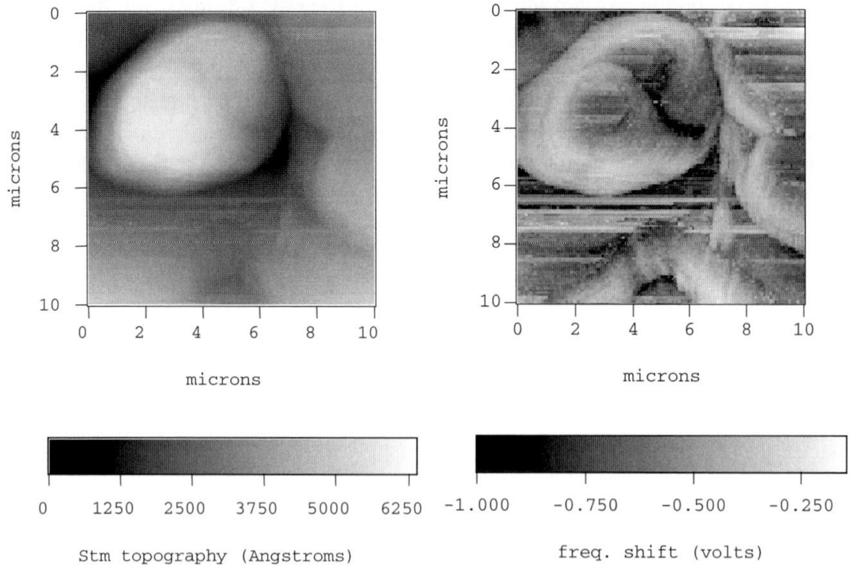

FIGURE 7. STM topography image and frequency shift image of the surface of a $La_{1.4}Sr_{1.6}Mn_2O_7$ single crystal, taken at 298 K and 7.47 GHz, with a bias of 1 V and tunnel current of 0.5 nA. Sample provided by Y. Mukovskii.

by combining the frequency shift and Q-factor measurements [14,93,121]. Other groups developed microscopes with similar capabilities [68,89,169,171–176], including measurements of thin-film conductivity in ferromagnetic combinatorial libraries [140]. An interesting use of the millimeter-wave waveguide microscope was the measurement of local Hall effect signals [177], and polarization contrast [39,178].

Also of significance is the search for nm-scale inhomogeneity in correlated electron metals such as cuprates and manganites [57,179]. Figure 7 shows an example result from an STM-assisted near-field microwave microscope. The frequency shift image shows distinct contrast from sample property variations that are not present in the STM topography image. These local variations contain information about the local sheet resistance and the microscopic physics of the manganite materials.

5.3.3 Ferromagnets/Permeability

The earliest microwave microscope was used to image ferromagnetic resonance [3,4]. Later generations of microscopes returned to the problem of permeability contrast and magnetic resonance. The scanning SQUID microscope was used to generate microwave and millimeter-wave signals in close proximity to the SQUID. The sample response to the high frequency magnetic fields modified the properties of the SQUID, giving rise to contrast [69,70]. Ferromagnetic resonance probes were employed for eddy current microscopy [180]. Later, reference [16] employed a scanned loop probe to make quantitative images of magnetic permeability in ferromagnetic and paramagnetic metals. This microscope was also used to perform imaging of ferromagnetic resonance fields in a bulk manganite sample. A resonant slit microscope was also used to measure ferromagnetic resonance in nanoparticles [181]. Ferromagnetic resonance force microscopy was also developed for investigation of very small localized magnetic moments [182,183].

5.3.4 Electron Spin Resonance Microscopy

An exciting use of microwave microscopes is in electron spin resonance (ESR) microscopy [184]. The pioneering work of Ikeya *et al.* on various ESR microscopes covered many applications in biology, medicine, dentistry, and materials [185–190]. An important objective is to detect spin resonance from a single spin in a sample. This ultimate limit has been pursued through an STM tunnel junction [188,189,191,192].

5.4 Superconductors

The interest in high-temperature superconductors since their discovery in 1986 has spawned research on local investigations of their linear and nonlinear response. These microscopes involve the use of a cryogenic environment for the sample, complicating the operation of the instrument. An early microscope could measure variations in the transition temperature of a superconducting film, as well as some temperature dependence of the local microwave losses [193]. Other work was done imaging high-T_c superconductors [73,194–196], including T_c imaging [172], as well as imaging bulk Nb [171,172].

The nonlinear properties of superconductors are important for both fundamental and practical reasons. On the fundamental side, the measurement of an intrinsic nonlinear Meissner response is an important goal. Most measurements of this small nonlinearity are overwhelmed by extrinsic nonlinearities, most of which come from edges and corners of samples where large current densities build up. Early experiments were done to identify the local sources of intermodulation distortion in resonant superconducting devices [197]. Later experiments used a scanned magnetic loop probe to apply a focused local current distribution at points far from edges and defects in the superconducting sample [71–75,198]. These experiments successfully resolved a localized source of nonlinearity (a single grain boundary Josephson tunnel junction) and found results for spatial and power dependence of the nonlinear signals in good agreement with models of the junction.

5.5 Biological Samples

Microwave techniques are attractive for biological applications because of their sensitivity to water and dielectric contrast [95]. Early dielectric measurements on biological samples were carried out by [124,199–201]. Scanned microwave measurements to determine water content and mineral content were carried out by

Tabib-Azar *et al.* [66,115,202]. Images were also made of dried DNA [116]. Living biological samples often have to be imaged inside saline solutions. Here, as in semiconductors, there are two parallel conduction channels: displacement currents through the water dielectric (with $\varepsilon_r \sim 80$), and conduction current through the salt ions. On a related note, an early use of the STM-assisted microwave microscope was molecular spectroscopy measurements. It was predicted that harmonics due to molecular rotations on a surface would provide unique signatures of molecules [61].

5.6 Parametric Microwave Microscopy

It is of interest to explore the properties of materials at microwave frequencies as a function of other variables or parameters. The dependence of conductivity or dielectric constant on other parameters can give insights into the physical origin of the microwave properties of materials, for example.

5.6.1 Electric Field

Low-frequency electric fields can be used to modify the structure of polar dielectric materials. These changes will in turn modify the dielectric properties of the material. The measurement of nonlinear dielectric response has been investigated by several groups [64,65,110,117,129,130,147]. The low-frequency electric field is typically applied between a sharp tip in contact with the dielectric and a counter electrode. Field strengths on the order of 10^6–10^8 V/cm can be easily applied in thin-film dielectrics. Further, if the material is polarized, it will contain internal depolarization fields. These fields act in a way similar to external bias and shift the nonlinear dielectric versus external bias curves left or right on the bias voltage axis. Such shifts indirectly measure the local polarization of the material underneath the tip [102,131]. Figure 6 shows an example of local polarization imaging in a ferroelectric crystal.

5.6.2 Magnetic Field

A low-frequency magnetic field can be applied during microwave microscopy to modify the magnetic properties of a sample. For example the magnetic field can be used to induce ferromagnetic resonance (FMR) in the sample at a frequency that matches that of the microwave microscope [3,4,16,28,29]. The work of [16] acquired images of the FMR resonant field as a function of position over a $La_{0.8}Sr_{0.2}MnO_3$ (LSMO) sample in a fixed external magnetic field.

5.6.3 Temperature

The microwave properties of materials can vary dramatically with temperature. For example, phase transitions can occur which cause a change of state and ordering in a material. One example is the ferroelectric to paraelectric phase transition at the Curie temperature, T_C. Steinhauer *et al.* have imaged the development of ferroelectric domains out of a paraelectric background as a deuterated triglycine sulfate sample is cooled through $T_C \approx 340$ K (Figure 6). They observed a peak in the dielectric constant at T_C, and a scaling of the domain sizes (coarsening) using a microscope that can vary the sample temperature from room temperature to 400 K [102,131].

Microwave microscopes have also been developed to study the low temperature microwave properties of materials [57,193,195,196]. Most of this work has investigated superconducting properties, including the transition temperature [193,195], the surface resistance [193], and local nonlinear electrodynamic properties [71–75]. One challenge with these microscopes is motion control at cryogenic temperatures. Solutions include generating the motion at room temperature and transmitting it to the low-temperature environment through mechanical coupling [94,126,171,172,195,197,198,203,204], or to use piezoelectric tubes in the cryogenic environment (see Figure 7) [57].

5.6.4 Frequency

Some physical properties measured at microwave frequencies can be frequency dependent. One example is dispersion of dielectric properties due to the contribution of different polarization mechanisms to the total dielectric response. One approach is to build a tunable resonant detection system [117,175,205] by varying the physical size of the microwave detector resonator. This can also vary the coupling between the microwave source and the microscope. One also has the option with transmission line resonators to excite higher-order resonant modes of the system to perform broadband microwave imaging [76]. This has been employed to perform broadband dielectric imaging of Ba-Sr-Ti-O thin-film samples [137,206].

6 Future Directions

The future of near-field microwave microscopy has many promising directions. We discuss a few of these below.

6.1 *Theory*

Much remains to be learned about generating near-zone or evanescent fields, how they interact with the sample, and the most efficient ways to control the confining volume and recover the information contained in their high spatial frequency harmonics (see also section 6.7). Another topic of theoretical study is the relationship between the near-zone and evanescent description of the fields generated by the near-field probes. In the area of quantitative measurements the theory of probe-sample interaction for a multilayer, stratified, and/or anisotropic sample is of great importance and such work is already underway [95]. We also believe that approaches developed in conventional antenna theory (see for example [112,207]) as well as the Poynting theorem (see Eqs. (1) and (2)) could be leveraged for this purpose. For example, they may be employed to determine the conditions when the far-field radiation from the tip can be neglected, which is important for measuring losses in samples.

6.2 Instrumentation

Development of novel probe geometries is necessary to reduce the tip stray capacitance found in many microscopes. This will inevitably improve the sensitivity and spatial resolution of microwave microscopes [208]. Advanced shielding structures for AFM probes are being fabricated [66]. Yet another goal is reduction of the parasitic far-field radiation from the probe, which can be achieved by using balanced quadrupole-like (and higher order) tip geometries [114]. One more open question is the influence of the tip-sample ponderomotive forces on the probe-sample distance control.

6.3 Nano-Imaging

It is clear that microwave microscopes can measure contrast on nanometer length scales, perhaps even quantitatively, and many new scientifically and technologically interesting materials have features on nm scales. For example, magneto-electric nano-composites have very large interface areas between the ferroelectric and ferromagnetic constituents [209]. This can lead to an enhanced magneto-electric effect. Other electrical and mechanical phenomena can be enhanced through nano-engineering of materials, and these would be interesting to investigate locally at microwave frequencies. There is also scientific interest in generating materials contrast on nanometer scales. For example, colossal magneto-resistive materials (see Figure 7) have phase separation into ferromagnetic metallic and charge-ordered insulating regions on nanometer length scales. There are other correlated electron materials that are expected to have similar nano-structure, such as cuprate superconductors. For example, it may also be possible to perform domain wall resonance imaging [210] with microwave microscopes. One concern with nano-imaging is the eventual breakdown of the macroscopic constitutive properties (ε and μ) concept [211,212], and the emergence of quantum properties.

6.4 Single-Spin Microwave Microscopy

Further progress on detecting the spin of a single electron through localized electron spin resonance [41,213] will be made. If this capability can be extended to spin-polarized tunneling, then new applications in quantum computing [192,214] and nano-microwave oscillators [215] may be forthcoming.

6.5 Coupling to High-Speed Nanodevices

The microwave microscope offers the opportunity to couple high-speed signals locally to nanodevices such as nano-electromechanical devices (quantum-limited

GHz oscillators) and carbon nanotube transistors. Multi-walled nanotubes are being investigated for use as interconnects in integrated circuits [216]. The microscope is well suited for measurements of the quantum of resistance, inductance, and capacitance in single nanotubes [217]. The microscope can also be used to measure Coulomb blockade tunneling at RF and microwave frequencies in a mesoscopic device [61].

6.6 Biological Imaging

Much remains to be learned about the capabilities of near-field microscopes in biological imaging [95,115]. There is a great challenge/opportunity to image living cells in their saline solutions. It is possible to design a microscope that has equal current flow through the dielectric and conduction channels in such solutions. These may be very sensitive to small variations in ion concentration or membrane dielectric properties. There is also interest in applying near-field microwave techniques to localized heating of biological media [218].

6.7 Metamaterial Near-Field Lenses

The development of left-handed metamaterials [219] has opened the possibility of new applications in near-field microscopy. One possible new application is that of a near-field flat lens [22,220]. Such lenses take a source in the near-field of a slab of negative index of refraction and convert it into a super-resolution image on the other side, and have been demonstrated at microwave frequencies [221]. In other words, the superlens retains high spatial frequency information in constructing an image. The lens can amplify evanescent waves in a range of wavenumbers, depending on its loss [222]. Integrating such a lens with a near-field microwave microscope may allow further improvements in spatial resolution, or may facilitate new measurements on systems that cannot be studied directly with scanned probes, such as buried features [223]. Another possibility is the manipulation of surface plasmons at microwave and millimeter-wave frequencies [224] to produce enhanced transmission through sub-wavelength apertures [225] for near-field micro- or nano-spectroscopy.

Acknowledgments. We are grateful to R. C. Black, C. Canedy, H. M. Christen, S. Dutta, B. J. Feenstra, T. Hartman, W. Hu, A. Imtiaz, J. Kanner, S.-C. Lee, J. Melngailis, D. Mircea, R. Moreland, Y. Mukovskii, M. Pollak, A. K. Prasad, Y. Qi, R. Ramesh, G. Ruchti, A. Scherz, I. Smolyaninov, D. Steinhauer, A. S. Thanawalla, A. Tselev, C. P. Vlahacos, and F. C. Wellstood for their assistance and fruitful discussions. We also thank M. Golosovsky, A. Reznik, and A. Tselev for critical reading of the manuscript. This work is supported by the National Science Foundation through the UMD MRSEC NSF/DMR-00-80008 and its Microwave Microscope SEF, NSF/GOALI DMR-0201261, and through NIST-ATP No. 70NANB2H3005.

References

1. M. Dressel and G. Grüner, *Electrodynamics of Solids*, Cambridge University Press, Cambridge, 2002.
2. M. Tinkham, *Introduction to Superconductivity*, 2nd Edition, McGraw-Hill, New York, 1996, p. 87.
3. Z. Frait, V. Kambersky, Z. Malek, and M. Ondris, Czeck. J. Phys. B **10**, 616 (1960).
4. R. F. Soohoo, J. Appl. Phys. **33**, 1276 (1962).
5. R. F. Soohoo, *Microwave Magnetics*, Harper & Row Publishers, New York, 1985.
6. E. H. Brandt and G. P. Mikitik, Phys. Rev. Lett. **85**, 4164 (2000).
7. S. M. Anlage, D. E. Steinhauer, B. J. Feenstra, C. P. Vlahacos, and F. C. Wellstood, "Near-Field Microwave Microscopy of Materials Properties," in *Microwave Superconductivity*, ed. by H. Weinstock and M. Nisenoff, (Kluwer, Amsterdam, 2001), p. 239. Also available at http://arXiv.org/abs/cond-mat/0001075.
8. X.-D. Xiang, C. Gao, Mat. Character. **48**, 117 (2002).
9. B. T. Rosner, D. W. van der Weide, Rev. Sci. Instrum. **73**, 2505 (2002).
10. R. Zoughi, *Microwave Non-Destructive Testing and Evaluation*, Kluwer, The Netherlands, 2000.
11. A. J. Bahr, R. Zoughi, N. Qaddoumi, "Microwave," in *Nondestructive Evaluation*, ed. by P. J. Shull (Marcel Dekker, New York, 2002), p. 645.
12. C. Gao, T. Wei, F. Duewer, Y. Lu and X.-D. Xiang, Appl. Phys. Lett. **71**, 1872 (1997).
13. D. W. van der Weide, Appl. Phys. Lett. 70, 677 (1997).
14. D. E. Steinhauer, C. P. Vlahacos, S. Dutta, F. C. Wellstood, S. M. Anlage, Appl. Phys. Lett. **71**, 1736 (1997).
15. V. V. Talanov, R. L. Moreland, A. Scherz, A. R. Schwartz, and Y. Liu, Mater. Res. Soc. Symp. Proc. **815**, F5.11.1 (2004).
16. S.-C. Lee, C. P. Vlahacos, B. J. Feenstra, A. Schwartz, D. E. Steinhauer, F. C. Wellstood, and S. M. Anlage, Appl. Phys. Lett. **77**, 4404 (2000).
17. J. D. Jackson, *Classical Electrodynamics*, 3rd Edition, Wiley, New York, 1999.
18. S. Schelkunoff, *Antennas, Theory and Practice*. Wiley and Sons, Inc., New York, 1952.
19. S. S. Osofsky, S. E. Swarz, IEEE Trans. Microwave Theory Tech. **40**, 1701 (1992).
20. D. M. Pozar, *Microwave Engineering*. John Wiley & Sons, Inc., New York, 1998.
21. E. A. Synge, Phil. Mag. C **6**, 356 (1928).
22. J. B. Pendry, Phys. Rev. Lett. **85**, 3966 (2000).
23. J. B. Pendry, Phys. Rev. Lett. **87**, 249704-1 (2001).
24. H. A. Bethe, Phys. Rev. **66**, 163 (1944).
25. C. J. Bouwkamp, Philips Res. Rep. **5**, 401 (1950).
26. G. A. Massey, Appl. Optics **23**, 658 (1984).
27. R. D. Grober, T. Rutherford, T. D. Harris, Appl. Optics **35**, 3488 (1996).
28. S. E. Lofland, S. M. Bhagat, H. L. Ju, G. C. Xiong, T. Venkatesan, and R. L. Greene, Phys. Rev. B **52**, 15058 (1995).
29. S. E. Lofland, S. M. Bhagat, Q. Q. Shu, M. C. Robson and R. Ramesh, Appl. Phys. Lett. **75**, 1947 (1999).
30. E. A. Ash and G. Nicholls, Nature **237**, 510 (1972).
31. C. A. Bryant and J. B. Gunn, Rev. Sci. Instrum. **36**, 1614 (1965).
32. R. G. Bosisio, M. Giroux, and D. Couderc, J. Microwave Power **5**, 25 (1970).

33. D. W. Pohl, W. Denk, and M. Lanz, Appl. Phys. Lett. **44**, 651 (1984).
34. E. Betzig, M. Isaacson and A. Lewis, Appl. Phys. Lett. **51**, 2088 (1987).
35. M. Fee, S. Chu and T. W. Hänsch, Optics Communications **69**, 219 (1989).
36. M. Golosovsky and D. Davidov, Appl. Phys. Lett. **68**, 1579 (1996).
37. M. Golosovsky, A. Galkin, and D. Davidov, IEEE Micro. Theor. Tech. **44**, 1390 (1996).
38. J. Bae, T. Okamoto, T. Fujii, K. Mizuno, T. Nozokido, Appl. Phys. Lett. **71**, 3581 (1997).
39. T. Nozokido, R. Iibuchi, J. Bae, K. Mizuno, Rev. Sci. Instrum. **76**, 033702 (2005).
40. G. P. Kochanski, Phys. Rev. Lett. **62**, 2285 (1989).
41. Y. Manassen, R. J. Hamers, J. E. Demuth, A. J. Castellano Phys. Rev. Lett. **62**, 2531 (1989).
42. R. J. Hamers and D. G. Cahill, Appl. Phys. Lett. **57**, 2031 (1990).
43. M. Völcker, W. Krieger, H. Walther, Phys. Rev. Lett. **66**, 1717 (1991).
44. W. Seifert, E. Gerner, M. Stachel, K. Dransfeld, Ultramicroscopy **42**, 379 (1992).
45. U. Ch. Fischer, M. Zapletal, Ultramicroscopy **42–44**, 393 (1992).
46. B. Michel, W. Mizutani, R. Schierle, A. Jarosch, W. Knop, H. Benedickter, W. Bächtold, H. Rohrer, Rev. Sci. Instrum. **63**, 4080 (1992).
47. W. Mizutani, B. Michel, R. Schierle, H. Wolf, H. Rohrer, Appl. Phys. Lett. **63**, 147 (1993).
48. G. Nunes, M. R. Freeman, Science **262**, 1029 (1993).
49. S. J. Stranick and P. S. Weiss, Rev. Sci. Instrum. **64**, 1232 (1993).
50. S. J. Stranick and P. S. Weiss, Rev. Sci. Instrum. **65**, 918 (1994).
51. L. A. Bumm and P. S. Weiss, Rev. Sci. Instrum. **66**, 4140 (1995).
52. A. Kramer, F. Keilmann, B. Knoll, R. Guckenberger, Micron **27**, 413 (1996).
53. F. Keilmann, D. W. van der Weide, T. Eickelkamp, R. Merz, and D. Stöckle, Optics Commun. **129**, 15 (1996).
54. T. Leinhos, O. Rudow, M. Stopka, A. Vollkopf, E. Oesterschulze, J. Microscopy **194**, 349 (1999)
55. F. Demming, J. Jersch, S. Klein, K. Dickmann, J. Microscopy **201**, 383 (2001).
56. Š. Lányi, M. Hruškovic, Rev. Sci. Instrum. **73**, 2923 (2002).
57. A. Imtiaz and S. M. Anlage, Ultramicroscopy **94**, 209 (2003).
58. L. A. Valiente, A. A. P. Gibson, A. D. Haigh, Electron. Lett. **40**, 1483 (2004).
59. A. Imtiaz , M. Pollak , S. M. Anlage, J. D. Barry and J. Melngailis , J. Appl. Phys. **97**, 044302 (2005).
60. F. Keilmann, Infrared Phys. Technol. **36**, 217 (1995).
61. S. J. Stranick, L. A. Bumm, M. M. Kamna, P. S. Weiss, "Linear and nonlinear spectroscopy with the tunable AC scanning tunneling microscope," in *Photons and Local Probes*, ed. by O. Marti and R. Möller (Kluwer, Dordrecht, 1995), p. 221.
62. H. Yokoyama, M. J. Jeffery and T. Inoue, Jpn. J. Appl. Phys. **32**, L1845 (1993).
63. H. Yokoyama and M. J. Jeffery, Colloids and Surfaces A **93**, 359 (1994).
64. Y. Cho, A. Kirihara and T. Saeki, Rev. Sci. Instrum. **67**, 2297 (1996).
65. Y. Cho, S. Kazuta, and K. Matsuura, Appl. Phys. Lett. **75**, 2833 (1999).
66. M. Tabib-Azar, Y. Wang, IEEE Trans. Microwave Theory Tech. **52**, 971 (2004).
67. H. G. Frey, F. Keilmann, A. Kriele, R. Guckenberger, Appl. Phys. Lett. **81**, 5030 (2002).
68. W. Park, J. Kim, K. Lee, Appl. Phys. Lett. **79**, 2642 (2001).
69. R. C. Black, F. C. Wellstood, E. Dantsker, A. H. Miklich, D. T. Nemeth, D. Koelle, F. Ludwig, and J. Clarke, Appl. Phys. Lett. **66**, 99–101 (1995).

70. R. C. Black, F. C. Wellstood, E. Dantsker, A. H. Miklich, D. Koelle, F. Ludwig, and J. Clarke, IEEE Trans. Appl. Supercon. **5**, 2137 (1995).
71. S.-C. Lee and S. M. Anlage, Appl. Phys. Lett. **82**, 1893 (2003).
72. S.-C. Lee and S. M. Anlage, IEEE Trans. Applied Supercond. **13**, 3594 (2003).
73. S.-C. Lee and S. M. Anlage, Physica C **408–410**, 324 (2004).
74. S.-C. Lee, M. Sullivan, G. R. Ruchti, S. M. Anlage, B. Palmer, B. Maiorov, E. Osquiguil, Phys. Rev. B **71**, 014507 (2005).
75. S.-C. Lee, S.-Y. Lee, and S. M. Anlage Phys. Rev. B **72**, 024527 (2005).
76. C. P. Vlahacos, R. C. Black, S. M. Anlage and F. C. Wellstood, Appl. Phys. Lett. **69**, 3272 (1996).
77. M. Tabib-Azar, N. Shoemaker and S. Harris, Meas. Sci. Tech., **4**, 583 (1993).
78. A. Tselev, S. M. Anlage, H. Christen, R. L. Moreland, V. V. Talanov, and A. R. Schwartz, Rev. Sci. Instrum. **74**, 3167 (2003).
79. E. Betzig, P. L. Finn, J. S. Weiner, Appl. Phys. Lett. **60**, 2484 (1992).
80. R. Toledo-Crow, P. C. Yang, Y. Chen, M. Vaez-Iravani, Appl. Phys. Lett. **60**, 2957 (1992).
81. K. Karrai, R. D. Grober, Appl. Phys. Lett. **66**, 1842 (1995).
82. M. S. Kim, S. Kim, J. Kim, K. Lee, B. Friedman, J.-T. Kim, J. Lee, Rev. Sci. Instrum. **74**, 3675 (2003).
83. C.-G. Gao, B. Hu, I. Tackeuchi, K.-S. Chang, X.-D. Xiang, and G. Wang, Meas. Sci. Tech. **16**, 248 (2005).
84. H. Odagawa and Y. Cho, Jpn. J. Appl. Phys. **39**, 5719 (2000).
85. D. W. van der Weide, P. Neuzil, J. Vac. Sci. Technol. B **14**, 4144 (1996).
86. *Scanning Tunneling Microscopy*, ed. by J. A. Stroscio and W. J. Kaiser, Academic Press, New York (1993).
87. R. Wiesendanger, *Scanning Probe Microscopy and Spectroscopy: Methods and Applications*, Cambridge (1994).
88. B. Knoll, F. Keilmann, A. Kramer, R. Guckenberger, Appl. Phys. Lett. **70**, 2667 (1997).
89. M. Tabib-Azar, D.-P. Su, A. Pohar, S. R. LeClair, G. Ponchak, Rev. Sci. Instrum. **70**, 1725 (1999).
90. F. Duewer, C. Gao, I. Takeuchi, X.-D.Xiang, Appl. Phys. Lett. **74**, 2696 (1999).
91. F. Duewer, C. Gao, I. Takeuchi, X.-D.Xiang, Rev. Sci. Instrum. **71**, 2414 (2000).
92. B. Knoll and F. Keilmann, Appl. Phys. Lett. **77**, 980 (2000).
93. D. E. Steinhauer, C. P. Vlahacos, S. K. Dutta, B. J. Feenstra, F. C. Wellstood, S. M. Anlage, Appl. Phys. Lett. **72**, 861 (1998).
94. A. S. Thanawalla, S. K. Dutta, C. P. Vlahacos, D. E. Steinhauer, B. J. Feenstra, S. M. Anlage, and F. C. Wellstood, Appl. Phys. Lett. **73**, 2491 (1998).
95. A.N. Reznik, N.V. Yurasova, Tech. Phys., **49**, 485 (2004).
96. S. Ramo, J. R. Whinnery, T. V. Duzer, *Fields and Waves in Communication Electronics*. John Wiley & Sons, Inc., New York, 1994.
97. Z.-Y. Shen, *High-Temperature Superconductive Microwave Circuits*. Artech House, Inc., Boston, 1994.
98. R. E. Matick, *Transmission Lines for Digital and Communication Networks*, IEEE Press, New York, 1995.
99. M. Hein, *High-Temperature-Superconductor Thin Films at Microwave Frequencies*, Springer Tracts in Modern Physics, Vol. 155, Springer-Verlag, 1999.
100. S.-C. Lee, Ph.D. thesis, University of Maryland, 2004.
101. C. P. Vlahacos, D. E. Steinhauer, S. K. Dutta, B. J. Feenstra, S. M. Anlage, and F. C. Wellstood, Appl. Phys. Lett., **72**, 1778 (1998).

102. D. E. Steinhauer, Ph.D. Thesis, University of Maryland (2000).
103. V. V. Talanov, L. V. Mercaldo, S. M. Anlage, and J. H. Claassen, Rev. Sci. Instrum. **71**, 2136 (2000).
104. T. Tran, D. R. Oliver, D. J. Thomson, and G. E. Bridges, Rev. Sci. Instrum. **72**, 2618 (2001).
105. C. Gao, and X.-D. Xiang, Rev. Sci. Instrum. **69**, 3846 (1998).
106. M. Tabib-Azar, D. Akinwande, G. Ponchak, and S. R. LeClair, Rev. Sci. Instrum. **70**, 3381 (1999).
107. M. Tabib-Azar, D. Akinwande, Rev. Sci. Instrum., **71**, 1460 (2000).
108. M. Tabib-Azar, T. Zhang, and S. R. LeClair, IEEE Trans. Instrum. Meas. **51**, 1126 (2002).
109. *Waveguide handbook*, ed. N. Marcuvitz. McGraw-Hill Book Co., Inc., New York, 1951.
110. D. E. Steinhauer, C. P. Vlahacos, C. Canedy, A. Stanishevski, J. Melngailis, R. Ramesh, F. C. Wellstood, and S. M. Anlage, Appl. Phys. Lett. **75**, 3180 (1999).
111. X.-D. Xiang, C. Gao, P. G. Schultz, T. Wei, US Patent # 6,532,806 (2003).
112. R. W. P. King, M. Owens, T. T. Wu, *Lateral Electromagnetic Waves*, Springer-Verlag, New York, 1992.
113. A. Imtiaz and S. M. Anlage, J. Appl. Phys. (in press), August 1, (2006).
114. V. V. Talanov, A. Scherz, R. L. Moreland, A. R. Schwartz, Appl. Phys. Lett. **88**, 134106 (2006); *ibid* **88**, 192906 (2006); V. V. Talanov, A. Scherz, A. R. Schwartz, Appl. Phys. Lett. **88**, 262901 (2006).
115. M. Tabib-Azar, P. S. Pathak, G. Ponchak, and S. LeClair, Rev. Sci. Instrum. **70**, 2783 (1999).
116. J. Kim, M. Kim, H. Kim, D. Song, K. Lee, B. Friedman, Appl. Phys. Lett. **83**, 1026 (2003).
117. M. Kim, J. Kim, H. Kim, S. Kim, J. Yang, H. Yoo, S. Kim, K. Lee, B. Friedman, Rev. Sci. Instrum. **75**, 684 (2004).
118. V. S. Zuev and A. V. Frantsesson, J. Commun. Tech. and Electronics **43**, 1021 (1998).
119. Š. Lányi, M. Hruškovic, J. Phys. D **36**, 598 (2003).
120. J. H. Lee, S. Hyun, K. Char, Rev. Sci. Instrum. **72**, 1425 (2001).
121. S. M. Anlage, B. J. Feenstra, D. E. Steinhauer, US Patent # 6,376,836 (2002).
122. R. Hillenbrand, F. Keilmann, Phys. Rev. Lett. **85**, 3029 (2000).
123. E. Tanabe and W. T. Joines, IEEE Trans. Instrum. Meas. **IM-25**, 222 (1976).
124. M. A. Stuchly and S. S. Stuchly, IEEE Trans. Instrum. and Meas. **IM-29**, 176 (1980).
125. Y. Lu, T. Wei, F. Duewer, Y. Lu, N. Ming, P. G. Schultz and X.-D. Xiang, Science **276**, 2004 (1997).
126. B. J. Feenstra, C. P. Vlahacos, A. S. Thanawalla, D. E. Steinhauer, S. K. Dutta, F. C. Wellstood and S. M. Anlage, IEEE MTT-S Int. Microwave Symp. Digest, p. 965 (1998).
127. C. P. Vlahacos, D. E. Steinhauer, S. M. Anlage, F. C. Wellstood, S. K. Dutta, J. B. Feenstra, The Americas Microscopy and Analysis, January, 13 (2000).
128. Y. G. Wang, M. E. Reeves, F. J. Rachford, Appl. Phys. Lett. **76**, 3295 (2000).
129. Y. Cho, S. Kazuta, K. Ohara, H. Odagawa, Jpn. J. Appl. Phys. **39**, 3086 (2000).
130. D. E. Steinhauer, C. P. Vlahacos, F. C. Wellstood, S. M. Anlage, C. Canedy, R. Ramesh, A. Stanishevsky, and J. Melngailis, Rev. Sci. Instrum. **71**, 2751 (2000).
131. D. E. Steinhauer and S. M. Anlage, J. Appl. Phys. **89**, 2314 (2001).
132. Y. G. Wang, M. E. Reeves, W. J. Kim, J. S. Horwitz, and F. J. Rachford, Appl. Phys. Lett. **78**, 3872 (2001).

133. Y. J. Feng, L. Y. Wu, T. Jiang, L. Liu, L. Kang, S. Z. Yang, P. H. Wu, Supercond. Sci. Technol. **15**, 390 (2002).
134. Y.-C. Chen, H.-F. Cheng, G. Wang, X.-D. Xiang, C.-M. Lei, I.-N. Lin, Jpn. J. Appl. Phys. **41**, 7214 (2002).
135. Z. Wang, M. A. Kelly, Z.-X. Shen, G. Wang, X.-D. Xiang, J. T. Wetzel, J. Appl. Phys. **92**, 808 (2002).
136. C. Gao, B. Hu, P. Zhang, M. Huang, W. Liu, I. Takeuchi, Appl. Phys. Lett. **84**, 4647 (2004).
137. A. Tselev and S. M. Anlage, in preparation, 2006.
138. H. Chang, C. Gao, I. Takeuchi, Y. Yoo, J. Wang, P. G. Schultz, X.-D. Xiang, R. P. Sharma, M. Downes, and T. Venkatesan, Appl. Phys. Lett. **72**, 2185 (1998).
139. H. Chang, I. Takeuchi, X.-D. Xiang, Appl. Phys. Lett. **74**, 1165 (1999).
140. Y. K. Yoo, F. Duewer, T. Fukumura, H. Yang, D. Yi, S. Liu, H. Chang, T. Hasegawa, M. Kawasaki, H. Koinuma, X.-D. Xiang, Phys. Rev. B **63**, 224421 (2001).
141. H. Chang, K.-M. Yu, Y. Dong, X.-D. Xiang, Appl. Phys. Lett. **81**, 2062 (2002).
142. N. Okazaki, S. Okazaki, H. Higuma, S. Miyashita, Y. Cho, J. Nishimaru, T. Fukumura, M. Kawasaki, M. Murakami, Y. Yamamoto, Y. Matsumoto, H. Koinuma, T. Hasegawa, Applied Surface Science **223**, 196 (2004).
143. H.-W. Cheng, X.-J. Zhang, S.-T. Zhang, Y. Feng, Y.-F. Chen, Z.-G. Liu, G.-X. Cheng, Appl. Phys. Lett. **85**, 2319 (2004).
144. X.-D. Xiang, Appl. Surf. Sci. **223**, 54 (2004).
145. I. Takeuchi, R.B. van Dover, and H. Koinuma, Mater. Res. Soc. Bull. **27**, 301 (2002).
146. H. Koinuma, I. Takeuchi, Nature Materials **3**, 429 (2004).
147. Y. Cho and K. Ohara, Appl. Phys. Lett. **79**, 3842 (2001).
148. C. Hubert, J. Levy, Rev. Sci. Instrum. **70**, 3684 (1999).
149. C. Hubert, J. Levy, E. J. Cukauskas, S. W. Kirchoefer, Phys. Rev. Lett. **85**, 1998 (2000).
150. S. Hyun, J. H. Lee, S. S. Kim, K. Char, S. J. Park, J. Sok, E. H. Lee, Appl. Phys. Lett. **77**, 3084 (2000).
151. H. Odagawa and Y. Cho, Appl. Phys. Lett. **80**, 2159 (2002).
152. M. J. Werner and R. J. King, MRS Proc. **430**, 67 (1996).
153. Y. S. Xu and R. G. Bosisio, IEE Proc. H **139**, 500 (1992).
154. R. J. Gutman and J. M. Borrego, IEEE MTT Digest, 281 (1987).
155. H. Bhimnathwala and J. M. Borrego, J. Vac. Sci. Technol. B **12** , 395 (1994).
156. T. Nozokido, J. Bae, K. Mizuno, IEEE Trans. Microwave Theory Tech. **49**, 491 (2001).
157. Y. Huang, C. C. Williams, J. Slinkman, Appl. Phys. Lett. **66**, 344 (1995).
158. M. Tabib-Azar, D. Akinwande, G. Ponchak, S. R. LeClair, Rev. Sci. Instrum. **70**, 3083 (1999).
159. C. C. Watson, W. K. Chan, Appl. Phys. Lett. **78**, 129 (2001).
160. M. Tabib-Azar, R. Ciocan, G. Ponchak, and S. R. LeClair, Rev. Sci. Instrum. **70**, 3387 (1999).
161. J.-P. Bourgoin, M. B. Johnson, B. Michel, Appl. Phys. Lett. **65**, 2045 (1994).
162. B. Knoll and F. Keilmann, Optics Commun. **182**, 321 (2000).
163. A comprehensive discussion of the VideoDisc technology may be found in a collection of articles published in RCA Rev. **39** (1978).
164. J. R. Matey and J. Blanc, J. Appl. Phys. **57**, 1437 (1985).
165. C. D. Bugg and P. J. King, J. Phys. E **21**, 147 (1988).

166. H. P. Kleinknecht, J. R. Sandercock, H. Meier, Scanning Microscopy **2**, 1839 (1988).
167. Š. Lányi, J. Török, P. Řehůřek, Rev. Sci. Instrum. **65**, 2258 (1994).
168. Š. Lányi, J. Török, P. Řehůřek, J. Vac. Sci. Technol. B **14**, 892 (1996).
169. A. F. Lann, M. Golosovsky, D. Davidov, and A. Frenkel, Appl. Phys. Lett. **73**, 2832 (1998).
170. A. Kim, J. Kim, S. Hyun, S. S. Kim, T. Kim, K. Char, Rev. Sci. Instrum. **74**, 3164 (2003).
171. S. M. Anlage, C. P. Vlahacos, D. E. Steinhauer, S. K. Dutta, B. J. Feenstra, A. Thanawalla, and F. C. Wellstood, Particle Accelerators **61**, 321./57 (1998). Also available at http://www.arXiv.org/abs/cond-mat/9808195.
172. S. M. Anlage, D. E. Steinhauer, C. P. Vlahacos, B. J. Feenstra, A. S. Thanawalla, W. Hu, S. K. Dutta, and F. C. Wellstood, IEEE Trans. Appl. Supercond. **9**, 4127 (1999).
173. Y. J. Feng, L. Liu, Z. L. Fu, L. Y. Wu, Q. G. Liu, L. X. You, L. Kang, S. Z. Yang, P. H. Wu, S. Z. Wang, IEEE Trans. Applied Supercond. **11**, 123 (2001).
174. F. Sakran, M. Golosovsky, H. Goldberger, D. Davidov, A. Frenkel, Appl. Phys. Lett. **78**, 1634 (2001).
175. J. Kim, K. Lee, B. Friedman, D. Cha, Appl. Phys. Lett. **83**, 1032 (2003).
176. Z. Wang, M. A. Kelly, Z.-X. Shen, L. Shao, W.-K. Chu, H. Edwards, Appl. Phys. Lett. **86**, 153118 (2005).
177. M. Abu-Teir, F. Sakran, M. Golosovsky, D. Davidov, A. Frenkel, Appl. Phys. Lett. **80**, 1776 (2002).
178. A. F. Lann, M. Golosovsky, D. Davidov, A. Frenkel, Appl. Phys. Lett. **75**, 603 (1999).
179. S. Hyun, J. H. Cho, A. Kim, J. Kim, T. Kim, and K. Char, Appl. Phys. Lett. **80**, 1574 (2002).
180. B. A. Auld, and D. K. Winslow, "Microwave Eddy-Current Experiments with Ferromagnetic Resonance Probes," in *Eddy-Current Characterization of Materials and Structures* (ASTM Special Technical Publication 722, Philadelphia, 1981), p. 348.
181. F. Sakran, A. Copty, M. Golosovsky, and D. Davidov, P. Monod, Appl. Phys. Lett. **84**, 4499 (2004).
182. Z. Zhang, P. C. Hammel and P. Wigen, Appl. Phys. Lett. **68**, 2005 (1996).
183. Z. Zhang, P. C. Hammel, M. Midzor, M. L. Roukes, and J. R. Childress, Appl. Phys. Lett. **73**, 2036 (1998).
184. G. R. Eaton and S. S. Eaton, Bull. Mag. Res. **10**, 22 (1987).
185. M. Ikeya and T. Miki, Jap. J. Appl. Phys. **26**, L929 (1987).
186. M. Ikeya, M. Furusawa, and M. Kasuyai, Scanning Microscopy **4**, 245 (1990).
187. M. Ikeya, "ESR Microscopy," in *New Applications of Electron Spin Resonance* (World Scientific, Singapore, 1993), p. 427.
188. M. Hirai, C. Yamanaka, M. Ikeya, Appl. Radiation and Isotopes **44**, 385 (1993).
189. K. Wago, D. Botkin, C. S. Yannoni, and D. Rugar, Appl. Phys. Lett. **72**, 2757 (1998).
190. F. Sakran, A. Copty, M. Golosovsky, N. Bontemps, D. Davidov, A. Frenkel, Appl. Phys. Lett. **82**, 1479 (2003).
191. Y. Manassen, Adv. Mater. **6**, 401 (1994).
192. C. Durkan and M. E. Welland, Appl. Phys. Lett. **80**, 458 (2002).
193. I. Takeuchi, T. Wei, F. Duewer, Y. K. Yoo, X.-D. Xiang, V. Talyansky, S. P. Pai, G. J. Chen, and T. Venkatesan, Appl. Phys. Lett. **71**, 2026 (1997).
194. A. F. Lann, M. Abu-Teir, M. Golosovsky, D. Davidov, A. Goldgirsch, and V. Berlin, Appl. Phys. Lett. **75**, 1766 (1999).

195. S. M. Anlage, C. P. Vlahacos, S. Dutta, and F. C. Wellstood, IEEE Trans. Appl. Supercond. **7**, 3686 (1997).
196. A. F. Lann, M. Abu-Teir, M. Golosovsky, D. Davidov, S. Djordjevic, N. Bontemps, and L. F. Cohen, Rev. Sci. Instrum. **70**, 4348 (1999).
197. W. Hu, B. J. Feenstra, A. S. Thanawalla, F. C. Wellstood, and S. M. Anlage Appl. Phys. Lett. **75**, 2824 (1999).
198. S. M. Anlage, W. Hu, C. P. Vlahacos, D. Steinhauer, B. J. Feenstra, S. K. Dutta, A. Thanawalla, and F. C. Wellstood, J. Supercond. **12**, 353 (1999).
199. E. C. Burdette, F. L. Cain, and J. Seals, IEEE Trans. Microwave Theory Tech. **MTT-28**, 414 (1980).
200. T. W. Athey, M. A. Stuchly and S. S. Stuchly, IEEE Trans. Microwave Theory and Tech. **MTT-30**, 82 (1982).
201. M. A. Stuchly, T. W. Athey, G. M. Samaras and G. E. Taylor, IEEE Trans. Microwave Theory and Tech. **MTT-30**, 87 (1982).
202. M. Tabib-Azar, J. L. Katz, and S. R. LeClair, IEEE Trans. Instrum. Meas. **48**, 1111 (1999).
203. S. M. Anlage, A. S. Thanawalla, A. P. Zhuravel', W. Hu, C. P. Vlahacos, D. E. Steinhauer, S. K. Dutta, and F. C. Wellstood, "Near-Field Scanning Microwave Microscopy of Superconducting Materials and Devices," in *Advances in Superconductivity XI*, ed. by N. Koshizuka and S. Tajima, (Springer-Verlag, Tokyo, 1999), pp. 1079.
204. A. S. Thanawalla, W. Hu, D. E. Steinhauer, S. K. Dutta, B. J. Feenstra, S. M. Anlage, F. C. Wellstood, and R. B. Hammond, IEEE Trans. Appl. Supercond. **9**, 3042 (1999).
205. S. Hong, J. Kim, W. Park, K. Lee, Appl. Phys. Lett. **80**, 524 (2002).
206. K. S. Chang, M. Aronova, O. Famodu, I. Takeuchi, S. E. Lofland, J. Hattrick-Simpers, H. Chang, Appl. Phys. Lett. **79**, 4411 (2001).
207. L. B. Felsen, N. Marcuvitz, *Radiaiton and Scattering of Waves*, IEEE Press, Piscataway, NJ, 1994.
208. C. Gao, F. Duewer, and X.-D. Xiang, Appl. Phys. Lett. 75, 3005 (1999).
209. H. Zheng, J. Wang. S. E. Lofland, Z. Ma, L. Mohaddes-Ardabili, T. Zhao, L. Salamanca-Riba, S. R. Shinde, S. B. Ogale, F. Bai, D. Viehland, Y. Jia, D. G. Schlom, M. Wuttig, A. Roytburd, R. Ramesh, Science **303**, 661 (2004).
210. E. Saitoh, H. Miyajima, T. Yamaoka, G. Tatara, Nature **432**, 203 (2004).
211. Y. Cho, K. Fujimoto, Y. Hiranaga, Y. Wagatsuma, A. Onoe, K. Terabe, K. Kitamura, Appl. Phys. Lett. **81**, 4401 (2002).
212. K. Fujimoto and Y. Cho, Jpn. J. Appl. Phys. **43**, 2818 (2004).
213. A. V. Balatsky, Y. Manassen, R. Salem, Phys. Rev. B **66**, 195416 (2002).
214. C. Durkan, Contemp. Phys. **45**, 1 (2004).
215. S. I. Kiselev, J. C. Sankey, I. N. Krivorotov, N. C. Emley, R. J. Schoelkopf, R. A. Buhrman, D. C. Ralph, Nature **425**, 380 (2003).
216. F. Kreupl, A. P. Graham, G. S. Duesberg, W. Steinhögl, M. Liebau, E. Unger and W. Hönlein, Microelec. Eng. **64**, 399 (2002).
217. P. J. Burke, IEEE Trans. Nanotech. **1**, 129 (2002).
218. A. Copty, M. Golosovsky, D. Davidov, A. Frenkel, IEEE Trans. Microwave Th. Tech. **52**, 1957 (2004).
219. R. A. Shelby, D. R. Smith, and S. Schultz, Science **292**, 77 (2001).
220. V. G. Veselago, Usp. Fiz. Nauk 92, 517 (1967) {English translation: Sov. Phys. Usp. **10**, 509 (1968)}.
221. A. A. Houck, J. B. Brock, I. L. Chuang, Phys. Rev. Lett. **90**, 137401 (2003).

222. D. R. Smith, D. Schurig, M. Rosenbluth, S. Schultz, S. A. Ramakrishna, J. B. Pendry, Appl Phys. Lett. **82**, 1506 (2003).
223. N. Fang, H. Lee, C. Sun, X. Zhang, Science **308**, 534 (2005).
224. J. B. Pendry, A. J. Holden, D. J. Robbins, W. J. Stewart, J. Phys. Condens. Matter **10**, 4785 (1998).
225. T. W. Ebbesen, H. J. Lezec, H. F. Ghaemi, T. Thio, P. A. Wolff, Nature **391**, 667 (1998).

I.8
Electromagnetic Singularities and Resonances in Near-Field Optical Probes

ALEXANDRE BOUHELIER AND RENAUD BACHELOT

1 Introduction

Over the last two decades scanning near-field optical microscopy (SNOM) has demonstrated its ability to provide optical resolution significantly better than the diffraction limit (<20 nm). The general principle of SNOM relies on the approach of a nanometer-sized object in the optical near-field of a sample to be studied. This nano-object (*NO*) is usually the extremity of a probe. Regardless of the nature of the observed SNOM signal (inelastic scattering, fluorescence, etc.), the detection of the light is achieved in the far-field regime where the *NO* acts as a mediator between the optical near-field and the detector. Figure 1 is a schematic illustration of the SNOM principle.

For a given position *M* of *NO* on the sample surface, the total field detected in the far-field (E_d) is given by the Huygens–Fresnel principle. It is the result of the interference between Huygens spherical wavelets issued from the sample detection zone *S*. In other words, the SNOM signal results from the integration of these wavelets over *S*. In the case of classical far-field microscopy, the minimum size of *S* is given by the diffraction limit. In this case E_d, describes only the sample optical properties averaged over *S*, which cannot be smaller than approximatively $\lambda/2n$, where λ is the light wavelength and n is the refraction index of the sample medium. Near-field optics provides the unique opportunity to move the *NO* laterally in a controllable way within the sample near-field. E_d becomes sensitive to E'_d, which represents the small contribution of the probe/sample interaction within the Huygens–Fresnel integration. E'_d corresponds in fact to the small contribution of the integration over *S'* (centered on *M*), which represents the spatial domain of the interaction defined by the size of the probe. Consequently, scanning the probe above the sample surface allows the optical properties of the surface to be described, pixel by pixel, with a resolution mainly affected by *S'*.

The challenge of SNOM is the optimization of the weight of E'_d respect to E_d, with respect to both dynamics and sensitivity of the detector. Two main approaches have been proposed, called "aperture" and "aperureless" SNOM.

In the first case the *NO* is a dielectric nano-aperture surrounded by metal which acts as an opaque screen confining the radiation into the aperture. This aperture

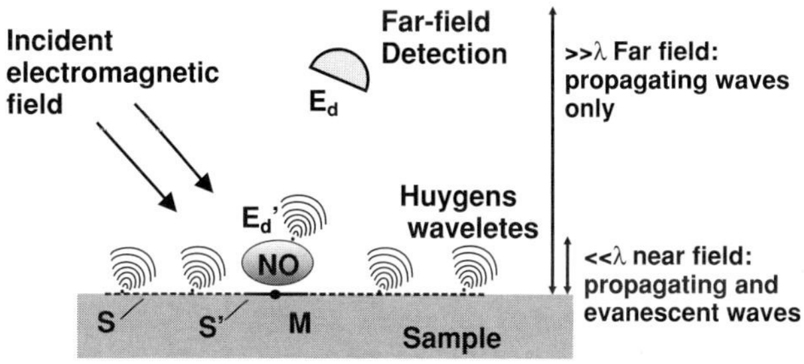

FIGURE 1. General principle of scanning near-field optical microscopy.

can be used indiscriminately as an optical near-field source or as a collector of the sample's near-field. The size of the aperture is an important parameter, if not the most important one, because the domain of integration S is drastically reduced. In this case $S \approx S'$, and $E_d \approx E'_d$ and high-contrast near-field imaging is made possible. This approach was first proposed in 1928 by Synge [1]. It was demonstrated experimentally in 1972 by Ash and Nichols in the microwave regime [2] and by Pohl et al. in the visible range in 1984 [3].

High-resolution imaging is thus achieved by minimizing the interaction volume between a probe and an object. Control over probe manufacturing is a prerequisite for routine high-resolution imaging. Unlike the successful batch microfabrication of AFM cantilevers, the reliable fabrication of near-field aperture probes remains a technological issue that has yet to be overcome. In its classical embodiment, sharply pointed optical fibers are usually utilized to reduce the optical interaction volume [4]. Optical fibers have the advantage of low fabrication costs and low propagation losses. The guiding properties of these optical waveguides fulfill the conditions needed for near-field applications to a large extent. For instance, polarization can be accurately controlled in the fiber [5]. The central step in the probe manufacturing is the formation of a taper region to form a nanometric glass tip terminating one end of the fiber. Such glass tips can be formed either by locally melting the glass and subsequent pulling of the fiber [6] or can be etched chemically [7, 8]. Both techniques can provide a relatively small tip end diameter (<100 nm) by carefully controlling the various fabrication parameters.

The guiding properties of an optical fiber are well understood. In the taper region, a propagating guided mode becomes increasingly delocalized, and the electromagnetic field reaches out of the fiber core. The consequence is that while the extremity of the fiber can be a few tens of a nanometer, the spatial extension of the mode significantly exceeds this dimension. To confine the mode and to guide it to the very tip, a metal coating is deposited on the outside surface of the fiber. The layer acts as a reflector and therefore prevents the light to spread out of the fiber. Aluminum, silver, and gold are the most commonly used materials, owing to their good optical properties at visible wavelengths (small skin depth and high reflectivity). If the metal coating covers the entire taper region, the fiber will be opaque and no light will be transmitted or collected. For the light to escape, a sub-wavelength hole or aperture is needed at the very apex of the coated tip. The fabrication of such an aperture with nanometer-sized dimensions poses a variety of technical difficulties. A large number of techniques have been tried to solve these issues with various degree of success. Squeezing of a completely metal-coated optical waveguide toward a hard surface in order to press the metal away from the foremost end [3, 9], the evaporation of the metal layer such that the very apex of a fiber tip is left uncoated, solid electro-erosion [10, 11], a self-terminated corrosion process [12], and focused ion beam modifications [13, 14] are the most common approaches used to produce a sub-50-nm aperture.

While aperture-based near-field optical probes have been successfully employed to produce sub-wavelength optical resolutions in a large number of studies, it is commonly accepted that these types of probe suffer from several fundamental limitations. For large-aperture (>50 nm), the lateral resolution of a probe is, at first approximation, given by the diameter of the aperture. However, when the diameter of the opening becomes comparable to the skin depth of the coating material, the spatial distribution of the electromagnetic field at the tip apex can be significantly larger than the physical size of the aperture. Furthermore, as the diameter of the aperture decreases, the intensity of the light transmitted through the opening also rapidly decreases following a sixth-order power law. Consequently, it becomes very difficult to sustain a good signal-to-noise ratio for very small apertures.

Alternative near-field optical methods have been recently developed to circumvent the limitations of aperture SNOM. These techniques, referred to as apertureless SNOM, are based on the local electromagnetic enhancement occurring in resonant nanostructures and/or the scattering properties of a homogeneous tip [15–19]. Metal surfaces, for instance, play an important role in spectroscopy, where they can greatly increase the yield of weak optical transitions through an electromagnetic enhancement. These increased responses originate from a complex interplay between surface plasmon resonances of the underlying structure and local geometric singularities. The use of nanometric metal particles locally to enhance an electromagnetic field and their potential use for optical microscopy was first suggested by John Wessel in 1985 [20]. In his pioneering article, Wessel envisioned a nanometric metal particle to act as a local enhanced source of photons that can be scanned above a sample, forming an optical image with sensitivity approaching the single molecule level. Among the increasing range of applications are Raman spectroscopy at a single molecule level [21], sensitive biorecognition [22], manipulation of light in nanostructures [23–25], and high-resolution optical imaging, which will be discussed below. To use the formalism used above, the *NO* acts as a homogeneous nano-antenna, scattering a field E'_d. Here, both illumination and detection are far-field in nature, and one has $S \gg S'$ and $E_d \gg E'_d$. E'_d can be extracted from the background field E_d only under specific conditions. The reader is referred to set a recent set of reviews for additional details [26–34]. The recent advances of field-enhanced near-field optical microscopy constitute the subject of the present chapter. The reader will find an introduction of the underlying physical mechanisms giving rise to the field-enhancement effect as well as some successful examples of high-resolution imaging, spectroscopy, and nanolithography.

2 The Field Enhancement Effect

The field enhancement occurring at the extremity of a homogeneous metal tip originates from two different mechanisms: surface plasmon oscillations and a lightning-rod effect. These two distinct phenomena can be combined for particular tip geometries and excitation conditions as discussed in the following paragraphs.

2.1 Surface Plasmon Resonances

Plasmon excitations, i.e., collective oscillations of the conduction electrons, strongly influence the optical properties of metal nanostructures and much interest and extensive works over the past 30 years have been devoted to their study [35–38]. In particular, plasmon resonance found applications in several areas of sciences and technology, including nanosensoring and plasmonics [39]. In this section, we will review some of the important plasmon properties and focus our attention on resonances occurring on SNOM probes. Like most physical systems, the electron density present in metals structures can be modulated by resonances. Resonance parameters such as peak wavelength and relaxation times depend mainly on charge density, the electron effective mass, and conductivity of the material. The Drude model is instrumental in predicting the dispersion of the permittivity $\epsilon(\omega)$. For bulk metal, $\epsilon(\omega)$ can be written as

$$\epsilon(\omega) = \epsilon_\infty - \frac{\omega_p^2}{\omega(\omega + i\gamma_o)} \approx \epsilon_\infty - \frac{\omega_p^2}{\omega^2} + i\frac{\gamma_o\omega_p^2}{\omega^3}, \tag{1}$$

where ω_p is the plasma frequency, γ_o is the electron relaxation rate (related to damping due to collisions), and ϵ_∞ is the dielectric constant at $\omega = \infty$ ($\epsilon_\infty = 1$ if only free electron are considered). γ_o is the inverse of τ, the electron relaxation time (basically, the average time between two succussive collisions), which depends on material properties

$$\tau = \frac{\sigma m^*}{ne^2}, \tag{2}$$

where n, m^*, σ, and e are the density, the effective mass of the conduction electrons, the material conductivity, and the electron charge, respectively.

The plasma frequency ω_p depends on material properties like

$$\omega_p = \sqrt{ne^2/\epsilon_o m^*}, \tag{3}$$

where ϵ_o is the dielectric constant of the vacuum. Many properties of real metals have been shown to be well predicted by the Drude model described by Eq. (1). It should be stressed that this model only takes into account free electrons within the conduction band and considers intraband electron transitions only. Correct

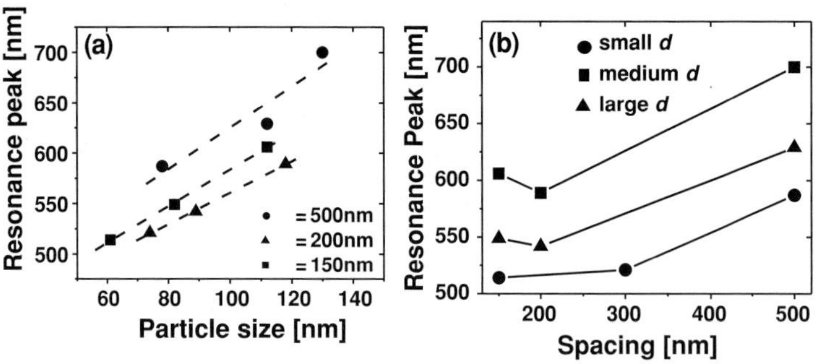

FIGURE 2. (a) Surface resonance peak λ_{max} plotted as a function of particles sizes d for three different lattice spacings Λ, 500 nm (circles), 200 nm (triangles) and 150 nm (squares). The lines represent the result of a linear fitting of the data for the three lattice spacings. (b) Spectral position of the peak plotted as a function of the lattice spacing.

description of nanometer-sized metal particles involves bounded electrons and interband transition from other bands to the conduction band. For bulk material, the electron resonance corresponds to a vanishing real part of the dielectric constant (see Eq. (1)). Plasmon modes actually exist in a number of geometries and in various metals, with the strongest responses in noble metals such as gold and silver. In confined geometries such as metallic nanoparticles, the plasmon resonance not only depends on materials parameters (σ, n, e, and m^*) but is also affected by the shape of the object and its local environment, as well as the condition of illumination such as polarization and incident angle [39]. As a significant example, Figure 2 shows the scattered plasmon resonance of a 2D arrays of gold disks for different particle sizes and lattice spacings [40]. The surface plasmon peak wavelength is drastically affected by the size of the particle (blue-shift of the resonance for decreasing particle diameter) and by the near-field interaction between neighboring particle (suggested by the red-shift of the resonance for spacing below 200 nm).

In metal nanoparticles, the resonance leads to a sudden increase of the extinction cross section over a relatively short spectral distribution. In the context of apertureless SNOM, a large effort has been dedicated to the development of local resonant objects that could serve as local optical probes. This approach was first discussed theoretically by Wessel in 1985 [20] and by Denk and Pohl in 1991 [41]. Experimentally, protrusions in a metallic film were successfully used to scatter surface plasmons providing thus near-field detailed information [42]. Following this approach, several groups were able to characterize and use isolated resonant metal nanoparticles to perform near-field microscopy [43–48]. In these studies, individual nanoparticles were optically trapped [48] or attached to the end of sharpened optical fibers by a pre-functionalization of the surface [44–47]. An example of such resonant particle is shown in Figure 3(a). Here, a single gold nanoparticle was attached at the end of an etched optical fiber by functionalizing the glass surface with silane vapor [47]. It is important to note that the magnitudes of the plasmon resonances occurring in the spherical metal particle are significantly smaller than

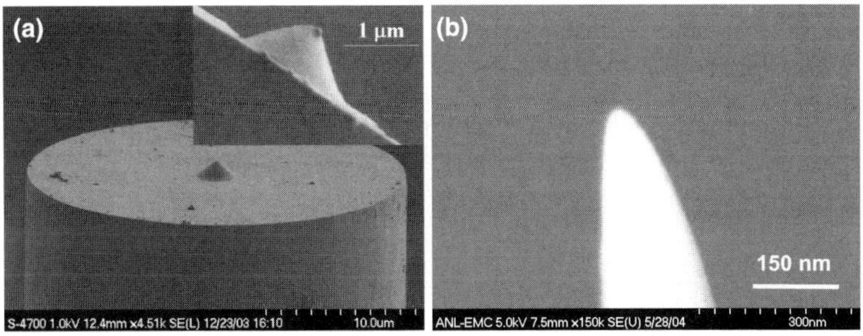

FIGURE 3. (a) Scanning electron micrograph of a single gold nanoparticle attached at the end of the etched fiber glass tip. Reprinted with permission from [47], Copyright 2005, American Institute of Physics. (b) Scanning electron micrograph of a gold tip etched in a Hydrochloric acid using an alternating current. The radius of curvature is approximatively 15 nm.

for nonspherical geometries [36]. In particular, structures such as ellipsoids or triangles were shown to have a considerable enhancement over the driving field and consequently may lead to better probe characteristics. For a conical tip geometry, where the metal structure can be considered as semi-infinite, there is a compromise between the establishment of large surface plasmon resonances and a local lightning-rod effect. Indeed, as discussed in the next section, an electromagnetic singularity at a tip end can be efficiently excited for an incident polarization parallel to the tip axis. However, for this polarization orientation, the surface plasmon resonance of the tip is probably located in the infrared region of the spectrum. Furthermore, because of the semi-infinite geometry, the main resonance is likely to be broad and contributes marginally to the overall field enhancement.

2.2 The Lightning-Rod Effect and Geometrical Singularities

The term *lightning-rod effect* refers to a phenomenon in which the electric charges on the surface of a conductive material are spatially confined by the shape of the structure. For a conductor with a nonspherical shape, the surface charge density, σ, will vary from point to point along the shape. In the regions of high curvature, σ is locally increased, resulting in a large electric field just outside the material. If the object is at a high electric potential, the field can be large enough to trigger a dielectric breakdown of the surrounding medium.

The confinement of the electric field around point of high curvature is routinely used to protect buildings from the destructive power of lightning. Sharply pointed poles ("lightning-rods") at an equipotential with the earth provide on one hand, a low-resistance electric path to a lightning discharge and, on the other hand, a means to reduce the surrounding local electrostatic potential [49–51].

Although this effect originates from macroscopic electrostatic considerations, a similar phenomenon can be observed in nanometer-sized metallic structures excited by electromagnetic radiation. The free electrons of the metal react to

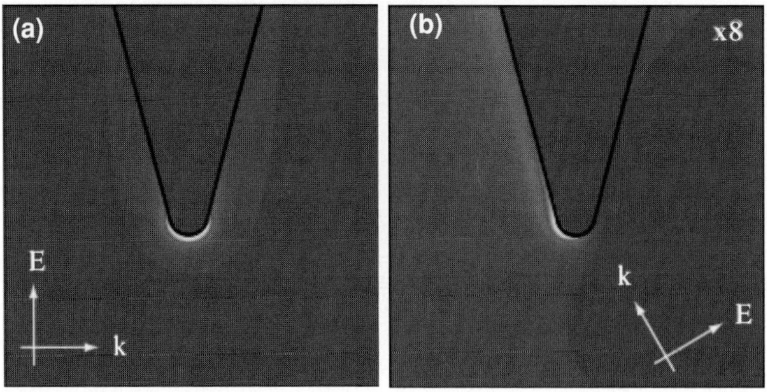

FIGURE 4. Calculated electric field intensity distribution around a 10 nm radius Au tip. (a) The incident polarization is oriented with the main axis of the tip. (b) Side illumination.

electromagnetic excitation by inducing oscillating surface charges. When the surface presents a geometrical singularity such as a tip apex, the local surface charge density is drastically increased in this region. As a consequence, the electromagnetic field outside the tip is not only locally enhanced over the driving field but is also highly confined near the tip apex.

This lightning-rod effect is illustrated in the calculations shown in Figure 4(a). The image shows the intensity of the electric field just outside a Au tip with a radius of 10 nm. The tip is illuminated by a monochromatic plane wave incident from the side of the image with a polarization vector aligned vertically. The intensity near the apex of the tip is increased with respect to the driving field, and is also largely confined within the tip radius. Figure 4(b) shows the same calculation for an illumination from the side where the polarization is no longer aligned with the tip main axis. The overall magnitude of the intensity is drastically reduced from the ideal illumination condition and the intensity is no longer localized at the very apex of the tip. This effect will be discussed in the next section in more details.

From the experimental point of view, several approaches have been used with various degree of success to produce nanometric metal tips. In many cases, the tips are fabricated using etching techniques originating from scanning tunnelling microscopy. Both direct (dc) and alternating current (ac) electrochemistry [52,53] produce tips with radius of curvature below 20 nm. For a listing of the various chemical techniques, the reader is referred to [54]. Figure 3(b) shows a scanning electron micrograph of a gold tip produced by ac electroerosion [53]. The radius of the apex is approximatively 15 nm. Focused ion beam (FIB) milling has been successfully used to tailor well-defined tip shapes [55]. The advantage of this top-down approach is that the technique can be employed regardless of the tip material, as opposed to classic wet etching techniques that are more material-specific. Another technological approach consists of taking advantage of the well-established batch fabrication of atomic force microscope cantilever tips. The bare silicon tips are usually subsequently coated with a thin metallic layer to increase their electromagnetic enhancement capability [56–59]. Recently, alternative probe fabrication techniques clearly demonstrated their potential for field-enhanced scanning near-field optical microscopy. For instance, properly engineered metallic nanotip grown near a sub-wavelength aperture can be used to control the physical phenomena responsible for a field enhancement effect [60–62]. Following the same idea of combining near-field aperture probes and the field enhancement effect, theoretical studies have shown that for completely metal-coated optical fiber tips and for particular guided modes, constructive interference at the extremity of the tip would provide the basis for a field enhancement effect while keeping well-established fabrication techniques [63–66].

2.2.1 Parameters Influencing the Electromagnetic Enhancement

The lightning-rod effect and the field localization at the apex of sharply pointed nanometric tips are strongly sensitive to experimental parameters. The polarization of the driving field is of primary importance, but the enhancement factor can also be varied by controlling the tip shape as well as the material used.

2.2.1.1 Polarization

The polarization of the incident field plays a crucial role in the lightning-rod effect. When the incident electric field is oriented perpendicular to the main axis of the tip, the oscillating charges have opposite signs on either side of the tip at any given time. As a consequence, the apex of the tip remains uncharged. If the polarization is rotated 90° to be aligned with the main axis of the tip, the surface charge density is maximum at its apex leading to an enhanced electromagnetic field (see Figure 4) [67]. The necessity to have a strong component of the electric field oriented axially in order to obtained a substantial enhancement of the intensity around the tip apex was confirmed by numerous theoretical studies [68–72]. An example of the polarization effect is shown in Figure 5 for three different orientations of the incident electric field. The total intensity in the gap between a tungsten tip and a silica substrate is maximized for an electric field oriented along the tip axis.

From the experimental point of view, there are a number of different options that can provide the necessary illumination polarization. The most common of them are sketched in Figure 6. Early schemes used a laser excitation illuminating the tip from the side as represented in Figure 6(a) [58, 73–79]. The polarization of the light is oriented in a way that a strong component of the electric field is projected along the main axis of the tip. The main advantage of this mode of operation is that there are few restrictions on the optical characteristics of the sample under investigation [80]. This could be of relevance when investigating opaque, thick, or buffered specimens. On the reverse side, the illumination beam is only relatively weakly focused at the tip extremity, mainly due to the geometrical constraints inherent to this approach (use of numerical aperture < 1, and long focal distance objective lens). The resulting relatively large excitation area covers several microns of the tip but also a substantial area of the sample itself. The far-field

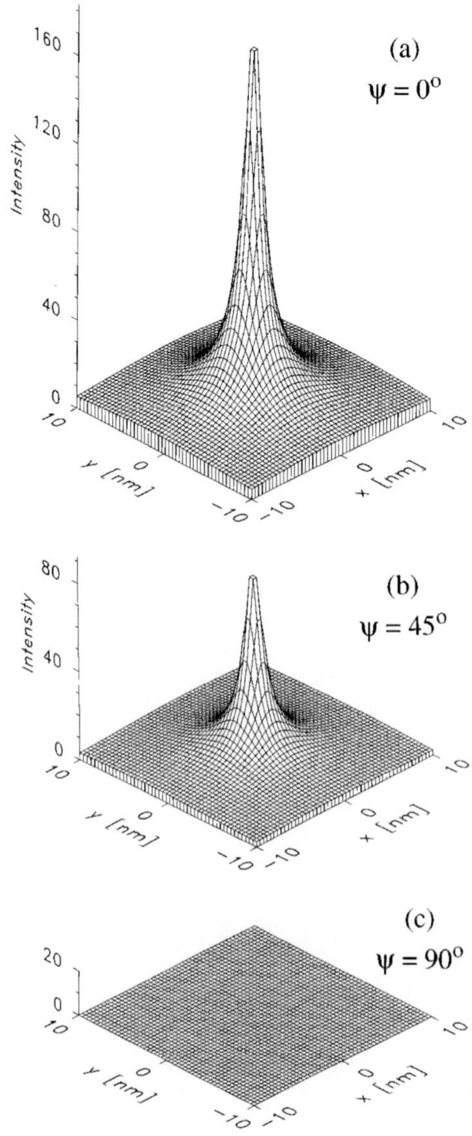

FIGURE 5. Intensity of the electric field at the junction between a tungsten tip and a glass substrate for p polarization ($\Psi = 0°$), intermediate polarization ($\Psi = 45°$), and s polarization ($\Psi = 90°$), respectively. The angle of incidence is 80° for a wavelength of 633 nm. Reprinted with permission from [68]. Copyright 1997, American Institute of Physics.

optical interactions originating from the sample give rise to a large background that is difficult to remove. For elastic processes, where illumination and detection are done at the same wavelength, the background can be partially suppressed by homodyne and heterodyne demodulation techniques [19,73,81]. For spectroscopic investigations, time-stamping background suppression can be applied to extract a near-field fluorescence molecular response [82].

Another illumination approach is depicted in Figure 6(b). Here, the tip is excited by an evanescent field created by total internal reflection inside a hemispherical lens

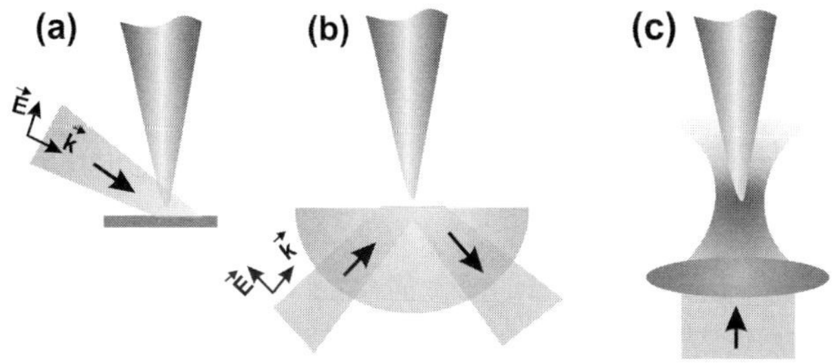

FIGURE 6. Schematic representation of the different experimental illumination schemes. (a) Side illumination with a strong polarization component aligned with the tip axis. The optical interactions are collected either via a side detection or by transmission. (b) Evanescent excitation through total internal reflection. p polarized light is necessary to excite the enhancement effect. The prism can be replaced by a immersion oil high numerical aperture objective. (c) Excitation of the tip by a longitudinal field created by highly focused laser beams.

or a prism. This configuration largely reduces the scattered background reflected off the surface, and has been implemented by several groups [83–85]. As expected, both theoretical and experimental studies revealed that only p polarized light generates a localized enhanced field at the tip apex. An interesting variant of this illumination scheme makes use of high numerical aperture oil immersion objectives (N.A. $>$ 1). A total internal reflection at a glass coverslip is achieved by using a beam mask covering the area corresponding to N.A. $<$ 1 [56, 80]. This geometry increases the electric field perpendicular to the interface [86, 87], providing better illumination conditions to create a field enhancement.

Figure 6(c) shows an alternative to the approaches described above. The concept here is to use the unusual electric field distributions found in tightly focused laser beam. For focal regions approaching the theoretical size limit imposed by diffraction, it has been shown that the electric field is composed of two transverse components (perpendicular to the wavevector k) and of longitudinal components oriented along the direction of propagation [88]. These particular components do not propagate and exist only in the focal region. For a fundamental Gaussian beam, these so-called longitudinal fields are relatively weak, and little enhancement can be expected [88, 89]. However, diffraction-limited focal spots of higher-order laser modes such as Hermite-Gaussian and radial beams present strong longitudinal field components exceeding their transverse counterparts. Because the polarization of the longitudinal fields is exactly aligned with the main axis of the tip (see Figure 6(c)), they provide the necessary illumination conditions to create a field enhancement localized at the apex of a tip [53, 90–93].

2.2.1.2 Probe Material

From an electrostatic point of view, the necessary condition on the type of material only requires a good electrical conductor. Aluminum, tungsten, iridium, silver and gold are good candidates since their electrical resistivities are below

$10^{-7}\Omega$ m at room temperature [94]. While some theoretical works calculated the field enhancement factor in the electrostatic limit (see, for instance, [41,95]), the presence of surface plasmon waves in the surface charge density [67], and possible phase retardation effects [69] suggest that a complete electrodynamics picture may be more accurate for the description of the phenomenon. In consequence, one has to consider not only good electrical conductors but also materials behaving like metals from the electrodynamics standpoint when the dielectric function ϵ is wavelength-dependent and size effects (retardation) are taken into account [36]. Aluminum, silver, and gold are again among the most likely metals, because in the visible region, the imaginary part of ϵ, i.e., the absorption loss, is relatively small compared to the real part of the permittivity ($\epsilon'' \ll \epsilon'$).

Various authors have theoretically investigated the enhancement factors obtained for different tip materials, including silver, gold, platinum, tungsten, iridium, and aluminum [41,67–69,83,95,96]. There is a general agreement between the published theoretical works that silver and gold are the best candidates for supporting a large resonant enhancement factor. Surface plasmon resonances occurring in the visible or in the near-infrared region are efficiently excited only for these noble metals. While silver has better optical properties in the visible range (reduced absorption losses), leading to higher enhancement and sharper resonance, gold seems a good choice of material from the point of view of tip manufacturing and because of its good chemical stability. On the reverse side, the d-band core electrons of gold play a substantial role in increasing the damping of the surface plasmon resonance. Additional mechanisms are responsible for limiting the strength of the enhancement in noble metal structures. Radiation damping and electron scattering at surfaces, particularly for structures smaller than the electron mean free path [97], impose an upper limit to the enhancement.

Localization and a relatively strong field gradient have been calculated for dielectric tips [67,70,95] as well. However, in all cases, the enhancement factors were considerably smaller than those for resonant materials and shapes. From the experimental point of view, enhancements were observed for non-metallic probes, mainly exploiting the antenna effect. Studies on dye-doped polystyrene nanospheres [98] and single quantum dots [99] showed that fluorescence intensity can be enhanced up to fivefold by silicon tips, giving rise to increased fluorescent contrast in the images.

2.2.1.3 Shape Effects

One of the necessary conditions for the lightning-rod effect is the presence of geometrical singularities at the structure that can locally increase the surface charge density. A significant number of theoretical works have been focused on understanding the mechanisms of field enhancement at a tip extremity when parameters such as tip radius and opening angle were systematically varied. For instance, studies of a perfectly conductive conical probe showed that the enhancement factor increases with cone angle [100]. By incorporating a complex permittivity, and solving the eigenvalue problem, high enhancement factors were achieved for acute angles, with a steep increase for angles below 30° [95]. However, in a parametric study including real excitation condition, Martin and coworkers calculated that wide gold tips (cone angle of 90°) have a significantly larger enhancement than tips with small opening angles. The authors argued that dephasing effects may be less pronounced in wide-cone tips than in narrow-cone tips [69]. In order to restore large field enhancement, the authors proposed to create a discontinuity in the dimensions and shape of the tip. They calculated the electric field for various geometries, including a small aluminum tip of predefined length protruding from a larger stub or part of a small gold ellipsoid positioned at the extremity of a wider nonmetallic tip. These ideas were later refined by Krug and colleagues. They simulated the optical response of truncated cones and trigonal pyramids by a three-dimensional finite-difference time-domain algorithm [101]. Field enhancement factors as high as 86 were estimated for gold right-angle trigonal pyramids with excitation in the near-IR. Some investigations included the proximity of a substrate and modeled the linear scattering properties for various parameters, including tip radius and sample materials [102]. The conclusions of these investigations revealed that the contrast can be sensitively influenced by the substrate composition, in agreement with early works [41, 73, 96, 103].

2.2.1.4 Excitation Wavelength

Although the lightning-rod effect can be explained by simple electrostatic considerations, the frequency dispersion of the material constituting the tip implies that the magnitude of the enhanced field will be influenced by the excitation wavelength. A few theoretical studies using finite-difference time-domain (FDTD) and multiple-multipole modeling (MMP) have addressed the wavelength dependence of the enhancement factor for various tip geometries. Elongated particles have resonant enhancement in the green-red part of the spectrum, depending of their relative dimensions [104]. FDTD calculations showed that electromagnetic resonance of trigonal pyramids with sharp edges occurs in the near-infrared [101] and the highest calculated enhancements for conical tips are in the visible region [72, 105].

Of course, parameters such as probe length, opening angle, material, and relative distance to the substrate play an important role in defining the spectral position of the resonance. The well-defined resonances described in these studies originate from the finite size of the probes and are most likely caused by the excitation of surface plasmon resonances. For semi-infinite tip geometries, it is not clear whether the surface plasmon resonance or the lightning-rod effect is responsible for the enhanced electromagnetic field. The optical scattering responses of the tungsten tip were measured experimentally using a white-light-incoherent illumination source [106]. The scattering efficiency did not show a resonance peak with wavelength. Instead, a monotonic increase of the scattered intensity was measured for increasing wavelength. This effect can be easily explained by the dispersion of the material where the imaginary part of the permittivity decreases with longer excitation wavelength. Although the material used in this work does not support surface plasmon in the visible region, it is reasonable to assume that for semi-infinite tip geometry composed of plasmonic materials (Au, Ag), the main plasmon resonance would be significantly broader than resonances occurring in finite-size structures. As a result, the field enhancement most likely originates from a complex interplay between a lightning-rod effect and weak electronic resonances and the largest factor would probably occur in the near-IR, where the material losses are minimal.

2.3 Experimental Evidence of a Field Enhancement Effect

Beyond the direct-imaging applications of the field-enhancement effect which are the subject of the last part of this chapter (localized multi-photon excitation, Raman imaging, and surface modifications), there are a number of physical phenomena that are triggered by large electromagnetic fields.

For instance, the scattering cross section of a metal nanoparticle is significantly increased at the surface plasmon wavelengths [35]. In the context of near-field optical microscopy, this effect was observed with surface plasmon-based probes. Strong scattered intensities were observed when the distance between a resonant probe and a surface of a glass slide was reduced to a few nanometers [42]. The excitation of a localized surface plasmon at the apex of a metal-coated glass tip was indirectly observed by measuring the increased transmission characteristic of the tip over a relatively small range of illumination wavelengths [59, 107]. More recently, some groups developed a technique where a well-characterized resonant nanoparticle was attached at the apex of a probe [44–47, 108]. Increased optical responses such as scattering and transmission were measured within the resonant wavelengths. The technique was further developed when approaches similar to tomography were used to characterize fully the resonant conditions of the nanoparticle as shown in Figure 7 [108]. The scattering spectra of a single gold particle were acquired for various incident polarizations and particle orientations as depicted in

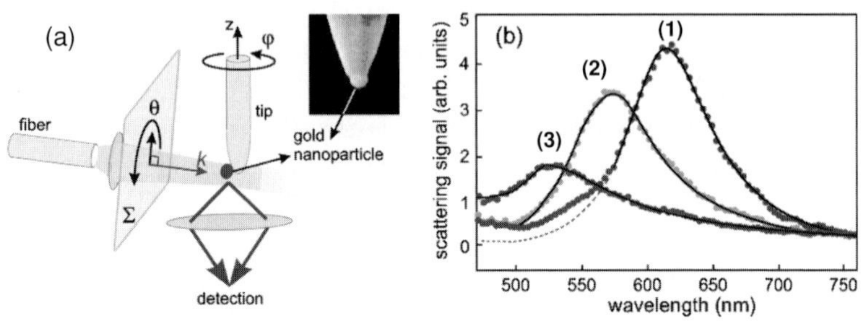

FIGURE 7. Optical tomography of a single gold nanoparticle. (a) Schematic of the measurement setup with the different orientation plans. The inset represents a scanning electron micrograph of a tip carrying a single gold nanoparticle. (b) Curves (1), (2), and (3) show the spectra surface plasmon scattering recorded at ($\varphi = 30°$, $\theta = 350°$), ($\varphi = 30°$, $\theta = 80°$), and ($\varphi = 120°$, $\theta = 290°$), respectively. The solid lines are theoretical fits. Reprinted with permission from [108]. Copyright 2004, American Chemical Society.

the schematic of Figure 7(a). Figure 7(b) shows the narrowest spectra observed for a combination of polarization angle and particle orientations, thus revealing the orientation of the particle and spectral inhomogeneities.

More indirectly, the effect of an enhanced field was observed in a tunneling junction. A rectification current was induced by the increased optical response of a resonant tip, and enhancement factors as large as 1,000 were evaluated [76]. The signature of a field enhancement at a tip was measured by Aigouy et al. by far-field measurement of the intensity of the light scattered by the tip as a function of the incident light polarization [78]. They observed maximum intensity in the case of p-polarization, in accordance with Figure 5. The field enhancement was also demonstrated by direct near-field photography on photopolymers, as discussed in more detail in section 3.2.

The localization of an efficient light source at the apex of the tip leads to very large field amplitude. When combined with ultrafast pulsed excitations, the local peak power can reach several 100 MW/cm^2 and trigger nonlinear optical responses of the tip material. The field enhancement effect can localize a second-harmonic generation to the very end of a metal tip, creating a highly confined source of photons [90, 93, 109, 110]. The sensitivity of the second-harmonic generation to the strength of the field enhancement was demonstrated by scanning the tip above the focal spot of a tightly focused laser beam at an glass/air interface. As discussed previously, the field distribution of a diffraction-limited focal spot is composed of transverse fields but also of regions where the polarization vectors are aligned with the propagation direction. If a tip is placed above these longitudinal regions, the excitation is such that a field enhancement effect can be created at the tip apex. Consequently, and as a result of the enhanced field, the generation of second-harmonic occurs only when the tip faces the longitudinal field regions as shown in Figure 8(a). Here, the image represents the intensity of the second-harmonic signal originating from the apex of the tip as a function of the tip position within the focal spot. The second-harmonic distribution is confined in a two-lobe region within the focal area indicative of the locations of the longitudinal field. Figure 8(b) represents the calculated distribution of these longitudinal fields in good agreement with the second-harmonic pattern. The sensitivity of the second-harmonic generation to the field polarization and magnitude result in a extreme localization of the emission. To illustrate this confinement, an intensity dependence of the second-harmonic signal was measured as a function of the distance between the tip and the glass sample where the laser beam was focused. The approach curve is shown in Figure 8(c). The second-harmonic intensity decreases dramatically as the tip is withdrawn from the focal area. An exponential decay length comparable to the tip-end diameter indicates that the second-harmonic emission is confined to regions near the tip extremity [90].

The sensitivity of the enhancement to regions of longitudinal polarizations was confirmed through the generation of a broad-spectral continuum occurring at a tip. White-light emission was observed from the tip extremity when the tip was excited by longitudinal fields [53]. The yield of the emitted continuum was shown to be very sensitive to the field enhancement factor.

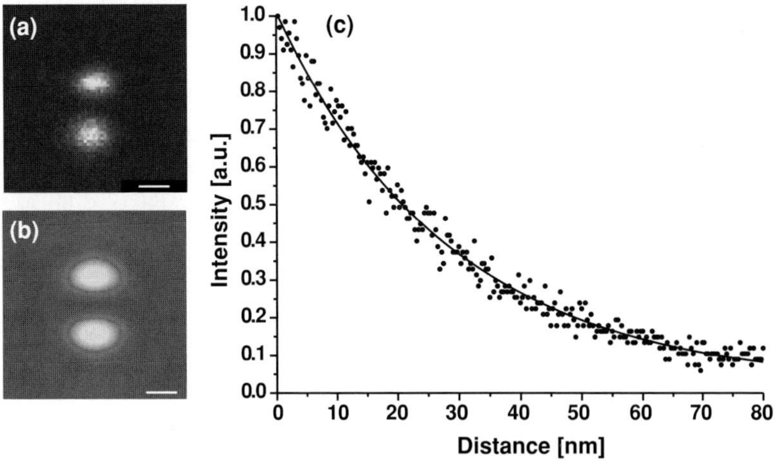

FIGURE 8. Optical response of a Au tip raster scanned in the focal region of a Gaussian laser mode. (a) Spatial distribution of the second-harmonic emitted by a Au tip. (b) Calculated spatial distribution of the longitudinal field. Scale bar is 250 nm. (c) Intensity of the second-harmonic as a function of the vertical distance between the tip and the focal spot region.

3 Applications to Nanosciences

3.1 Nanoscale Imaging and Spectroscopy

The recent advances in near-field optical microscopy enabled to image nanosystems with resolutions far below the conventional limit imposed by diffraction. Interestingly, some of the major breakthroughs in ultrahigh resolution imaging came from optical interactions with low to very low optical cross sections, such as multiphoton and Raman processes. In this context, field enhanced near-field optical microscopy is of particular interest because the magnitude of the interactions can be greatly increased by the proximity of a resonant probe.

In the next paragraphs, two examples of high-resolution imaging will be given. The ability to resolve optically nanometer-sized systems is of great interest to to a wide spectrum of scientists because it opens the way to local spectroscopies and chemical specifications.

3.1.1 Near-Field Two-Photon Excitation of *J*-Aggregates

It had been suggested that true optical confinement within nanoscopic volumes can be achieved by the combination of a field enhancement effect and nonlinear processes such as multiphoton transitions [95]. The existing substantial far-field localization of nonlinear processes due to their strong dependence on the field strength [111] would be further increased by the proximity of a metal probe. High-resolution images of the fluorescent responses of molecular aggregates and harvesting complexes were acquired by a combination of a tightly focused femtosecond laser beam and a modified gold tip [89]. The fluorescence of the specimens were triggered through a two-photon absorption process, and lateral optical resolutions approaching 20 nm were achieved. Similar results are shown in Figure 9, where pseudo-isocyanine (PIC) *J*-aggregates were excited using a similar approach [93]. J-aggregates are characterized by a red-shifted and sharp absorption band relative to the monomer band, a result of exciton delocalization over the molecular building blocks of the aggregate. The *J*-aggregates consists of a ribbon-like structure forming a mesh of interwoven strands as shown in the topography image of Figure 9 (left inset). The aggregates have a variety of sizes ranging from a few nanometers to several tens of nanometers. The fluorescence image (Figure 9, right inset), essentially reproduces the pattern seen in the topographic image with the advantage of providing an emission spectrum for each pixel of the image. The estimated optical resolution is 25 nm. The far-field contribution acts as a background in the fluorescent image. The effect is obvious in the right-hand side of the image where the density of tubules is larger. The fine optical features are surrounded by a background which is the far-field signature of the aggregates. It is very interesting to compare the resolution obtained in far-field (background) and in near-field (individual tubules). The near-field nature of the excitation source is readily seen when the distance between the tip and the sample is varied, as seen in the curve of Figure 9. A constant background representing the far-field fluorescence contribution dominates the signal when the tip is retracted. A steep increase of the fluorescence signal is detected when the tip approaches the last 10 nm.

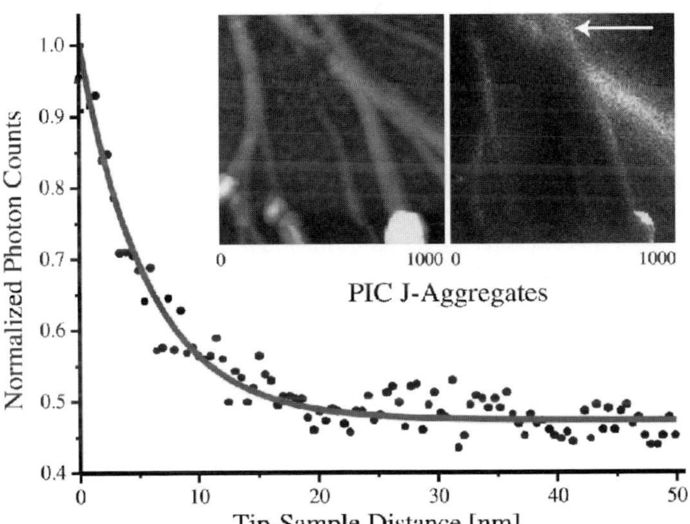

FIGURE 9. Tip-sample distance dependence of two-photon-excited fluorescence rate of PIC *J*-aggregates. The approach curve was taken in the location indicated by the arrow in the optical image (right) shown in the inset. On the left inset is the simultaneously acquired topographical image.

The fluorescent rate can be fitted by a single exponential decay with a typical length of ∼6 nm.

3.1.2 Chemical Imaging and Raman Spectroscopy

Compared to fluorescence spectroscopy, Raman scattering gives a much more detailed spectrum that can be uniquely linked to a molecular structure and chemical composition of a sample. While such insight is important, the Raman scattering cross sections are extremely small (10^{-31} to 10^{-28}), typically 15 orders of magnitude smaller than fluorescence cross sections. From the experimental point of view, the access of such weak signals relies greatly on the substantial enhancement capability of metal tips. Early experiments demonstrated a 30-fold increase in the Raman response of cresyl blue when a metallized cantilever tip was placed in the vicinity of the analyte [112]. Raman enhancements as large as 10^6 were estimated in an optical cavity formed by a STM gold tip and a Au (111) substrate [113, 114].

High-resolution, two-dimensional near-field Raman imaging stemmed from the Raman investigation of single-walled carbon nanotubes (SWNTs) [92, 115–117]. Near-field Raman images were acquired by collecting the overtone of a disorder-induced mode around 2,600 cm^{-1}, the so-called G' band. Optical resolution below 20 nm and single nanotube sensitivity were demonstrated. The authors also reported variations in the Raman spectrum revealing modifications of the molecular structure of the tubes.

Figure 10 shows simultaneous near-field Raman images of a single nanotube acquired over various active modes. The images show evidence for localized Raman scattering for all the recorded modes as well as localized spectral variations. The optical resolution obtained is about 15 nm, thus avoiding the averaging effect of the Raman spectrum along individual SWNTs.

Recently, the field enhancement capability of nanometer-size metal tips enabled workers to perform localized surface-enhanced coherent anti-Stokes Raman scattering (CARS) [118, 119]. CARS is a third-order nonlinear process providing spectroscopic information sensitive to molecular species and molecular orientations [120]. In this technique, a pump beam and a probe beam at a frequency ω_1 and a Stokes beam at ω_2 induce a nonlinear polarization $\omega_{CARS} = 2\omega_1 - \omega_2$ in the sample of interest. When the pulsation difference between the pump beam and the Stokes beam coincides with a molecular vibrational level, an anti-Stokes Raman signal is resonantly excited. The presence of a metal tip amplifies the CARS signal. The CARS polarization is generated within a very small volume near the tip, so high-resolution tip-enhanced CARS microscopy of a specific vibrational mode can be performed.

3.2 Nanoscale Lithography

One of the most promising features of near-field techniques is the ability to induce physical and chemical processes optically at the molecular scale. These applications have been implemented using different near-field sources. Since a complete

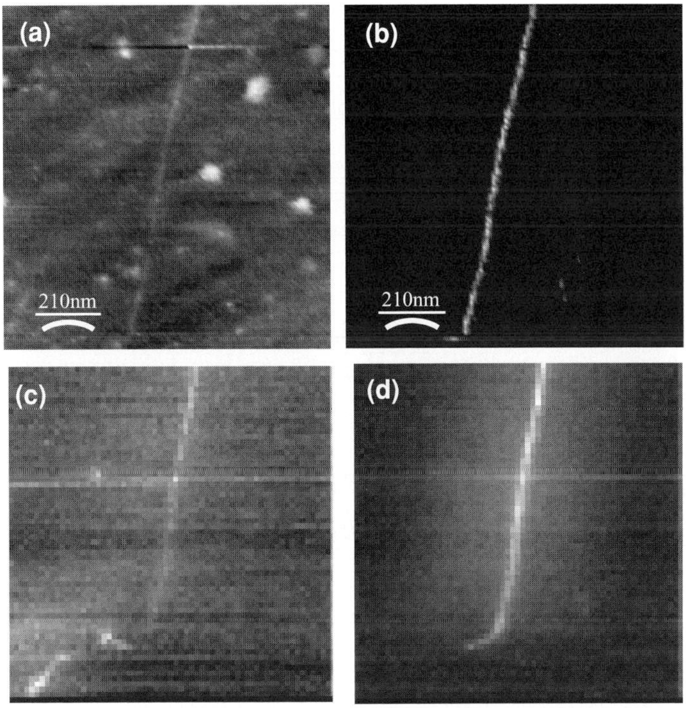

FIGURE 10. (a) Topographical image a single carbon nanotube. Near-field images integrated over various Raman-active modes: $\nu = 1{,}267\ \mathrm{cm}^{-1}$ (b), G band at $1{,}592\ \mathrm{cm}^{-1}$ (c), and G' band at $2{,}600\ \mathrm{cm}^{-1}$ (d). Courtesy of N. Anderson, University of Rochester.

review of near-field optical lithography is beyond the scope of this chapter, the reader is referred to a recent review by Royer et al. [27], where a survey of the literature on near-field optical lithography (materials, processes, and methods) is presented. In this section, we focus our attention on a particular approach of nanophotolithography based on the local field enhancement at the extremity of a metal tip. In the following, the acronym TEOL (tip-enhanced optical lithography) will be used to describe this approach. Several groups have successfully performed TEOL on both inorganic and organic materials. Nowadays, far-field optical lithography is the most widely used technique for patterning mass production devices in various fields ranging from micro-electronics and micro-optics [121]. This technology, based on patterning replica, or masks, is well-optimized and relatively easy to implement as compared to X-ray or electron-beam lithography. The photoresists are derived from well-known formulations and have been optimized for various applications. However, the bottleneck of optical lithography is the diffraction-limited spatial resolution of approximatively $\lambda/2$. Typical resolution with the new UV standard is currently slightly below 200 nm. The current trend is to implement "extreme UV" sources which would enable a resolution of $\approx$50 nm in the near future. However, this approach requires expensive development of new technologies:

new light sources, optics, and photopolymers. TEOL provides an alternative and elegant method for improving the resolution. Let us stress that TEOL essentially remains an optical technique as compared to other types of lithography (X-ray or electron-beam) with the distinct advantage of *not* being limited by diffraction. Indeed, for a near-field illumination, the spatial confinement of the light/matter interaction is not limited by the wavelength of the light but by the probe size and its distance to the surface. Additionally, TEOL avoids the technical challenges associated with aperture SNOM probes because the technique uses homogeneous metal tips widely used in atomic-force microscopy (AFM) or scanning tunneling microscopy (STM).

A survey of the literature shows that the first TEOL experiments were demonstrated on metal surfaces. Jersch and Dickman used laser field enhancement in the near-field of a STM tip to fabricate nanostrutures on gold surfaces [74]. The lithographic process was based on gold evaporation and surface grain reorganization. They achieved a 10-nm resolution with tungsten, silver, and platinum–iridium tips illuminated by 5-ns pulses at $\lambda = 532$ nm from a Nd:YAG laser. More recently, the experiment was confirmed by A. Chimmalgi and coworkers, who used AFM Si tips and femtosecond laser pulses, showing that an electromagnetic enhancement can be also excited at the end of a semiconductor singularity [122]. As an example, Figure 11 illustrates TEOL on an Au surface.

TEOL was experimentally demonstrated on light-sensitive resists by several groups. Wurtz and coworkers have shown that it is possible to induce one-photon absorption free-radical photopolymerization locally by a tip field enhancement [123]. Th enhancement allowed the polymerization threshold to be exceeded in a limited region around the tip extremity. The authors reported polymerized features with line-width of approximatively 70 nm. The local photolymerization was obtained with a tungsten tip illuminated by a *p*-polarized laser beam at a

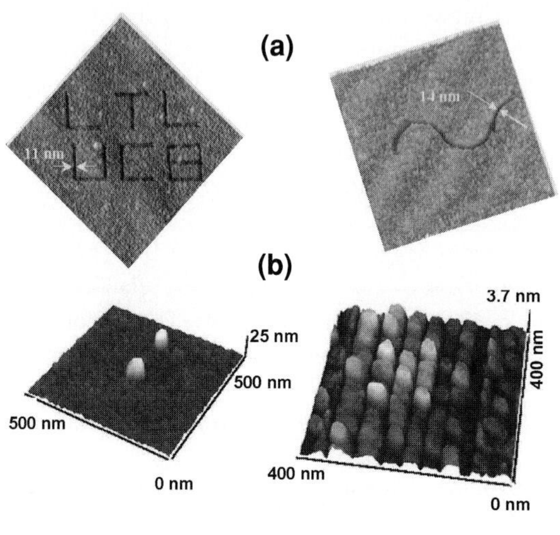

FIGURE 11. Example of nanometer sized pattern manufactured by TEOL on an Au surface. (a) AFM image of continuous letters and curve showing a TEOL resolution close to 10 nm. Reprinted with permission from [122]. Copyright 2003, American Physical Society. (b) STM image of hillocks obtained with a tungsten tip illuminated by femtolaser pulses. Reprinted with permission from [74]. Copyright 1995, American Physical Society.

wavelength λ of 532 nm. The experiments demonstrated the use of metal tips for inducing local photochemical process on polymers. Two years later, a similar resolution was achieved by near-field two-photon lithography on commercial negative tone photoresists [124]. The authors used a metal-coated Si AFM tip illuminated by 120-fs laser pulses at 790 nm to induce polymerization locally. Positive-tone photoresists were also considered. Tarun and coworkers successfully fabricated 100-nm-wide lines using metal-coated AFM tips illuminated by a laser beam ($\lambda = 402$ nm) through a high-numerical-aperture immersion objective lens. In particular, they demonstrated that for low incident energy, the local field enhancement effect occurring at the tip end can provide the necessary amount of energy required to induce photolysis [125].

Semiconductor samples were also modified by TEOL. Lu and coworkers used this approach locally to induce nanooxydation on a hydrogen-passivated Ge surface with an STM tip under laser illumination [126]. The process of lithography was based on the local photo-breaking of the Ge-H bond. The region illuminated by the tip extremity oxidizes instantaneously. The Ge-H/GeO_2 contrast was characterized by STM imaging. Twenty-nanometer resolution was achieved with tungsten tips. Similar results were obtained by Herndon and coworkers, who used illumination mode aperture SNOM to perform near-field optical lithography on H-passivated Si surface [127].

Before ending this section, let us stress that, from general point of view, TEOL experiments are motivated by three objectives.

1. Photochemistry at the nanometer scale: the use of sub-wavelength optical sources opens the possibility of local studies of photochemical reaction, allowing for a better understanding of light–matter interactions at the molecular scale.
2. Characterization of the local sources: each molecule of the photosensitive materials can be considered as a probe sensitive to the intensity of the electric field. A "snapshot" of the local optical source can thus be obtained by this approach.
3. High-resolution optical lithography.

3.3 Nanoscale Manipulation

Nanoscale manipulation using optical forces induced by a illuminated metal tips has been discussed theoretically [67, 71]. This approach provides a powerful tool for manipulation and engineering at the molecular scale. Compared to nanomanipulation based on STM, one could trap a large variety of objects, including dielectric and biologically active particles, using the field gradient associated with the local confinement. Additionally, the capture or release of the object could be easily selected by adjusting the polarization of the incident light polarization (TM for capture and TE for release). However, an experimental verification has yet to be demonstrated.

Related to this topic, matter migration induced by a field enhancement effect has been recently reported. H'Dhili and coworkers used metal tips to induce

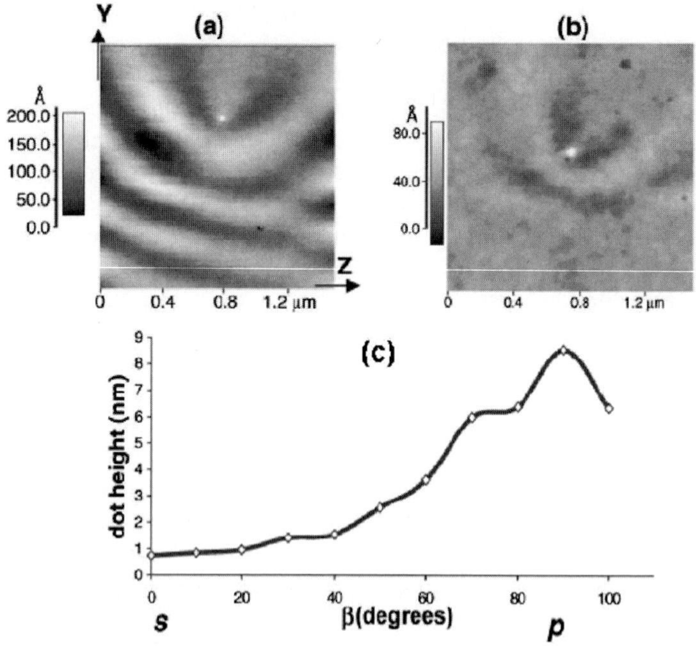

FIGURE 12. Example of study of parameters influencing tip field enhancement. (a,b) Influence of the tip illumination geometry. AFM images obtained after illumination (a) reflection mode illumination. Both central singularity and far-field type fringes are visible. (b) Illumination by total internal reflection. The fringes vanished, and only the central nanometer sized central spot is obtained. (c) Influence of the incident polarization. The height of the central dot as a function of the angle of polarization. The incident polarization was linear. β represents the polarization angle with regard to the sample surface.

photoisomerization of azobenzene dye molecules grafted to a polymer matrix [79, 128, 129]. The isomerization induces a migration of the matrix that is easily observable by AFM. The photopolymer acts as a self-developing material. The migration relied on successive trans-cis-trans isomerization cycles induced by the local field located at the tip extremity. This approach enabled workers to perform a parametric analysis, leading to valuable information about the field enhancement effect. In particular the enhancement was qualitatively evaluated as a function of illumination conditions (state of polarization, angle of incidence...) and characteristics of the tip (see section 2).

As an example, Figure 12 shows the influence of the illumination geometry and the polarization on the polymerization. Figure 12(a) shows an AFM image obtained after reflection mode illumination of a metallized Si tip in interaction with the azodye surface. The center of the image corresponds to the tip position during the exposure. The figure shows two different patterns: a extended fringe system, which corresponds to an interference of light scattered by the tip and

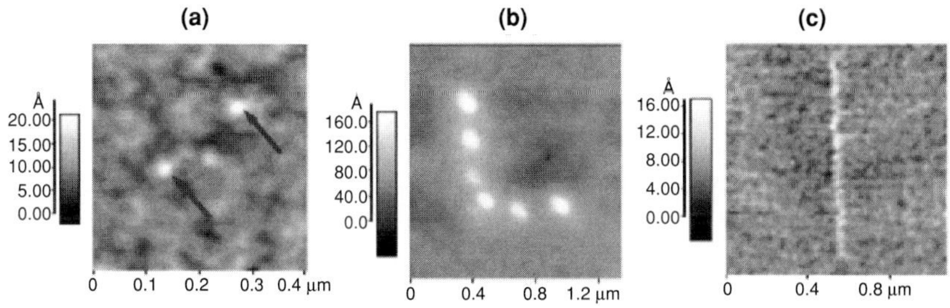

FIGURE 13. AFM images of patterns obtained by TEOL on self-developing azobenzene containing photosensitive surface: (a) two 20-nm dots, (b) letter "L," (c) 40-nm-wide continuous line.

the incident light, and a central nanometric dot due to the local enhancement of the electromagnetic field below the metallic tip. The fringes vanish if the tip is illuminated by total internal reflection (see Figure 12(b)). This type of study provides valuable information about the response of the tip to various experimental parameters. Figure 12(c) shows a plot of the height of the central near-field spot as a function of the incident angle of polarization, confirming that the intensity of the local tip field is enhanced gradually from a s-polarization to p-polarization. Additionally, the authors showed preliminary examples of nanopatterning based on this local enhancement effect (see Figure 13 as an illustration).

The approach introduced in [79, 129] was recently applied to other metal nanoobjects. The migration of the azo-molecule were shown to be possibly induced by a field locally enhanced in the vicinity of noble metal nanostructure produced by electron lithography [130]. This method may ultimately be applied to the nanooptical manipulation of photon-assisted molecular motors to produce nanoscale transport of a material. In that sense, data presented in references [79, 129, 130] can be considered as the original observations of matter manipulation induced by optical field enhancement.

4 Conclusions

Nanoscale metallic structures are extensively studied for their extraordinary optical properties. The local electromagnetic enhancement associated with the electronic resonances of the structure is the basis for an increasing range of discoveries. While the concept of tip-induced field-enhanced near-field optical microscopy was suggested in 1985, it took over ten years to demonstrate high-resolution near-field imaging based on this effect experimentally. This time lag reflects the experimental and theoretical efforts needed to understand the underlying physical mechanisms. Polarization and wavelength of the excitation field as well as the shape and the material of a tip were found to be important parameters in maximizing the enhancement factor. The electromagnetic enhancement effect is used in various aspects of near-field microscopy, and optical resolutions approaching 10 nm have been reported. Chemical imaging, as well as fluorescence spectroscopy, are routinely performed in increasingly complex systems. Surface photomodifications were shown to be localized below 40 nm, opening the possibility for optical control of matter migration in nanoscale dimensions.

References

1. E. H. Synge, Phil. Mag. **6**, 356 (1928).
2. E. A. Ash and G. Nichols, Nature **237**, 510 (1972).
3. D. W. Pohl, W. Denk, and M. Lanz, Appl. Phys. Lett. **44**, 651 (1984).
4. E. Betzig, P. L. Finn, and J. S. Weiner, Appl. Phys. Lett. **60**, 2484 (1992).
5. O. J. F. Martin, J. Microscopy. **194**, 235 (1999).
6. E. Betzig and J. K. Trautman, Science **257**, 189 (1992).
7. P. Lambelet, A. Sayah, M. Pfeffer, C. Philipona, and F. Marquis-Weible, Appl. Opt. **37**, 7289 (1998).
8. R. Stoeckle, Ch. Fokas, V. Deckert, R. Zenobi, B. Sick, B. Hecht, and U. P. Wild, Appl. Phys. Lett. **75**, 60 (1999).
9. T. Saiki and K. Matsuda, Appl. Phys. Lett. **74**, 2773 (1999).
10. D. Mulin, D. Courjon, J-P. Malugani, and B. Gauthier-Manuel, Appl. Phys. Lett. **71**, 437 (1997).
11. A. Bouhelier, J. Toquant, H. Tamaru, H.-J. Guentherodt, D. W. Pohl and G. Schider, Appl. Phys. Lett. **79**, 683 (2001).
12. D. Haefliger and A. Stemmer, Ultramicroscopy **100**, 457 (2004).
13. Th. Lacoste, Th. Huser, R. Prioli and H. Heinzelmann, Ultramicroscopy **71**, 333 (1998).
14. J. A. Veerman, A. M. Otter, L. Kuipers, and N. F. Hulst, Appl. Phys. Lett. **7**, 3115 (1998).
15. M. P. O'Boyle and H. K. Wickramasinghe, Appl. Phys. Lett. **65**, 1623 (1994).
16. R. Bachelot, P. Gleyzes, and A. C. Boccara, Microsc. Microanal. Microstruct. **5**, 389 (1994).
17. Y. Inouye and S. Kawata, Opt. Lett. **19**, 159 (1994).
18. F. Zenhausern, M. P. O'Boyle, and H. K. Wickramasinghe, Science **249**, 1083 (1995).
19. R. Bachelot, P. Gleyzes, and A. C. Boccara, Opt. Lett. **20**, 1924 (1995).
20. J. Wessel, J. Opt. Soc. Am. B **2**, 1538 (1985).
21. S. Nie, S. R. Emory, Science **275**, 1102 (1997).
22. A. D. McFarland and R. P. Van Duyne, Nano. Lett. **3**, 1057 (2003).
23. B. Lamprecht, G. Schider, R. T. Lechner, H. Ditlbacher, J. R. Krenn, A. Leitner, and F. Aussenegg, Phys. Rev. Lett. **84**, 4721 (2000).
24. S. A. Maier, P. G. Kik, and H. A. Atwater, Appl. Phys. Lett. **81**, 1714 (2002).
25. G. A. Wurtz, J. S. Im, S. K. Gray, and G. P. J. Wiederrecht, Phys. Chem. B **107**, 14191 (2003).
26. G. Wiederrecht, Eur. Phys. J. Appl. Phys. **28**, 3 (2004).
27. P. Royer, D. Barchiesi, G. Lerondel, and R. Bachelot, Phil. Trans. R. Soc. Lond. A **362**, 821 (2003).
28. S. Patan, G. P. Gucciardi, M. Labardi, and M. Allegrini, Riv. Nu. Cim. **27**, 1 (2004).
29. R. Bachelot, G. Lerondel, S. Blaize, S. Aubert, A. Bruyant, and P. Royer, Microsc. Res. Tech. **64**, 441 (2004).

30. D. Courjon, *Near-Field Microscopy and Near-Field Optics*, Imperial College Press (2003).
31. S. Kawata, M. Ohtsu, and M. Irie, *Nano-Optic*, Springer Berlin (2002).
32. W. E. Moerner and D. P. Fromm, Rev. Sc. Instr. **74**, 3597 (2003).
33. P. N. Prasad, *Nanophotonics*, John Wiley & Sons, Inc. Hoboken, New Jersey (2004).
34. A. Bouhelier, Mat. Res. Tech., **69**, 563 (2006).
35. C. Bohren and D. Huffman. *Absorption and Scattering of Light by Small Particles*. John Wiley & Sons, New York (1982).
36. U. Kreibig and M. Vollmer. *Optical Properties of Metal Clusters*, Vol. 25 of Springer Series in Materials Science. Springer, Berlin (1996).
37. J. Jackson, *Classical Electrodynamics*. John Wiley & Sons, New York, 2nd. edition (1975)
38. H. Raether, *Surface Plasmons on Smooth and Rough Surfaces and Gratings*. Springer, Berlin (1988).
39. E. Hutter and J. H. Fendlcr, Adv. Mater. **16**, 1685 (2004).
40. A. Bouhelier, R. Bachelot, J. S. Im, G. P. Wiederrecht, G. Lerondel, S. Kostcheev, and P. Royer, J. Phys. Chem. B **109**, 3195 (2005).
41. W. Denk and D. W. Pohl, J. Vac. Sci. Technol B **9**, 510 (1991).
42. U. Ch. Fisher and D. W. Pohl, Phys. Rev. Lett. **62**, 458 (1989).
43. T. J. Silva, S. Schultz, and D. Weller, Appl. Phys. Lett. **65**, 658 (1994).
44. O. Sqalli, M.-P. Bernal, P. Hoffmann, and F. Marquis-Weible, Appl. Phys. Lett. **76**, 2134 (2000).
45. Th. Kalkbrenner, M. Ramstein, J. Mlynek, and V. Sandoghdar, J. Microsc. **202**, 72 (2000).
46. O. Sqalli, I. Utke, P. Hoffmann, and F. Marquis-weible, J. Appl. Phys. **92**, 1078 (2002).
47. S.-K. Eah, H. M. Jaeger, N. F. Scherer, G. P. Wiederrecht, and X.-M. Lin, Appl. Phys. Lett. **86**, 031902 (2005).
48. M. Gu and P. C. Ke, Opt. Lett. **24**, 74 (1999).
49. J. Zeleni, Science **79**, 269 (1932).
50. D. J. Mallan, *Physics of Lightning*. London: English University Press (1963).
51. J. A. Chalmers, Weather **20**, 183 (1965).
52. L. Libioulle, Y. Houbion, and J.-M. Gilles, J. Vac. Sci. Technol. B **13**, 1325 (1995).
53. M. R. Beversluis, A. Bouhelier, and L. Novotny, Phys. Rev. B **68**, 115433 (2003).
54. B. Ren, G. Picardi, and B. Pettinger, Rev. Sci. Instr. **75**, 837 (2004).
55. E. J. Sànchez, J. T. Krug II, and X. S. Xie, Rev. Sci. Instr. **73**, 3901 (2002).
56. N. Hayazawa, Y. Inouye, Z. Sekkat, and S. Kawata, Chem. Phys. Lett. **335**, 369 (2001).
57. R. Eckert, J. M. Freyland, H. Gersen, H. Heinzelmann, G. Schürmann, W. Noell, U. Staufer, and N. F. de Rooij, Appl. Phys. Lett. **77**, 3695 (2000).
58. L. T. Nieman, G. M. Krampert, and R. E. Martinez, Rev. Sci. Instr. **72**, 1691 (2001).
59. T. Yatsui, K. Itsumi, M. Kourogi, and M. Ohtsu, Appl. Phys. Lett. **80**, 2257 (2002).
60. T. Matsumoto, T. Ichimura, T. Yatsui, M. Kourogi, T. Siaki, and M. Ohtsu, Opt. Rev. **5**, 369 (1998).
61. H. G. Frey, F. Keilmann, A. Kriele, and R. Guckenberger, Appl. Phys. Lett. **81**, 5030 (2002).
62. H. G. Frey, S. Witt, K. Felderer, and R. Guckenberger, Phys. Rev. Lett **93**, 200801 (2004).
63. A. Bouhelier, J. Renger, M. R. Beversluis, and L. Novotny, J. Microsc. **210**, 220 (2003).
64. L. Vaccaro, L. Aeschimann, U. Staufer, H. P. Herzig, and R. Dändliker, Appl. Phys. Lett. **83**, 584 (2003).

65. A. J. Babadjanyan, N. L. Margaryan, and Kh. V. Nerkararyan, J. Appl. Phys. **88**, 3785 (2000).
66. F. Keilmann, J. Microsc. **194**, 567 (1999).
67. L. Novotny, R. X. Bian, and X. S. Xie, Phys. Rev. Lett. **79**, 645 (1997).
68. O. J. F. Martin and Ch. Girard, Appl. Phys. Lett. **70**, 705 (1997).
69. Y. C. Martin, H. F. Hamann, and H. K. Wickramasinghe, J. Appl. Phys. **89**, 5774 (2001).
70. J. L. Bohn, D. J. Nesbitt, and A. Gallagher, J. Opt. Soc. Am. A **17**, 2998 (2001).
71. P. C. Chaumet, A. Rahmani, and M. Nieto-Vesperinas, Phys. Rev. Lett. **88**, 123601 (2002).
72. W.-X. Sun, Z.-X. Shen, J. Opt. Soc. Am. A **20**, 2254 (2003).
73. S. Kawata and Y. Inouye, Ultramicrosc. **57**, 313 (1995).
74. J. Jersch and K. Dickmann, Appl. Phys. Lett. **68**, 868 (1995).
75. A. Lahrech, R. Bachelot, P. Gleyzes, and A. C. Boccara, Appl. Phys. Lett. **71**, 575 (1997).
76. A. V. Bragas, S. M. Landi, and O. E. Martínez, Appl. Phys. Lett. **72**, 2075 (1998).
77. B. Knoll and K. Keilmann, Nature **399**, 134 (1999).
78. L. Aigouy, A. Lahrech, S. Grésillon, H. Cory, A. C. Boccara, and J. C. Rivoal, Opt. Lett. **24**, 187 (1999).
79. F. H'Dhili, R. Bachelot, G. Lerondel, D. Barchiesi, and P. Royer, Appl. Phys. Lett. **79**, 4019 (2001).
80. N. Hayazawa, A. Tarun, Y. Inouye, and S. Kawata, J. Appl. Phys. **92**, 6983 (2002).
81. R. Hillenbrand and F. Keilmann, Phys. Rev. Lett. **85**, 3029 (2000).
82. T. J. Yang, G. A. Lessard, and S. T. Quake, Appl. Phys. Lett. **76**, 378 (2000).
83. H. Furukawa and S. Kawata, Opt. Comm. **148**, 221 (1998).
84. H. F. Hamann, G. Gallagher, and D. J. Nesbitt, Appl. Phys. Lett. **73**, 1469 (1998).
85. R. G. Milner and D. Richards, J. Microsc. **202**, 66 (2000).
86. H. Kano, S. Mizugushi, and S. Kawata, J. Opt. Soc. Am. B **15**, 1381 (1998).
87. B. Sick, B. Hecht, and L. Novotny, Phys. Rev. Lett. **82**, 4482 (2000).
88. L. Novotny, E. J. Sànchez, and X. S. Xie, Ultramicrosc. **71**, 21 (1998).
89. E. J. Sánchez, L. Novotny, and X. S. Xie, Phys. Rev. Lett. **82**, 4014 (1999).
90. A. Bouhelier, M. R. Beversluis, A. Hartschuh, and L. Novotny, Phys. Rev. Lett. **90**, 13903 (2003).
91. A. Bouhelier, M. R. Beversluis, and L. Novotny, Appl. Phys. Lett. **82**, 4596 (2003).
92. A. Hartschuh, E. J. Sánchez, X. S. Xie, and L. Novotny, Phys. Rev. Lett. **90**, 95503 (2003).
93. A. Bouhelier, M. R. Beversluis, and L. Novotny, Ultramicrosc. **100**, 413 (2004).
94. *Handbook of Chemistry and Physics*, 72nd edition, CRC Press, (1992).
95. Y. Kawata, C. Xu, and W. Denk, J. Appl. Phys. **85**, 1294 (1999).
96. A. V. Zayats, Opt. Comm. 1**61**, 156 (1999).
97. A. Wokaum, J. P. Gordon, and P. F. Liao, Phys. Rev. Lett. **48**, 957 (1982).
98. H. F. Hamann, M. Kuno, A. Gallagher, and D. J. Nesbitt, J. Chem. Phys. **144**, 8596 (2001).
99. V. V. Protasenko, M. Kuno, A. Gallagher, and D. J. Nesbitt, Opt. Comm. **210**, 11 (2002).
100. H. Cory, A. C. Boccara, J. C. Rivoal, and A. Lahrech, Microwave. Opt. Techno. Lett. **18**, 120 (1997).
101. J. T. Krug II, E. J. Sánchez, and X. S. Xie, J. Chem. Phys. **116**, 10895 (2002).
102. M. B. Raschke and Ch. Lienau, Appl. Phys. Lett. **83**, 5089 (2003).

103. J. A. Porto, O. Johansson, S. P. Apell, and T. López-Ríos, Phys. Rev. B **67**, 085409 (2003).
104. J. Renger, S. Grafstrom, L. Eng, and V. Deckert, J. Opt. Soc. Am. A **21**, 1362 (2004).
105. F. Festy, A. Demming, and D. Richards, Ultramicrosc. **100**, 437 (2004).
106. L. Aigouy, F. X. Andréani, A. C. Boccara, J. C. Rivoal, J. A. Porto, R. Carminati, J. J. Greffet, and R. Mégy, Appl. Phys. Lett. **76**, 397 (2000).
107. M. Ashino and M. Ohtsu, Appl. Phys. Lett. **72**, 1299 (1998).
108. Th. Kalkbrenner, U. Hakanson, and V. Sandoghdar, Nano. Lett. **4**, 2309 (2004).
109. A. V. Zayats, T. Kalkbrenner, V. Sandoghdar, and J. Mlynek, Phys. Rev. B **61**, 4545 (2000).
110. A. V. Zayats, and I. I. Smolyaninov, Phil. Trans. R. Soc. Lond. A **362,** 843 (2004).
111. W. Denk, J. H. Strickler, and W. W. Webb, Science **248**, 73 (1990).
112. R. M. Stöckle, Y. D. Suh, V. Deckert, and R. Zenobi, Chem. Phys. Lett. **318**, 131 (2000).
113. B. Pettinger, G. Picardi, R. Schuster, and G. Ertl, J. Electro. Chem. **554**, 293 (2003).
114. B. Pettinger, B. Ben, G. Picardi, R. Schuster, and G. Ertl, Phys. Rev. Lett. **92**, 096101 (2004).
115. A. Hartschuh, M. R. Beversluis, A. Bouhelier, and L. Novotny, Phil. Trans. R. Soc. Lond. A. A **362**, 807 (2003).
116. A. Hartschuh, N. Anderson, and L. Novotny, J. Microsc. **201**, 234 (2003).
117. N. Anderson, A. Hartschuh, and L. Novotny, J. Am. Chem. Soc, **127**, 2533 (2005).
118. H. Hayazawa, T. Ichimura, M. Hashimoto, Y. Inouye, and S. Kawata, J. Appl. Phys. **418**, 2676 (2004).
119. T. Ichimura, H. Hayazawa, M. Hashimoto, Y. Inouye, and S. Kawata, Phys. Rev. Lett. **92**, 220801 (2004).
120. Y. R. Shen, *The Principles of Nonlinear Optics*, New-York: Wiley (1984).
121. F. M. Schellenberg, *Selected Papers on Resolution Enhancement Techniques in Optical Lithography*, SPIE Press (2004).
122. A. Chimmalgi, T. Y. Choi, C. P. Grigoropoulos, and K. Komvopuolos, Appl. Phys. Lett. **82**, 1146 (2003).
123. G. Wurtz, R. Bachelot, F. HDhili, P. Royer, C. Triger, C. Ecoffet, and D. J. Lougnot, Jpn. J. Appl. **39**, 98 (2000).
124. X. Yin, N. Fang, X. Zhang, I. B. Martini, and B. J. Schwartz, Appl. Phys. Lett. **81**, 3663 (2002).
125. A. Tarun, M. R. H. Daza, N. Hayazawa, Y. Inouye, and S. Kawara, Appl. Phys. Lett. **80**, 3400 (2002).
126. Y. F. Lu, Z. H. Mai, G. Qiu, and W. K. Chim, Appl. Phys. Lett. **75**, 2359 (1999).
127. M. K. Herdon, R. T. Collins, R. E. Hollingswouth, P. R. Larson, and M. B. Johnson, Appl. Phys. Lett. **74**, 41 (1999).
128. F. H'Dhili, R. Bachelot, A. Rumyantseva, G. Lerondel, and P. Royer, J. Microsc. **209**, 214 (2002).
129. R. Bachelot, F. H'Dhili, D. Barchiesi, G. Lerondel, R. Fikri, P. Royer, N. Landraud, J. Peretti, F. Chaput, G. Lampel, J. P. Boilot, and K. Lahlil, J. Appl. Phys. **94**, 2060 (2003).
130. C. Hubert, A. Rumyantseva, G. Lerondel, J. Grand, S. Kostcheev, L. Billot, A. Vial, R. Bachelot, P. Royer, S. Chang, S. K. Gray, G. P. Wiederrecht, and G. C. Schatz, Nano Lett. **5**, 615 (2005).

I.9
Electrochemical SPM
Fundamentals and Applications

T. J. SMITH AND K. J. STEVENSON

The application of scanning probe microscopy (SPM) techniques to the study of electrochemical interfaces is discussed in this review. Experimental requirements for performing electrochemical scanning probe microscopy (EC-SPM) experiments are briefly discussed as well as their successful merger with conventional scanning probe microscopy instrumentation. Methods highlighted include electrochemical-scanning tunneling microscopy (EC-STM), electrochemical-atomic force microscopy (EC-AFM), and hybrid variants of AFM, including scanning electrochemical microscopy-atomic force microscopy (SECM-AFM) and local electrochemical impedance spectroscopy-atomic force microscopy (LEIS-AFM). Applications of these modern EC-SPM techniques in emerging areas of energy storage and conversion, corrosion, catalysis, and electrochemical deposition processes are underscored to emphasize the resolving power of these methods.

1 Introduction

Many technological and industrial processes result from electrochemical oxidation and reduction reactions occurring at electrified solid/liquid interfaces. Examples include refining [1] and deposition [2] of metals, synthesis [3] and remediation [4] of chemicals, and storage [5] and conversion [6] of chemical to electrical energy. These electrochemical reactions depend predominantly on (1) mass transport of reactants from the bulk solution to a region near a charged surface and (2) the rate of electron transfer. Specifically, the rates of the electrochemical reactions are strongly influenced by the applied electric potential, interfacial structure and the propensity of an electrochemical reactant, intermediate, or product to adsorb on the electrode surface. Therefore, a detailed understanding of the complex relationships between adsorption, surface reactivity (e.g., catalysis, corrosion, and passivation) and interfacial structural features (e.g., morphology, crystallinity, and chemical defects) is essential for establishing efficient and productive control of electrochemical processes. Furthermore, such information is important for establishing material design parameters in the optimization of future technological devices such as in microelectronics [7], miniaturized chemical sensors [8], advanced power sources [9], and improved multifunctional coatings [10].

Over the past two decades, the ability to probe the local character of solid/liquid interfaces spatially and temporally while under potential or current control has helped advance the knowledge of several electrochemical processes. Prior to the advent of scanning probe microscopy (SPM), the surface characterization methods available to the electrochemist were either *ex situ*, typically requiring high vacuum (e.g., scanning electron microscopy and low-energy electron diffraction) or ensemble averaging techniques with large analytical spot sizes (e.g., UV-vis and IR spectroscopies), which provided only an averaged macroscopic picture of interfacial structure and reactivity. As a result, data acquired by these methods was sometimes misinterpreted, yielding oversimplified or misidentified reaction mechanisms. This is especially true when studying heterogeneous materials and interfaces since electrochemical behavior is characteristically dominated by localized reactivity occurring at energetically distinct surface sites and structural features. Better comprehension of controlling factors can be significantly aided by spatially resolved measurements to allow for direct correlation of structure/composition/reactivity relationships. With this motivation in mind, the field of electrochemical SPM (EC-SPM) continues to advance at a rapid pace. Many improvements and modifications to standard SPM techniques, including those involving electrochemical applications, have been reviewed recently. Recent monographs on electrochemical applications of SPM include those by Gewirth and Niece [11], Ward and White [12]; electrochemical-scanning tunneling microscopy (EC-STM) by Bard and Fan [13] and Moffat [14]; electrochemical-atomic force microscopy (EC-AFM) by Macpherson [15]; and scanning electrochemical microscopy (SECM) by Bard and Fan [13] and Horrocks [16]. Comparisons between UHV and *in situ* techniques have been reviewed by Weaver [17]. We limit this review to applications of electrochemical-scanning tunneling microscopy (EC-STM), electrochemical-atomic force microscopy (EC-AFM), scanning electrochemical microscopy-atomic force microscopy (SECM-AFM), and local electrochemical impedance-atomic force microscopy (LEIS-AFM), published since the year 2000 that involve *in situ* characterization of electrode surfaces under potential or current control.

Our review is divided into two main sections: Fundamentals and Applications. The first section introduces the new methods of SPM that have been extended to analyze electrified interfaces within an actively controlled electrochemical cell; it discusses the essential equipment modifications or additions needed to extend standard SPM to the study of electrochemical systems. The second section reviews recent applications to several fundamental electrochemical problems which illustrate the resolving power of these newly developed EC-SPM techniques.

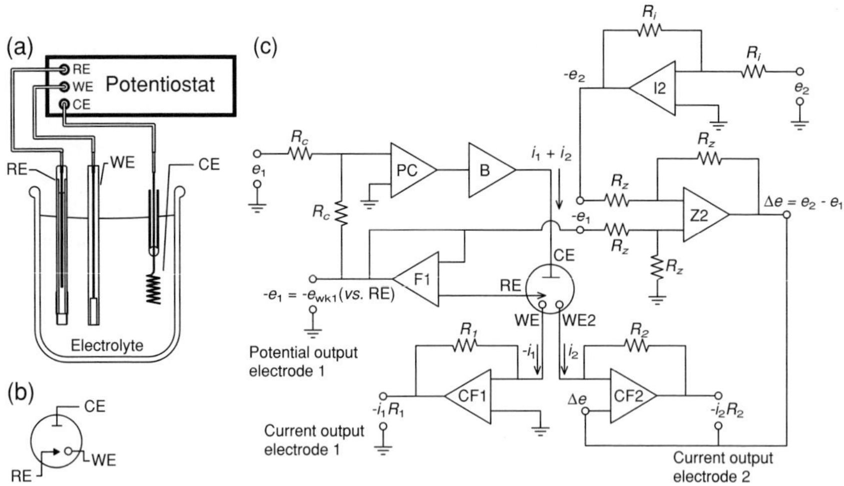

FIGURE 1. (a) Standard three-electrode electrochemical cell, (b) potentiostat connection diagram, and (c) four-electrode electrochemical cell using a bipotentiostat configuration. Reprinted with permission from [18]. Copyright 2001, John Wiley & Sons, Inc.

2 Fundamentals

2.1 *Electrochemical Cells*

The successful integration of an electrochemical cell into a scanning probe microscope requires an understanding of the necessary control elements, including both electrodes and electronics, as well as how the presence of the electrolyte influences the SPM imaging conditions. Electrochemical cells utilize two to four electrodes, depending on the purpose of the cell and the desired level of control over the reactions desired (Figure 1). Two electrode cells are sometimes used to evaluate the potential of an electrified interface without attempting to actively control its potential. They are also used where precision control is unnecessary, e.g., metal plating [19]. A three-electrode system facilitates accurate control of the electrified interface potential by providing a fixed reference potential from a standard reference electrode. A four-electrode cell maintains the accuracy of the three-electrode cell, and adds the ability to control concurrently the potential of a second electrode with respect to the same reference potential.

The electrodes used in an electrochemical cell (Figure 1 (a,b)) are named by their function in the cell design. The working electrode (WE), or indicator electrode, is the electrified interface (substrate) under study and is normally under active potential or current control. The reference electrode (RE) maintains a well-defined reference potential from which the potential of the working electrode is both determined and controlled. It is important that negligible current passes to/from the RE as this would change its potential, thus sacrificing the accuracy it contributes to the potential control. The third electrode is the counter or auxiliary electrode (CE) and is used as a current source or sink, enabling the potentiostat electronics to control the WE potential with respect to the RE. The CE is typically spatially isolated from the WE and sometimes also the RE to prevent contamination from any byproducts generated at the CE during maintenance of the cell environment. When a fourth electrode is incorporated into the cell, it is usually a second working electrode (WE2) where its potential is also determined, and controlled, in relation to the RE using a bipotentiostat setup. A schematic showing the internal circuitry of a basic bipotentiostat connected to a four-electrode electrochemical cell is shown in Figure 1(c). In the case of EC-SPM, the sample or substrate under study is the WE and, when necessary, the SPM probe tip is WE2. Utilizing these electrode configurations in combination with a potentiostat (or galvanostat), the electrostatic potential of the substrate is controlled with an applied potential (or current), which allows electrochemical processes to be evaluated with the techniques discussed in this chapter.

2.2 *Electrochemical SPM*

In general, EC-SPM can be separated into two basic groups depending on whether the SPM tip requires potential control. Figure 2 illustrates the various electrochemical configurations for EC-SPM. In *passive-probe* EC-SPM, the probe tip is unbiased, requiring only a three-electrode potentiostat. In experiments where the SPM probe tip must either be held at a specific potential or subjected to some type of potential perturbation, designated *active-probe* EC-SPM, a bipotentiostat is typically required. The first type of passive-probe EC-SPM, electrochemical AFM (EC-AFM), requires a three-electrode potentiostat and utilizes an unbiased probe to map the topographical features (height, phase) of an electrified substrate under potentiostatic (or galvanostatic) control (Figure 2(a)). When an insulating substrate

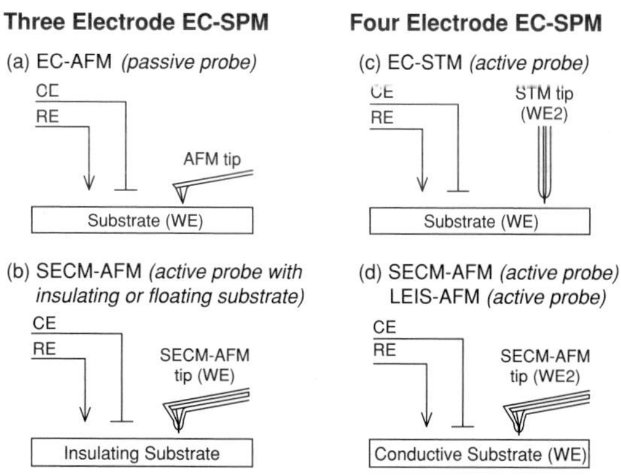

FIGURE 2. EC-SPM configurations.

or floating potential (unbiased) electrode is probed such as with SECM-AFM, only three electrodes are needed, although this technically can still be considered active probe EC-SPM since the potential of the probe tip is actively controlled, but not the substrate (Figure 2(b)). All other types of electrochemical SPM discussed in this monograph require a bipotentiostat to control a four-electrode electrochemical cell (Figure 2(c,d)). In these experiments, both the electrified substrate and the probe tip are biased and must be controlled by the bipotentiostat. As such, electrochemical SPM requires a fluid cell designed to incorporate both a reference and counter-electrode submersed within it or incorporated into the flow path of the electrolyte into or out of the electrochemical cell.

2.2.1 Three-Electrode Experiments

2.2.1.1 EC-AFM

In EC-AFM experiments, either the potential or current to/from the tip or substrate is controlled, but not both. In most cases, the potential of the substrate is manipulated and the topography (height) of the substrate is analyzed (Figure 2(a)). Many phenomena, including surface adsorption of ions and molecules [20], underpotential deposition (UPD) [21], electroplating [22], and corrosion [23–25], can be studied in this fashion. One major concern when acquiring data in this configuration is how the probe tip may influence results. The presence of the probe tip in the substrate's vicinity can hinder diffusion of solution species in the neighborhood of the tip and can result in artifacts in these studies. This aspect is discussed in more detail below. When imaged in contact mode the probe tip can wear, damage, or otherwise change the surface being studied. This effect can be induced deliberately to aid in the study of certain applications, such as in the study of film dissolution or corrosion.

2.2.1.2 SECM-AFM

SECM-AFM on either an insulating substrate or an unbiased conductor can be used to study transport through membranes [26,27], dissolution of ionic crystals [28], or a proximity-blocking effect [29] as the electrified probe tip (WE) approaches the surface (Figure 2(b)). SECM-AFM can also be used to determine the presence of conductive structures in an insulating substrate [30,31] by analyzing the response of the tip current during the approach to the substrate, where the current is plotted as a function of vertical distance from the surface. This is discussed in more detail in section 3.2.1. An insulating substrate exhibits negative feedback, where the probe tip current drops as it approaches the surface, and a conductive substrate shows a positive feedback current as the distance between tip and sample decreases.

2.2.2 Four-Electrode Experiments

2.2.2.1 EC-STM

All electrochemical-scanning tunneling microscopy (EC-STM) experiments must be controlled by a bipotentiostat permitting the independent potential control of both the substrate (WE) and the STM tip (WE2) (Figure 2(c)). The potential of the STM tip is typically held at a fixed offset from the substrate to maintain similar tunneling conditions for the duration of an experiment. Probes used for EC-STM must be insulated, with only the tip exposed, to reduce the Faradaic current that results from biasing it in an electrochemical cell [18]. This Faradaic current must be minimized such that it is one to several orders of magnitude less than the tunneling current; otherwise, the feedback of the STM signal will be difficult to maintain. In particular, the effects of coadsorbates on UPD [32–37] on single-crystal metal surfaces have been studied with EC-STM.

2.2.2.2 SECM-AFM

Scanning electrochemical microscopy-atomic force microscopy (SECM-AFM) with active control of substrate potential requires a bipotentiostat (Figure 2(d)). This configuration can be used to evaluate diffusion patterns emanating from a small electrified interface [30] (i.e., an ultramicroelectrode), monitor the products of a surface-catalyzed reaction [38], or monitor enzymatic activity [29,39]. One factor limiting implementation of SECM-AFM has been the ability to fabricate conductive probes of nanoscale dimensions repeatedly and consistently as discussed in section 3.2.1.

2.2.2.3 LEIS-AFM

Local electrochemical impedance spectroscopy-atomic force microscopy (LEIS-AFM), also known as scanning impedance microscopy (SIM), performed *in situ*, is a technique that probes frequency-dependent electrical characteristics (capacitance and resistance) of electrochemical interfaces. LEIS-AFM is configured like Figure 2(d), but employs a frequency-response analyzer in addition to a bipotentiostat. In this configuration, the conductive AFM tip acts as a local impedance probe that has been used to study electrochemical catalysis [40]. The advantage of this configuration is that with some knowledge of the process being studied both kinetic and mechanistic information can be obtained, along with spatial distributions and topographic information. Very little work has been done in this area, as it is difficult to combine these two techniques and still collect data fast enough for experimentally significant processes to be studied with a scanning probe technique. One main reason for this is that each frequency is measured sequentially so most impedance measurement schemes involve lengthy signal acquisition protocols.

2.3 Incorporating Electrochemical Cells With SPM

The addition of an electrochemical cell to SPM requires consideration for electrolyte retention as well as counter- and reference-electrode incorporation. Each of these components can hinder the free movement of the SPM scan head. Several different designs of electrochemical cells are available from commercial SPM manufacturers, although some research groups prefer to fabricate their own cells due to additional experimental constraints such as inert atmosphere requirements

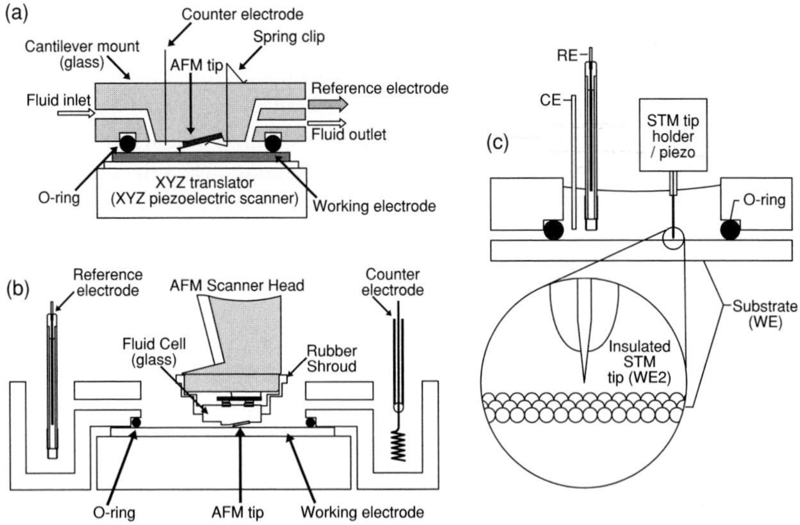

FIGURE 3. (a) Schematic model of the cross section of an electrochemical AFM cell with a scanning stage. Reprinted with permission from [47]. Copyright 2003, Elsevier B.V. (b) Electrochemical AFM with a scanning head. Adapted from [48]. (c) Schematic of EC-STM.

or atypical substrate geometry. Several EC-SPM cell designs are illustrated in Figure 3. The AFM cell designs (Figure 3(a,b)) can be used for SECM-AFM or LEIS-AFM by insulating all but the apex of a conductive AFM tip and connecting this tip as WE2 to a bipotentiostat. A generic EC-STM design is shown in Figure 3(c). Whether the tip or substrate is scanned in EC-SPM is predetermined by the SPM design. It is important, regardless of the technique and the cell design, that the SPM electronics be protected from the electrolyte; otherwise, corrosion or short circuits can reduce the lifetime of the SPM.

2.3.1 Probe Modifications/Special Requirements

SPM experiments that use probe tips as WEs require special tip-preparation procedures to minimize or localize Faradaic currents at the tip in STM and SECM-AFM, respectively. Methods have been published where the probe tip is insulated with nail polish [41], apeizon wax [42], or electrophoretic paint [43]. Other methods utilized include the chemical vapor deposition (CVD) of SiN_3 [44], SiO_2 [45], or a polymer such as Parylene [46]. Each of these methods has its benefits and limitations. Commercial sources for insulated STM tips are available; however, to our knowledge, there are currently no commercial sources for SECM-AFM probe tips.

2.3.2 Liquid Cells

The *in situ* characterization of solid/liquid interfaces requires that the SPM scan head remain in close proximity of an electrolyte solution for the duration of an experiment. The design of a liquid cell must consider the isolation of the electrochemical cell from SPM electronics, preferably immersing as little of the SPM head as possible. Typically, only the probe tip and probe holder are immersed. The design must also protect the SPM from any corrosive solution or vapor components. The position of the CE and RE also need to be considered. As the distance between the RE and WE (and WE2 if used) increases, so does the uncompensated solution resistance between the RE and WEs. This will adversely affect the actual potential at the electrode surface. Placement of the CE is also important as it can change the current density throughout the cell and introduce unwanted species in solution due to the counter reactions that take place at the CE during accurate potential control of the WE.

2.3.3 Diffusion Interference

SPM analyses of electrochemical interfaces suffer from interference to the free diffusion of the solution components in the vicinity of the probe tip. Additionally, when the potential of the probe tip is changed, as in EC-STM, the tip can cause redox reactions on the solution components that can interfere with the interface studied. This is typically overcome by biasing the potential of the tip at a non-reactive potential, as well as insulating the tip to minimize the exposed area where reactions can occur, as discussed above.

AFM instruments that measure cantilever deflection with a laser beam can also suffer from diffusion effects when used in an electrochemical cell. If the concentration of the electrolyte differs along the path of the laser, the resultant change in refractive index of the electrolyte will alter the trajectory of the laser beam, and as a result, interfere with measurements of cantilever position at an active electrified interface. Stappers and Fransaer [49] report two major effects seen in EC-AFM: concentration-gradient-dependent beam deflection and beam deflection oscillations. Both effects show a dependence on the geometry of the electrochemical cell as well as the nature of the electrified interface under study; and while both cause beam deflections, a special case arises when the least concentrated portion of the electrolyte is also in the lowest spot of the electrochemical cell. They determined that if the concentration gradient becomes large enough, specifically when the Rayleigh number is greater than ~1,700, Rayleigh-Bénard convection [50] will cause oscillations in laser beam trajectory. Changes in the geometry of the cell, such as placing the electrified interface at the top or bottom of the cell, can alleviate the oscillation problems but are typically limited by instrumental constraints. Other possible solutions to these problems include reducing the current flow, minimizing the electrode dimensions, using fluid flow to reduce the thickness of the diffusion layer, or allowing time for the solution at the electrified interface to reach equilibrium with the bulk solution.

3 Applications

3.1 Conventional Electrochemical SPM

Conventional *in situ* SPM experiments include both EC-STM and EC-AFM, where electrochemical currents or potentials are measured at the electrified interface, rather than the probe tip. Typically, these techniques are designed to elucidate structural and morphological changes of the sample as a function of time, applied potential or current flow through the sample. One goal of conventional EC-SPM is to ensure that analyzing a sample with the SPM has a negligible effect on the surface under study. Applications of conventional EC-SPM are illustrated below.

3.1.1 Copper Electrodeposition

Electrodeposited copper is used extensively in ultra-large-scale integration (ULSI) microelectronics and micro-electro-mechanical systems (MEMS), particularly in chip metallization such as contacts and interconnects and the filling of vias [51]. Several reviews in this area are available [52]. As these processes and devices are continually driven toward the nanoscale, the need for more research into conditions affecting the electrochemical plating process arises. Electrochemical SPM techniques have been extremely beneficial in understanding the dynamics of nucleation and growth processes, and how additives and adsorption can inhibit or enhance these processes.

The effects of sulfate adsorption on copper underpotential deposition (UPD) on platinum and rhodium single crystal surfaces have been studied with EC-STM. Wu et al. [32] demonstrated the strength of adsorption has an effect not only on the potential required for copper UPD but also on the nucleation sites and surface diffusion of these sites once deposition begins. In particular, time-dependent *in situ* EC-STM experiments showed that copper UPD on Pt(111) proceeds via a three-stage process. Initially, a copper ion/sulfate ion bilayer structure is formed, followed by a disordered phase, and finally a complete monolayer of Cu on the Pt(111) surface. This monolayer is covered with adsorbed sulfate, exhibiting a $(\sqrt{3} \times \sqrt{7})$ oblique structure with 20% coverage. In contrast, on a Rh(111) surface, copper UPD proceeds via nucleation at step edges, followed by growth to near a monolayer from these sites. The same $(\sqrt{3} \times \sqrt{7})$ oblique sulfate adsorption layer is observed throughout this process. Wu proposed that the higher potential required for the UPD of copper onto Rh(111) is due to the increased strength of sulfate adsorption onto rhodium, resulting in less surface diffusion and rearrangement of the copper deposit.

An additional effort to probe the complex nature of sulfate adsorption during the UPD of copper has been reported by Okada et al. [33]. EC-STM images of the sulfate adlayer on Pd(111), shown in Figure 4, are proposed to have a bridge-like water layer bound between sulfates (Figure 4(b)). This group further evaluated the effects of adsorption from the supporting electrolyte on the UPD of copper on a Pt(111) electrode, noting that in hydrochloric acid, a copper/chloride ionic bilayer is also formed, similar to that with sulfate discussed above as well as a disorder-to-order transition in the adlayer structure as a submonolayer of copper is formed. Wu et al. [34] also reported on the effect of scanning parameters on the imaging of a surface with an adsorbate. By adjusting the STM tip voltage and the current setting, they were able to image either the chloride adsorbate or the underlying platinum crystal, as shown in Figure 5. This further demonstrates the

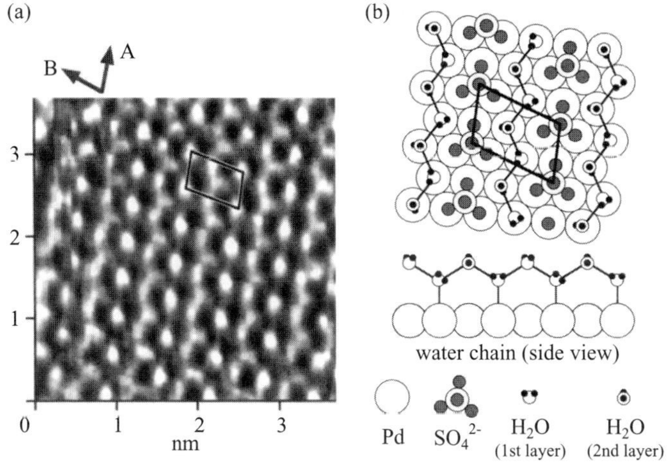

FIGURE 4. (a) High-resolution STM image of sulfate adlayer on Pd(111) in 0.05 M H_2SO_4 + 1 mM $CuSO_4$ before UPD of Cu; scan area, 3.7 × 3.7 nm^2; electrode potential, 0.65 V; tip potential, 0.75 V; tunneling current, 15 nA; scanning rate, 21.70 Hz. (b) Real space model for the Pd(111)-($\sqrt{3} \times \sqrt{7}$) sulfate adlayer with hydrogen-bonded water chains shown between adsorbed sulfate rows. Reprinted with permission from [33]. Copyright 2001, the PCCP Owner Societies.

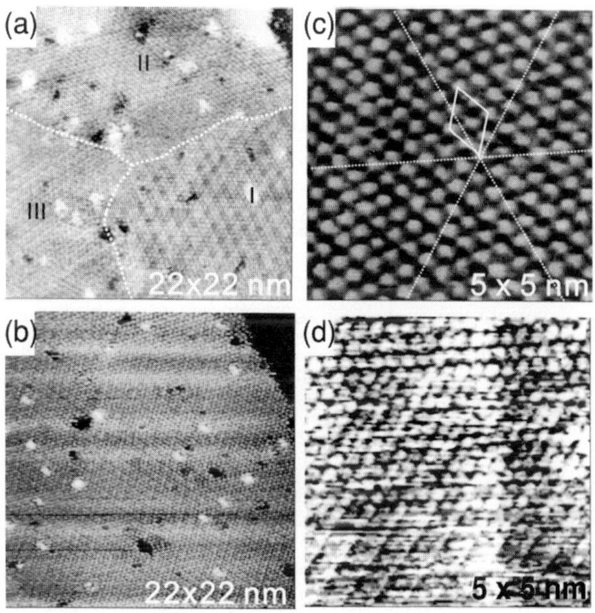

FIGURE 5. *In situ* STM images of Pt(111) at 0.65 V in 0.01 M HCl + 1 mM $Cu(ClO_4)_2$. (a) Domain boundaries between (4 × 4) and the new structure, resulting from continuous deposition of Cu. Panels (b) and (c) show the long-range ordering and internal atomic arrangements of the ($\sqrt{7} \times \sqrt{7}$) R19.1° structure. Adjusting the imaging conditions from −200 mV and 3 nA to 50 mV and 5 nA reveals the substrate lattice (d). Reprinted with permission from [34]. Copyright 2001, American Chemical Society.

utility of accurate potential control in the *in situ* EC-STM study of electrodeposition processes.

The adsorption of an iodine monolayer on a Pt(100) electrode and its effects on the UPD of copper have been reported by Shue and Yau [35]. The iodine monolayer makes it more difficult to deposit copper, shifting the UPD potential ~350 mV more negative, while causing peak broadening for the UPD. Adsorbed iodine is displaced by the copper as it is deposited, with the least strongly bound iodine being replaced first. Shue and Yau concluded that the *in situ* STM experiments confirmed the accepted view of copper UPD in the presence of a halide monolayer.

Many types of additives are used in the electroplating industry to help control the morphology and overall quality of deposits. These include blocking additives, accelerating additives, surfactants, complex forming or ion-pairing additives, and film forming additives [36]. Recently Petri et al. [37] discussed the effects of mercaptopropionic acid (MPA) adsorption on copper electrodeposition at a Au(111) electrode. They showed that when MPA is permitted to form a complete monolayer, it inhibits copper deposition onto the electrode surface, resulting in a defect rich copper UPD layer. At sub-monolayer MPA coverage the copper deposit was shown to have a smaller grain size than either the additive free system or a surface with a full monolayer of MPA. MPA is demonstrated to be a blocking or grain refining additive depending on the level of coverage on the electrode surface prior to deposition.

EC-SPM has also been used to probe the dynamics of copper electrodeposition using constant height mode STM at rates of 5–25 frames per second, termed video-STM. Zitzler et al. [53], Polewska et al. [54], and Magnussen et al. [55,56] have employed this technique to study fast kinetic processes at single crystal copper electrodes. Zitzler examined the adsorption of chloride to a single crystal copper electrode as well as terrace growth processes, and demonstrated copper deposition/dissolution fluctuations at kink sites in HCl. Polewska [54] reported that video-STM is able to capture the fluctuations of atoms on the interior or exterior corner edge of a terrace at equilibrium (Nernst potential), shown in Figure 6. The ability to probe fast, dynamic processes with EC-STM is proving to be an invaluable tool in the understanding of electrochemical deposition processes.

3.1.2 Energy Storage

3.1.2.1 Lead Acid Battery

Lead acid batteries have been the secondary (rechargeable) energy storage solution of choice for over 100 years [57]. In that time, the design of the battery has undergone many modifications, including alloying support grids, organic coatings on the negative electrode, and gelled electrolytes. The lead acid battery is still the most popular battery in use today due to its high power output. In an effort to extend the use of lead acid batteries, EC-AFM has been used better to understand degradation and failure mechanisms that occur at both electrodes

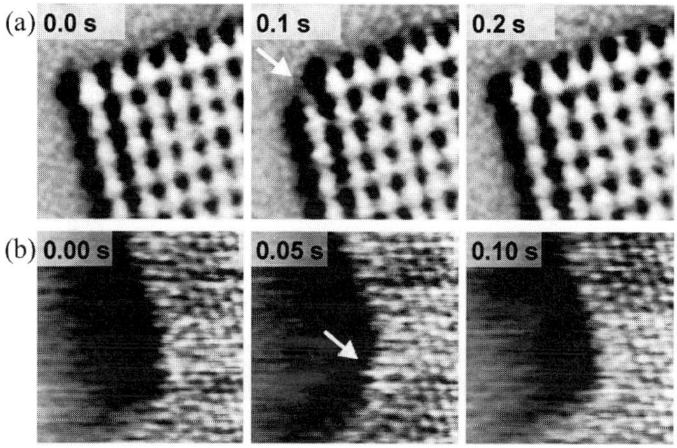

FIGURE 6. Two series of *in situ* video-STM images recorded on Cu(100) in 0.01 M HCl showing fluctuations of atoms at (a) an outer ($-230\,mV_{SCE}$, 25×25 Å^2) and (b) an inner (-190 mV_{SCE}, 45×45 Å^2) terrace corner. Fluctuating atom at (a) outside and (b) inside corner of terrace highlighted by arrows. Reprinted with permission from [54]. Copyright 2003, Elsevier Ltd.

in the process of charging and discharging each half cell. EC-AFM is better suited to study these systems than EC-STM due to the roughness of the surfaces and the poor conductivity of samples covered by semiconducting or insulating films.

Yamaguchi et al. [58] were the first to utilize EC-AFM to investigate lead-acid battery electrodes. Initial observations at open circuit potential of a lead electrode in a sulfuric acid solution showed quick coverage with lead sulfate as chemical equilibrium is established. Controlling the rate of oxidation through potential control of the lead electrode resulted in a smoother sulfate deposit than the spontaneously formed sulfate at open circuit. These results led Yamaguchi to hypothesize that the discharge reaction at the negative electrode occurs via a two-step process with the lead electrode initially being oxidized to saturate the electrolyte with Pb^{2+} ions. The level of saturation increases until lead sulfate precipitates onto the lead electrode surface. If left at open circuit, an Oswald ripening process produces larger lead sulfate crystals, reducing the exposed surface area, and results in a slower reverse reaction. A further study of the negative electrode by the same group [59] imaging with EC-AFM during a cyclic voltammogram (CV) showed the quick precipitation of lead sulfate immediately after the sharp oxidation peak in the CV, but displayed a much slower dissolution during the reverse cycle. Both high sulfate concentrations and fast potential sweep rates were observed to result in smaller crystallites being formed during the oxidation process.

Additional studies of the negative electrode reaction by Yamaguchi et al. [60] evaluated the influence of standing time at open circuit on the ability to recharge

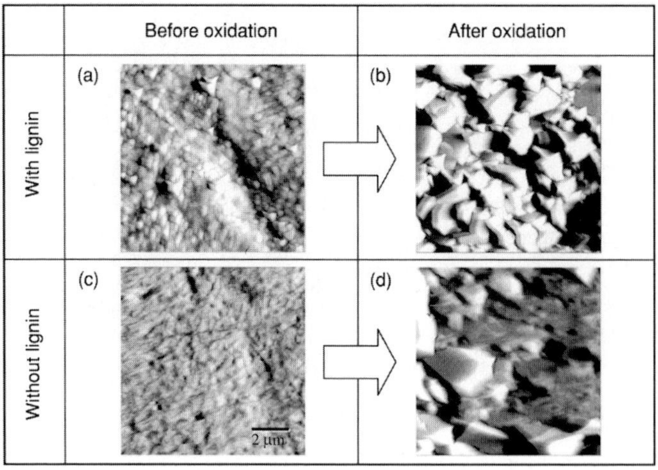

FIGURE 7. EC-AFM topographic images of lead electrode surfaces before/after potentiostatic oxidation. (a) Image before oxidation with lignin; (b) image after oxidation with lignin; (c) image before oxidation without lignin; and (d) image after oxidation without lignin. Reprinted with permission from [20]. Copyright 2002, Elsevier B.V.

the anode and correlated this data with changes in the morphology of the electrode surface. The lead sulfate on the electrode surface was observed to undergo Oswald ripening as a function of the time spent at open circuit potential. This resulted in a smoother lead sulfate layer with lower surface area, as well as reduced charge acceptance. The effect of idle time prior to charging was explored on commercial batteries and was found to agree with potentiostatic transient experiments on the electrodes used for EC-AFM.

Additional studies reported by Ban et al. [20] showed the effects of additives such as lignosulfonate (lignin) on the lead sulfate deposition at the negative electrode. The lead sulfate deposits were smaller and more evenly dispersed across the electrode surface in the presence of lignin than in its absence (Figure 7). Hirai et al. [61] also explored the effects of oxidation at -20°C and reduction at room temperature in dilute (50 mM) and concentrated (5 M) sulfuric acid in the presence or absence of lignin. They concluded that irrespective of the temperature, the precipitation of lead sulfate upon oxidation of the electrode occurred immediately following the current peak. The coverage density of the lead sulfate decreased in the presence of lignin, while the size of individual crystallites increased. The lignin still slowed the dissolution of the precipitate upon oxidation of the electrode.

3.1.2.2 Lithium Ion Batteries

Lithium ion batteries, introduced commercially in the 1990s by Sony [62], are a relatively new battery technology. Research into more efficient, higher-capacity, longer-life, and safer electrode materials is still an active topic in the scientific community. The lithium ion secondary storage battery has very high energy density (~400 W-hours/liter) [62], enabling consumer devices such as cell phones and portable computing to be reduced in size and weight with long charge cycles. EC-SPM has been used to evaluate lithium intercalation and its effects on the morphology and stability of new electrode materials.

Negative Li Ion Electrodes. The majority of EC-SPM analysis of Li-ion negative electrode materials has been EC-AFM on model carbon materials, such as highly oriented pyrolytic graphite (HOPG). Additional work has been performed to confirm the use of HOPG as a model surface compared to composite graphite electrodes which are composed of graphite particles bound in a polymer matrix.

Solvent instability in lithium ion batteries at the potential of the negative electrode leads to solvent decomposition, forming an ionically conductive separation layer between the electrode material and the solvent, termed the solid electrolyte interphase (SEI). Edström and Herranen [63] recently reported on the thermal stability of the SEI on a HOPG electrode, showing topographical changes to the HOPG surface as the SEI is destroyed thermally while imaged *in situ* with AFM. Comparison of topography images before insertion to post-Li insertion in 1M $LiBF_4$ in 2:1 ethylene carbonate (EC)/γ-butyrolactone with 0.51% water impurity showed a clear indication of the SEI layer formation. Heating the cell to 40°C produced no change in the topography of the SEI layer, yet from 50°C to 70°C there was an increase in blister formation and at the highest temperature, HOPG becomes exposed through the SEI layer.

The effects of electrode potential on the formation of a surface film on HOPG electrodes in 1 M $LiClO_4$ in EC/diethyl carbonate (DEC) have been reported by Jeong et al. [64]. Two processes can be differentiated using AFM and CV data: the subsurface decomposition of co-intercalated solvent molecules and the decomposition of solvent on the surface of the HOPG. The first process is observed to occur only during the initial potential sweep, resulting in distinct cathodic peaks in the CV and 1–2-nm-tall "hill-like structures" on the surface of the HOPG due to the intercalation of solvated lithium ions. Immediately following the appearance of these structures, at higher potentials, a 15–25-nm- tall swelling occurs where the cointercalated solvent molecules decompose below the surface of the HOPG. Finally, the decomposition of solvent directly on the surface of the HOPG is observed as a precipitate of approximately 40 nm particles that increase in size to 70 nm on subsequent potential sweeps. As can be expected of poorly adhered deposits, continuous scans with the AFM resulted in particle-like solvent decomposition products being cleared from the scan area by the sweeping motion of the AFM probe used in contact mode.

The addition of a co-solvent to EC on the formation of the SEI on HOPG has been reported by Jeong et al. [65]. In the presence of EC + DEC (1:1), co-intercalation of solvated lithium ions was observed as blisters on the surface of the HOPG after the first potential sweep and subsequent use of the AFM to clear the surface of the solvent decomposition product. These blisters were not observed to the same extent in either EC or EC + dimethyl carbonate (DMC) solutions (Figure 8). When DMC

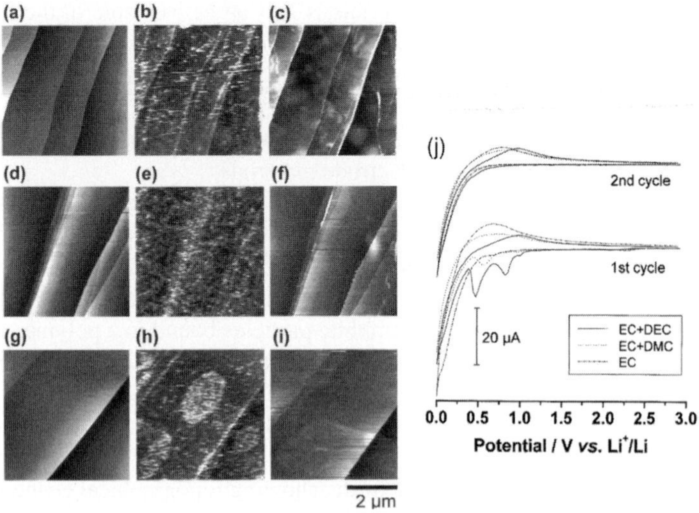

FIGURE 8. AFM topographic images (5×5 mm^2) of the HOPG basal plane surface obtained at 2.9 V (a,d,g) before and (b,e,h,) after the first cycle of CV and (c,f,i) after the precipitate layer was removed by repeated scanning. Solution: 1 M $LiClO_4$ dissolved in (a–c) EC + DEC (1:1); (d–f) EC + DMC (1:1); and (g–i) EC. (j) Cyclic voltammograms of the HOPG basal plane in 1 M $LiClO_4$ dissolved in EC + DEC (1:1), EC + DMC (1:1), and EC. Sweep rate: 0.5 mV s^{-1}. Reprinted with permission from [65]. Copyright 2002, Elsevier Ltd.

was used as the main solvent, similar blister formation is observed. Examination of the solvated lithium ion species with Raman spectroscopy showed that in EC + DEC lithium is solvated by three EC molecules and one DEC molecule, whereas in EC and EC + DMC the lithium is solvated solely by EC molecules. Due to the similarity in chemical structure between DEC and DMC, an additional *in situ* AFM experiment was performed, with DMC as the main solvent. In the absence of EC, lithium ions also co-intercalated with the DMC solvent, exhibiting a blistered topography similar to the EC + DEC sample. The authors suggest that solvation of lithium ions by alkyl carbonates, such as DMC and DEC, enhance the co-intercalation and the subsurface decomposition of these species, although more studies are needed to confirm this hypothesis.

In order to take advantage of the higher conductivity of propylene carbonate (PC) solutions in place of EC in lithium ion batteries, the solvent decomposition that causes exfoliation of HOPG must be suppressed to allow the formation of an SEI. Jeong et al. [66] used EC-AFM to evaluate vinylene carbonate (VC), fluoroethylene carbonate (FEC), and ethylene sulfite (ES) as additives (3% wt.) in PC and their effects on the formation of an SEI on HOPG. Each additive was successful in forming an SEI through its decomposition, and the layer formed inhibited the decomposition of PC and the resultant exfoliation of the HOPG as well as the co-intercalation of solvated Li. The HOPG in VC + PC showed the highest reversible capacity and capacity retention due to the smallest SEI thickness.

Campana and coworkers [67] reported step height changes between individual graphene layers of an HOPG electrode during Li^+ intercalation/deintercalation (charge/discharge) cycles as determined with EC-AFM measurements. The first intercalation/deintercalation cycle resulted in a permanent increase in the step height between graphene layers. Additional cycles are reversible and show a $17 \pm 5\%$ change in distance between graphene layers.

EC-AFM has also been enlisted to probe the surface of composite graphite electrodes in an effort to study commercially available electrode designs and to supplement the data obtained studying HOPG electrodes. The increased roughness of this type of electrode material heightens the difficulty of EC-SPM, requiring the careful preparation of an electrode for study. Koltypin et al. [68] report their efforts to follow the morphological changes of the same particle in a composite graphite electrode during several lithium insertion/deinsertion cycles in 1 M $LiAsF_6$ or $LiPF_6$ in EC + DMC. The irreversible formation of an SEI layer was observed on the composite graphite electrode during the initial lithium insertion, similar to that observed on HOPG. Subsequent insertion/deinsertion cycles exhibit reversible morphological changes seen as expansion and contraction of inter particle gaps. Due to the higher exposure of edge plane graphite on these particles compared to HOPG, the SEI layer on this portion of the particle is stressed during the insertion cycles and continues to be reformed as the graphite is exposed to solvent. The fade in capacity of these electrodes is attributed to this process.

Using EC-AFM to better understand the surface evolution that occur upon lithium insertion, Jeong et al. [69] detail three types of morphological changes observed on a composite graphite electrode during the initial insertion in 1M $LiClO_4$ in EC + DEC. The AFM images and diagrams of the mechanistic changes are shown in Figure 9. This suggests similar mechanisms of solvent decomposition between the graphene sheets of the graphite particles to that observed on HOPG electrodes. In PC, the solvent decomposition caused exfoliation of the graphene sheets severe enough to cause rupture of the particles.

Positive Li Ion Electrodes. A host of materials have been studied as potential replacements for lithium metal as the positive electrode material in lithium ion secondary storage batteries [62]. Typical commercial materials are alloys of lithium and include layered materials such as $LiCoO_2$, $LiNiO_2$, and all stoichiometries in between ($LiNi_{1-x}Co_xO_2$), as well as spinel type materials such as $LiMn_2O_4$. Each has its own strengths and weaknesses; an ideal electrode material has not been found. Both EC-AFM and EC-STM have been used to evaluate materials in an effort to better understand the effects of alloying lithium on the morphology of the electrode surface. These attempts to elucidate lithium incorporation mechanisms and isolate material failure points are essential to the development of higher quality positive electrode materials.

The effect of lithium reduction and oxidation in 1 M $LiPF_6$ in EC + DMC + DEC (2:2:1) on the topography of a $LiMn_2O_4$ electrode has been reported by Vidu and coworkers [70]. *In situ* lateral force microscopy (LFM) was used in addition to

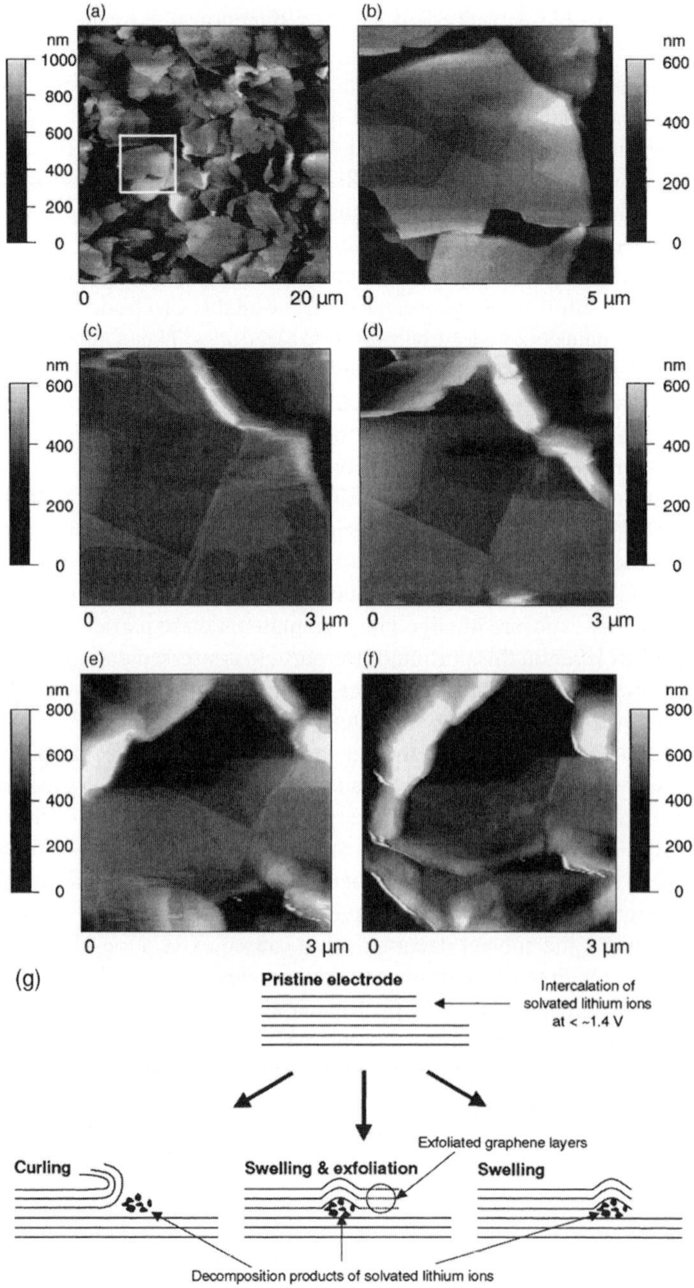

FIGURE 9. AFM topographic images of the composite graphite electrode surface obtained at the open circuit potential (~3.3 V) (a,b), 1.4 V (c), 1.1 V (d), 1.0 V (e), and 0.8 V during constant current charging at 30 μA cm^{-2} in 1 M $LiClO_4$/(EC + DEC) (f). The square in (a) shows the 5 × 5 μm area in (b). (g) Schematic models for the three topographical features observed in (a–f). Reprinted with permission from [69]. Copyright 2003, Elsevier B.V.

standard topographical AFM to distinguish between polymer binder and acetylene black from the more rigid $LiMn_2O_4$ component. Comparison between *ex situ* and *in situ* AFM and LFM showed an increase in roughness at the electrode surface in the presence of the organic solvent that was attributed to electrolyte penetration. The dissolution of surface particles during the second discharge cycle was found to be kinetically controlled by fitting the particle volume calculated using the height distribution function from the topographical data to $V \sim t^{1/2}$, not $t^{3/2}$ as would be suggested using a film dissolution model. The formation of new particles was observed to occur quickly, and was suggested to follow a concentration limited growth model.

The possibility of a high-capacity secondary storage battery using a vanadium oxide electrode due to the four oxidation states of vanadium has elicited interest in the investigation of V_2O_5 for use as a lithium-ion-positive electrode material. Cohen and Aurbach [71] reported the effects of the lithium counter-ion on lithium insertion into V_2O_5 thin film electrodes, observed using EC-AFM. When a vanadium oxide electrode in $LiPF_6$ in PC is cycled, the passivation layer that forms slowed the subsequent insertion of lithium ions and was seen as an irreversible change in the morphology at the surface of the electrode. This effect was not observed when $LiClO_4$ was used in PC as there was no change in the morphology of the electrode surface over several charging cycles. Slower lithium insertion with $LiPF_6$ was attributed to the blocking nature of the passivation layer.

In addition to the study of the metal oxides as positive electrode materials, metals and metal alloys have also been investigated using EC-AFM by several research groups. Beaulieu et al. [72] recently reported on "colossal" volume changes ($\geq$100%) during the electrochemical alloying of lithium into SiSn thin films. These films experience anisotropic growth during the initial deposition, followed by a more isotropic shrinkage during the first charge, when lithium is removed from the film. During this initial shrinkage, lateral stresses are relieved by cracking of the film, and curling at the newly formed edges. When lithium is deposited again, the film expands isotropically with the edges contacting seamlessly due to the lateral expansion, with virtually no detectable traces of the cracks, as shown in Figure 10. The next lithium cycle is accompanied by another isotropic shrinkage; the film contracts at the original crack lines, though the gaps are larger during this and successive cycles. Additional cycling results in reversible, isotropic expansion and contraction in these films.

In later work, Beaulieu et al. [73] reported on the catalytic electrolyte decomposition on pure Sn electrodes that results in an irreversible loss of capacity. It was found that if lithium is deposited at a fast enough initial rate, alloying occurs and as a result, decomposition is suppressed. The LiSn alloy is not catalytically active for decomposition of the electrolyte, and no additional decomposition occurs if the electrode is alloyed at the solvent electrode interface by this fast initial discharge. Cycling this thin film electrode results in an overall increase in thickness as lithium becomes trapped within the film due to the electrical disconnection in portions of the thin film. This also results in continuous loss of capacity throughout the cycling of the electrode.

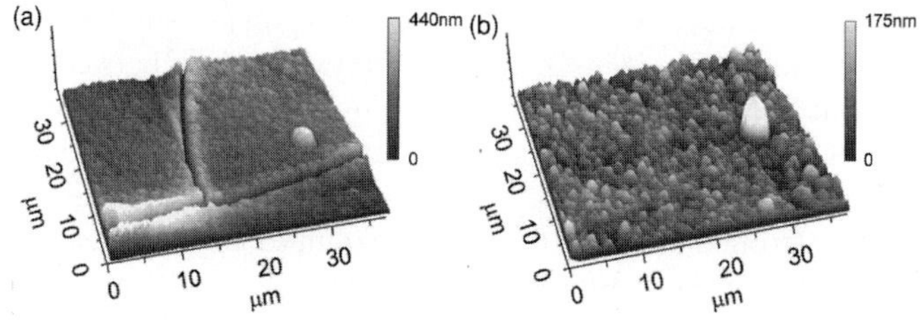

FIGURE 10. Three-dimensional topographic views of (a) unlithiated and (b) lithiated SiSn films. The height contrast scale is shown on the right of each image. Reprinted with permission of [72]. Copyright 2001, The Electrochemical Society.

Comparison of crystalline versus amorphous thin film electrodes by Beaulieu et al. [74], in which crystalline Al or Sn, or amorphous Si or $Si_{0.64}Sn_{0.36}$ alloy thin films were studied with EC-AFM, determined that the amorphous alloys are much better suited to lithium positive electrodes than their crystalline counterparts. The amorphous materials exhibit reversible, homogeneous expansion, whereas the crystalline materials expand inhomogeneously and exhibit a high degree of irreversibility. This irreversibility results in poor capacity retention when used as a positive electrode in a lithium ion battery.

Mogi and coworkers [75] examined the effects of additives on the formation of an SEI on a Ni positive electrode. The cycling of Ni electrodes in 5 wt% FEC, VC, or ES in 1 M $LiClO_4$ in PC was examined with EC-AFM to determine how each additive affected the morphology of the electrode surface through repeated cycles of the electrode. The authors showed that while FEC improved the efficiency, both VC and ES significantly reduced the cycling efficiency of the electrode. Analyzing the morphology during cycling in the presence of each additive displayed a uniform layer of close packed particles with FEC, but a more solid or an inhomogeneous layer with VC and ES, respectively. Impedance analysis showed the film formed in FEC had the lowest impedance, even lower than with an additive free solution. This was in agreement with the cycling efficiency data presented, where the FEC exhibits the best performance of the additives studied.

3.1.3 Corrosion and Anodic Coatings

The anodization of aluminum has been used as a method to provide a protective coating on aluminum since the 1930s [76]. Through the control of anodization parameters, namely, the nature of the acid electrolyte and the potential or current used to oxidize the aluminum, the size and surface coverage of hexagonally arranged pores can be produced. This anodic aluminum has been used as a template for the growth of metallic [77] and semiconducting nanorods [78].

Wu et al. [23] monitored the formation of pores on high-purity aluminum foil under galvanostatic control in a phosphoric acid solution with *in situ* AFM. They compared the behavior of as received foil to cathodically cleaned foil prior to the anodization procedure. As shown in Figure 11, the cathodically precleaned

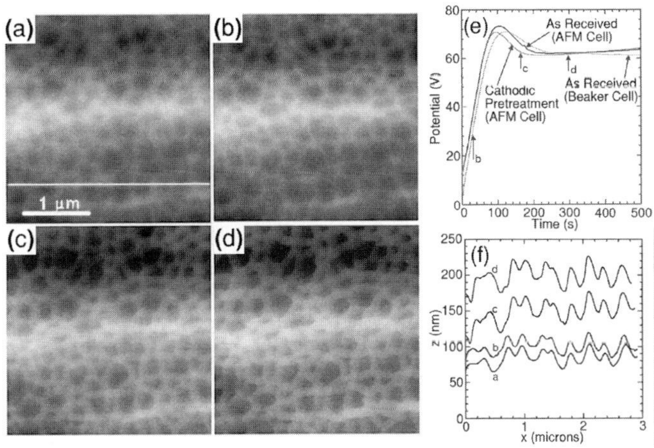

FIGURE 11. AFM topographic view images during oxidation of Al foil pretreated by cathodic dissolution. Image (a) captured at open circuit just before current application, and images (b–d) were obtained during oxidation, at mean times indicated in plot (e). Maximum height contrast 200 nm for all images. (e) Potential transients measured during anodic oxidation of aluminum samples in the AFM fluid cell. Arrows mark the average times during the capture of images (b–d). (f) Elevation profiles during oxidation of cathodically treated Al foil, at the location of the horizontal line drawn in image (a). Profiles are offset for clarity of presentation; the z scale indicates relative heights along individual profiles, not absolute heights. Reprinted with permission from [23]. Copyright 2000, The Electrochemical Society.

foil forms a hexagonally arrayed pore structure that is characteristic of anodic aluminum oxide (AAO). Through the analysis of the potential transients and the topographical data collected with the AFM, a mathematical model was used to describe the mechanism of oxide growth on an aluminum surface.

Leblanc and Frankel [24] have reported on the corrosion of AA2024-T3 aluminum in chloride and chromate solutions using a combination of *ex situ* Kelvin probe force microscopy (KPFM), and *in situ* contact mode AFM to expose desired areas of the aluminum substrate to the solution by scratching away a protective ink. The comparison of Volta potential maps acquired with KPFM to topography images allowed studies to focus on specific inclusion sites on the aluminum substrate, such as Al-Cu-Fe-Mn or Al_2CuMg intermetallic phases. A potential dependent AFM experiment evaluating this alloy in 0.5 M NaCl and 0.1 mM $Na_2Cr_2O_7$, resulted in sustained local hydrogen evolution at the AFM probe tip when the potential of the aluminum was between −0.65 and −0.63 V versus SCE. This prevented the collection of topography images during the potential sweep of the aluminum substrate. After the potential sweep was completed, topography showed that pits formed only at Al-Cu-Fe-Mn inclusions. These experiments show the utility of the EC-AFM in its ability to selectively expose an area by scratching the surface while the potential is controlled. Also seen is the ability of the tip to

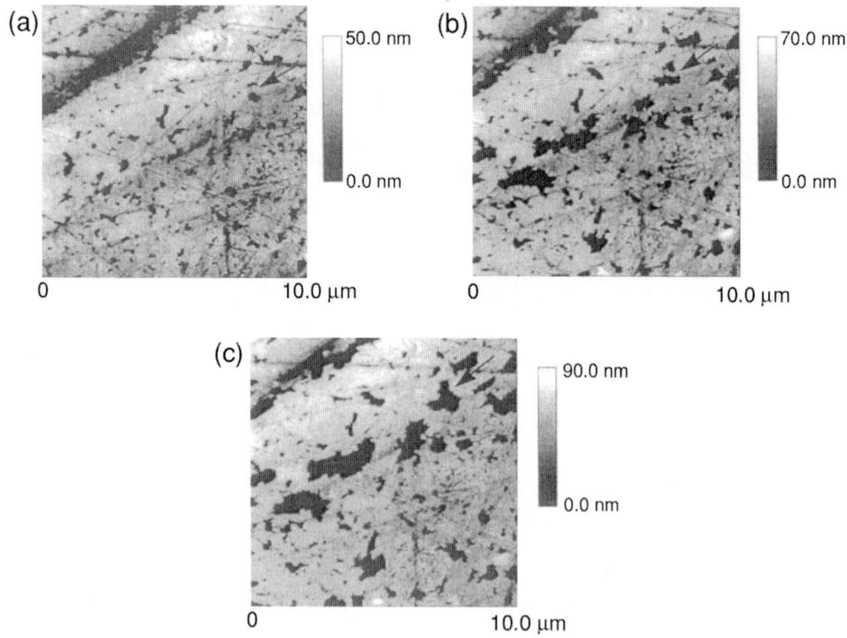

FIGURE 12. Time-dependent *in situ* AFM images (topography vs. time) of the copper surface polarized at +30 mV (SCE) in 0.1 M H_2SO_4: (a) t = 0 min, (b) t = 10 min, (c) t = 20 min. Reprinted with permission from [25]. Copyright 2000, Elsevier B.V.

act as a nucleation center for the evolution of a gaseous byproduct of a surface reaction, resulting in poor imaging conditions.

As discussed above, copper is invaluable in numerous technologies; hence, its corrosion behavior has been studied extensively. Studies on the corrosion of copper and the effects of different inhibiting agents on corrosion are a good topic for EC-SPM evaluation. Bertrand et al. [25] reported on the morphological changes that occur to polycrystalline copper electrodes when exposed to H_2SO_4 and $NaSO_4$ solutions, as well as the passivation of copper with either Borax ($Na_2B_4O_7$) or sodium heptanoate, ($CH_3(CH_2)_5COONa$). Figure 12 illustrates the time-dependent growth of corrosion pits on the copper substrate while exposed to H_2SO_4. When exposed to $NaSO_4$ at open circuit potential (OCP, no applied bias), dissolution occurred creating pits on the electrode and the dissolved Cu precipitated as oxide, forming a nodular coating on the surface. This precipitate could be reduced with applied potential, but the morphology of the surface was irreversibly changed. When oxidized in a borax solution, a thin passivation layer forms, the morphology of which is independent of the potential that is applied to form the layer. A poorly adhering passivation layer is formed on the copper surface if an oxidizing potential is applied in sodium heptanoate solution, making AFM imaging difficult. When this layer is reduced, no pits are observed on the copper surface, unlike each of the other cases discussed, showing the protective nature of this passivation coating.

3.2 Advanced Electrochemical SPM Techniques

These techniques probe specific electrical signals at the electrochemical interface near the electrode in addition to looking at the topography before, during, and/or after an electrochemical process. This additional channel of information compliments the topography by helping to elucidate the local environment from the solution side of the interface, probing the effects of diffusion and, in some cases, the chemical changes taking place at the electrode. These are the most powerful electrochemical SPM methods, since interfacial structure can be obtained as a function of tip bias voltage, tip-to-sample separation, and applied substrate potential.

3.2.1 SECM-AFM

SECM utilizes an ultramicroelectrode (UME) locally to probe the electrochemical properties in the vicinity of a substrate. Two types of feedback, positive or negative, can be observed when an SECM probe tip approaches either a conductive or insulating substrate in the presence of a redox couple. By recording the response of current from the redox couple at the tip versus tip-substrate separation distance, an approach curve can be collected. Three types of electrochemical responses due to different diffusion conditions of a redox active mediator are shown in Figure 13. The first case (Figure 13(a)) shows hemispherical diffusion of an oxidized species that is reduced at a disk-shaped SECM probe tip that is held away from a substrate. This is used to characterize the baseline current ($i_{T,\infty}$) at the SECM probe tip electrode. Figure 13(b) shows how approaching an insulating substrate

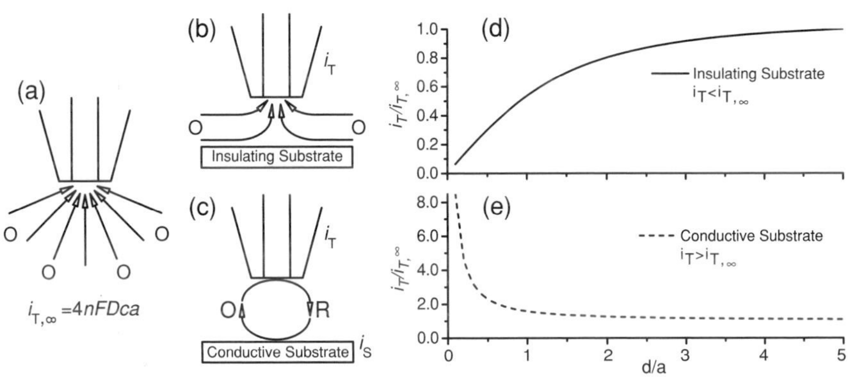

FIGURE 13. (a) Hemispherical diffusion at an SECM probe tip with a planar disk shaped UME, and the resulting current observed; (b) inhibition of diffusion as an SECM probe tip approaches an insulating substrate; (c) redox mediator regeneration between an SECM probe tip and a conductive substrate; (d) ideal approach curve for a disk shaped SECM probe tip approaching an insulating substrate (negative feedback); (e) ideal approach curve for a disk-shaped SECM probe tip approaching a conductive substrate (positive feedback). Adapted from [18].

can inhibit diffusion of the oxidized species, O, to the electrode, termed negative feedback. When the SECM probe tip is brought close to a conductive substrate in the presence of a redox couple, the redox species is regenerated via oxidation at the substrate, resulting in positive feedback (Figure 13(c)). Figure 13(d,e) illustrates ideal (dimensionless) approach curves normalized by $i_{\mathrm{T},\infty}$ and a (tip radius), collected during the approach toward an insulator (negative feedback) and a conductive substrate (positive feedback) using a redox active mediator, respectively. Another ability of SECM experiments is to operate in a generation/collection mode where the tip (or substrate) is used to generate a moiety in solution via oxidation or reduction and the products are interrogated by measuring the substrate (or tip) current, resulting from reduction or oxidation of the species produced at the tip (or substrate). This mode can be used to probe local differences in reactivity by generating reactants at the SECM tip. A thorough discussion of the various SECM modes is provided by Bard and Mirkin [80].

There has been much work with SECM since the technique was introduced [80], with several reviews written recently [16,81] as well as a comprehensive text edited by Bard and Mirkin [79] covering the theory, instrumentation, numerical analysis, and various applications of SECM. Interested readers are directed to these sources for a more through discussion on SECM principles and applications. The combination of scanning electrochemical microscopy with atomic force microscopy (SECM-AFM) combines the topographical resolution of AFM with the electrochemical characterization capabilities of SECM. The integration of these two techniques promises to improve the resolution of SECM through shrinking the scanning electrode to nanometer dimensions, as well as providing a means to collect topographical information concurrently.

3.2.1.1 SECM-AFM Probe Development

One vital area of research toward improving SECM-AFM has been the development of tip-fabrication techniques that can yield high-resolution topographical and electrochemical information. Efforts have included both modifications of commercial probes and "ground up" fabrication processes. In addition to creating AFM probes that incorporate a nanoscale electrode while retaining a sharp tip, the control or repeatability of the electrode geometry has also been a focus of these efforts. The main driver in this area is the ability to develop new probe technologies, some of which are highlighted in Figure 14 and discussed in more detail below.

Probes developed for combined SECM-AFM experiments can be classified as either *intrinsic* or *extrinsic* electrode probes. Intrinsic electrode probes utilize a conductive tip to probe both the topography and the current, whereas extrinsic probes have isolated or spatially separated tip/electrode arrangements.

Intrinsic tips require multiple scans for any given area: one to measure the topography followed by another with the probe tip offset vertically from the substrate for the SECM scan to acquire additional signal (e.g., current) at a fixed tip-to-sample distance. While this type of tip is typically easier to fabricate, the fabrication process is less controlled resulting in inconsistencies that require thorough characterization

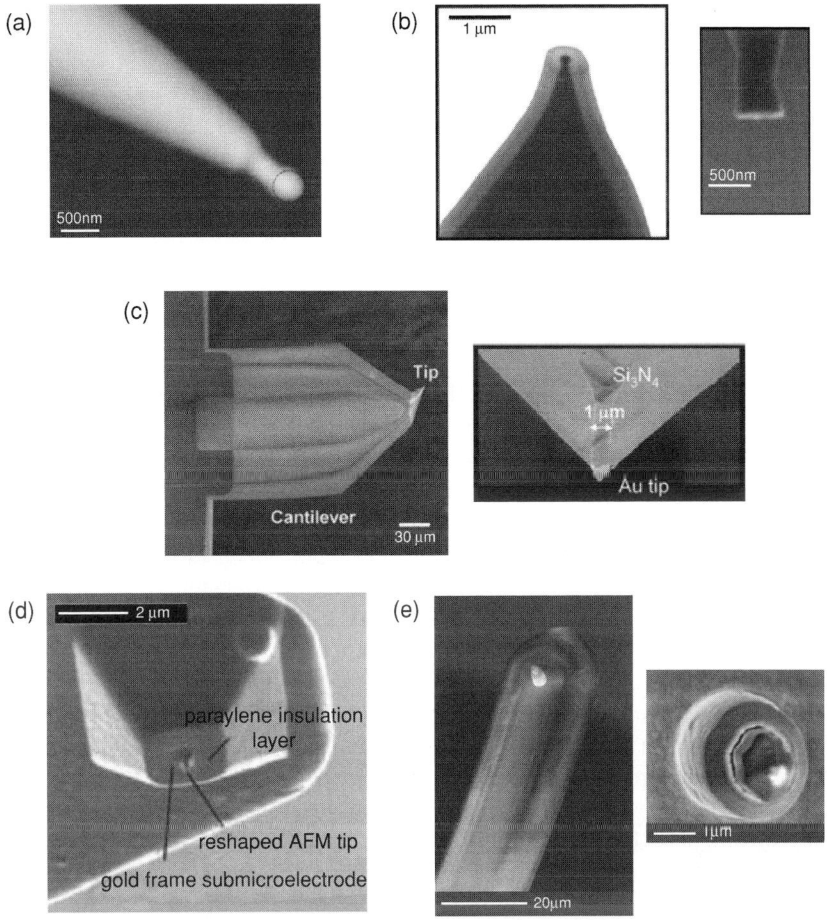

FIGURE 14. Recent SECM-AFM tips developed: (a) spherical Au tip, (b) Au-plated carbon nanotube, (c) batch-fabricated SECM-AFM tip with triangular Au electrode. (d) FIB-modified commercial probe with square frame Au electrode, and (e) high aspect ratio tip with Pt ring electrode. Reprinted in part with permission from (a) [83b], copyright 2004, American Chemical Society; (b) [84], copyright 2005, American Chemical Society; (c) [85], copyright 2005, American Chemical Society; (d) [39], copyright 2003, Wiley-VCH Verlag GmbH & Co.; and (e) [44b], copyright 2004, The Society for Photo-Optical Instrumentation Engineers.

(calibration) of each tip. Intrinsic electrode tips can suffer from wear effects that cause changes in electrode geometry and current response, as well as a loss of topographical resolution.

Because the conductive and topographical probe tip are either isolated or spatially separated, extrinsic electrode probes can be used to concurrently measure the topography and the current or potential. The fabrication of extrinsic electrode probes is more complex and typically requires several steps and expensive equipment, often including focused ion beam (FIB) milling. One distinct advantage of extrinsic electrode probes is that they provide consistent electrode geometry and tip/sample separation that can both reduce the time needed to image a surface and reduce the need to extensively characterize a tip before use.

The first report [28] of SECM-AFM was in 1996 when the dissolution of a KBr crystal in acetonitrile was investigated using a commercial SiN cantilever that was coated with a Pt layer, and insulated using a nonconductive polymer. Only the base of the probe was insulated, with both the apex and the cantilever beam left exposed. This prevented high-resolution imaging in the electrochemical portion of the experiment due to the large exposed electrode area.

Hirata et al. [82] reported efforts to modify commercial AFM tips to enable SECM-AFM scanning. They metallized Si_3N_4 cantilevers by physical vapor deposition of gold with a thin chromium adhesion layer. The probes were insulated at all but the very apex of the probe using a digital projector coupled to a microscope to pattern photoresist. This method for modifying commercial tips is simple and easily implemented with standard laboratory equipment.

Macpherson and Unwin [26] have developed SECM-AFM probes by etching Pt wire, similar to making an STM tip, followed by bending the tip at a 90-degree angle and compressing a portion of the wire to form a cantilever beam. This wire is then insulated with electrophoretic paint and glued to an AFM chip to complete the fabrication process. While these tips do improve on the electrochemical resolution achieved over other metallized and insulated commercial probes, this method is labor intensive and each tip must be characterized individually due to the variability of the fabrication process.

To produce tips with predictable geometry and electrochemical response, Abbou and coworkers [83] used an arc discharge method to modify the tip of a gold wire electrode fabricated by electrochemical etching and electrophoretic paint deposition, similar to that described by Macpherson and Unwin [26]. This produced an exposed electrode on the tip with a spherical geometry (Figure 14(a)) that could then be used in the SECM-AFM apparatus. Although diffusion of electroactive species toward a spherical electrode is well characterized [18], simplifying the interpretation of experimental data, the lateral topographical resolution is poor.

Single-wall carbon nanotubes (SWNT) have also been used to build high-resolution SECM-AFM tips by Burt et al. [84] (Figure 14(b)). These tips are fabricated by gold plating a SWNT AFM probe and insulating with an electrodeposited polymer over coated with a layer of Si_3N_4. A focused ion beam mill was then used to remove the end of the insulated SWNT, leaving a disk shaped electrode with a SWNT core and a gold outer layer. The well-defined electrode geometry produced simplifies data treatment, as the diffusion toward a disk electrode is well studied and understood [18].

Dobson et al. [85] reported the use of electron beam lithography to batch fabricate intrinsic electrode SECM-AFM probes. These probes (Figure 14(c)) exhibit a high degree of reproducibility due to the automated nature of the fabrication methods. This reproducibility allows the authors to use finite element modeling software to simulate diffusion profiles of these tips, simplifying interpretation of experimental data. Both approach curves and SECM-AFM data are presented demonstrating the capabilities of these tips.

The first SECM-AFM extrinsic electrode probes were demonstrated by Kranz and coworkers [31] in 2001. These probes (Figure 14(d)) have a frame shaped electrode geometry that is separate from the Si_3N_4 tip and can simultaneously collect topographical and current data, simplifying SECM-AFM and reducing the time needed to complete one scan. Fabrication is complicated and labor intensive, since this process requires first coating a commercial Si_3N_4 cantilever with gold, then with Si_3N_4. These coated tips are then extensively modified in a FIB mill, producing a SECM-AFM probe with a sharp Si_3N_4 tip for topographical imaging and a separate gold electrode for electrochemical analysis. Additionally, square-frame electrodes are not common, and have not been exhaustively treated and no theoretical models exist to describe their mass transport characteristics. Sklyer et al. [86] used numerical simulations to evaluate diffusion patterns at the electrode and showed that experimental results agree quantitatively with numerical simulations. These tips have been used for tapping mode imaging without damaging or otherwise changing the electrode geometry [46].

Another design has been reported by Fasching and coworkers [44] for an extrinsic electrode SECM-AFM probe that has a ring-shaped electrode fabricated onto a high aspect ratio silicon cantilever. The resultant tips (Figure 14(e)) are reproducible, but, as with other extrinsic probe designs, fabrication is complex and requires elaborate equipment. The electrode was examined in 10 mM $Ru(NH_3)_6^{3+}$ with electrochemical impedance spectroscopy (EIS) and cyclic voltammetry and displayed characteristic UME behavior. The probes were undamaged after scanning a grating structure.

3.2.1.2 SECM-AFM Applications

With the increased resolution of newly developed SECM-AFM probes (section 3.2.1.1) *in situ* analysis of both topography and local electrochemical behavior has reached new limits. It is now technologically feasible to evaluate diffusion through 100-nm pores [27] and the performance of sub-micron clusters of catalyst particles [38]. Several applications of this technique are discussed below.

Kucernak et al. [38] reported on SECM-STM analysis of commercial platinum catalyst particles for the hydrogen evolution reaction by both tip generation/substrate collection (TG/SC) and substrate generation/tip collection (SG/ TC) modes. STM was performed *ex situ*, followed by tip positioning, and tip/substrate immersion prior to performing the SECM experiment. Kucernak demonstrated the capability of using STM hardware for SECM experiments as well as the utility of SECM to analyze the catalytic current at single catalyst particles during the TG/SC experiment. The SG/TC experiment also demonstrated the ability to observe the hydrogen generation from the substrate with the ability to separate the contribution from each catalyst particle.

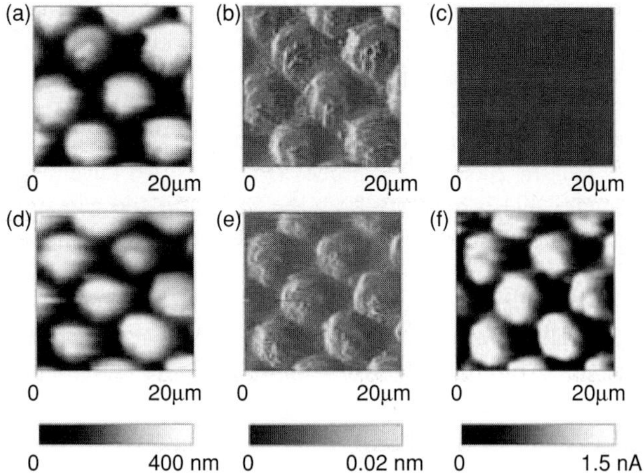

FIGURE 15. Simultaneously recorded topography of the enzyme-containing periodic polymer structure (height image) and current image of GOx activity in AFM tapping mode with free drive amplitude of 5 V and drive frequency of 32.4 kHz. Height (a,d), corresponding amplitude (b,e), and simultaneously recorded Faradaic current (c,f) images recorded in air-saturated phosphate buffer (0.1 mol L^{-1}, pH 7.4) in the absence (a–c) and in the presence (d–f; 50 mM) of glucose. The tip was held at a potential of 0.75 V (versus Ag/AgCl quasi-reference electrode). Electrode edge length: 770 nm; tip height: 700 nm; scan rate: 3 Hz. Reprinted with permission from [39]. Copyright 2003, Wiley-VCH Verlag GmbH & Co.

Kueng et al. [39] utilized the ability of tapping mode AFM to image soft samples coupled with the electrochemical characterization of SECM to image the enzymatic redox activity of glucose oxidase (GOx) entrapped within a polymer. The polymer was contained within micropatterened pores in a SiN layer over a gold electrode. This results in topography of this pattern, and a Faradaic current in the vicinity of the GOx. A similar sample without GOx interspersed in the polymer shows no redox activity (Figure 15). The SECM was used in a SG/TC mode, where the redox byproduct of GOx, H_2O_2, is detected by the tip as it passes over the polymer entrapped enzyme. FIB patterned probes (Figure 14(d)) were used for these experiments, allowing for the concurrent collection of height and current rather than the sequential acquisition of topography and current associated with SECM-AFM tips.

Kranz et al. [29] further demonstrated the power of SECM-AFM to analyze the activity of horseradish peroxidase (HRP) that is immobilized via thiol linkage to a patterned gold electrode. When scanned with H_2O_2 in solution, enzymatic activity could be detected by the current channel in the immediate vicinity of the spots containing immobilized HRP. The researchers also noted that when the tip to electrode height was comparable to the feature height on the substrate, the electrochemical signal is convoluted due to geometric interference at feature edges.

Given the flexibility of FIB fabricated probes, the researchers suggest that a probe with a larger tip to electrode separation would reduce this effect.

Jones et al. [87] used a SECM-AFM probe as a local proton generator to study dissolution processes of calcite in aqueous solutions. A Pt-coated commercial probe was insulated with polystyrene up to the base of the cantilever, providing for a relatively large electrode area incorporating the tip and the entire cantilever beam. By controlling the rate of proton flux at the calcite surface through tip bias, dissolution processes can be monitored in different regimes. At low proton fluxes of 1 nmol cm^{-2} s^{-1}, they reported that monolayer deep pits are formed that are less than 250 nm wide at point defect sites and proceeding along step edges. At 10 nmol cm^{-2} s^{-1} flux, these pits are larger than 250 nm in width and grew fast enough that nucleation at point defects is not seen due to the time scale of the SECM-AFM experiment. These measurements demonstrate the utility of SECM-AFM to exploit small local perturbations in the study of dissolution processes.

Monitoring flux through micrometer and nanometer dimension pores, whether in a membrane or on a cell, is another area where the unique abilities of SECM-AFM can contribute. Initial studies by Macpherson and Unwin [26] displayed this capability by mapping diffusion profiles for $IrCl_6^{3-}$ diffusion through polycarbonate track etched membranes with pore sizes of 0.6 μm and 1.2 μm. They also reported on the use of these membranes as microcavities, were diffusion was limited due to blocking by the probe itself. Also demonstrated was the ability to pattern a surface through selective dissolution of the substrate using electrochemistry at the probe tip.

Macpherson et al. [27] further studied diffusion of $IrCl_6^{3-}$ through polycarbonate track etch membranes with nanometer size (100 nm) pores and presented theoretical discussions to describe the effects of tip shape on diffusion. As with standard SECM, approach curves can be used to elucidate the tip geometry. The actual tip shape is required to fully analyze data collected in any SECM-AFM experiment.

3.2.2 LEIS-AFM

Electrochemical impedance spectroscopy (EIS) is a powerful electrochemical technique that utilizes small ac electrical perturbations to a system held at a steady state potential. By analyzing how the substrate responds to these small electrical perturbations, it is possible to extract meaningful kinetic data even for complex reactions with multiple processes occurring with differing time constants [88]. This technique is also useful for the extraction of double layer capacitance and uncompensated resistance in an electrochemical system. This is an ensemble average technique and the localized response of a heterogeneous electrode surface is difficult to extract. For a thorough discussion of this technique, the reader is directed to the monograph edited by Macdonald [86]. When EIS is implemented as a scanning probe technique, termed local electrochemical impedance spectroscopy (LEIS), only one frequency can be analyzed for each scan due to limitations on the acquisition time of the impedance signal. For a frequency-dependent spectrum, the probe is held stationary and the frequencies are scanned with the aid of a frequency response analyzer.

An impedance-based scanning probe technique was introduced by Lillard et al. [89] that used a bielectrode to collect local electrochemical impedance spectroscopy (LEIS). This bielectrode consisted of two microelectrodes separated vertically by less than 1 mm, scanned across a surface. The difference in potential measured between the two electrodes is used along with the solution conductivity to determine a local current density that, when compared to the potential waveform applied, can be used to calculate the local impedance. This method has been used to examine coating failure on a urea-formaldehyde modified epoxy coating on cold-rolled steel [90] and has also been used to study aluminum and aluminum/molybdenum electrodes [89]. Baril et al. [91] reported a bielectrode LEIS system to study corrosion of an AZ91 magnesium alloy in a low concentration Na_2SO_4 solution.

In an effort to improve the resolution of LEIS, Bayet and coworkers [92] introduced a modified scanning vibrating electrode technique (SVET), where ac perturbations to the substrate are analyzed using a SVET probe, and the complex waveform data deconvoluted to give both local current density and the local voltage. This permits a single microelectrode or even an UME to be used in the collection of LEIS, improving the resolution of the technique. Zou and Thierry [93] used both experimentation and computer simulations to show that LEIS data collected within the diffusion layer may differ from conventional EIS due to heterogeneous concentration gradients. Bayet et al. [94] further analyzed this SVET based LEIS system with computer simulations to compare conventional EIS to both the bielectrode system, where only a local current density is acquired, and the SVET based LEIS system, where both local current density and local potential are acquired. They concluded that LEIS requires both current and potential to be measured locally, close to the electrode surface, to minimize errors associated with the different environments that can be encountered over an electrified substrate. These researchers also note that large deviations from expected impedance spectra can be obtained if the probe is within the diffusion layer at the surface. Katemann et al. [95] have taken a similar approach, and introduced an alternating current-SECM (ac-SECM) technique to collect LEIS data.

Experiments using *ex situ* AFM to analyze impedance data across the interface of solid state devices have been reported by Kalinin and Bonnell [96] and termed scanning impedance microscopy (SIM). In this report an ac signal was applied across the grain boundary of an Nb-doped $SrTiO_3$ crystal and the local phase angle of an oscillating AFM tip scanned across the grain boundary and was used in conjunction with scanning surface potential microscopy (SSPM) to quantify both the interface resistance and capacitance of the grain boundary.

Additional developments of AFM-based impedance measurements have been reported by Shao et al. [97], which they term nanoimpedance microscopy/spectroscopy (NIM). This technique is used to study the transport behavior within a grain as well as across grain boundaries on a polycrystalline ZnO substrate. This experiment also introduced the AFM tip as an electrode through

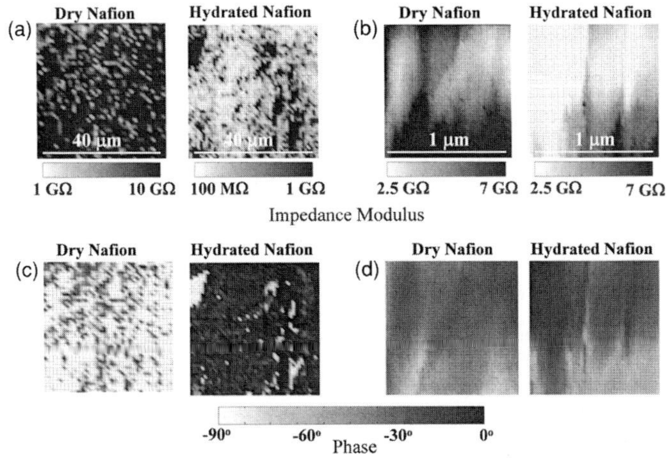

FIGURE 16. Impedance modulus (a,b) and impedance phase (c,d) images of the Nafion electrolyte membrane as a function of humidity. The image sets on the left compare the impedance response under dry (0% relative humidity) vs ambient (30% relative humidity) conditions for a 40-μm area of the Nafion membrane. The image sets compare the impedance response under dry vs. ambient humidity conditions for a 1-μm area of the Nafion membrane. Images acquired at 1 Hz with a 30 mV excitation signal, 0.4 V dc cathodic bias. The hydration level of the Nafion electrolyte significantly influences its H_3O^+ concentration and ionic conductivity, as is readily apparent from the impedance images (note change in the impedance modulus scale for the left image set). Reprinted with permission from [40]. Copyright 2004, American Institute of Physics.

which the ac potential was applied to the substrate, while monitoring the current through the tip. The researchers could distinguish between grain boundary and tip-sample contributions by analyzing the impedance spectra.

O'Hayre et al. [40] reported the first electrochemical LEIS-AFM experiment, evaluating the impedance through a protonated Nafion solid polymer electrolyte at various stages of hydration. A reversible hydrogen electrode was used as the counter-electrode on the opposite side of the Nafion membrane and coupled with a platinum-coated AFM tip held at a potential to reduce H^+ to water. This formed a local hydrogen fuel cell that was scanned spatially across the membrane with a small ac bias applied to produce an impedance image of the membrane surface. (Figure 16) The tip position was also held stationary (Figure 17(a)) and the frequency scanned producing a local impedance plot (Figure 17(b)).

Local electrochemical impedance spectroscopy-atomic force microscopy (LEIS-AFM) can expand on the abilities of SECM-AFM by incorporating ac perturbations to analyze local reaction time constants. This technique offers great promise to probe corrosion processes on coated metals, heterogeneities in the reactivity of alloys, and catalytic reactions. While little has been reported on *in situ* LEIS-AFM, the LEIS technique has been commercialized, and research has been reported in using SVET apparatus to probe LEIS at a higher resolution, as well

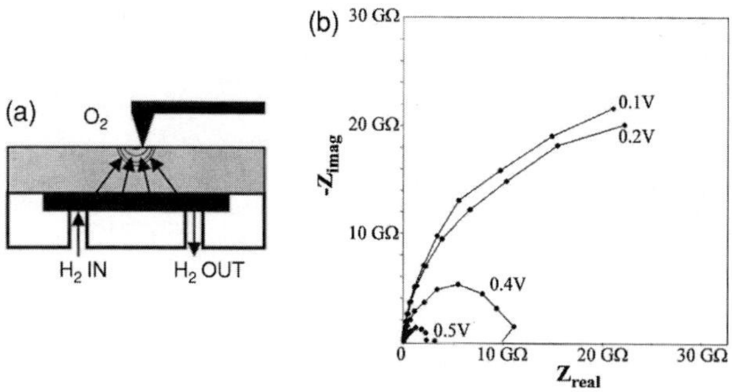

FIGURE 17. (a) Configuration for Nafion characterization experiment. The impedance is measured between the Pt-coated AFM tip on the Nafion top surface and a bulk bottom electrode (a reversible hydrogen electrode) which is hermetically sealed and supplied with hydrogen gas. (b) Nyquist spectra of the Pt-coated AFM tip/Nafion impedance response as a function of applied dc bias. Impedance spectra are measured according to the experimental configuration in (a) over the frequency range 100 kHz–100 mHz with a 30 mV excitation signal. H_2 gas is provisioned to the anode, the AFM tip is in hard contact with the membrane. The impedance loop shrinks with increasing dc bias, characteristic of a Faradaic charge transfer process. Reprinted with permission from [40]. Copyright 2004, American Institute of Physics.

as SIM and NIM experiments with the AFM. Much of the development of special tips for SECM-AFM can be harnessed for application to *in situ* LEIS-AFM, further improving the resolution of the technique. An oscillating cantilever could be used much as SVET to probe the local voltage and current density in the vicinity of an electrode surface. Coupling the topographical analysis capability of AFM along with this oscillating cantilever based LEIS measurement, will allow precise placement of a local impedance probe, increasing the resolution of the LEIS technique.

4 Summary

Although electrochemical SPM is a relatively young field, a wide range of systems have been examined ranging from studies of surface adsorption, film growth and dissolution, and membrane transport to catalysis. The summarized work in electrochemical-SPM demonstrates the impressive resolving power of this approach, and the results of these studies provide an atomic and molecular basis for understanding the kinetics and thermodynamics of electrified interfaces. The field of electrochemical SPM is anticipated to expand significantly as new technical advances in tips and cantilevers and hybrid methods and measurements emerge. Application of new electrochemical SPM methods will impact research in several disciplines, including biology, physics, chemistry, materials science, engineering, and medicine. As have conventional scanning tunneling and atomic force microscopies, electrochemical SPM will prove itself an exceptional and indispensable tool for direct high-resolution studies of physicochemical processes, especially since many technological and industrial processes are based on electrochemical reactions and principles.

Acknowledgments. Financial Support from the National Science Foundation (grant CNE-0134884) is gratefully acknowledged.

References

1. B. J Welch, JOM, **52**, 24 (1999).
2. M. Paunovic and M. Schlesinger, *Fundamentals of Electrochemical Deposition*, (Wiley-Interscience, New York, 1998).
3. A. J. Fry, *Synthetic Organic Electrochemistry*, (John Wiley, New York, 1988).
4. A.M.O. Mohamed and H.E. Antia, *Geoenvironmental Engineering,* (Elsevier, Amsterdam, 1998).
5. D. Linden, *Handbook of Batteries*, 3rd ed., edited by D. Linden and T. B. Reddy (McGraw-Hill, New York, 2002), Chap. 1, pp. 1.1–1.18.
6. (a)D. Linden and T.B. Reddy, *Handbook of Batteries*, 3rd ed., edited by D. Linden and T. B. Reddy (McGraw-Hill, New York, 2002), Chap. 22, pp. 22.3–22.24; (b) A. Kaufman and H. F. Gibbard, *ibid.,* Chap. 43, pp. 43.1–43.20.
7. M. Datta, *Microelectronic Packaging*, edited by M. Datta, T. Osaka, and J.W. Schultz (CRC Press, London, 2004).
8. R.J. Forster, *Principles of Chemical and Biological Sensors; Chemical Analysis Series,* edited by D. Diamond (Wiley-Interscience, New York, 1998), Vol. 150.
9. Y. Yamazaki, Electrochim. Acta **50**, 663 (2004).
10. N.N. Voevodin, V.N. Balbyshev, M. Khobaib, and M.S. Donley, Prog. Org. Coat. **47**, 416 (2003).
11. A.A. Gewirth and B.K. Niece, Chem. Rev. **97**, 1129 (1997).
12. M.D. Ward and H.S. White, *Modern Techniques in Electroanalysis, Chemical Analysis Series*, edited by P.Vanýsek (Wiley-Interscience, New York, 1996), Vol. 139.
13. A.J. Bard and F.F. Fan, *Scanning Probe Microscopy and Spectroscopy*, 2nd ed., edited by D.A. Bonnell (Wiley-VCH, New York, 2001), Chap. 12, pp. 421–452.
14. T.P. Moffat in *Encyclopedia of Electrochemistry,* edited by A.J. Bard, M. Stratmann, and P.R. Unwin (Wiley-VCH, Weinheim, 2003), Vol. 3, pp. 393–414.
15. J.V. Macpherson, *Encyclopedia of Electrochemistry,* edited by A.J. Bard, M. Stratmann, and P.R. Unwin (Wiley-VCH, Weinheim, 2003), Vol. 3, pp. 415–434.
16. B.J. Horrocks *Encyclopedia of Electrochemistry,* edited by A.J. Bard, M. Stratmann, and P.R. Unwin (Wiley-VCH, Weinheim, 2003), Vol. 3, pp. 444–490.
17. M.J. Weaver, J. Phys. Chem. **100**, 13079 (1996).
18. A.J. Bard and L.R. Faulkner, *Electrochemical Methods Fundamentals and Applications*, 2nd ed. (John Wiley, New York, 2001).
19. *Modern Electroplating,* 4th ed., edited by M. Schlesinger and M. Paunovic (John Wiley, New York, 2000).
20. I. Ban, Y. Yamaguchi, Y. Nakayama, N. Hirai, and S. Hara, J. Power Sources **107**, 167 (2002).
21. R. Vidu and S. Hara, J. Electroanal. Chem. **475**, 171 (1999).

22. (a) J. Zak and M. Kolodziej-Sadlok, Electrochim. Acta 45 2803 (2000); (b) R. Vidu and S. Hara, Scripta Mat. **41**, 617 (1999).
23. H. Wu, X. Zhang, and K.R. Herbert, J. Electrochem. Soc. **147**, 2126 (2000).
24. P. LeBlanc and G.S. Frankel, J. Electrochem. Soc. **149**, B239 (2002).
25. G. Bertrand, E. Rocca, C. Savall, C. Rapin, J.-C. Labrune, and P. Steinmetz, J Electroanal. Chem. **489**, 38 (2000).
26. J.V. Macpherson and P.R. Unwin, Anal. Chem. **72**, 276 (2000).
27. J.V. Macpherson, C.E. Jones, A.L. Barker, and P.R. Unwin, Anal. Chem. **74**, 1841 (2002).
28. J.V. Macpherson, P.R. Unwin, A.C. Hillier, and A.J. Bard, J. Am. Chem. Soc. **118**, 6445 (1996).
29. C. Kranz, A. Kueng, A. Lugstein, E. Bertagnolli, and B. Mizaikoff, Ultramicroscopy **100**, 127 (2004).
30. J.V. Macpherson and P.R. Unwin, Anal. Chem. **73**, 550 (2001).
31. C. Kranz, G. Friedbacher, B. Mizaikoff, A. Lugstein, J. Smoliner, and E. Bertagnolli, Anal. Chem. **73**, 2491 (2001).
32. Z.-L. Wu, Z.-H. Zang, and S.-L. Yau, Langmuir **16**, 3522 (2000).
33. J. Okada, J. Inukai, and K. Itaya, Phys. Chem. Chem. Phys. **3**, 3297 (2001).
34. Z.-L. Wu and S.-L. Yau, Langmuir **17**, 4627 (2001).
35. C.-H. Shue and S.-L. Yau, J. Phys Chem. B **105**, 5489 (2001).
36. D. Landolt, J. Electrochem. Soc. **149**, S9 (2002).
37. M. Petri, D.M. Kolb, U. Memmert, and H. Meyer, Electrochim. Acta **49**, 183 (2003).
38. A.R. Kucernak, P.B. Chowdhury, C.P. Wilde, G.H. Kelsall, Y.Y. Zhu, and D.E. Williams, Electrochim. Acta **45**, 4483 (2000).
39. A. Kueng, C. Kranz, A. Lugstein, E. Bertagnolli, and B. Mizaikoff, Angew. Chem. Ind. Ed. **42**, 3238 (2003).
40. R. O'Hayre, M. Lee, and F.B. Prinz, J. Appl. Phys. **95**, 8382 (2004).
41. S.-L. Yau, C.M. Vitus, and B.C. Schardt, J. Am. Chem. Soc. **112**, 3677 (1990).
42. L.A. Nagahara, T. Thundat, and S.M. Lindsay, Rev. Sci. Instrum. **60**, 3128 (1989).
43. C.J. Slevin, N.J. Gray, J.V. Macpherson, M.A. Webb, and P.R. Unwin, Electrochem Commun. **1**, 282 (1999).
44. (a) R.J. Fasching, Y. Tao and F.B. Prinz, Chem. Sens. **20**(Suppl. B), 318 (2004); (b) R.J. Fasching, Y. Tao and F.B. Prinz, Proc. SPIE **5342**, 53 (2004).
45. G. Zhao, D.M. Giolando, and J.R. Kirchhoff, J. Electranal. Chem. **379**, 505 (1994).
46. A. Kueng, C. Kranz, B. Mitzaikoff, A. Lugstein, and E. Bertagnolli, Appl. Phys. Lett. **82**, 1592 (2003).
47. K. Kubo, N. Hirai, T. Tanaka, and S. Hara, Surf. Sci. **546**, L785 (2003).
48. *Dimension 5000 Instruction Manual*, v.4.23ce (Digital Instruments, Veeco Metrology Group, Santa Barbara, 1999).
49. L. Stappers and J. Fransaer, J. Appl. Phys. **92**, 5543 (2002).
50. A.V. Getling in *Advanced Series in Nonlinear Dynamics*, edited by R.S. MacKay (World Scientific, New Jersey, 1998), Vol.11.
51. V.M. Dubin, H.S. Simka, S. Shankar, P. Moon, T. Marieb, and M. Datta in *New Trends in Electrochemical Technology* edited by M. Datta, T. Osaka, and J.W. Schultz (CRC Press, London, 2004), Vol. 3.
52. (a) Y.-H. Joung and M.G. Allen, *Proceedings – 53rd Electronic Components & Technology Conference* (IEEE, 2003), pp. 630–646; (b) M. X. Yang, D. Mao, C. Yu, J. Dukovic, and M. Xi, Sol. St. Tech. **46**, 37 (2003); (c) M. Rousseau, Semiconductor Int. **26**, 58 (2003).

53. L. Zitzler, B. Gleich, O.M. Magnussen, and R.J. Behm, Electrochem. Soc. Proc. **99-28**, 38 (2000).
54. W. Polewska, R.J. Behm, and O.M. Magnussen, Electrochim. Acta **48**, 2915 (2003).
55. O.M. Magnussen, W. Polewska, L. Zitzler, and R.J. Behm, Faraday Discuss. **121**, 43 (2002).
56. O.M. Magnussen, L. Zitzler, B. Gleich, M.R. Vogt, and R.J. Behm, Electrochim. Acta **46**, 3725 (2001).
57. A.J. Salkind, A.G. Cannone, and F.A. Trumbure in *Handbook of Batteries*, 3rd ed., edited by D. Linden and T.B. Reddy (McGraw-Hill, New York, 2002), Chap. 23, pp. 23.1–23.88.
58. Y. Yamaguchi, M. Shiota, Y. Nakayama, N. Hirai, and S. Hara, J. Power Sources **85**, 22 (2000).
59. Y. Yamaguchi, M. Shiota, Y. Nakayama, N. Hirai, and S. Hara, J. Power Sources **93**, 104 (2001).
60. Y. Yamaguchi, M. Shiota, M. Hosokawa, Y. Nakayama, N. Hirai, and S. Hara, J. Power Sources **102**, 155 (2001).
61. N. Hirai, D. Tabayashi , M. Shiota, and T. Tanaka, J. Power Sources **133**, 32 (2004).
62. G.M. Ehrlich in *Handbook of Batteries*, 3rd ed., edited by D. Linden and T.B. Reddy (McGraw-Hill, New York, 2002), Chap. 35, pp. 35.1–35.94.
63. K. Edström and M. Herranen, J. Electrochem. Soc. **147**, 3628 (2000).
64. S.-K. Jeong, M. Inaba, T. Abe, and Z Ogumi, J. Electrochem. Soc. **148**, A989 (2001).
65. S.-K. Jeong, M. Inaba, Y. Iriyama, T. Abe, and Z. Ogumi, Electrochim. Acta **47**, 1975 (2002).
66. S.-K. Jeong, M. Inaba, R. Mogi, Y. Iriyama, T. Abe, and Z Ogumi, Langmuir **17**, 8281 (2001).
67. F.P. Campana, R. Kötz, J. Vetter, P. Novák, and H. Siegenthaler, Electrochem. Commun. **7**, 107 (2005).
68. M. Koltypin, Y.S. Cohen, B. Markovsky, Y. Cohen, and D. Arbach, Electrochem. Commun. **4**, 17 (2002).
69. S.-K. Jeong, M. Inaba, Y. Iriyama, T. Abe, and Z Ogumi, J. Power Sources **119–121**, 555 (2003).
70. R. Vidu, F.T. Quinlan, and P. Stroeve, Ind. Eng. Chem. Res. **41**, 6546 (2002).
71. Y.S. Cohen and D. Aurbach, Electrochem. Commun. **6**, 536 (2004).
72. L.Y. Beaulieu, K.W. Eberman, R.L. Turner, L.J. Krause, and J.R. Dahn, Electrochem. Solid-State Lett. **4**, A137 (2001).
73. L.Y. Beaulieu, S.D. Beattie, T.D. Hatchard, and J.R. Dahn, J. Electrochem. Soc. **150**, A419 (2003).
74. L.Y. Beaulieu, T.D. Hatchard, A. Bonakdarpour, M.D. Fleischauer, and J.R. Dahn, J. Electrochem. Soc. **150**, A1457 (2003).
75. R. Mogi, M. Inaba, S.-K. Jeong, Y. Iriyama, T. Abe, and Z. Ogumi, J. Electrochem. Soc. **149**, A1578 (2002).
76. V.F. Henley, *Anodic Oxidation of Aluminum & Its Alloys* (Pergammon Press, Oxford, 1982).
77. (a) G. Meng, A. Cao, J.-Y. Cheng, A. Vijayaraghavan, J. Aravind Y.J. Jung, M. Shima, and P.M. Ajayan, J. App. Phys. **97**, 064303 (2005); (b) Q. Zhang, Y. Li, D.-S. Xu, and Z. Gu, J. Mat. Sci. Lett. **20**, 925 (2001).
78. (a) C.-M. Shen, X.-G Zhang, and H.-L Li, Mat. Sci. & Eng. A: Structural Materials: Properties, Microstructure and Processing **A303**, 19 (2001); (b) Y.F. Mei, Z.M. Li,

R.M. Chu, Z.K. Tang, G.G. Siu, R.K.Y. Fu, P.K. Chu, W.W. Wu, and K.W. Cheah, App. Phys. Lett. **86**, 021111 (2005).
79. *Scanning Electrochemical Microscopy*, edited by A.J. Bard and M.V. Mirkin (Marcel Dekker, New York, 2001).
80. R.C. Engstrom, M. Weber, D.J. Wunder, R. Burgess, and S. Winquest, Anal. Chem. **58**, 844 (1986).
81. (a) D.O. Wipf in *Characterization of Materials* edited by E.N. Kaufmann (Wiley, Hoboken, 2003), Vol. 1, p. 636–653; (b) R.E. Gyurcsanyi, G. Jagerszki, G. Kiss, and K. Toth, Bioelectrochemistry **63**, 207 (2004); (c) G. Wittstock in *Topics in Applied Physics*, edited by K. Wandelt and S. Thurgate (Springer-Verlag, New York, 2003), Vol. 85, pp.335–364; (d) H. Shiku, H. Ohya, T. Matsue in *Encyclopedia of Electrochemistry,* edited by A.J. Bard, M. Stratmann, and G.S. Wilson (Wiley-VCH, Weinheim, 2002), Vol. 9, pp. 257-275.
82. Y. Hirata, S. Yabuki, and F. Mizutani, Biochemistry **63**, 217 (2004).
83. (a) J. Abbou, C. Demaille, M. Druet, and J. Moiroux, Anal. Chem. **74**, 6355 (2002). (b) J. Abbou, A. Anne, and C. Demaille, J. Am. Chem. Soc.,**126**, 10095 (2004).
84. D.P. Burt, N.R. Wilson, J.M.R. Weaver, P.S. Dobson, and J.V. Macpherson, Nano Lett. **5**, 639 (2005).
85. P.S. Dobson, J.M.R. Weaver, M.N. Holder, P.R. Unwin, and J.V. Macpherson, Anal. Chem. **77**, 424 (2005).
86. O. Sklyer, A. Kueng, C. Kranz, B. Mizaikoff, A. Lugstein, E. Bertagnolli, and G. Wittstock, Anal. Chem. **77**, 764 (2005).
87. C.E. Jones, P.R. Unwin, and J.V. Macpherson, Chemphyschem **4**, 139 (2003).
88. J.R. Macdonald, *Impedance Spectroscopy Emphasizing Solid State Materials and Systems* (Wiley, New York, 1987).
89. R.S. Lillard, P.J. Moran, and H.S. Isaacs, J. Electrochem. Soc. **139**, 1007 (1992).
90. R.S. Lillard, J. Kruger, W.S. Tait, and P.J. Moran, Corrosion **51**, 251 (1995).
91. G. Baril, C. Blanc, M. Keddam, and N. Pébère, J. Electrochem. Soc. **150**, B488 (2003).
92. E. Bayet, F. Huet, M. Keddam, K. Ogle, and H. Takenouti, J. Electrochem. Soc. **144**, L87 (1997).
93. F. Zou and D. Thierry, J. Electrochem. Soc. **146**, 2940 (1999).
94. E. Bayet, F. Huet, M. Keddam, K. Ogle, and H. Takenouti, Electrochim. Acta **44**, 4117 (1999).
95. B.B. Katemann, A. Schulte, E.J. Calvo, M. Koudelka-Hep, and W. Schuhmann, Electrochem. Comm. **4**, 134 (2002).
96. S.V. Kalinin and D.A. Bonnell, App. Phys. Lett. **78**, 1306 (2001).
97. R. Shao, S.V. Kalinin, and D.A. Bonnell, App. Phys. Lett. **82**, 1869 (2003).

I.10 Near-Field High-Frequency Probing

C. A. PAULSON AND D. W. VAN DER WEIDE

We review a range of near-field probes that have been used for measuring the localized high-frequency electromagnetic properties of materials. First, near-field microscopy is briefly discussed, thereby establishing a background for describing near-field probing experiments. The basic dielectric properties of metals, insulators, and dielectrics are briefly reviewed since these provide the basis for image contrast. Finally, a wide range of near-field probes are described, including optical, infrared, microwave, etc. The literature on high-frequency near-field probes is surveyed throughout the text.

1 Introduction

Microscopic investigation of optical phenomena (such as luminescence, infrared absorption, microwave reflection, etc.) suffers from an inherent frequency-dependent resolution limit that is much larger than the nanoscale, owing to a phenomena called the diffraction limit. This paper discusses near-field probes that circumvent the resolution limitations of conventional optics, thereby allowing for nanoscale imaging.

Many of the topics discussed here have been reviewed previously. High-frequency microscopy was recently reviewed by Rosner and van der Weide [1]. There is also a useful text on near-field optics by Paesler and Moyer [2]. Near-field optics has been reviewed by Hecht [3], Pohl [4, 5], Lewis [6], Hsu [7], and Courjon and Bainier [8]. Optical and infrared frequency scattering techniques were reviewed by Keilmann [9]. Microwave microscopes have been reviewed by Tabib-Azar [10, 11] and briefly by Xiang and Gao [12]. Finally, evanescent fields and a wide range of microscopies have been reviewed by de Fornel [13]. The history of near-field microscopy is given in many of these references and is therefore not repeated here. Resolution and contrast have great importance for the microscopist, so we begin by discussing these quantities.

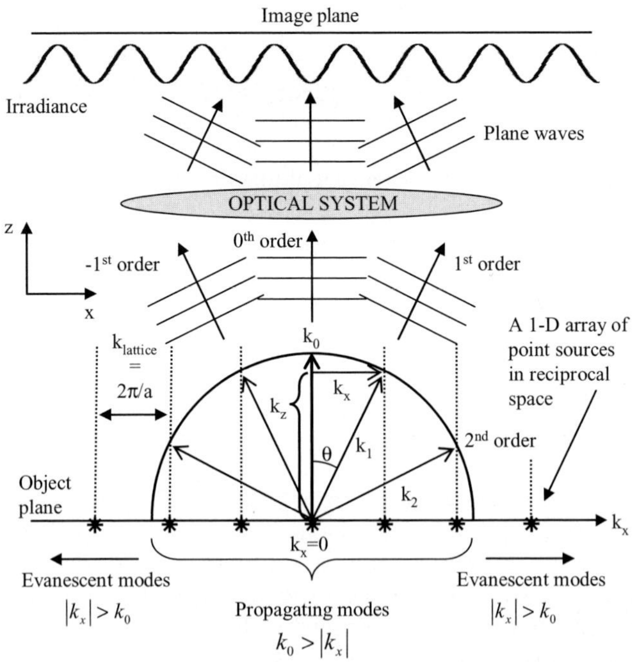

FIGURE 1. The bottom shows a 1-D lattice of coherent sources, in reciprocal space. Above, collection and focusing of propagating fields is illustrated, in real space. The spatial extent (the size) and the focal length of the optical system place an upper limit on the angle θ of diffracted waves that can be collected and brought to focus at the image plane.

2 Evanescent Fields and Resolution Limits

There is a mathematical approach to image formation called the "angular spectrum of plane waves" approach, wherein plane waves are focused by an optical system onto a screen in an image plane. Consider plane waves that have their propagation direction slightly tilted with respect to one another, that are brought together at the image screen. The resulting irradiance (W/m^2) at the image screen will therefore have a harmonic wave spatial profile across the image screen, similar to the pattern seen at the top of Figure 1. Evidently, in order to have a high spatial frequency sine pattern at the image plane, one must pass plane waves with high angles relative to one another. Figure 1 is discussed in greater detail below.

In a more rigorous discussion, once a propagating electric field is completely specified in a source plane, $E_s(x, y, z)$, with z being normal to the plane, it can be specified at any point in any distant spectrum plane, by the Fourier transform shown in Eq. 1 [3,14–17]:

$$E_p(k_x, k_y, z) = \iint E_s(x, y, z)e^{i(xk_x+yk_y)}dxdy. \tag{1}$$

Here k_x and k_y are the angular spatial frequencies,

$$k_x = \frac{kX}{r_0}, \quad k_y - \frac{kY}{r_0} \tag{2}$$

where X and Y are real space coordinates in the image plane, and the distance between planes is r_0. An inverse transform of E_p yields the real space electric field in the spectrum plane. From this description, components in $E_s(x, y, z)$ that have small values for their spatial periods (small objects in real space) are associated with large valued reciprocal space components, having large k_x or k_y, in the spectrum (image) plane. However, all fields traveling along z from source plane to spectrum plane, must obey the scalar wave equation,

$$\nabla^2\psi - \frac{1}{c^2}\frac{\partial^2\psi}{\partial t^2} = 0, \tag{3}$$

and therefore must also obey the vector Helmholtz equation,

$$(\nabla^2 + k^2)E(x, y, z) = 0, \tag{4}$$

where

$$k = \sqrt{\mu\varepsilon}\frac{2\pi}{\lambda} = \sqrt{k_x^2 + k_y^2 + k_z^2}. \tag{5}$$

Hence, large values for k_x or k_y in the plane of the sample are associated with imaginary k_z values, and hence with non-propagating waves. This description is very general; next we try to simplify this description using a simple 1-D analog for this description of imaging.

Figure 1 shows a 1-D array of coherent point sources of light in reciprocal space along the k_x axis, having a reciprocal lattice spacing given by

$$k_{lattice} = \frac{2\pi}{a}, \tag{6}$$

corresponding to real space period a. Constructive interference occurs in the directions of reciprocal space vectors k_1, k_2, etc, for real space angles θ that satisfy the grating equation,

$$m\lambda = a\sin(\theta). \tag{7}$$

Here, m is an order number, and λ is the wavelength of the radiation being emitted by the sources. This wavelength has a propagation constant (k_0), where

$$k_0 = \frac{2\pi}{\lambda}. \tag{8}$$

Rearranging Eq. (7) and multiplication by 2π yields

$$m\frac{2\pi}{a} = \frac{2\pi}{\lambda}\sin(\theta) = k_0\sin(\theta) = \sqrt{k_0^2 - k_z^2} = k_x. \tag{9}$$

Here, $k_0\sin(\theta)$ has been expressed in terms of the moduli (lengths) of components k_0, k_x, and k_z, as seen in the geometry of Figure 1. Note that, $k_0^2 = k_1^2$. The Ewald construction described above provides a simple geometrical interpretation of the real space angle of diffraction from the object, as given by Eq. (9), and as seen in Figure 1.

A wave along k_1 in Figure 1 can be described by

$$\exp(i(k_z z + k_x x - \omega t)), \tag{10}$$

where x and z are real space coordinates, with $z = 0$ being the object plane. Here, k_x and k_z are spatial frequency components of k_1. The number k_z is a real when $k_0 > |k_x|$. If $|k_x| > k_0$, then k_z is imaginary (from Eq. (9)), producing a wave that is damped in space along the $z-$direction (an evanescent wave). This wave is proportional to

$$\exp(i(k_x x - \omega t)) \exp(-k_z z). \tag{11}$$

Only the fundamental wave k_0 (with $k_x = 0$) travels through the optical system when k_x becomes larger than k_0 (as a becomes less than λ, in this model). In this case, the irradiance distribution in the image plane is uniform (plane wave illumination), and the point sources are not resolved. Real samples have a wide range of k_x and k_y values, with some values above and some below cutoff (greater than k_0) for a given optical system.

The lateral spatial resolution (Δx) in far-field imaging is inversely proportional to the highest lateral spatial frequency (k_x in Figure 1) that is brought to the image plane. The numerical aperture ($N.A.$) determines the value for the highest k_x waves that are not filtered from the image plane. The $N.A.$ is the index of refraction times the largest sine θ (see Figure 1) that an optical system can collect. The Rayleigh minimum resolvable distance uses the $N.A.$ [15,18],

$$\Delta x_{\min} \approx \frac{1.22\lambda}{N.A.}. \tag{12}$$

Here, the real space resolution $(\Delta x)_{\min}$ becomes worse as the wavelength being imaged gets longer, and as the $N.A.$ gets smaller.

The resolution of near-field imaging techniques is not limited by diffraction (and the $N.A.$) since near-field techniques utilize the high spatial frequency waves (those with high k_x values). For instance, Figure 2(a) illustrates a sharp fiber optic probe that can collect localized high spatial frequency fields around point sources by optical frustration. Therefore, by scanning the probe in the near-field the two point sources can be resolved. In contrast, in the far-field, the spherical waves from the point sources in Figure 2 merge. This makes the points appear as a single source. The experiment in Figure 2(a) is called scanning (or photon) tunneling optical microscopy (STOM) [19–23]. Rigorous treatments for the near-fields of gratings [24] and for near-field imaging are cited here [25].

Alternately, a tapered fiber optic can be coated with metal while leaving a small aperture at the apex. Light emanating from the tip can *excite* a sample locally, allowing for spectroscopy. This experiment is depicted in Figure 2(b), and is called near-field scanning optical microscopy (NSOM). An example of a coated probe, with ~100 nm aperture is shown in Figure 2(c). In NSOM, the aperture size imposes a limit on resolution.

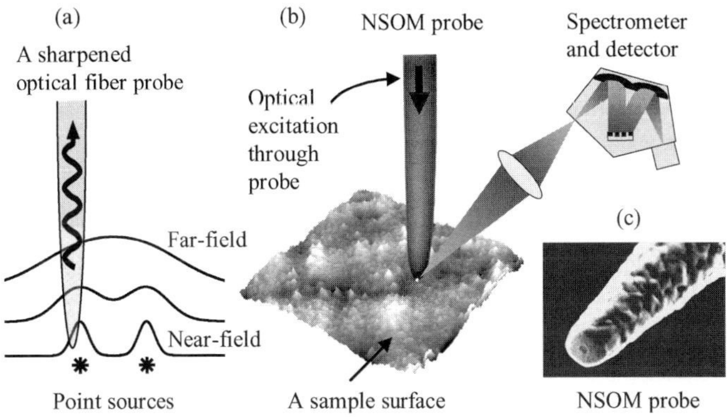

FIGURE 2. (a) An illustration of scanning tunneling optical microscopy (STOM), where the probe collects near-fields from sub-wavelength objects. (b) Near-field scanning optical microscopy (NSOM) in excitation mode, and (c) an NSOM probe with 100-nm aperture.

Simple Fresnel intensity profiles [15–17] have been calculated for a range of aperture sizes in Figure 3. Note that the intensity is confined to a region that is about the size of the aperture when the profile is calculated close to the aperture. Note the absence of Fresnel interference fringes near the aperture and that the profile diverges considerably at a distance of only 1λ, for the smallest aperture case. These profiles are common to many types of near-field probes.

At this point, we make a few general observations. In near-field microscopy the fast parallel imaging of optical systems (lenses) is replaced with a point by point

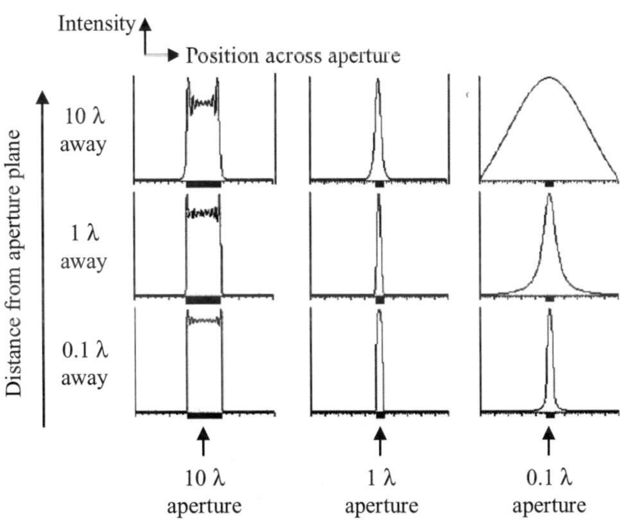

FIGURE 3. Fresnel intensity profiles, calculated for a range of aperture sizes and for a range of distances from the apertures.

scanning (serial acquisition). The scanned probes have a spatially localized or confined element, such as an aperture, which approximates a delta function in real space coordinates, and which is often coupled to a waveguide. Finally, the lateral scan speeds are slow (compared with the light speed performance of the optical microscope). Methods to improve scan speed are an ongoing topic of research and development [26–31]. For instance, Seo [32] used a 2-MHz quartz crystal for fast (1.2 mm/s) acquisition, with high resolution (~140 nm).

We mention here that image artifacts pose a serious challenge to near-field microscopy [2, 12, 33]. These artifacts are due, in part, to the rapid fall-off and divergence of the near-field, since changes in tip to sample separation can cause changes in the near-field coupling with the sample [34]. This may result in artificial contrast (which is not due to the intrinsic properties of the sample). Many techniques minimize the convolution of topographic and electrical properties by scanning at constant height or at constant gap, or both, using a topographic channel that is auxiliary to the optical/electrical channel [33,35]. Comparing data between channels allows identification of image artifacts. Intensity imaging is particularly prone to topographic artifacts. A related well-known effect involving the scan height is that resolution decreases when scanning further from the sample [36].

We are going to discuss briefly microwave, infrared, and optical near-field microscopies, which are all interrelated. The optical/electrical response of the sample is source of contrast in imaging. Next we discuss the basic optical properties of materials.

3 Frequency Response of Materials—Examples of Contrast

Image contrast originates from spatial changes in the optical properties of the sample. Here we discuss the linear response of (isotropic) metals, insulators, and dielectrics. The real (n) and imaginary (k, extinction coefficient) parts of the index of refraction [15–17] are given by Eqs. (13) and (14):

$$n^2 = \frac{1}{2}\left[\sqrt{\mu^2\varepsilon^2 + 4\mu^2\sigma^2/\nu^2} + \mu\varepsilon\right], \tag{13}$$

$$k^2 = \frac{1}{2}\left[\sqrt{\mu^2\varepsilon^2 + 4\mu^2\sigma^2/\nu^2} - \mu\varepsilon\right], \tag{14}$$

respectively. Here, μ and ε are the permeability and permittivity, respectively; and σ and ν are the conductivity and frequency, respectively. All of these quantities can be complex, and they are all functions of frequency. The complex dielectric function ($\varepsilon = \mathrm{Re}(\varepsilon) + \mathrm{Im}(\varepsilon)$) is related to n and k by $\mathrm{Re}(\varepsilon) = n^2 + k^2$ and $\mathrm{Im}(\varepsilon) = nk$. $\mathrm{Re}(\varepsilon)$ and $\mathrm{Im}(\varepsilon)$ are related through a Kramers-Kronig transform [37], and thus they are not independent.

The absorption coefficient is given by

$$\alpha = \frac{4\pi\nu k}{c} = \frac{4\pi k}{\lambda}. \tag{15}$$

This determines power attenuation as a wave travels through a material,

$$\frac{P(x)}{P(0)} = e^{-\alpha x}, \tag{16}$$

where $P(x)$ is the power at a distance x into the material. The inverse of α is skin depth, d, a measure of radiation penetration,

$$d = \frac{c}{4\pi \nu k} = \frac{\lambda}{4\pi k}. \tag{17}$$

As a material becomes insulating ($\sigma = 0$), n becomes equal to $\sqrt{\mu\varepsilon}$, while k and α tend to 0, making the material transparent. When a material becomes highly conducting, and $(\mu\sigma/\nu) \gg \mu\varepsilon$, then $n \sim k = \sqrt{\mu\sigma/\nu}$ and the skin depth equation becomes

$$d = \frac{c}{4\pi\sqrt{\mu\sigma\nu}}. \tag{18}$$

Also, as n becomes large, the reflectance (R) of the material,

$$R = \frac{(n-1)^2 + k^2}{(n+1)^2 + k^2}, \tag{19}$$

approaches 1. Hence, good conductors become highly reflective, with a very shallow penetration depth (and therefore have low power loss).

Dielectrics are materials that can maintain an internal electric field. The dielectric constant is related to the polarization P (C/m^2) in a material by

$$\varepsilon = \frac{\varepsilon_0 E + P}{\varepsilon_0 E} = 1 + \chi, \tag{20}$$

where E is the externally applied field and χ is the susceptibility. The polarization is related to the polarizability, α(Cm^2/V), of an atom or molecule by

$$P = \sum N\alpha E_{loc}, \tag{21}$$

where the sum is over all atoms or molecules, N is a concentration, and E_{loc} is the local field at the atom or molecule. Note, E is only one term in E_{loc}, and $E_{loc} \neq E$, since depolarization (shape effects) and the geometric distribution of polarizable elements must also be taken into account.

The polarizability, α is a property of the atom or material, providing a link that relates microscopic properties to macroscopic measurements such as reflectivity and absorption. Polarization as a function of frequency is often modeled as a harmonic oscillator with the following relation:

$$n^2(\omega) = \varepsilon(\omega) = 1 + \frac{Ne^2}{m\varepsilon_0}\sum_j \frac{f_j}{\omega_j^2 - \omega - i\gamma_j\omega}. \tag{22}$$

Here, f_i is an oscillator strength and γ_j is a damping constant for a particular resonance frequency ω_j. A material may have many resonance frequencies and absorptions which can be used for chemical identification in the infrared fingerprint

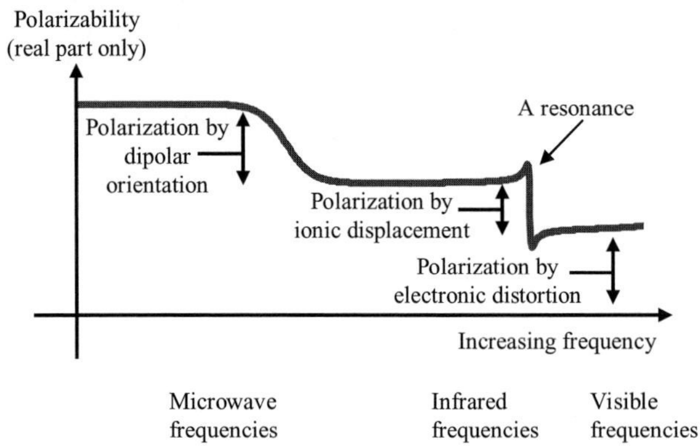

FIGURE 4. The polarizability versus frequency for a hypothetical material. Note the frequency dependence of the microscopic contributors to polarization.

region. Oscillator strengths and damping constants can be determined by quantum mechanical calculation [38].

The polarizability of a material is often composed of dipolar, ionic, electronic contributors [39]. The dipolar refers to molecules reorienting in response to an electric field, ionic refers to ionic displacement, and the electronic refers to electron cloud motions relative to the nuclei in a material. Generally, all three mechanisms can contribute to polarizability at low frequencies (below ~1 GHz), whereas only the electronic response is fast enough to respond in the optical frequencies (above ~100 THz). An example of the polarization versus frequency, for a hypothetical material, is shown in Figure 4. Plots of dielectric function versus frequency often show similar features.

Contrast in an image can result from spatial changes in the electrical or optical properties of a sample (it can also result from aberrations in the imaging system). The contrast (C) across an image may be defined by

$$C = \frac{S_{\max} - S_{\min}}{S_{\max} + S_{\min}}, \tag{23}$$

where S refers to a signal that is being imaged, such as reflection [2]. Note that imaging at frequencies near a polarization resonance offers the opportunity for high contrast, since one area may absorb strongly, while another area does not. However, this is not a requirement for image contrast.

Owing to the spectrum of resonant absorptions in the fingerprint region, THz imaging promises to revolutionize applications in many important research fields from biological imaging to semiconductor device inspection [40,41]. A challenge to this field is the lack of powerful sources and/or detectors of THz radiation [42, 43]. Despite this fact, THz imaging is an active area of near-field research [44,45], as will be shown in subsequent sections.

4 Apertured Probes, Throughput and Cutoff

The idea for the NSOM experiment was shown in Figure 2(b). NSOM has traditionally been an optical frequency experiment, although examples abound for IR and microwave investigations as well [2,3]. As an example of aperture NSOM we cite the work of Masaki et al. [46]. They presented infrared measurements with an aperture on cantilever probe. The probe was hollow, and fabricated by Si anisotropic etching and focused ion beam milling [47]. They measured line/space structures of poly(methylmethacrylate) (PMMA) films, for which far-field IR spectra revealed a C=O stretching absorption at 5.78 μm. Imaging at 5.78 μm revealed high contrast from changes in the real and the imaginary parts of the index of refraction. Weaker contrast was found at 5.55 μm, where there is no absorption and contrast arises from only the real part of the index of refraction [48].

The principal challenge for aperture probes is severe power loss due to the cutoff phenomenon [49,50], where the waveguide becomes too narrow to support a propagating wave. The same cutoff occurs for modes in cylindrical and rectangular waveguides [51], as a consequence of Eq. (4). In the case of a NSOM probe, a cylindrical waveguide, the last mode present before cutoff is the HE_{11} mode [3,51]. Other modes are less effective in delivering power to the aperture, they contribute to heating and back reflection, in greater proportion, when compared to the HE_{11} mode.

The power transmission (T) through the cutoff region is given by

$$T = \exp\left(4\pi\sqrt{(f_c^2 - f^2)t}\right), \tag{24}$$

where f_c is the cutoff frequency, f is the frequency of the radiation, and t is the penetration depth into the cutoff region [52]. Cutoff was described analytically by Knoll and Keilmann [53]. For TE_{01} modes in a square metal waveguide, cutoff occurs for edge lengths (of the square waveguide) less than a_c,

$$a_c = \lambda_0/2\sqrt{\varepsilon}, \tag{25}$$

where λ_0 is the wavelength of the radiation in vacuum, and ε is the dielectric constant of the filling material. Minimizing power loss caused by cutoff can be achieved by using large taper angle for the apex region, thus decreasing the distance (t) between the onset of cutoff and the aperture of the probe [53]. Also, the transmission through a sub-wavelength circular aperture is proportional to $(d/\lambda)^4$, where d is the aperture diameter and λ is the wavelength of the radiation [53–55]. As a result, the transmitted power through sub-wavelength NSOM probes is extremely small.

For some experiments the aperture size sets the resolution of the probe. However, as the aperture is made smaller and smaller, eventually one is left with no signal. This is an obvious but important phenomenon in near-field microscopy—as the resolution increases, the signal strength decreases. Therefore, in some cases the sensitivity of the detector can limit the resolution. Alternatively, the skin depth can sometimes set resolution; when an aperture becomes smaller than about twice the skin depth, light penetrating into the metal coating around the aperture results in an increased "effective aperture." The skin depth is only ~6 nm for green light in Al [2]. However, skin depth increases with wavelength. Experimentally, most NSOM experiments report resolutions that are not better than about 50 nm. However, special aperture probes with plasmon effects [56–58], non-circular-shaped apertures [59–61], and coaxial probes (see below) can increase the throughput, resulting in a stronger signal and potentially improved resolution.

Many features common to aperture probes were demonstrated by the finite element modeling (FEM) of rectangular aperture near-fields done by Golosovsky and coworkers [62]. In this work, the field is well collimated outside of the slot, in the proximity zone (where $k_z z \ll 1$). The collimation is lost as z increases, over a distance referred to as the collimation length. The field magnitude is also enhanced, relative to the excitation field, in the vicinity of the slot. There is an edge singularity (an enhancement) at the sharp conducting edges of the aperture (the lightening rod effect). Related effects had been predicted and observed in optical NSOM [63,64]. They proposed that the field strength enhancement (proportional to $1/w$, where w is the slot width) is due to the capacitance of the slot. A similar argument was reported by Ash and Nicholls [49]. Golosovsky and coworkers also report the field strength decays faster for narrower slots, in the direction normal to the slot. They also derive an analytic model [62], and point out that experimental studies [65] agree well with numerical simulations. The main challenge for FEM modeling is the disparate length scales, requiring complicated meshing. Microwave aperture probes were also investigated experimentally and theoretically by Symons and coworkers [66]. They discuss an impedance-based shape effect, where coupling impedance (for a wire sample) is inversely proportional to transmission. Consequently, the image features were distorted by spatial variation in sample impedance. Hence, the frequency employed for imaging affects the features that are observed.

Now we discuss a few specific examples of apertured probe experiments. Imaging of living leaves at 12 cm^{-1} (0.36 THz) was demonstrated with 130-μm resolution using a conical aperture scanning near-field infrared microscope (SNIM) probe at a synchrotron [67]. Higher resolution was realized with visible regime imaging, but in the THz regime, the dynamics of leaf rehydration could be imaged since THz is sensitive to water concentration. Interesting contrast mechanisms of this sort are a driving force for near-field microscopy in the THz and microwave regimes. In this case, the synchrotron provided a brilliant coherent source of THz radiation, compensating for losses in the aperture probe.

Highly collimated radiation from 12 different backward-wave oscillators, covering the 30 GHz to 1.5 THz range, was used to excite samples through subwavelength-sized pinholes in copper foil [68]. Comparisons were made to pulsed radiation sources, which are broadband but have relatively poor energy resolution. Since they had a wide range of frequencies available, quantitative spectroscopy of small samples was possible, as long as the probe-surface distance was stable. They mention that apertures made from finite thickness metals may have a faster decay than infinitely thin apertures. They found the decay (in the $z-$direction) to be independent of the frequency for the near-field of the probe (providing that the wavelength was much greater than the aperture size). The decay was proportional to $\exp(-z/ta)$, where z is tip-sample distance, a is the aperture radius and t ($\sim$1.1) is a fitting factor.

Limited transmissivity, especially of the fiber, can also be an issue for infrared NSOM experiments. To overcome this, tapered Ag-halide fiber probes with apertures, with high transmission in the mid-infrared from 0.6 to 10 μm, have been developed [69]. In this report, radiation distribution was measured by a specially designed knife-edge technique, and images of 5-μm-wide stripes on Si were acquired. The Ag-halide transmission window (from 0.6 to 30 μm) is wider than other those for chalcogenide [70] (2–12 μm) and fluoride [71] (0.4–6 μm) fiber materials.

5 Scattering SNOM

An experiment that does not suffer from waveguide cutoff, which is also broadband, and has high resolution, is the (scattering) s-SNOM [9]. This experiment is illustrated in Figure 5 (a). In s-SNOM, a vibrating tip at a sample interface modulates the polarizability of the interface at the frequency of vibration, Ω. The tip is

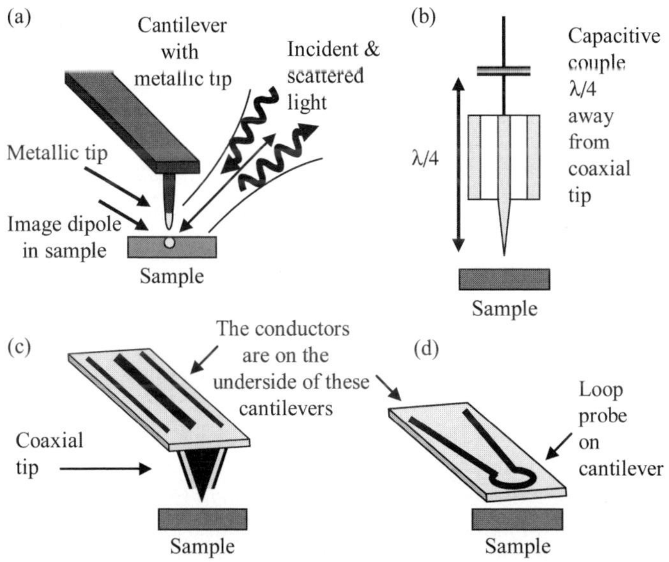

FIGURE 5. High-frequency near-field probes. (a) s-SNOM, which is mainly used in the visible and IR. Resonant (b) and broadband (c) coaxial probes for microwave electric fields and (d) a loop probe for microwave magnetic fields.

modeled as a polarizable sphere, with polarizability

$$\alpha = 4\pi a^3 \frac{\varepsilon_t - 1}{\varepsilon_t + 2}, \tag{26}$$

where ε_t is the dielectric constant of the tip, and a is the tip radius. The dielectric surface response function of the sample is

$$\beta = \frac{\varepsilon_s - 1}{\varepsilon_s + 1}, \tag{27}$$

where ε_s is the dielectric constant of the sample, and the effective polarizability of the tip-sample system is

$$\alpha_{eff} = \frac{\alpha(1+\beta)}{1 - \alpha\beta/(16\pi(a+z)^3)}, \tag{28}$$

where z is the distance between the tip and sample. Scattering of the incident field depends on the effective polarizability of the tip-sample region, $E_s \propto \alpha_{eff} E_i$, where E_s and E_i are the scattered and incident fields, respectively [72]. The sample influences the scattered light through its dielectric function in the equation for β. If z and ε_t are kept constant, then imaging measures contrast which is only due to changes in ε_s.

When the interface is illuminated, the modulated polarizability can be detected by the amplitude and phase of the scattered radiation. Because the effective polarizability is a highly nonlinear function of distance, the scattered signal can be demodulated at higher harmonics of the cantilever's tapping frequency Ω. This is important, since it allows discrimination of the scattered signal at higher harmonics, versus direct scattering from the cantilever body at Ω. However, this suppression technique also relies on the absence of higher harmonics in the mechanics of the cantilever itself [73]. The mechanical higher harmonics are negligible for very weak interactions between tip and sample (such as with high feedback set-points that result in gentle feedback). The resolution of the experiment is related to the radius of the tip, which can be very small (20 nm or less). The resolution can even be better than the tip radius since the z-extent of the near-field decreases with demodulation order.

The antenna function of the metallic s-SNOM probe has been investigated by rotating the polarization of light impinging on the tip-sample interface [74]. The contrast was diminished when the polarization was not parallel to the axis of the metallic probe, because the long axis of the antenna determines the polarizability direction for the tip sample system (the scattering center).

Materials specific s-SNOM was demonstrated, with 10 nm spatial resolution, using a sample consisting of Au, Si, and polystyrene [75]. The authors used scattering at a single wavelength (633 nm). Even though the dielectric functions for these materials do not exhibit any resonant absorption at this wavelength, the effective polarizability, which depends on the dielectric constants for each material, was used as a source of contrast via the scattering magnitude. These experiments were also performed at 9.7 μm. However, a complication in the single wavelength scattering experiments when used for materials identification [76] is that dielectric contrast is affected by vertical inhomogenieties in the material beneath the probe, on the length scale of the penetration depth for the radiation in question.

Polymer recognition was also demonstrated using vibrational IR lines in a s-SNOM incorporated into a Michelson interferometer [77]. This work is also motivated by chemical contrast identification in the vibrational fingerprint region. A primary conclusion was that polymer films that are far thinner than the IR penetration depth can be imaged using the scattering technique.

In another s-SNOM experiment, 150-nm resolution was demonstrated from a gold stripes on semi-insulating Si sample at 2 THz (150 μm) [78]. The THz was generated from a pulsed n-InAs semiconductor sample. Detection was by either electro-optic sampling (with sensitivity to $\sim$ 2.5 THz) or bolometry (with sensitivity to $\sim$ 10 THz, detecting power, not amplitude).

The effect of antenna geometry on the s-SNOM spectroscopy has been investigated by Wang et al. [79], where a metallic tip was modeled as an oscillating dipole,

$$E_{dipole} = \frac{\mu_0 c^2}{4\pi}\left[\hat{\theta}\sin\theta\left(\frac{(p)}{(r^3)} + \frac{(p')}{(cr^2)} + \frac{(p'')}{(c^2 r)}\right) + \hat{r}\cos\theta\left(\frac{2p}{r^3} + \frac{2p'}{cr^2}\right)\right]. \tag{29}$$

Here, μ_0 is the permeability of free space, c is the speed of light, p is dipole moment and the primes indicate derivatives in time, and r is the radial distance. The near-field component of the electric field, which dominates at short distances ($1/r^3$), was separated out, and the dipole moment, $p(t)$, was related to current in the antenna, $I_{THZ}(t)$, to the through the continuity equation,

$$p(t) \propto \int_{-\infty}^{t} I_{THZ}(t)\,dt \tag{30}$$

The antenna was modeled as a simple RLC network, and therefore the current could be related to the incident electric field,

$$I_{THZ}(\omega) \propto \frac{E_{THZ}(\omega)}{(Rr + j\omega L - j(\omega C)^{-1})}. \tag{31}$$

Here, E_{THZ} is the terahertz electric field. The authors calculated the spectral power for this network and compared it with the input and the measured near-field power spectra (up to 6 THz). They experimentally found a three-fold bandwidth reduction in the s-SNOM near-field spectrum when compared to the input spectrum. They attribute this to the inductive component of the metallic antenna, which dominates at higher frequency. Because of this, one cannot assume that the input radiation spectrum for the s-SNOM experiment is the same as the actual near-field spectrum, especially at high frequencies.

Si AFM probes for s-SNOM have been compared with metal-based scattering probes by Haefliger and coworkers [80]. The Si probes can be advantageous due to increased wear resistance, smaller tip radius, and, for experiments where metal could cause sample photobleaching. The performance of the Si probes compares favorably with metal probes, as long as an oxide layer on their surface is removed, thus increasing the polarizability of the probe. The authors describe a (fortuitous) tip crash event which removed the oxide of a Si probe while performing s-SNOM. This event decreased the topographic resolution but enhanced the optical resolution. This observation fits well with the idea that the Si core ($\varepsilon = 15$) is more polarizable than the oxide ($\varepsilon = 2.1$).

The fact that the s-SNOM exhibits strong field enhancement directly under the metallic tip has been experimentally verified [81]. In this work, a metallic tip was scanned over single fluorophores imbedded in a PMMA matrix. An enhanced fluorescence was observed, indicative of an enhanced field strength, when the probe was directly over a fluorophore. We also mention here that carbon nanotubes have also been used in s-SNOM experiments [82].

6 Coaxial Probes

Coaxial probes support transmission of fundamental TEM_{00} waves without cut-off effect [51, 83]. Coaxial probes show dramatic increases for IR transmission (320 μm radiation) versus near-field apertures, but it is a challenge to fabricate coaxial probes in the size regime required for near-field optical probes [52]. The coaxial probe can be flat at its end, or it can have a sharp probe protruding from its end. The probe can be resonant when capacitively coupled (as in Figure 5(b)) or non-resonant. A resonant probe increases quality factor (Q) and sensitivity, but decreases measurement bandwidth (BW).

Kantor and Shvets [84] have reported on a coaxial antenna microwave probe (a monopole). They note that the length of the protruding center conductor must not exceed the desired spatial resolution, and the diameter of the shielding must be kept small, since surface currents in the shield can induce signal in the center conductor. They also note that the monopole probe is usually considered to be immersed in a sample field that is homogeneous. However, as the center conductor shrinks in size, the fields around it become increasingly distorted, and measurements made with the probe are less amenable to quantitative modeling. The authors present a solution to these challenges: the position/signal difference (PSD) method. In this method, two scans are recorded at different scan heights above the sample, displaced by a small amount. An image is produced from the difference between the images at the two heights. The resulting image is predominately due to the displaced region of the apex, where the field changes most rapidly. PSD images therefore reduce the background from the rest of the antenna, including the shielding. The authors calibrated their measurement using an air suspended transmission line sample, and report a minimum detectible field of 15 V/m (for a displacement of 20 μm). They also demonstrated a lateral resolution of 30 μm (while operating at 3.8 GHz). In a later contribution [85], they use the PSD method with antennas that are inclined (45°) to the surface normal. This allowed measurement of tangential fields (E_x and E_y).

Several authors [86–90] have integrated coaxial probes onto AFM cantilevers (similar to Figure 5(c)). As a particular example, Wang and Tabib-Azar et al. [90]

developed a coaxial microwave probe on an AFM cantilever. They demonstrate microwave reflection imaging of a Si_3N_4 grid pattern, and report a microwave resolution of 50 nm at 2.8 GHz.

7 Microcoils—Magnetic Dipole Probes

There is a large amount of literature on microcoils for magnetic resonance (MR) spectroscopy, from which a few representative publications are cited here [91–95]. In a particular work, the signal to noise for microcoils has been derived, from which the on-axis MR sensitivity varies inversely with the radius of the microcoil [96].

A 6-μm-diameter loop probe on cantilever design (similar to Figure 5(d)) was published by Agrawal et al. [97]. The authors demonstrated magnetic imaging of coplanar waveguide structures, with 10-GHz sensitivity that is comparable to that for superconducting quantum interference detectors (SQUIDs).

A 100-μm-diameter planar microcoil for electron spin resonance (ESR) was developed by Boero et al. [98]. They note that as the microcoil diameter decreases, the spin number sensitivity increases but the concentration sensitivity decreases. They demonstrated ESR on $\sim$10 μm^3 of DPPH (1,1-diphenyl-2-picrylhydrazyl, a paramagnetic molecule), with a spin sensitivity of 2×10^{10} spins/$Hz^{1/2}$. They also measure 2,2,6,6-tetramethylpiperidine-N-oxyl, which has a melting temperature of 310 K. At low powers delivered to the coil, the sample was solid, with one absorption line in the ESR spectrum. At higher powers the sample melted, and exhibited a triplet spectrum. From these experiments they estimate the temperature change of the coil is less than 20 K.

Lee and Anlage [99] presented a semicircular loop, 200 μm in diameter, made by shorting the center conductor of a coaxial probe to its shield. They can identify nonlinear edge effects, related to screening currents, in a superconductor sample. They also observe strong second and third harmonics above the center of a Josephson junction, in a YBCO bi-crystal film. These harmonics were absent in an area that is far from the grain boundary.

8 Slot Resonators

As mentioned previously, resonant probes increase sensitivity by their quality factor, Q, but this comes with a loss in bandwidth (proportional to $1/Q$). The quantities usually measured with resonant probes are Q and frequency shift. A perturbation can cause these quantities change simultaneously, which makes it necessary to give careful consideration of the source of signal during imaging. For instance, tip-sample distance changes can cause a change in the resonance frequency, as can changes in the sample's dielectric constant. Slot resonators (similar to Figure 6(a)) exhibit different resolution in orthogonal scan directions. The E-field lines for slot probes (parallel to the sample) are orthogonal to those for coaxial-type probes (neglecting fringe fields).

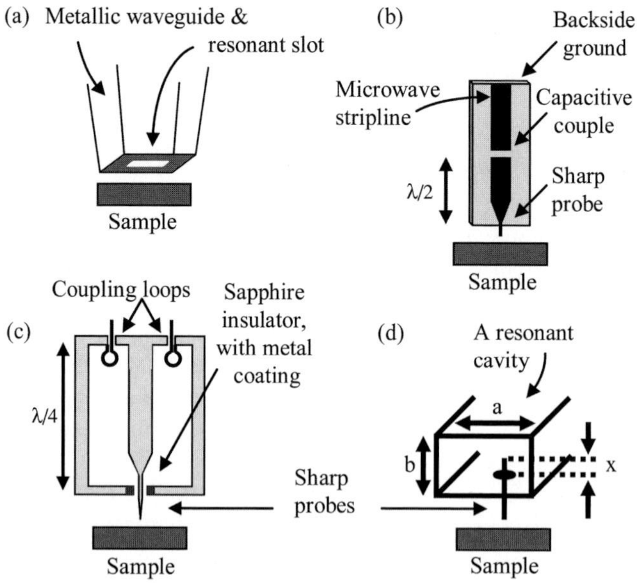

FIGURE 6. (a) Resonant slot with waveguide feed. (b) Microstripline resonator. (c) Coaxial resonator with loops for excitation and detection, and (d) sharp probe coupled to a cavity resonator (which could be filled with a dielectric in order to increase field strength).

Lann et al. [100] describe a mm-wave microscope with a slot aperture (10 μm wide) in the convex end plate of a rectangular waveguide. The capacitance between the probe and the sample depends on the distance between them. This capacitance was incorporated into an LC oscillator at 5 MHz. Changes in the resonant frequency of the oscillator was used to control the tip to sample distance. Simultaneously, they imaged sheet resistance ($R_x = \rho/t$) of silver films on glass at 82 GHz. They use transmission line theory for calculation, and related reflectivity to the impedance of the sample-substrate assembly. By monitoring amplitude (Q) the microscope could discriminate resistance (energy dissipation) values below 2 Ω.

Sakran et al. [101] discuss an electrically open dielectric resonator with a 10-μm-wide slot aperture. In this case, the slot length and the dielectric resonator size were chosen such that they both resonate at the same frequency. They used their microscope for imaging ESR from of DPPH grains on glass. They calculate a sensitivity limit of $\sim 10^{11}$ spins. They note that the slot probe should be curved, for highest resolution [102]. In a related contribution, resonant slots were also used for localized heating, coagulation, and melting of soft matter by Copty and coworkers [103].

9 Stripline and Transmission Line Resonators

Microwave probes based on a microstrip resonator (similar to Figure 6(b)) with a sharp tip attached have been demonstrated by Tabib-Azar and coworkers [104], where the microstrip was capacitively coupled to a feedline. An electrically open probe tip resulted in an electric field probe. Alternatively, a loop shorted to ground resulted in a magnetic probe, where currents can be induced in the probe. They demonstrated microwave probing of skeletal tissues, including human femur, a tooth, and balsa wood.

Self-oscillating probes, where the oscillator was a microstripline resonator in a feedback loop across a microwave amplifier ($f_0 \sim 2.25$ GHz) have been demonstrated [105]. The circuit is compact, making it convenient for incorporation into scanning probe systems. Tabib-Azar and Wang [106] also discuss operating parallel microstrip line resonators, each at different frequencies, for hyperspectral analysis and quality control in manufacturing.

Most papers considered the sharp tip to be modeled by a perfect sphere. Imtiaz et al. [107] discuss the effects of a conical protrusion on a sphere, using models of a coaxial transmission line resonator with sharp probe. When the probe-sample distance is less than 10 nm, the protrusion produces deviations from the logarithmic capacitance versus distance expected for a sphere-infinite plane geometry, which modifies the interpretation of frequency shifts that are observed during imaging.

Artificial contrast in microwave imaging that can be related to changes in tip to sample distance was discussed by Kim et al. [108]. Their solution is to scan with the tip using a topography-following method that is independent of the microwave probe. They use tuning fork shear-force feedback [109] where an etched tungsten tip on microstrip line was attached to one prong of the tuning fork. Vlahacos and coworkers also discuss intrinsic information (dielectric constant, surface resistivity) and extrinsic information (topography and artifacts) in microwave imaging [110]. The authors describe an open-ended transmission line resonator. By measuring frequency shifts they image topography changes across US quarter dollar coin, and they report height sensitivity of 55 nm, at a tip to sample distance of 30 μm.

Tselev and colleagues present a height-modulated distance following (HMDF) method using an open-ended coaxial transmission line resonator [111]. Here, the tip to sample distance is modulated so that the tip is in contact with the sample during part of its motion (similar to tapping mode AFM). The reflected microwave signal is modulated by the tip to surface separation, and changes in the resonance frequency can be used to follow the topography with high fidelity. They demonstrate microwave reflection images of doped lines in Si (free of topography artifacts), with about 2-μm resolution.

Microwave probes have been used to measure ferroelectric samples [112]. In this work, a sharp probe tip protrudes from the open end of a resonant coaxial cable, and makes gentle contact with the sample. A sample crystal of duterated triglycine sulfate (DTGS) was heated above $T_c \sim 335$ K (the Curie temperature, where ferroelectric polarization disappears), and then allowed to cool. Periodic imaging showed the evolution of large ferroelectric domains at the expense of small domains. Ferroelectric relaxation times were calculated from histograms.

10 Coaxial and Cavity Resonators

A sketch of a common type of coaxial resonator is shown in Figure 6(c). Here, a sharp probe protrudes through a hole in a thin (~skin depth) metallic shield supported on a ceramic insulator. The shield around the tip keeps the propagating far-field components within the cavity. Thus this probe emits little energy, resulting in high values for Q. The metallic probe tip acts as a point-like evanescent wave emitter. At least two versions of this approach exist. An insulator can separate the sample from the tip (the insulator can be coated either on tip, or on sample). Otherwise, a tip-independent topography following scheme can be implemented. Two measurement techniques are most common, a total response technique (reflectivity detection) or the phase-sensitive technique in which the resonator's phase is compared with that of a reference oscillator.

Gao et al. reported quantitative coaxial resonator microwave microscopy of dielectric materials [12, 113] and insulators and conductors [114]. The resonant system was modeled as a RLC resonator. Analytic expressions were given for the unloaded (Q_u) and sample coupled (Q_c) quality factors. From these the Johnson noise limited dielectric sensitivity is derived as

$$\frac{\delta\varepsilon}{\varepsilon} = \frac{1}{A}\sqrt{\frac{k_B T B}{4P Q_c Q_u}}, \tag{32}$$

where A is a function of the geometry of the tip-sample region, and P is the power delivered to the cavity. From Eq. (32), they calculate the minimum detectable change in dielectric constant $(\delta\varepsilon/\varepsilon)_{\text{min}} \sim 1 \times 10^{-5}$. Gao and colleagues [113,114] also calculated the field distribution inside the probe tip (E_1). They use E_1 with resonant cavity perturbation theory [115], in order to calculate frequency shift. They comment that, for sharp probes (and thick films without substrates), the penetration depth for fields into the tip is approximately equal to the sphere's radius, limiting the spatial resolution. Xiang and Gao also note that for dielectrics of moderate values (~50), the resolution can be finer than the tip radius, because of field concentration effects [12]. However, metals, having very large dielectric constants, are not appropriate for determining resolution that would be achieved on dielectric or insulating samples. In a later publication, Gao and colleagues [116] present a recursive image charge approach for calculating frequency shifts, taking into account film thickness and the dielectric constants for the film and substrate. Key in their analysis is an analytic expression for the capacitance between a sphere and a conducting plane.

The operation principles and implementation of a $\lambda/4$ wave coaxial resonator were published by Kim et al. [117]. They discuss the importance of impedance matching to maximize sensitivity. Using cavity tuning [118] they demonstrated changes in the spatial resolution and sensitivity during imaging. For superconducting yttrium barium copper oxide (YBCO) films on MgO they report a resolution of 4 μm at a probe frequency of 1–1.5 GHz.

Gao and coworkers [119] reported 100 nm resolution at 2.3 GHz results using their $\lambda/4$ coaxial resonator for measurements on a high dielectric constant "relaxor material." They had incorporated their resonator into a phase-locked loop (PLL) configuration, where phase and Q were monitored by the PLL.

Wang and colleagues [120] used a 2-GHz microwave coaxial resonator with a protruding SiO_2 coated metallic tip for the examination of 25 different low-k dielectrics. These samples require greater sensitivity as compared with high-k dielectrics [113, 114]. FEM calculations, and simple analytic models, show good agreement between the measured and known dielectric constants.

Feng and coworkers [121] have reported a low-temperature $\lambda/4$ wave coaxial resonator [122] with an electrochemically etched niobium tip for investigation of superconducting YBCO films on $LaAlO_3$. Ag electrodes allowed for four point measurements of resistance. The sample was covered by 250 nm of CaF_2 (an insulator). The Nb tip made soft contact with the CaF_2, maintaining a constant tip to YBCO distance. During cooling, the four-point probe identified a drop in resistance at T_c where, concurrently, the microwave Q increased significantly (since the sample became much less lossy). The resonance frequency did not shift significantly. Q scans crossing the YBCO to Ag boundary showed higher surface resistance for YBCO when $T > T_c$. For $T < T_c$, the YBCO surface resistance fell below that for Ag.

A topography following microwave microscope, using a coaxial resonator, which minimizes topographic artifacts was discussed by Kim and coworkers [123]. Kim and coworkers also investigated probe tip shape effects [124]. The largest taper angle tip gave images with low resolution (3 μm) and the smallest taper angle tip had good resolution (0.8 μm) but low sensitivity. Kim and coworkers [125] have also reported on a 60-GHz microscope, consisting of an etched tungsten probe attached to rectangular waveguide (similar to Figure 6(d)). They use a moveable plunger within the resonator, which facilitates impedance matching (thus increasing sensitivity). They operated the waveguide in TE_{10} mode, and when the probe shaft (similar to the short "x" in Figure 6(d)) was aligned with the electric field inside the waveguide, fields could be coupled efficiently into or out of the waveguide. Further examples of waveguide coupling to E- and H-field probes are discussed by Ingram [126]. Kim and coworkers [127] also presented a microwave microscope that uses a small cylindrical cavity which incorporated a dielectric resonator ($Ba(ZrTa)O_3$ ($\varepsilon_r = 29$)). This resonator increased the Q of the cavity. A sharp metallic probe protruded out of the resonant cavity. They calculated the 4.5-GHz E- and H-field lines that exist within the cylindrical cavity using finite element modeling software. They report a resolution of 1 μm at ~4.5 GHz during imaging of Cr on glass. Transmission line theory was used to model S_{11} (the wave reflection coefficient for an impedance of Z_0).

Park and colleagues [128] reported on a resonant cavity coupled to a sharp metallic probe, and spatial resolution better than 2 μm at 36 GHz on a patterned superconductor sample. Park and coworkers also used a quarter-wave resonator design for observation of biological systems [129]. They performed imaging on cells from a plant leaf, where water (a lossy dielectric at microwave frequencies) distribution delineated the cell wall, cytoplasm, and the stoma. They also measured several NaCl aqueous solutions. They used a blunt tip, which enhanced sensitivity, but reduced resolution, and also allowed probing through a thin glass slide barrier to the solution, since the z-direction probing depth is set by the tip radius. With increasing NaCl concentration, the real and imaginary parts of the dielectric constant increased, thus decreasing the resonant frequency (which is inversely proportional to $\sqrt{C}$ and $\sqrt{\varepsilon_{\text{real}}}$) and also decreased Q (which is inversely proportional to loss).

Many authors have used microwave probes for dielectric measurements. Chang and colleagues [130] have used a coaxial resonator for exploring the local dielectric and loss tangent in a ternary oxide sample. They prepared a triangular sample, where the composition of the sample was varied along each edge of the triangle, much like a ternary phase diagram. Localized 1-GHz probing quickly surveyed the dielectric values of the ternary, and identified regions (and compositions) with anomalously high dielectric values (above 600). In related work, Zhang and colleagues [131] have used a coaxial resonator for determining the ε_{33} dielectric constants for samples of $Sr_{m-3}Bi_4Ti_mO_{3m+3}$ (m = 3, 4, 5, and 6). Later, Zhang and colleagues [132] studied $Sr_3Bi_4Ti_6O_{21}$, for which they report a dielectric constant of 211 ± 20.

In other microwave microscopy work, Hughes and colleagues [133] have used evanescent microwave probes for subsurface detection of corrosion precursor pitting underneath paint, an important problem in aircraft maintenance [134–136]. Studies of cracks in metal were using microwave probes have also been reported [137–139], as have corrosion investigations using SQUID microscopy (see below) [140].

11 Multifunctional Probes

There are many hybrid probes presented in the literature, a few examples of which are discussed here. Lann and colleagues [141] use a resonant slot concurrently with a low-resolution, fast, eddy current system to image smooth superconductor films. The oxygen-depleted areas (lower conductivity) were clearly identified by microwave reflectivity images. The slot probe also found localized T_c values that differ with bulk (eddy current) measurements.

An open-ended coaxial resonator with a NSOM probe center conductor was published in work by Aga and coworkers [142]. During simultaneous operation, NSOM images exhibited higher resolution than the microwave image. Imaging of dielectric resonators revealed evidence for the spatial coincidence of non-uniform microwave absorption in regions with increased optical transmission, possibly indicating the presence of pinholes.

Frey and coworkers [143] have fabricated an optical NSOM probe in which a metallic tip protrudes from the aperture, thus combining apertured and apertureless SNOM designs. They claim that near-field illumination, with their probe, reduces photobleaching, and that the fixed source-tip geometry improves stability and compactness. They demonstrated 25-nm resolution on a sample of fluorescent beads.

12 Scanning Tunneling Microscopy

Atomic resolution can be achieved in microwave imaging with an experiment called ac-scanning tunneling microscopy (STM), which was originally developed in order to permit measurement of insulating samples [144]. In ac-STM, alternating current causes cyclic charging and discharging of the surface, and thus avoids net transport of charge through the tip-sample junction. The tunneling current in STM is nonlinear in the potential energy difference, ϕ, between the tip and sample, where $I_{tunneling} \propto \exp(-\kappa z)$, and $\kappa = \sqrt{2m\phi/\hbar}$. m is the electron mass and $\hbar$ is the reduced Planck's constant [145]. Because of the nonlinearity, a high-frequency voltage, applied across the tip-sample interface, results in generation of harmonic frequency components. Higher harmonics can be used as a source of feedback, and can allow for atomic resolution images [146, 147]. The motivation for much of the recent effort in STM is chemical identification at the atomic level. McCarty and Weiss [148] have used higher harmonics in ac-STM for differentiation between chemical species on surfaces.

In a STM experiment reported by Stipe and coworkers [149] and later by Wang and coworkers [150], the vibrational absorptions of individual molecules can be identified during STM inelastic electron tunneling spectroscopy (IETS). In this experiment, the tip-sample dc voltage is swept while monitoring d^2I/dv^2. Peaks can be observed in the tunneling current at potentials that equal the vibrational energies of molecules.

Durkan and Welland [151] reported STM tunneling into individual molecules at the Larmor precession frequency during STM-ESR (electron spin resonance) spectroscopy. They showed that tunneling, at the resonance frequency, changes in response to changes in the external magnetic field. In other work, ferromagnetic resonance (FMR) in single Ni dots (70 nm wide) was detected by thermoelastic expansion of the dots in a STM [152]. Here, the STM-FMR lineshape matched the conventional FMR lineshape. FMR with 15-μm resolution from Fe_3O_4 nanoparticles was detectable by miniature thermocouple or an infrared thermometer as reported by Sarkan et al. [153]. Subsurface imaging capability was also demonstrated. Finally, a variety of NSOM combined with STM techniques are reviewed by Grafstrom [154], and a recent paper by Colton reviewed STM, STS (scanning tunneling spectroscopy, which measures dI/dv as a function of v), and related techniques [155].

13 SQUID Microscopy

Kirtley has presented an excellent introduction to superconducting quantum interference device (SQUID) microscope [156]. The main advantage of the SQUID, is its excellent magnetic field sensitivity. For example, a sensitivity of 50 $pT/\sqrt{Hz}$ with spatial resolution better than 2 μm was reported by Anderberg and coworkers [157]. The disadvantages include a relatively low resolution (μm level and above), the saturation magnetic field limit for SQUID operation, and the required cryogenic cooling of the SQUID sensor.

As the SQUID loop (or its inductively coupled pick-up loop) is made smaller, it offers higher resolution, but the loop should not be separated from the sample by more than the diameter of the loop. For room-temperature samples, this necessitates high thermal gradients, across some window material; and therefore, a tradeoff between the spatial resolution and the temperature of the sample. As an example of a room-temperature experiment, SQUID microscopy was used for measuring ionic currents across cell membranes by Tanaka and coworkers [158]. In a later contribution, Kirtley reviews [159] several techniques for coupling magnetic flux to the SQUID. These include using a superconducting pick-up loop for inductive coupling to the SQUID, and ferromagnetic flux guides for focusing fields to the SQUID [160–162].

Gudoshnikov and coworkers [163] used a permalloy needle as a "flux guide," which increases distance of separation to the sample, and can also increase the resolution of the microscope. However, they note that permalloy can distort the resulting images. A mathematical model for the flux guide operation was presented in a subsequent publication [164].

Chatraphorn and coworkers presented an analytic formula for the noise and spatial resolution in SQUID microscopy [165,166]. Their experiment used a scanning SQUID for exploring their room temperature samples, from which they report a resolution of 50 μm on a test IC. They note that SQUID microscopy permits subsurface imaging, which is important for metrology through die packaging. SQUID devices have also been integrated onto tuning forks, thus adding a topography channel to the microscope [167].

Black and coworkers discuss the biasing of a dc SQUID device just to its critical current point (where a voltage appears across the SQUID) [168]. In this situation, small changes in magnetic flux, at high frequencies, induce an oscillating voltage across the SQUID. For very high frequencies, the associated electronics cannot follow the voltage waveform. However, owing to nonlinearity, high frequencies become rectified in the device. Using this effect, the amplitude of fields up to 12.5 GHz can be measured with spatial resolution of 15 μm. In a subsequent report, Black and colleagues report on a critically biased SQUID which undergoes self-oscillation at microwave frequencies [169]. The SQUID junction serves as source and detector during scanning. Changes in the mutual inductance of the SQUID-sample system thus induce changes in the flux through the SQUID, which changes the modulation depth of the voltage versus flux. The resulting self-oscillation frequency is tunable up to about 200 GHz, and allowed imaging with 30-μm resolution.

14 Pump-Probe Techniques

Electrooptic measurements have been a very successful method for measurement of localized high-speed waveforms and were recently reviewed by Grafstrom [154].

As an example, Takeuchi and colleagues [170, 171] used a photoconductive semiconductor switch (PCSS), placed near the end of a STM probe. This switch toggles the tunneling current through the probe on and off by an optical pulse train at frequency f. When a waveform that is sampled by the probe differs from f by Δf, then the a tunneling current replica of the device under test, stretched out in the time domain by an amount $f/\Delta f$, is passed through the probe and read by an oscilloscope. By setting the bandwidth for the STM feedback low, the high-frequency components of the optical sampling circuit have no effect on the ordinary operation of the STM. They reported time resolution of 160 pS. Later publications reported 2 pS resolution [172].

Oesterschulze and colleagues presented a coplanar waveguide (CPW) structure, integrated with a GaAs cantilever [173]. The exposed GaAs between the electrodes of the waveguide acts as a photoconductive switch. The waveguide can be matched to the sample impedance, providing an advantage over in-line switching schemes. Also, the cantilever is used in a scanning force microscope, which can be advantageous over STM, since insulating samples can be imaged at dc frequencies. The PC switch is turned on with a laser pulse, temporarily introducing carriers and increased conductivity between the electrodes of the cantilever-based CPW. While the pulse is on, a voltage on a device under test (DUT) will drive current between the electrodes of the waveguide. The duration of the pulse, and the relaxation time of carriers determines the fastest sample signals that can be measured. They demonstrate spatio-temporal imaging (with 1.9 pS resolution) of a voltage pulse on a coplanar strip line sample. In another reports, by Khusnatdinov and coworkers [174], pS transients were measured with 1-nm spatial resolution [174].

15 Electrostatic Force Microscopy

Finally, we discuss electrostatic force microscopy, a technique which is capable of measuring high-frequency voltage waveforms quantitatively. A heterodyne technique for measuring high-frequency potentials on integrated circuits (ICs) by sensing the force of interaction between sample and a biased scanning probe was described by Bridges and coworkers [175]. Their probe was a thin, bent molybdenum wire, and a fiber optic interferometer was used to sense the deflection of the wire. The deflection is given by $\Delta z = QF_z/k$, where Q is mechanical quality factor, and k is the spring constant. F_z is the electrostatic force between probe and circuit, and is given by

$$F_z = \frac{1}{2}\frac{\partial}{\partial z}C(x, y, z)[v_{probe}(t) - v_{circuit}(x, y, t)]^2. \tag{33}$$

Here, z is the direction normal to the IC surface. The square law causes mixing and conversion of high-frequency components, into dc, intermediate, and very high-frequency force components on the cantilever.

In their experiment, a modulated signal was sent to the probe,

$$v_{probe}(t) = [A + K\cos(\omega_r t)]v_s(t). \tag{34}$$

Here, the user determines A, K, and v_s, which is called the sampling voltage. For determining sinusoidal waveforms, such as $v_{circuit}(x, y, t) = V_c \sin(\omega_0 t + \phi_c)$, a sinusoidal sampling voltage was used, $v_s(t) = V_s \sin(\omega_0 t + \phi_s)$. A lock-in was used for signal recovery at the mechanical resonance frequency of the wire, ω_r. Therefore, the force component at ω_r is given by,

$$F_z|_{\omega \simeq \omega_r} = -\frac{\partial}{\partial z}C(x, y, z) \times \left[\frac{AV_s^2}{2} - \frac{V_s V_c \cos(\phi_s - \phi_c)}{2}\right] K\cos(\omega_r t). \tag{35}$$

From Eq. (35), the user controlled quantities AV_s^2 and ϕ_s could be used to null the force, at which point the quantities V_c and ϕ_c of the IC's waveform could be determined.

This technique can also be used for arbitrary waveforms. A more general form of Eq. (35) is

$$F_z|_{\omega \simeq \omega_r} = -\frac{\partial}{\partial z}C(x, y, z) \times [A\langle v_s(t), v_s(t)\rangle - \langle v_s(t), v_{circuit}(x, y, t)\rangle]K\cos(\omega_r t). \tag{36}$$

Here,

$$\langle a, b\rangle = \int a(t)b(t)dt/T \tag{37}$$

is the inner product. To measure arbitrary waveforms, a narrow pulse, with adjustable delay, was used for the sampling pulse. The pulse is written $v_s(t) = G_\delta(t - \tau)$, and consisted of a pulse of width δ, at time τ. A narrow pulse results in a force given by

$$F_z|_{\omega \simeq \omega_r} = -\frac{\partial}{\partial z}C(x, y, z) \times {}^{\delta}/_{T}[A - v_{circuit}(x, y, t = \tau)]K\cos(\omega_r t). \tag{38}$$

The magnitude of the signal $v_{circuit}(x, y, t)$ is measured by adjusting A to null the force. By repeating the nulling procedure, for several values of τ across one period, an arbitrary waveform can be measured. In related publications, angle modulation methods have been demonstrated [176, 177].

Later, a quantitative null-force technique was described by Weng and coworkers [178]. They noted that owing to work function differences, a dc offset ($\Delta\Phi$) should be present in Eq. (38), making the bracketed term $[A - v_{circuit}(x, y, t = \tau) + \Delta\Phi]$. To deal with this, they chose a sampling pulse of the form, $v_s(t) = G_\delta(t - \tau) - \langle G_\delta(t)\rangle$. Here, $v_{circuit}(x, y, t)$ is the new dc offset, and $v_{circuit}(x, y, t = \tau)$ is measured relative to $v_{circuit}(x, y, t)$, where t is an arbitrary reference time for measuring the waveform. The signal is nulled when $A = v_{circuit}(x, y, t = \tau) - v_{circuit}(x, y, t)$. Again, by measuring the null condition, at many values of τ , the wavefrom can be measured quantitatively. In addition to being quantitative, another advantage of this technique is that it unnecessary to calibrate dC/dz, Q, and k, in separate experiments. They demonstrate absolute voltage measurements at 250 Mbit clock rates on an IC (a wideband distributed amplifier), and reported voltage error of less than 30 mV in a dynamic range of 2.5 V.

We also note related techniques involving frequency down conversion of magnetic fields, at 1-GHz rates, using magnetic force microscopy probes [179]. These studies are relevant for magnetic storage and readout devices.

16 Conclusions

We have discussed the spatial resolution limit for far-field optics, which is on the order of λ. Many phenomena, including magnetic resonance (FMR, ESR), IR resonance absorptions, and several modern-day devices (such as microprocessors and cell phones) are based on high-frequency signals in the GHz through THz frequency ranges. Conventional microscopic imaging of these signals is subject to coarse spatial resolution in the micron to centimeter range. However, imaging at high frequencies is attractive owing to unusual image contrast, such as the leaf rehydration observed in the microwave regime [67], and many other examples that we have mentioned. In order to image these signals with a resolution better than the diffraction limit, one can use the near-field probing techniques that were discussed in this chapter.

To review, we discussed near-field aperture probes and described throughput cutoff that limits the ultimate resolution for aperture probes to the 10-nm range. We also discussed scattering and coaxial-type probes, which do not suffer throughput cutoff, and have achieved high spatial resolution for IR and visible frequencies, in the 1 nm range. A variety of microwave probes were discussed, some of which can achieve nanometer resolution; and STM techniques were described which can achieve atomic resolution at microwave frequencies. We briefly discussed SQUID and microcoil measurements of localized magnetic fields. Finally, we discussed pump probe and electrostatic force techniques which are useful for high-frequency circuit probing. For all of these microscopies, as the resolution increases, the absolute signal level decreases, thus necessitating careful impedance matching and signal recovery techniques.

An important area of research in near-field high-frequency probing is development of quantitative measurement techniques and analysis [9, 113, 114, 116, 119, 178]. For the scattering and microwave reflection measurements, quantitative measurement allows for discrimination of chemical species, even at frequencies removed from resonant absorptions [9, 77]. In the area of IC testing, quantitative measurements permit absolute voltage measurements, which are important for modeling IC transfer functions and signal distortion [178, 179].

Acknowledgments. This work was supported by the Air Force Office of Scientific Research, MURI 2003 (Award number F49620-03-1-0420). We also acknowledge long-standing collaborations with Fritz Keilmann.

References

1. B. T. Rosner, D. W. van der Weide, Rev. Sci. Instrum. **73**, 2505 (2002).
2. M. A. Paesler, P. J. Moyer, *Near-Field Optics, Theory Instrumentation and Applications* (John Wiley and Sons, NY, 1996).
3. B. Hecht, B. Sick, Urs. P. Wild, V. Deckert, R. Zenobi, O. J. F. Martin, J. Chem. Phys. **112**, 7761 (2000).
4. D. W. Pohl, Phil. Trans. R. Soc. Lond. A **362**, 701 (2004).
5. D. W. Pohl, Thin Solid Films **264**, 250 (1995).
6. Lewis, in *Current Trends in Optics*, ed. by J. C. Dainty (Academic Press, Cambridge, UK, 1994).
7. J. W. P. Hsu, Mar. Sci. Eng. Rep. **33**, 1 (2001).
8. D. Courjon, C. Bainier, Rep. Prog. Phys. **57**, 989 (1994).
9. F. Keilmann, Phil. Trans. R. Soc. Lond. A **362**, 787 (2004).
10. M. Tabib-Azar, AIP Proc. **20**, 400 (2001).
11. M. Tabib-Azar, *Evanescent Microwave Microscopy for High-Speed and High-Resolution Material Characterizations* (Kluwer Academic, Boston, 1999).
12. X.-D. Xiang, C. Gao, Materials Char. **48**, 117 (2002).
13. F. de Fornel, *Evanescent Waves from Newtonian Optics to Atomic Optics* (Springer, Berlin, DE, 2000).
14. J. W. Goodman, *Introduction to Fourier Optics* (McGraw-Hill, San Francisco, 1968).
15. M. Born, E. Wolf, *Principles of Optics* (Cambridge University Press, Cambridge, UK, 1998).
16. E. Hecht, *Optics*, 3rd Ed. (Addison Wesley Longman, Reading, MA, 1998).
17. F. L. Pedrotti, L. S. Pedrotti, *Introduction to Optics*, 2nd Ed. (Prentice Hall, Englewood Cliffs, NJ, 1993).
18. Lord Rayleigh, Philos. Mag. **5**, 167 (1896).
19. D. Courjon, K. Sarayedine, M. Spajer, Opt. Commun. **71**, 23 (1989).
20. R. C. Reddick, R. J. Warmack, T. L. Ferrell, Phys. Rev. B **39**, 767 (1989).
21. R. Quidant, J. C. Weeber, A. Dereux, D. Peyrade, Y. Chen, C. Girard, Europhys. Lett. **57**, 191 (2002).
22. A. Dereux, C. Girard, J. C. Weeber, J. Chem. Phys. **112**, 7775 (2000).
23. J. R. Krenn, J.-C. Weeber, Phil. Trans. R. Soc. Lond. A **362**, 739 (2004).
24. Y. Chen, R. K. Kupka, Ultramicroscopy **57**, 153 (1995).
25. P. S. Carney, R. A. Frazin, S. I. Bolzhevolnyi, V. S. Volkov, A. Boltasseva, J. C. Schotland, Phys. Rev. Lett. **92**, 163903-1 (2004).
26. T. R. Albrecht, P. Grutter, D. Horne, D. Rugar, J. Appl. Phys. **69**, 668 (1991).
27. W. A. Atia, C. C. Davis, Appl. Phys. Lett. **70**, 405 (1997).
28. H. Edwards, L. Taylor, W. Duncan, A. J. Melmed, J. Appl. Phys. **82**, 981 (1997)
29. F. J. Giessibl, Appl. Phys. Lett. **73**, 3956 (1998).
30. A. Kikukawa, S. Hosaka, Y. Honda, R. Imura, Jpn. J. Appl. Phys. **33**, L1286 (1994).
31. S. J. Bukofsky, R. D. Grober, Appl. Phys. Lett. **71**, 2749 (1997).
32. Y. Seo, J. H. Park, J. B. Moon, W. Jhe, Appl. Phys. Lett. **77**, 4274 (2000).
33. B. Hecht, H. Bielefeldt, Y. Inouye, D. W. Pohl, L. Novotny, J. Appl. Phys. **81**, 2492 (1997).
34. C. A. Paulson, A. B. Ellis, L. McCaughan, B. Hawkins, J. Sun, T. F. Kuech, Appl. Phys. Lett. **77**, 1943 (2000).
35. B. Knoll, F. Keilman, A. Kramer, R. Guckenberger, Appl. Phys. Lett. **70**, 2667 (1997).

36. E. Betzig, J. K. Trautman, Science **257**, 189 (1992).
37. C. Kittel, *Introduction to Solid State Physics*, 7th ed. (John Wiley and Sons, Inc., New York, 1996).
38. W. A. Harrison, *Electronic Structure and the Properties of Solids, the Physics of the Chemical Bond*, 2nd ed. (Dover, Toronto, CN 1989).
39. P. W. Atkins, *Physical Chemistry*, 4th ed. (Freeman and Company, NY, 1990).
40. D. Mittelman, *Sensing with Terahertz Radiation* (Springer, Berlin, 2002).
41. X.-C. Zhang, Phys. Med. Biol. **47**, 3667 (2002).
42. J. F. Federici, O. Mitrofanov, M. Lee, J. W. P. Hsu, I. Brener, R. Harel, J. D. Wynn, L. N. Pfeiffer, K. W. West, Phys. Med. Biol. **47**, 3727 (2002).
43. T. Kiwa, H. Murakami, M. Tonouchi, Supercond. Sci. Technol. **16**, 1540 (2003).
44. M. K. Hong, A. Swan, S. Erramilli, Optics and Photonics News, p. 30, July (2004).
45. N. C. J. van der Valk, P. C. M. Planken, Appl. Phys. Lett. **81**, 1558 (2002).
46. T. Masaki, Y. Unouye, S. Kawata, Rev. Sci. Instrum. **75**, 3284 (2004).
47. T. Masaki, A. Ono, K. Goto, Y. Inouye, S. Kawata, H. Kuroda, Jpn. J. Appl. Phys., Suppl. **41–1**, 71 (2002).
48. B. Dragnea, J. Preusser, W. Schade, S. R. Leone, J. Appl. Phys. **86**, 2795 (1999).
49. E. A. Ash, G. Nicholls, Nature **237**, 510 (1972).
50. R. F. Soohoo, J. Appl. Phys. **33**, 1276 (1962).
51. D. K. Cheng, *Field and Wave Electromagnetics* (Addison Wesley, Reading, MA, 1989).
52. F. Keilmann, Infrared Phys. Technol. **36**, 217 (1995).
53. B. Knoll, F. Keilmann, Opt. Comm. **162**, 177 (1999).
54. C. J. Bouwkamp, Phillips Res. Rep. **5**, 401 (1950).
55. G. Massey, J. Davis, S. Katnick, E. Omon, Appl. Opt. **4**, 1498 (1985).
56. T. Thio, K. M. Pellerin, R. A. Linke, H. J. Lezec, T. W. Ebbesen, Opt. Lett. **26**, 1972 (2001).
57. J. R. Krenn, J.-C. Weeber, Phil. Trans. R. Soc. Lond. A **362**, 739 (2004).
58. T. Yatsui, K. Itsumi, M. Kourogi, M. Ohtsu, Appl. Phys. Lett. **80**, 2257 (2002).
59. X. Shi, L. Hesselink, R. L. Thornton, Opt. Lett. **28**, 1320 (2003).
60. T. Thio, K. M. Pellerin, R. A. Linke, H. J. Lezec, T. W. Ebbesen, Optics Letters **26**, 1972 (2001).
61. M. Mansuripur, A. R. Zakharian, J. V. Moloney, Optics and Photonics News, p. 38, March (2004).
62. M. Golosovsky, E. Maniv, D. Davidov, A. Freqnkel, IEEE Trans. Instrum. Meas. **51**, 1090 (2002).
63. E. Betzig, J. K. Trautman, J. S. Weiner, T. D. Harris, R. Wolfe, Appl. Opt. **31**, 4563 (1992).
64. O. J. F. Martin, M. Paulus, J. Micros. **205** part 2, 147 (2002).
65. M. Golosovsky, A. F. Lann, D. Davidov, A. Frenkel, Rev. Sci. Instrum. **71**, 3927 (2000).
66. W. C. Symons, K. W. Whites, R. A. Lodder, IEEE Trans. Microwave Theo. Tech. **51**, 91 (2003).
67. U. Schade, K. Holldack, P. Kuske, G. Wustefeld, H. W. Hubers, Appl. Phys. Lett. **84**, 1422 (2004).
68. S. Mair, B. Gompf, M. Dressel, Appl. Phys. Lett. **84**, 1219 (2004).
69. P. Ephrat, K. Roodenko, L. Nagli, A. Katzir, Appl. Phys. Lett. **84**, 637 (2004).
70. D. Schaafsma, R. Mossadegh, J. Sanghera, I. Aggarwal, M. Luce, R. Generosi, P. Perfetti, A. Cricenti, J. Gilligan, N. Tolk, Opt. Eng. **38**, 1381 (1999).

71. C. A. Michaels, S. J. Stranick, L. J. Richter, R. R. Cavanagh, J. Appl. Phys. **88**, 4832 (2000).
72. R. Hillenbrand, T. Taubner, F. Keilmann, Nature **418**, 159 (2002).
73. R. Hillenbrand, M. Stark, R. Guckenberger, Appl. Phys. Lett. **76**, 3478 (2000).
74. B. Knoll, F. Keilmann, J. Microsc. **194**, 512 (1999).
75. R. Hillenbrand, F. Keilmann, Appl. Phys. Lett. **80**, 25 (2002).
76. M. B. Raschke, C. Lienau, Appl. Phys. Lett. **83**, 5089 (2003).
77. T. Taubner, R. Hillenbrand, F. Keilmann, Appl. Phys. Lett. **85**, 5064 (2004).
78. H. T. Chen, R. Kersting, G. C. Cho, Appl. Phys. Lett. **83**, 3009 (2003).
79. K. Wang, D. M. Mittleman, N. C. J. van der Valk, P. C. M. Planken, Appl. Phys. Lett. **85**, 2715 (2004).
80. D. Haefliger, J. M. Plitzko, R. Hillenbrand, Appl. Phys. Lett. **85**, 4466 (2004).
81. A. Kramer, W. Trabesinger, B. Hecht, U. P. Wild, Appl. Phys. Lett. **80**, 1652 (2002).
82. R. Hillenbrand, F. Keilmann, P. Hanarp, D. S. Sutherland, J. Aizpurua, Appl. Phys. Lett. **83**, 368 (2003).
83. J. Takahara, S. Yamagishi, H. Taki, A. Morimoto, T. Kobayashi, Opt. Lett. **22**, 475 (1997).
84. R. Kantor, I. V. Shvets, J. Appl. Phys. **93**, 4979 (2003).
85. R. Kantor, I. V. Shvets, IEEE Transactions Microwave Theo. Tech. **51**, 2228 (2003).
86. D. Drews, W. Ehrfeld, M. Lacher, K. Mayr, W. Noell, S. Schmitt, M. Abraham, Nanotechnology **10**, 61 (1999).
87. B. T. Rosner, T. Bork, V. Agrawal, D. W. van der Weide, Sensors and Actuators A **102**, 185 (2002).
88. D. W. van der Weide, P. Neuzil, J. Vac. Sci. Technol. B **14**, 4144 (1996).
89. D. W. van der Weide, Appl. Phys. Lett. **70**, 677 (1997).
90. Y. Wang, M. Tabib-Azar, IEEE Transactions Microwave Theo. Tech. **52**, 971 (2004).
91. J. E. Stocker, T. L. Peck, A. G. Webb, M. Feng, R. L. Magin, IEEE Trans. Biomed. Eng. **44**, 1122 (1997).
92. S. C. Grant, L. A. Murphy, R. L. Magin, G. Friedman, IEEE Trans. on Mag. **37**, 2989 (2001).
93. R. Subramanian, W. P. Kelley, P. D. Floyd, Z. J. Tan, A. G. Webb, J. V. Sweedler, Anal. Chem. A **71**, 5335 (1999).
94. D. L. Olson, T. L. Peck, A. G. Webb, R. L. Magin, J. V. Sweedler, Science **270**, 1967 (1995).
95. Y. Li, A. M. Wolters, P. V. Malawey, J. V. Sweedler, A. G. Webb, Analytical Chemistry **71**, 4815 (1999).
96. T. Peck, R. Magin, P. Lauterbur, J. Magn. Reson. Ser. B **108**, 114 (1995).
97. V. Agrawal, P. Neuzil, D. W. van der Weide, Appl. Phys. Lett. **71**, 2343 (1997).
98. G. Boero, M. Bouterfas, C. Massin, F. Vincent, P.-A. Besse, R. S. Popovic, A. Schweiger, Rev. Sci. Instrum. **74**, 4794 (2003).
99. S-C. Lee, S. M. Anlage, IEEE Trans. Appl. Superconductivity **13**, 3594 (2003).
100. A. F. Lann, M. Golosovsky, D. Davidov, A. Frenkel, Appl. Phys. Lett. **73**, 2832 (1998).
101. F. Sakran, M. Copty, M. Golosovsky, N. Bontemps, D. Davidov, A. Frenkel, Appl. Phys. Lett. **82**, 1479 (2003).
102. M. Golosovsky, A. F. Lann, D. Davidov, A. Frenkel, Rev. Sci. Instrum. **71**, 3927 (2000).
103. A. Copty, F. Sakran, M. Golosovsky, D. Davidov, A. Frenkel, Appl. Phys. Lett. **84**, 5109 (2004).

104. M. Tabib-Azar, J. L. Katz, S. R. LeClair, IEEE Trans. Instrum. Meas. **48**, 1111 (1999).
105. M. Tabib-Azar, T. Zhang, S. R. LeClair, IEEE Trans. Instrum. Meas. **51**, 1126 (2002).
106. M. Tabib-Azar, R. Wang, AIP Proc. **20**, 446 (2001).
107. A. Imtiaz, M. Pollak, S. M. Anlage, J. D. Barry, J. Melngailis, J. Appl. Phys. **97**, 044302 (2005).
108. M. S. Kim, S. Kim, J. Kim, K. Lee, B. Friedman, J. T. Kim, J. Lee, Rev. Sci. Instrum. **74**, 3675 (2003).
109. K. Karrai, R. D. Grober, Appl. Phys. Lett., **66**, 1842 (1995).
110. C. P. Vlahacos, D. E. Steinhauer, S. K. Dutta, B. J. Feenstra, S. M. Anlage, F. C. Wellstood, Appl. Phys. Lett. **72**, 1778 (1998).
111. A. Tselev, S. M. Anlage, H. M. Christen, R. L. Moreland, V. V. Talanov, A. R. Schwartz, Rev. Sci. Instrum. **74**, 3167 (2003).
112. D. E. Steinhauer, S. M. Anlage, J. Appl. Phys. **89**, 2314 (2001).
113. C. Gao, F. Duewer, X.-D. Xiang, Appl. Phys. Lett. **75**, 3005 (1999).
114. C. Gao, X.-D. Xiang, Rev. Sci. Instrum. **69**, 3846 (1998).
115. R. F. Harrington, *Time-Harmonic Electromagnetic Fields* (Wiley-IEEE Press, NY, 2001).
116. C. Gao, B. Hu, P. Zhang, M. Huang, W. Liu, I. Takeuchi, Appl. Phys. Lett. **84**, 4647 (2004).
117. J. Kim, M. S. Kim, K. Lee, J. Lee, D. Cha, B. Friedman, Meas. Sci. Technol. **14**, 7 (2003).
118. S. Hong, J. Kim, W. Park, K. Lee, Appl. Phys. Lett. **80**, 524 (2002).
119. C. Gao, T. Wei, F. Duewer, Y. Lu, X.-D. Xiang, Appl. Phys. Lett. **71**, 1872 (1997).
120. Z. Wang, M. A. Kelly, Z.-X. Shun, G. Wang, X.-D. Xiang, J. T. Wetzel, J. Appl. Phys. **92**, 808 (2002).
121. Y. L. Feng, L. Y. Wu, K. L. Wang, T. Jiang, L. Kang, S. Z. Yang, P. H. Wu, IEEE Trans. Appl. Superconductivity **13**, 2901 (2003).
122. L. Liu, Y. J. Feng, L. Y.Wu, Q. G. Liu, E. H. Zhao, Z. L. Fu, L. X. You, L. Kang, S. Z. Yang, and P. W. Wu, Solid State Commun. **119**, 133 (2001).
123. A. Kim, J. Kim, S. Hyun, S. S. Kim, T. Kim, K. Char, Rev. Sci. Instrum. **74**, 3164 (2003).
124. J. Kim, M. S. Kim, H. Kim, D. Song, B. Friedman, Appl. Phys. Lett. **83**, 1026 (2003).
125. M. Kim, J. Kim, H. Kim, S. Kim, J. Yang, H. Yoo, S. Kim, K. Lee, B. Friedman, Rev. Sci. Instrum. **75**, 684 (2004).
126. D. J. E. Ingram, *Spectroscopy at Radio and Microwave Frequencies* (Plenum Press, NY, 1967).
127. J. Kim, K. Lee, B. Friedman, D. Cha, Appl. Phys. Lett. **83**, 1032 (2003).
128. W. Park, J. Kim, K. Lee, Appl. Phys. Lett. **79**, 2642 (2001).
129. J. Park, S. Hyun, A. Kim, T. Kim, K. Char, Ultramicroscopy **102**, 101 (2005).
130. H. Chang, K.-M. Yu, Y. Dong, X.-D. Xiang, Appl. Phys. Lett. **81**, 2062 (2002).
131. S. T. Zhang, Y. F. Chen, H. P. Sun, X. Q. Pan, Z. G. Liu, N. B. Ming, Appl. Phys. Lett. **81**, 5009 (2002).
132. S. T. Zhang, Y. F. Chen, H. P. Sun, X. Q. Pan, W. S. Tan, Z. G. Liu, N. B. Ming, J. Phys. Condens. Matter **15**, 1223 (2003).
133. D. Hughes, R. Zoughi, R. Austin, N. Wood, R. Englebart, AIP **22**, 462 (2003).
134. N. Qaddoumi, A. Shroyer, R. Zoughi, Research in Nondestructive Evaluation **9**, 201 (1997).
135. N. Qaddoumi, L. Handjojo, T. Bigelow, J. Easter, A. Bray, R. Zoughi, Materials Evaluation **58**, 178 (2000).

136. D. Hughes, N. Wang, T. Case, K. Donnel, R. Zoughi, R. Austin, M. Novack, Proceedings of the Twenty-seventh Annual Review of Progress in Quantitative Nondestructive Evaluation, **20A**, 460 (2000).
137. Y. Wang, R. Zoughi, Materials Evaluation **58**, 1228 (2000).
138. C. Huber, S. I. Ganchev, R. Mirshahi, J. Easter, R. Zoughi, Nondestructive Testing and Evaluation **13**, 227 (1997).
139. C. Yeh, R. Zoughi, Research in Nondestructive Evaluation **6**, 35 (1994).
140. M.v. Kerutzbruck, K. Allweins, G. Gierelt, H.-J. Krause, S. Gartner, W. Wolf, Physica C **368**, 85 (2002).
141. A. F. Lann, M. Abu-Teir, M. Golosovsky, D. Davidov, S. Djordjevic, N. Bontemps, L. F. Cohen, Rev. Sci. Instrum. **70**, 4348 (1999).
142. R. S. Aga, Jr., J. Brookman, J. Dizon, J. Wu, Appl. Phys. Lett. **84**, 1979 (2004).
143. H. G. Frey, F. Keilmann, A. Kriele, R. Guckenberger, Appl. Phys. Lett. **81**, 5030 (2002).
144. S. J. Stranick, P. S. Weiss, Rev. Sci. Instrum. **65**, 918 (1993).
145. J. C. Chen, *Introduction to Scanning Tunneling Microscopy* (Oxford University Press, Oxford, UK, 1993).
146. J. Schmidt, D. H. Rapoport, H.-J. Frohlich, Rev. Sci. Instrum. **70**, 3377 (1999).
147. B. Michel, W. Mizutani, R. Schierle, A. Jarosch, W. Knop, H. Benedickter, W. Bachthold, H. Rohrer, Rev. Sci. Instrum. **63**, 4080 (1992).
148. G. S. McCarty, P. S. Weiss, Chem. Rev. **99**, 1983 (1999).
149. B. C. Stipe, M. A. Rezaei, W. Ho, Science **280**, 1732 (1998).
150. W. Wang, T. Lee, I. Kretzschmar, M. A. Reed, Nano. Lett. **4**, 643 (2004).
151. C. Durkan, M. E. Welland, Appl. Phys. Lett. **80**, 458 (2002).
152. R. Meckenstock, D. Spoddig, D. Dietzel, J. Pelzel, J. P. Bucher, Rev. Sci. Instrum. **74**, 789 (2003).
153. F. Sakran, A. Copty, M. Golosovsky, D. Davidov, Appl. Phys. Lett. **84**, 4499 (2004).
154. S. Grafstrom, J. Appl. Phys. **91**, 1717 (2002).
155. R. J. Colton, J. Vac. Sci. Technol. B **22**, 1609 (2004).
156. J. Kirtley, IEEE Spectrum, p. 40, Dec. (1996).
157. J. Anderberg, M. S. Colclough, D. B. Crum, Y. Tokura, D. N. Paulson, R. L. Fagaly, IEEE Trans. Appl. Supercond. **13**, 231 (2003).
158. S. Tanaka, N. Tanaka, M. Natsume, M. Uchida, M. Oshita, Z. Aspanut, T. Eki, S. Yoshida, Supercond. Sci. Technol. **16**, 1536 (2003).
159. J. R. Kirtley, Physica C **308**, 55 (2002).
160. H. Itozaki, T. Kondo, T. Nagaishi, Physica C **392**, 1392 (2003).
161. T. Nagaishi, K. Minamimura, H. Itozaki, IEEE Trans. Appl. Supercond. **13**, 227 (2003).
162. T. Kondo, H. Itozaki, R. Koetitz, Supercond. Sci. Technol. **16**, 1528 (2003).
163. S. A. Gudoshnikov, Y. V. Deryuzhkina, P. E. Rudenchik, Y. S. Sitnov, S. I. Bondarenko, A. A. Shablo, P. P. Pavlov, A. S. Kalabukhov, O. V. Snigirev, P. Seidel, IEEE Trans. Appl. Supercond. **11**, 219 (2001).
164. S. A. Gudoshnikov, B. Ya. Liubimov, L. V. Matveets, A. P. Mikhailenko, Y. V. Deryuzhkina, Y. S. Sitnov, O. V. Snigirev, Physica C **368**, 66 (2002).
165. S. Chatraphorn, E. F. Fleet, F. C. Wellstood, L. A. Knauss, T. M. Eiles, Appl. Phys. Lett. **76**, 2304 (2000).
166. S. Chatraphorn, E. F. Fleet, F. C. Wellstood, L. A. Knauss, IEEE Trans. Appl. Supercond. **11**, 234 (2001).
167. K. Hasselbach, C. Veauvy, D. Mailly, Physica C **332**, 140 (2000).

168. R. C. Black, F. C. Wellstood, E. Dantsker, A. H. Miklich, D. Koelle, F. Ludwig, J. Clarke, Appl. Phys. Lett. **66**, 1267 (1995).
169. R. C. Black, F. C. Wellstood, E. Dantsker, A. H. Miklich, D. T. Nemeth, D. Koelle, F. Ludwig, J. Clarke, Appl. Phys. Lett. **66**, 99 (1995).
170. K. Takeuchi, A. Mizuhara, Y. Kasahara, IEEE Trans. Instrum. Meas. **44**, 815 (1995).
171. K. Takeuchi, Y. Kasahara, Appl. Phys. Lett. **63**, 3548 (1993).
172. K. Takeuchi, A. Mizuhara, Electron. Lett. **32**, 1709 (1996).
173. E. Oesterschulze, S. Heisig, W. M. Steffens, J. Vac. Sci. Technol. B **19**, 107 (2001).
174. N. N. Khusnatdinov, T. J. Nagle, G. Nunes, Appl. Phys. Lett. **77**, 4434 (2000).
175. G. E. Bridges, R. A. Said, D. J. Thompson, Electron. Lett. **29**, 1448 (1993).
176. A. S. Hou, F. Ho, D. M. Bloom, Electron. Lett. **30**, 560 (1994).
177. C. Bohm, F. Saurenbach, P. Taschner, C. Roths, E. Kubalek, J. Phys. D **26**, 1801 (1993).
178. Z. Weng, C. J. Falkingham, G. E. Bridges, D. J. Thompson, J. Vac. Sci. Technol. A **20**, 999 (2002).
179. S. Li, S. Stokes, Y. Liu, S. Foss-Schroder, W. Zhu, D. Palmer, J. Appl. Phys. **91**, 7346 (2002).

II
Electrical and Electromechanical Imaging at the Limits of Resolution

II.1 Scanning Probe Microscopy on Low-Dimensional Electron Systems in III-V Semiconductors

MARKUS MORGENSTERN

Since the discovery of the quantum Hall effect [1], a lot of insight into the behavior of quasi-free electron systems housed in III-V-semiconductors has been gained by macroscopic measurements such as transport, magnetization, or optics [2,3]. However, theoretical descriptions of the observed phenomena mostly rely on a distinct local arrangement of the corresponding electron phases [4–8]. For example, it is believed that the quantum Hall plateaus require localization of the electrons everywhere within a sample except close to the edge. The electrons inside the sample should reside at equipotential lines in valleys and at hills of the potential disorder. This specific description, as well as many others explaining other effects, triggered the effort to observe the local arrangement of the electrons in real space by scanning probe techniques. In this chapter the current state of the art of the different scanning probe methods as applied to low-dimensional III-V semiconductors is reviewed and the specific advantages and drawbacks are discussed.

1 Introduction

The investigation of electron systems in solid state still provides puzzling results with many waiting for decades to be explained. This is due to the intricate quantum mechanics describing the behavior of interacting electrons. A systematic approach to understand this behavior requires tuning of the thermodynamic parameters guiding the electron systems. This is usually done by modifying solid-state samples chemically or physically.

Another, rather elegant approach uses III-V-semiconductors. There, most of the parameters are externally accessible [2,3]. In addition, the relevant band structure is quite simple. It consists of a nearly parabolic conduction band with dominating s-character; i.e., the wave functions are largely isotropic. Thus, III-V-materials are considered to be the paradigmatic case of interacting electron systems. Indeed, a large number of different electron phases have been found using these materials [1,4,9–14].

The electron phases are guided by several interactions, which compete with the kinetic energy. In particular, the electron–electron interaction energy or the electron-disorder interaction energy can dominate the behavior of the system. Since these interactions are tunable in III-V-semiconductors, one can realize a wealth of different electron phases covering a large part of the general phase diagram for electrons. Only some phases are not accessible. These are phases driven by a strong electron–phonon interaction as the Peierls phase [15] or the superconducting phase [16]. Moreover, strong local electronic correlations as they exist, e.g., for d-like electrons [17] cannot be realized.

All other relevant cases of interaction-induced electron phases can be studied systematically. The following parameters are tuned in order to change the strength of the different interactions. The dimensionality of the system is modified from 3D down to 0D by simple growth techniques or electrostatics [2,3]. The electron density is tuned rather accurately using gates. The temperature is reduced down to about 10 mK. Relatively large magnetic fields are applied resulting in Landau energies and Zeeman energies larger than the kinetic energy. Even the relative importance of the two magnetic energies could be tuned by using different III-V materials, i.e., g-values from $g = 0$ to $g = 50$ can be realized [18]. Finally, the residual disorder in the sample is tuned by using advanced growth techniques. Currently, samples with a scattering length of the electrons close to 1 mm are available [19]. Thus, the most relevant thermodynamic parameters are accessible over a wide range, which provides the base for systematic studies.

The paradigmatic III-V samples are usually studied by macroscopic techniques such as electrical transport or optics, which yielded an enormous amount of information on different phases and phase transitions [1,4,9–14]. However, the techniques cannot access the underlying physical arrangement of the electrons. Consequently, scanning probe techniques [20], which measure the electronic properties down to the atomic scale, appear to be a good complementary approach to investigate these samples.

Here, the most recent approaches based on a variety of different scanning probe techniques are reviewed. The most direct approach uses scanning tunneling spectroscopy (STS). It images the local density of states, i.e., the wave functions of the system, down to the atomic scale [21]. However, this technique suffers from the fact that it can only access electrons very close to the sample surface. Consequently, one is restricted to special samples, which typically do not posses the high mobilities available nowadays in heterostructures.

Other techniques couple electrostatically to the sample, which allows the use of high-mobility heterostructures. There, the 2D systems are typically buried about 100 nm below the surface. Electrostatic force microscopy (EFM) has been used to measure the potential distribution within such samples [22]. The same quantity has been measured with higher energy resolution using a single electron transistor (SET) deposited directly on the scanning tip (SSET: scanning SET) [23]. Even the current induced potential has been mapped by correlating the induced current with the measured potential by lock-in technique [24]. Finally, combining SSET or EFM with a backgate allows us to measure the local compressibility, giving direct access to localization phenomena [25]. This local compressibility has also been measured using scanning capacitance microscopy (SCM) within the quantum Hall regime [26,27]. Another approach uses the tip as an electrostatic perturbation for the current flow within the sample [28]. This technique visualizes the preferential current paths, e.g., through a quantum point contact [29,30]. All these electrostatic techniques have the advantage that they can be applied to high mobility samples due to the long-range character of the electrostatic fields. Unfortunately, the lateral resolution is currently limited to about 100 nm, which is above the characteristic length scales of electron waves in these samples.

Finally, scanning near-field optical microscopy (SNOM) [30] has been used to address optical properties of individual quantum dots [31]. Moreover recombination of injected electrons has been investigated on the local scale within 1D and 2D systems [32–34]. Since the SNOM investigations are only indirectly related to phase properties of the III-V-systems, they are discussed rather briefly at the end of this article.

2 Scanning Tunneling Spectroscopy (STS)

STS measures the differential conductivity *dI/dV* of the tunneling gap between a metallic tip and a sample [20,21]. Parameters are the applied voltage V and the position of the tip with respect to the sample $\mathbf{x}$. At low voltages ($V < 200$ mV) and a spatial homogeneous electron decay into vacuum, $dI/dV(\mathbf{x}, V)$ is directly proportional to the local density of states (LDOS$(\mathbf{x}, E)$), where energy E is identified with voltage. More precisely, one measures the probability of finding an electronic state at the position $\mathbf{x}$ and at the energy eV with respect to the Fermi level. If one describes the system by single particle wave functions, this corresponds to the sum of all squared single particle wave function amplitudes at the desired energy and position. In real experiments, the finite energy resolution (1 meV at 4 K) and spatial resolution (about 0.1 nm) has to be taken into account [21].

Individual wave functions can be measured in quantum dots, if the level spacing is larger than the energy resolution. For III-V semiconductors, this has been realized on strain-induced InAs quantum dots on GaAs(001) [35,36] as well as on chemically prepared InAs clusters having a ZnSe shell [37].

Moreover, completely confined states in stacking fault pyramids on InAs(111) have been studied [38]. Figure 1 shows an example of a strain-induced quantum dot [36]. The *dI/dV* curve in Figure 1(a) is measured on a particular position of the quantum dot and shows two peaks belonging to two different single particle energies within the dot. The lateral distribution of the peak intensity (Figure 1(b,c)) shows the symmetry of the corresponding wave functions to be s- and p-like. Figure 1 (d–g) shows the squared wave functions of another InAs quantum dot.

The same technique can be applied to systems of higher dimensionality. Since the energy levels are dense in such systems, one usually measures not only one but several squared wave functions at the same voltage [21]. More precisely, one measures all wave functions having an energy within the energy resolution of the experiment. The major problem is to provide the respective samples exhibiting 1D or 2D behavior. They have to be close to the surface in order to get a tunneling

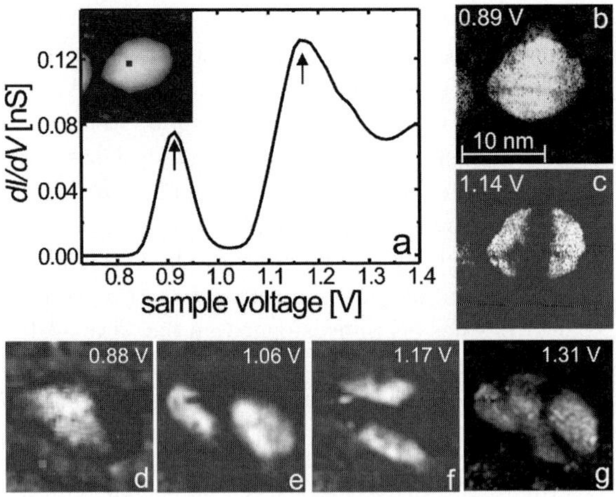

FIGURE 1. **(a)** *dI/dV*-curve measured on an InAs quantum dot; arrows mark the voltages where the *dI/dV* maps in **(b)** and **(c)** are taken; inset: STM image of the quantum dot; black square marks the position of the spectrum in the main image; (b–c) *dI/dV* maps taken at the voltages indicated and showing s- and p-like symmetry of the corresponding wave functions; **(d–g)** *dI/dV*-maps taken on a different InAs quantum dot and showing s-, p_x-, p_y- and d-like symmetry; $T = 6$ K [36].

current between the system and the STM tip. A possible access is to use naturally formed systems of lower dimension. For example, a 1D system exists below positively charged [112]-step edges on InAs(110). The corresponding LDOS has been measured and could be related to the underlying electrostatic disorder potential [39]. Figure 2 shows the disorder potential and the resulting LDOS at different energies. Basically, the LDOS can be reproduced by a single particle calculation. This is disappointing, since one of the most interesting effects in 1D systems appears if the electron-electron interaction guides the system leading to the so-called Luttinger properties [11]. But indications of these intriguing properties have not been found.

Two-dimensional systems can be either prepared by the growth of InAs on GaAs(111) [40] or by depositing minute amounts of adsorbates on InAs(110) [41]. In the former case, the interface to the GaAs realizes the 2D confinement, while in the latter case a surface accumulation layer is formed. Both systems provide relatively large electron densities and a moderate disorder strength [42]. However, the interaction of the electrons with the disorder, which is known to lead to weak localization in 2D can be measured on a local scale [43]. Figure 3(a) shows an example. A complex wave pattern with a preferential wavelength can be seen.

The Fourier transform of the real space data is shown in the inset, which is a representation of the wave vectors contributing to the image. The apparent ring structure demonstrates that the majority of the contributing wave vectors has the same length. Figure 3(c) shows a histogram of all *dI/dV* values obtained in the

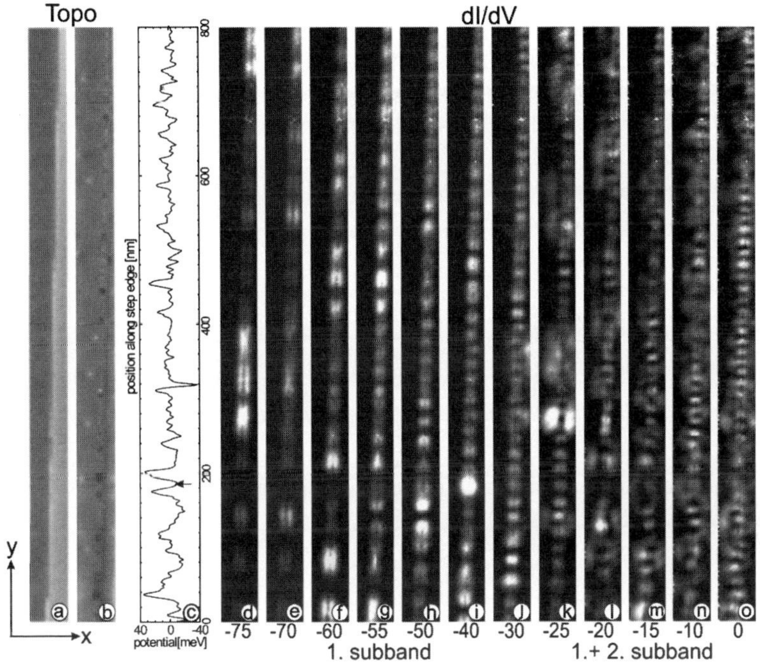

FIGURE 2. (a) STM image of a [112]-step edge on InAs(110); (b) same image after substracting the average apparent height of the step edge; dopants are visible as bright and dark spots; (c) deduced disorder potential along the step edge; (d–o) *dI/dV*-maps measured at the voltages indicated in mV; 1D standing waves indicating localized states in a 1DES are visible; $T = 6$ K [39]

image. The histogram is broad, i.e., one finds many different LDOS-values within the investigated sample area. The LDOS spatially fluctuates between large values and small values, i.e., the LDOS corrugation is large. Again, a single-particle calculation using the measured disorder potential can reproduce the features of the *dI/dV*-image [43].

Since the pristine 3D InAs(110) system can also be measured [44], a direct comparison between 2D and 3D behavior of electrons is possible. The 3D-LDOS and its histogram are shown in Figs. 3 (b,d). Note that both systems, 2D and 3D, are measured at the same temperature, with the same disorder strength and at the same energy with respect to the band edge. Thus, the comparison is rather direct. Obviously, the 3D pattern is much more regular, consisting of self-interference rings around individual scatterers, and the corrugation is an order of magnitude less. This is a direct visualization of the fact that 2D systems tend to localize weakly, while 3D systems do not [45]. In 3D the pattern results from scattering processes at individual dopants, while in 2D more complex scattering paths including several dopants lead to the observed pattern. The fact that the scattering paths are closed in 2D leads to the observed strong corrugation and also to weak localization.

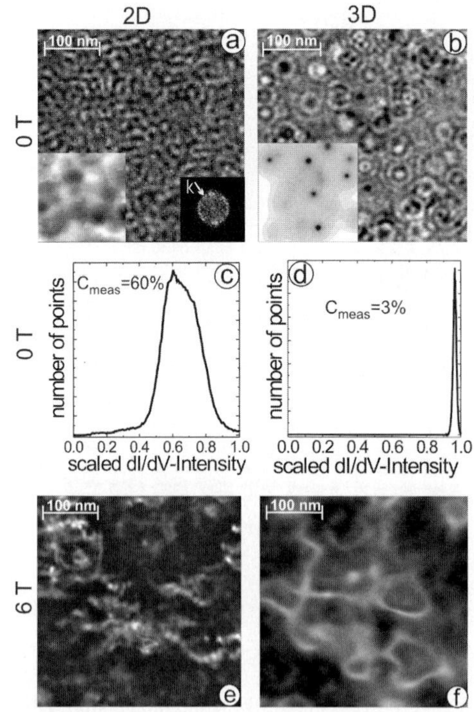

FIGURE 3. (a) Measured LDOS of a 2DES at $B = 0$ T. Insets: (left) measured disorder potential of the 2DES; (right) Fourier transform of the LDOS with dominating k-vector length indicated [43]. (b) LDOS of a 3DES measured at the same kinetic energy as the 2DES at $B = 0$T. Inset: measured disorder potential of the 3DES at the surface [44]; (c,d) histograms of the data of (a) and (b) with indicated values of relative LDOS corrugation (C_{meas}); (e) LDOS of the 2DES at $B = 6$ T [46]; (f) LDOS of the 3DES at $B = 6$ T [48]; $T = 6$ K.

As an important thermodynamic parameter, a magnetic field up to 6 T has been applied to systems from 0D to 3D. Landau and spin quantization have been observed for 2D and 0D systems (the latter only for 0D) [46,47]. More interestingly, the transformation of the wave functions in magnetic field could be observed in real space. While 0D and 1D systems did not show pronounced changes due to the relatively large confinement, a distinct change has been observed for 2D [46] and 3D [48]. In both cases, serpentine structures exhibiting strong corrugation appear as shown in Figs. 3(e,f). The full width of half-maximum of the structures is exactly the magnetic length $l_B = \sqrt{\hbar/eB}$. Moreover, in 2D, the patterns are periodic in energy having a periodicity of the Landau energy. Thus, the theoretically predicted drift states [6] which arise due to the interaction of Lorentz forces and electrostatic disorder have been measured. The states run along equipotential lines of the disorder potential as predicted [6]. The fact that drift states appear also in 3D was surprising and could be linked to the appearance of a Coulomb gap at the Fermi level [49]. Probably, a partial localization of the electrons parallel to the magnetic field arises in the extreme quantum limit prior to magnetic freeze out. This results in local 2D properties of the electrons [50]. Thus, electrons running along equipotential lines of the disorder potential also exist in 3D at sufficiently high magnetic fields.

The results provided by STS demonstrate the possibilities of the technique to image the wave functions directly in the model systems of III-V-semiconductors.

However, it appears necessary to extend this technique to higher-mobility samples, lower temperatures, and higher magnetic fields in order to get access to a larger part of the phase diagram, including phases which are guided by electron–electron interaction.

3 Electrostatic Force Microscopy (EFM)

Since an inhomogeneous distribution of electrons leads to long-range electrostatic forces, such a distribution can also be measured in the far-field. An atomic force microscope can detect electrostatic forces if the tip is charged [22]. The electrostatic force leads to a shift in the resonance frequency of the cantilever, on which the tip is mounted. In case of an inhomogeneous charge distribution within the sample, the frequency shift depends on the position of the tip with respect to the sample and the map of the frequency shift $\Delta f(x,y)$ is roughly proportional to the electrostatic potential distribution of the sample. To increase the signal-to-noise ratio, one can use lock-in technique with respect to the applied voltage between sample and tip.

The technique is called electrostatic force microscopy (EFM) [51]. The spatial resolution of EFM is limited to about 10 nm, since the whole area of the tip is charged and interacts with the sample and since electrostatic forces decay only by $1/r$. The big advantage of EFM is the possibility of investigating samples which are buried below the surface [24]. Thus, a typical high-mobility 2DES, which is buried about 100 nm below the surface, can be studied.

EFM has been applied to a 2DES within a GaAs heterostructure exhibiting an electron mean free path of 6 μm [52–54]. Figure 4(a) shows the experimental setup with the upper part being the usual EFM setup. It measures the frequency shift of the cantilever via lock-in technique. The Hall bar sample contains the 2DES which is marked in black. It is connected via current pads (crossed boxes). The electronics shown in the lower part of Figure 4(a) allows to measure the current induced potential within the 2DES. Therefore, an ac current is guided through the 2DES, which is linked to the measured EFM signal by lock-in technique [52]. A magnetic field is applied in order to detect the current-induced potential profile within the quantum Hall regime. An additional difficulty is to eliminate all electrostatic fields, which are not related to the 2DES. This can be done by performing adequate calibration measurements without the 2DES present. Therefore, a back gate is used to deplete the 2DES [55].

Figure 4(b) shows the local Hall potential as a function of the magnetic field measured across the micro-structured Hall bar. At most magnetic fields, one can see three areas of different potential. Two smaller, black-and-white areas are visible at both rims of the sample and a large, gray area is visible in the center of the Hall bar. Thus, the Hall potential drops in small areas close to the rim between black and gray and gray and white, respectively. Only around certain magnetic fields (e.g., 4 T or 6 T) do these rims move toward the center and the potential drops nearly homogeneously across the whole bar [56]. They are marked by black lines

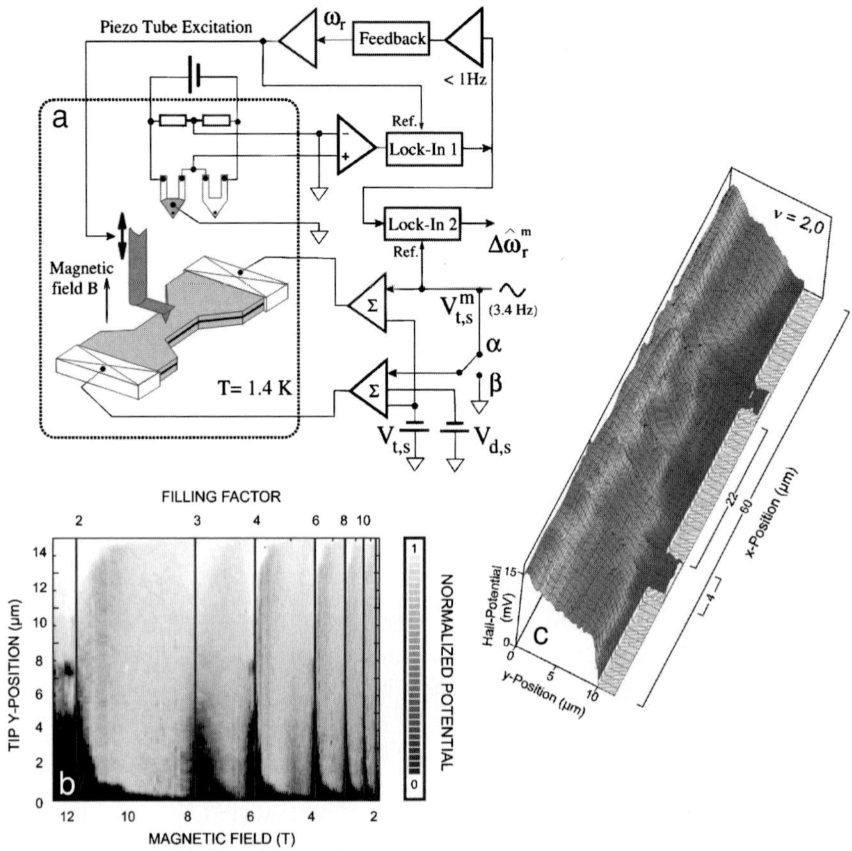

FIGURE 4. (a) Experimental setup of EFM applied to a 2DES in B-field: cantilever with tip is positioned above a sample in Hall bar geometry, which includes a 2DES (black) and two current pads connecting the 2DES. The upper part of the electronic circuit including feedback, and lock-in 1 detects electrostatic forces. The lower part injects an ac current into the 2DES, and lock-in 2 allows us to measure the local potential change induced by this current [55]. (b) Gray-scale plot of the Hall voltage across the Hall bar depicted in (a) at different magnetic fields; corresponding filling factors of the 2DES are marked on the upper axis [56]. (c) Gray-scaled 3D plot of the local Hall potential at filling factor 2.0; the flat area on the right is the area of the contact pads [56] (courtesy of J. Weiss).

in Figure 4(b). These magnetic fields correspond exactly to the fields, where a Landau level is completely filled with electrons (integer filling factor). Consequently, the pattern is periodic in $1/B$, as expected for the periodicity of integer filling factors.

The reason for the measured Hall profile is the well-known fact that a 2DES in magnetic field forms Landau levels. All Landau levels have to be above the Fermi level at the rim of the sample, since the electron density must become zero outside the sample. Thus, each Landau level which is below the Fermi level in the center of the Hall bar has to cross the Fermi level somewhere between the center and the rim. At the crossing point, electrons are available at the Fermi level and can screen the electrostatic potential. Consequently, the potential is constant in a certain area around the crossing point. This leads to so-called alternating incompressible and compressible stripes close to the rim of the sample [57]. Within the compressible stripes the Fermi level is located within a Landau level, while within the incompressible stripes, the Fermi level is located between Landau levels. Importantly, compressible stripes are characterized by a constant potential, while incompressible stripes are characterized by a spatially changing potential.

The large current-induced voltage drop observed in the EFM experiment appears to be in the incompressible stripe closest to the center of the sample. This behavior is attributed to the fact that the inner incompressible stripe has the largest width [58,59] and, thus, can decouple the compressible center of the Hall bar most effectively from the compressible regions of the rim, where the current flows. The other incompressible stripes do not necessarily exist, since the Thomas–Fermi approach of screening is not valid in the resulting relatively small areas [59]. Exactly at integer filling factor the center of the Hall bar itself is incompressible and thus the Hall voltage drops across the whole bar. This picture rationalizes the experimental observations.

Interestingly, the voltage drop at integer filling factor is not linear. The nonlinearity depends critically on the position of the cross section as shown in Figure 4(c), where a 2D map of the Hall potential at an integer filling factor is displayed in a 3D representation. The visible strong potential fluctuations are caused by the potential disorder seen by the 2DES, i.e., a map of the disorder potential appears, if the center of the Hall bar is (largely) incompressible and cannot screen the disorder (completely).

The most interesting point of the EFM measurements is that the Hall potential map can be linked to the current distribution within the 2DES. More exactly, one knows that the excess current runs along the gradients in Hall potential [59]. Consequently, the EFM results show that the excess current in quantum Hall samples mostly runs in the innermost incompressible stripe.

4 Scanning Single-Electron Transistor (SSET)

Another scanning technique which is sensitive to the electrostatic potential of a 2DES is the scanning single-electron transistor (SSET) [23]. A sketch of the experimental setup is shown in Figure 5(a). The single-electron transistor is fabricated on the apex of a glass fiber tip using shadow evaporation techniques. At the apex resides a small Al island (Al), which is connected via tunneling junctions to source (S) and drain (D). The island is small enough to show Coulomb blockade at low temperatures (300 mK) [60]. Thus, current between source and drain can only flow if an energy level of the island resides at the Fermi level. The position of the energy

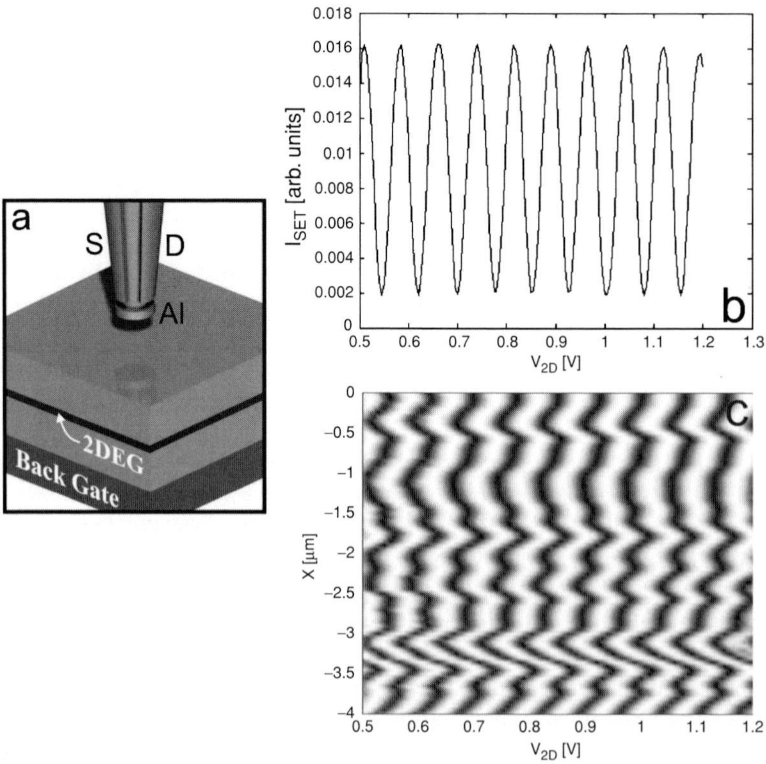

FIGURE 5. (a) Experimental setup of the scanning SET: source, S; drain, D; and Al island, Al, of the SET are marked. The sample contains the 2DES at about 90 nm below the surface and a back gate. (b) SET current as a function of applied voltage to the 2DES at a constant position of the SET. (c) Gray-scale plot of the SET current as a function of position and 2DES voltage (courtesy of A. Yacobi) [60].

levels depends on the electrostatic potential; and, thus, in turn, the electrostatic potential of the sample can be measured by this technique. Applying a voltage to the 2DES one can move several energy levels across the Fermi level, leading to the so called Coulomb blockade oscillations with a periodicity of 80 mV (Figure 5(b)). The potential fluctuations across the sample shift the Coulomb oscillations in voltage. This is visible in Figure 5(c), which is a gray-scale representation of the SET-current as a function of position x and 2D potential V_{2D}. The shift of the coulomb oscillations is a direct measure for the local electrostatic potential within the 2DES. The technique is extremely sensitive, and it has been estimated that electrostatic potential differences caused by 0.01 electron can be detected [60]. However, the lateral resolution is only 100 nm due to the size of the SET and the distance of the 2DES from the probe.

Again different lock-in-techniques can be used in order to measure different sample properties. Firstly, the electrostatic potential map basically caused by donors

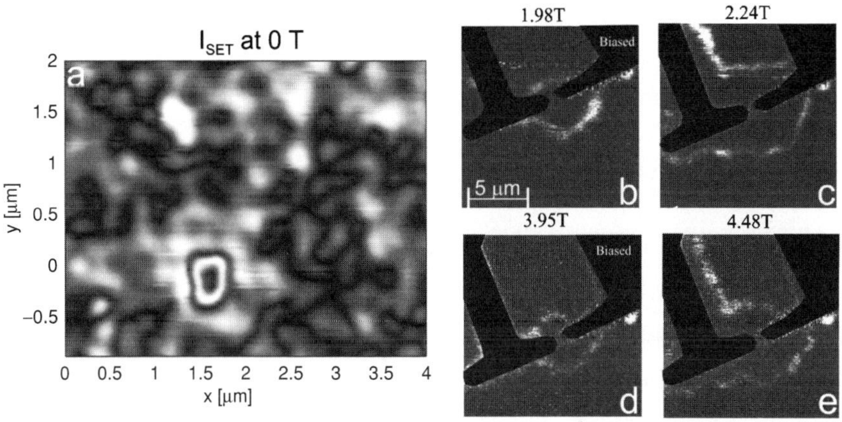

FIGURE 6. (a) SET current as a function of position measured at $B = 0$ T. The dark lines represent equipotential lines of the 2DES [60]; (b–e) transparency of the 2DES measured at different B-fields close to filling factor 2 (b,c) and filling factor 1 (d,e). Black areas are contact pads; transparency is the response of the SET to an applied back gate voltage [61] (courtesy of A. Yacobi).

within the sample can be measured by the technique as shown in Figure 5(c). Secondly, the current-induced potential map can be measured using the same method as described in the previous section, i.e., applying a current to the 2DES and measuring the current induced shift of the Coulomb oscillations. Finally, the local compressibility of the 2DES can be studied by applying an ac voltage to the back gate of the 2DES and measuring the resulting energy shift of the Coulomb oscillations [61].

Figure 6(a) shows the electrostatic potential map of the 2DES at $B = 0$ T [60]. The current of the SET is displayed as a function of position. Thus, equidistant equipotential lines are imaged. The visible potential fluctuations are caused by the donor distribution in the sample. Figure 6(b–e) show the inverse compressibility (transparency) of the 2DES at the rim of the sample [61]. Two voltage pads are visible as black areas of high compressibility. The 2DES area displayed at different magnetic fields exhibits a meandering bright stripe about 500 nm wide. The bright stripe is an incompressible region, separating the rim of the sample from the sample center. The stripe moves inward with increasing field, e.g., from 1.98 T to 2.24 T. Moreover, the stripe patterns are repeatable with filling factor. Thus, Figure 6(b,c) is imaged close to filling factor 2, while Figure 6(d,e) is measured at the doubled B-field, i.e., close to filling factor 1.

Figure 7(a–c) shows an inner part of the sample at a magnetic field close to filling factor one [25]. The area includes about 400 electrons. By changing the back-gate voltage, the density in the 2DES is slightly changed and, thus, the filling factor. Here, the measured compressibility, which is the inverse transparency, is

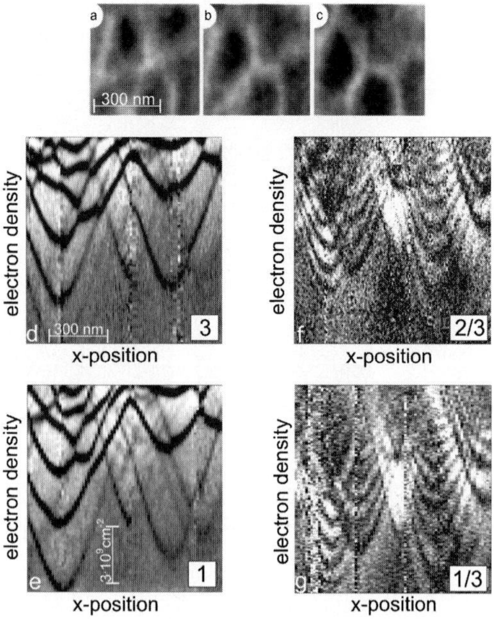

FIGURE 7. (a–c) SET images of the compressibility of a 2DES close to filling factor 1; adjacent images are measured at slightly different back gate voltage corresponding to different densities of 0.16 electrons per image [25]; (d–g) gray-scale plot of transparency as a function of electron density and position close to different filling factors as indicated (3,1,2/3,1/3); $B = 7.25$ T; a small potential is applied to the tip, leading to the curved shape of the charging lines [62] (courtesy of A. Yacoby).

gray scaled. Thus, areas of high compressibility appear bright. The electrons are added in these bright areas. Obviously, the compressible regions form a network which undergoes morphological changes with electron density. Surprisingly, the difference in density between adjacent images corresponds to only 0.16 added electrons within the image area. Additional measurements reveal that the positions of the compressibility stripes are not directly related to the electrostatic potential of the sample.

The reason for these patterns is most likely a small bias of the SET island with respect to the sample. The following scenario can explain the data, which have been recorded as a function of density, magnetic field, and position [62]. If the Fermi level is close to the center of a Landau level, the electrons screen the electrostatic disorder efficiently. Thus, each added electron sees a spatially flat potential and is delocalized. However, if the Fermi level is close to the rim of a Landau level, certain areas of the 2DES exhibit a completely filled Landau level and become incompressible, leaving some compressible droplets in between. Since these droplets are relatively small in size, they exhibit Coulomb blockade. Thus, the electrons can only be filled into these droplets, if the external (backgate) potential matches the charging energy. This leads to compressibility spikes, which are nearly periodic with applied bias. Figures 7(d,e) shows these spikes as dark lines in the voltage-position plane. Here, the back-gate voltage has been converted to electron density. The fact that the lines are not parallel to the x-axis is due to the fact that the potential of tip and sample are not exactly matched. Consequently, a spatially parabolic potential originating from the tip is added to the screened electrostatic potential of the 2DES. Moving the tip changes the effective confinement potential of a certain droplet and, thus, continuously changes the charging voltage of the droplet. This leads to the parabolic shape of the charging lines in Figure 7(d,e), respectively, to the movement of the network lines in Figure 7(a–c).

The interesting point of the SSET measurements is that they clearly show that localization in high-mobility samples is not related to drift states as one expects for non-interacting particles [6], but is simply due to incomplete screening close to integer filling factors. Since this incomplete screening arises at the same electron deficiency of each Landau level, the patterns are exactly reproducible close to each integer filling factor as evidenced by Figure 7(d) (filling factor 3) and Figure 7(e) (filling factor 1). Drift states appear only to be relevant if complete screening of the disorder is impossible as in the larger disorder case discussed in section 2 [46].

The same SSET has been used in the fractional quantum Hall regime, i.e., at filling factors smaller than 1 [63]. Here, the Coulomb interaction between the electrons leads to quasi particle gaps preferring certain fractional filling factors. Interestingly the relevant quasi-particles are fractionally charged. For example, at filling factor 1/3 or 2/3, the quasi-particles have a charge of e/3 due to dressing of the electrons by magnetic flux [64].

The fractional charge, which has earlier been evidenced by shot noise measurements [65,66], can be directly seen in the charging patterns of the compressible droplets close to the gapped fractional filling factors. The patterns of the same sample area in Figure 7(d,e) have been measured in Figure 7(f,g) close to filling factor 1/3 and 2/3 [63]. About three times more charging lines appear; and, indeed, the signal strength of each line is reduced by a factor of three. Thus, the images show, directly, the addition of portions of e/3 to each droplet.

5 Scanning Capacitance Microscopy (SCM)

Another approach to investigate a heterostructure 2DES in a magnetic field uses scanning capacitance microscopy [67]. A sketch of the experiment is shown in Figure 8(a) [26,27]. A metallic tip is coupled to a sensitive charge detector (Q_{ac}) and an ac voltage of 100 kHz and 1–4 mV is applied at the 2DES contact. The locally induced charge into the sample is capacitively coupled to the tip and is, thus, measured. Typically, one records in-phase and out-of-phase signal of the charge detector [26]. The in-phase signal gets low if an incompressible stripe separates the measured 2DES region from the contact.

A simple model assumes a capacitive coupling between the tip and the particular 2DES region below the tip and a resistive coupling between this 2DES region and the 2DES contact [68]. They are connected in series, resulting in a roll-off frequency of the circuit. This roll-off frequency changes from point to point. Only if the applied 100 kHz are smaller than the roll-off frequency at a particular point, can one measure a signal at the charge detector. If the roll-off frequency is exactly identical to the excitation frequency, the out-of-phase signal measured at the charge detector becomes large. Since the resistivity increases dramatically at the position

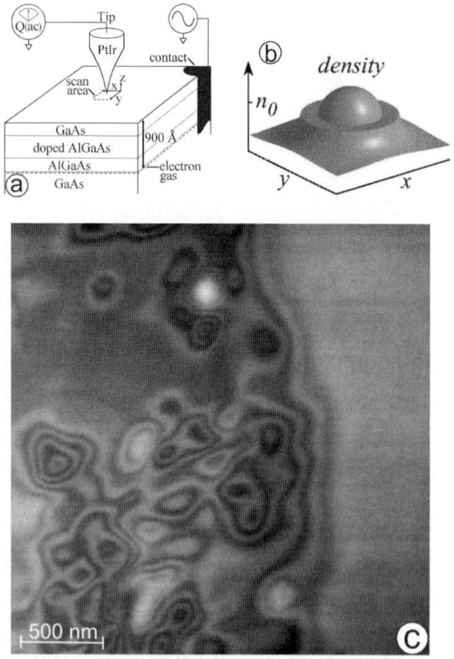

FIGURE 8. (a) Sketch of the setup for SCM measurements of a 2DES with the charge detector Q(ac) indicated; (b) sketch of the electron density within the electron bubble induced into thc 2DES by the applied tip voltage V_{tip}; the ring of constant density marks the incompressible stripe separating the bubble from the rest of the 2DES; (c) in-phase image of the local compressibility; the bright stripes within the image mark the positions, where an electron is added to the bubble and correspond to equipotential lines within the 2DES; the right part without lines is a gate electrode; $B = 5.8$ T, $T = 300$ mK, $V_{\text{tip}} = 1$ V, [27] (courtesy of R. Ashoori).

of an incompressible stripe, the out-of-phase signal measures the position of the outermost incompressible stripe having sufficient resisitivity to drive the roll-off frequency below 100 kHz [69,70]. Note that the excitation frequencies in STS, EFM and SSET are several orders of magnitude lower than in SCM, thus, roll-off is only important in SCM measurements.

Nevertheless, SCM can also measure incompressible stripes at the rim of a 2DES [69] or, close to integer filling factors, in the center of a 2DES [70]. With a reasonable assumption on the capacitance, the resistance of this stripe can be estimated by directly measuring the frequency, where the stripe starts to separate the sample center from the rim [69]. Stripe resistances up to a GΩ/μm have been found.

In order to measure the charging of the unperturbed 2DES, one has to compensate for work function differences between sample and tip. However, one can also use a potential difference between sample and tip to measure the screened electrostatic potential within the 2DES [27]. If the tip potential is attractive, a small electron bubble with diameter of about 100 nm is formed below the tip. This bubble can be moved through the potential landscape of the sample, and each time an electron moves in or out of the bubble a signal will be detected by the charge sensor. If the 2DES potential is relatively smooth on the 100-nm length scale, the charging always appears at the same value of the local potential. Thus, charging lines are equipotential lines. The technique can be used in the magnetic field, if the filling factor is slightly lower than integer. Then, the bubble has a larger filling factor than an integer and is separated from the unperturbed 2DES by an incompressible stripe as sketched in Figure 8(b). This ensures charge quantization within the bubble very similar to the charge quantization seen within the SSET measurements of section 4. Thus, one can measure the screened disorder potential of a high-mobility 2DES close to integer filling factors. Figure 8(c) shows an example of such a map measured close to filling factor 1 at the rim of a 2D sample [27]. The bright and dark lines are charging and decharging lines, respectively, of the moving bubble and, thus, represent equipotential lines. One can quantify the data by changing the bias applied to the tip. Potential fluctuations of about 2 meV are found on a length scale of 0.5 μm for a sample mobility of 150 m^2/Vs. These values are in agreement with the SSET measurements.

6 Scanning Gate Microscopy (SGM)

Scanning probes can act as a perturbation to the current flow within a sample [28]. As described above, a potential difference between tip and sample leads to a local band bending within the sample. As a consequence, either an electron bubble (accumulation layer) can be formed below the tip or a depletion region is formed, leading to back-scattering of electrons at the rim of the tip induced depletion region. If the tip is placed at a position where current is flowing within the sample, it reduces the total current due to back-scattering. If the tip is placed at a region without current flow, the total current is unchanged. Thus, by measuring the resistance of the sample as a function of the position of the negatively biased tip, one can detect the pattern of the current flow. The technique is called scanning gate microscopy (SGM). The most spectacular SGM experiment has been performed on a quantum point contact as shown in Figures 9 and 10 [29,76]. The conductance of an ideal quantum point contact is known to be quantized in multiples of $2e^2/h$ [3,77]. Such quantum point contacts can be built within a 2DES by using two gates to deplete the 2DES locally. Conductance plateaus at multiples of $2e^2/h$ are indeed observed, if the conductance is measured as a function of the voltage applied to the two gates [78]. A sketch of the quantum point contact used in the SGM experiment is shown in Figure 9(a) [71]. The 2DES is buried 57 nm below the surface, and the two gates forming the quantum point contact are indicated. The current flow pattern is sketched schematically on the surface and the tip is visible above the surface. The current through the point contact is measured as a function of the lateral position of the repelling tip. If one places the tip as the third gate in between the other two gates, one blocks the channel and the current goes down to zero [28]. Thus, the current obviously runs through the constriction. More interestingly, the outer parts of Figure 9(b–d) shows the current flow pattern further away from the constriction at conductances $1 \cdot 2e^2/h$, $2 \cdot 2e^2/h$ and $3 \cdot 2e^2/h$, respectively. Obviously, one, two, and three branches are observed at the first, second, and third conductance plateau [29]. This result makes perfect sense, since the conductance plateaus are related to the first, second, and third quantized state within the constriction area having zero, one, and two nodes. These

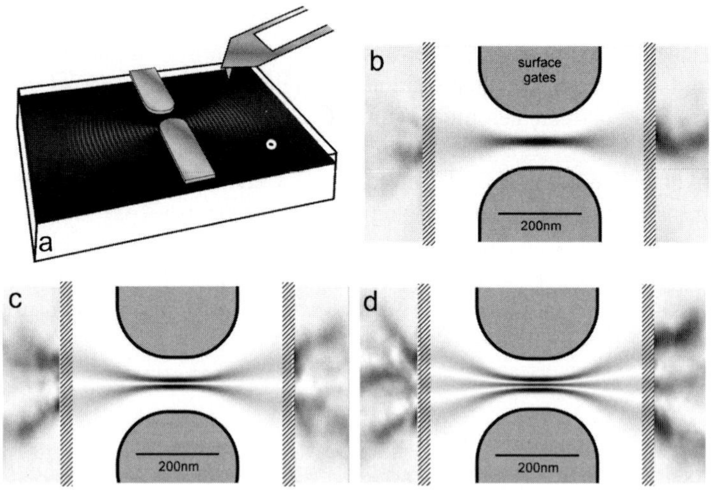

FIGURE 9. (a) Sketch of SGM experiment using a quantum point contact as the sample; the two lateral gates confine the 2DES located 57 nm below the surface; the expected current flow pattern of an ideal point contact is indicated on the sample surface; (b–d) inner part—sketch of the two gates with inverted gray-scale plot of the calculated $|\Psi|^2$ of the corresponding first (b), second (c), and third (d) wave function within the constriction; outer part—gray-scale plot of the point contact conductance as a function of the position of the tip charged by −3 V. The gates are tuned to give a point contact conductance of $1 \cdot e^2/h$ in (b), $2 \cdot e^2/h$ in (c), and $3 \cdot e^2/h$ in (d) in the absence of the tip; the contrast in (b–d) is about 0.6 e^2/h [29] (courtesy of M. Topinka).

states have been calculated and are displayed in the inner part of Figure 9(b–d). They indeed exhibit one, two, and three maxima of $|\Psi|^2$ perpendicular to the current flow. The angular spread is simply due to the shape of the electrostatic potential in the area of the constriction. Thus, branching close to the point contact simply reflects the symmetry of the current carrying wave function. However, if one goes further apart from the constriction, the pattern of the first conductance plateau becomes more complicated, consisting of a number of branches as shown in Figure 10(a) [76]. The visible fringe pattern is due to self-interference of the electrons. It has also been observed closer to the constriction, but has been filtered out in Figure 9(b–d) in order to improve the visibility of the different branches. The appearance of the fringes indicates coherent electron flow up to distances of several μm. The complex branching away from the constriction could be modelled by considering the disorder potential of the sample, which has a strength of 1.3 meV and a correlation length of 25 nm. Figure 10(b) shows a classical simulation of electron flow within an adequate disorder potential, which indeed reproduces the data qualitatively. The simple classical focusing of electrons within a potential valley together with the irregular structure of valleys explains the complex pattern except the fringes. Thus, the branching itself is a pure classical disorder effect. However, the fringes are quantum mechanical interference effects. Interestingly,

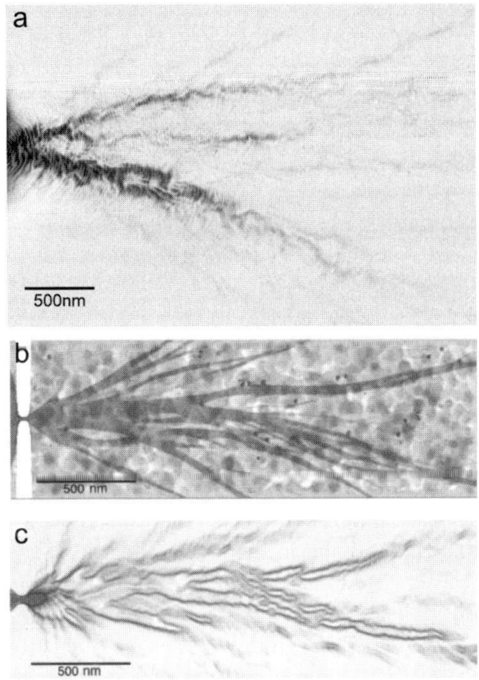

FIGURE 10. (a) SGM conductance map of the point contact of Figure 9(a) at $1 \cdot e^2/h$; same conditions as Figure 9(b), but larger scale; the point contact is on the left of the image; contrast: about 0.25 e^2/h. (b) Calculated classical flux through a point contact; the gray-scale image in the background shows the assumed disorder potential resulting from the donors located 22 nm below the 2DES; the gray, transparent pattern shows the flux distribution of an ensemble of classical trajectories. (c) Flux density of a quantum mechanical wave function in the disorder potential of (b) [76] (courtesy of M. Topinka and E. Heller).

fringes are observed far away from the constriction at distances much larger than the thermal length. If the interference results from the scattering of the electrons between the tip and the constriction only, this would be impossible. But, of course, self-interference appears also due to scattering between the tip and scatterers in its immediate environment. Figure 10(c) shows the calculated pattern of quantum mechanical flux through the point contact including disorder. It indeed shows some fringes up to large distances.

The SGM technique can be applied to many other structures [79]; e.g., it has been used to detect the weak links between edge channels in a quantum Hall device [72], it has been used to follow the electron path in a focusing experiment of a 2DES in small magnetic fields [73], and it has been used to manipulate charge levels of a quantum dot [74,75]. An interesting experiment has been performed on a large and slightly irregular quantum dot coupled to two leads [80]. The quantum dot has been fabricated by the tip in contact with the sample surface using voltage pulses in order to charge the surface locally [81]. The local charge acts as a barrier within the 2DES, confining the electrons to the quantum dot region marked in Figure 11(a). The size of the quantum dot was 2×3 μm^2, and the phase coherence length of the sample was 24 μm. Consequently, several reflections at the boundaries could occur coherently. Due to the curvature and irregularities of the boundary, the dot is in the chaotic regime. Indeed, the analysis of the G (conductance) versus B curves reveals a fractal dimension of 1.44. However, it is known for chaotic quantum dots

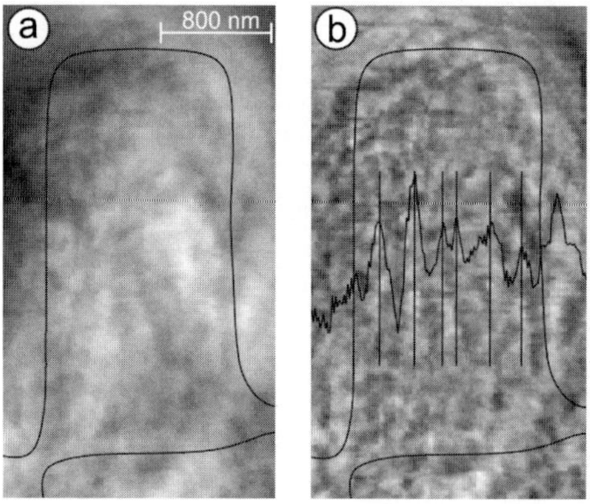

FIGURE 11. (a) SGM image of an open, chaotic quantum dot at $B = 39.1$ mT and a conductance of about 40 μS. The shape of the electrostatically defined quantum dot is marked. (b) The same data as (a) after high-pass filtering; the black line across the image is a cross section of measured conductance values spatially averaged along the dot; the vertical lines mark the parallel areas of high intensity presumably belonging to the so-called scarred wave function [80] (courtesy of R. Crook).

that they exhibit striking regularities as for example Aharonov–Bohm oscillations. These regularities are usually attributed to so-called scarred wave functions being relatively regular in shape [80].

The SGM images of the dot exhibit fluctuations on all length scales, as shown in Figure 11(a). However, if a high-pass is used to filter the images (Figure 11(b)), a certain structure exhibiting lines parallel to the rim of the dot appears. The authors believe that this pattern corresponds to a scarred wave function.

The big advantage of SGM is that one modifies exactly the same current that one usually measures in transport of mesoscopic devices [2,3]. This gives a very direct link to existing theory as well as to applications.

7 Scanning Near-Field Optical Microscope (SNOM)

A very popular technique is the scanning near-field microscope (SNOM), which can be used to perform optical experiments with sub-wavelength resolution [30]. Mostly, one uses a coated glass fiber with a small aperture at the end. The aperture is brought close enough to the sample to detect the near-field optical signal. Consequently, spatial resolution down to about 30 nm [82,83] can be combined with high energy resolution in the μeV-range [31] and temporal resolution in the 100-fs range [33].

The technique has been largely used to investigate single quantum dots in detail [31, 84,86], however reports on spatial structures within the quantum dot are rare [87]. Other measurements investigating quantum wires could map the confinement potential [88], could distinguish between localized and delocalized excitons [34,89] and could give evidence for level repulsion of spatially overlapping localized states [90]. The use of femtosecond-time resolution in these experiments provided insight into the diffusion dynamics from a surrounding quantum well into the wire [91,92] and the relaxation of excitons within the wire [33]. Measurements on quantum wells have been used to identify electron density fluctuations on larger length scales, which partly show long-range spatial correlations probably caused by corrugation patterns [93,94]. Evidence for electron localization has also been found by studying the spatial distribution of the charged excitons [32,95]. The resulting level repulsion has been detected by analyzing two-energy autocorrelation functions of the local exciton peaks [96,97]. Time resolution of individual local excitons clearly exhibits different decay times of different localized excitons [98]. The dynamics of carrier trapping around defects and local Schottky-barriers has also been investigated [99,100].

These measurements show the large potential of the SNOM for the investigation of disordered III-V semiconductors and might be used in the future for the detection of phase properties of these materials by optical means on the local scale, especially if they are combined with magnetic field [101].

8 Conclusion

The different scanning probe techniques applied to III-V semiconductors provide a great deal of information on different length scales. Consequently, SPMs are a good complementary approach with respect to macroscopic techniques. While it is easier to go to extreme areas of the phase diagram (lower temperature, higher magnetic field) with the latter in order to find new electron phases, only the spatial resolution of the scanning probes can give detailed insight into the electron arrangement within a particular phase, which partly leads to rather unexpected results.

As a good example, the localization behavior within the quantum Hall regime detected by SSET clearly showed that the partially neglected screening is of crucial importance in high-mobility samples. Also, the disorder-induced branching of the current flow behind quantum point contacts was never considered, before it was measured by SGM. Moreover, the current flow in quantum Hall systems revealed by EFM to be concentrated in the innermost incompressible stripe is contradictory to the often used picture of current flow in all compressible stripes. Finally, the existence of drift states in disordered 3D systems could only be deduced from directly imaging them by STS.

The ongoing development of scanning probe techniques toward higher resolution, lower temperature, higher magnetic fields, higher-mobility samples, or higher complexity of the investigated structures will lead to additional surprises in the understanding of the complex behavior of electrons in the paradigmatic III-V samples.

Acknowledgments. The images from J. Weiss, A. Yacobi, R. Ashoori, M. Topinka, E. Heller, and R. Crook are greatly appreciated. Moreover, I would like to acknowledge gratefully that the measurements described in section 2 were mostly performed by T. Maltezopolous, C. Meyer, J. Klijn, D. Haude, C. Wittneven, and R. Dombrowski.

References

1. K. v. Klitzing, G. Dorda, and M. Pepper, Phys. Rev. Lett. **45** (1980) 494.
2. T. Ando, A.B. Fowler, and F. Stern, Rev. Mod. Phys **54** (1982) 437.
3. C. W. J. Beenakker, Rev. Mod. Phys. **69** (1997) 731.
4. E. P. Wigner, Phys. Rev. **46** (1934) 1002.
5. T. Jungwirth and A. H. MacDonald, Phys. Rev. Lett. **87** (2001) 216801.
6. R. Joynt and R. E. Prange, Phys. Rev. B **29**, (1984) 3303.
7. A. H. MacDonald and M. P. A. Fisher, Phys. Rev. B **61** (2000) 5724.
8. B. Kramer and A. MacKinnon, Rep. Prog. Phys. **56** (1993) 1469.
9. S. V. Kravchenko, G. V. Kravchenko, J. E. Furneaux, V. M. Pudalov, and M. D'Iorio, Phys. Rev. B **50** (1994) 8039.
10. D. C. Tsui, H. L. Störmer, and A. C. Gossard, Phys. Rev. Lett. **48** (1982) 1559.
11. J. M. Luttinger, J. Math. Phys. **4** (1963) 1154.
12. J. H. Smet, R. A. Deutschmann, W. Wegscheider, G. Abstreiter, and K. von Klitzing, Phys. Rev. Lett. **86** (2001) 2412.
13. V. W. Scarola, K. Park, and J. K. Jain, Nature **406** (2000) 863.
14. R. Willett, J. P. Eisenstein, H. L. Störmer, D. C. Tsui, A. C. Gossard, and J. H. English, Phys. Rev. Lett. **59** (1987) 1776.
15. S. Kagoshima and H. Nagasawa: One-dimensional conductors (Springer, Berlin 1988).
16. M. Tinkham: Introduction to superconductivity (McGraw-Hill, New York 1975).
17. N. Tsuda, K. Nasu, A. Yanase, and K. Siratori: Electronic conduction in oxides (Springer, Berlin 1983).
18. G. Salis, Y. Kato, K. Ensslin, D. C. Driscoll, A. C. Gossard, and D. D. Awschalom, Nature **414** (2001) 619.
19. L. Pfeiffer, K. W. West, H. L. Störmer, and K. W. Baldwin, Appl. Phys. Lett. **55** (1989) 1888.
20. R. Wiesendanger: Scanning Probe Microscopy and Spectroscopy (Cambridge University Press, Cambridge 1994)
21. M. Morgenstern, Surf. Rev. Lett. **10** (2003) 933.
22. M. Nonnenmacher, M. P. O'Boyle, and H. K. Wickramashinge, Appl. Phys. Lett. **58** (1991) 2921.
23. M. J. Yoo, T. A. Fulton, H. F. Hess, R. L. Willet, L. N. Dunkelberger, R. J. Chichester, L. N. Pfeiffer, and K. W. West, Science **276** (1997), 579.
24. K. L. McCormick, M. T. Woolside, M. Huang, M. Wu, and P. L. McEuen, Phys. Rev. B **59** (1999) 4654.
25. N. B. Zhitenev, T. A. Fulton, A. Yacoby, H. F. Hess, L. N. Pfeiffer, and K. W. West, Nature **404** (2000) 473.
26. S. H. Tessmer, P. I. Glicofridis, R. C. Ashoori, L. S. Levitov, and M. R. Melloch, Nature **392** (1998) 51.
27. G. Finkelstein, P. I. Glicofridis, R. C. Ashori, and M. Shayegan, Science **289** (2000) 90.
28. M. A. Eriksson, R. G. Beck, M. Topinka, J. A. Katine, R. M. Westervelt, K. L. Campman, and A. C. Gossard, Appl. Phys. Lett **69** (1996) 671.
29. M. A. Topinka, B. J. LeRoy, S. E. J. Shaw, E. J. Heller, R. M. Westervelt, K. D. Maranowski, and A. C. Gossard, Science **289** (2000) 2323.
30. D. W. Pohl, W. Denk, and M. Lanz, Appl. Phys. Lett **44** (1984) 651.
31. R. J. Guest, T. H. Stievater, G. Chen, E. A. Tabak, B. G. Orr, D. G. Steel, D. Gammon, and D. S. Katzer, Science **293** (2001) 2224.

32. G. Eytan, Y. Yayon, M. Rappaport, H. Shtrikman, and I. Bar-Joseph, Phys. Rev. Lett. **81** (1998) 1666.
33. T. Guenther, V. Emiliani, F. Intonti, C. Lienau, T. Elsaesser, R. Nötzel, and H. Ploog, Appl. Phys. Lett. **75** (1999) 3500.
34. M. Achermann, B. A. Nechay, U. Siegner, A. Hartmann, D. Oberli, E. Kapon, and U. Keller, Appl. Phys. Lett. **76** (2000) 2695.
35. B. Grandidier, Y. M. Niquet, B. Legrand, J. P. Nys, C. Priester, D. Stiévenard, J. M. Gérard, and V. Thierry-Mieg, Phys. Rev. Lett. **85** (2000) 1068.
36. T. Maltezopoulos, A. Bolz, C. Meyer, C. Heyn, W. Hansen, M. Morgenstern, and R. Wiesendanger, Phys. Rev. Lett. **91** (2003) 196804.
37. O. Millo, D. Katz, Y. Cao, and U. Banin, Phys. Rev. Lett. **86** (2001) 5751.
38. K. Kanisawa, M. J. Butcher, Y. Tokura, H. Yamaguchi, and Y. Hirayama, Phys. Rev. Lett. **87** (2001) 196804.
39. Chr. Meyer, J. Klijn, M. Morgenstern, and R. Wiesendanger, Phys. Rev. Lett. **91**, (2003) 076803.
40. K. Kanisawa, M. J. Butcher, H. Yamaguchi, and Y. Hirayama, Phys. Rev. Lett. **86** (2001) 3384.
41. V. Yu Aristov, G. LeLay, P. Soukiassian. K. Hricovini, J. E. Bonnet, J. Osvald, and O. Olsson, Europhys. Lett. **26** (1994) 359.
42. M. Morgenstern, Physik Journal **8** (2004) 83.
43. M. Morgenstern, J. Klijn, Chr. Meyer, M. Getzlaff, R. Adelung, K. Rossnagel, L. Kipp, M. Skibowski, and R. Wiesendanger, Phys. Rev. Lett. **89** (2002) 136806.
44. Chr. Wittneven, M. Morgenstern, R. Dombrowski, and R. Wiesendanger, Phys. Rev. Lett. **81** (1998) 5616.
45. E. Abrahams, P. W. Anderson, D. C. Licciardello, and T. V. Ramakrishnan, Phys. Rev. Lett. **42** (1979) 673.
46. M. Morgenstern, J. Klijn, Chr. Meyer, and R. Wiesendanger, Phys. Rev. Lett. **90** (2003) 056804.
47. M. Morgenstern, V. Gudmundsson, R. Dombrowski, Chr. Wittneven, and R. Wiesendanger, Phys. Rev. B **63** (2001) 201301.
48. D. Haude, M. Morgenstern, I. Meinel, and R. Wiesendanger, Phys. Rev. Lett. **86** (2001) 1582.
49. M. Morgenstern, D. Haude, J. Klijn, and R. Wiesendanger, Phys. Rev. B **66** (2002) 121102.
50. M. Morgenstern, D. Haude, Chr. Meyer, and R. Wiesendanger, Phys. Rev. B **64** (2001) 205104.
51. Y. Martin, D. W. Abraham, and H. K. Wickramasinghe, Appl. Phys. Lett. **52** (1988) 1103.
52. P. Weitz, E. Ahlswede, J. Weis, K. von Klitzing, and K. Eberl, Physica E **6** (2000) 247.
53. E. Ahlswede, P. Weitz, J. Weis, K. von Klitzing, and K. Eberl, Physica B **298** (2001) 562.
54. E. Ahlswede, J. Weis, K. v. Klitzing, and K. Eberl, Physica E **12** (2002) 165.
55. P. Weitz, E. Ahlswede, J. Weis, K. v. Klitzing, and K. Eberl, Appl. Surf. Sci. **157** (2000) 349.
56. K. Güven and R. R. Gerhardts, Phys. Rev. B **67** (2003) 115327.
57. D. B. Chklovskii, B. I. Shklovskii, and L. I. Glazman, Phys. Rev. B **46** (1992) 4026.
58. A. Siddiki and R. R. Gerhardts, Phys. Rev. B **68** (2003) 125315.
59. A. Siddiki and R. R. Gerhardts, Phys. Rev. B **70** (2004) 195335.
60. H. F. Hess, T. A. Fulton, M. J. Yoo, and A. Yacoby, Sol. St. Com. **107** (1998) 657.

61. A. Yacoby, H. F. Hess, T. A. Fulton, L. N. Pfeiffer, and K. W. West, Sol. St. Com. **111** (1999) 1.
62. S. Ilani, J. Martin, E. Teitelbaum, J. H. Smet, D. Mahalu, V. Umansky, and A. Yacobi, Nature **427** (2004) 328.
63. J. Martin, S. Ilani, B. Verdene, J. H. Smet, V. Umansky, D. Mahalu, D. Schuh, G. Abstreiter, and A. Yacoby, Science **305** (2004) 980.
64. R. E. Prange and S. M. Girvin: The quantum Hall effect (Springer, New York 1987).
65. R. de Picciotto, M. Reznikov, M. Heiblum, V. Umansky, G. Bunin, and D. Mahalu, Nature **389** (1997) 162.
66. M. Reznikov, R. de Picciotto, T. G. Griffiths, M. Heiblum, and V. Umansky, Nature **399** (1999) 238.
67. C. C. Williams, W. P. Hough, and A. Rishton, Appl. Phys. Lett. **55** (1989) 203.
68. S. H. Tessmer, G. Finkelstein, P. I. Glicofridis, and R. C. Ashoori, Phys. Rev. B **66** (2002) 125308.
69. P. I. Glicofridis, G. Finkelstein, R. C. Ashoori, and M. Shayegan, Phys. Rev. B **65** (2002) 121312.
70. G. Finkelstein, P.I. Glicofridis, S. H. Tessmer, R. C. Ashoori, and M. R. Melloch, Phys. Rev. B **61** (2000) R16323.
71. M. A. Topinka, R. M. Westervelt, and E. J. Heller, Physics Today (Dec 2003) 47.
72. M. T. Woodside, C. Vale, P. L. McEuen, C. Kadow, K. D. Maranowski, and A. C. Gossard, Phys. Rev. B **64** (2001) 041310.
73. R. Crook, C. G. Smith, M. Y. Simmons, and D. A. Ritchie, Phys. Rev. B **62** (2000) 5174.
74. R. Crook, C. G. Smith, W. R. Tribe, S. J. O'Shea, M. Y. Simmons, and D. A. Ritchie, Phys. Rev. B **66** (2002) 121301.
75. A. Pioda, S. Kicin, T. Ihn, M. Sigrist, A. Fuhrer, K. Ensslin, A. Weichselbaum, S. E. Ulloa, M. Reinwald, and W. Wegscheider, Phys. Rev. Lett. **93** (2004) 216801.
76. M. A. Topinka, B. J. LeRoy, R. M. Westervelt, S. E. J. Shaw, R. Fleischmann, E. J. Heller, K. D. Maranowski, and A. C. Gossard, Nature **410** (2001) 183.
77. E. Scheer, N. Agraït, J. C. Cuevas, A. Levy Yeyati, B. Ludoph, A. Martín-Rodero, G. Rubio Bollinger, J. M. van Ruitenbeek, and C. Urbina, Nature **394** (1998)154.
78. B. J. van Wees, H. van Houten, C. W. J. Beenakker, and J. G. Williamson, Phys. Rev. Lett. **60** (1988) 848.
79. M. T. Woodside and P. L. McEuen, Science **296** (2002), 1098.
80. R. Crook, C. G. Smith, A. C. Graham, I. Farrer, H. E. Beere, and D. A. Ritchie, Phys. Rev. Lett. **91** (2003) 246803.
81. R. Crook, A. C. Graham, C. G. Smith, I. Farrer, H. E. Beere, and D. A. Ritchie, Nature **424** (2003) 751.
82. M. N. Islam, X. K. Zhao, A. A. Said, S. S. Mickel, and C. F. Vail, Appl. Phys. Lett. **71** (1997) 2886.
83. K. Matsuda, T. Saiki, S. Nomura, M. Mihara, and Y. Aoyagi, Appl. Phys. Lett. **81** (2002) 2291.
84. L. Landin, M. S. Miller, M. E. Pistol, C. E. Pryor, and L. Samuelson, Science **280** (1998) 262.
85. Y. J. Yu, W. Jhe, and Y. Arakawa, Appl. Phys. Lett. **83** (2003) 3024.
86. A. Chevez-Pirson, Y. Temmyo, H. Kamada, H. Gotoh, and H. Ando, Appl. Phys. Lett. **72** (1998) 3494.
87. K. Matsuda, T. Saiki, S. Nomura, M. Mihara, Y. Aoyagi, S. Nair, and T. Takagahara, Phys. Rev. Lett. **91** (2003) 177401.

88. Ch. Lienau, A. Richter, G. Behme, M. Süptitz, D. Heinrich, T. Elsaesser, M. Ramsteiner, R. Nötzel, and K. H. Ploog, Phys. Rev. B **58** (1998) 2045.
89. V. Emiliani, F. Intonti, C. Lienau, T. Elsaesser, R. Nötzel, and K. H. Ploog, Phys. Rev. B **63** (2001) 075313.
90. A. Feltrin, R. Idrissi Kaitouni, A. Crottini, M.-A. Dupertuis, J. L. Staehli, B. Deveaud, V. Savona, X. L. Wang, and M. Ogura, Phys. Rev. B **69** (2004) 205321.
91. A. Richter, G. Behme, M. Süptitz, Ch. Lienau, T. Elsaesser, M. Ramsteiner, R. Nötzel, and K. H. Ploog, Phys. Rev. Lett. **79** (1997) 2145.
92. A. Richter, M. Süptitz, D. Heinrich, Ch. Lienau, T. Elsaesser, M. Ramsteiner, R. Nötzel, and K. H. Ploog, Appl. Phys. Lett. **73** (1998) 2176.
93. Y. Yayon, A. Esser, M. Rappaport, V. Umansky, H. Shtrikman, and I. Bar-Joseph, Phys. Rev. Lett. **89** (2002) 157402.
94. Y. Yayon, M. Rappaport, V. Umansky, and I. Bar-Joseph, Phys. Rev. B **66** (2002) 033310.
95. Y. Yayon, M. Rappaport, V. Umansky, and I. Bar-Joseph, Phys. Rev. B **64** (2001) 081308.
96. F. Intonti, V. Emiliani, Ch. Lienau, T. Elsaesser, V. Savona, E. Runge, R. Zimmermann, R. Nötzel, and K. H. Ploog, Phys. Rev. Lett. **87** (2001) 076801.
97. Ch. Lienau, F. Intonti, T. Guenther, T. Elsaesser, V. Savona, R. Zimmermann, and E. Runge, Phys. Rev. B **69** (2004) 085302.
98. U. Neuberth, G. Walter, G. v. Freymann, B. Dal Don, H. Kalt, M. Wegener, G. Khitrova, and H. M. Gibbs, Appl. Phys. Lett. **80** (2002) 3340.
99. M. Achermann, B. A. Nechay, F. Morier-Genoud, A. Schertel, U. Siegner, and U. Keller, Phys. Rev. B **60** (1999) 2101.
100. M. Achermann, F. Morier-Genoud, W. Seifert, L.-E. Wernersson, U. Siegner, and U. Keller, Phys. Rev. B **65** (2002) 045322.
101. T. Tokizaki, T. Onuki, T. Tsuchya, and H. Yokoyama, J. of Micr. **210** (2003) 315.

II.2
Spin-Polarized Scanning Tunneling Microscopy

WULF WULFHEKEL, UTA SCHLICKUM, AND JÜRGEN KIRSCHNER

We present an overview of spin-polarized scanning tunneling microscopy (Sp-STM). As in STM, the electron density near the sample surface can be imaged. In addition, Sp-STM allows us to map the spin polarization. Thus, information on the magnetic configuration of the sample surface can be gathered. Three imaging modes are currently being used: the constant-current mode, the spectroscopic mode, and the differential magnetic mode. The principles of the three modes are explained, and their advantages and limitations are discussed in the framework of imaging ferromagnetic and antiferromagnetic surfaces of bulk materials and thin film systems. Further, two approaches for controlling the spin direction of the tip apex, i.e., the sensitive spin component, are discussed. Surface or interface magnetic anisotropies at the tip apex may be used to align the axis of sensitivity or alternatively, the shape anisotropy of the whole tip may determine the spin direction. Finally, it is demonstrated that Sp-STM can be used beyond magnetic imaging. Valuable information on the spin resolved electronic structure or on the fundamental processes of spin-polarized tunneling may be obtained.

1 Introduction

In scanning tunneling microscopy (STM), the electron charge is used to carry information in the imaging process. The small tunneling current between the tip and the conductive surface is used as a feedback parameter to move the tip. During scanning in the constant current mode, the apex of the tip is held on equi-current lines several Å above the sample surface by the current feedback mechanism. In first approximation, i.e., in the Tersoff–Hamann model [1,2], these lines correspond to lines of constant charge density of the sample surface probed by the tip apex. A plot of the z-coordinate, i.e., the coordinate perpendicular to the overall sample surface, as a function of the position x and y in the surface plane is therefore called a topographic STM image and reflects the spatial distribution of the density of states of the electrons [3]. In the case of a sharp apex, individual atoms can be resolved in topographic images [4]. In the above picture, the spin of the electron has been neglected. For most materials, this simplification is well justified as the electron density does not depend on the spin of the electron. For ferromagnetic or antiferromagnetic materials, however, the density of states is spin-split into majority and minority states. A net spin polarization is present in the atoms and the individual atoms carry a magnetic moment. In spin-polarized STM (Sp-STM) the tip itself is a source of spin-polarized electrons. Information on the spin polarization of the sample surface may be obtained via the spin-dependent tunneling process between tip and sample. In the case that the spin-dependent part of the tunneling current can be separated from the spin-independent part, it is possible to obtain information on the spin- or magnetic configuration of a sample surface with the same lateral resolution as topographic information, i.e., with atomic resolution. Thus, Sp-STM is the magnetic imaging technique with ultimate lateral resolution, which for the first time allows the study of the magnetic configuration of antiferromagnets in real space. Examples for imaging both ferromagnetic and antiferromagnetic surfaces are given, after the principles of spin-polarized tunneling are introduced in detail. Three different approaches to separate spin information from the tunneling current are discussed, and examples are given for all three imaging modes. The potential of the different techniques is illustrated and their advantages and disadvantages are discussed. Further, we focus on tip preparation and finally show that Sp-STM can be used beyond magnetic imaging to learn more about spin-split density of states and the mechanism of spin-polarized tunneling, per se.

1.1 Spin-Polarized Tunneling

The principle of operation of Sp-STM is based on the fundamental property of ferromagnets and antiferromagnets—that their magnetic moment is related to an imbalance in occupation of electrons of different spins. Due to the spin-sensitive exchange interaction, the density of states splits up into minority and majority densities (see Figure 1(a)). The imbalance causes a spin polarization. This is in contrast to paramagnetic substances, where the distributions of spin-up and spin-down electrons are identical and no spin polarization is present. The splitting

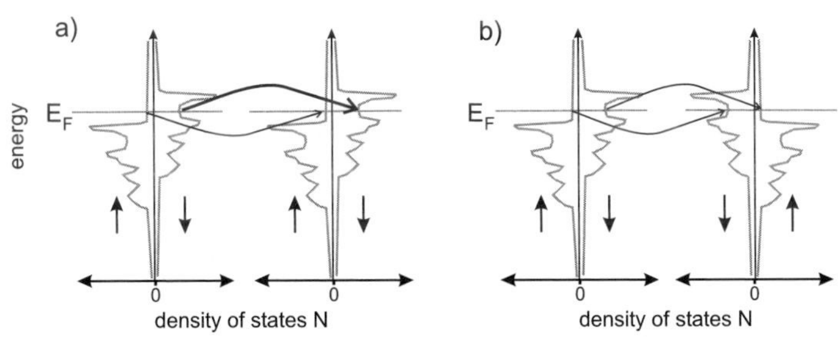

FIGURE 1. Tunneling between two identical ferromagnetic electrodes that show a spin split density of states N (↑/↓ indicate majority/minority states). In (a) and (b) the magnetization of the two electrodes is parallel and antiparallel, respectively. The conductivities for tunneling from the left to the right electrode are indicated by arrows.

of the density of states has immediate consequences on the tunneling current as pioneering field-emission experiments with ferromagnetic tips have shown [5, 6]. During the tunneling process from the ferromagnetic tip into the free states of the vacuum, the spin polarization of the density of states in the tip is partly transferred to the emitted electrons and a spin-polarized current was observed. This can easily be understood using a simple model. As the tunneling barrier, i.e., the potential step of the work function in front of the tip, is not spin dependent, the transmission probability through the barrier does not depend on the electron spin. The observed polarization is just a consequence of the different number of states for minority and majority electrons in accordance with Fermi's golden rule: the more states are allowed to tunnel, the higher the resulting tunneling current. Therefore, the tunneling current from a spin-polarized tip is spin-polarized according to the imbalance in spin of the involved states.

This effect alone is not enough to obtain spin information in STM, since the spin polarization of the tunneling current is not directly accessible. The only measured parameter is the magnitude of the current. It was Jullière [7] who discovered that when electrons tunnel between two ferromagnets, not only the current is spin-polarized but also the magnitude of the current is influenced. The phenomenon was therefore referred to as the tunneling magnetoresistance (TMR) effect. In his experiment, two magnetic films were separated by a thin insulator film to form a planar tunnel junction. The two magnetic films were chosen to have different coercive fields. This permitted to align their magnetization parallel or antiparallel as a function of an applied magnetic field. Jullière found that the tunneling conductance G (and by this the resistance) depends on the relative orientation of the magnetization of the two layers. For parallel orientation, G was higher than for antiparallel orientation. This finding can be explained on the basis of a simple model for tunneling. As above, we neglect any spin dependence in the transmission through the barrier and focus solely on the electronic properties of the two electrodes. Under the assumption of a small bias voltage across the junction and in the absence of spin-flip scattering during the tunneling process, the electrons in the ferromagnets near the Fermi energy determine the tunneling conductance of the junction. For a parallel orientation, the majority/minority electrons of the first electrode tunnel into the majority/minority states in the second electrode, respectively, as sketched in Figure 1(a). Using Fermi's golden rule, the conductance G is proportional to the density N of initial (i) and final (f) states at the Fermi edge. Combining both spin channels, the conductance for parallel oriented magnetizations is given by

$$G_{\uparrow\uparrow} \propto N^i_\uparrow N^f_\uparrow + N^i_\downarrow N^f_\downarrow . \tag{1}$$

For the antiparallel orientation, electrons of majority character in one electrode tunnel into states of minority character in the other electrode and the conductance is given by a mixed product:

$$G_{\uparrow\downarrow} \propto N^i_\uparrow N^f_\downarrow + N^i_\downarrow N^f_\uparrow . \tag{2}$$

These two conductivities are in general not identical, leading to a variation of the tunneling current with the magnetic configuration of the electrodes. Slonczewski treated the problem of spin-polarized tunneling more rigorously [8]. Neglecting higher-order spin effects like spin accumulation, he calculated the dependence of G on the angle θ between the magnetization of the two electrodes. With the spin polarization $P = (N_{\uparrow} - N_{\downarrow})/(N_{\uparrow} + N_{\downarrow})$, the conductance is given by

$$G = G_0(1 + P^i P^f \cos\theta). \tag{3}$$

The $\cos\theta$ dependence is rigorous, reflecting the quantum mechanical rotation behavior of the spin $\frac{1}{2}$ tunneling electrons. Slonczewski's prediction for the angular dependence of the TMR effect was later experimentally confirmed [9].

In general, if a finite bias is applied, all states between the two Fermi levels are involved in tunneling. They have to be weighted according to their tunneling probability through the barrier, which is, e.g., energy dependent. This scenario is more complex but G can be expressed using effective, i.e., correctly weighted, densities or polarizations.

2 Imaging Modes

Already in 1988, Pierce suggested setting up a STM that uses the TMR effect to image the sample magnetization with high lateral resolution [10]. He suggested two different approaches to create a spin-polarized current—the use of ferromagnetic tips and the possibility to photo-excite spin-polarized carriers in GaAs tips. While we focus on the first, the latter was realized by Suzuki et al. [11]. In the latter approach, circularly polarized light is used to excite spin-polarized carriers into the conduction band of the tip that then tunnel into the sample. The spin polarization of the electrons can be selected by the helicity of the light. By modulating the helicity of the light, modulations in the tunneling current are induced due to spin dependent tunnelling. The modulations were detected with a lock-in amplifier to separate spin information from topographic information. The optical modulation technique, however, suffers from a rather low contrast and an unintended additional magneto-optical contrast of low resolution. Only a few studies on domain patterns have been published using this technique. Note that the inverse effect, i.e., the tunneling form a ferromagnetic Ni tip into GaAs followed by the emission of polarized light, was experimentally demonstrated as well [12]. Much more successful was the development of Sp-STM using ferromagnetic tips. In his pioneering publication, Pierce suggested three different imaging modes of Sp-STM which have been all realized experimentally and which are discussed in detail below. A theoretical description of the three imaging modes was given recently by Heinze et al. [13].

2.1 The Constant-Current Mode

In the constant-current mode of Sp-STM, a spin-polarized tip is scanned over a spin-polarized surface in the conventional constant-current mode of STM. In this

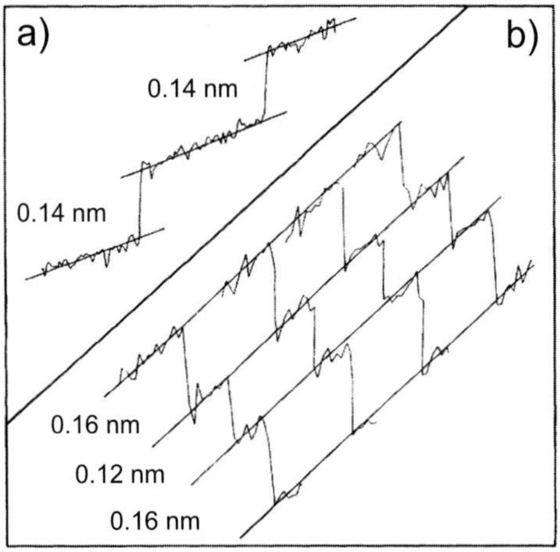

FIGURE 2. Constant current line scans of the Cr(001) surface with (a) a W tip and (b) a CrO_2 tip. The variation of the step height in (b) is a result of spin-polarized tunneling. Figure used with kind permission of R. Wisendanger [14].

case, the images contain mixed topographic and spin information. Wiesendanger et al. were the first to use this imaging mode. They reported on spin-polarized vacuum tunneling at room temperature between a ferromagnetic CrO_2 tip and the (100) surface of the layer-wise antiferromagnetic bcc Cr [14]. Using a non-magnetic W tip, topographic line-scans revealed atomic steps on the Cr surface of the expected step height of 0.14 nm (see Figure 2(a)). When, however, the experiment was repeated with ferromagnetic CrO_2 tips, alternating step heights of 0.16 and 0.12 nm were observed (see Figure 2(b)). This was attributed to the TMR effect between the ferromagnetic tip and the antiferromagnetically ordered Cr terraces. When the spin polarizations of the tip and the Cr terrace atoms are parallel, the tunneling current is enhanced (see Eq. (3)). In the constant current mode of the STM, the tip is retracted by a small amount (0.02 nm). Due to the layer-wise antiferromagnetic order of Cr(001) [15], the spin polarization of the tip and the Cr atoms is antiparallel on the adjacent atomic terrace. The TMR effect leads to a reduction of the current and the STM tip approaches. This mechanism results in alternating step heights seen with a spin-polarized tip and allowed an indirect observation of the sample spin polarization. No separation of topography and spin information could, however, be obtained in this approach and reference measurements had to be acquired with non-magnetic tips. Further, the interpretation of the data was based on the assumption that both spin polarizations of the tip apex and the Cr sample were oriented along the surface normal. As we know now, this assumption was wrong, at least for the Cr surface [16].

Recently, the constant-current mode was used to obtain information on the sample spin on the atomic scale. Heinze et al. used Fe-coated W tips to resolve the antiferromagnetic order in Mn films on W(110) [17]. When imaging the Mn film with nonmagnetic tips, the bcc(110) 1×1 unit cell could be observed. When using a spin-polarized tip, a spin contrast is added in the topographic image and the unit cell expanded to a $c(2 \times 2)$ cell, reflecting the magnetic unit cell. Note,

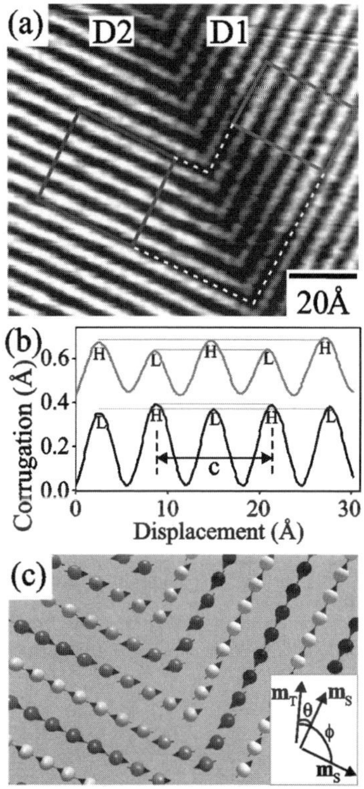

FIGURE 3. (a) Constant current images of the $Mn_3N_2(010)$ taken with a spin-polarized tip and (b) averaged line scans of the corrugation in the domain D1 (top) and D2 (bottom). (c) Model of the surface spins. Reprinted figure with permission of A.R. Smith [18]. Copyright 2002 by the American Physical Society.

however, that the additional corrugation is damped, as the structures are smaller than the lateral resolution of the tunneling process [17].

Since in the constant current mode, the contrast is largely due to the topography and only a small variation of the tip-sample distance is related to the spin polarization, it has mostly been used on surfaces of constant electronic structure like simple metallic surfaces with small corrugation. It may, however, also be applied to complex materials with large topographic corrugation in the unit cell. Yang et al. have used Fe- and Mn-coated tips to study the spin configuration of the $Mn_3N_2(010)$ surface [18]. Figure 3 shows constant current images with two crystallographic domains. Clearly a row structure can be seen that was related to Mn planes in Mn_3N_2. With spin-polarized tips, a modulation of the height of the rows was detected on both domains (see Figure 3(b)) that reflects the antiferromagnetic coupling between the Mn planes. Due to the different orientation of the Mn planes and by this the different alignment of the Mn moments, the modulation in the topography is different on the two domains. The higher the component of the Mn spins along the tip magnetization, the larger the modulation of the corrugation. By taking the average and the difference of two line scans that are shifted by half the magnetic periodicity, Yang et al. managed to separate the spin independent and spin dependent part of the corrugation.

This method, however, is only suitable for systems with a known magnetic unit cell.

2.2 The Spectroscopic Mode

A mode that under certain circumstances allows separation of topographic information from spin information is spin-polarized scanning tunneling spectroscopy. It was initially suggested by Pierce [10] and Stroscio et al. [19] and was first realized by Bode et al. [20]. It uses the fact that the spin polarization of the sample density of states is a function of the energy.

For simplicity, we assume that the density of states in the tip is featureless, i.e., constant. The spin polarization of the tip states is constant as well. When a finite bias is applied between tip and sample, a set of states may contribute to the tunneling. As illustrated in Figure 4(a), a positive sample bias U leads to the tunneling of all tip states between $E_{F_t} - U$ and E_{F_t} to empty states in the sample between E_{F_s} and $E_{F_s} + U$. Using the same simple approximation for the tunneling current as above, the tunneling current is proportional to the density of states of the sample integrated from E_{F_s} to $E_{F_s} + U$ for both spin channels and weighted by the (constant) density of states of the tip. In this approximation, the differential conductance dI/dU as function of the sample bias U is simply given by the spin split density of states of the sample weighted by the constant tip density of states [2, 13]. In the example of Figure 4(a), where the tip and the sample magnetization are parallel, the measured dI/dU is dominated by the minority density of states of the sample and contains only little information on the majority density of states, as the tip states are mainly of minority character. When the sample is magnetized in opposite

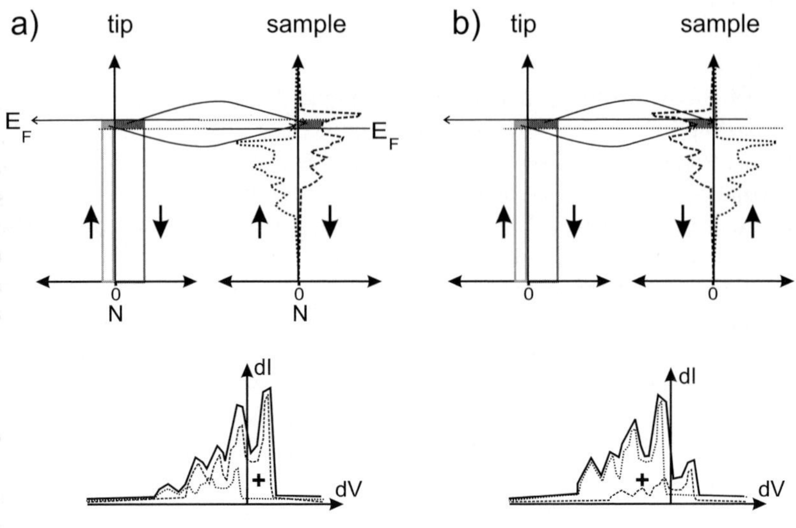

FIGURE 4. Tunneling between a ferromagnetic tip and sample for (a) parallel and (b) antiparallel magnetization of both. The corresponding differential conductivities dI/dU as function of bias U are sketched below.

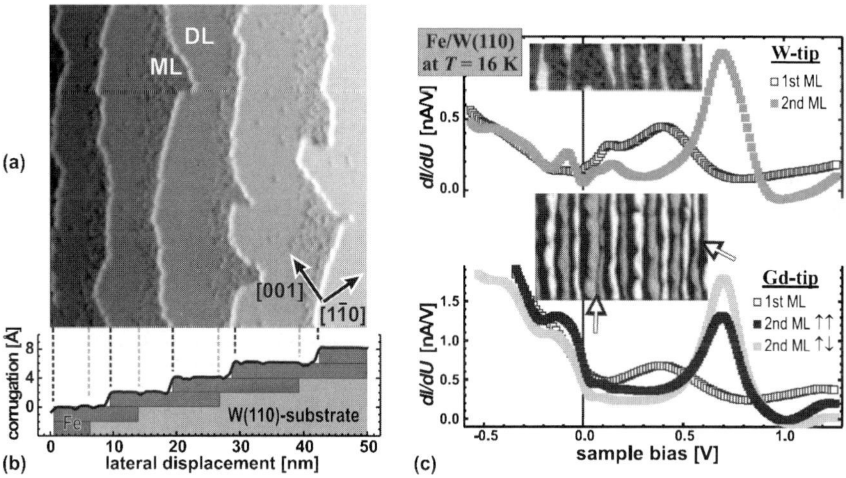

FIGURE 5. Topographic image (a) and line scan (b) of ≈ 1.5 ML Fe on W(110). (c) Differential conductance of ML and DL areas taken with a W tip (top) and a Gd tip (bottom) as function of the sample bias. The insert shows maps of the differential conductance revealing magnetic information in case of Gd tips. Figure kindly provided by M. Bode.

direction to the tip, the balance is changed and the contribution of majority density of states in the differential conductance dominates (see Figure 4(b)). The resulting differential conductance is simply a linear combination of the minority and majority density of states depending only on the relative orientation of tip and sample magnetization. Therefore, the characteristic energies, e.g. the position of peaks in the dI/dU spectra, do not change, whereas the peak heights vary with the sample magnetization.

In the pioneering experiment by Bode et al., the height of the spin-split surface state of Gd(0001) in laterally resolved dI/dU spectra was used to obtain magnetic information [20].

Figure 5 shows an example for magnetic contrast obtained with Sp-STM using Gd-coated W tips. The tips show a sensitivity for the out-of-plane magnetization, which suggest that they are magnetized along the tip axis. When Fe is deposited onto W(110) at 300 K followed by annealing to elevated temperatures, the Fe films grow in the step-flow growth mode [21]. For coverages between one and two mono-layers (ML), this results in alternating mono-layer and double-layer (DL) stripes as depicted in the topographic image of Figure 5(a) and the schematic sketch of Figure 5(b). As has been shown by Elmers et al., the ML stripes have an in-plane magnetization, while the DL stripes are magnetized out-of-plane [21]. When performing tunneling spectroscopy on these samples with W tips, two different dI/dU spectra were observed (see Figure 5(c)), reflecting the different electronic structure of the ML and the DL areas. When plotting the differential conductance at 0.68 V as a function of the tip position, a map of the local differential conductance (see inset in Figure 5(c)) shows the DL areas as bright and the ML areas as dark patches, since dI/dU of the DL is much higher than that of the ML at the chosen voltage. When using a ferromagnetic, Gd-coated tip, the overall shape of the dI/dU spectra changes little, but the DL stripes display two distinct types of spectra. Due to spin-polarized tunneling, one spectrum has a higher differential conductance around the peak at 0.68 V while the other has a higher differential conductance at around -0.2 V. The first is the spectrum for antiparallel tip and sample magnetization and the latter for parallel. When plotting dI/dU at 0.68 V as a function of position, a magnetic contrast is visible in the maps of the local differential conductance. Besides the contrast between DL and ML stripes mentioned above, an additional contrast between different DL stripes appears (see inset of Figure 5(c)). Neighboring DL stripes have an alternating contrast reflecting an antiferromagnetic alignment of neighboring DL stripes. This confirms the dipolar antiferromagnetic coupling that reduces the stray field energy of the sample similar to stripe domains in perpendicular magnetized films [21, 22].

The magnetic configuration of antiferromagnets can be investigated as well, as Sp-STM is sensitive to the spin polarization at the surface [16]. Figure 6 shows

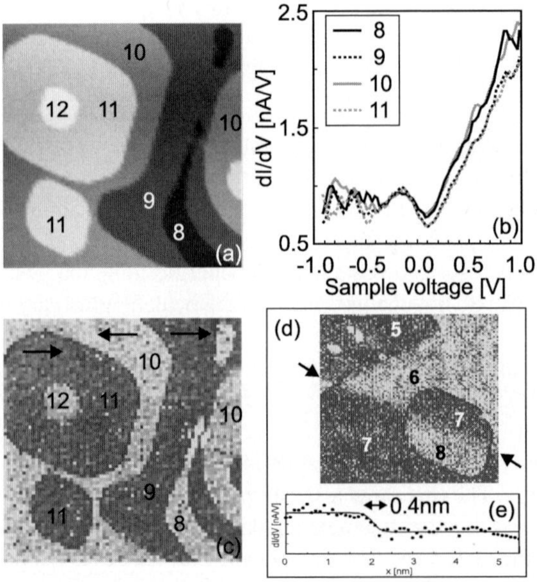

FIGURE 6. STM and SP-STM measurements of 10 ML Mn deposited on Fe(001) at 370 K. These measurements were performed with an Fe- coated W tip. (a) $100 \times 100\text{nm}^2$ topographic image ($U = -0.5$ V, $I = 0.5$ nA). Numbers in (a) and (c) denote the numbers of the Mn layers. (b) dI/dV curves as a function of the sample bias voltage obtained on the Mn layers in (a). dI/dV curves obtained on even (odd) Mn layers are shown as solid (dashed) curves, which were numerically obtained from $I(V)$ curves measured at a set point of $U = -0.5$ V, $I = 0.5$ nA. (c) *dI/dV* map at +0.2 V measured at the same area as a). (d) *dI/dV* map at +0.4 V of 7.2 ML Mn on Fe(001) ($U = -0.6$V, $I = 0.5$ nA, 70×70 nm^2). The magnetic contrast reverses across the hidden step. (e) An averaged line profile across the hidden step in the dI/dV map is shown. Figure kindly provided by T.K. Yamada.

the example of thin Mn films grown on Fe(001). Mn grows in the body centered tetragonal (bct) configuration and shows ferromagnetic coupling within an atomic layer while the different layers couple antiferromagnetically [23–25]. Figure 6(a) shows the topography of the Mn film. Due to imperfect growth, several layers are exposed at the surface. Sp-STM spectra taken with Fe-coated tips on the different layers are shown in Figure 6(b). While the differences between spectra taken on odd (even) layers are minute, the spectra of odd and even layers vary strongly, i.e., the spectra taken on odd and even layers fall into two classes. This reflects the layer-wise antiferromagnetic order of the Mn film, changing the tunneling probability between sample and spin-polarized tip due to the TMR effect. When plotting the differential conductivity at 200 mV as a function of position (see Figure 6(c)), the layer-wise antiferromagnetic order results in alternating bright and dark terraces. Interestingly, the antiferromagnetic order of the Mn film can be disturbed at buried Fe step edges, as will be discussed in detail below. As a result, a magnetic frustration similar to a domain wall is induced in the Mn-film (see Figure 6(d, e)). These walls can be rather sharp. A lateral resolution of 0.4 nm has been claimed by Yamada et al. [25], which is close to the theoretical limit for s-electron tunneling [2].

Sp-STM in the spectroscopic contrast mode has been used in a variety of systems (Gd/W(110), Fe/W(110), Fe/W(100), Fe/Mo(110), Cr(001), Fe/Cr(001), Co/Cu(111), Mn/Fe(001)), with ferromagnetic and antiferromagnetic tips that are sensitive to one in-plane or the out-of-plane component of the spin polarization [16, 20, 22, 25–30].

2.3 The Differential Magnetic Mode

In the differential magnetic mode, the tip magnetization is modulated to rigorously separate the spin information of a sample surface from the topography. In Sp-STM, the nonmagnetic tip is replaced by a bulk ferromagnetic STM electrode. This method was proposed by Pierce [10] as well. First attempts using Ni-tips were of limited success [31]. Only recently this mode was realized to obtain magnetic contrast [32].

The basic concept of this mode is related to Eq. (3). The tip magnetization is periodically reversed, which is equivalent to changing the sign of the spin polarization of the tip apex. In the experimental setup, the magnetization of the tip is reversed by an alternating current through a small coil that is fixed to the tip. The frequency of the alternating current is chosen above the cutoff frequency of the feedback loop of the STM [32]. Thus, the feedback loop only detects the averaged tunneling current $\overline{I} = I_0$ for the two spin polarizations (positive and negative) of the tip apex. The spin-dependent part in the averaged tunneling current cancels out, so that the topographic image contains no magnetic information. With a phase-sensitive lock-in amplifier, the alternating part of the tunneling current ΔI is detected which is within the Jullière model proportional to $P^i P^f \cos\Theta$. It contains all the spin information. This way, topographic and spin information are

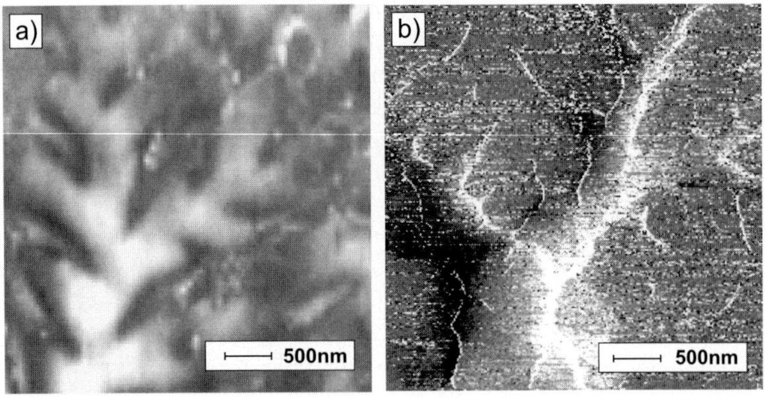

FIGURE 7. (a) MFM and (b) Sp-STM image of the fractal domain pattern of Co(0001). The scans were performed on the same sample but not on the same area.

separated and an image of the spin component along the magnetization axis of the tip magnetization can be recorded simultaneously with the topography [32].

In the experiment, the alternating magnetic field induced within the coil has to be large enough fully to reverse the magnetization of the tip. The alternating field, however, may also create induction currents in the tunneling loop. For this reason, only extremely soft magnetic materials may be used as tips. Moreover, magnetostriction of the tip during the reversal must be avoided. The initial experiments of Johnson et al. suffered from large magnetostriction of the Ni tip, so that no stable magnetic or topographic imaging was possible [31].

When using tips that are magnetized perpendicular to the sample plane, the out-of-plane component of the spin polarization may be imaged [32]. As an example, the closure domain pattern of Co(0001) is shown in Figure 7. In magnetic force microscopy (MFM) images of the sample magnetization, the fractal domain pattern of Co(0001) is visible. It originates from the partial flux closure of the perpendicular bulk domains at the sample surface [33]. The limited lateral resolution of MFM becomes apparent at the magnification of the image. In contrast to MFM, Sp-STM reveals the full detail of the fractal structure as depicted in Figure 7(b). Ultranarrow 20°-domain walls of 1.1-nm width were observed [34, 35], indicating a lateral resolution better than 1 nm.

Bulk in-plane magnetized STM electrodes also allow the imaging of one in-plane component of the spin polarization [36]. The differential magnetic mode was applied to layer-wise antiferromagnetic Mn films on Fe(001), which allows direct comparison to the spectroscopic mode shown above. Figure 8(a) shows a schematic picture of the alignment of the magnetic moments in the ferromagnetic Fe substrate and the antiferromagnetic Mn film. Within the Mn film, the magnetization is reversed in every atomic layer, while the coupling at the Mn-Fe interface is ferromagnetic [37]. When several Mn layers are exposed at the surface, their

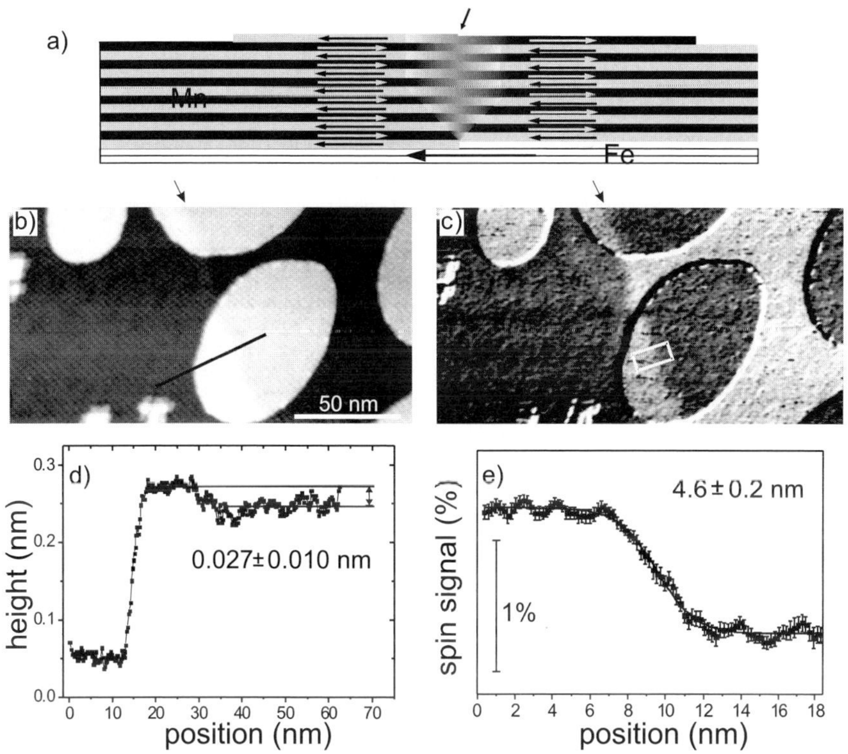

FIGURE 8. (a) Schematics of Mn layers overgrowing an Fe substrate step. Above the buried Fe step edge a magnetically frustrated region occurs. Sp-STM image of (b) the topography and (c) the corresponding spin signal of 11.9 ML Mn on Fe(001). A buried Fe step edge is running almost vertically through the center of the images, indicated by arrows. (d) Line profile taken along the line in (b) showing a monatomic Mn step and a step of subatomic height formed by a buried Fe step. (e) Line profile (averaged over 70 lines) across the magnetically frustrated region in the Mn film at the position of the box in (c). The solid line represents a fit.

magnetization points into opposite directions on adjacent Mn terraces due to the layer-wise antiferromagnetic order. The situation of the magnetic order above a buried Fe step edges is more complicated. The thickness of the Mn layers on both sides of a monatomic Fe step differs by one ML. Due to the vertical lattice mismatch (bcc Fe versus bct Mn), subatomic steps are formed at the Mn film surface at the position of Fe step edges. An undisturbed layer-wise antiferromagnetic order within the Mn film is not possible when the Mn moments at the interface on both sides of the step edge are aligned in the same direction by the Fe substrate. Instead, Mn layers which meet at the position of the Fe step edge have an opposite spin polarization. This leads to a magnetic frustration similar to a 180° domain wall in the Mn film [25, 38–40].

A region above a buried Fe step edge is shown in Figure 8(b) for the topography and (c) for the corresponding spin signal. A buried Fe step edge is running almost vertically through the center of the imaged area (see black arrows). The line profile in Figure 8(d) taken along the line in Figure 8(b) shows a step of monatomic height between two different Mn terraces ($\approx$0.16 nm) and a step of subatomic height ($\approx$0.018 nm) at the position of a buried Fe step edge. The latter is caused by the different lattice constants of Fe and Mn. In Figure 8(c) clearly the layer-wise antiferromagnetic order between the Mn islands and the Mn layer underneath is visible similar to Figure 6(c). Following the way of the buried Fe step edge, a magnetically frustrated region is formed in the spin signal, and a reversal of contrast appears. Figure 8(e) presents an averaged line profile across the topologically enforced magnetic frustration at the position of the box in Figure 8(c). The measurements indicate that the magnetic frustration has a certain lateral extension. To estimate the wall width at the surface, the experimental profile is fitted with a standart wall profile $\tanh(x/w)$. This function is the analytical solution for a one-dimensional domain wall neglecting dipolar energies. The curve is plotted as a solid line in Figure 8(e) and fits the shape of the transition region well. The resulting width ($2w$) was in this case 4.6 ± 0.2 nm.

A systematic study of the wall width as a function of the Mn film thickness has shown a widening of these regions with increasing Mn film thickness (see Figure 9) [40]. Neglecting the magnetostatic energy in the antiferromagnet, the width of a 180° domain wall in the Mn film is determined by a competition between the exchange energy and the magnetic anisotropy energy. Assuming an antiferromagnetic exchange proportional to the Néel temperature [41] and a Mn anisotropy similar to that of the 3D ferromagnets, one expects between 20 and 60 nm for the bulk domain wall width in Mn. The pinned domain walls at Fe step edges result, however, in narrow widths at the interface. The driving force for widening of the wall for thicker Mn films is the exchange energy which is gained by approaching the bulk domain wall configuration. Thus, the width of the wall should asymptotically approach its bulk wall width with increasing film thickness. As expected, we observed a widening, but it is linear with no sign of saturation. This is in agreement with the relatively narrow walls when compared to the expected bulk wall width.

To calculate the width of the wall in the Mn film on Fe(001), an effective Heisenberg model with classical spins was used and a fourfold magnetic anisotropy was included. While the values for the exchange coupling constants and the anisotropy constant are well known for bulk bcc Fe [41] the values for bct Mn were estimated. The coupling constants are depicted in the inset of Figure 9. When taking into account only the nearest neighbor interaction, the modelled walls are too narrow (open squares). When the nearest- and next-nearest-neighbor exchange interaction were considered, the simulations (open triangles) are closer to the experimental data, but still a small offset occurs. For both calculations the coupling across the interface between Fe and Mn was identical to the one within the Mn film. When reducing the interface coupling to 25%, the agreement to the experimental data is

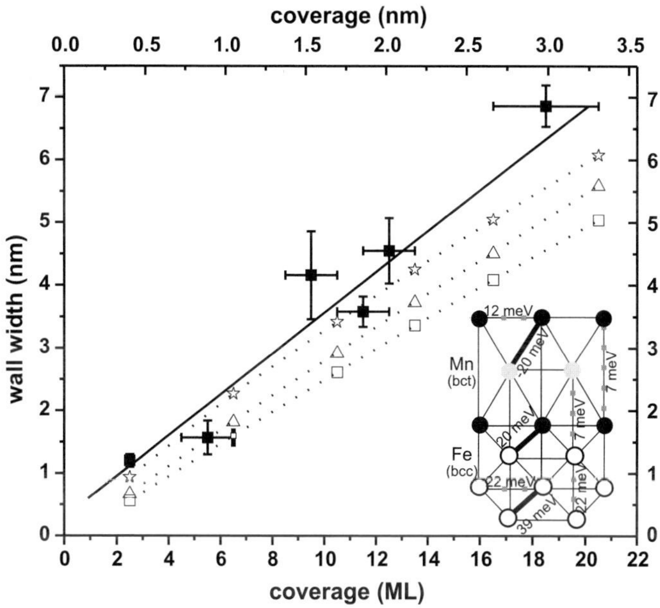

FIGURE 9. The measured wall width as a function of Mn film thickness with linear fit (solid line) together with calculated wall widths of magnetically frustrated regions using the Heisenberg model. The calculated data shown in open squares are obtained by considering only the nearest-neighbor exchange interaction and in open triangles by considering the nearest- as well as the next-nearest-neighbor exchange interaction and the bct structure of Mn. Using the same parameter as in the latter case and reducing the exchange at the interface to 25%, one obtains the values presented in open stars. The insert shows the exchange constants used in the Heisenberg model.

better (open diamonds) which suggest that the exchange interaction at the interface between Fe and Mn may be reduced. This demonstrates that Sp-STM may also provide insight into fundamental mechanisms in magnetism.

The differential magnetic contrast mode was used in polycrystalline Ni, Co(0001), Fe(001) and Mn/Fe(001) [35, 36, 40, 42].

3 Tip Preparation

For spin-polarized STM, at least the last atom at the tip apex must have a spin polarization stable in magnitude and orientation. Two different approaches have been used in the past to reach these conditions. First, the whole tip may consist of a material with a certain spin polarization or a non-magnetic tip is coated with spin-polarized material. These two approaches are discussed below.

3.1 Tuning of the Axis of Sensitivity in Thin Film-Coated Tips

Coated tips are suitable for the constant current and the spectroscopic imaging mode. Most frequently, nonmagnetic W tips are used. To obtain spin contrast, these tips are coated in situ with a thin film of either a ferromagnet or an antiferromagnet. Best results have been reported if the W tips are flashed to high temperatures before coating. This cleans the tip apex from the natural oxide of W and leads to a reproducible but rather blunt tip [43]. When depositing, e.g., an Fe film of 10 ML on the tip followed eventually by gentle annealing, the tip is covered by either a continuous film or patches of Fe, that show a magnetization along the curved surface of the tip. This direction of magnetization is a consequence of the shape anisotropy of the thin Fe film. Therefore, the tunneling part of the tip has most probably a magnetization in the sample plane and one in-plane component of the magnetization can be imaged [20]. When, however, the tip is coated with a $\approx$10 ML film of Gd, the tips exhibit sensitivity for the out-of-plane component of the spin polarization [22]. This is most likely due to the interfacial magnetic anisotropy of Gd on W that favors magnetization perpendicular to the interface, i.e., at the apex along the tip axis. Both Fe- and Gd-coated tips produce a small magnetostatic stray field at the tip apex that in some cases may influence the magnetic object under investigation [26]. Alternatively, antiferromagnetic Cr [26] or Mn [25] coatings have been used. While the orientation of the spin polarization of the apex is more difficult to control, the tips are mostly free of stray fields. Yamada et al. showed that in antiferromagnetically coated Mn tips, voltage pulses on the tip may even be used to change the direction of sensitivity [44].

3.2 Controlling the Magnetization Direction in Bulk Magnetic tips

Bulk tips are suitable for all three imaging modes. Here, however, we focus on the special necessities of the differential magnetic imaging mode.

The direction of magnetization in bulk magnetic tips is primarily determined by the shape of the tip. For tips that are sensitive to the out-of-plane component of the spin polarization, sharp tips have produced best results [35]. Tips were electrochemically etched from thin CoFeSiB wires of 130 μm diameter. As etching agent, a dilute mixture of HCl and HF was used that was suspended by surface tension as a thin liquid membrane in a Pt ring during etching. The pH value was adjusted so that the formation of silica from the Si in the amorphous wire was prevented. Using low etching currents of the order of 250 μA, pointed tips of cone angles typically between 8° and 15° were created. Due to the large shape anisotropy, a tip magnetization along the tip axis is ensured [45]. When using the differential magnetic contrast mode, it is important to avoid vibrations due to magnetic switching. The chosen material offers extremely low coercivities in the range of 50 μT and negligible magnetostriction. This ensures virtually vanishing vibrations [45]. While magnetic stray fields cannot be avoided in this configuration, hard magnetic samples could be imaged without problem.

For imaging one in-plane component of the spin polarization, ring-shaped STM electrodes can be used. In these ring electrodes, the magnetization direction lies tangential to the outer perimeter of the ring. Thus, at the bottom of the ring where the tunneling occurs, the magnetization lies in the plane of the sample surface. Because the magnetic flux in an ideal ring is closed, the magnetic stray field is zero. Experimentally, even soft magnetic materials could be imaged [36] using ring electrodes. By choosing the plane in which the ring is oriented, the magnetization direction of the ring is defined and thus the direction of the sensitivity in the surface plane for the measured spin signal is known. Although the rings used as STM electrodes are not sharp, a lateral resolution below 1 nm could be achieved [36,40]. Most likely nanotips exist at the apex which give a high lateral resolution.

To obtain spin contrast, the STM electrodes need to be cleaned in situ, e.g., by Ar sputtering. The contrast can be increased by coating the CoFeSiB tips with several ML of Fe.

4 Advantages and Disadvantages of the Different Modes

The different contrast modes all have their specific advantages and drawbacks.

The constant current imaging mode is the simplest imaging mode. It has the advantage that it does not need any sophisticated spectroscopy. It can be carried out with any STM in vacuum that shows a high enough stability. The nonmagnetic tip is replaced either by a ferromagnetic or an antiferromagnetic tip. Both bulk and coated tips can be used. The mode in principle offers atomic resolution. It is well suited to study antiferromagnets and the spin structure in more complex materials. For itinerant ferromagnets, it is, however, less appropriate. As the image consists of mixed topographic and spin information, an unfolding of the two channels is necessary to gain information on the spin configuration. On the atomic scale, this can be done by studying the corrugation with nonmagnetic and spin-polarized tips. The additional corrugation due to the spin is of the order of several picometers. In itinerant ferromagnets, the magnetic domains often extend over the range of a micron. Currently, STM is not able to resolve an additional picometer scale corrugation on these lateral length scales. Further, a small variation of the density of states by contaminations, statistical variation in alloy composition, etc., leads to a corrugation that is hard to distinguish from a spin-induced corrugation in the constant current mode. Therefore, this mode has been used in chemically ordered and single crystalline systems. The use of antiferromagnetic tips in this mode allows avoiding magnetic dipole interaction of the tip and the sample.

The spectroscopic mode is a more sophisticated mode. In samples with a homogeneous density of states, the magnetic signal can be separated from the topographic information. However, additional studies of the differential conductivity with an unpolarized tip are required. In case the spin polarization is not constant with bias voltage, the differential conductance recorded with a spin-polarized tip contains spin information. In case of a constant spin polarization with bias voltage, the topographic tip stabilization levels out any variations of the tunneling current, and no contrast can be observed. As a spurious effect, the topographic images contain spin information similar to the constant current mode. Samples with strong variation of the density of states and spin polarization with bias are most suited for this technique. These are surfaces with surface states or thin films with quantized states. The imaging mode may be used with bulk and coated ferromagnetic or antiferromagnetic tips. The latter allows the influence of the magnetic stray field to be minimized and even allows operation in high magnetic fields. Its application to systems with a very inhomogeneous or unknown electronic structure is problematic. The technique offers high lateral resolution and has been used to image the magnetic structure of simple metallic ferromagnets and antiferromagnets.

The differential magnetic mode allows rigorous separation of topographic and spin information. The average tunneling current contains no spin information, and therefore constant current topographic images reflect the spin averaged density of states. The differential magnetic signal, only contains spin information. It represents the local TMR between the tip and the sample surface, which is within the Julière model related to the integrated spin polarized density of states between the Fermi edge and the bias voltage. The technique can be applied to inhomogeneous samples. It has been used, however, only with bulk ferromagnetic tips for technical reasons. Further, tip preparation is more demanding, but the tip geometry fixes the direction of sensitivity. As the tip magnetization has to be switched periodically, thin film tips with high coercivities are not suited. A further limitation is that the application of a large magnetic filed affects the tip switching so that only studies in limited applied fields are possible. Using the above-mentioned simple Jullière model, the differential magnetic mode allows the measurement of the sample spin polarization independently of the existence of domains and is therefore well suited to study the TMR effect across vacuum barriers.

5 Beyond Magnetic Imaging

Spin-polarized STM has much more to offer than high-resolution imaging of the spin configuration. Like STM, it can be used to gain information of the density of states and fundamental mechanisms of tunneling. In contrast to STM, Sp-STM may give spin-resolved information. Two examples of such studies are discussed below.

5.1 Vacuum Tunneling Versus Tunneling Across an Insulator

Spin-polarized tunneling has been of particular interest in the last years due to its application to magnetic tunnel devices. In the pioneering work of Jullière on planar junctions, the dependence of the tunneling resistance on the relative orientation of the electrode magnetization was related to the spin polarization of the density of states at the Fermi energy [7]. The TMR effect, however, is still far from being completely understood. A particular aspect, namely, that the TMR in planar junctions with oxide barriers decreases with increasing bias voltage, is still puzzling. Due to its technological importance, many studies have been devoted to this. In the early work of Jullière, a bias as small as 3 mV was needed to halve the value of the TMR. With increased control of the preparation, this value increased up to 700 mV over the years [46]. Many models have been proposed to explain this behavior. First, biasing of the junctions leads to elastic tunneling of electrons mostly from the Fermi energy of the negative electrode into empty states of the positive electrode. Since the spin polarization of the empty states depends in general on the energy, it should cause variations of the TMR due to density-of-states effects. Via this mechanism, information on the spin-dependent density of states can be obtained. Second, hot electrons from the positive electrode might be scattered in a spin-dependent way at defects in the amorphous barriers [47] or might create magnons [48]. The latter mechanisms reduce the spin polarization and, consequently, the TMR.

To elucidate the mechanism, difficulties related to the complex structure of the tunnel junctions (polycrystalline electrodes, poorly characterized amorphous barriers) have to be overcome to perform well-defined experiments and to allow comparison with theoretical results. Here, Sp-STM measurements of the voltage dependence of the TMR across the vacuum barrier, where effects due to impurities in the spacer are obviously not present, are helpful. Magnon creation and the density-of-states effect are in principle still present. Further, theoretical modeling of spin-polarized tunneling across a well-defined vacuum barrier is feasible. Measurements of the TMR were carried out in ultrahigh vacuum with a Sp-STM between an amorphous tip and a Co(0001) sample. During the measurement, the bias voltage was varied while keeping the tip at a fixed position and measuring the averaged tunneling current I_t and the modulated current ΔI. The TMR, defined as the asymmetry δ of the tunneling currents observed for parallel and antiparallel magnetization alignment (P and AP), was obtained from the ratio of these two currents. The measured TMR (Figure 10(a)) obtained with the tip stabilized at 1 V, 1 nA, is almost constant with bias voltage. This is in contrast to the case of planar tunnel junctions with amorphous spacers. If spin-dependent scattering at magnons was the dominant mechanism for the drop of the TMR, a similar decrease of the TMR with bias voltage should also be present in our case. Its absence, however, indicates that this mechanism is not dominant.

Additional insight into the mechanisms of spin-polarized tunneling can be gained by first-principle calculations of the TMR. Several modern approaches beyond the simple Jullière model exist. Especially successful is the scheme proposed by MacLaren and coworkers [49] for planar tunnel junctions. In this model, the electronic structures of the electrode and the barrier are treated by first principle calculations and the current is modelled as a coherent scattering process of electrons from one electrode into the other. These calculations for Co(0001) junctions reveal a similar behavior as the Sp-STM experiments. Using semi-infinite

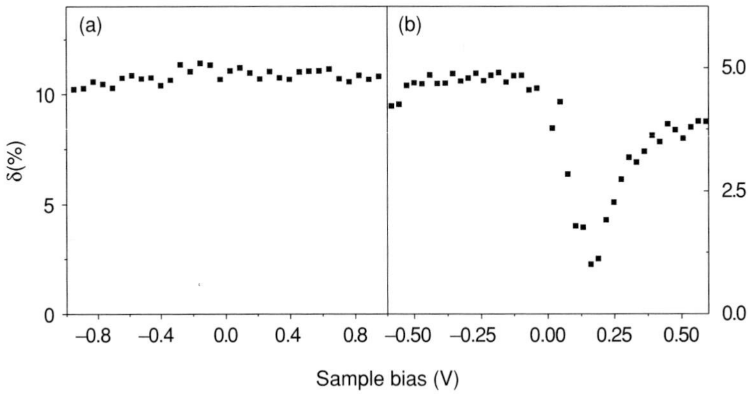

FIGURE 10. Tunnel magnetoresistance δ of a clean Co(0001) surface versus bias voltage, obtained with a magnetic tip stabilized at (a) 1 V, 1 nA, and (b) 100 mV, 1 nA.

Co(0001) electrodes, conductances G for parallel and antiparallel alignment were computed [50]. As usual, the calculated conductances (Figure 11(a)) for parallel alignment are larger than those for antiparallel alignment of the electrode magnetizations. Further, the tunneling conductances at $d = 2d_0$ separation is almost two orders of magnitude larger than those for $3d_0$ ($d_0 = 2.035$ Å, the Co interlayer distance), as expected for vacuum tunneling. In agreement with the experimental results for large separations, the TMR is almost constant [50].

For small barrier widths, tunneling via surface states could become important in STM experiments [19]. This might lead to pronounced changes in the TMR. Figure 10(b) presents the experimental TMR versus bias voltage obtained at a smaller tip-sample separation (feedback conditions: 100 mV, 1 nA). For negative bias, a

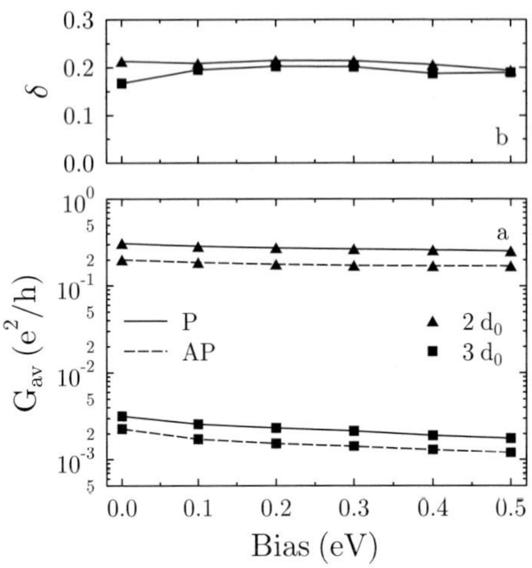

FIGURE 11. Theoretical spin-dependent tunneling through Co(0001) planar junctions. (a) Conductances G for parallel (P) and antiparallel (AP) magnetization alignment at various barrier widths d *vs* bias voltage. (b) Tunnel magneto-resistance δ obtained from the data shown in (a).

constant TMR is still observed. For positive bias, however, a strong dip at 200 mV is present. According to Jullière's model, one can speculate that the dip is related to majority states which reduce the minority-dominated spin polarization at this energy. Indeed, inverse photoemission measurements revealed a majority surface state of Co(0001) at ≈ 0.2 eV [51] which is also found in the calculations. At small tip-sample separations, the tunneling probability through this surface state can be enhanced and thus would decrease significantly the TMR. The density-of-states mechanism qualitatively explains the constant TMR for large tip-sample separations and the minimum at small tip-sample separations. The dip at 200 mV does not show up in our transport calculations because surface states do not contribute to the conductance in that model. It is, however, conceivable that the surface state contributes to the tunneling current through scattering at steps or defects [50].

5.2 Spin-Orbit Mixing as a Function of Magnetization Direction

Sp-STM in the spectroscopic mode may also be used to study the electronic structure of ferromagnets as a function of the magnetic configuration. Due to the spin-orbit interaction, the band structure of a ferromagnet depends on the direction of magnetization [52] and the degeneracies of the 3d-bands can be lifted [53]. Sp-STM in the spectroscopic mode allows to compare the local density of states in areas of different sample magnetization. By this the minute changes in the density of states induced by the spin-orbit interaction may be resolved. Figure 12(a), shows the topographic image of $\approx$1.75 ML of Fe on W(110). The DL stripes show a stripe domain pattern of domains magnetized up and down out of the stripe plane, as Figure 12(b) reveals [54]. Fe coated tips image the in-plane magnetized domain walls between the stripe domains as alternating bright and dark lines (see Figure 12(c)). Maps of the differential conductance taken at certain bias voltages with bare W tips, however, also show these in-plane domain walls (see Figure 12(d)). They appear as dark lines, i.e. do not contain information on the in-plane direction of the magnetization in the walls, but indicate whether the magnetization is locally out-of-plane (gray) or in-plane (black). Figure 12(e) shows in all detail the minute changes of the differential conductivity upon reorientation of the magnetization from in-to out-of plane. Ab initio calculations revealed that the differences are due to changes of the mixing of the d_{xy} and the d_{z^2} states depending on the magnetization direction [54]. This example shows the high potential of Sp-STM to study the density of states in ferromagnets. The accuracy is high enough to resolve the weak impact of the spin-orbit interaction on the density of states.

6 Outlook

With the successful operation of Sp-STM in all three modes proposed by Pierce, Sp-STM has become an established technique. It is now widely used by several

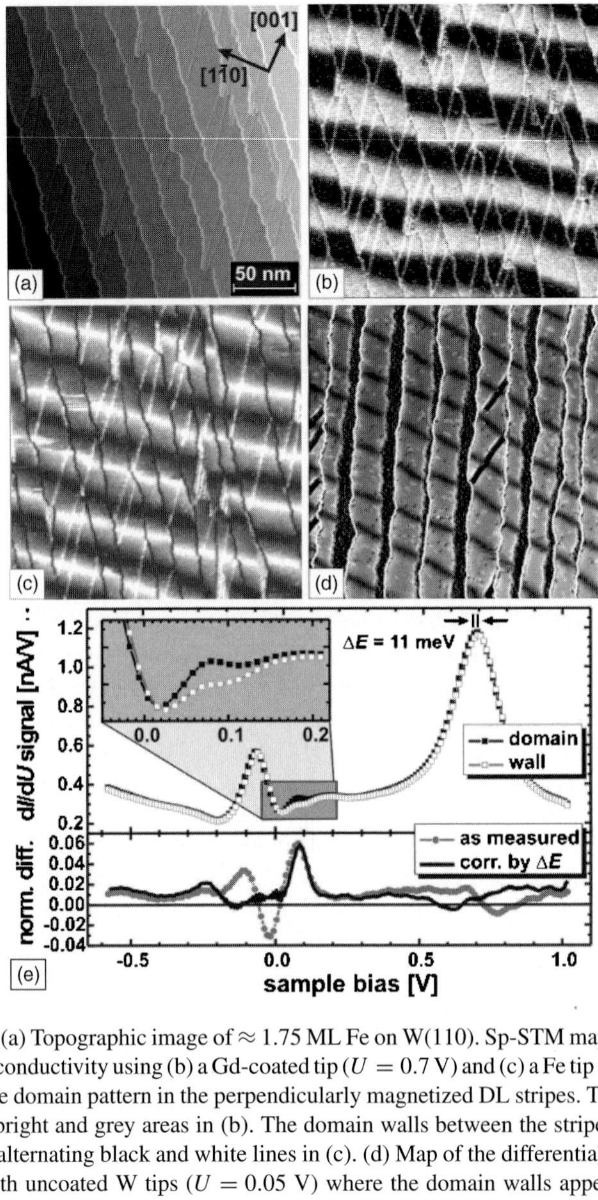

FIGURE 12. (a) Topographic image of ≈ 1.75 ML Fe on W(110). Sp-STM maps of the local differential conductivity using (b) a Gd-coated tip ($U = 0.7$ V) and (c) a Fe tip ($U = 0.05$ V) show a stripe domain pattern in the perpendicularly magnetized DL stripes. They appear as alternating bright and grey areas in (b). The domain walls between the stripe domains are resolved as alternating black and white lines in (c). (d) Map of the differential conductance recorded with uncoated W tips ($U = 0.05$ V) where the domain walls appear darker. (e) Comparison of the differential conductance recorded on domains and domain walls as a function of sample bias. Figure kindly provided by M. Bode.

groups in Europe, Asia, and America. Its potential for high-resolution magnetic imaging has only been exploited with a small number of systems. More studies on all kinds of systems will give a deeper understanding of magnetism on the nanometer scale. Sp-STM is also the method of choice to investigate the spin structure of antiferromagnetic surfaces. The real-space information gained is complementary to the information already available from magnetic scattering techniques like neutron scattering. Sp-STM allows us to tackle new classes of problems in antiferromagnets: frustrations and aperiodic structures. Finally, Sp-STM may be used to investigate the density of states and their spin polarization at the same time. Because of this, it has the potential to become a valuable tool for electron spectroscopy—beyond high-resolution magnetic imaging.

References

1. J. Tersoff and D. R. Hamann, Phys. Rev. Lett. **50**, 1998 (1983).
2. J. Tersoff and D. R. Hamann, Phys. Rev. B **31**, 805 (1985).
3. G. Binning, H. Rohrer, Ch. Gerber, and E. Weibel, Appl. Phys. Lett. **40**, 178 (1982).
4. G. Binning, H. Rohrer, Ch. Gerber, and E. Weibel, Phys. Rev. Lett. **49**, 57 (1982).
5. N. Müller, W. Eckstein, W. Heiland, and W. Zinn, Phys. Rev. Lett. **29**, 1651 (1972).
6. M. Landolt and Y. Yafet, Phys. Rev. Lett. **40**, 1401 (1978).
7. M. Jullière, Phys. Lett. **54A**, 225 (1975).
8. J. C. Slonczewski, Phys. Rev. B **39**, 6995 (1989).
9. T. Miyazaki and N. Tezuka, J. Magn. Magn. Mater. **139**, L231 (1995).
10. D. T. Pierce, Physica Scripta **38**, 291 (1988).
11. Y. Suzuki, W. Nabhan, and K. Tanaka, Appl. Phys. Lett. **71**, 3153 (1997).
12. S. F. Alvarado and P. Renaud, Phys. Rev. Lett. **68**, 1387 (1992).
13. S. Heinze, P. Kurz, D. Wortmann, G. Bihlmayer, and S. Blügel, Appl. Phys. A **75**, 25 (2002).
14. R. Wiesendanger, H. J. Güntherodt, G. Güntherodt, R.J. Gambino, and R. Ruf, Phys. Rev. Lett. **65**, 247 (1990).
15. S. Blügel, D. Pescia, and P.H. Dederichs, Phys. Rev. B **39**, 1392 (1989).
16. M. Kleiber, M. Bode, R. Ravlic, and R. Wiesendanger, Phys. Rev. Lett. **85**, 4606 (2000).
17. S. Heinze, M. Bode, A. Kubetzka, O. Pietzsch, X. Nie, S. Blügel, and R. Wiesendanger, Science **288**, 1805 (2000).
18. H. Yang, A. R. Smith, M. Prikhodko, and W.R.L. Lambrecht, Phys. Rev. Lett. **89**, 226101 (2002).
19. J. A. Stroscio, D. T. Pierce, A. Davies, R. J. Celotta, and M. Weinert, Phys. Rev. Lett. **75**, 2960 (1995).
20. M. Bode, M. Getzlaff, and R. Wiesendanger, Phys. Rev. Lett. **81**, 4256 (1998).
21. H. J. Elmers, J. Hauschild, and U. Gradmann, Phys. Rev. B **59** 3688 (1998).
22. O. Pietzsch, A. Kubetzka, M. Bode, and R. Wiesendanger, Phys. Rev. Lett. **84**, 5212 (2000).
23. B. Heinrich, A. S. Arrott, C. Liu, and S. T. Purcell, J. Vac. Sci. Technol. A **5**, 1935 (1987).
24. T.G. Walker and H. Hopster, Phys. Rev. B **48**, 3563 (1993).
25. T.K. Yamada, M.M.J. Bischoff, G. M. M. Heijnen, T.Mizoguchi, and H. van Kempen, Phys. Rev. Lett. **90**, 056803 (2003).
26. A. Kubetzka, M. Bode, O. Pietzsch, and R. Wiesendanger, Phys. Rev. Lett. **88**, 057201 (2002).
27. R. Ravlic, M. Bode, and R. Wiesendanger, Journ. Phys., Cond. Matter **15**, 2513 (2003).
28. A. Yamasaki, W. Wulfhekel, R. Hertel, S. Suga, and J. Kirschner, Phys. Rev. Lett. **91**, 127201 (2003).

29. M. Bode, O. Pietzsch, A. Kubetzka, and R. Wiesendanger, Phys. Rev. Lett. **92**, 067201 (2004).
30. O. Pietzsch, A. Kubetzka, M. Bode, and R. Wiesendanger, Phys. Rev. Lett. **92**, 057202 (2004).
31. M. Johnson and J. Clarke, J. Appl. Phys. **67**, 6141 (1990).
32. W. Wulfhekel and J. Kirschner, Appl. Phys. Lett. **75**, 1944 (1999).
33. J. Unguris, M. R. Scheinfein, R. C. Celotta, and D.T. Pierce, Appl. Phys. Lett. **55**, 2553 (1989).
34. H. F. Ding, W. Wulfhekel, C. Chen, J. Barthel, and J. Kirschner, Mater. Sci. and Engineering B **84**, 96 (2001).
35. H. F. Ding, W. Wulfhekel, and J. Kirschner, Europhys. Lett. **57**, 100 (2002).
36. U. Schlickum, W. Wulfhekel, and J. Kirschner, Appl. Phys. Lett. **83**, 2016 (2003).
37. S. Andrieu, E. Foy, H. Fischer, M. Alnot, F. Chevrier, G. Krill, and M. Piecuch, Phys. Rev. B **58** 8210 (1998).
38. A. Berger and H. Hopster, Phys. Rev. Lett. **73**, 193 (1994).
39. A. Berger A and E. E. Fullerton, J. Magn. Magn. Mater **165**, 471 (1997).
40. U. Schlickum, N. Janke-Gilman, W. Wulfhekel, and J. Kirschner, Phys. Rev. Lett. **92**, 107203 (2004).
41. M. Pajda, J. Kudrnovský, I. Turek, V. Drchal, and P. Bruno, Phys. Rev. B **64**, 174402 (2001).
42. W. Wulfhekel, H. F. Ding, G. Steierl, M. Vazquez, P. Marin, A. Hernando, and J. Kirschner, Appl. Phys. A **72**, 463 (2001).
43. M. Bode and R. Wiesendanger, In: *Magnetic Microscopy of Nanostructures*, ed. by H. Hopster, H.P. Oepen (Springer, Berlin Heidelberg New York, 2004), p. 205.
44. T.K. Yamada, M.M.J. Bischoff, T. Mizoguchi, and H. van Kempen, Appl. Phys. Lett. **82**, 1437 (2003).
45. W. Wulfhekel, R. Hertel, H. F. Ding, G. Steierl, and J. Kirschner, J. Magn. Magn. Mater. **249**, 368 (2002).
46. H. Boeve, E. Girgis, J. Schelten, J. De Boeck, and G. Borghs, Appl. Phys. Lett. **76**, 1048 (2000).
47. J. Zhang and R. White, J. Appl. Phys. **83**, 6512 (1998).
48. J. S. Moodera, J. Nowak, and R. J. M. van de Veerdonk, Phys. Rev. Lett. **80**, 2941 (1998).
49. J. M. MacLaren, X.- G. Zhang, W. H. Butler, and X. Wang, Phys. Rev. B **59**, 5470 (1999).
50. H. F. Ding, W. Wulfhekel, J. Henk, P. Bruno, and J. Kirschner, Phys. Rev. Lett. **90**, 116603 (2003).
51. C. Math, J. Braun, and M. Donath, Surf. Sci. 482–485, 556 (2001).
52. R. Wu, and A. J. Freeman, J. Magn. Magn. Mater. **200**, 498 (1999).
53. A. Lessard, T. H. Moos, and W. Hübner, Phys. Rev. B **56**, 2594 (1997).
54. M. Bode, S. Heinze, A. Kubetzka, O. Pietsch, X. Nie, G. Bihlmayer, S. Blügel, and R. Wiesendanger, Phys. Rev. Lett. **89**, 237205 (2002).

II.3
Scanning Probe Measurements of Electron Transport in Molecules

KEVIN F. KELLY AND PAUL S. WEISS

The ability to control the placement of molecules is essential for the patterning and fabrication of nanoscale electronic devices. We apply selective chemistry and self-assembly to reach higher resolution, greater precision, and chemical versatility in the nanostructures that we create. These methods demonstrate the possibilities of patterning films by exploiting the intrinsic properties and interactions of the molecules. We employ self-assembled monolayers as a means to isolate molecules with various electronic properties to determine the fundamental transport mechanisms, and the relationships between molecular structure, environment, connection, coupling, and function. Using self-assembly techniques in combination with scanning tunneling microscopy (STM), we are able to study candidate molecular switches individually and in small bundles. Alkanethiolate self-assembled monolayers on gold are used as a host two-dimensional matrix to isolate and to insulate electrically the molecular switches. We then individually address and electronically probe each molecule using STM. The conjugated molecules exhibit reversible conductance switching, manifested as a change in apparent topographic height in STM images. The origins of switching and the relevant aspects of the molecular structure and environment required are discussed below.

1 Introduction

Control and stabilization of molecular assemblies at the nanometer scale are crucial steps in the fabrication of molecular-scale devices. Current techniques such as photolithography, electron beam lithography [1] and "soft lithography" [2,3] are limited in their resolution and cannot reproducibly achieve patterns with dimensions at the nanometer scale. At the other end of the spectrum, single molecule manipulation has been successfully demonstrated using scanning probe microscopy but is unable to produce devices in parallel and is still too time-consuming to be practical as a fabrication technique [4–8]. We anticipate the need to combine the speed and versatility of lithographic techniques with single-molecule resolution and control in order to create viable molecular devices.

2 Self-Assembled Monolayers

We and others have developed and utilized methods that exploit the inherent chemical, physical, and thermodynamic properties of molecules to pattern surfaces at the molecular level using self-assembly techniques in order to investigate molecular-based electronic devices. Self-assembly is a natural phenomenon that can be observed in many biological, chemical, and physical processes [9,10]. This method has been explored recently to produce supramolecular assemblies in a straightforward manner [9–14]. The most commonly studied and best characterized systems are alkanethiolate self-assembled monolayers (SAMs) on Au{111} [15,16]. Alkanethiolate SAMs form spontaneously on Au{111} through chemisorption of the S head group to the Au surface. The monolayers interact on the surface through van der Waals interactions between adjacent alkyl chains. The origin of the stability of SAMs is thus twofold, due to the covalent S–Au bond and the attractive van der Waals forces between the methylene groups. As a result of the attractive interactions imparting stability and order to these systems, SAMs have low defect densities and resist degradation in air.

The process of self-assembly lends itself naturally to controlling the local placement of molecules. In particular, SAMs have been used as model systems for fabricating structures with controlled geometries [3,17–21], as well as essential components in device structures [22–25]. Multicomponent SAMs formed by codeposition of two or more adsorbates from solution have been investigated for their patterning potential [13,14,26–31]. When more than one adsorbate is considered, it is necessary to account for the interactions between the different adsorbates and whether these favor mixing or separation.

Insofar as molecular-level patterning is concerned, where the size regime for structures at the nanometer scale is desired, it is necessary to use techniques that are able to distinguish molecular level characteristics. With the development of scanning probe microscopies, such as scanning tunneling microscopy (STM) [32] and atomic force microscopy (AFM) [33], it has become possible to gain local information at the atomic scale and obtain real-space images of surfaces [34–36]. This has proven to be enormously beneficial in studies of phenomena such as phase separation of SAMs, where previous analytical methods were able to measure only ensemble averages.

We are concerned with the design and control of systems such as these, and have been successful in demonstrating a number of methods for patterning SAMs using self-assembly to form the structures and STM (along with ensemble-averaging techniques) to determine the success of the method. This paper outlines recent research conducted by our group involving patterning of SAMs, and controlled placement of individual molecules by insertion.

3 Single Molecule Conductance Switching

Microelectronics technology has focused on producing ever smaller and faster devices, which has led to formidable technological challenges in fabricating and characterizing devices at the nanometer and sub-nanometer scales. If miniaturization of electronic components and devices is to reach its ultimate limits, new methods and strategies will need to be developed. One promising alternative to conventional technology is molecular electronics, where single molecules may be used as active components in electronic devices. Although the field is still in its infancy, there have been demonstrations of functional single molecule electronics. Two-terminal devices have been constructed that examine the viability of individual and bundled molecules as wires [47–49], switches [50–53], and diodes [54,55].

Conjugated phenylene-ethynylene oligomers have been identified as a family of molecules with favorable characteristics for molecular electronic design; they are rigid, fully conjugated, and have tunable functionality [46,48,53–60]. Nanopore experiments have demonstrated the viability of these molecules as electronic devices. In these experiments, a monolayer of molecules was sandwiched between two evaporated gold electrodes in a small pore, <30 nm in diameter. The nanopore devices were used to test groups of thousands of these molecules, elucidating interesting functional properties including negative differential resistance (NDR), two persistent conductance states, and controlled switching via electric field [53–55]. As a complement to the nanopore work, theoretical studies have suggested a variety of mechanisms that produce switching effects. These mechanisms range from reduction and conformational twisting of the molecules [56] to rotation of substituents [61].

We seek experimental data that will determine the factors that play important roles in the switching of these molecules. To achieve this goal we study molecules in this family on an individual or collective basis using STM [47,48,51]. Unlike the aforementioned nanopore experiments, here the molecules are bound only to one gold electrode, and the STM tip acts as the second electrode that can individually address each molecule to determine its electronic and structural properties. Systems for study must be carefully designed, as precise control at the single molecule level is often a difficult venture. To this end, we utilize self-assembly to control the placement, orientation, and local environment of single molecules.

3.1 Isolation of Single Molecules

Whereas codeposition of a multicomponent SAM has been demonstrated to result in the formation of nanometer-scale domains, it is also possible to control the insertion of *individual* molecules into an alkanethiolate matrix. We have previously shown that monolayers of alkanethiolates are well suited for the two-dimensional matrix isolation of individual molecules [47,48,51]. The monolayers are well understood, as discussed above, so they are easily manipulated to serve as host matrices for the insertion of molecules of interest. The host monolayers contain a variety of defect sites, including domain boundaries, step edges, and substrate vacancy islands. At these locations, guest molecules have access to the underlying gold, and may be able to bind to the substrate via a sulfur head group. Varying the deposition and processing conditions of the host matrix film can manipulate the type and densities of these defects [47]. Due to the limited space in the defect sites of the SAM matrix, the guest molecules are forced to stand up from the gold surface, and a portion of the conjugated molecules then protrude from the host matrix. This arrangement allows us to locate the molecules with the STM, and to probe the molecules end-to-end. In our configuration, the STM tip serves as one electrode and the gold substrate serves as the other. Because STM images are a convolution of the tip and surface structures, it is not always possible to distinguish between individual and bundled molecules. For surface features with low corrugation, such as the molecular lattice of the matrix, STM images are good representations of the lateral surface structure. Details of features with higher aspect ratios, such as the protruding switch molecules, can be more difficult to resolve with the STM, and thus more difficult to interpret. As in our earlier work, we are able to infer that many of the similar protruding features at structural domain boundaries on substrate terraces, where access for insertion is limited, are individual molecules [47,48].

When co-deposited films of molecular wire candidates and alkanethiolates were prepared, they were found to be highly disordered compared to pure alkanethiolate films. Even at low concentrations of the molecular wire in solution (11% by weight), no crystalline order was observed. It was concluded that the molecular wires disturb the ordering process during deposition. However, the insertion process allows us to produce well-ordered SAMs before exposure to the molecular wires in solution. Prior to and immediately following the insertion, the crystalline order of the dodecanethiolate matrix could easily be observed. Substrate defects and domain boundaries that are indicative of alkanethiolate SAMs also remain before and after insertion. After insertion, the films are interspersed with protrusions appearing $<$4–6 Å higher than the film in STM images. These were attributed to the molecular wires, as they were not present in the films imaged immediately before insertion. The molecular wires insert at domain boundaries, substrate defects and step edges. It is thus assumed that the molecular wire candidates chemisorb to the Au surface through a S–Au bond at sites that are particularly accessible to the molecules.

This same insertion strategy has been used to bind tethers designed and synthesized by Weck and Grubbs to serve as the anchor and initiation points for ring-opening metathesis polymerization (ROMP) [62,63]. By controlling the initial SAM defect density, we controlled the number of inserted tethers and their separation. Polymerization from these norbornene-terminated phenylene-ethynylene thiolates then resulted in individual, isolated polymer chains in the low-SAM-defect limit, or micron-wide polymer brushes where large defects had been opened to allow substantial insertion. Since ROMP yields an active polymer, the chain length could be controlled via monomer supply and polymerization time.

The insertion process is an effective way to measure single molecules. Further control over the placement of the molecules can be achieved by controlling the substrate defect density by solution annealing (detailed in above section), or by carefully selecting the matrix molecule to control the uniformity of the environment of the inserted molecules [64].

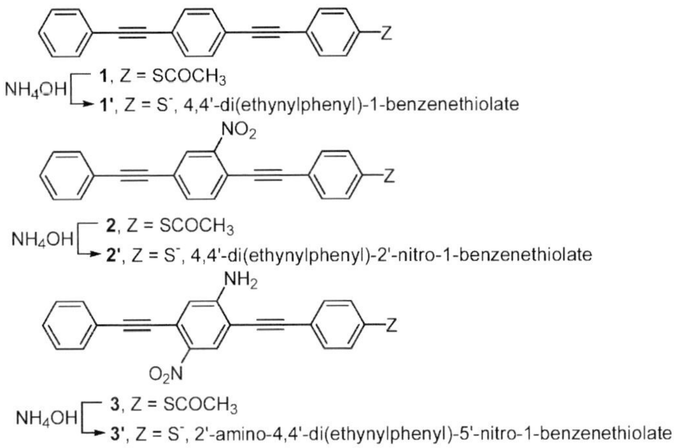

FIGURE 1. Phenylene ethynylene oligomers (**1, 2, 3**) that were inserted into alkanethiolate matrices. The scheme for base-promoted hydrolysis is show beneath each molecule.

In more recent work, we have compared molecules (shown in Figure 1) that have shown switching behavior – **2′** and **3′** in nanopore experiments and in calculations, with those that have not – 1′ [55,57]. We have found that all three of these molecules exhibit stochastic switching between at least two conductance states, as do oligothiophene thiols [65]. The amount of switching directly correlates to the amount of order in the insertion matrix, leading us to conclude that the mechanism of conductance switching involves motion of the molecules. Finally, we describe vapor annealing, a novel technique for modification of the local structure of SAMs [66].

For these experiments SAM matrices were prepared on commercially obtained Au{111} on mica [67]. The Au substrates were annealed and cleaned using a hydrogen flame immediately prior to film preparation. SAMs for the host matrices were deposited from a 1 mM solution of dodecanethiol in ethanol. Three types of SAMs were made: (1) SAMs were adsorbed by exposing the gold surface to the alkanethiol solution for 24 h; (2) SAMs were adsorbed for 5 min deposition time, to allow less time for self-assembly of the matrix and to leave more and larger defect sites for insertion; (3) following the insertion step (described below), typical (deposited from solution for 24 h) SAMs were annealed in dodecanethiol vapor at 80°C for 2 h in a sealed vial. The third type of SAM processing was utilized to improve SAM quality, and reduce defect density. This process allows us to add matrix molecules from the gas phase, while reducing exchange processes that are inevitable during solution phase annealing [66]. Following all SAM depositions, the substrates were rinsed in ethanol and dried with nitrogen.

To insert guest molecules into the host SAM matrices, solutions were prepared containing 0.1 μM of **1, 2,** or **3** (defined in Figure 1) in dry tetrahydrofuran. Aqueous ammonia was added to hydrolyze the thioacetyl protecting group, generating the thiol in situ to adsorb as the thiolate **1′**, **2′**, or **3′** on the surface. The SAMs were

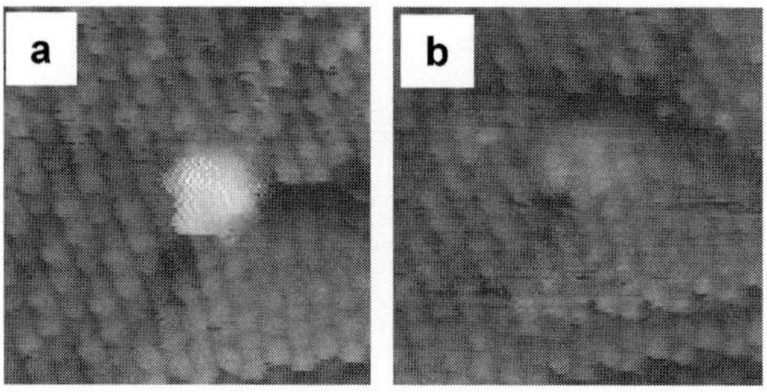

FIGURE 2. STM images of a 5 nm × 5 nm area showing a single molecule of **3′** protruding from a matrix of dodecanethiolate, in the ON (a) and OFF (b) states (a 0.3 nm height difference). Imaged at $V_{sample} = -1.0$ V, I = 1 pA.

exposed to this solution for times ranging from 0.5 to 3 h. After rinsing and drying, the samples were stored at room temperature in a desiccator until imaging.

Figure 2(a) is a STM image obtained under ambient conditions of a single molecule of **3′** inserted in a dodecanethiolate monolayer, located at a domain boundary. Although the host matrix is generally very stable over long periods of time, inserted molecules are observed to change conductance state frequently over time, or to move within the site. The inserted molecules have at least two states that differ in their apparent height in STM topography by ~3 Å. Since topographic STM images represent a convolution of the electronic and topographic structure of the surface, the apparent height change observed in the STM images can be due to a change in physical height of the molecule, a change in the conductance of the molecule, or both. We refer to the more conductive state (greater apparent height) as ON (Figure 2(a)), and the less conductive state as OFF (Figure 2(b)). The molecule in the OFF state is still visible as a smaller protrusion, indicating that it is still present in the same defect site. The conductance state switching observed between Figure 2(a) and Figure 2(b) occurs spontaneously during STM imaging.

Stochastic switching between the ON and OFF states is reversible. Time-lapse series of images acquired over several (up to 26) hours recorded the temporal behavior of the molecules. Drift arising from thermal fluctuations and creep in the STM piezoelectric imaging mechanism makes it difficult to observe a single small feature on the surface for long times so relatively large scan areas ([1,000–2,500 Å]^2) were used to ensure that several molecules of interest remained in the area sampled for the full duration of imaging. Additionally, large scan areas allowed several molecules to be monitored simultaneously, enabling statistical analysis of the multiple switching events. Figure 3 shows a series of four images selected from a larger series of time-lapse images, taken over a span of ~10 hours. Four substrate terraces are visible in each 770 × 770 Å image, separated in height by about 2.4 Å. The host matrix monolayer contains a variety of defect sites, including domain boundaries and substrate vacancy islands. These defect sites served as adsorption sites for the inserted molecules of **2′**, which are visible as 3 Å protrusions in the images.

Using an automated digital tracking and extraction algorithm discussed in detail below, the three molecules labeled A–C in Figure 3 were individually monitored for a complete series of 98 images. For each image in sequence, a smaller image of each molecule was extracted, and assembled into the sequences in Figure 4(a–c). Because each frame contains only a single molecule, the molecular height above

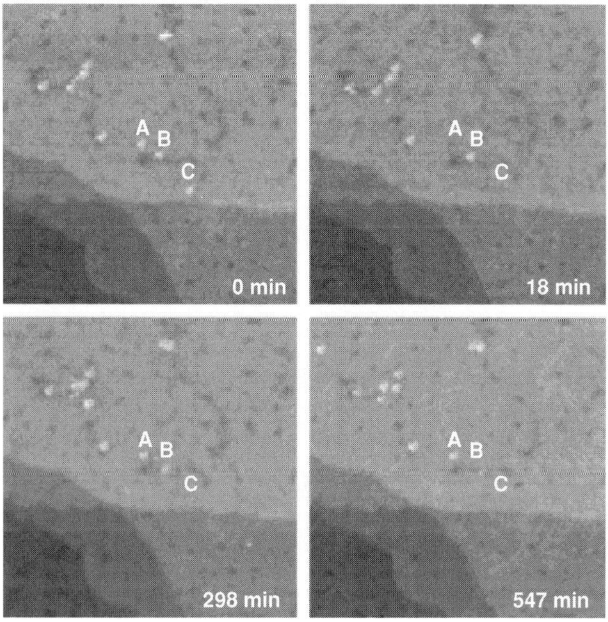

FIGURE 3. Four representative images of **2′** isolated in a SAM matrix of dodecanethiolate on Au{111} from a time-lapse series of images. Each 77 nm× 77-nm image was acquired at $V_{sample} = -1.0$ V, I = 1 pA. The three molecules labeled A–C were found to switch conductance states several times between the four images.

FIGURE 4. (a–c) Extracted areas around each **2′** molecule labeled A–C in Figure 3 for the full time-lapse series of images. These images were used to calculate the height profiles shown in Figure 5.c.

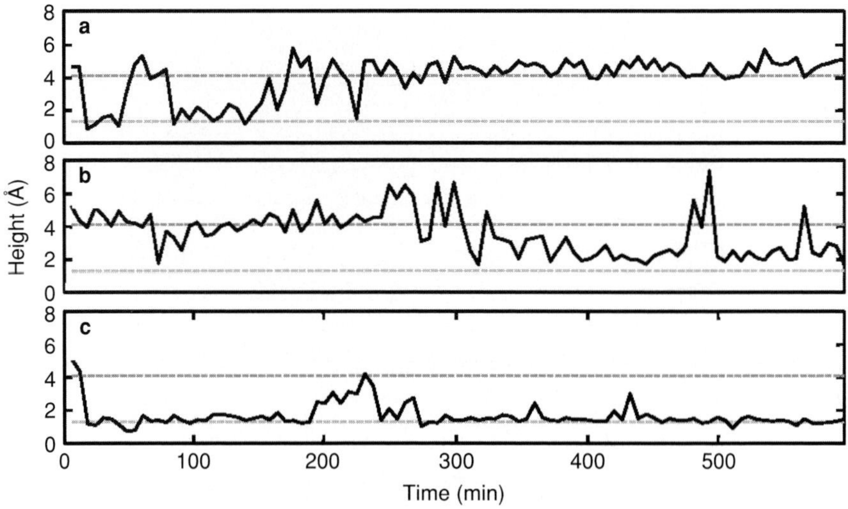

FIGURE 5. Height versus time profiles for the molecules shown in Figs. 3 and 4. Heights were calculated as described in [68].

the SAM can be automatically measured for each frame, generating the height versus time plots shown in Figure 5(a–c) [68]. Molecule A was relatively active in switching between the ON and OFF states during the first 220 min of acquisition, but then stabilized in the ON state for the remainder of the images. Molecule B exists predominantly in the ON state for the first 300 min, before switching predominantly into the OFF state for the last half of the sequence. Molecule C switched OFF after just two images and remained mainly in the OFF state for the rest of the images. Several important observations can be made from these three representative molecules: (1) In all cases, switching between states is reversible. (2) Both the ON and the OFF states can persist for hours, or can change rapidly, persisting for just one image. (3) In the OFF state, the molecules still appear to protrude slightly from the host film, indicating that they remain bound to the same defect site, and that they do not desorb or diffuse out of the imaging area.

While the molecules shown in Figs. 3 and 4 are all **2**′, analogous switching behavior was also observed for **1**′ inserted into matrices of dodecanethiolate. Additionally, we have determined that switching is not unique to phenylene ethynylene oligomers. We have also analyzed SAM matrices with oligothiophenes inserted. Like the phenylene ethynylenes, the thiophene molecules are fully conjugated, and consist of three or four thiophene units. However, they are not completely rigid and have more possible conformations. The oligothiophenes exhibit similar two-state reversible switching [65], demonstrating that the switching we observe may not depend on the specific functionality of the phenylene ethynylene oligomers.

3.2 Feature Tracking in STM

A common problem with many scanning probe microscope (SPM) techniques is drift caused by the relative motion of the sample and the probe. In applications using piezoelectric transducers for raster scanning common in most SPMs, nonlinear responses such as hysteresis and creep [69,70] can cause the probe to drift across the surface [71]. When attempting long-term temporal data acquisition (i.e., several hours), the drift can be severe enough to cause the scan window to drift beyond the original region of interest. In an effort to compensate for this drift, we have developed a fast image cross-correlation (XC) technique that is used for post-acquisition drift climination to enable several types of temporal data analyses, including temporal topographic analysis and diffusion of individual molecules.

This method of drift elimination is a computational solution; others have implemented instrumentation-based solutions such as tracking tunneling microscopy [72] where the tip is moved in a circle laterally (x, y) and the x and y topographic gradients are measured to keep the tip located over one molecule. This method has been used to track diffusion [73,74] and rotation [75] of individual molecules on surfaces. While instrumentation-based tracking has the capability of measuring events on the millisecond time scale, it is limited in that it can track only one molecule at a time. In contrast, tracking diffusion or other events in an image allows monitoring of many molecules with the disadvantage of a slower time scale limited by the image acquisition rate. However, the use of fast scanning instruments (e.g., 20 frames/s) [76] can bring image tracking close to the time scale of instrumentation-based tracking. One of our interests is in monitoring topographic changes, such as the change in apparent height or even disappearance of a protrusion. In this latter case, the instrumentation-based tracking method, would fail, as a protrusion (or depression) is required for tracking. With XC tracking, the entire image is tracked such that the behavior of individual molecules (topographic or spatial changes) does not play a major role in the overall tracking algorithm; these features are extracted and analyzed after tracking and data acquisition.

Image tracking is a vital factor in video compression [77] and machine vision, and as such, is currently an area of intense research [78–81]. Several algorithms have recently been developed to track three-dimensional (3D), two-dimensional (2D), and rotational motions for various microscopies [82–85]. In this work, we assume that SPM techniques will produce image sequences of constant image size and resolution. Because of the orthogonal geometry of most scanning probe actuators, such as the beetle-style STM [86], only lateral, non-rotational drift should occur. Therefore, a simple XC approach can be used to find the best alignment, and thus the drift between successive images. The maximum of the 2D XC represents the best alignment, also referred to as registration, of two images. The deviation of this maximum from the center of the image represents the drift (in pixels) between the two images.

By determining the drift from a sequence of images, we have been able to select and to extract specific regions from these images for further analysis. In this chapter, we discuss methods to resolve temporally the conductance behavior of conjugated molecular switches inserted into alkanethiol self-assembled monolayers [51] and to analyze the diffusion of individual benzene molecules adsorbed on Ag{110} [87]. Through these analyses, statistical information on conformational switching and hopping distances, respectively, are obtained.

The real-space cross-correlation $C(x,y)$ for two digital images, $a_1(I,J)$ and $a_2(I,J)$, can be calculated from a comparison of the subsets of each image [$a_1(i,j)$ and $a_2(i,j)$] at multiple offsets (x and y) specified by [88,89]:

$$C(x,y) = \frac{\sum_i \sum_j \left[a_1(i+x, j+y)a_2(i, j)\right]}{\left[\left(\sum_i \sum_j a_1(i+x, j+y)^2\right)\left(\sum_i \sum_j a_2(i, j)^2\right)\right]^{1/2}} \quad (1)$$

Variations on this technique have been used previously for several applications including molecular tagging velocimetry [89], stability measurements of STMs [90], and atomic force microscopy image alignment [84]. While this approach yields the desired XC, large images can present an exceptionally demanding computational load. In addition, this technique is limited in the available search window (limits on x and y), determined by the size of the subset images such that $i + x < I$ and $j + y < J$ (Figure 6), thus limiting the maximum detectable shift from the origin. One significant disadvantage to this technique is that doubling the size of the search window would result in a fourfold increase in the number of calculations described by Eq. (1).

Using the discrete correlation theorem [91], one can calculate the XC of the entire image in the Fourier domain with significantly less computational expense. In this case, the XC computation is reduced to

$$C_n(I, J) = \mathcal{F}^{-1}[A_{n-1}(I, j)A_n^*(I, J)], \quad (2)$$

$$A_n(I, J) = \mathcal{F}[h(I, J)a_n(I, J)], \quad (3)$$

where n is the frame number, C is the XC image, $\mathcal{F}$ denotes the Fourier-transform operator [92], $\mathcal{F}^{-1}$ denotes the inverse Fourier-transform operator, A is the Fourier

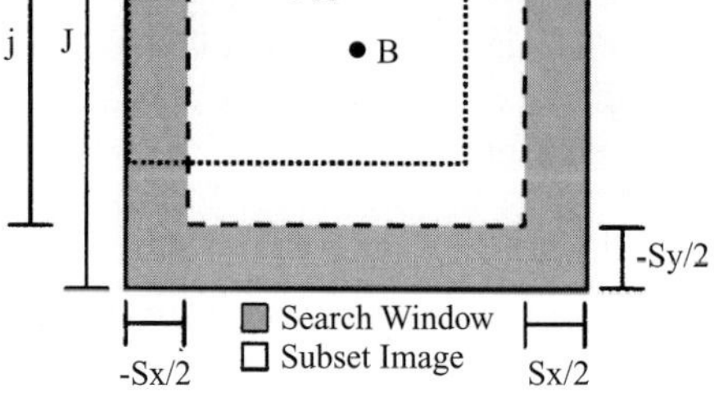

FIGURE 6. Dimensions for real-space calculation using Eq. (1). I and J are the number of pixels in the image, i and j are the number of pixels in the subset image, S_x and S_y determine the size of the search window. The location of the subset image at point A (dotted line) corresponds to the offset positions $x = -S_x/2$, $y = S_y/2$, and the location of the subset image at point B (dashed line) corresponds to offset positions $x = y = 0$.

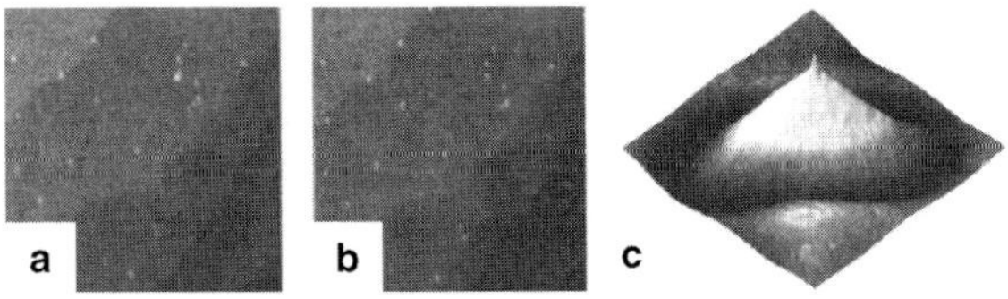

FIGURE 7. STM images of the conjugated oligomers inserted into a host SAM of dodecanethiol on Au{111}, frame 1 (a) and frame 15 (b). The cross-correlation of these two images results in (c) where, the maxima represents best alignment of images.

domain image, A^* denotes the complex conjugate of the Fourier domain image, h is a Hanning window [93], and a is the original image. In performing the Fourier domain XC, we found that a 2D Hanning window function was required to obtain the true XC maximum as the non-windowed Fourier transform generates a white-noise response along the horizontal and vertical center of the image resulting from the lack of periodicity at the image edges. This generates an overwhelming peak in the center of the XC image not related to the registration of the two images. An example XC image is shown in Figure 7, where (a) and (b) are the images to be cross-correlated, and the maximum in Figure 7(c) represents the best alignment of the two images.

Using Fourier domain XC maximum tracking, van Noort et al.were able to track the diffusion of DNA in sequential AFM images [83]. While our interests are in tracking adsorbates similar to this experiment, compensation for lateral tip-sample drift [71] must be performed before adsorbate analysis can be accomplished. The drift between each image is obtained by recording the deviation of the XC maxima pixel coordinates from the center of the image. We apply this procedure to track the relative movement of the surface and the probe tip to generate a drift track. The drift track is used to calculate the location and to extract a specific region from a sequence of images for further analysis.

While individual adsorbates could be used as the tracking images, this has the disadvantages that the adsorbate could change over time and lead to poor tracking, and potentially the XC calculation would have to be computed for each adsorbate. For these reasons, we calculate the XC for an entire image where large stationary features such as step edges and other surface defects act as the major alignment features. The drift track calculated from a full image should be applicable to all adsorbates, assuming no mid-image tip jumps occur in the image sequence. A method to correct for mid-image tip jumps is discussed below. Purely periodic surfaces such as well-resolved atomic lattices would be difficult to track with the Fourier transform (or any other) XC method because the periodic lattices would generate many XC maxima. However, the presence of any nonperiodic features such as defects or step edges eliminates this problem.

All of our calculations were implemented in Matlab [94]. Image sizes ranged from 256 $\times$ 256 to 400 $\times$ 400 pixels. While this calculation was done serially, the XC calculations are independent from frame to frame; therefore, a parallel implementation could be used for large-scale calculations. The computational expense of real-space convolution of an image of size N^2 and search window of size M^2 ($S_x = S_y = M$) yields approximately $N^2(M - N + 1)^2$ integer multiplications and the same number of integer additions. Whereas the Fourier domain method uses approximately $12M^2 \log_2 M$ floating point multiplications and $18M^2 \log_2 M$ floating point additions [79]. When M is much larger than N, the spatial correlation is more computationally advantageous. For our drift track calculations, the Fourier domain method is the most efficient.

After using a drift track to determine the region of interest to extract for a given adsorbate, we found that there was a cumulative error in our tracking algorithm such that as more images were tracked, the track became less accurate. For example, when a single stationary molecule was extracted with the calculated drift track, it should appear stationary in the extracted images; however, in this case, the molecule would appear to move or even to drift out of the extraction region. We attribute this error to fractional pixel drift. Because successive frames are compared and the Fourier-transform technique is limited to 1 pixel resolution, non-integer pixel drift cannot be detected. As each successive correlation is performed, the error propagation is additive. While this problem could be addressed with the more computationally expensive method of increased resolution via image interpolation [84], or the seventh-order polynomial fit to the XC peak used by Zheng and Klewicki [89], we have developed a key frame technique that has *no* impact on the computational load.

A key frame is an image with a frame number $(n + 1)$ that is a multiple of the key frame interval (k). The concept of key frames is that successive images are correlated, with the exception that every key frame is correlated with the previous key frame, thus altering Eq. (2) to $C_n(I, J) = \mathcal{F}^{-1}[A_{n-k}(I, J)A_n^*(I, J)]$ for key frames only. The effect of key frame interval size can be seen in Figure 8, where $k = 1$ represents no key frames (sequential XC), and $k = 5$ would compare frame

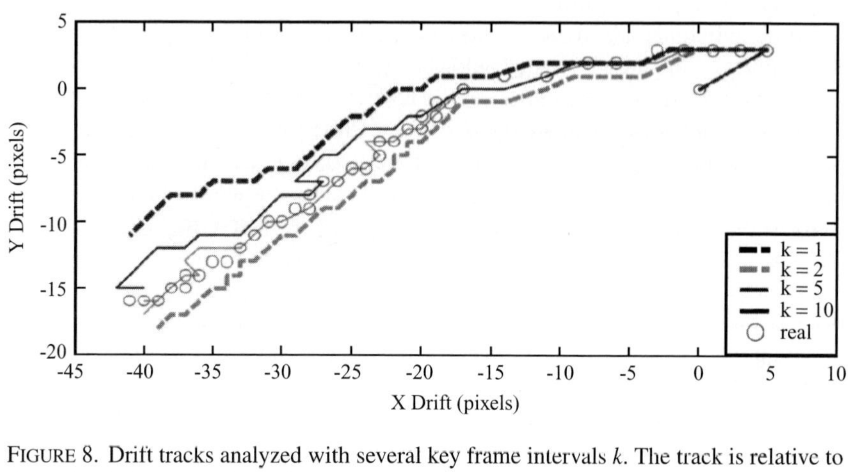

FIGURE 8. Drift tracks analyzed with several key frame intervals k. The track is relative to frame 1, which starts at (0,0). Each data point thereafter represents the pixel offset relative to frame 1.

2 to 1, 3 to 2, 4 to 3, 5 to 4, 6 to 1, and then 7 to 6. The circles in Figure 8 represent the "real" track, calculated by methods discussed in Sec. IV. The error with and without key frames can be seen by comparing the drift track for $k = 1$ and the real drift track in Figure 8, where the $k = 1$ track differs from the real track by 5 pixels for the last frame. For this series of images, $k = 5$ agrees well with the real track in the last frame and appears to give the optimal fit.

To determine the optimal key frame interval, a simulated image of a Gaussian peak was positioned on a flat background with a specified orthogonal velocity (pixels/frame) and tracked with this algorithm. Because the drift is known, a track error can be calculated. For integer velocities all values of k yielded a perfect track; however, for non-integer velocities the optimal k was found to be the inverse of the fractional component of the velocity, the number of frames required for the fractional component to propagate a full pixel of drift. For example, the optimal key frame interval for drift velocities of 0.2, 1.2, or 2.2 pixels/frame was $k = 5$. While this would imply that selection of k relies on knowing the drift, we have found that slower drifts (0–1 pixels/frame) track better with larger k values (2–10), and most all tracks are more accurate for $k > 2$. As a future improvement to this algorithm, an average velocity could be monitored to adjust k as the tracking progresses. Correlating all images to the first image was not done because subsequent images may have enough drift or change over time to the extent that they do not resemble the initial image.

3.3 Single-Molecule Switching

Using an insertion process described in detail above, single switch molecules were inserted at defect sites in self-assembled monolayers of n-dodecanethiolate, such as step edges and substrate vacancies, resulting in a low density of isolated switch molecules adsorbed on the surface (Figs. 7 and 9). When imaged, switches would spontaneously switch from their ON state, appearing in STM images as a 4–5 Å protrusion, to their OFF state appearing as a slight protrusion out of the SAM [48,51]. Persistence times of both states range from fractions of seconds to many hours.

Constant-current STM images represent a convolution of topography and electronic state (conductivity) of the surface. Therefore, a change in either or both will result in an apparent change in height. To measure this change in height, a movie spanning several hours of data acquisition would be tracked. Using the drift track, individual switch molecules are extracted and processed to calculate the apparent height of a switch molecule above the SAM surface throughout the course of the movie. Large scan areas on the order of $1{,}500 \times 1{,}500$ to $2{,}500 \times 2{,}500$ Å^2 [2] were acquired over a period of up to 25 h (400 images) to observe as many molecules and switching events as possible. While this tracking algorithm may be implemented in the acquisition software, we have currently used it only as a post acquisition tool to extract and to monitor the topography of each single switch.

The topographic height of a switch was obtained by extracting a 16 pixel $\times$ 16 pixel region from each image, and calculating the difference of the median of

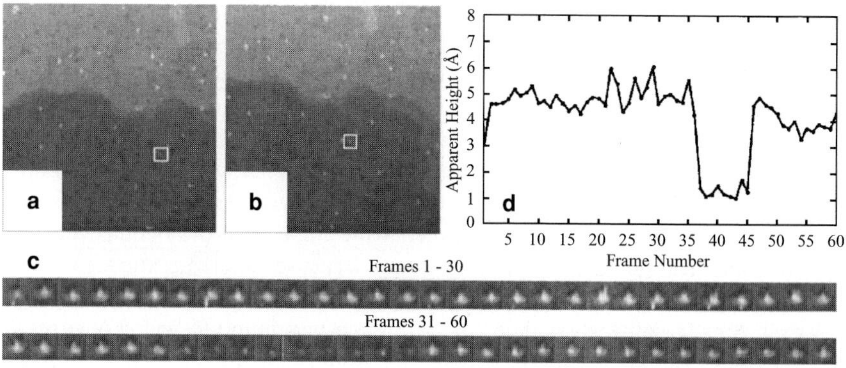

FIGURE 9. Topographic analysis of switch molecules. (a) and (b) are frames 1 and 60, respectively, from a series of time-lapse STM images. The white protrusions are individual switch molecules, the darker large spots are substrate vacancies or SAM domain boundaries. The white boxes represent the extracted region for one switch. Imaging conditions: $V_{sample} = -1.0$ V, $I = 0.8$ pA, 1,000 Å × 1,000 Å, 256 pixels × 256 pixels, frame acquisition = 2.07 min/frame, delay between frames 556 s. (c) Using the real track from Figure 8, a region was extracted from 60 frames of a movie, bicubicly interpolated (for improved display, height calculations do not use interpolated data) to twice the original pixel density, and median filtered. The 60 frames displayed were extracted from a 200-frame series acquired over 10 h; further switching was not observed after the displayed frames by this particular switch. Note that each frame has been rotated by 90° relative to (a) in the frame view. (d) The calculated height of the molecule above the SAM.

the 9 highest pixels in the extracted region and the median of the lowest fourth of the points from the extracted region. This method was used because of the different adsorption sites in which each switch could reside, such as in or beside a substrate vacancy, or on or next to a step edge. We found that this method gave reproducible topographic heights for most of the molecules analyzed, independent of the insertion site. An example of the time-resolved switching can be seen in Figure 9(c) and the corresponding calculated topographic height difference in Figure 9(d).

A consequence of this method is that an apparent height of zero will occur only if the image is flat and featureless. Because our images have corrugation from the superlattice of matrix molecules and noise, the minimum calculated height of a switch molecule was typically 2 Å. In addition, this method works best for well-isolated molecules; if adjacent switch molecules are inside the extracted region, their topographic height can influence the calculated apparent height of the switch molecule of interest, especially in the case of measuring a switch in an OFF state. Another complication for the height calculation is that the difference in height between the SAM and the switch can be sufficient to drive the STM feedback mechanism to overcompensate and to "overshoot" the real topographic height of the molecule and possibly to oscillate while scanning over the molecule, yielding falsely large apparent heights. To minimize these problems, the extracted region is median-filtered before the height calculation to eliminate any noise spikes or gross overcompensation of the feedback mechanism. To characterize the reversible conductance switching, a statistical analysis of the calculated height distributions was performed. The distribution shows the bimodal characteristic of the switches and the ON/OFF ratios of the molecules. From these data we were able to conclude that the switching observed by the STM was induced by a change in molecular conformation [51].

3.4 Controlling Matrix Order

While the functionality of the oligomer may not be essential for stochastic switching, the insertion matrix plays a critical role in controlling the guest molecules. By changing the defect density and film quality, we changed the rates and numbers of molecules that switched. As described in the experimental section, we compared three methods for controlling the defect density of our matrices: vapor annealing to decrease defect density, typical SAM deposition (24 h), and decreased deposition time to increase defect density (5 min).

Vapor-phase annealing was shown to be an effective means to fill defects in the monolayer [66]. If insertion of guest molecules is done prior to vapor-phase annealing, the vapor molecules have the opportunity to "backfill" around inserted molecules, as they cannot desorb into solution. It is assumed that desorption of the chemisorbed molecules from the film into the vapor is significantly reduced at 80°C, thus allowing us to tailor the local environment of inserted molecules.

To demonstrate the viability of the vapor annealing process, a decanethiolate film was prepared, and molecules of **2′** were inserted. This film was then annealed in dodecanethiol [$CH_3(CH_2)_{11}SH$] vapor. Introduction of dodecanethiol [$CH_3(CH_2)_9SH$] from the vapor phase allows us to place these molecules selectively at monolayer defect sites, domain boundaries and substrate step edges. The apparent difference in height of the dodecanethiolate and decanethiolate molecules is ~1.1 Å, as observed with STM [66], making it easy to distinguish between the two molecules. This type of mixed monolayer provides a convenient method to assess the effectiveness of the vapor annealing process.

An image of the inserted decanethiolate film vapor, annealed in dodecanethiol, is shown in Figure 10. The vapor-phase molecules have adsorbed along existing monolayer domain boundaries, as evidenced by the networked structure of the dodecanethiolate domains. Additionally, dodecanethiolate molecules (topographically higher areas) occupy most of the space along the step edges and around substrate vacancy islands. An inserted switch molecule is visible on the lower substrate terrace (shown in greater detail in the inset). Dodecanethiolate molecules supplied from the vapor are located around this molecule. Surrounding the guest molecule with vapor phase molecules may restrict its movement, if dodecanethiolate molecules occupy lattice sites previously unoccupied by the host matrix. This ability to "backfill" the host monolayer with vapor molecules is confirmation of our ability to tailor the local environment of inserted molecules in an attempt to regulate their behavior.

FIGURE 10. STM image of an 80 nm × 80 nm area of a decanethiolate (less protruding matrix regions) monolayer with **2′** inserted and vapor annealed in dodecanethiol (more protruding matrix regions). Inset is a 13 nm× 13 nm image of the same molecule. Both images were acquired at $V_s = -1$ V.

Vapor annealing is used to reduce defect density; the converse method is the use of a short monolayer deposition time. With a 5-min SAM solution deposition, the monolayer is chemically attached, essentially complete, but not fully organized. However, slower adsorption processes that serve to increase order in the film have not yet completed. This is because initial monolayer formation occurs rapidly, but film restructuring and further ordering occurs over long times during exposure to the alkanethiol solution. The result of the shorter adsorption time is a generally ordered film with high defect density.

4 Switching Mechanisms

With the successful observation by STM of conductance switching in a single molecule, the questions turn to which mechanisms are responsible for the observed changes in conductance, and how can the chemical configuration of the molecule and its surrounding matrix influence the electronic properties of the system.

4.1 Molecular Motion

To test the ability of the matrix to affect switching, a pure dodecanethiolate vapor annealed film with **1′** inserted was compared to an unannealed SAM with **1′** inserted. Similarly, **2′** inserted in a SAM prepared over 5 min was compared to a SAM produced from the standard 24 h solution deposition. The results of these experiments are summarized in Figure 11 and Table 1.

Each histogram in Figure 11 shows the distribution of heights for a representative sample of molecules extracted from a sequence of 200 images taken over 20 h

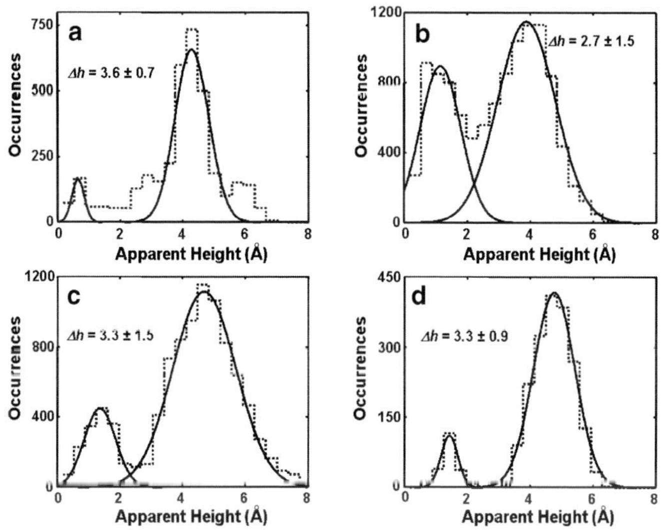

FIGURE 11. Height distributions of **1′** (c,d) and **2′** (a,b). All data were digitally extracted from series of images acquired over 20 h.

(summarized in Table 1). The apparent heights were calculated in the same fashion as for Figure 4(a–c). The bimodal distribution observed for the molecular heights in each sample is caused by bistability of the conductances of the inserted molecules; the peak at lower apparent height results from molecules in the OFF state, while the other peak corresponds to the ON state. Figure 11(a,b) show the distributions for **2′** in the (more-ordered) SAM deposited for 24 h and in the (more defective) SAM deposited for 5 min, respectively. The apparent height difference between the ON and the OFF states is approximately the same, regardless of the host matrix order. However, the measured fraction of molecules in the OFF state increases dramatically in the less-ordered film. Gaussian fits to the peaks reveal ON/OFF ratios of 12.3 for the more-ordered SAM and 2.1 for the less-ordered SAM (Table 1). The low ratio in the less-ordered SAM is a direct result of increased switching activity in this sample. The high ratio in the SAM adsorbed for 24 h may be caused by undersampling of molecules in the OFF state, caused by the less frequent switching in the more-ordered films; there are not necessarily fewer molecules in the

TABLE 1. Summary of data for the bimodal distributions shown in Figure 11

Matrix preparation	Insertion coverage (molecules/Å^2)	Inserted molecule	Number samples	% that switched	On/off radio
5 min	33×10^{-6}	**2′**	74	47	2.1
24 h	12×10^{-6}	**2′**	28	32	12.3
24 h	25×10^{-6}	**1′**	57	11	5.2
24 h, vapor anneable	6.7×10^{-6}	**1′**	15	7	9.2

OFF state. This undersampling problem arises because molecules continually in the OFF state are difficult to locate in large STM images, leading us to overlook switches that remain OFF through the many hours of data acquisition. Another result of the shorter SAM deposition time is an increased coverage of inserted molecules. The measured coverage of the inserted molecules in the less-ordered film was found to be 2.6 times that of the more-ordered film. The higher coverage is likely caused by the higher defect density of the less-ordered film, as there are more sites for insertion. Not only are there more defects, but some are likely larger, allowing more freedom of movement for the inserted molecules. This, in turn, leads to a higher switching activity.

The results presented in Figure 11(c,d) demonstrate that vapor annealing of the host monolayer also affects switching. The number of molecules observed in the OFF state decreased in the vapor-annealed film (Figure 11(d)) when compared with the standard film (Figure 11(c)). ON/OFF ratios were 5.2 for the unannealed film versus 9.2 for the vapor-annealed film (Table 1). Again, the higher ratio in the vapor-annealed film is indicative of lower switching activity. Thus, the host matrix can hinder switching when it is well ordered with low defect density, by obstructing motion of the guest molecules.

Details of the mechanism of the switching observed here are not yet fully understood. However, based on our observations and the data of others, we can make some inferences regarding the origins of switching.

The results for **1′** and **2′** demonstrate that the molecules are very similar in their stochastic switching behavior; switching of both molecules does not appear to depend on the specific chemical nature of the molecule but is strongly correlated to the order of the host matrix. While the data presented in Table 1 may imply that **1′** has a higher switching activity, a full analysis of several data sets (not presented here) suggests that there is no significant difference between **1′** and **2′** in their stochastic switching activity. This result eliminates the possibility that rotation of functional groups, such as the nitro group in **2′** [61], is responsible for the switching we observe. We conclude that the nitro functionality plays little role in the stochastic switching observed here, while the local environment plays a much more critical role in mediating stochastic switching. As mentioned above, further evidence of the non-specificity of the molecular functionality to switching is that we have also observed reversible switching for other families of molecules, including oligothiophenes [65].

It has been suggested that large conductance changes in molecular films of **2′** and **3′** may be due to an internal twist in the molecule that alters the conjugation of the molecule and thus change the conductance of the molecule [56]. For analogous systems, it has been calculated that the phenyl rings can freely rotate with respect to each other at temperatures as low as 30 K [57]. However, our experiments were performed at room temperature, and the states we observe can persist for several hours. It is unlikely that an unrestricted molecule would be able to maintain a static twisted or planar conformation for these time periods. For this mechanism to contribute significantly to the stochastic switching it is necessary for the host matrix to inhibit intramolecular rotation enough to stabilize each state. The viability of this mechanism requires the ability of the matrix to be somewhat fluid (or at least more conformationally relaxed) at defect sites where active molecules reside. They would have to be large enough to permit initial adsorption, but small enough to hinder intramolecular rotations.

To examine the possibility of an electronic or electrochemical mechanism for switching, we have studied this system over a range of voltages and currents. The magnitude of the tunneling current determines the number of electrons that are available to interact with local electronic states to induce such phenomena as local heating, electron charging, or electrolytic cleavage of chemical bonds [95]. The applied voltage creates large local electric fields that can interact with molecular or induced dipoles to produce molecular motion [96]. Time-lapse series of STM images have been acquired at bias voltages ranging from −1.4 to +1.4 V (with magnitudes as low as 250 mV), and at tunneling currents ranging from 0.1 to 4.0 pA. In this range of tunneling conditions, we have not observed significant correlation with the amount of switching. This observation diminishes the possibility of an electrochemical mechanism, in which changes in conductance of the molecules may be caused by a twist in the molecules that is induced by reduction of the molecules [56]. Because such charging effects are intimately related to the applied voltages and magnitude of tunneling currents in STM experiments, it is difficult to reconcile our observations with a mechanism that is dominated by electronic effects such as reduction or charging.

The switching behavior observed here is consistent with tilting of the inserted molecules or a change in the orientation of the molecule with respect to the STM tip. This type of motion can cause a change in the physical height of the molecule above the host matrix, which would result in a change in the apparent height of the molecule as observed with STM. This type of motion of the molecule would result from a change in the angle of the gold-sulfur-molecule bonding geometry. Experimental [97] and theoretical [98] studies have suggested two energetically similar geometries for thiolate monolayers bound to gold, due to varying hybridization of the sulfur atom (sp vs. sp^3). The energy barrier for the transition between these two states may be as low as 2.5 kcal/mol [98]; switching between these states is feasible for molecules at room temperature. The speed of such a switching mechanism would be limited by the rotational frequencies of the molecules involved to a few GHz. Physical changes such as tilting can also result in large changes in molecular conductance. A change in the molecular orientation with respect to the electrodes (the substrate and the STM tip) can cause a variation in the effective conductance of the molecule by two orders of magnitude [99]. This effect is caused by a change in the overlap of the π molecular orbitals with the orbitals of the metal electrodes. Of course, a tilting mechanism would require a large defect site in the matrix to allow significant movement of the molecule. Additionally, because there are only two observed states for these molecules, this implies that there are just two preferred tilts of the molecule, independent of defect type and size. A concerted motion involving attached substrate atoms is also possible.

As to the roles of the functionality in **2′** and **3′**, we speculate that the permanent dipole in the molecules caused by the functional groups may be useful in deliberately switching the molecules [51,53]. Large electric fields applied with the STM tip or in nanopores may interact with these dipoles, providing a "handle" for applying torque and changing the tilt of the inserted molecules. As a final note, we have recently observed switching behavior in these systems using a UHV STM, operating at room temperature. Conducting the experiments in vacuum enables us to confirm that the switching is not related to interactions with surface contaminants or the ambient atmosphere.

4.2 Internal Molecular Rotation

To understand what effect the geometry of molecules has on their conductance states, 2-thiophenanthrene and 4-thiobiphenyl molecules were inserted into octanethiolate SAMs and were studied on individual bases with the STM [100]. Insertion of the molecules into SAMs reproducibly isolates each molecule (or small bundle of molecules) from one another and restricts them to the SAM defects and gold step edges, thus controlling the placement of the molecule within the SAM. At these sites, the guest molecules insert nearly normal to the surface with the headgroups adhering to the exposed gold and the conjugated region of the molecule sticking up out of the SAM. Similar to the methods described above, large areas (1,200–1,700 Å on a side) were imaged repeatedly for 10 or more hours. The data for each molecule were extracted from those images and the statistics for many individual molecules were extracted simultaneously. Using tracking software, the data extracted from each image were correlated to the next image to piece together the conductance history (as observed by changes in apparent height) over the duration of the time-lapse sequence. An example apparent height versus time plot for the duration of the movies is given for each type of molecule in Figure 12. As discussed elsewhere, the apparent height changes of the molecules throughout the duration of the movie are not correlated to any changes in tip geometry, changes of the gold substrate, or movement across substrate step edges. It was also possible to distinguish single molecules from bundles of molecules and to interpret separately each molecule's conductance state. It should be noted again that this sampling method favors molecules that started in or switch to a high conductance state, as molecules in a high conductance state were more easily discernible from the surrounding matrix and were therefore more likely to be included.

As shown by the histogram (extracted from a 250-frame sequence taken over a 12-h period) in Figure 12(a), the 4-thiobiphenyl demonstrates two conductance states. From Gaussian fits to the distributions, the difference in apparent height of the ON and OFF states is 4.5 ± 0.9 Å. The apparent height of the molecule was measured from the apparent height of the SAM and changed with the height of the SAM. Although an OFF state was usually still visible protruding from the SAM, any molecule that fell below the level of the SAM would have an apparent height of zero resulting in narrower distributions for the OFF states.

Thc 2-thiophcnanthrene also exhibited two conductance states, with an apparent height difference between them of 5.1 ± 1.2 Å. The topographic height of these two molecules normal to the surface is comparable, so the deviations in apparent

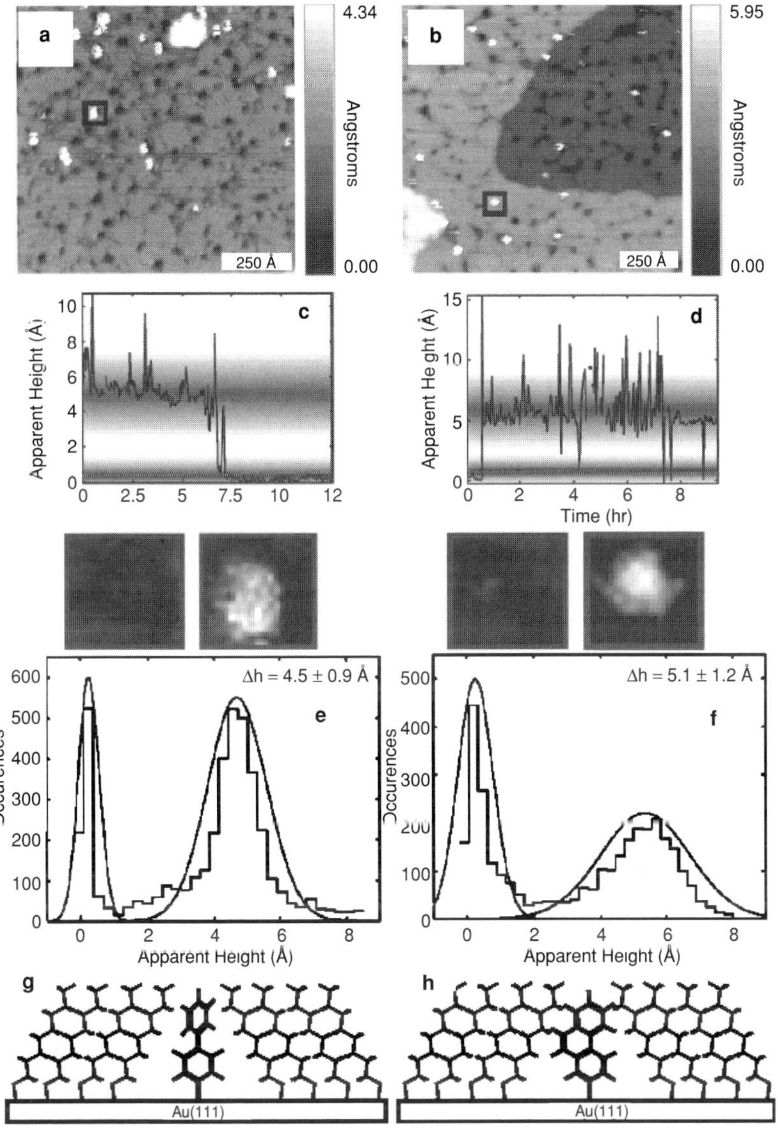

FIGURE 12. Height distributions for 4-thiobiphenyl (a) and 2-thiophenanthrene (b). For each distribution the ON and OFF heights are shown with typical correlating images and a height versus time plot (c,d) for one representative molecule (black circle). The distribution data (e,f) were digitally extracted from sequential images taken over 10 h. Tunneling conditions: 1,200 Å $\times$ 1,200 Å, sample bias: 1 V, current: 3 pA. (a) Octanethiol SAM with 4-thiobiphenyl inserted $h_{\text{on}} = 4.7 \pm 0.9$ Å, $h_{\text{off}} = 0.3 \pm 0.2$ Å (b) Octanethiol 24 h SAM with 2-thiopehnanthrene inserted $h_{\text{on}} = 5.4 \pm 1.1$ Å $_{\text{off}} = 0.4 \pm 0.2$ Å. Schematics (g,h) of the corresponding molecules inserted into an octanethiolate matrix are shown in the bottom of the figure [100].

height were associated with differences in conductance. It is expected that the 2-thiophenanthrene molecule would be more conductive because of the increased conjugation of the middle ring extending the overlap of the π orbitals. The existence of switching in the 2-thiophenanthrene molecule eliminates internal ring rotation as the sole mechanism responsible for the change in conductance states. In addition, internal rotation of biphenyl at room temperature would give states that are too short-lived to be observed by these methods.

Qualitatively, the 2-thiophenanthrene molecules switch states less frequently and favor the OFF state. This is quantitatively verified by the calculated ON/OFF ratios (calculated from the areas under the Gaussian fits) of 2.45 for the 2-thiophenanthrene and 4.61 for the 4-thiobiphenyl molecules. Considering that both molecules are in similar environments, this difference can be explained by the degrees of freedom available to each molecule. In both cases, the alkanethiolate SAM is the surrounding matrix. Due to size and rotation constraints, the 2-thiophenanthrene molecule needs larger defect sites to be able to insert and to have the same degrees of freedom afforded to the 4-thiobiphenyl. For the same defect site, one would expect that the biphenyl would be more likely to switch, but we do not know if the distributions of sites at which the two molecules insert are equivalent. Theory predicts that a tilted thiolate bond is energetically favorable and a more conductive (ON) molecular conformation (vide intra), although it is not yet possible to measure experimentally only the thiolate-substrate bond angle with the STM. It is expected that a molecule that is increasingly constrained by the host SAM is less likely to undergo motion that will result in a dramatic change of conductivity.

4.3 Molecular Latching

While not only the functionality of the molecular wire candidate affects the stochastic switching, the insertion matrix also plays a critical role in controlling the electronically active guest molecules. Toward this end, we have designed a system that mediates stochastic switching of **2′** using localized electric fields and hydrogen bonding between the inserted **2′** and its host matrix [101].

Here, we report a method to mediate the switching of inserted OPE molecules by varying their local chemical environment. We have inserted **2′** into host amide-containing alkanethiol SAMs that are composed of 3-mercapto-N-nonyl-propionamide (1ATC9). The embedded amide moiety in 1ATC9 leads to extended hydrogen bonding throughout the monolayer matrix, thereby increasing the crystallinity of the film and limiting the conformational motion of inserted **2′**. The internal amide group can also hydrogen bond with the -NO_2 substituent of inserted **2′**, resulting in bias-dependent switching as shown by the schematic in Figure 13.

The polarity of the applied electric field between the STM tip and the substrate strongly influence the switching behavior of **2′** and leads to bias-dependent switching. At positive sample bias, **2′** has a preference for the ON state, by the large ON/OFF ratio calculated from the histograms generated by the molecular

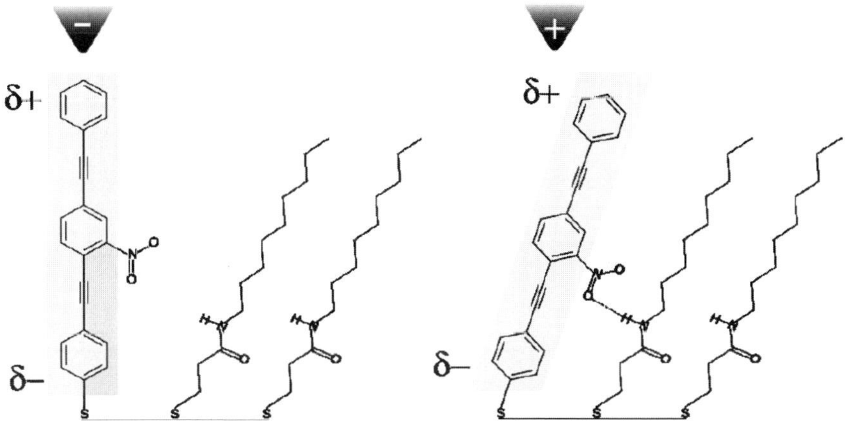

FIGURE 13. Schematic showing the possible mechanism for bias-induced switching via a polar host matrix.

tracking algorithms. Conversely, we observe a strong preference for the OFF state at negative sample bias. The trend in the height distribution of **2′** for a series of images taken over a 1,500 Å× 1,500 Å area of 2′/1ATC9 for ~8.3 h at negative sample bias. The OFF and ON peaks occur at approximately the same apparent topographic heights as at positive sample bias (1.0 ± 0.6 Å and 6.7 ± 1.3 Å, respectively; $\Delta h = 5.7 \pm 1.5$ Å) [101]. However, it is apparent from the histograms that the OFF state was significantly populated at negative sample bias (ON/OFF peak area ratio of 1.10) [101]. This low ratio is indicative of a high number of molecules initially in the ON state switching to the OFF state for the duration of imaging. It is important to note that the switching activity is still limited by the hydrogen-bonded matrix at negative sample bias; that is, molecules typically remain in the OFF state and do not undergo subsequent switching events. This phenomenon is evident in Figure 14(a–d), which shows four selected 1,400 Å × 1,400 Å areas extracted from the images used to generate the histogram. The images were taken ~165 min apart. In the first image, there are many **2′** molecules in the ON state (indicated as protrusions). As the imaging progressed, the **2′** molecules in the imaging field were affected by the applied electric field for a longer period of time, which resulted in an overall decrease in the number of **2′** molecules in the ON state. At the end of the series of images, almost all of the **2′** molecules initially in the ON state switch to and remain in the OFF state.

The bias dependence observed for **2′** in 1ATC9 SAMs is reversible in that the conductance state of the molecules can be changed by reversing the polarity of the applied bias. Subsequent images acquired after changing to positive sample bias show that it is possible to drive the **2′** molecules back into the ON state, see Figure 14(e,f). The reversible switching of **2′** molecules excludes the possibility that they are removed from the surface by the STM tip. In addition, it further demonstrates control over the state of the inserted NPPB molecules using the applied electric field.

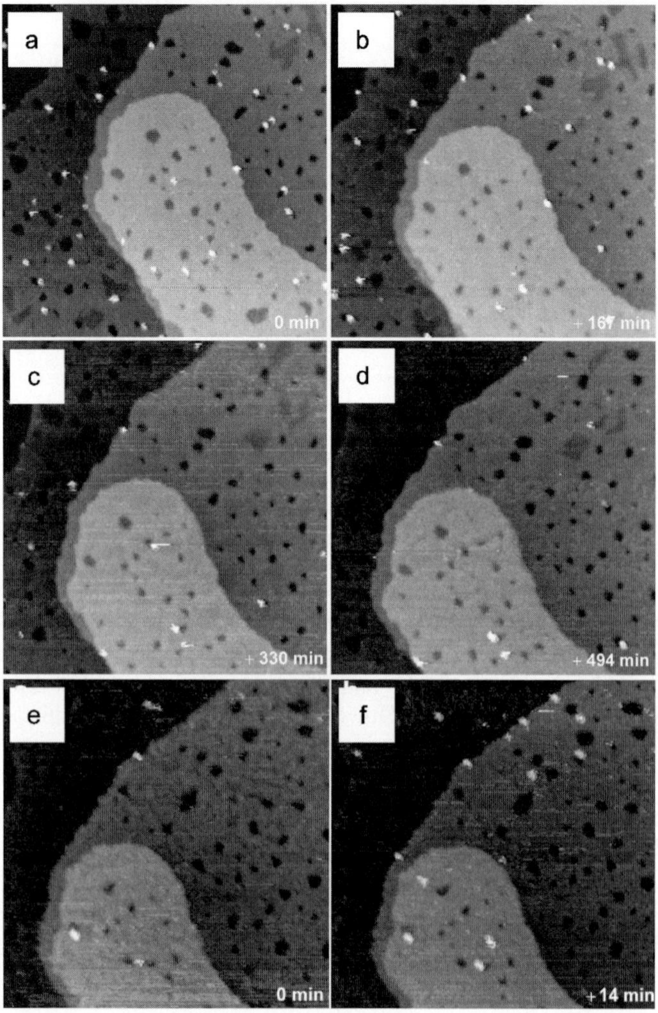

FIGURE 14. (a-d) Selected representative images taken at ~165 min intervals throughout a series of time-lapse images at −1.0 V sample bias. The bright protrusions are NPPB molecules in the ON state. 1400 Å × 1400 Å, $V_{sample} = -1$ V, I = 2 pA. (e) STM image taken at the end of a series of time-lapse images showing most of the NPPB molecules in the OFF state at negative sample bias. $V_{sample} = -1$ V, I = 2 pA. (f) STM image taken in the same area 14 min after switching polarity to positive sample bias. Some of the NPPB molecules have switched back to the ON state. $V_{sample} = +1$ V, I = 2 pA. [from ref. 101]

This result demonstrates a method for mediating conformational switching of **2′** by modifying the chemical functionality of the host matrix. Using this strategy, we can induce bias-dependent switching of individual **2′** molecules, which was formerly not possible based on the intrinsic characteristics of **2′**. This shows that the chemical and physical environments of proposed molecular devices are crucial to their function and can be exploited to impart tunable electronic properties. Currently, we are investigating the bias-dependence mechanism and tilt angles by incorporating substituents on the inserted OPE that alter the degree and location of hydrogen bonding between the inserted molecules and the host matrix.

5 Conclusions

We have used our control over monolayer formation, dynamics, and interactions to observe switching of single molecules with STM, manifested as changes in the apparent heights of molecules inserted into host matrices of alkanethiols. We have demonstrated that the nature of defects in the host films can control the motion of inserted molecules and affect the amount of switching. We have also shown that the local environment plays an extremely important role in determining the behavior of single molecules, and may be an essential consideration for single molecule electronic design.

Acknowledgments. We would like to acknowledge the contributions to this work by our highly productive and stimulating collaborators: Profs. Dave Allara, Bob Grubbs, Jim Hutchison, Jim Tour, and Marcus Weck and their research groups. We also thank and are deeply indebted to Lloyd Bumm, Arrelaine Dameron, Zach Donhauser, Penelope Lewis, Brent Mantooth, Sanjini Nanayakkara, Tom Pearl, and Rachel Smith for their experimental assistance. The continuing financial support of the Army Research Office, Defense Advanced Research Projects Agency, National Institutes of Standards and Technology, National Science Foundation, Office of Naval Research, and the Semiconductor Research Corporation is gratefully acknowledged.

References

1. P. Rai-Chaudhury, *Handbook of Microlithography, Micromachining, and Microfabrication*; SPIE: London, (1997).
2. Y. N. Xia and G. M. Whitesides, Ang. Chem. Int. Ed. **37**, 551 (1998).
3. X. M. Zhao, Y. N. Xia, and G. M. Whitesides, J. Mat. Chem. **7**, 1069 (1997).
4. R. S. Becker, J. A. Golovchenko, and B. S. Swartzentruber, Nature **325**, 419 (1997).
5. D. M. Eigler and E. K. Schweizer, Nature **344**, 524 (1990).
6. P. S. Weiss and D. M. Eigler, in NATO ASI Series E: Applied Sciences **235**, 213 (1993).
7. J. K. Gimzewski and C. Joachim, Science **283**, 1683 (1999).
8. S.-W. Hla, L. Bartels, G. Meyer, and K.-H. Rieder, Phys. Rev. Lett. **85**, 2777 (2000).
9. A. Ulman, *An Introduction to Ultrathin Organic Films: From Langmuir-Blodgett to Self-Assembly*; Academic: San Diego, (1991).
10. A. Ulman, Chem. Rev. **96**, 1533 (1996).
11. R. G. Nuzzo and D. L. Allara, J. Am. Chem. Soc. **105**, 4481 (1983).

12. D. L. Allara, Biosens. Bioelec. **10**, 771 (1995).
13. C. D. Bain and G. M. Whitesides, J. Am. Chem. Soc. **111**, 7164 (1989).
14. C. D. Bain and G. M. Whitesides, Langmuir **5**, 1370 (1989).
15. L. H. Dubois and R. G. Nuzzo, Ann. Rev. Phys. Chem. **43**, 437 (1992).
16. G. E. Poirier, Chem. Rev. **97**, 1117 (1997).
17. S. Hong, J. Zhu, and C. A. Mirkin, Science **286**, 523 (1999).
18. G.-Y. Liu, S. Xu, and Y. L. Qian, Acc. Chem. Res. **33**, 457 (2000).
19. M. Zharnikov, S. Frey, K. Heister, and M. Grunze, Langmuir **16**, 2697 (2000).
20. A. Gölzhäuser, W. Geyer, V. Stadler, W. Eck, M. Grunze, K. Edinger, T. Weimann, and P. Hinze, J. Vac. Sci. Technol. B **18**, 3414 (2000).
21. K. Heister, M. Zharnikov, M. Grunze, L. S. O. Johansson, and A. Ulman, Langmuir **17**, 8 (2001).
22. J. Collet, O. Tharaud, A. Chapoton, and D. Vuillaume, Appl. Phys. Lett. **76**, 1941 (2000).
23. J. Collet, and D. Vuillaume, Appl. Phys. Lett. **73**, 2681 (1998).
24. C. Boulas, J. V. Davidovits, F. Rondelez, and D. Vuillaume, Phys. Rev. Lett. **76**, 4797 (1996).
25. S. A. Lee, Y. Yoshida, M. Fukuyama, and S. Hotta, Syn. Metals **106**, 39 (1999).
26. C. D. Bain and G. M. Whitesides, J. Am. Chem. Soc. **110**, 6560 (1988).
27. C. D. Bain, J. Evall, and G. M. Whitesides, J. Am. Chem. Soc. **111**, 7155 (1989).
28. J. P. Folkers, P. E. Laibinis, G. M. Whitesides, and J. Deutch, J. Phys. Chem. **98**, 563 (1994).
29. R. K. Smith, S. M. Reed, P. A. Lewis, J. D. Monnell, R. S. Clegg, K. F. Kelly, L. A. Bumm, J. E. Hutchison, and P. S. Weiss, J. Phys. Chem. B **105**, 1119 (2001).
30. S. J. Stranick, A. N. Parikh, Y.-T. Tao, D. L. Allara, and P. S. Weiss, J. Phys. Chem. B **98**, 7636 (1994).
31. S. J. Stranick, S. V. Atre, A. N. Parikh, M. C. Wood, D. L. Allara, N. Winograd, and P. S. Weiss, Nanotechnology **7**, 438 (1996).
32. G. Binnig, H. Rohrer, C. Gerber, and E. Weibel, Appl. Phys. Lett. **40**, 178 (1982).
33. G. Binnig, C. F. Quate, and C. Gerber, Phys. Rev. Lett. **56**, 930 (1986).
34. E. Delamarche, B. Michel, C. Gerber, D. Anselmetti, H.-J. Güntherodt, H. Wolf, and H. Ringsdorf, Langmuir **10**, 2869 (1994).
35. D. Anselmetti, A. Baratoff, H.-J. Güntherodt, E. Delamarche, B. Michel, C. Gerber, H. Kang, H. Wolf, and H. Ringsdorf, Europhys. Lett.**27**, 365 (1994).
36. N. Camillone, P. Eisenberger, T. Y. B. Leung, P. Schwartz, G. Scoles, G. E. Poirier, and M. J. Tarlov, J. Chem. Phys. **101**, 11031 (1994).
37. M. T. Cygan, T. D. Dunbar, J. J. Arnold, L. A. Bumm, N. F. Shedlock, T. P. Burgin, L. Jones, D. L. Allara, J. M. Tour, and P. S. Weiss, J. Am. Chem. Soc. **120**, 2721 (1998).
38. L. A. Bumm, J. J. Arnold, M. T. Cygan, T. D. Dunbar, T. P. Burgin, L. Jones II, D. L. Allara, J. M. Tour, and P. S. Weiss, Science **271**, 1705 (1996).
39. V. J. Langlais, R. R. Schlittler, H. Tang, A. Gourdon, C. Joachim, and J. K. Gimzewski, Phys. Rev. Lett. **83**, 2809 (1999).
40. F. Moresco, G. Meyer, K.-H. Rieder, H. Tang, A. Gourdon, and C. Joachim, Phys. Rev. Lett. **86**, 672 (2001).
41. Z. J. Donhauser, B. A. Mantooth, K. F. Kelly, L. A. Bumm, J. D. Monnell, J. J. Stapleton, D. W. Price, A. M. Rawlett, D. L. Allara, J. M. Tour, and P. S. Weiss, Science, **292**, 2303 (2001).
42. C. Joachim, J. K. Gimzewski, R. R. Schlittler, and C. Chavy, Phys. Rev. Lett. **74**, 2102 (1995).

43. M. A. Reed, J. Chen, A. M. Rawlett, D. W. Price, and J. M. Tour, Appl. Phys. Lett. **78**, 3735 (2001).
44. J. Chen, W. Wang, M. A. Reed, A. M. Rawlett, D. W. Price, and J. M. Tour, Appl. Phys. Lett. **77**, 1224 (2000).
45. J. Chen, M. A. Reed, A. M. Rawlett, and J. M. Tour, Science, **286**, 1550 (1999).
46. J. M. Seminario, A. G. Zacarias, and J. M. Tour, J. Am. Chem. Soc. **122**, 3015 (2000).
47. J. M. Seminario, A. G. Zacarias, and J. M. Tour, J. Am. Chem. Soc. **120**, 3970 (1998).
48. J. M. Tour, Chem. Rev. **96**, 537 (1996).
49. J. M. Tour, M. Kozaki, and J. M. Seminario, J. Am. Chem. Soc. **120**, 8486 (1998).
50. J. M. Tour, W. A. Reinerth, L. Jones II, T. P. Burgin, C. W. Zhou, C. J. Muller, M. R. Deshpande, and M. A. Reed, in Molecular Electronics: Science and Technology **852**, 197 (1998).
51. M. Di Ventra, S. G. Kim, S. T. Pantelides, and N. D. Lang, Phys. Rev. Lett. **86** 288 (2001).
52. M. Weck, J. J. Jackiw, P. S. Weiss, and R. H. Grubbs, Proc. Poly. Mat. Sci. Eng. **79**, 72 (1998).
53. M. Weck, J. J. Jackiw, R. R. Rossi, P. S. Weiss, and R. H. Grubbs, J. Am. Chem. Soc. **121**, 4088 (1999).
54. L. F. Charles, Masters thesis, The Pennsylvania State University: University Park (1999).
55. H. Sakaguchi, K. F. Kelly, Z. J. Donhauser, P. A. Lewis, and P. S. Weiss, manuscript in preparation.
56. Z. J. Donhauser, D. W. Price II, J. M. Tour, and P. S. Weiss, J. Am. Chem. Soc. **125**, 11462 (2003).
57. Molecular Imaging Corp., Phoenix, AZ, USA.
58. B. A. Mantooth, Z. J. Donhauser, K. F. Kelly, and P. S. Weiss, Rev. Sci. Instrum. **73**, 313 (2002).
59. J. F. Jorgensen, L. L. Madsen, J. Garnaes, K. Carneiro, and K. Schaumburg, J. Vac. Sci. Technol. B **12**, 1698 (1994).
60. J. F. Jorgensen, K. Carneiro, L. L. Madsen, and K. Conradsen, J. Vac. Sci. Technol. B **12**, 1702 (1994).
61. Supplemental material: An example set of movies demonstrating drift and image extraction can be found at http://www.nano.psu.edu/supplemental/tracking.html
62. D. W. Pohl and R. Moller, Rev. Sci. Instrum. **59**, 840 (1988).
63. M. Aketagawa, K. Takada, Y. Minao, Y. Oka, and J. D. Lee, Rev. Sci. Instrum. **70**, 2053 (1999).
64. B. S. Swartzentruber, Phys. Rev. Lett. **76**, 459 (1996).
65. L. J. Lauhon, and W. Ho, J. Chem. Phys. **111**, 5633 (1999).
66. S. Renisch, R. Schuster, J. Wintterlin, and G. Ertl, Phys. Rev. Lett. **82**, 3839 (1999).
67. G. Tziritas, and C. Labit, *Motion Analysis for Image Sequence Coding*, Elsevier Science B. V., New York (1994).
68. C. Stiller, and J. Konrad, IEEE Signal Process. Mag. **16**, 70 (1999).
69. J. P. Lewis, http://www.idiom.com/;zilla/Papers/nvisionInterface/nip.html, Industrial Light & Magic.
70. D. I. Barnea, and H. F. Silverman, IEEE Trans. Comput. **21**, 179 (1972).
71. *Handbook of Image and Video Processing,* edited by A. Bovik; Academic Press, San Diego, CA (2000).
72. W. D. Lockwood, and A. P. Reynolds, Mater. Charact. **42**, 123 (1999).

73. S. J. T. van Noort, K. O. van der Werf, B. G. de Grooth, and J. Greve, Biophys. J. **77**, 2295 (1999).
74. J. Romer, M. Plaschke, and J. I. Kim, Ultramicroscopy **85**, 99 (2000).
75. G. C. Rosolen, and W. D. King, Scanning **20**, 495 (1998).
76. L. A. Bumm, and P. S. Weiss, Rev. Sci. Instrum. **66**, 4140 (1995).
77. J. J. Jackiw, Doctoral thesis, The Pennsylvania State University: University Park (2001).
78. J. C. Russ, *The Image Processing Handbook,* 2nd ed. (Chemical Rubber, Ann Arbor) 1995.
79. Q. X. Zheng, and J. C. Klewicki, Meas. Sci. Technol. **11**, 1282 (2000).
80. W. H. Huang, W. W. Wang, A. D. Xia, N. Jin, and Z. Q. J. Hu, J. Vac. Sci. Technol. B **18**, 2027 (2000).
81. W. H. Press, B. P. Flannery, S. A. Teukolsky, and W. T. Vetterling, *Numerical Recipes, the Art of Scientific Computing* (Cambridge University, Cambridge) 1986.
82. Fast Fourier transform libraries often return a Fourier domain image where the dc peak is in the corners of the image; for this calculation the dc peak must be shifted to the center of the image. In addition, note that the multiplication of Fourier domain images is elemental, and not matrix multiplication. When computing the inverse Fourier transform, the correlation image is flattened into only real numbers by calculating the absolute value of the real and imaginary (resulting from rounding errors) components returned by a Fourier transform.
83. A. V. Oppenheim, and R. W. Schafer, *Discrete-Time Signal Processing* (Prentice Hall, Englewood Cliffs, NJ), pp. 63–67, 447, 746–747, 839–842 (1989).
84. The MathWorks, Inc., 3 Apple Hill Drive, Natick, MA 01760-2098.
85. C. F. Herrmann and J. J. Boland, J. Phys. Chem. B **103**, 4207 (1999).
86. H. C. Akpati, P. Norlander, L. Lou, and P. Avouris, Surf. Sci. **372**, 9 (1997).
87. Y.-T. Tao, C. C. Wu, J. Y. Eu, W. L. Lin, and K. C. Wu, Langmuir **13**, 4018 (1997).
88. H. Sellers, A. Ulman, Y. Shnidman, and J. E. Eilers, J. Am. Chem. Soc. **115**, 9389 (1993).
89. P. E. Kornilovitch, and A. M. Bratkovsky, Phys. Rev. B **64**, 5413 (2001).
90. A. A. Dameron, J. W. Ciszek, J. M. Tour, and P. S. Weiss, J. Phys. Chem. B **108**, 16761 (2004).
91. P. A. Lewis, C. E. Inman, Y. Yao, J. M. Tour, J. E. Hutchison, and P. S.Weiss, J. Am. Chem. Soc. **126**, 12214 (2004).

II.4 Scanning Probe Microscopy of Individual Carbon Nanotube Quantum Devices

C. STAII, M. RADOSAVLJEVIC, AND A. T. JOHNSON

Since the discovery of carbon nanotubes, considerable insight into their electronic properties has been gained by electrical transport characterization of individual nanotube devices. At the same time, both experimental and theoretical investigations found that many of the phenomena observed in transport studies of these devices are not due to bulk properties of nanotubes, but depend instead on local atomic-scale details including defects and impurities within the nanotube itself and interfaces between the nanotube and microfabricated electrodes. This prompted application of a number of scanning probe microscopies (SPM), where local probes are used to both stimulate the nanotube and its response to localized electrostatic potentials. In this chapter, we review the state of the art in several methods of SPM, including scanning gate microscopy (SGM), electrostatic force microscopy (EFM), and current-sensitive and thermal SPM, as they are applied to study physical phenomena in carbon nanotube devices at both room and low temperatures.

1 Introduction to Carbon Nanotubes

Carbon is unique among chemical elements in its ability to assume a wide variety of different structures and forms. In addition to the sp^3-bonded diamond, sp^2-bonded graphite, and fullerenes [1], the fourth form of carbon, the sp^2-bonded carbon nanotubes, was first discovered in 1991 by S. Iijima [2], in the form of multiple coaxial carbon fullerene shells (multi-wall carbon nanotubes, mwCN). Later, in 1993 single fullerene shells (single-wall carbon nanotubes, swCN) were synthesized using transitional metal catalysts [3,4]. A swCN can be simply described as a sheet of graphene coaxially rolled to create a cylindrical surface [5]. In this way the two-dimensional (2D) hexagonal lattice of graphene is mapped onto a cylinder, leading to a unique situation where geometric arrangement of carbon atoms around the CN cylinder determines the electronic properties of this material [5]. Indeed, experimental measurements [6–10] have confirmed the theoretical predictions [11,12] that, depending on their diameter and chirality, swCNs could be either metallic or semiconducting.

The discovery of CNs has sparked new research areas in physics, chemistry, and materials science with observation of new phenomena and important implications in the development of unique technological devices. From the fundamental point of view CNs provide a beautiful model system for studying one-dimensional (1D) physics. Such diverse phenomena as Luttinger liquid behavior [13,14], single-electron tunneling [10,15], Aharonov–Bohm oscillations [16], and photoluminescence [17–19] have been observed in CN devices. On the other hand, their outstanding mechanical, electronic, and thermal properties offer great potential for applications in areas as diverse as materials design [20], molecular electronics [8,21], nanoelectromechanical systems [22], optoelectronics [19,23], and biodetection and chemical sensing [24–29].

For example, direct measurements using atomic force microscopy (AFM) have shown that the Young modulus of mwCNs is approximately 1.28 TPa [30], which is about one order of magnitude larger than that of steel. These excellent mechanical characteristics suggest that CNs have potential for applications in fabricating superstrong composites, and that they could be unique force transducers to the molecular world.

During the past few years, individual molecular electronic devices based on CNs, such as field-effect transistors [8,21], nanodiodes [31], memory elements [32–35], and even color display [36] have been fabricated. In addition, Coulomb blockade phenomena detected in transport measurements demonstrate that the nanotubes are suitable building blocks for single-electron transistors [9,15,37,38]. Structural and mechanical deformations also have important effects on the electronic properties of swCNs. Experimental studies have shown the possibility of tuning the electronic structure of swCNs by inducing mechanical deformations via an AFM tip [39]. Conversely, the electrostatically driven mechanical response of CNs shows their utility in nanomechanical applications such as tweezers [22] and actuators [40].

There is already a large number of excellent reviews in the literature dedicated to this subject [41–44]. For an excellent general introduction into the field of CNs we refer the reader to ref. [5].

Successful implementation of swCNs as nanoelectronic circuit elements requires reliable techniques for measuring their local electronic properties and structure. The atomic force microscope (AFM) [45–47] is the most versatile scanning probe available, and all the electrostatic scanning probe techniques (SPM) described in this chapter rely on the AFM to position the tip close to substrate, and at a number of well defined locations along the nanotube device.

The remainder of this chapter consists of four sections: (II) scanning gate microscopy (SGM) for locating defects within nanotubes, (III) scanning impedance microscopy (SIM) for potential profiling along nanotube devices and scanning conductance microscopy (SCM), (IV) tunneling AFM (TUNA) and movable current injecting electrodes, and (V) novel AFM-based techniques, including single-electron electrostatic force microscopy (e-EFM) and scanning thermal microscopy (SThM). In each case we begin by briefly describing the AFM-based technique, followed by its applications to nanotube devices.

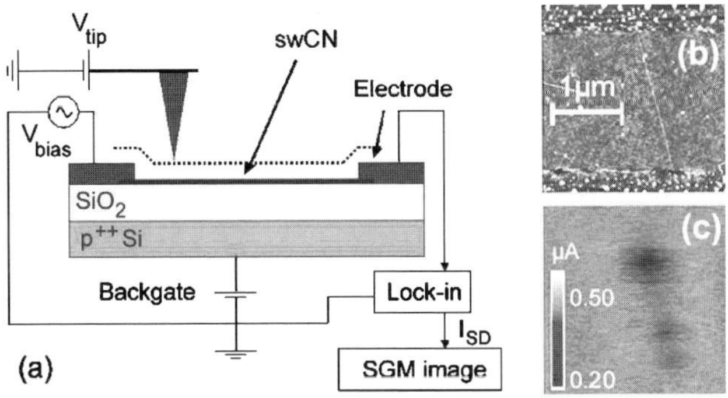

FIGURE 1. (a) Schematic of SGM. (b) AFM image of a swCN-FET circuit. The diameter of the swCN is 1 nm. (c) SGM image for the same device, with $V_{tip} = +8$ V, $V_{bias} = 100$ mV, source-drain ac frequency $f = 10$ kHz, backgate voltage $V_G = 0$ V, and scan height $h = 15$ nm. The positive tip voltage depletes holes locally; the current I_{SD} decreases when the tip is above defect sites. Reproduced with permission from [35].

2 Scanning Gate Microscopy for Locating Defects Within Nanotubes

Scanning gate microscopy (SGM) [48–52] is a very powerful technique for measuring the local electronic properties of nanoscale circuits. In SGM imaging mode (Figure 1(a)), a conducting tip with applied voltage V_{tip} is scanned at a fixed height over an electrically biased sample, and the transport current through the sample: I_{SD} is recorded as a function of tip position. In contrast to a macroscopic backgate that couples capacitively to the entire sample, the tip is a spatially localized gate whose position can be varied. For example, in Figure 1(c) (taken from [35]) a positively biased tip depletes the carrier (hole) concentration locally, and three spots (defects) are seen along the nanotube length.

In the following subsections, we discuss the use of SGM to image Schottky barriers at metal contacts to a single-walled carbon nanotube field effect transistor (swCN-FET) and at junctions between metallic and semiconducting nanotubes, to study gate-susceptible defects along the nanotube length, and to demonstrate the local nature of the memory effect normally observed in swCN-FET [33,34]. We will also present an extension of SGM that can be used to get quantitative information about the behavior of swCN-FET at high frequencies [35,53].

2.1 SGM for Studying Schottky Barriers in 1D Systems

It has recently been demonstrated [54–58] that Schottky barriers formed at the contact between the swCN and the metallic electrode play a very important role in the field effect behavior of the semiconducting swCN-based devices. SGM allows

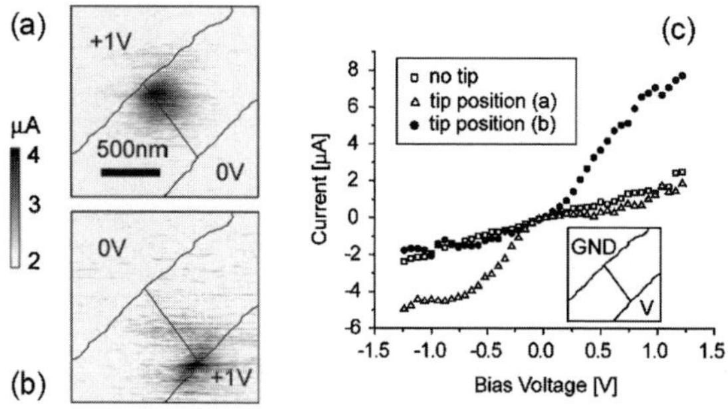

FIGURE 2. Control creation of a carbon nanotube diode (reproduced with permission from [48]): (a,b) SGM images of the swCN-FET with the top-left electrode at +1 V bias and the bottom electrode grounded (a), and with the bias and ground connections reversed (b). The tip potential is −2 V, and the backgate is grounded. The electrodes and nanotube are sketched for clarity. In each case, the current is strongly enhanced when the tip gate is near the positively biased contact. (c) I_{SD}-V_{bias} characteristics of the nanotube with and without tip gating. The bias is applied to the bottom-right electrode while the top left electrode is grounded. Without tip gating (open squares), the I_{SD}-V_{bias} is symmetric. With tip gating near the top left or bottom-right contact (open triangles or filled circles), the I_{SD}-V_{bias} is rectifying, with a sharp current increase when the electrode near the tip gate is positively biased.

us to address this issue directly by using the voltage-biased AFM tip as a spatially localized gate in order to probe the local field effect in a swCN-FET [48,59].

This is illustrated in Figure 2 [48], which shows SGM data for a CVD-grown swCN-FET. In Figure 2(a,b) the SGM-tip is biased at −2 V and a dc-bias of +1 V is applied to the left (Figure 2(a)) and right (Figure 2(b)) electrodes, respectively, while the other electrode is grounded. A pronounced gating action (current enhancement shown by the dark spots) is seen in each case, close to the contact between the swCN and the positively biased electrode. Equally important, this current enhancement is seen only when a negative voltage is applied to the tip.

Figure 2(c) shows the transport current I_{SD} vs. source-drain bias V_{bias} with the SGM-tip gate placed at particular locations above the device, and −2 V applied to the tip. With no tip gate (open squares) the I_{SD}-V_{bias} is symmetric. In the presence of the tip gate the I_{SD}-V_{bias} becomes strongly asymmetric (open triangles or filled circles), with a pronounced current increase when the electrode near the tip is positively biased.

These observations can be explained by considering the Schottky barriers that form at the contact between the p-type swCN and the metal electrodes (Figure 3, [48]). At zero bias both barriers have approximately equal height and depletion region width (Figure 3(a)). An applied bias voltage is increasing the depletion region at the positively biased electrode while is decreasing the same region at the other electrode (Figure 3(b)). In this way a reverse biased Schottky barrier is formed at the positive electrode, which acts as a bottleneck for the majority carriers (holes). A negatively biased tip positioned near this barrier induces holes

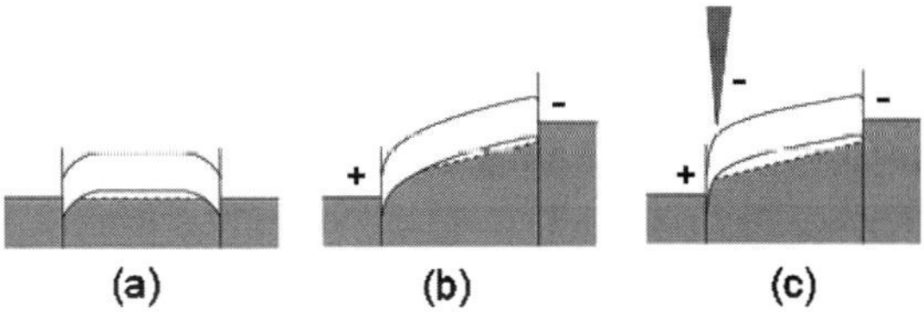

FIGURE 3. Energy band diagrams for the nanotube FET: zero bias (a), biased FET (b), biased and tip-gated FET (c). Reproduced with permission from [48].

in the area beneath the tip, and reduces the width of the Schottky barrier, therefore increasing the transport current through the device (Figure 3(c)). Based on the SGM data, the upper limit for the size of the depletion region is estimated to be 50 nm [48].

In a similar way, SGM was used to show unambiguously that the current rectification normally observed at the metallic-semiconducting nanotube cross-junctions [60] is due to a localized Schottky barrier at the junction [59,61]. It has previously been suggested [54,60] that if a Schottky barrier is responsible for the rectification behavior observed in such devices then the gating action should be confined to an area where the *p*-type swCN is depleted due to charge transfer from the metallic swCN. The SGM tip was used to show that indeed the size of the gate active region is confined to a small area at the junction [61].

2.2 Role of single defects in electronic transport and local memory effect of swCN-FET

SGM can be used to probe the nature of conduction directly in both metallic and semiconducting swCN. For example, it has been shown [62,63] that transport properties in metallic nanotubes are influenced by resonant electron scattering at defects sites in the nanotube as the Fermi level is varied by the SGM tip gate. In this case the SGM data can be used to measure the electron reflection coefficients of the defects.

SGM can also resolve local fluctuations in the valence band [50–52] and conduction band [33,59] edge in semiconducting swCNs. Tans and Dekker have shown for the first time that the edge of the valence band in swCN-FETs does not vary in a smooth fashion but is instead modulated by a sequence of barriers for the hole carriers [50]. Similar modulations were found for the conduction band of the *n*-type swCN-FETs [33,59]. Quite generally, SGM experiments show that the resistance of the swCN-FET is dominated by a few strong barriers (defects) [33,35,50–52].

We illustrate these general remarks with data taken from UPenn group [52]. The sample shown in Figure 4 is a *p*-type swCN-FET grown by CVD. A positively biased SGM-tip resolves valence band modulations by depleting carriers (holes) locally (Figure 4(c)). SGM images taken at different positive tip voltages show four strong defects (Figure 4(a)). Electrostatic considerations [52] show that the SGM-imaged diameter D_i of the *i*-th defect increases linearly with the applied tip voltage V_{tip}. More precisely, if R is the tip radius and k the dielectric constant of

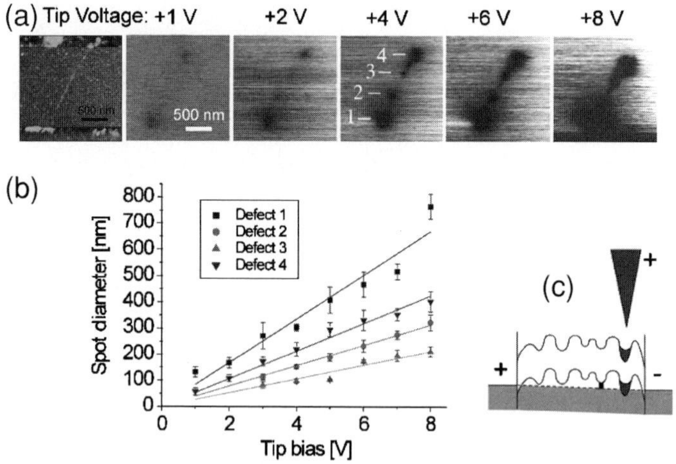

FIGURE 4. (a) Defects imaged by SGM with tip voltage varying between +1 and +8 V. (b) SGM imaged diameter increases linearly with tip voltage. The slope of the line gives the depletion surface potential of each defect. (c) Schematic of valence band energy modulation due to defects and Schottky barriers. The tip depletes the shaded part of the tube below it. Reproduced with permission from [52].

the substrate then

$$D_i = \frac{V_{tip}}{V_i^*}\frac{4\alpha R}{\kappa + 1}, \tag{1}$$

where α is the ratio of the tip-surface capacitance to that of a sphere of radius R, and V_i^* is the surface depletion potential (the surface potential at which the current through the device is suppressed).

Experimentally, the linear dependence presented in Figure 4(b) can be used to measure the depletion surface potential for each defect. From these measurements the local Fermi energy near the defect can be determined using the following considerations: adding carriers to a depleted region will increase both the electrostatic energy $e\Phi$, and the Fermi energy E_F of that region; with N carriers added to a region of length l, these become $e\Phi = Ne^2/Cl$, and $E_F = N/Dl = Ne^2/C_Q l$, where C is the total capacitance of the region per unit length, D is the density of states per unit length, and C_Q is the "quantum capacitance" [64]. The capacitance per unit length can be estimated from a co-axial geometry $C \sim 2\pi\varepsilon\varepsilon_0/\ln(2h/r) \sim 30$ pF/m $= 0.2$ e/V − nm (with $h = 200$ nm the oxide thickness and $r = 1$ nm the nanotube radius). The quantum capacitance (per unit length) is: $C_Q = e^2 D \sim 2.4$ e/V− nm, assuming eightfold level degeneracy and Fermi velocity $\hbar v_F = 0.5$ eV − nm. Then the ratio $C/C_Q = 1/12$ gives an estimated Fermi energy [52] of 20, 45, 65, and 35 meV for the defects 1 – 4 in Figure 4(a).

Recent theoretical calculations have addressed explicitly the interaction between SGM-tip, tube and substrate by using a combination of first-principles density

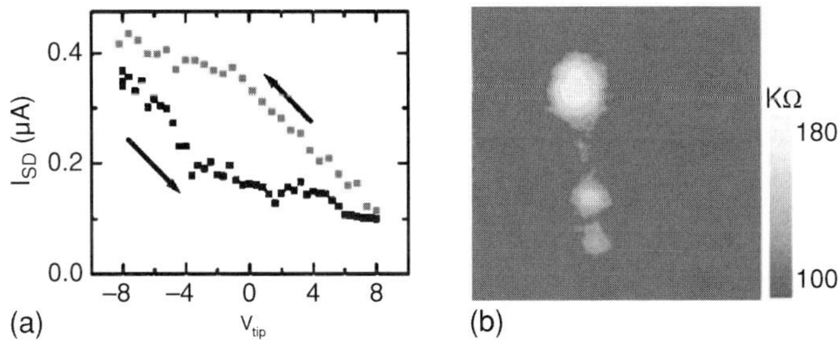

FIGURE 5. (a) Transport current vs. tip voltage with the SGM tip positioned above the top defect in Figure 1c. The bias voltage is 30 mV, and the backgate voltage is $V_G = 0$ V. A local hysteresis (memory effect) is seen as the tip voltage is swept from positive to negative and back. The memory effect is seen only with the tip above the SGM defect, which indicates that it can be induced at well-defined locations along the nanotube length. Reproduced with permission from [35]. (b) High frequency scanning gate (HF-SGM) image of the sample impedance modulus |Z| vs. tip position; data are shown for the same device presented in Figure 1.

functional calculations and continuum electrostatics modeling [65]. The result of this rigorous approach shows that the potential on the swCN is only a small fraction (10^{-2} to 10^{-3}) of the applied tip bias. This leads to a shift in the Fermi energy of tens of meV at defect sites, in good agreement with the experimental results presented above.

SGM can also be used to study the nature of the "memory effect" normally observed in swCN-FETs [33,34]. This phenomenon consists in a large hysteresis in the transport current vs. backgate voltage curves that is observed when the backgate voltage is swept back and forth between negative and positive values. The observed hysteresis was attributed to charge injection into stable trap states in the SiO_2 substrate due to the large electric field at the surface of the swCN-FET [33,34]. Later experiments showed this hysteresis was very sensitive to details of the surface chemistry [66]. When the backgate voltage is swept, electric field-induced charge injection occurs all along the swCN-FET length. In contrast, it has been recently shown [35] that an SGM-tip gate induces a local memory effect only in a region of size roughly equal to the tip radius (30 nm). For example, Figure 5(a) shows the transport current I_{SD} as a function of V_{tip} with the tip positioned at a fixed height ($h = 15$ nm) above the top defect in Figure 1(c). As the tip gate voltage is swept back and forth between +8 V and −8 V, a modulation in the transport current by a factor of 4 and a pronounced hysteresis is seen in $I_{SD}(V_{tip})$. There should be no memory effect if charge injection is localized to a region of the swCN-FET that shows no local gating response. In agreement with this prediction, the memory effect of Figure 5(a) was observed only when the tip was located over a defect region and never with the tip located elsewhere over the swCN-FET [35]. This is strong evidence that the active region for the memory effect can be as small as a few tens of nanometers. It should therefore be possible to miniaturize a nanotube-based memory element to less than 100 nm in size.

Finally, we mention that all the SGM measurements presented above provide information on variation of the low-frequency (DC–10 kHz) resistance of the sample in response to the tip-gate bias. A recent extension of this technique, called high-frequency scanning gate microscopy (HFSGM), allows to investigate the role of single defects in electronic transport through swCN-FET at high frequencies [35,53]. In high-frequency SGM, the voltage-biased SGM tip is scanned over the swCN-FET at a fixed height as in Figure 1(a), and the circuit impedance modulus $|Z|$ and phase θ, measured at a particular frequency f, are recorded as a function of tip position. The HFSGM data (Figure 5(b)) show that at high frequencies the device behaves as a simple parallel *RC* circuit, where *R* is the resistance set by the swCN-FET and *C* is the stray capacitance. The effect of the tip-gate is simply to increase the sample resistance, in agreement with the results of SGM measurements at low frequency.

3 Scanning Impedance Microscopy for Potential Profiling Along Nanotube Devices and Scanning Conductance Microscopy

As described in the previous section, images in "depletion mode" SGM are used to quantify the depletion surface potential and Fermi level associated with individual defects. However information about the local potential distribution along the swCN-FET for different backgate voltages is also very important in order to further characterize defect related transport properties of the nanotube. This goal was achieved by using another dual-pass technique called scanning impedance microscopy (SIM) [51,67]. In SIM (Figure 6(a) [59]), the tip is held at a constant dc voltage V_{tip} and an ac lateral potential $V_{lat} = V_{dc} + V_{ac}\sin(\omega t)$ is applied across the sample. This produces an ac electrostatic force on the cantilever whose first harmonic is proportional to the local potential oscillation amplitude $V_{ac}(x)$ beneath

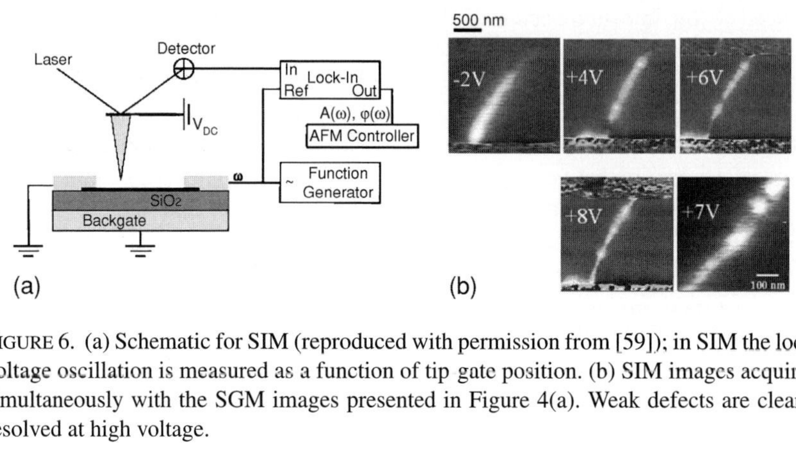

FIGURE 6. (a) Schematic for SIM (reproduced with permission from [59]); in SIM the local voltage oscillation is measured as a function of tip gate position. (b) SIM images acquired simultaneously with the SGM images presented in Figure 4(a). Weak defects are clearly resolved at high voltage.

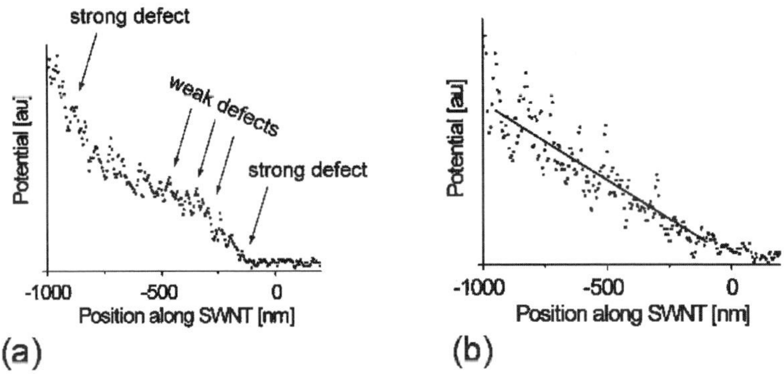

FIGURE 7. (a) Potential distribution in the swCN-FET when the device is in the "off" state. SIM shows voltage steps at the position of strong defects. The backgate is 0.7 V, and the lateral bias is 1 V. (b) Uniformly increasing potential along the nanotube FET in the "on" state (bias voltage 0.5 V, backgate voltage 0 V, and tip voltage −1 V). Reproduced with permission from [59].

the tip:

$$F_{\omega}(z) = \frac{\partial C_z}{\partial z}(V_{tip} - V_s)V_{ac}(x), \tag{2}$$

where V_s is the dc surface potential, C_z is tip-surface capacitance, and z is tip-surface separation. The oscillation amplitude of the cantilever is measured using a lock-in amplifier, and the resulting signal is proportional to $V_{ac}(x)$.

Figure 6(b) [52] shows SIM images acquired simultaneously with the SGM data presented in Figure 4a. For small negative tip bias, the local carrier density is only slightly enhanced, and the tip acts as a non-invasive probe of the potential distribution. This is shown in Figure 7 [52], which shows the potential distribution along the swCN-FET length. When the device is in the OFF state (poorly conducting, transport current 40 nA at backgate voltage $V_G = +0.7$ V), SIM shows clearly resolved potential steps (Figure 7(a)) at defects labeled 2 and 4 in Figure 4(a). A series of weaker scattering centers give a uniform voltage increase in the neighborhood of defect 3. The nanotube is divided into multiple conducting segments separated by distinct barriers at stronger defects. On the other hand, Figure 7b indicates a *uniform potential drop* along the device when the nanotube FET is ON (conducting, transport current 350 nA, with $V_G = 0$ V). All defects are far from depletion, and each acts as a weak scattering site: transport along the nanotube FET is diffusive.

When biased at a positive voltage the tip gate strongly alters the potential distribution beneath it, resulting in a sawtooth like signal [59,68] as the tip position is changed from one defect to the next. If the tip voltage is sufficient to deplete a defect, a potential step is created at the defect site, creating a large SIM signal when the tip is positioned above the defect, with a sharp signal reduction as the tip moves away. The width of the transition region is determined by the tip-surface transfer function and provides a direct measurement of the SIM resolution [69]. This "tip-gating" SIM mode enables high resolution imaging of weak defects, even if they are in close proximity to strong defects. It could also be used as an extremely effective method for measuring the local Fermi energies associated with weak defects [59].

SIM was also used recently to measure frequency-dependent transport through conducting networks of swCN embedded in polymers [70]. It has been shown that SIM can measure transport along nanotubes inside the polymer matrix up to 100 nm depth below the surface, illustrating the applicability of SIM for noninvasive transport imaging in nanocomposite materials.

We end this section with a discussion of scanning conductance microscopy (SCM), a mode of electrostatic force microscopy, which is extremely useful for investigating the electrical properties of nanoscale samples without electrical contacts [71,73]. Like SIM, SCM is a dual-pass technique. In the first line scan, the tip acquires a topography profile in tapping mode. In the second (interleave) line scan, the tip travels at a defined height above the surface. A dc voltage is applied to the tip, and the cantilever is mechanically driven at its free oscillation resonant frequency. The SCM image records the phase of the cantilever oscillation as a function of tip position [72]. The SCM cantilever can be modeled as a driven harmonic oscillator with resonant frequency ω_0 and spring constant k. When the tip is scanned at height h above the bare insulating substrate, the electrostatic force between metallic tip and surface leads to a decrease of the resonant frequency of the cantilever and therefore to a negative value Φ_0 for the *background* phase lag: $\tan(\Phi_0) = -(Q/2k)C_1''(h)V_{tip}^2$, where $Q = \gamma\omega_0$ is the quality factor of the cantilever and $C''_1(h)$ *is the second derivative of the tip-substrate capacitance.* The value of Φ_0 is independent of the tip horizontal position and is used as the reference zero. When the tip is at height h above the sample (swCNs, nanowires etc.) the total capacitance of the system changes to $C_2(h)$. Again, assuming that the electrostatic forces are small, the phase shift relative to that over the bare substrate is

$$\tan(\Phi - \Phi_0) \approx \frac{Q}{2k}\left(C_1''(h) - C_2''(h)\right) \cdot V_{tip}^2. \tag{3}$$

Quite generally, explicit calculations of the capacitances called for in Eq. (3) predict that in SCM swCNs and other small-diameter conducting nanowires always show a negative phase shift, while insulating nanowires show a positive phase shift that increases with fiber diameter. Equation (3) also predicts that conducting nanowires with diameter larger than 30nm have a negative-positive-negative ("double dark line") contrast in SCM [72]. These quantitative predictions are in excellent agreement with data collected on swCNs (Figure 8(a)), insulating polyethylene oxide (PEO) nanofibers (Figure 8(b)), λ-DNA, and conducting doped polyaniline/polyethylene oxide (PAn/PEO) nanofibers (Figure 8(c)) [71,72,74]. The method can also be used to measure the dielectric constant of nanoscale objects such as insulating nanowires, polymer nanofibers and nanospheres [72].

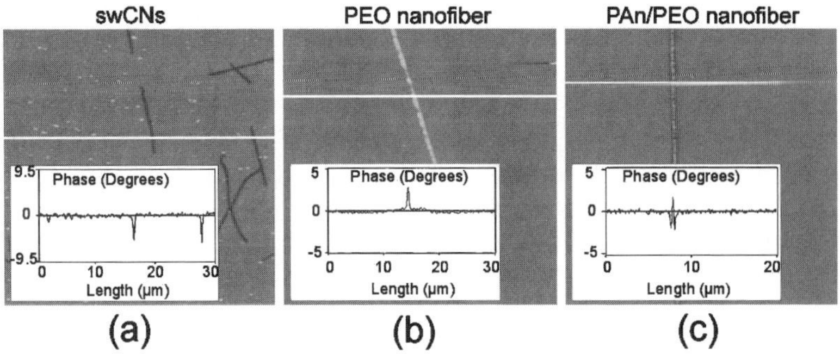

FIGURE 8. SCM images of carbon nanotubes (a), insulating PEO nanofiber (diameter: 10–100 nm) (b), and conducting PAn.HCSA/PEO nanofiber (diameter 100 nm) (c). Insets show line scans along the black lines. Differences in phase shift are explained in the text. Reproduced with permission from [72].

4 Tunneling AFM (TUNA) and Movable Current Injecting Electrodes

In this section we discuss AFM-based techniques which use current measurements to gather additional information about the material system under study. In particular, such an approach, called tunneling AFM (TUNA) combines the best of conventional AFM and scanning tunneling microscopy (STM). It utilizes conventional AFM (force) feedback to guide the tip along non-conductive surfaces, while relying on sensitivity of a tunneling current for enhanced spatial resolution.

The first demonstration of TUNA in CN system was done by Clauss et al. [75]. In this study, CNs were deposited either on conducting substrates or on insulating substrates with pattern electrodes for local contact. While the samples are imaged using force feedback the tunnel current is observed as the tip is positioned above CNs. Interestingly, the tunnel current is easier to observe over CNs than over metallic electrodes (such as Au), presumably due to very different elastic responses of the two materials to the AFM tip, and to the fact that carbon is hydrophobic, while Au gets contaminated in air. Moreover, the exponential dependence of tunnel current on distance allows an unprecedented atomic scale resolution of CNs in force feedback mode as shown in Figure 9 [75].

Enhanced spatial resolution of TUNA is particularly useful when studying circuits of CNs [76]. An example is shown in Figure 10, where electrochemical potential of a CN bundle is resolved down to an individual CN level. In this setup the TUNA tip is used as a weakly coupled voltage probe locating the position in the circuit which has the same electrochemical potential as the tip. The location of equal electrochemical potential is identified as the position where no current flows between the tip and the sample. The electrochemical potential of the CN is lower when the current is positive: i.e., it flows from tip to the sample, and vice versa. What is observed in typical CN bundle samples is that there are locations,

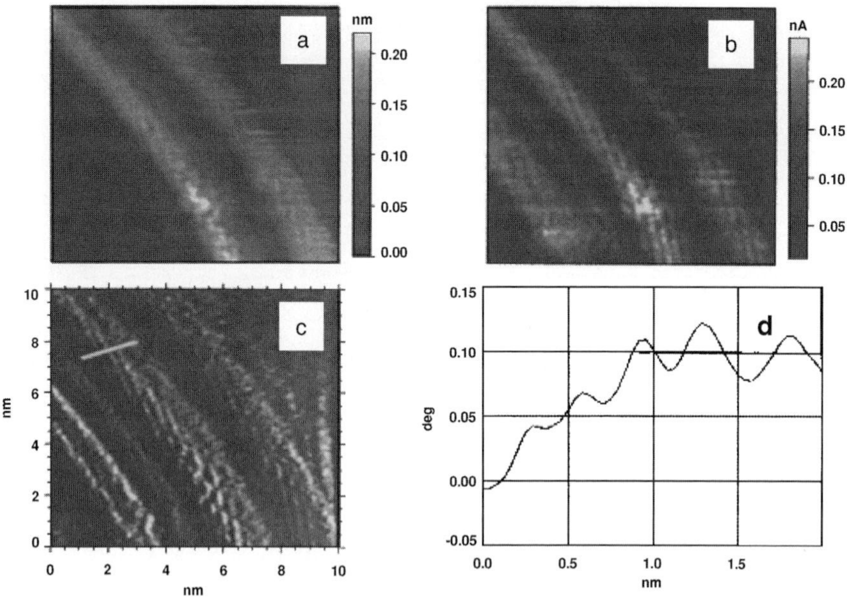

FIGURE 9. Enhanced resolution of CN bundle using tunneling AFM (TUNA). (a) Topographic image of a CN bundle. (b) Tunnel current image. (c) Feedback error image showing atomic imaging within a CN. (d) Line scan from (c). Reproduced with permission from [75].

such as shown in Figure 10, where the electrochemical potential varies along the cross section of the bundle [76]. This indicates that the coupling between CNs in the bundle is weak, such as tunneling and that electrons are well localized on individual CNs within the bundle as predicted by theory [77]. More recently, similar techniques have been used to study coupling in nanotube mats [78].

We close this section by describing a technique which uses AFM tip as a movable current injecting electrodes as shown in Figure 11(a) [79]. Instead of using force

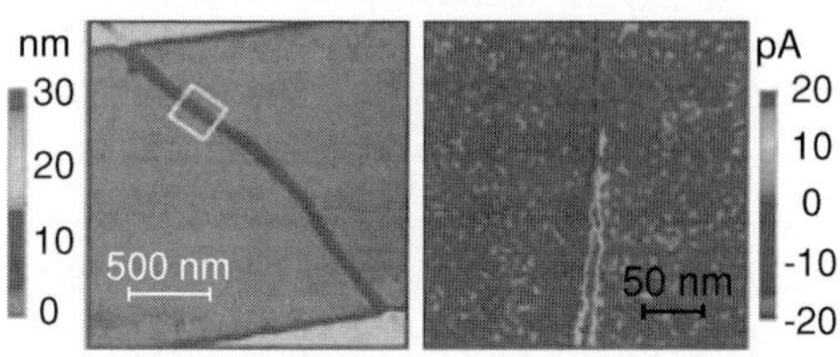

FIGURE 10. TUNA imaging of CN circuits in operation. **(left)** Topographic image of a 8 nm diameter CN bundle contacted by two electrodes. The top contact is grounded while the lower one is biased to 0.6 V. **(right)** Image of tunneling current with tip biased at 0.45 V. Coexistence of positive and negative current at same bundle cross section indicates weak coupling within CNs in a bundle. Reproduced with permission from [76].

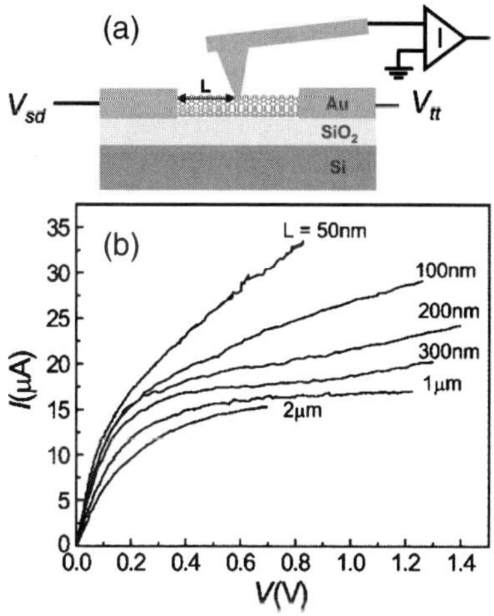

FIGURE 11. (a) Schematic of three terminal measurement setup. The voltage drop is set up between the left electrode and the gold-coated AFM tip, while the right electrode is measuring the voltage drop across the CN. The effective length of the sample can be varied by moving the AFM tip. (b) Current-voltage (I-V) characteristics of a same metallic CN as a function of length. The data can be used to extract mean free path in the CN as discussed in the text. Reproduced with permission from [79].

feedback and relying on tunnel current, the tip-sample force is increased until the tip is in contact with the device and large currents can be passed through the tip/CN junction. Once reliably implemented this technique is used to study length depended resistance of metallic swCNs, which is not easily achieved due to sample-to-sample variations in these devices. Using an AFM tip as the movable current injecting electrodes, current-voltage (I-V) characteristics of the metallic CN can be obtained on segments as short as 50 nm. The results shown in Figure 11(b) are used to extract the two characteristic resistivities of the material, $\rho = 4$ KΩ/μm at low bias, and 800 KΩ/μm at high bias. These two values correspond to electronic mean free paths of roughly 2 μm at low bias, limited by elastic phonon scattering, and 10 nm at high bias limited by optical phonons.

5 Novel AFM-Based Techniques

In this last section we describe some of the more exotic AFM based techniques which have been demonstrated only recently, and their full potential remains unexplored. One such example is scanning thermal microscopy (SThM), which allows accurate sample temperature measurements in local ~50-nm radius. In order to accomplish this goal, specialized cantilever tips have been fabricated whose thermal resistances (see Figure 12(a)) are carefully designed to respond preferentially to the sample temperature rather than that of the ambient [80].

SThM using such probes has been used to investigate temperature profiles in CNs carrying large current densities [80]. Figure 12(b,c) shows the topographic and thermal images of a mwCN contacted by four electrodes. Qualitatively it is evident that the temperature increases with voltage bias that is applied to the CN, as expected. Observed thermal images are due to phonon coupling between the tip and the sample, and can be used to study phonon population and dynamics in

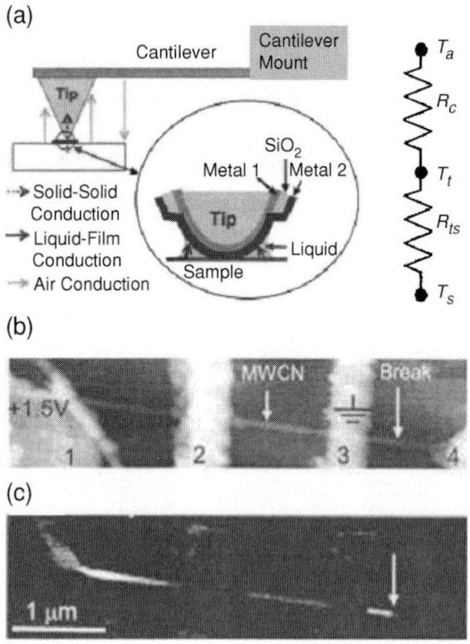

FIGURE 12. Scanning thermal microscopy (SThM) of CNs. (a) Schematic diagram of the cantilever probe used for SThM. (b) Topographic and (c) Thermal images of a multi-wall CN with voltage applied between contacts 1 and 3. Reproduced with permission from [80].

CNs. For example, SThM shows additional rise in temperature between contacts 3 and 4 along the segment that does not carry any current, implying a rather efficient phonon transport. The fact that the temperature profile changes abruptly between the two contacts implies a break in the CN which is also observed electrically [80]. Currently, this technique is used to study phonon dynamics as well as electron-phonon coupling in nanotubes complementing purely electrical studies as described in section 4.

Lastly, we revisit SGM and EFM techniques as they are scaled to single-electron level by applying them at cryogenic temperatures where CN devices exhibit Coulomb-blockade phenomena [13]. At low temperatures, the tunnel barriers form within CN devices (either at the contacts to the macroscopic electrodes or at the defects within the CN) which serve to confine a well-define number of electrons on the CN quantum dot. This number of electrons is quantized when temperature is much lower than the electrostatic cost (also known as charging energy) necessary to add additional electron on the CN dot. Figure 13 shows how single electron EFM (e-EFM) can be used to measure the charge state of the CN quantum dot as a function of tip position. The single electron force is measured by applying a ac voltage in resonance with the tip frequency to the CN contacts. Concentric rings of force peaks are centered on the CN and correspond to the change of a charge state of the quantum dot by one electron. In this specific case the two sets of concentric rings suggest that the CN under study is segmented into two quantum dots [81]. This type of technique has important implications both for understanding the operation of single electron devices and for measuring how these devices affect nanoscale mechanical oscillators operating in their proximity.

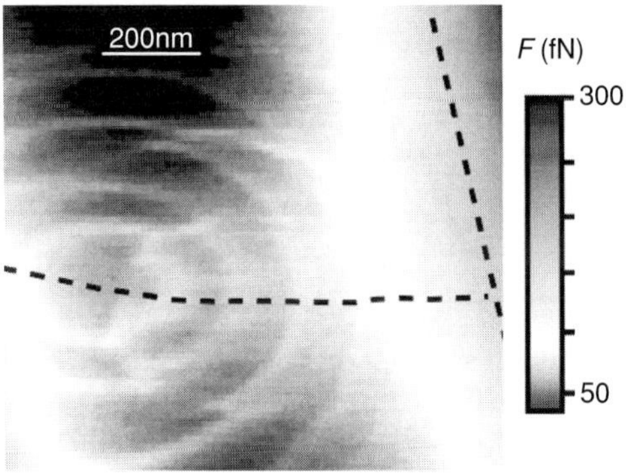

FIGURE 13. Single electron electrostatic force microscopy (e-EFM) image in CN device at T = 0.6 K. Concentric rings of force are seen enclosing two quantum dots in series. Reproduced with permission from [81].

References

1. H. W. Kroto, J. R. Heath, S. C. O'Brien, R. F. Curl, and R. E. Smalley, Nature **318**, 162 (1985).
2. S. Ijima, Nature **354**, 56 (1991).
3. S. Ijima, and T. Ishibashi, Nature **363**, 603 (1993).
4. D. S. Bethune, C. H. Klang, M. S. de Vries, G. Gorman, R. Savoy, J. Vazquez, and R. Beyers, Nature **363**, 605 (1993).
5. R. Saito, G. Dresselhaus, and M. S. Dresselhaus, *Physical properties of carbon nanotubes*, World Scientific, Singapore, (1998).
6. J. W. G. Wildöer, L. C. Venema, A. G. Rinzler, and R. E. Smalley, and C. Dekker, Nature **391**, 59 (1998).
7. T. W. Odom, J. L. Huang, P. Kim and C. M. Lieber, Nature **391**, 62 (1998).
8. S. J. Tans, R. M. Verschueren, and C. Dekker, Nature **393**, 49 (1998).
9. S. J. Tans, M. H. Devoret, H. Dai, A. Thess, R. E. Smalley, L. J. Geerligs, and C. Dekker, Nature **386**, 474 (1997).
10. Ph. Avouris, Acc. Chem. Res. **35**, 1026 (2002).
11. N. Hamada, S. Sawada, and A. Oshiyama, Phys. Rev. Lett. **68**, 1579 (1992).
12. R. Saito, M. Fujita, G. Dresselhaus, and M. S. Dresselhaus, Appl. Phys. Lett. **60**, 2204 (1992).
13. M. Bockrath, K. H. Cobden, J. Lu, A. G. Rinzler, R. E. Smalley, L. Balents, and P. L. McEuen, Nature **397**, 598 (1999).
14. B. Gao, A. Komnik, R. Egger, D. C. Glattli, and A. Bachtold, Phys. Rev. Lett. **92**, 216804 (2004).
15. M. Bockrath, D. H. Cobden, P. L. McEuen, and N. G. Chopra, A. Zettl, A. Thess, and R. E. Smalley, Science **275**, 1922 (1997).
16. A. Bachtold, C. Strunk, J. P. Salvetat, J. M. Bonard, L. Forro, T. Nussbaumer, and C. Schönenberger, Nature **397**, 673 (1999).
17. M. J. O'Connel, S. M. Bachilo, C. B. Huffman, V. Moore, M. S. Strano, E. Haroz, K. Rialon, P. J. Boul, W. H. Noon, C. Kittrell, J. Ma, R. H. Hauge, R. B. Weisman, and R. E. Smalley, Science **297**, 593 (2002).
18. S. M. Bachilo, M. S. Strano, C. Kittrell, R. H. Hauge, R. E. Smalley, and R. B. Weisman, Science **298**, 2361 (2002).

19. J. Lefebvre, Y. Homma, and P. Finnie, Phys. Rev. Lett. **90**, 217401 (2003).
20. B. I. Yakobson and R. E. Smalley, American Scientist **85**, 324 (1997).
21. R. Martel, T. Schmidt, H. R. Shea, T. Hertel, and Ph. Avouris, Appl. Phys. Lett. **73**, 2447 (1998).
22. P. Kim and C. M. Lieber, Science **286**, 2148 (1999).
23. J. Lefebvre, P. Finnie, and Y. Homma, Phys. Rev. B **70**, 045419 (2004).
24. R. J. Chen, Y. Zhang, D. Wang, and H. Dai, J. Am. Chem. Soc. **123**, 3838 (2001).
25. K. A. Williams, P. T. M. Veenhuizen, B. G. de la Torre, R. Eritja, and C. Dekker, Nature **420**, 761 (2002).
26. K. Besteman, J. Lee, F. G. M. Wiertz, H. A. Heering, and C. Dekker, Nano Letters **3**, 727 (2003).
27. J. Kong, N. R. Franklin, C. Zhou, M. G. Chapline, S. Peng, K. Cho, and H. Dai, Science **287**, 622–625 (2000).
28. E. S. Snow, F. K. Perkins, E. J. Houser, S. C. Badescu, and T. L. Reinecke, Science **307**, 1942–1945 (2005).
29. C. Staii, M. Chen, A. Gelperin, and A. T. Johnson, Nano Letters **5**, 1774 (2005).
30. E. W. Wong, P. E. Sheehan, C. M. Lieber, Science **277**, 1971 (1997).
31. R. D. Antonov and A. T. Johnson, Phys. Rev. Lett. **83**, 3274 (1999).
32. T. Rueckes, K. Kim, E. Joselevich, G. Y. Tseng, C. L. Cheung, and C. M. Lieber, Science **289**, 94 (2000).
33. M. Radosavljevic, M. Freitag, K. V. Thadani and A. T. Johnson, Nano Letters **2**, 761 (2002).
34. M. S. Fuhrer, B. M. Kim, T. Durkop, and T. Brintlinger, Nano Letters **2**, 755 (2002).
35. C. Staii, R. Shao, D. A. Bonnell, and Alan T. Johnson, Nano Letters **5**, 893 (2005).
36. W. N. Choi, D. S. Chung, J. H. Kang, H. Y. Kim, Y. W. Jin, I. T. Han, Y. H. Lee, J. E. Jung, N. S. Lee, G. S. Park, and J. M. Kim, Appl. Phys. Lett. **75**, 3129 (1999).
37. S. J. Tans, M. H. Devoret, R. J. Groeneveld, and C. Dekker, Nature **394**, 761 (1998).
38. J. Lefebvre, M. Radosavljevic, and A. T. Johnson, Appl. Phys. Lett. **76**, 3828 (2000).
39. E. D. Minot, Y. Yaish, V. Sazonova, J. Park, M. Brink, and P. L. McEuen, Phys. Rev. Lett. **90**, 156401 (2003).
40. R. H. Baughman, C. Cui, A. A. Zakhidov, Z. Iqbal, J. N. Barisci, G. M. Spinks, G. G. Wallace, A. Mazzoldi, D. De Rossi, A. G. Rinzler, O. Jaschinski, S. Roth, and Miklos Kertesz, Science **284**, 1340 (1999).
41. S. Ciraci, A. Buldum, I. P. Batra, J. Phys.: Condens. Matter **13**, R537 (2001).
42. T. W. Odom, J. L. Huang, P. Kim, and C. M. Lieber, J. Phys. Chem. B **104**, 2794 (2000).
43. M. Ouyang, J. J. Huang and C. M. Lieber, Acc. Chem. Res. **35**, 1018 (2002).
44. P. L. McEuen and J. Y. Park, MRS Bulletin, **April 2004**, 272 (2004).
45. G. Binning, C. F. Quate, and C. Gerber, Phys. Rev. Lett. **56**, 930 (1986).
46. R. Wiesendanger, *Scanning Probe Microscopy and Spectroscopy: Methods and Applications*, Cambridge University Press, Cambridge, (1994).
47. D. A. Bonnell, ed. *Scanning Probe Microscopy and Spectroscopy, Theory, Techniques and Applications*. 2nd ed. 2001, John Wiley and Sons: New York.
48. M. Freitag, M. Radosavljevic, Y. X. Zhou, W. F. Smith, and A. T. Johnson, Appl. Phys. Lett. **79**, 3326 (2001).
49. M. A. Eriksson, R. G. Beck, M. Topinka, J. A. Katine, R. M. Westervelt, K. L. Campman, and A. C. Gossard, Appl. Phys. Lett. **69**, 671 (1996).
50. S. J. Tans and C. Dekker, Nature **404**, 834 (2000).
51. A. Bachtold, M. S. Fuhrer, S. Plyasunov, M. Forero, E. H.Anderson, A. Zettl, and P. L. McEuen, Phys. Rev. Lett. **84**, 6082 (2000).

52. M. Freitag, A. T. Johnson, S. V Kalinin, D. A. Bonnell, Phys. Rev. Lett. **89**, 216801 (2002).
53. C. Staii, R Shao, N. J. Pinto, D. A.Bonnell, and A. T. Johnson, Jr, M. R. S. Proceedings of the Fall Meeting 2004 **838E**, 09.9.1 (2004).
54. J. Leonard and J. Tersoff, Phys. Rev. Lett. **83**, 5174 (1999).
55. V. Derycke, R. Martel, J. Appenzeller, and Ph. Avouris, Nano Letters **1**, 453 (2001).
56. R. Martel, V. Derycke, C. Lavoie, J. Appenzeller, K. K. Chan, J. Tersoff, and Ph. Avouris, Phys. Rev. Lett. **87**, 256805 (2001).
57. S. Heinze, J. Tersoff, R. Martel, V. Derycke, J. Appenzeller, and Ph. Avouris, Phys. Rev. Lett. **89**, 106801 (2002).
58. Y. Xue, M. A. Ratner, Appl. Phys. Lett. **83**, 2429 (2003).
59. M. Freitag, *Local electronic functionality in carbon nanotube devices*, PhD thesis, University of Pennsylvania, (2002).
60. M. S. Fuhrer, J. Nygård, L. Shih, M. Forero, Y. G. Yoon, M. S. C. Mazzoni, H. J. Choi, J. Ihm, S. G. Louie, A. Zettl, and P. L. McEuen, Science **288**, 494 (2000).
61. M. Freitag and A. T. Johnson, M. R. S. Proceedings of the Fall Meeting 2001 **706**, 371 6 (2001).
62. M. Bockrath, W. Liang, D. Bozovic, J. H. Hafner, C. M. Lieber, M. Tinkham, H. Park, Science **291**, (2001).
63. C. Staii, *Local electronic, sensing and optical properties of carbon nanotubes*, PhD thesis, University of Pennsylvania (2005).
64. S. Luryi, Appl. Phys. Lett. **52**, 501 (1988).
65. V. Meunier, S. V. Kalinin, J. Shin, A. P. Baddorf, and R. J. Harrison, Phys. Rev. Lett. **93**, 246801 (2004).
66. W. Kim, A. Javey, O. Vermesh, Q. Wang, Y. Li, and H. Dai, Nano Letters **3**, 193 (2003).
67. S. V. Kalinin and D. A. Bonnell, Appl. Phys. Lett. **78**, 1306 (2001).
68. S. V. Kalinin, D. A. Bonnell, M. Freitag, A. T. Johnson, Appl. Phys. Lett. **81**, 5219 (2002).
69. S. V. Kalinin, D. A. Bonnell, M. Freitag, A. T. Johnson, Appl. Phys. Lett. **81**, 754 (2002).
70. S. V. Kalinin, S. Jesse, J. Shin, A. P. Baddorf, M. A. Guillorn, and D. B. Geohegan, Nanotechnology **15**, 907 (2004).
71. M. Bockrath, N. Markovic, A. Shepard, M. Tinkham, L. Gurevich, L.P. Kouwenhoven, M. W. Wu, and L. L. Sohn, Nano Letters **2**, 187 (2002).
72. C. Staii, N. J. Pinto, and A. T. Johnson, Nano Letters **4,** 859 (2004).
73. J. Heo and M. Bockrath, Nano Letters **5**, 853 (2005).
74. Y. Zhou, M. Freitag, J. Hone, C. Staii, A. T. Johnson, N. J. Pinto, and A. G. MacDiarmid., Appl. Phys. Lett. **83**, 3800 (2003).
75. W. Clauss, M. Freitag, D. J. Bergeron and A. T. Johnson, Carbon **38**, 1735 (2000).
76. M. Freitag, M. Radosavljevic, W. Clauss and A. T. Johnson, Phys. Rev. B. **62**, R2307 (2000).
77. A. A. Maarouf, C. L. Kane, and E. J. Mele, Phys. Rev. B. **61**, 11156 (2000).
78. M. Stadermann, S. J. Papadakis, M. R. Falvo, J. Novak, E. Snow, Q. Fu, J. Liu, Y. Fridman, J. J. Boland, R. Superfine, and S. Washburn, Phys. Rev. B. **69**, 201402 (2004).
79. J. Y. Park, S. Rosenblatt, Y. Yaish, V.Sazonova, H. Üstünel, S. Braig, T. A. Arias, P. W. Brouwer, and P.L. McEuen, Nano Letters **4**, 517 (2004).
80. L. Shi, S. Plyasunov, A. Bachtold, P. L. McEuen, and A. Majumdar, Appl. Phys. Lett. **77**, 4295 (2000).
81. M. T. Woodside and P. L. McEuen, Science **296**,1098 (2002).

II.5
Conductance AFM Measurements of Transport Through Nanotubes and Nanotube Networks

M. STADERMANN AND S. WASHBURN

Conducting scanning probe microscopy provides a powerful tool for measuring electric transport through small surface features. In this chapter, carbon nanotubes and carbon nanotube networks are analyzed with a scanning probe microscopy method that employs solid metal tips to provide improved electrical contact to the nanotubes. The study reveals paths of electrical transport through carbon nanotubes and provides a means to differentiate between semiconducting and metallic nanotubes. Finally, high-resolution images provide insight into the conductance decay around nanotube junctions.

1 Introduction

Since the discovery of carbon nanotubes (CNT) by Iijima in 1991 [1], their electrical properties have been an area of active research. Conducting atomic force microscopy (AFM) has been established as an ideal method to measure the electronic properties of CNT. Not only is the sample preparation for conducting AFM easier than for standard two- or four-probe measurements, but the measurement type also allows the study of a larger amount of samples in a given time frame and gives information on the length dependence of a given electrical property. Only few studies have been performed with mobile probes [2–5], and fewer still had high resolution [2, 3, 5]. In this chapter, we describe conducting AFM studies on carbon nanotubes and present results of measurements on nanotube networks, as well as high-resolution images of nanotube junctions. With conducting tips, we obtain not only local topographic information, but also local conductance data, including the important contact resistance between different pieces of material and electrical decay lengths within individual elements of material.

One-dimensional localization has been studied theoretically for decades [6,7] and some excellent transport experiments have been performed [8]. These disordered samples have been related theoretically to percolating systems [9]. At the opposite end of the disorder spectrum, ballistic conductors also have been studied in both theoretical [10] and experimental [11–14] investigations. Both characters of transport have been observed in carbon nanotubes and in networks of CNT. On the border between the two extremes of disorder, Coulomb blockades have been observed in single electron transistor devices comprising carbon nanotubes [15]. In both limits (strong and weak disorder) screening and the decay of carrier wave functions have been characterized. In this chapter, we will introduce conductance imaging atomic force microscopy (CIAFM) and attempt to convey it's utility for experiments on carbon nanotubes in two dimensional random mats.

2 Sample Preparation and Method

Typically, conductance AFM measurements are either performed with metal coated tips [17,18] or coatings of NbN, doped diamond, or tungsten carbide [3,19]. None of these coating types provides optimal performance for measurements on carbon nanotubes. The metal-coated tips suffer from wear, making them unreliable. The other coatings are typically harder and less prone to wear, but have a higher resistance than the metal coatings. In conductance imaging atomic force microsopy, we use solid metal tips that are more durable and reliable than metal-coated tips and have a lower resistance than the other coatings. Most importantly, they also provide ohmic contact to carbon nanotubes [18, 20, 21]. The solid metal tip is implemented by using a tuning fork as the force sensor on which the tip is mounted [22]. The application of tuning forks as force sensors in AFM techniques is thoroughly described elsewhere [23–25]. We have chosen this method for our measurements because it is the only method that permits the use of a solid metal tip. The resonant oscillation of the fork and the attached tip is damped slightly when the tip contacts a surface and this slight damping can be read as a frequency or phase shift which can in turn be used as the error signal for the servo circuit that controls the tip holder via piezo actuation.

The tuning fork setup is shown in Figure 1. The current signal is processed with a lock-in amplifier and fed into the computer. Due to averaging within the software, the true magnitude of the signal is usually lost, and we can only infer a minimum conductance. The proportion of the amplitudes is retained, however, so that our setup can give us valuable information on the changes in conductance on the sample. The minimum resistance measured with the tip on a gold sample is 200 Ω; the method is not reliable for absolute measurement of smaller resistances.

For our experiments, we typically used an oscillation amplitude of about 30 nm, and scanning is performed at a frequency shift $\Delta f = 1.5$ Hz. The force between tip and sample was estimated from frequency shift/distance curves to be on the order of 30 nN. Voltages of up to ± 1 V were applied to the sample, while the tip remained grounded.

While the probe moves across the sample, the force between tip and sample is held constant. As a result, the pressure applied to the sample varies with its smoothness. Sharp features have more pressure applied to them, which results in a better contact, so edges of an electrode, e.g., always appear to have a higher conductance than the top surface of the electrode. When imaging nanotubes, this effect is highly advantageous: as the contact area between the nanotube and the

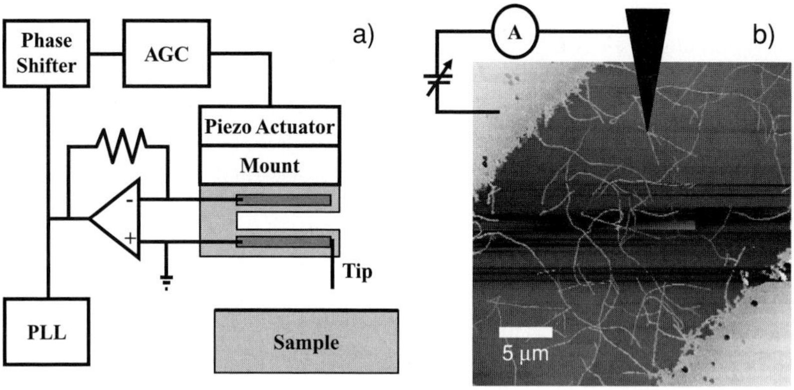

FIGURE 1. (a) Experimental setup: the tuning fork is driven at resonance frequency using a self-excitation circuit consisting of an amplifier, a phase shifter, an automatic gain control (AGC), and a piezo actuator. The amplifier output is fed into a phase-locked loop frequency demodulator (PLL). (b) Voltage is applied to the sample. When the tip touches a part of the sample that is connected to the electrode, a current can be measured and recorded simultaneously with topography.

tip is always very small, the pressure is correspondingly high, and thus the contact between tip and tube is improved. On conducting substrates, such as graphite, the nanotube has the same width on the conductance image as is does on the topography image. On insulating surfaces, however, the nanotube always appears broader on the conductance image than it does on the topography image. The degree of broadening is somewhat dependent on the tip sharpness. We do not know what causes the broadening, but assume that charging of the substrate may play a role.

The single-wall carbon nanotube (SWNT) samples in our experiments have to fulfill three requirements: the nanotubes have to form a well-dispersed network, the macroscopic contact to the nanotubes should have a low resistance, and the sample should be patterned, so that specific areas on the sample can be revisited. Three different sample types were made, each one compromising a different requirement. The first sample consisted of a silicon dioxide surface, upon which 20 nm of gold were deposited by sputtering. A sharpened wood applicator was used to scratch patterns into the film. Then, tubes synthesized by laser ablation and purified with oxidizing agents [26] were spun onto the surface from a 1-methylpyrrolidinone suspension. These samples provide good revisitability and contact to the nanotubes, but the nanotubes are typically deposited in bundles and form only poorly dispersed networks. The second type of sample used nanotubes grown directly onto silicon dioxide substrates using chemical vapor deposition (CVD) [27]. Gold was deposited onto the surface by thermal evaporation and patterned by shadow-masking. These networks are well-dispersed and well-contacted, but the shadowmasking provides only crude patterning capabilities that limit revisitability.

The third sample type also utilizes CVD-grown nanotubes. Chrome (10 nm) and gold (50 nm) were patterned onto the front by photolithography. This type of sample was annealed at 600°C for 10 min in vacuum after the metal deposition. Again, the SWNT networks are well dispersed, and the photolithographically defined features make it easy to revisit a specific part of the sample, but the contact resistance is much higher than on the other two samples. On samples with back gates, aluminum was deposited on the back before the gold deposition, and then the sample was sintered for 30 minutes at 400°C in vacuum. The oxide thickness was 1 μm. Etches performed on the samples are done with freshly prepared piranha (1:3 H_2O_2:H_2SO_4), with a typical etch taking 60–120 sec.

3 Carbon Nanotube Networks

Carbon nanotube field effect transistors (CNFET) have been shown to be very sensitive to ambient conditions [28–32], which makes them ideal sensors for gases.

There are, however, significant challenges associated with fabrication of devices out of individual CNT. Exactly one semiconducting SWNT needs to be placed between two electrodes. This kind of precision with placement is becoming more and more routine, as it is possible to define catalyst sites lithographically on the substrate and then to grow nanotubes with chemical vapor deposition in a controlled manner. As a second lithography step, electrodes are deposited. The principal challenge arises from the fact that the chirality of tubes grown in this fashion cannot be controlled. Statistically, the ratio of semiconducting to metallic tubes is 2:1 [33]. Since a metallic tube will not function as a sensor, a third of the sensors produced in this fashion will not work, which makes this kind of device unsuitable for mass production.

As an alternative, random networks of carbon nanotubes have been explored [31,32]. The minimum dimensions of a device based on a nanotube network are greater than those of one based on an individual SWNT, but the network devices are much easier to fabricate and show great promise for simple mass fabrication: instead of growing a single nanotube, the network of nanotubes is grown on the substrate, and electrodes are patterned upon it. This will create a device that consists of both semiconducting and metallic SWNT. When a gate voltage is applied, the semiconducting tubes are depleted of carriers, and they no longer contribute to conductance. As a result, the current drops. The ratio between "on" and "off" current for this type of device is, of course, much lower than for an individual CNFET, but they have the advantage that they always work, since there will always be semiconducting tubes present in the device. The macroscopic behavior of these networks has been investigated [31], and their usefulness as sensors was demonstrated [32]. The ratio between on- and off-current typically is 10:1, even though devices with a ratio of up to 10^4:1 have been observed for devices with a channel length of 10 μm and a width of 35 μm.

It might appear that these networks could be readily treated by percolation theory [9]. The percentage of metallic nanotubes is known, and it should be possible to calculate a threshold coverage at which connections consisting only of metallic tubes form from one electrode to the other for a given electrode configuration. It should also be possible to estimate how much of the total current is carried by such metallic connections. In practice, it turns out that the mat sizes for devices applications have too few elements for this kind of analysis to be instructive. Thus, to determine the percolation length for semiconducting and metallic tubes in a nanotube network of a given density of tubes per area, a nanoscale investigation of the current paths through nanotube network would give useful information.

3.1 Sparsely Connected Networks

It was found that the conductance of nanotube networks behaves very differently from that of thin films. The conductance of the network is not inversely proportional to the distance of the probe from the electrode. Instead, we find a conductance behavior that is strongly dependent on the connectivity of the network, as would be expected for a system whose behavior is determined by a percolating set of connections.

We define sparsely connected networks as networks in which each branch extends linearly away from the electrode, and there is no contact between the branches. Figure 2 shows one branch of such a network: there is only one connection to the electrode for the entire branch, and some parts of the branch are connected to the other parts by only one or two tubes. This type of network usually is the result of depositing SWNT from suspension onto the surface.

The conductance throughout these networks stays at the same order of magnitude throughout large parts of the network, but then decreases dramatically at certain points in the network, as shown in Figure 3. The resistance increase at these points is on the order of $10^8\,\Omega$, and they coincide with nanotube junctions.

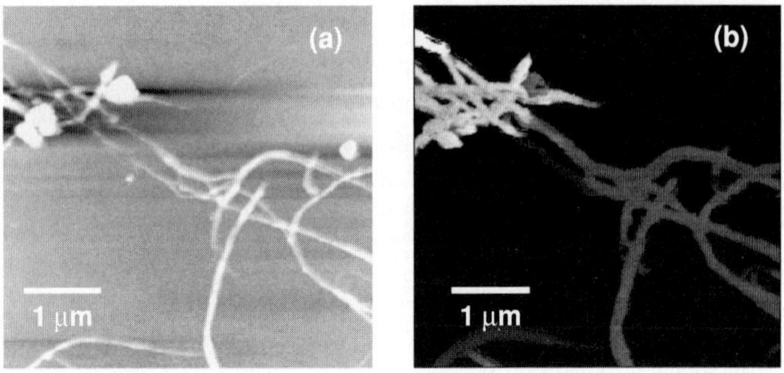

FIGURE 2. (a) Topography image of a nanotube network with low interconnectivity. The network contains bundles as well as individual tubes. These tubes were deposited from suspension and form clusters. The maximum feature height is 7.8 nm. (b) Conductance image of the same network taken simultaneously. The sample is biased at +0.5 V, the tip is grounded. The maximum conductance is 7.6 nS.

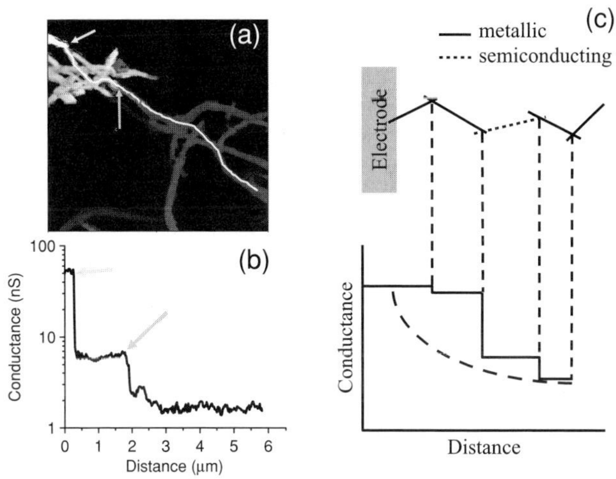

FIGURE 3. (a) Current image of the sparsely connected network from Figure 2. The white line indicates the path of the conductance trace in (b), and the arrows show the location of the conductance drops. (b) Conductance trace through the network. There is no measurable change in conductance throughout most of the network, but there are dramatic decreases at certain points. (c) The network is modelled as a chain of metallic and semiconducting nanotubes. The expected conductance change over distance is shown on the graph below the model. The dashed line in the conductance graph shows the conductance drop for a linear resistor in comparison.

Since the total resistance of the network even on the higher plateau (between the two arrows in Figure 3) is on the order of 100–200 MΩ and the noise on the current signal is larger than 1%, we do not expect to be able to resolve resistance changes below 1 MΩ. The literature values for resistances across junctions of nanotubes with the same electronic properties is about 200 kΩ for metallic tubes and 400 kΩ for semiconducting tubes [34]. The resistance of the SWNT themselves should be smaller still [33]. Thus, we expect the conductance drops caused by either to be negligible in this measurement. We believe that the conductance drops are associated with junctions formed by crossed *p*-type semiconducting and metallic nanotubes. This type of junction forms a Schottky barrier [34,35] and would have a significantly higher resistance than metallic/metallic or semiconducting/semiconducting junctions. When a positive voltage is applied to the sample, the Schottky barrier is reverse-biased, resulting in the large conductance drops we see. In almost all samples except for acid-treated ones, the semiconducting tubes behave like *p*-type semiconductors, as nanotubes in air are expected to be. In the following examples, we assume our semiconducting tubes to be *p*-type unless noted otherwise.

The network in Figure 2 consists of areas where we have a cluster of tubes all in contact with each other, and stretches in between the clusters where the network is connected by only one or two tubes. In consequence, we can model networks with poor interconnection as chains of SWNT, as shown in Figure 3.

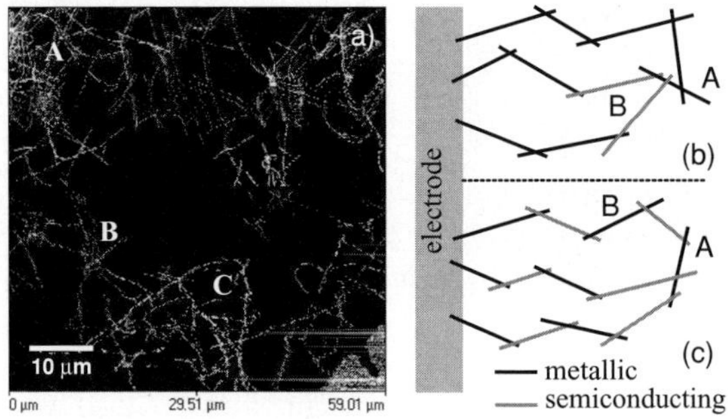

FIGURE 4. (a) Conductance image of a nanotube network with high interconnectivity. These tubes were directly grown by CVD onto the sample. The electrode is in the bottom right corner. The conductances at the points A, B, and C are 700 nS, 200 nS, and 700 nS. (b, c) Models of well-connected networks consisting of semiconducting and metallic SWNT that demonstrate how conductance at long distances from the electrode may be higher than at short distances. Explanations are in the text.

The resistance of these chains should increase monotonically with the distance from the electrode. If the distribution of metallic and semiconducting nanotubes is statistical (1:2), 2/9 of the connections between clusters should be a connection between a semiconducting and metallic tube and result in a conductance increase.

3.2 Well-Connected Networks

Well-connected networks are networks in which all parts of the network are connected to each other and no distinct branches can be identified. Such a network is shown in Figure 4. These networks are obtained when the nanotubes are grown on the surface from well-dispersed catalyst particles.

In these networks, we again find that the conductance is not inversely proportional to the distance, as would be the case for an ohmic conductor. The conductance of a given point in the network does not even decrease monotonically with increased distance from the electrode (Figure 4). Instead, we can identify areas of high conductance more than 80 μm away from the electrode. Local conductance is more dependent on the number of connections to the specific area: clusters of nanotubes with many paths to the electrode have significantly higher conductances than those parts of the network connected to the rest through fewer paths. Not only do multiple paths reduce resistance through the presence of multiple parallel resistors, but they also increase the probability that there is a chain of nanotubes stretching to the electrode that is purely metallic. Areas with low conductance typically only have two or three connections to the network, thus it is likely that these connections are dominated by the presence of highly resistive metallic/semiconducting junctions, just as the sparsely connected networks are.

Figure 4(b, c) demonstrates how it is possible to obtain high conductance areas in a network that have a larger distance to the electrode than low conductance areas closer to the electrode. In both examples, we assume that a positive bias is applied to the sample and the important drops occur only across reverse biased Schottky barriers, i.e. contacts between metallic and semiconducting nanotubes where the semiconducting tube is closer to the tip than the metallic tube. Figure 4(b) shows an example of the significance of purely metallic connections for the conductance of a given point in the network: the point that is furthest from the electrode (point A) has a higher conductance than point B (which could be anywhere on the semiconducting tubes), because it is part of a completely metallic connection with (for our purposes) negligible resistance. When the tip is placed anywhere on the semiconducting tubes, the connection to the electrode involves at least one reverse-biased Schottky barrier, which means the conductance measured on these tubes is significantly lower than the one at point A.

Figure 4(c) shows how multiple connections increase the conductance of a point at a large distance from the electrode. Assuming that the Schottky barriers all have about the same resistance R_S, the conductance at point A is

$$G_A = \frac{1}{R_A} = \frac{1+1+1}{2R_S} = \frac{3}{2R_S}, \tag{1}$$

since we have three parallel branches with 2 R_S each. At point B (any point on one of the metallic tubes closer to point A), we again have multiple parallel resistors: we have a single R_S to the left of B, and to the right a series of a single R_S and two parallel branches with a total resistance of R_S as well, yielding for G_B

$$G_B = \frac{1}{R_B} = \frac{1}{R_S} + \frac{1}{2R_S} = \frac{3}{2R_S} \tag{2}$$

as well. So G_A and G_B are equal, even though B is much closer to the electrode. Using this example, it is easy to imagine that the resistance at point A might actually become lower than the resistance at some points in the branches, depending on the number of branches and the actual resistance of each branch.

3.3 Characterizing the Contacts and Identifying the Extent of Metallic Connections

In the previous discussion, it was assumed that the major conductance drops occur across semiconducting/metallic junctions. The straightforward way to verify this appears to be taking several images at different voltages to see how the conductance drop across the junction changes with applied voltage. When the sign of the bias is changed, the Schottky barriers that were originally reverse biased are now forward biased, and vice versa. As a result, some of the conductance drops should decrease or disappear, whereas other will increase and new barriers may appear. This does not take into account the fact that the contact between the metal tip and a semiconducting tube might also form a Schottky barrier, and while the actual experiment shows differences for changed biases, these differences are not pronounced enough to clearly distinguish the contact types.

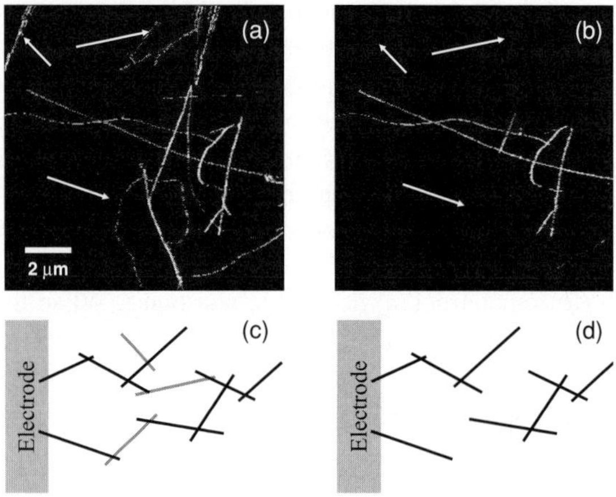

FIGURE 5. A back-gate voltage blocks current flow through semiconducting tubes. (a) Conductance image taken at $V_g = 0$ V. (b) Conductance image taken at $V_g = 10$ V. Parts of the network visible in (a) disappear because they are connected through semiconducting tubes. (c) and (d) show models of the network for (a) and (b). The semiconducting tubes (gray) are depleted of carriers and no longer carry current, which effectively removes them from the current image and severs the connections to the parts of the network connected through semiconducting tubes.

Instead, we use a different method to identify semiconducting tubes: when a sufficiently large positive back gate voltage is applied to the sample, current flow through the semiconducting tubes in the network is suppressed (Figure 5). The p-doped semiconducting tubes act like field effect transistors that are depleted of carriers by the positive gate voltage. Additionally, the width of the Schottky barrier is increased as the voltage increases [36]. This means that the semiconducting tubes and all parts of the network connected by semiconducting tubes are no longer visible in the conductance image. This allows the junctions between semiconducting and metallic tubes to be identified, and verifies that conductance drops indeed occur at these junctions.

This technique also enables the visualization of the purely metallic parts of the network and identify their spatial extent, which was our original goal. Our data demonstrate that the majority of high-conductance connections between the two metal electrodes consist of metallic tubes. When connecting the two electrodes on the sample macroscopically and measuring the change in source-drain current with applied gate voltage, we find for the sample in Figure 5, at a source-drain voltage of $V_{sd} = 100$ mV applied between the two electrodes, a current of $I_{sd} = 2.94 \cdot 10^{-7}$ A at a gate voltage $V_g = 0$ V, and $I_{sd} = 2.54 \cdot 10^{-7}$ A at $Vg = 10$ V. There is a decrease in current of only 14%, even though the percentage of tubes that is turned off by the back gate is much larger. Other samples show conductance changes between 1% and 50%. These ratios are much lower than those in the literature [31], likely due to differences in electrode geometry and nanotube coverage. This means that, in a mat of SWNT between two electrodes, a large part of the current is carried by a small number of highly conductive metallic connections, and the connections involving semiconducting tubes, even though much more numerous, carry a comparable or smaller portion of the current. Unless a way can be found to eliminate the metallic tubes, e.g., by burning them off by passing high currents through the mat while the semiconducting tubes are turned off with a back gate [37], the sensitivity of SWNT mat sensors will remain limited.

4 Measuring Decay Lengths in Carbon Nanotubes

We have analyzed the transport properties of the nanotubes by performing two-point measurements with a mobile probe on the nanotube [38]. Intuitively, the decay length is coupled to the amount of defects in the nanotube. To verify this assumption, we used untreated chemical vapor deposition (CVD) tubes that should be in pristine condition. After each measurement step, we briefly etched the the sample with piranha solution (1:3 H_2O_2:H_2SO_4) to increase the amount of defects in the tubes, and then measure again. Measurements were also performed on tubes synthesized by laser ablation that were purified by refluxing in highly oxidative media.

To investigate changes that occur directly after the contact, the analysis of current change was always performed around junctions between two nanotubes. The tip size prevented analysis of the metal-tube contacts, where the metal layer is 30 nm thick. Thus, resistances measured are typically upward of 100 kΩ, and changes of magnitudes below the noise level of about 5% were not measurable. Again, all numbers should be considered lower bounds of the conductance due to the averaging in signal processing.

The unetched CVD-grown samples show no change in current along their lengths, only sharp changes at the junctions, as shown before in the previous section. After a piranha etch, the changes in conductance around some junctions are no longer sharp, but decay gradually, as shown in Figure 6.

The conductance change was extracted from the image by tracing the backbone of the nanotube and then plotting the largest measured conductance versus the length (Figure 7).

It can be ruled out immediately that the change in conductance is a tip artifact. The actual length scale on which the change in conductance takes place is

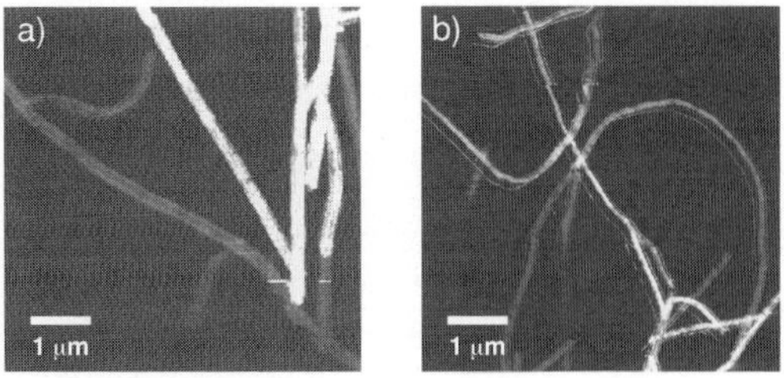

FIGURE 6. (a) Conductance image of untreated carbon nanotubes grown by chemical vapor deposition. (b) Conductance image of a nanotube network that was etched for 210 s in Piranha.

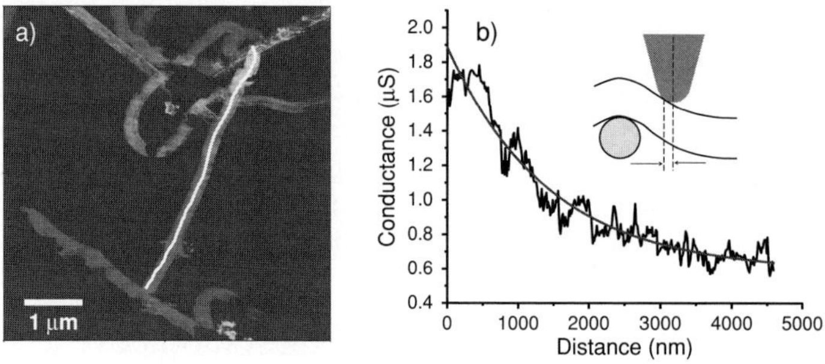

FIGURE 7. (a) The conductance image shows a network of SWNT. The white line indicates where the current trace on the right was taken. The tube close to the traced tube at the middle of the trace is not actually in electrical contact with the traced tube. (b) Trace along the tube. The smooth line is an exponential decay fit to the data (see equation 3, with $A = 2.7\ \mu S$, $B = 0.56 \pm 0.1\ \mu S$, and $l = 1{,}430 \pm 50$ nm). The inset shows how a sloped nanotube causes a tip offset.

significantly larger than any estimate of the tip radius, even for short length scales. The tip radius can, however, distort the relationship between conductance image and topography image, making it difficult to locate the point in the topographical image precisely where the change in conductance begins. A large tip can broaden the conductance signal near the junction where the tube slopes, since the point of contact between tip and tube changes as the tip moves off the junction. This problem does not exist far from the junction where the CNT lies flat on the substrate.

To extract a decay length l from these data, the following equation was fit:

$$G = B + Ae^{-x/l}. \tag{3}$$

In a first set of experiments, we compared the decay lengths measured on samples after etching for 120 s and and then etching again for an additional 90 s. We found that after the first etch, the decay lengths initially appear, and their length is reduced after the second etch step.

While the decay behavior shown in Figure 7 is most common, we also observed two other types of decay that are shown in Figure 8 and Figure 9. The former is found only at nanotube junctions which meet end-to-end. That the nanotubes are touching this way becomes apparent when the junction is pushed apart, but it is not certain whether the nanotubes merely touch at the ends, or whether they overlap over a certain length. The latter curve type was only observed rarely and occurs exclusively at the end of the tube. To extract values for l from these curves, we fit

$$G = \frac{G_1 - G_2}{1 + e^{(x-x_0)/l}} + G_2, \tag{4}$$

which is a sigmoidal function, to the data in Figure 8, and

$$G = B - Ae^{x/l} \tag{5}$$

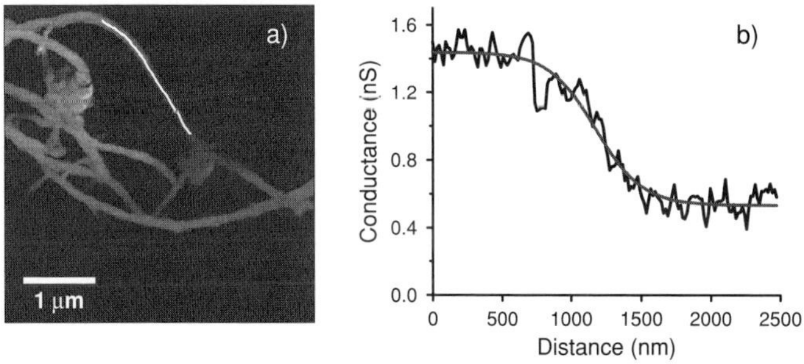

FIGURE 8. **Left:** Conductance image of a decay that follows a tanh curve at an end-to-end junction between two nanotubes. The white line indicates where the conductance trace shown was taken. **Right:** Trace along the tube. The smooth line is a tanh fit to the data (see Eq. (4), with $G_1 = 1.44 \pm 0.1$ nS, $G_2 = 0.53 \pm 0.1$ nS, and $l = 167 \pm 13$ nm).

to the data in Figure 9. On the same sample, the different decay types produce values for l that are within an order of magnitude. A summary of the decay lengths l that were measured is given in Table 1.

In a second set of experiments, we determined the behavior of the decay lengths when the sample bias is changed and when a back gate voltage is applied. Back-gating the samples introduces two difficulties. The noise in the topographical image increases, which makes it more difficult to take clean conductance image. The amount of noise seems to be connected to the tip size, but the cause for this noise could not been identified. Second, the decay length changes gradually towards an equilibrium value after every time the gate bias is changed (see Figure 10). The equilibrium value is typically reached after about 80 mins. We believe that this is due to injection of charge into the substrate from the nanotube [39,40]. The charge screens the gate voltage to some degree and reduces the effective applied gate voltage. This memory effect makes it difficult to determine the exact

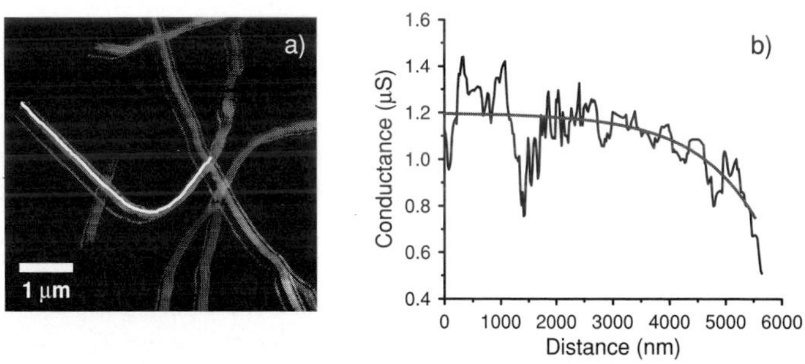

FIGURE 9. (a) Conductance image of an exponentially increasing resistance toward the end of the tube. The white line indicates where the conductance trace shown was taken. (b) Trace along the tube. The smooth line is a fit to the data using equation 5, with A $= 0.67\,\mu$S, B $= 1.20 \pm 0.1\,\mu$S, and $l = 1{,}110 \pm 125$ nm.

TABLE 1. Summary of the decay lengths extracted from measurements on etched samples without back gate

Sample	Shortest decay length (nm)	Longest decay length (nm)
CVD, no etch	—	—
CVD, 120 s etch	1,180	2,250
CVD, 210 s etch	180	890
Laser ablation	27	160

gate voltage for each measurement. We can, however, say with certainty that our measured decay lengths depend strongly on the applied gate voltages: positive gate bias increases the decay length, and negative bias decreases the decay length, as expected for charge density in a p-doped channel.

The decay length also changes if the source-drain bias (the bias between tip and sample) is changed. In all of the above measurements, the sample bias was $+1$ V, and the tip was held at 0 V. When the sample bias was changed to -1 V, again with the tip at 0 V, the decay length was reduced to approximately half of the length it had at positive bias. There appears to be no change in decay length over time after a change of the sample bias. This implies that the glassy effects mentioned above and in the literature are the result of charge motion in the oxide—not the CNT itself.

Conductance behavior as shown in Figure 7 has been observed previously [4,41]. Cumings [41] attributed this behavior to localization of electron wave functions. While the data in Table 1 suggest that the decay lengths in our experiments may be caused by the number of defects in our tubes, we almost never observe that the conductance goes to zero at long ranges. Instead, it usually reaches a finite value, indicating that localization is not the cause for the observed decay. Moreover, localization would also not explain the decay types shown in Figure 8 and Figure 9.

More consistent with our data would be that the observed decreases in conductance are caused by depletion lengths in semiconducting nanotubes around metallic/semiconducting contacts (Schottky barriers). For p-type semiconductors,

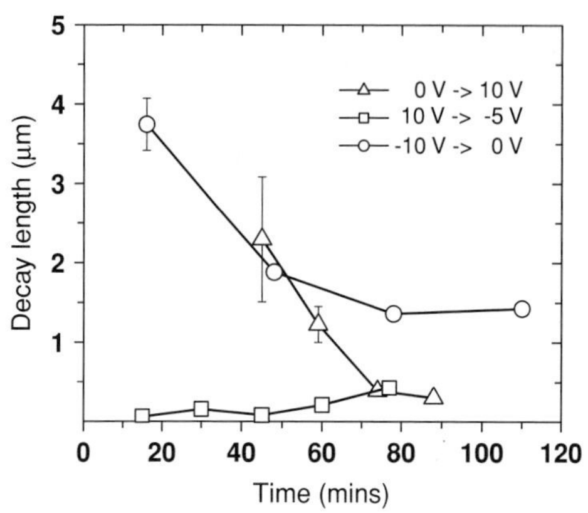

FIGURE 10. Change of decay length after the gate bias is changed. In case of the squares and triangles, the gate bias was changed from a lower value to a higher value. The circles show a change from a high value to a lower value. The decay length seems to equilibrate after about 80 min. The squares and the circles were measured sequentially on the same tube.

which nanotubes in air are assumed to be, these depletion lengths should be larger when positive bias is applied to the back gate or the electrodes, and they should be smaller when negative biases are applied. An increase in depletion length would correspond to an increase in decay length. Both of these dependencies can be observed in our data. If we assume that the decays are caused by different doping levels and screening lengths, then we can offer an explanation for the decay types shown in Figure 8 and Figure 9. The decay in Figure 8 follows the same shape as the potential decay in a junction between two semiconductors of the same dopant type, but different dopant densities, and the decay in Figure 9 looks like the potential level of a semiconductor toward the surface. In bulk semiconductors, the surface acts as a defect, causing this potential change. We assume that in the one-dimensional nanotube, the end of the tube could act as a similar defect.

It is not certain what effect the piranha etch has on the electronic properties of tube for the purpose of our measurement. The etching with piranha has been shown to introduce carboxyl groups [42]. We speculate that the high electronegativity of the oxygen in these grpups reduces the electron density on the carbon atoms, and thus the carboxyl-groups act as p-dopants. This assumption is consistent with the reduction of the decay length arising from increased carrier density if the decay is a result of screening.

5 Conclusion

Conductance imaging AFM is shown to be a powerful tool for measuring the electrical properties of carbon nanotubes and carbon nanotube mats. The connectivity of large networks can be analyzed within minutes, and the proportion of semiconducting tubes in the network can be readily identified. On the small scale, conductance imaging AFM enables high-resolution imaging of changes in electrical properties and let us track the decay lengths in nanotubes.

References

1. S. Iijima, Nature **354**, 56 (1991).
2. A. Bachtold, M. S. Fuhrer, S. Plyasunov, M. Forero, E. H. Anderson, A. Zettl, and P. L. McEuen, Phys. Rev. Lett. **84**, 6082 (2000).
3. M. Freitag, M. Radosavljević, W. Clauss, and A. T. Johnson, Phys. Rev. B **62**, R2307 (2000).
4. P. J. de Pablo, C. Gómez-Navarro, J. Colchero, P. A. Serena, J. Gómez-Herrero, and A. M. Baró, Phys. Rev. Lett. **88**, 036804 (2002).
5. M. Stadermann, S. Papadakis, M. Falvo, J. Novak, E. Snow, Q. Fu, J. Liu, Y. Fridman, J. Boland, R. Superfine, and S. Washburn, Phys. Rev. B **69**, 201402(R) (2004).
6. D. J. Thouless, Phys. Rev. Lett. **39**, 1167 (1977).
7. N. Mott and W. Twose, Adv. Phys. **10**, 107 (1961).
8. A. B. Fowler, A. Hartstein, and R. A. Webb, Phys. Rev. Lett. **48**, 196 (1982).
9. A. Miller and E. Abrahams, Phys. Rev. **120**, 745 (1960).
10. J.Voit, Rept. Prog. Phys. **58**, 977 (1996).

11. A. Yacoby, H. L. Stormer, N. S. Wingreen, L. N. Pfeiffer, K. W. Baldwin, and K. W. West, Phys. Rev. Lett. **77**, 4612 (1996).
12. F. P. Milliken, C. P. Umbach, and R. A. Webb, Solid State Comm. **97**, R10653 (1996).
13. M. Bockrath, D. Cobden, A. Rinzler, R. Smalley, L. Balents, and P. McEuen, Nature **397**, 598 (1999).
14. H. W. C. Postma, M. de Jonge, Z. Yao, and C. Dekker, Phys. Rev. B **62**, R10653 (2000).
15. C. White and T. Todorov, Nature **393**, 240 (1998).
16. P. A. Lee and T. V. Ramakrishnan, Rev. Mod. Phys. **57**, 287 (1985).
17. F. Houzé, R. Meyer, O. Schneegans, and L. Boyer, Appl. Phys. Lett. **69**, 1975 (2000).
18. M. Lantz, S. O'Shea, and M. Welland, Phys. Rev. B **56**, 15 345 (1997).
19. H. Dai, E. Wong, and C. Lieber, Science **272**, 523 (1996).
20. M. Lantz, S. O'Shea, and M. Welland, Rev. Sci. Instr. **69**, 1757 (1998).
21. S. O'Shea, R. Atta, and M. Welland, Rev. Sci. Instr. **66**, 2508 (1995).
22. M. Stadermann, H. Grube, J. Boland, S. Papadakis, M. Falvo, R. Superfine, and S. Washburn, Rev. Sci. Instr. **74**, 3653 (2003).
23. H. Edwards, L. Taylor, W. Duncan, and A. J. Melmed, J. Appl. Phys. **82**, 980 (1997).
24. F. Giessibl, Appl. Phys. Lett. **73**, 3956 (1998).
25. F. Giessibl, Appl. Phys. Lett. **76**, 1470 (2000).
26. J. Liu, A. Rinzler, H. Dai, J. Haffner, R. Bradley, P. Boul, A. Lu, T. Iverson, K. Shelimov, C. Huffman, F. Rodriguez-Macias, Y. Shon, T. Lee, D. Colbert, and R. Smalley, Science **280**, 1253 (1998).
27. Q. Fu, C. Lu, and J. Liu, Nano Letters **2**, 329 (2002).
28. J. Kong, N. Franklin, C. Zhou, M. Chapline, S. Peng, K. Cho, and H. Dai, Science **287**, 622 (2000).
29. P. Qi, O. Vermesh, M. Grecu, A. Javey, Q. Wang, H. Dai, S. Peng, and K. Cho, Nano Letters **3**, 347 (2003).
30. T. Someya, J. Small, P. Kim, C. Nuckolls, and J. Yardley, Nano Letters **3**, 877 (2003).
31. E. Snow, J. Novak, P. Campbell, and D. Park, Appl. Phys. Lett. **82**, 2145 (2003).
32. J. Novak, E. Snow, E. Houser, D. P. and. J.L. Stepnowski, and R. McGill, Appl. Phys. Lett. **83**, 4026 (2003).
33. M. Dresselhaus, G. Dresselhaus, and P. Avouris, *Carbon Nanotubes: Synthesis, Structure, Properties, and Applications*, Vol. 80 of *Springer Topics in Applied Physics* (Springer Verlag, Berlin Heidelberg New York, 2001).
34. M. Fuhrer, J. Nygård, L. Shih, M. Forero, Y. Yoon, M. Mazzoni, H. Choi, J. Ihm, S. Louie, A. Zettl, and P. McEuen, Science **288**, 494 (2000).
35. F. Léonard and J. Tersoff, Phys. Rev. Lett. **84**, 4693 (2000).
36. J. Appenzeller, J. Knoch, V. Derycke, R. Martel, S. Wind, and P. Avouris, Phys. Rev. Lett. **89**, 126801 (2002).
37. P. Collins, M. Arnold, and P. Avouris, Science **292**, 706 (2001).
38. M. Stadermann, S. Papadakis, M. Falvo, Q. Fu, J. Liu, Y. Fridman, .1. Boland, R. Superfine, and S. Washburn, Phys. Rev. B, **72**, 245406 (2005).
39. M. Fuhrer, B. Kim, Dürkop, and T. Brintlinger, Nano Lett. **2**, 755 (2002).
40. M. Radosavljević, M. Freitag, K. Thadani, and A. Johnson, Nano Letters **2**, 761 (2002).
41. J. Cumings and A. Zettl, Phys. Rev. Lett. **93**, 086801 (2004).
42. A. Kuznetsova, D. Mawhinney, V. Naumenko, J. Yates, J. Liu, and R. Smalley, Chem. Phys. Lett. **321**, 292 (2000).

II.6
Theory of Scanning Probe Microscopy

VINCENT MEUNIER AND PHILIPPE LAMBIN

1 Introduction

One of the central difficulties of devising new nanostructures is to monitor characterization and manipulation processes at extreme dimensions. In that respect, scanning probe microscopes (SPMs) constitute an important class of experimental tools for imaging the local atomic, mechanical, electronic, and transport properties of samples at dimensions ranging from the mesoscale to the nanoscale.

There are a number of challenges to be met in order to utilize the full capabilities of SPMs in nanoscale chemistry, nanobiology, and nanoelectronics. Aside from experimental difficulties, which will not be covered here, one of the most important challenge remains the interpretation and prediction of the actual measurements. It is necessary to get a good understanding of the interaction between the probe, the sample, and the substrate, and, in turn, of the electronic and mechanical processes governing the image formation. In particular, much care must be taken to separate the intrinsic properties of the sample under investigation from the effects of the probe on the overall response.

SPM techniques can be distinguished according to the way the probe interacts with the sample [1, 2]. Atomic force microscopy (AFM) is a beautiful example where different operation modes of the microscope (the most common modes being contact, noncontact, and tapping), associated with different strengths of interaction of the tip, provide different kinds of information. SPMs that work with a conducting tip held at a given electric potential have also three major nonexclusive interaction modes: strong (mechanical), weak or tunneling, and electrostatic (probe far from the sample).

While any measurement intrinsically modifies the state of a nanoscopic system, salient information concerning the sample property can usually be deconvoluted from the raw experimental data. This deconvolution, which unfortunately and most often depends on sample and experimental conditions, is sometimes tedious and therefore constitutes an important source of error in data interpretation. The interpretation and prediction of actual measurements should ideally rely on accurately simulating images, while faithfully keeping simulation parameters as close as possible to experimental ones. Starting from a structural model of the tip and the sample, and from a modeled or ad hoc tip-sample interaction, images can usually be obtained for various experimental conditions and sample modifications. In AFM for instance, there are quite accurate theories of the force between tip and sample in different contact regimes [3], on which image simulation can rely.

In scanning tunneling microscopy (STM), a common approach for image simulation is based on the Tersoff–Hamann approximation [4], where the electron flow from the tip to the sample is given by the convolution of the electronic densities of states of the tip and the sample within the potential bias window centered around the common Fermi level. In this approach, the coupling of the STM tip to the sample is treated as a perturbation of the sample Hamiltonian and first-order perturbation theory usually yields accurate comparison with experiment. Although successful, it must be remembered that only weak couplings (tunneling regime) can be treated in this manner, which rules out contact modes, field-effects, and strong interaction in general. At the same time, the majority of SPMs apart from AFM are based on the detection of ac or dc current induced by a bias applied to the sample and/or to the tip. Thus, the spatial resolution of these techniques is determined by the contact separation, and the exact nature of the probe-sample interaction. All these features, some of which are inherent to a given SPM operating mode, call for dedicated theoretical modeling. A successful theoretical approach must include a precise description of the electronic properties of the sample, an unambiguous depiction of the transmission of electrons in the sample and the tip, and the more challenging understanding of the tip-sample coupling for the various interaction modes discussed above.

In the following sections, a brief introduction of the principal theory of several SPM techniques is presented. Scanning tunneling microscopy and spectroscopy in the elastic regime are first discussed. Here, information on the local electronic density of states of the sample can be obtained. In the next section, inelastic scanning tunneling microscopy is briefly described. The technique has the capability of providing information on the local vibrational structure of the sample. Atomic force microscopy and its major operation (static and dynamic) modes are covered next. The technique is particularly useful for exploring structural and mechanical properties of a sample. Each technique is illustrated with examples chosen from the characterization of carbon nanotubes. The last part of the chapter discusses the image formation mechanism in scanning gate microscopy, a variant of the AFM in which a biased conducting tip acts as a local gate to modify the electrical transport properties of a sample.

There are many SPM techniques other than those listed above that can be used for characterization and modification at the nanoscale. For an extensive list, refer to the classification of scanning probe microscopies published by the international union of pure and applied chemistry (IUPAC) [5]. Many SPMs are derivatives of the AFM [6]. With a magnetic tip, the AFM becomes a magnetic force microscope (MFM) that can be used for imaging the structure of magnetic domains in magnetic materials and super-current vortices in superconductors, for instance. The electrostatic force microscope (EFM) and Kelvin force microscope (KFM) are variants of the AFM working in a noncontact mode with an electrically biased conducting tip that can be used for imaging surface charges and contact potential differences (local variations of the difference between the Fermi levels of the tip and the sample). Optical probe techniques, such as scanning near-field optical microscopy (SNOM), use light or infrared radiations scattered by the surface of the sample as a signal. Interestingly, the Bardeen formalism that forms the basis of the Tersoff–Hamann description of the STM, can be adapted to the SNOM in the weak-coupling regime [7].

2 Scanning Tunneling Microscopy

Since its invention in 1982 by Binnig, Röhrer, Gerber and Weibel [8], scanning tunneling microscopy has proven to be a particularly well adapted local technique for the atomic scale examination of numerous surface types and nanoscopic systems. In STM, a weak potential difference is applied between the sample and a metallic tip, which ideally has an atomically sharp end. The flow of a tunneling current is possible as long as the tip-sample distance is smaller than a few Ångströms. Depending on the bias polarity, electron transfer can either be established from occupied states of the sample into unoccupied states of the tip (positive tip potential) or, conversely from occupied states of the tip into unoccupied states of the sample (negative tip potential). In topographic mode, a feedback mechanism permits tuning of the tip vertical position, in such a way to keep a constant tunneling current. In other words, an STM image displays the variations of the tip vertical position while horizontally scanning the sample. Alternatively, at constant height mode, maps of the tunneling current over the sample are obtained by varying the horizontal position of the tip. For a typical applied bias up to about 1 V, the tunneling rate is very sensitive to the electronic states close to the Fermi level. In many systems, the charge distribution of those states turns out to be a signature of the total charge, and it follows that a typical STM image reflects the atomic arrangements of the sample. Therefore, scanning tunneling microscopy can provide a direct determination in real space of 3D atomic arrangements, regardless of whether they are periodic or not. As a step beyond, besides the topographic mode, STM techniques can also be used in spectroscopic mode (STS). In this mode, the feedback loop is turned off, and the tunneling current is recorded at a given tip position while varying the applied potential. This allows for an accurate determination of the local electronic properties of the sample [9, 10]. Tersoff and Hamann made a key contribution to the quantitative understanding of STM operation by showing that the tunneling rate is proportional to the local electronic density of states at the center of curvature of the tip [4]. Below, we outline the Tersoff–Hamann approach and show how it allows for an accurate interpretation and prediction of STM and STS experiments.

Consider a physical system described by a time-independent Hamiltonian H_0. The Hamiltonian represents a system constituted by the sample and the STM tip, where the tip is positioned far away from the sample. In a linear combination of atomic orbital (LCAO) description, H_0 would be a matrix made up of two independent blocks H_0^s and H_0^t, corresponding to the sample and the tip, respectively.

Suppose that at time $t = 0$, the tip is brought in close contact (typically, a few Ångströms) with the sample. Formally, this can be expressed as a perturbation $W(t) = \lambda \widehat{W}(t)$ to the original system, where the matrix elements of $\widehat{W}(t)$ are comparable in amplitude to those of H_0. After $t = 0$, the system is therefore described by the following time-dependent Hamiltonian

$$H(t) = H_0 + W(t). \tag{1}$$

For $t < 0$, one assumes that the system is described by the state $|\varphi_i\rangle$, which is an eigenstate of H_0. Transitions between the eigenstates of the system can be induced by turning on the perturbation $W(t)$. Quantitatively, the transitions can be computed as the probability $\mathcal{P}_{if}(t)$ to find, at any time t, the system in an eigenstate $|\varphi_f\rangle$ of H_0, knowing that it was initially in the state $|\varphi_i\rangle$

$$\mathcal{P}_{if}(t) = |\langle\varphi_f|\psi(t)\rangle|^2, \tag{2}$$

where $|\psi(t)\rangle$ describes the system at time t. In the time interval between 0 and t, the state of the system obeys the time-dependent Schrödinger equation

$$i\hbar\frac{d}{dt}|\psi(t)\rangle = [H_0 + \lambda\widehat{W}(t)]|\psi(t)\rangle \tag{3}$$

with the following initial condition:

$$|\psi(t = 0)\rangle = |\varphi_i\rangle. \tag{4}$$

In general, an analytic solution to Eq. (3) does not exist. However, if the perturbation W is weak (i.e., if λ is small), an approximate solution can be determined and $\mathcal{P}_{if}(t)$ can be calculated.

In the representation of the basis constituted by the eigenstates of H_0 ($\{|\varphi_n\rangle\}$), the LCAO coefficients $c_n(t)$ of $|\psi(t)\rangle$ and the matrix elements $\widehat{W}_{nk}(t)$ can be written as

$$|\psi(t)\rangle = c_n(t)|\varphi_n\rangle \tag{5}$$

and

$$\widehat{W}_{nk}(t) = \langle\varphi_n|\widehat{W}|\varphi_k\rangle. \tag{6}$$

Since the basis is complete ($\sum_k |\varphi_k\rangle\langle\varphi_k| = 1$) and orthogonal ($\langle\varphi_i|\varphi_j\rangle = \delta_{ij}$), Eq. (3) can be written as

$$i\hbar\frac{d}{dt}c_n(t) = E_n c_n(t) + \lambda\sum_k \widehat{W}_{nk}(t)c_k(t). \tag{7}$$

E_n is an eigenvalue of H_0 corresponding to the eigenstate $|\varphi_n\rangle$. When $\lambda \ll 1$, the solutions of H and H_0 are very similar, and can be written as

$$c_n^0(t) = b_n e^{-iE_n t/\hbar}, \tag{8}$$

where b_n only depends on the initial conditions. It follows that the unknown $c_n(t)$ can be expressed as

$$c_n(t) = b_n(t)e^{-iE_n t/\hbar}, \tag{9}$$

where $b_n(t)$ are functions that are slowly varying with the time coordinate. Defining the Bohr pulsation as $\omega_{nk} = (E_n - E_k)/\hbar$, Eqs. (7) and (9) yield

$$i\hbar \frac{d}{dt} b_n(t) = \lambda \sum_k e^{i\omega_{nk}t} \widehat{W}_{nk}(t) b_k(t). \tag{10}$$

This system of equations, formally equivalent to the initial problem, in general cannot be solved exactly, and to obtain a tractable solution, an approximation is necessary. The approximation consists in looking for a solution taking the form of a power series in λ. Since $\lambda \ll 1$, it is reasonable to assume the series is convergent and then, $b_n(t)$ can be expressed as

$$b_n(t) = b_n^{(0)}(t) + \lambda b_n^{(1)}(t) + \frac{\lambda^2}{2!} b_n^{(2)}(t) \ldots \tag{11}$$

Introducing the above expression into Eq. (10), we have

$$\begin{cases} b_n^{(0)} = \text{constant} \\ i\hbar \dfrac{d}{dt} b_n^{(r)}(t) = \sum_k e^{i\omega_{nk}t} \widehat{W}_{nk}(t) b_k^{(r-1)}(t). \end{cases} \tag{12}$$

This recursion relation leads to the evaluation of $b_n^{(r)}$ for any order, r, provided that the solution at order $r = 0$ can be obtained.

For $t < 0$, the system is in state $|\varphi_i\rangle$ and all coefficients $b_n(t)$ are zero, except for $b_i(t)$. At $t = 0$, the Hamiltonian undergoes a discontinuity, but the solution of the Schrödinger equation is continuous and it follows that $b_n(t = 0) = \delta_{ni}$ for any value of λ. Hence,

$$\begin{cases} b_n^{(0)}(t = 0) = \delta_{ni} \\ b_n^{(r)}(t = 0) = 0 \text{ if } r \geq 1. \end{cases} \tag{13}$$

For $t > 0$, $b_n^{(0)}(t)$ can be obtained from Eq. (12):

$$b_n^{(0)}(t) = \delta_{ni}. \tag{14}$$

Finally, to first order we find,

$$b_n^{(1)}(t) = \frac{1}{i\hbar} \int_0^t dt' e^{i\omega_{ni}t'} W_{ni}(t). \tag{15}$$

$\mathcal{P}_{if}(t)$ can be calculated from Eq. (2) as

$$\mathcal{P}_{if}(t) = |c_f(t)|^2 = |b_f(t)|^2, \tag{16}$$

where

$$b_f(t) = \sum_{r=0}^{\infty} \lambda^r b_f^{(r)}(t). \tag{17}$$

For differing initial and final states, we have to the first order,

$$\mathcal{P}_{if}(t) = \lambda^2 |b_f^{(1)}(t)|^2, \tag{18}$$

and in turn,

$$\mathcal{P}_{if}(t) = \frac{1}{\hbar^2} \left| \int_0^t dt' e^{i\omega_{fi}t'} W_{fi}(t') \right|^2 . \tag{19}$$

In the case of a constant perturbation, we see that

$$\mathcal{P}_{if}(t) = \frac{|W_{fi}|^2}{\hbar^2} \left[\frac{\sin(\omega_{fi}t/2)}{\omega_{fi}/2} \right]^2 . \tag{20}$$

Using a representation of the delta function,

$$\lim_{t\to\infty} \frac{\sin(\omega_{fi}t/2)^2}{\omega_{fi}/2^2} = \pi t \delta(\omega_{fi}), \tag{21}$$

the transition probability can be written as

$$\mathcal{P}_{if}(t) = 2\pi t \frac{|W_{fi}|^2}{\hbar^2} \delta(\omega_{fi}) \tag{22}$$

or

$$\mathcal{P}_{if}(t) = 2\pi t \frac{|W_{fi}|^2}{\hbar} \delta(E_f - E_i). \tag{23}$$

Since the probability increases linearly with time, the transition probability per unit of time from state $|\varphi_i\rangle$ to state $|\varphi_f\rangle$ is readily computed:

$$\mathcal{W}_{if} = \frac{d}{dt} \mathcal{P}_{if}(t) = 2\pi \frac{|W_{fi}|^2}{\hbar} \delta(E_f - E_i). \tag{24}$$

This result can be used to derive a general expression of the tunneling current to the first order. It is clear that the tunneling current will only be established if the initial state is occupied and the final state is unoccupied (the reverse argument is true for hole transitions). Depending on the initial conditions, the electron flow will either take place from the tip into the sample or from the sample into the tip. The current can be seen as the sum of all possible transitions between pairs of initial (μ) and final (ν) states $I_{\mu\nu}$:

$$I = \sum_{\mu,\nu} I_{\mu\nu}. \tag{25}$$

μ and ν label the eigenstates of the sample and tip associated with energies E_μ and E_ν, respectively. Therefore

$$I_{\mu\nu} \propto f_s(E_\mu)(1 - f_t(E_\nu)) - f_t(E_\nu)(1 - f_s(E_\mu)) = f_s(E_\mu) - f_t(E_\nu),$$

where $f_{t(s)}(E)$ is the Fermi-Dirac distribution on the tip (sample) side. Thus, from equation (24),

$$I = \frac{2\pi e}{\hbar} \sum_{\mu\nu} \left[f_s(E_\mu) - f_t(E_\nu) \right] \left| \langle \varphi_\mu | W | \varphi_\nu \rangle \right|^2 \delta(E_\mu - E_\nu), \tag{26}$$

where $e = |e|$ is the electron charge. When a positive (negative) potential V is applied to the tip while the sample is set to the ground potential, all the tip states

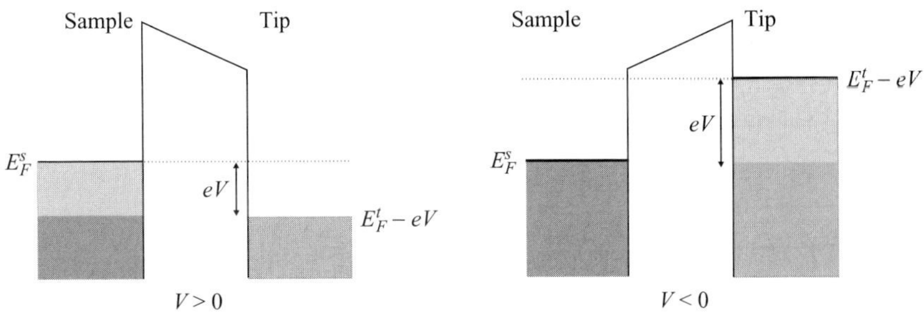

FIGURE 1. Schematic representation of the electronic processes governing the tunneling current in the presence of an external potential V between the tip and the sample. The electronic states at the Fermi level must overcome a potential barrier equal to the work function in order to tunnel. In the absence of the external potential, this barrier makes it impossible for the electrons to escape the sample or the tip. Note that the (un-)occupied states of the sample contribute to the tunneling current when the external tip potential is positive (negative).

corresponding to E_ν are lowered (raised) by $-eV$. At low temperature, tip states are occupied up to the Fermi level $E_F^t - eV$, which has also been shifted by eV (Figure 1).

Under these conditions, using the properties of the Fermi-Dirac distribution, in the low temperature limit, one finally obtains

$$I = \frac{2\pi e}{\hbar} \int_{-eV}^{0} dE \sum_{\mu\nu} \left|\langle\varphi_\mu|W|\varphi_\nu\rangle\right|^2 \delta(E - E_\mu + E_F^s)\delta(E - E_\nu + eV + E_F^t). \tag{27}$$

The summation is performed over all eigenstates of the sample (μ) and of the tip (ν). Note that the eigenstates are evaluated before connecting the system to the external potential V. The states must be rigidly shifted to account for the contact potential $(E_F^t - E_F^s)/e$ (related to the difference in work functions) and for the applied potential V, while W represents here the explicit tip-sample interaction.

A tight-binding version of Eq. (27) will now be derived to provide a practical way to compute the tunneling current from first order perturbation theory. Tight-binding theory offers a straightforward and intuitive way to evaluate electronic properties of a large spectrum of materials in the solid state [11]. To begin, an LCAO expression for the tunneling current will be derived. For the sake of simplicity, one atomic orbital per atomic center will be assumed. This limitation can easily be generalized to multiple atomic orbitals per atom. In the following the index μ (ν) will refer to an eigenstate of the sample (tip) and index I (J) will be used for an atom in the sample (tip). Therefore, $|\varphi_\mu\rangle$ and $|\varphi_\nu\rangle$ can be expressed as

$$|\varphi_\mu\rangle = \sum_{I\in s} \chi_I^\mu |I\rangle \tag{28}$$

$$|\varphi_\nu\rangle = \sum_{J\in t} \psi_J^\nu |J\rangle. \tag{29}$$

The functions $|J\rangle$ and $|I\rangle$ are atomic orbitals centered on the tip site J and sample site I. The scalars $\chi_I^\mu = \langle I|\mu\rangle$ and $\psi_J^\nu = \langle J|\nu\rangle$ are LCAO coefficients of the eigenstates of the tip and sample, respectively.

In the LCAO basis, the matrix elements $\langle\varphi_\mu|W|\varphi_\nu\rangle$ can be represented as

$$\langle\varphi_\mu|W|\varphi_\nu\rangle = \sum_{I\in s}\sum_{J\in t}\chi_I^{\mu*}\psi_J^\nu\langle I|W|J\rangle. \tag{30}$$

The generalized tight-binding density of states is given by

$$n_{II'}(E) = \sum_\mu \chi_I^{\mu*}\chi_{I'}^\mu \tag{31}$$

for the sample and

$$n_{JJ'}(E) = \sum_\nu \psi_J^{\nu*}\psi_{J'}^\nu \tag{32}$$

for the tip. By combining Eqs. (27) and (30), the tunneling current can be expressed as

$$I = \frac{2\pi e}{\hbar}\int_{-eV}^{0} dE \sum_{I,I'\in s}\sum_{J,J'\in t} W_{IJ}W^*_{I'J'}n_{II'}(E+E_F^s)n_{JJ'}(E+eV+E_F^t) \tag{33}$$

where $W_{IJ} = \langle I|W|J\rangle$ are the tip-sample coupling matrix elements expressed in the atomic orbital basis. There exists a number of approaches to compute the LCAO density of states in a variety of different approximations and levels of theory [12].

In the spectroscopic (STS) mode, when the feedback loop is turned off, the tip position is fixed. The applied tip potential is varied and an $I(V)$ curve is plotted. It is generally accepted that the first derivative of the recorded $I(V)$ curve is a measure of the local density of states of the sample below the tip. To check this, let's suppose the tip is placed just above an atomic center of the sample. In first approximation, only atomic center $I = i$ will contribute to the tunneling current. In addition, we will assume that the tunneling current is recorded by a single $J = j$ atom of the tip. By taking the first derivative of Eq. (33), we find, for a weak applied potential:

$$\begin{aligned}\frac{dI}{dV}(V) &= \frac{d}{dV}\left[\frac{2\pi e}{\hbar}\int_{-eV}^{0} dE\,|W_{ij}|^2 n_{ii}(E+E_F^s)n_{jj}(E+eV+E_F^t)\right]\\ &\simeq \frac{2\pi e^2}{\hbar}|W_{ij}|^2 n_{ii}(E_F^s - eV)n_{jj}(E_F^t).\end{aligned} \tag{34}$$

This shows that the first derivative of the current with respect to the applied potential is proportional to the local density of states of the sample below the tip. Note that (un-)occupied states of the samples are probed at (negative) positive tip potential. This means that the $dI/dV(V)$ curve should be compared to $n(E_F^e - E)$, not to $n(E_F^e + E)$, with $E = eV$. Equation (34), though correct, is an approximation: the current also depends on off-diagonal terms of the generalized density of states. Since they correspond to the description of bonding or anti-bonding states, they can be asymmetric with respect to the Fermi level and it follows that the first derivative of the current can display some asymmetry, even if the density of states is symmetric [13].

We see from the above that the tunneling current, and in turn the STM image, is formed from a convolution of the density of states of the tip and the sample to the first order. The tip-sample interaction is exclusively contained in the coupling operator W, while the intrinsic properties of the sample are included in the density of states. Computing an STM image as a plot of the density of states on a plane is very tempting since it is a natural output of most electronic structure codes. Most of the time, this basic approach yields good results, especially in the case of flat conducting surfaces. It is sometimes important to avoid neglecting W, especially in the case of a very sharp tip and a clean molecular sample where the coupling between orbitals may lead to somewhat nonintuitive features to the image. For example, we illustrate that in the case of the interaction of a point-like tip with a curved graphitic surface (i.e., a carbon nanotube).

It should be mentioned that *ab-initio* calculations have shown that the substrate on which a nanotube is deposited plays only a small part in the STM image [14]. For the applications illustrated below, the tip is described by a single atom with an s orbital. The corresponding density of states $n^t(E)$ is assumed to have a Gaussian shape. The nanotube can be described by a π-electron tight-binding Hamiltonian with constant first-neighbor interactions ($\gamma_0 = -2.9\,\text{eV}$). The matrix elements coupling the tip apex atoms J to the atoms I of the nanotube are Slater-Koster sp-like hopping interactions decaying exponentially with separation distance d_{IJ}:

$$W_{IJ} = W_0 v_{IJ} e^{-d_{IJ}/\lambda} \cos\theta_{IJ} \tag{35}$$

$$v_{IJ} = e^{-ad_{IJ}^2} \Big/ \sum_{J'} e^{-ad_{IJ'}^2}. \tag{36}$$

W_0 is a scaling factor (of the order of 1 eV), θ_{IJ} is the angle between the π-orbital lobe on site I and the local direction of the tip atom, and W_{IJ} is a weighting factor used to select the carbon atoms closest to the tip (within a disk of radius 2 Å centered on the projection of the tip apex on the nanotube). The parameters used were $\lambda = 0.85$ Å, $a = 0.6$ Å^{-2}. With these values, the tunneling current above a flat sheet of graphite decreases by an order of magnitude when increasing tip–sample distance by 1 Å. It has also been verified that the current asymmetry brought about by the nonequivalence of the so-called A and B atoms in multi-layered graphite comes out correctly from the calculations [15].

As an example, Figure 2 shows constant-current images of four single-walled carbon nanotubes computed for a tip potential equal to −0.4 V and +0.4 V. In the following, we will make use of the (n, m) Hamada notation to label different carbon nanotubes [16]. The first three nanotubes are semiconductors (the difference between their wrapping indices n and m is not a multiple of 3, written thereafter $M(3)$), with a band gap about 0.6 eV. There is a clear effect of the tip polarity on the images of the semiconducting nanotubes. In the case of (17,0), the bonds parallel to the tube axis (horizontal direction in the figure) appear as strong protrusions for $V = +0.4$ V. With a negative tip, by contrast, those bonds have a dip at their center, whereas the bonds inclined at 60° with respect to the axis have increased

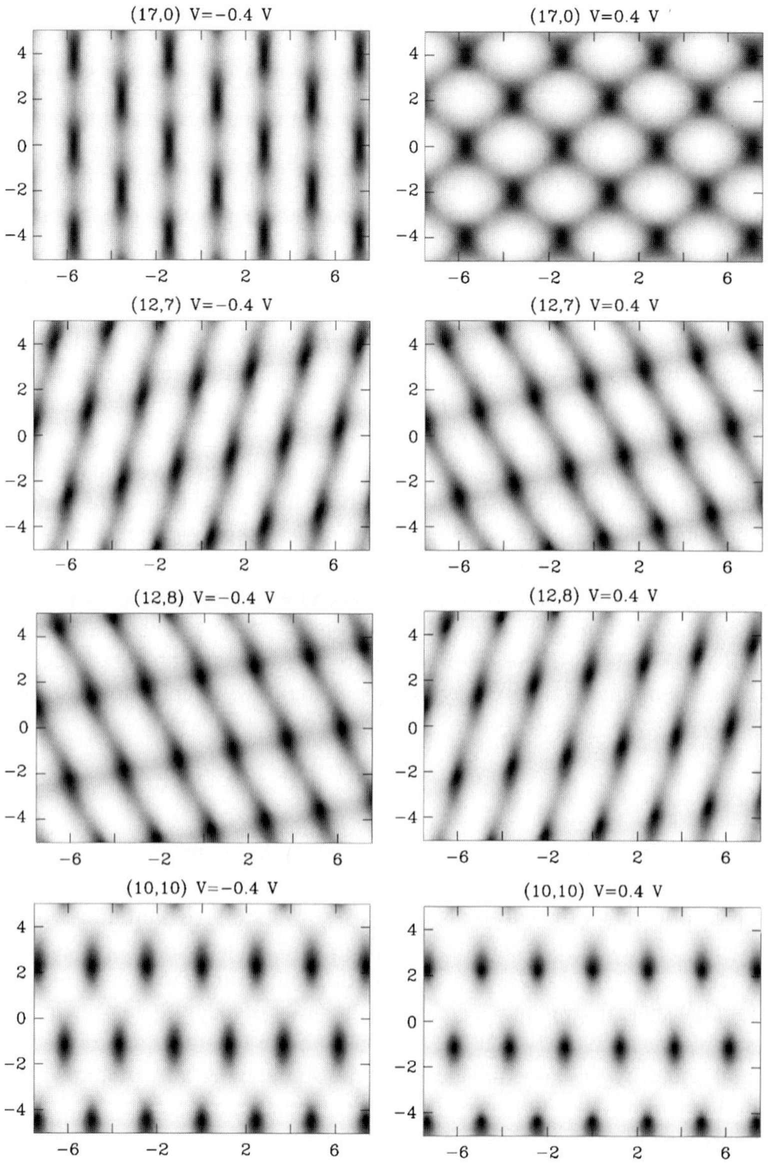

FIGURE 2. STM images of three semiconducting nanotubes and one metallic nanotube computed for $V = -0.4$ V (**left**) and $V = +0.4$ V (**right**). The nanotubes are labelled following Hamada's indexing system [16]: (17,0), (12,7), (12,8) (semiconducting), and (10,10) (metallic). The centers of the honeycomb hexagons appear as sharp dips validating thereby the interpretation of the experimental observations [17, 18]. The hexagon holes of the nanotubes are elongated along the vertical direction (perpendicular to the tube axis) due to the geometrical distortion that is inherent to the cylindrical shape of the nanotube [19]. A consequence of the distortion is that the apparent angle of a zigzag chain of atoms with respect to the tube axis is overestimated, an effect that is easily corrected in an experimental image [20].

strength. The strongest bonds are those which have a bonding character. In the above formalism, this means a positive value of the non-diagonal element $n_{II'}(E)$ corresponding to the II' bond for energies E in the bias window. These elements are odd functions of $E - E_F$, which means that the same bond may have a bonding character near the top of the valence band (probed by a positive tip) and a anti-bonding character near the bottom of the conduction band (probed by a negative tip). As a consequence, the STM image of a nanotube exhibits a marked bond anisotropy: the bonding orbitals form strips around the nanotube and the direction of the stripped pattern changes by reversing the sign of the tip potential, a peculiarity that has been observed experimentally [21]. Interestingly, (n, m) nanotubes with $n - m = M(3) + 1$ and those with $n - m = M(3) - 1$ have opposite behaviors [22] (compare (12,7) with (12,8) in Figure 2: the image of one tube with a positive bias is similar to that of the other tube with a negative tip). Metallic nanotubes also have a bond anisotropy in the STM images, displaying a maximum for the zig-zag orientation and a minimum for the armchair geometry. Here, the origin of the bond anisotropy is geometric [23], at least for a small potential bias. As long as the tip potential remains within the metallic plateau that contains the Fermi level in the density of states, the sign of the potential has only a small effect on the STM image. This invariance is illustrated with the case of the (10,10) metallic nanotube at the bottom of Figure 2.

In an alternative description of the tunneling regime, it is possible to consider the tip as a moving terminal and the sample as a fixed terminal in a Landauer/Buttiker formalism [24–26]. In this case, the transport properties of the probe–sample system can be evaluated in the open system using a non-equilibrium Green functions. The actual calculation of the electron transmission in multi-terminal devices (i.e., the probe coupled to the sample) does require generalized Landauer/Buttiker formalism. The role of the probe (i.e., the nature of the contact and its invasiveness in the electronic process) must be accurately described to determine its effects on the computed SPM image. While some of the techniques are special purpose (e.g., non-equilibrium Green functions), most of the methods are cross-cutting, especially for systems requiring highly accurate large-scale simulations. These methods include high-order quantum methods for quantitative calculations on large systems such as multi-grid density functional theory methods based on localized orbital basis sets.

3 Inelastic Scanning Tunneling Microscopy

In the STM tip-sample junction, an inelastic process may take place when the tunneling electrons have lost the energy $\hbar\omega$ of a vibrational mode—or possibly another low-energy excitation—of the system. By opening a new tunneling channel at the corresponding energy $\hbar\omega$, this process increases the tunneling differential conductance dI/dV of the junction compared to that arising from a pure elastic transfer of the electrons. As a result, the second derivative d^2I/dV^2 of the tunneling current plotted against the bias potential V displays narrow peaks at the energies of the vibrational modes of the sample, and possibly the tip [27]. Because the

STM has atomic spatial resolution, the Inelastic Electron Tunneling Spectroscopy (IETS) techniques combined with STM makes it possible to probe local vibrational properties of the sample, including the vibrations of a single molecule adsorbed on a surface [28]. This technique can be used to excite surface phonons of an extended sample, like graphite [27,29], and to scan the phonon structure of a nanostructure, such as a carbon nanotube [30].

The differential conductance of the STM junction can be estimated by the Tersoff–Hamann theory (see previous section)

$$dI(V)/dV \propto n(\boldsymbol{r}_t, E_F - eV) \tag{37}$$

where V denotes the tip-sample potential as above, E_F is the Fermi level of the sample, and $n(\boldsymbol{r}_t, E)$ is the local density of states (LDOS) of the sample at the location $\boldsymbol{r}_t$ of the tip,

$$n(\boldsymbol{r}_t, E) = \sum_{\mu} |\varphi_\mu(\boldsymbol{r}_t)|^2\, \delta(E_\mu - E). \tag{38}$$

The sum runs over all electronic states μ of the sample, with energies E_μ. $\varphi_\mu(\boldsymbol{r}_t)$ denotes the corresponding wave-function at the location of the tip. The change of LDOS induced by the electron–phonon coupling (assumed to be small) can be shown to have a discontinuity, Δn, when the tip bias, eV, matches the energy, $\hbar\omega$, of a phonon or a molecular vibration of the sample [31]. If it is further assumed that there is no interference between the elastic and inelastic scattered STM wave-functions, Eq. (37) indicates that there will be a discontinuity of the differential conductance when $eV = \hbar\omega$, induced by the phonon, and directly proportional to Δn, that can be approximately written as

$$\Delta n = \sum_{\mu} \left[\left| \sum_{\mu'} \frac{\langle \mu|\delta v|\mu'\rangle\, \varphi_{\mu'}(\boldsymbol{r}_t)}{E_\mu - E_{\mu'} + \hbar\omega + i0} \right|^2 - 2\pi^2 \left| \sum_{\mu'} \langle \mu|\delta v|\mu'\rangle\, \varphi_{\mu'}(\boldsymbol{r}_t)\, \delta(E_\mu - E_{\mu'} + \hbar\omega) \right|^2 \right] \delta(E_\mu - E_F). \tag{39}$$

The first term in brackets originates from the emission of a real phonon whereas the second term results from virtual phonon emission-absorption processes. In this expression, $\delta v = \sqrt{\hbar/2M\omega}\, \partial v_{eff}/\partial Q$ is the deformation potential, where v_{eff} is the one-electron effective potential in the Schrödinger equation of the sample, Q is the normal coordinate and M is the mass associated with the phonon. In certain cases, the contribution of the virtual phonons can be neglected when compared to the inelastic part that involves real phonon excitations. If it is further assumed that for most of the intermediate states μ', $\hbar\omega \ll |E_\mu - E_{\mu'}|$, then

$$\Delta n \approx \frac{\hbar}{2M\omega} \sum_{\mu} |\partial\varphi_\mu(\boldsymbol{r}_t)/\partial Q|^2\, \delta(E_\mu - E_F). \tag{40}$$

IETS-STM spectra of carbon nanotubes obtained recently [30] revealed spectral features located between 10 and 200 meV that can be attributed to phonons. At low energy, in particular, the signature of the radial breathing mode was clearly identified in the inelastic spectra. This particular mode, for which the normal coordinate Q is proportional to the change of radius of the nanotube, is inversely proportional to the diameter. This dependence with frequency could be checked by measuring the nanotube diameter in STM image mode for a series of tubes. It was also shown that nanotubes can have a radial breathing vibration only if its length exceeds a few nanometers. In the case of isolated C_{60} fullerene adsorbed on an Ag surface, the Hg(2) mode (at about 50 meV) is excited by inelastic tunneling electrons [32].

4 Atomic Force Microscopy

An atomic force microscope (AFM) is a widely used instrument that can be operated in many modes. AFM can work on nonconducting materials and it can image the surface of many different types of materials, such as glass, polymers, nanomaterials, and biological samples. AFM consists of a hard tip mounted on a cantilever. The tip is brought in close proximity to the surface to be analyzed, the tip-sample interaction produces a deformation of the cantilever, which is measured, often by the deflection of a laser beam reflected from the back of the AFM lever [1,2,6].

When the tip is very close to the surface, the force F_t exerted on it is dominated by the repulsive interaction between the tip apex and the nearest atoms of the surface. This is the repulsive regime corresponding to the so-called contact mode. In the constant-force mode, the tip-sample distance is adjusted to keep the same deflection of the lever, which is proportional to the force F_t. With the lateral force operating mode, the torsion of the cantilever related to in-plane force components is measured.

AFM is most often used in dynamical modes where the cantilever-tip assembly is oscillating at a frequency close to its resonance. The sample is scanned while the feedback loop adjusts the tip-sample separation to keep the oscillation frequency (non-contact mode) or the amplitude of the oscillations (tapping mode) at a constant set-point value [33]. In the non-contact mode, the tip does not touch the sample. In semi-contact or tapping mode, the oscillations bring the tip closer to the surface, in such a way that the tip partly probes the repulsive part of the interaction.

First-principles theories are more and more frequently applied to study the interaction between the AFM tip and the sample [34]. This interaction includes a repulsive part coming from the overlap of the core electron wave-functions, the decrease of the kinetic energy of the electrons due to their delocalization over tip and sample, the London dispersion interaction, and possibly some other forces (electrostatic, capillar...) [3, 35]. The first two contributions decrease exponentially with increasing tip distance. They can be evaluated using *ab-initio* methods provided the tip-sample system is restricted to a few atoms [36]. The London dispersion force, simply called van der Waals interaction, may prevail over the first two contributions and certainly dominates at large distance. Density functional theory may fail in reproducing the dispersion force correctly, since it oiginates in electron correlations. The alternative approach is to sum r^{-6} pair potentials with parameters fitted to experimental data.

Evaluating the force exerted on the tip in the repulsive regime is difficult, because both the sample and the tip are deformed. This problem is often addressed by molecular dynamics, most generally with empirical atomic potentials because it involves too many atoms. The simplest description of the tip-sample interaction relies on pair-wise potentials between the tip atoms (label j) and those (label i) of the sample,

$$V_{ts} = \sum_{i \in s} \sum_{j \in t} V_{ij}(r_{ij}), \tag{41}$$

where V_{ij} can be, for instance, the Lennard–Jones 12-6 potential

$$V_{ij}(r_{ij}) = 4\epsilon_{ij}[(\sigma_{ij}/r_{ij})^{12} - (\sigma_{ij}/r_{ij})^6]. \tag{42}$$

The net force acting on the tip is readily deduced in the form

$$\boldsymbol{F}_t = -\sum_{i \in s} \sum_{j \in t} \frac{dV_{ij}}{dr_{ij}} \frac{\boldsymbol{r}_{ij}}{r_{ij}}. \tag{43}$$

This expression may be sufficient for a qualitative calculation of the AFM image of a sample [37].

It is instructive to consider a continuous medium approximation in which the sums in the above expressions are replaced by integrals over simple shaped bodies. If, for instance, the sample is represented by a half-space medium with a flat surface and if the tip is modeled by a paraboloid of revolution having its axis perpendicular to the surface, then

$$F_{tz} = \frac{B}{180} \frac{R_t}{z^8} - \frac{A}{6} \frac{R_t}{z^2}, \tag{44}$$

where R_t is the tip radius at the apex, z is the height of the tip apex above the sample surface, $B = 4\pi^2\epsilon\sigma^{12}/\Omega_t\Omega_s$ and $A = B/\sigma^6$. Here, Ω_t and Ω_s are the atomic volumes of the tip and the sample, respectively. The second term is the London dispersion force. Rather than integrating the r^{-6} part of the Lennard-Jones potential over the tip and sample, the London dispersion force can be evaluated by application of the Lifchitz theory, based on zero-point fluctuations of the electromagnetic field. Alternatively, the constant A in Eq. (44) can be taken as the Hamacker constant usually defined for two semi-infinite bodies with parallel faces [38]. It is related to the dielectric functions of the tip and sample materials and also depends on the dielectric properties of the medium between tip and sample (if different from vacuum like in immersed AFM experiments) [3].

The implications of Eq. (44), though approximate, are twofold. First, the attractive part decreases slowly with increasing distance and it is rather long-ranged (Figure 3). Second, the repulsive part of the tip-sample force is short-ranged. This part, however, is not well described by Eq. (44) at short distance. It is indeed important to take the atomic structure of the tip and sample into account when the distance between them is below one nanometer. In addition, the deformation of the

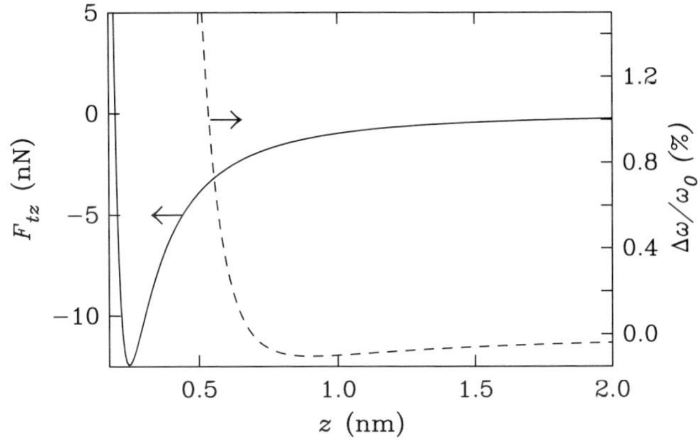

FIGURE 3. The vertical force exerted on an AFM diamond tip (Eq. (44), solid line) and the relative change of resonant frequency (Eq. (47), dashed line) versus the distance z of the tip apex above a graphite sample. The parameters used are $\epsilon = 2.97$ meV, $\sigma = 0.341$ nm, $R_t = 10$ nm, $k = 40$ N/m, $a = 10$ nm.

sample and the tip softens their repulsive interaction. In strong contact (indentation), most of the interaction is governed by the elastic, or even plastic, deformation of the two bodies. In the elastic domain, Hertz theory establishes that the force increases like the power 3/2 of the penetration $-\Delta z$ of the tip into the sample,

$$F_{tz} = (4/3)K_{ts}\sqrt{R_t}\,(-\Delta z)^{3/2}, \tag{45}$$

where $K_{ts} = [(1-\nu_t^2)/Y_t + (1-\nu_s^2)/Y_s]^{-1}$ with ν and Y respectively the Poisson coefficients and Young moduli of the tip and sample materials.

In dynamic operation modes of the AFM, the important quantity is the change of oscillation frequency $\Delta\omega$ with respect to the eigenfrequency ω_0 of the tip-cantilever assembly. The change of frequency is related to the interaction force through the following convolution integral, valid for the harmonic approximation [39]

$$\Delta\omega/\omega_0 = \frac{1}{\sqrt{2}\pi k a^{3/2}} \int_z^{\infty} \frac{F_{tz}(\zeta)}{\sqrt{z-\zeta}}\, d\zeta, \tag{46}$$

where a is the amplitude of oscillation, k is the effective spring constant of the cantilever, and z denotes the closest approach distance of the tip. With the expression (44) of the force, the above equation yields

$$\Delta\omega/\omega_0 = \frac{1}{\sqrt{2}ka^{3/2}} \left(\frac{3003\,B}{512} \frac{R_t}{z^{15/2}} - \frac{A}{12} \frac{R_t}{z^{3/2}} \right). \tag{47}$$

The result obtained from this last equation is illustrated in Figure 3 together with the force (Eq. (44)) for the case of a diamond tip on graphite. While F_{tz} becomes repulsive at very short distance ($0.57\,\sigma$), $\Delta\omega$ becomes positive at distances about 4 times as large ($2.03\,\sigma$).

Nanomaterials can be analyzed by AFM when they are deposited on a flat surface (glass, SiO_2, MoS_2 ...). In this respect, carbon nanotubes have been observed and characterized by AFM soon after their discovery. Due to the tip-shape convolution effect, a nanotube has always an apparent width larger than its height in topographic AFM images [40]. In tapping mode, the apparent height of a single-walled nanotube seems to be smaller than its actual diameter, due to mechanical compression [41]. By pressing the AFM tip onto the substrate that holds the nanotubes and dragging it along the surface, AFM is a perfectly adapted tool for moving or bending a nanotube [42], and for all types of manipulation [43]. The strength and Young modulus of a nanorod, including a nanotube, can be estimated by measuring the displacement versus the force applied by an AFM tip on a suspended portion of the nanostructure [44]. In the indentation mode, Hertz theory (Eq. (45) with coefficients adapted to the contact between a sphere and a cylinder) makes it possible to measure the radial compressibility of multi-walled nanotubes [45].

The electromechanical properties of single-walled carbon nanotubes can be probed by bending and deforming a suspended nanotube with an AFM tip, while simultaneously measuring its conductance. The experiment, confirmed by atomistic calculations for an armchair (5,5) nanotube, indicates a reversible drop of conductivity by two orders of magnitude when the bending angle reaches 14° [46]. As shown by density functional theory (DFT) calculations, the deformation of the nanotube by the AFM tip induces some sp^3 bonding to the carbon network [47], which reduces the conductivity of the nanotube. An electromechanical device operated by an AFM tip was used later on to measure the current-voltage characteristics of single-walled nanotubes under axial stretching [48]. The experiments indicate a strain gauge effect, i.e., a linear dependence of the resistance versus mechanical stress, the largest effect being observed with quasi-metallic nanotubes.

An interesting study of the electromechanical properties of *coiled* nanotubes has appeared recently [49]. These coils are multi-walled nanotubes produced by catalytic vapor deposition (CVD) that adopt the shape of a regular helix during their growth. After adsorption of the coiled nanotube on a flat surface, each of its windings has a free-standing part that can oscillate at its specific resonance frequency (200 MHz, typically). These oscillations can be excited mechanically by applying an ac electric field to a nearby conducting AFM tip and they can be detected by measuring the electrical response of the coil through contact electrodes deposited on it (microphone-like effect taking place in the contact resistances). Adsorption of nanoparticles on the resonating windings lowers their frequencies, making it possible to measure masses as small as a few tens of attograms.

The elastic properties of a nanostructure can be probed by an operating mode of the AFM known as force modulation, where a small-amplitude vertical modulation of sample vertical position is generated while the tip scans the surface of the object. The slope of the force versus distance curve is measured by synchronous detection, which is correlated to the local elasticity of the sample. This technique is particularly useful for samples that present important variations of their apparent stiffness. For instance, coiled nanotubes deposited on a flat substrate have been studied in this way, giving access to the Young modulus of the nanotubes [50]. In each case, measurements have to be be accompanied by theoretical simulations and modelling, in order to reach a qualitative and accurate interpretation of the experimental data.

5 Scanning Gate Microscopy

When dealing with low-dimensional systems, it is possible to modify the AFM apparatus to use it in conjunction with transport measurements. In this case, the nanostructure is connected to two fixed electrodes and the current along the sample is recorded as a function of an external potential. A biased AFM tip is then moved toward the structure, and the conductance of the device is measured as a function of tip position. When the tip is positioned over a portion of the structure where no current is flowing, the current distribution is not affected. Conversely, the tip can strongly affect regions of high current density, and a map of the current paths in the structure can be obtained. Since the tip acts as a local gate, this SPM technique is called scanning gate microscopy (SGM) [51]. In a modified approach, the tip-sample bias is fed back in the AFM tip (in noncontact mode) to cancel out the potential difference between the tip and the sample. Thus, when current is flowing through the structure, one can map the position and magnitude of the voltage drops between the electrodes [52]. In SGM the AFM tip is used to perturb the electrostatic potential of the sample. The tip is no longer a force sensor as in AFM mode but rather a movable local gate. Transport measurements are then performed in the same way as with a planar back-gate to change the electrostatic potential, except that now the tip affects a specific part of the sample, since the gate only changes the electrostatic potential on the part of the sample below the tip [53]. By scanning the tip over the sample with a fixed tip voltage, SGM images can reveal the locations that are particularly sensitive to changes in the electrostatic potential. For example, SGM can be used to image electronic features such as molecular orbitals [51,54–56], or scattering centers such as defects [57–61].

SGM can be used in a spectroscopic mode where quantum energy levels or modulation of the valence and conduction bands are probed. For example, when probing electronic defects, the local gate potential shifts energy levels of the target area of the sample, which turns on or off the passage of the tunneling current though the sample. When the current is canceled, the applied tip potential corresponds to a depletion surface potential which can be directly converted into information concerning the local electronic properties of the sample under investigation. The main difficulty is to link quantitatively the tip potential into an estimate of the local Fermi energy in the vicinity of the defect. The origin of that complication comes from the fact that the effective electrostatic potential induced on the sample below the tip depends on many variables, including geometric ones (tip-sample, substrate system seen as a series of capacitors), electronic ones (screening due to defects,), and intrinsic properties of the sample itself (local density of states).

The growing number of reported results in the literature shows the success and relative ease of use of this new technique to study the *local* transport properties of one-dimensional and zero-dimensional nanostructures. Despite this outstanding progress, transport SPMs have largely been qualitative techniques since the complexity of the electrostatic probe-sample interactions and geometry of the system has precluded rigorous interpretation of SPM data in terms of atomic-scale properties [61].

The main challenge in the interpretation of SGM data is to address the formal connection between the tip potential and the effective electrostatic potential shift on the sample. For example, the perturbation to the local electrostatic potential as a function of the tip bias can be evaluated from semi-classical arguments, using electrostatic theory and equivalent circuit model [61]. As carriers are added to the depleted region, both the electrostatic potential, ϕ, and the Fermi energy, E_F, increase. When N carriers are added to a region of length l, the electrostatic potential and the Fermi energy become

$$e\Delta\phi = \frac{Ne^2}{Cl} \tag{48}$$

$$\Delta E_F = \frac{N}{Dl}, \tag{49}$$

where C is the total capacitance of the region per unit length, D is the local density of states per unit length, assumed constant at the vicinity of E_F. We therefore find that

$$\Delta E_F = \frac{\Delta\phi C}{e^2 D}, \tag{50}$$

and we see that the ratio between the quantum [62] ($e^2 D$) and geometric (C) capacitance links the defect depletion surface potential and actual Fermi level shift. For example the ratio has been evaluated to about $1/12$ in the case of a carbon nanotube field effect transistor (FET), in the experimental setup of Freitag et al. [61].

It is desirable to go beyond purely electrostatic modelling and introduce an atomistic approach for the understanding of the tip-nanostructure interaction. Intensive and large scale DFT calculations of the interaction between a point charge and a 1D-nanostructure, in the present illustration a carbon nanotube, have been performed. DFT calculations yield a spatially resolved electrostatic potential distribution, V_a, in the point charge—nanotube system and include charge screening and electron redistribution effects. The present molecule-in-a-box calculation corresponds to a *floating* neutral system that differs from a *grounded* one in that the total charge on the nanotube remains constant. In actual experiments the one-dimensional system is contacted to the macroscopic world by electrodes and charge flows from the external circuit to equalize the electrochemical potential throughout the system. To address this difference in electrostatic boundary conditions, the potential distribution, V_b, must be evaluated in the point charge—nanotube system with additional charge (typically one electron) placed on the nanotube. The difference in potentials, $V_d = V_b - V_a$, describes the behavior of a free charge on the finite system in the presence of external charge.

The external charge induces a polarization on the nanotube which decays slowly along the tube, in agreement with electrostatics of one-dimensional systems [63]. The uniformity of potential inside the nanotube and symmetry with respect to the tube axis, suggest that the charge dynamics of the point charge deposited on the nanotube can be well described by a continuum model for the finite charged cylinder. In this case, the potential distribution is given by $V_{c0}(r) = \lambda/2\pi\epsilon_0 \ln(a_2/a_1)$ inside and $V_c(r) = \lambda/2\pi\epsilon_0 \ln(a_2/r)$ outside the cylinder, where a_1 is the effective electrostatic radius of the nanotube, λ is the charge density, and a_2 is the external cut-off radius related to the nanotube length.

For all nanotubes longer than 11 Å, the effective electrostatic radius is 4.1 Å, as compared to the geometrical radius of 3.4 Å, a signature of the fact that the excess charge resides on the outward lobes of the π^* orbitals, which on average extend 0.7° outside the atomic centers [64]. The external cut-off radius scales linearly with the number of atoms, as expected for a finite cylinder. The potential at the center of the nanotube can also be compared with the approximate expression for the finite conductive cylinder of length L with charge q, $V_0(L) = \lambda/2\pi\epsilon_0 \ln(L/a_1) = q/(2\pi\epsilon_0 L)\ln(L/a_1)$. The excellent agreement between continuum electrostatics and DFT theory, despite the lack of free parameters, shows that the free charge dynamics on a finite metallic carbon nanotube is well described using a conductive cylinder model, $V_d \sim V_c$, independent of external charge conditions. Thus, the potential distribution in the floating and grounded systems can be related as $V_a = V_g + \alpha V_c$, where α is a coefficient that accounts for the magnitude of the induced charge on the nanotube. The correspondence between DFT and continuum theory descriptions can be established by noting that $V_c = V_{c0}$ inside the nanotube. The electric field, $E = dV/dx|_{x=0}$, at the center of the nanotube, provides the measure of deviation from an ideal conductive cylinder behavior due to the finite density of states on the nanotube. This behavior does not depend on the nanotube length. Note that while the potential on the nanotube decreases as a function of its length, the electrostatic field at the center remains constant.

Since DFT calculations are limited to finite tube lengths and small tube-charge distances, it is imperative to construct a continuum electrostatic model to extend the *ab-initio* theory. Any description of the interaction between a point charge and a one-dimensional system must account for the finite density of states of the latter [62, 66, 67]. In our extension to continuum electrostatics, the quantum contribution to the electrostatic properties is described using the inverse screening length, k_s, which links the surface charge density and the surface potential as $\sigma_s = k_s\epsilon_0 V_s$. In cylindrical coordinates where the z-axis coincides with the tube axis, the point charge is located at distance ρ in the $\varphi = 0$ direction at $z = 0$. The potential distributions inside the nanotube, $V_n(r, \rho, \varphi, z)$, and in vacuum, $V_v(r, \rho, \varphi, z)$, for $r < \rho$ are then given by

$$V_n = \sum_{m=-\infty}^{\infty} \int_0^{\infty} dk \cos(kz) e^{im\varphi} A_{mk} I_m(kr) \tag{51}$$

$$V_v = \sum_{m=-\infty}^{\infty} \int_0^{\infty} dk \cos(kz) e^{im\varphi} \times [B_{mk} I_m(kr) + C_{mk} K_m(kr)], \tag{52}$$

where I_m and K_m are Bessel functions of the second kind, A_{mk} and C_{mk} are coefficients determined by boundary conditions [68], and B_{mk} is derived from the bare potential of the charge. The boundary conditions for a dielectric cylinder (of radius R) with finite surface density of states reflect the continuity of the potential and the linear screening approximation:

$$V_n = V_v, \text{ and } \quad \epsilon \frac{dV_n}{dr} - \frac{dV_v}{dr} = -k_s V_s, \tag{53}$$

where ϵ is the bulk dielectric constant. The potentials in vacuum and inside the nanotube are obtained from Eqs. (52) and (53):

$$B_{mk} = \frac{q}{2\pi^2 \epsilon_0} K_m(k\rho) \tag{54}$$

$$A_{mk} = B_{mk} \frac{I'_m(kR)K_m(kR) - I_m(kR)K'_m(kR)}{D_{mk}} \tag{55}$$

$$C_{mk} = -B_{mk} I_m(kR) \frac{k_s I_m(kR)/k + (\epsilon - 1)I'_m(kR)}{D_{mk}}, \tag{56}$$

where $D_{mk} = \epsilon I'_m(kR)K_m(kR) + I_m(kR)(k_s K_m(kR)/k - K'_m(kR))$. The above approximation is a priori only applicable to infinite 2D systems. This is not the case for a single-walled carbon nanotube, where the energy is quantized around its circumference. A more rigorous solution would include distinct longitudinal and circumferential screening lengths. This can be addressed by considering k_s as an effective parameter that depends on the charge-tube separation (Figure 4(a)). Practically, the effective k_s value for different charge-tube separations is determined so that the electric field at the center of the tube matches that of the DFT calculation. We find that the distance dependence can be approximated by $k_s(x) = k_{si} + b/x$, where $k_{si} = 5.0 \pm 0.1$ Å^{-1} and $b = 12.4 \pm 0.3$. The distance dependence originates from the fact that the electrostatic screening becomes less effective when the charge is moved away from the tube, since for a small charge-tube distance there is a large tangential component of the field which induces longitudinal polarization in the sp^2 bonded electron system of the nanotube. For large charge-tube separations

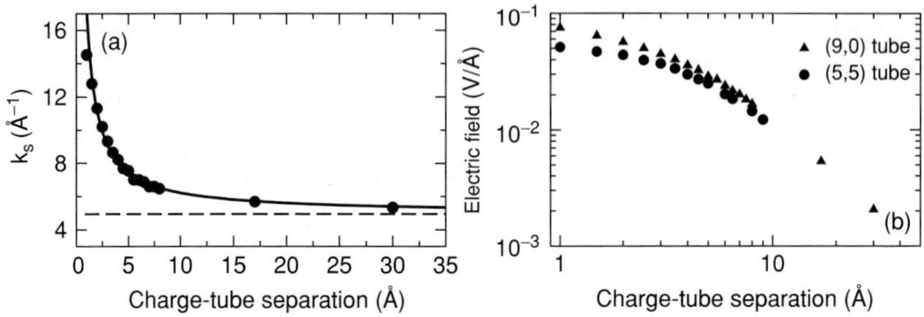

FIGURE 4. (a) Effective screening length for a (9,0) nanotube as a function of the charge-tube separation. (b) Electric field at the center of the (5,5) and (9,0) nanotubes as a function of charge-tube separation, as computed by DFT. Reproduced from Meunier et al [65].

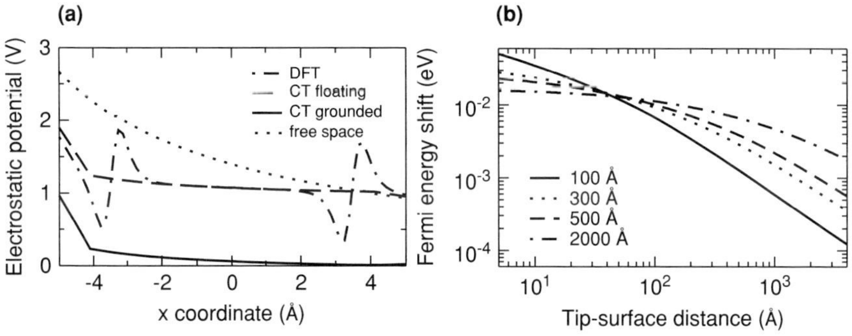

FIGURE 5. (a) Electrostatic potential perpendicular to the nanotube from the DFT model, compared to the electrostatic model for grounded and floating nanotubes. (b) Fermi energy shift in the nanotube as a function of tip-surface separation for $V_{\rm tip} = 1$ V. Reproduced from Meunier et al. [65].

the field is essentially normal to the surface and the screening is dominated by the p_z orbitals.

To demonstrate the universality of the model, we have computed the charge-tube separation dependence of the electrostatic field at the center of (5,5) and (9,0) nanotubes (Figure 4(b)). There is a significant difference in the electrostatic behavior of the nanotubes for small charge surface separations, which results from the different atomic structures and, in turn, of the atomic mechanism of screening. At large charge-tube separations, the screening behaviors become virtually identical. To account for the difference in bias conditions, the potential of the nanotube based on continuum theory (CT) is corrected as $V_{\rm eff} = V_g + \alpha V_c$ with $\alpha = 0.56$ being the induced charge on the nanotube per electron present on the tip. With this correction, the potential distributions in the DFT and CT models are very similar, the main difference remaining at the locations of the atoms due to the atomic polarizabilities (Figure 5(a)).

With the above-described method to evaluate the interaction between a point charge and a given sample as a function of the distance between the charge and the sample, it is now possible to extend that knowledge to realistic experimental situations. Under typical SGM experimental conditions, the tip radius of curvature and the tip-substrate separation (ranging from 10 to 100 nm and 3 to 500 nm, respectively) are significantly larger than the typical 0.5–3 nm radius of a 1D nanostructure. This implies that the nanostructure-tip capacitance is negligible compared to the tip-substrate capacitance and the charge states of the tip are only weakly affected by the sample. Thus, the point-charge interaction can be extended to the spherical tip model as follows: in the sphere-plane approximation, the image charge distribution on the tip can be represented by a set of charges Q_i located at distances r_i from the center of the sphere such that

$$Q_{i+1} = \frac{\kappa - 1}{\kappa + 1} \frac{r_{i+1}}{R_{\rm tip}} Q_i, \quad r_{i+1} = \frac{R_{\rm tip}^2}{2(R_{\rm tip} + d) - r_i}, \tag{57}$$

where R_{tip} is the tip radius, κ is the dielectric constant of the substrate, d is the tip-surface separation, $Q_0 = 4\pi\epsilon_0 R_{\rm tip} V$, and V is the tip bias. Hence, the tip induced potential on the nanotube is, with $\rho_0 = R_{\rm tip} + d - r_i$

$$V_t = \sum_{i=0}^{\infty} \left(Q_i v_s(\rho_0 - 2R) - \frac{\kappa - 1}{\kappa + 1} Q_i v_s(\rho_0 + 2R) \right), \tag{58}$$

where the first term is due to the image charges in the tip and the second term comes from the image charges in the substrate. The potential induced by one electron, $v_s(\rho)$, is a rapidly decaying function of tip-surface separation (faster than $1/\rho$) and, for large tip-tube separations, $d \gg R_{tip}$, Eq. (58) is simplified to $V_t = 2v_s(R_{\rm tip} + d)CV/(\kappa + 1)$ where C is the tip-surface capacitance that depends only on geometry, for a dielectric substrate. Note that for a metallic substrate, the potential is approximately given by $V_t = 4RCV\partial v_s(R_{\rm tip} + d)/\partial z$. Given that different limiting behaviors are expected for large separations and large dielectric constants, the use of Eqs. (57) and (58) is required for quantitative evaluations.

We can analyze the distance dependence of the Fermi energy shift in the nanotube corresponding to a given bias applied to the tip using the multiscale model outlined above (Figure5(b)). The potential on the nanotube is only a fraction (10^{-3} to 10^{-2}) of the applied tip bias. Thus, the application of ~10 V to the SGM tip shifts the Fermi level on the nanotube by tens of meV, comparable with the energy levels of defects obtained experimentally using STS [69]. It has long been realized that the shift of the Fermi level of the nanotube is significantly smaller than the tip bias. Notably, Woodside has used a simple electrostatic model to describe the decay of electrostatic potential at large separations from the tip [53]. A similar approach taking into account quantum capacitance of the nanotube was employed by Freitag et al. to describe experimental SGM data on semiconductive nanotubes [61]. However, both these approaches are phenomenological and do not explicitly address the electrostatics of tip, tube, and substrate interactions as developed here.

In summary, we see that a combination of first-principles and continuum theory addressing the interaction between a probe and a one-dimensional system on a substrate leads to a quantitative interpretation of the SGM data in terms of relevant electronic properties of the system which ultimately permits a full description of defects. Only such a rigorous analysis of tip-surface interactions taking into account quantum capacitance of the system and localized properties of the tip allows us to quantitatively relate tip bias and Fermi level in reduced-dimensionality systems.

Conclusion

Nanoscale materials are being extensively investigated as a new frontier beyond conventional systems. As materials are scaled to smaller and smaller sizes within the nano-regime, a number of technological and economical problems arise, and the rules of classical physics must give way to the ones of quantum mechanics.

In 1981, a new revolutionary method for characterization at the nanoscale, now known as scanning probe microscopy (SPM), was invented and its applications have been blooming in many fields of materials sciences, chemistry, biology, and technology. SPM is also emerging as a key technique for R&D, not only for the characterization but also for the manipulation of nano-objects, and tremendous potential exists for development of new SPM-based techniques. In parallel with these instrumental developments, more and more care must be taken to interpret and understand experimental findings. It is indeed true that the nanotechnology revolution calls for new computational models that account for fundamental physical interactions at the quantum mechanical level. The need for new highly accurate and large-scale modeling and computational tools has evolved tremendously.

Local probe microscopes constitute a very important class of experimental techniques for investigating various local properties of a sample at the nanoscale, as briefly reviewed in the above sections and detailed elsewhere in this book. The capabilities of the SPM techniques can only be fully exploited via an accurate theoretical description of the interaction between the local probe and the specimen. This is needed for the interpretation and sometimes the prediction of experimental data. In that respect, the development of tip-sample interaction models within a precise theoretical framework is already well advanced, as demonstrated with a few examples in this chapter.

Acknowledgments. This research is supported in part by the Division of Materials Science, U.S. Department of Energy under Contract No. DEAC05-00OR22725 with UT-Battelle, LLC, at Oak Ridge National Laboratory (V.M.) and in part by the IUAP project P5/01 "Quantum size effects in nanostructured materials" of the Belgian Science Policy Programming (Ph.L). The research presented here used in part resources of the Center for Computational Sciences at Oak Ridge National Laboratory.

References

1. R. Wiesendanger, ed., *Scanning Probe Microscopy: Analytical Methods* (Springer Berlin, Heidelberg, New York, 1998).
2. E. Meyer, H. J. Hug, and R. Bennewitz, *Scanning Probe Microscopy. The Lab on a Tip* (Springer, 2003).
3. B. Cappella and G. Dietler, Surf. Sci. Rep. **34**, 1 (1999).
4. J. Tersoff and D. Hamann, Phys. Rev. B **31**, 805 (1985).
5. G. Friedbacher and H. Fuchs, Pure Appl. Chem. **71**, 1337 (1999).
6. D. Bonnell, *Scanning Probe Microscopy and Spectroscopy: Theory, Techniques, and Applications* (Wiley-VCH, Inc., New York, 2001).
7. R. Carminati and J. Saenz, Phys. Rev. Lett. **84**, 5156 (2000).
8. G. Binnig, H. Röhrer, C. Gerber, and E. Weibel, Phys. Rev. Lett. **49**, 57 (1982).
9. M. Crommie, C. Lutz, and D. Eigler, Science **262**, 218 (1993).
10. M. Crommie, C. Lutz, and D. Eigler, Nature **363**, 424 (1993).

11. W. Harrison, *Electronic Structure and the properties of solids: The Physics of the Chemical Bond* (W. H. Freeman and Company, San Fransisco, 1980).
12. P. Lambin, *The tight-binding hamiltonian and its diagonalization*, Laboratoire de Physique du Solide, FUNDP (1983).
13. V. Meunier, P. Senet, and P. Lambin, Phys. Rev. B **60**, 7792 (1999).
14. A. Rubio, D. Sanchez-Portal, E. Artacho, P. Ordejon, and J. Soler, Phys. Rev. Lett. **82**, 3520 (1999).
15. D. Tomanek and S. Louie, Phys. Rev. B **37**, 8327 (1988).
16. N. Hamada, S. Sawada, and A. Oshiyama, Phys. Rev. Lett. **68**, 1579 (1992).
17. J. Wildoer, L. Venema, A. Rinzler, R. Smalley, and C. Dekker, Nature **391**, 59 (1998).
18. T. W. Odom, J. L. Huang, P. Kim, and C. M. Lieber, Nature **391**, 62 (1998).
19. V. Meunier and P. Lambin, Phys. Rev. Lett. **81**, 5888 (1998).
20. L. C. Venema, V. Meunier, P. Lambin, and C. Dekker, Phys. Rev. B **61**, 2991 (2000).
21. W. Clauss, D. Bergeron, M. Freitag, C. Kane, E. Mele, and A. Johnson, Europhys. Lett. **47**, 601 (1999).
22. C. Kane and E. Mele, Phys. Rev. B **59**, R12759 (1999).
23. V. Meunier and P. Lambin, Phil. Trans. R. Soc. Lond. A **362**, 2187 (2004).
24. P. Sautet and C. Joachim, Chem. Phys. Lett. **185**, 23 (1991).
25. J. Cerda, M. A. V. Hove, P. Sautet, and M. Salmeron, Phys. Rev. B **56**, 15885 (1997).
26. M. Toerker, T. Fritz, H. Proehl, R. Gutierrez, F. Grossmann, and R. Schmidt, Phys. Rev. B **65** (2002).
27. W. Olejniczak, Z. Klusek, and M. Bieniecki, Appl. Phys. A **66**, S191 (1998).
28. B. Stipe, M. Rezaei, and W. Ho, Science **280**, 1733 (1998).
29. L. Vitali, M. A. Schneider, K. Kern, L. Wirtz, and A. Rubio, Phys. Rev. B **69**, 121414 (2004).
30. L. Vitali, M. Burghard, M. A. Schneider, L. Liu, S. Wu, C. Jayanthi, and K. Kern, Phys. Rev. Lett. **93**, 136103.1 (2004).
31. N. Lorente and M. Persson, Phys. Rev. Lett. **85**, 2997 (2000).
32. J. Pascual, J. Gómez-Herrero, D. Sánchez-Portal, and H. Rust, J. Chem. Phys. **117**, 9531 (2002).
33. R. Garcia and R. Perez, Surf. Sci. Rep. **47**, 197 (2002).
34. D. Drakova, Rep. Prog. Phys. **64**, 205 (2001).
35. M. Sitti and H. Hashimoto, IEEE/ASME Trans. Mechatronics **5**, 199 (2000).
36. A. Buldum, S. Ciraci, C. Fong, and J. Nelson, Phys. Rev. B **59**, 5120 (1999).
37. H. Holscher, W. Allers, U. Schwarz, A. Schwarz, and R. Wiesendanger, Phys. Rev. B **62**, 6967 (2000).
38. M. Bordag, G. Klimchitskaya, and V. Mostepanenko, Surf. Sci. **328**, 129 (1995).
39. U. Dürig, Appl. Phys. Lett. **75**, 433 (1999).
40. M. Gallagher, D. Chen, B. Jacobsen, D. Sarid, L. Lamb, F. Tinker, J. Jiao, D. Huffman, S. Seraphin, and D. Zhou, Surf. Sci. **281**, L335 (1993).
41. H. Postma, A. Sellmeijer, and C. Dekker, Adv. Mater. **12**, 1299 (2000).
42. M. Falvo, G. Clary, R. T. II, V. Chi, F. Brooks, S. Washburn, and R. Superfine, Nature **389**, 582 (1997).
43. J. Muster, G. Duesberg, S. Roth, and M. Burghard, Appl. Phys. A **69**, 261 (1999).
44. E. Wong, P. Sheehan, and C. Lieber, Science **277**, 1971 (1997).
45. W. Shen, B. Jiang, B. Han, and S. Xie, Phys. Rev. Lett. **84**, 3634 (2000).
46. T. Tombler, C. Zhou, L. Alexseyev, J. Kong, H., Dai, L. Lei, C. Jayanthi, M. Tang, and S. Wu, Nature **405**, 769 (2000).
47. A. Maiti, Chem. Phys. Lett. **331**, 21 (2000).

48. J. Cao, Q. Wang, and H. Dai, Phys. Rev. Lett. **90**, 157601 (2003).
49. A. Volodin, D. Buntinx, M. Ahlskog, A. Fonseca, J. B. Nagy, and C. V. Haesendonck, Nano Lett. **4**, 1775 (2004).
50. A. Volodin, M. Ahlskog, E. Seynaeve, C. V. Haesendonck, A. Fonseca, and J. B.Nagy, Phys. Rev. Lett. **84**, 3342 (2000).
51. M. A. Eriksson, R. G. Beck, M. Topinka, J. A. Katine, R. M. Westervelt, K. L. Campman, and A. C. Gossard, Appl. Phys. Lett. **69**, 671 (1996).
52. S. Kalinin and D. A. Bonnell, Appl. Phys. Lett. **78**, 1306 (2001).
53. M. T. Woodside, Ph.D. thesis, schoolUC Berkeley (2001).
54. R. Crook, C. G. Smith, C. H. W. Barnes, M. Y. Simmons, and D. A. Ritchie, J. Phys. C **12**, L167 (2000).
55. M. A. Topinka, B. J. LeRoy, S. E. Shaw, E. J. Heller, R. M. Westervelt, K. D. Maranowski, and A. C. Cossard, Science **289**, 2323 (2000).
56. M. A. Topinka, , B. J. LeRoy, R. M. Westervelt, S. E. J. Shaw, R. Fleischmann, E. J. Heller, K. D. Maranowski, and A. C. Gossard, Nature **410**, 183 (2001).
57. A. Bachtold, M. S. Fuhrer, S. Plyasunov, M. Forero, E. H. Anderson, A. Zettl, and P. L. McEuen, Phys. Rev. Lett. **84**, 6082 (2000).
58. S. J. Tans and C. Dekker, Nature **404**, 834 (2000).
59. M. Bockrath, W. J. Liang, D. Bozovic, J. H. Hafner, C. M. Lieber, M. Tinkham, and H. K. Park, Science **291**, 283 (2001).
60. M. T. Woodside and P. L. McEuen, Science **296**, 1098 (2002).
61. M. Freitag, A. T. Johnson, S. V. Kalinin, and D. A. Bonnell, Phys. Rev. Lett. **89**, (2002).
62. S. Luryi, Appl. Phys. Lett. **52**, 501 (1998).
63. A. A. Odintsov, Phys. Rev. Lett. **85**, 150 (2000).
64. L. X. Benedict, S. G. Louie, and M. L. Cohen, Phys. Rev. B **52**, 8541 (1995).
65. V. Meunier, S. Kalinin, J. Shin, A. P. Baddorf, and R. J. Harrison, Phys. Rev. Lett. **93** (2004).
66. M. Krcmar, W. M. Saslow, and A. Zangwill, J. Appl. Phys. **93**, 3495 (2003).
67. M. R. Krcmar and W. M. Saslow, Phys. Rev. B **66** (2002).
68. J. Jackson, *Classical Electrodynamics* (John Wiley, New York, 1998).
69. P. Kim, T. W. Odom, J. L. Huang, and C. M. Lieber, Phys. Rev. Lett. **82**, 1225 (1999).

II.7 Multi-Probe Scanning Tunneling Microscopy

SHUJI HASEGAWA

A multi-tip scanning tunneling microscope (STM) can be used as a very versatile tool for electrical transport measurements at the nanometer scale. The STM tips are employed as current sources, voltage pick-up probes, and field-gate electrodes. This tool is very useful not only for fundamental physics research, but also for testing electrical characteristics of nanometer-scale electronic devices. Several groups are developing different types of multi-tip STM's [1–7]. Some companies have begun to deliver products combining multi-probe scanning probe microscopes with conductive atomic force microscopes [8], but the apparatus and operating system are not yet fully developed and still have much room for evolution in many aspects.

In this chapter, our homemade four-tip STM operating in ultrahigh vacuum is introduced—four STM tips are driven independently with the aid of piezo-actuators under a scanning electron microscope (SEM) for positioning the tips precisely with arbitrary arrangements on specified areas on the sample surface. The spacing between the tips can be changed from 200 nm to 1 mm. The SEM electron beam is used also for confirming the sample atomic structure in reflection-high-energy electron diffraction mode. With this machine, the surface sensitivity in electrical conductivity measurements has been greatly enhanced, resulting in successful measurements of electrical conduction through the topmost atomic layers (surface-state conductivity) on crystal surfaces. Anisotropic conductivity of an array of metallic atom chains is also measured by the "square-four-point probe method," in which the four tips are arranged in a square on the sample surface. The influence of atomic steps on conductivity is directly measured. The machine is also useful for measuring conductivity of individual nanometer-scale objects, such as silicide nanowires and carbon nanotubes.

This chapter is divided into four parts that focus on the four-tip STM apparatus and its applications, highlighting work performed at our laboratory in the University of Tokyo. The first section introduces the idea of the microscopic four-point probe (μ-4PP) method to measure the topmost atomic layers of three-dimensional bulk crystals. The second section is devoted to the description of the apparatus. The following four sections describe some applications of the four-tip STM (surface sensitivity versus probe spacing, anisotropic surface conductivity, conductivity across an atomic step, conductivity of nanowire and nanotubes). In the final section some future prospects, including a recent result aimed at developing a real nanometer-scale multi-tip STM, are described, and metal-covered carbon nanotube tips are introduced.

1 Four-Tip STM for μ-4PP Conductivity Measurements

The topmost layers of crystal surfaces are known to have characteristic electronic band structures that are sometimes quite different from that of the bulk crystal. While such *surface states* have been well studied so far by, e.g., photoemission spectroscopy and scanning tunneling spectroscopy, the electrical conduction through them (*surface-state conduction*) has been little studied because of lack of experimental tools [9–12]. Due to the thinness of the surface atomic layers, surface-state conductance is usually much lower than the conductance through the underlying bulk crystal. Furthermore, surface defects like steps and domain boundaries greatly perturb the electron conduction through the surface states. These facts have prohibited the direct detection and quantitative measurement of intrinsic surface-state conductivity. Since, however, surface-state conductance, i.e., electron conduction through only one or two atomic layers, is a key issue in the study of electronic transport in nanometer-scale regions or objects, it has recently attracted much interest, and great efforts are now being directed at detecting and measuring it. Here, we introduce a novel tool, a microscopic four-point probe (μ-4PP) conductivity measurement method with a four-tip STM, and demonstrate its effectiveness for such purposes [13,14].

A 4PP method is the most direct and common approach to measure the electrical conductivity of samples. We first briefly introduce the principles of 4PP method together with electrical conduction near a semiconductor surface, and we next explain why a μ-4PP is needed to detect and measure the surface-state conductance.

When, as shown in Figure 1(a), the outer pair of probes is placed in contact with a sample surface and a voltage is applied between them, a current I flows through the sample. The inner pair of probes, which are also in contact with the sample, detect a voltage drop V along the surface due to the resistance of the sample. Thus, one can obtain a 4PP resistance $R = V/I$ (strictly speaking, it is multiplied by a correction factor depending on the specimen shape and probe arrangement). In this configuration, one can correctly measure the resistance of a sample without any influence of contact resistance at the probe contacts, irrespective of whether the probe contacts are ohmic or Schottky-type. This is because virtually no current flows through the voltage pick-up contacts, so that no voltage drops occur at the probe contacts, thus minimizing the contact resistance contribution to the measured signal.

When the specimen is a semiconductor crystal, the current can flow through three independent channels in the sample [15]:

1. Surface states on the topmost atomic layers (when a well-ordered surface superstructure is developed).

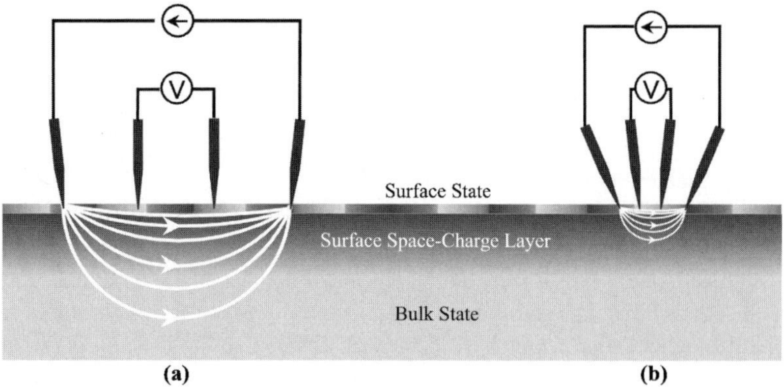

FIGURE 1. (a) Macro- and (b) micro-four-point probe methods to measure electrical conductivity. The distribution of current flowing through a semiconductor specimen is also schematically drawn.

2. Bulk states in the surface space-charge layer beneath the surface (when the bulk bands bend beneath the surface, the carrier concentration can be different from in the inner bulk).
3. Bulk states in the interior of crystal (which do not depend on the surface structures and states).

In general, the resistance measured by the 4PP method contains the contributions from all of these channels, and it is difficult to separate them. In the case of measurements in air, however, the sample surface is usually contaminated and does not have a well-ordered surface structure, so that the measured resistance is interpreted as a bulk value only. But under special conditions where the bands bend sharply under the surface to produce a carrier accumulation layer, or in ultrahigh vacuum (UHV) where the sample crystal has a well-ordered surface superstructure having a conductive surface-state band, the contributions from the surface layers cannot be ignored. However, even under such conditions, the surface contributions have been considered to be very small, because, as shown in Figure 1(a), the measurement current flows mainly through the underlying bulk in the case of macroscopic probe spacing. But these expectations include over simplified pictures in some cases; there can be a Schottky barrier between the surface state and underlying bulk state, and a *p*-*n* junction between the space-charge layer and inner bulk region. In these cases, the measurement current is confined to the surface states and surface space-charge layer; it does not leak into the inner region of the substrate so that the measurement can be surface-sensitive even with the macroscopic probe spacing.

If, as shown in Figure 1(b), one makes the probe spacing as small as the thickness of the space-charge layer or less than it, the measurement current will mainly flow through the surface region, which diminishes the bulk contribution in conductance measurement even when there are no Schottky barrier nor *p*-*n* junction between

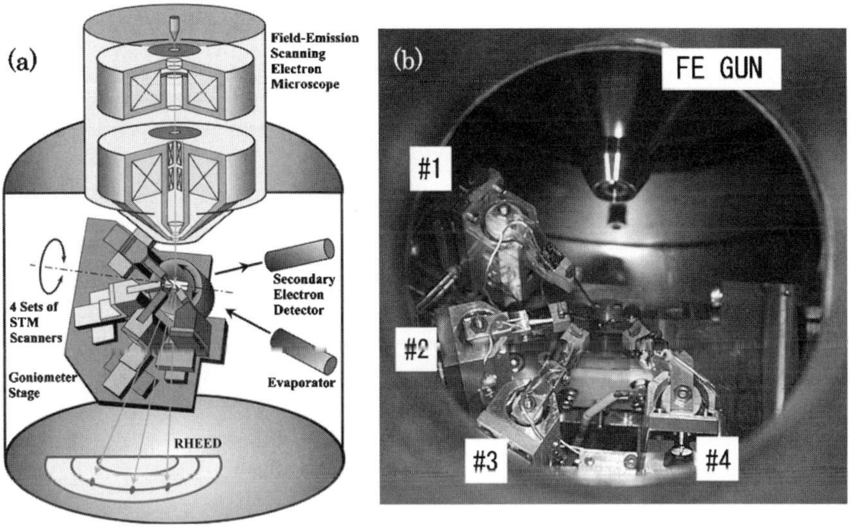

FIGURE 2. (a) Schematics of the independently driven four-tip STM, installed in an UHV-SEM-RHEED system. (b) A photograph of the goniometer stage on which a sample and the fourtip-scanners are mounted under the SEM column [17,18].

the surface region and the inner bulk region. Thus, this μ-4PP method has a higher surface sensitivity. The experimental results described below qualitatively show the validity of this intuitive picture in Figure 1.

The μ-4PP method has other advantages. It enables local measurements by selecting the area under concern with the aid of microscopes, so that the influence of observable defects can be avoided or intentionally included. Furthermore, by scanning the probes laterally over the sample surface, one can obtain a map of conductivity with a high spatial resolution [16]. In many cases, since the specimen size ($\sim$mm) is much larger than the probe spacing ($\sim$ μm) in μ-4PP measurements, the sample can be regarded as infinitely large, which makes the analysis of measurement results considerably easier. We can also measure the conductance of not only the surface, but also individual nanometer-scale objects.

2 Apparatus

Figure 2(a) shows a schematic of the four-tip STM apparatus in an ultrahigh vacuum (UHV) chamber developed at the University of Tokyo, and Figure 2(b) shows a photograph of the goniometer stage on which the sample and four sets of scanners are mounted [17,18]. Each tip points to the sample at 45° from the sample surface and is driven by a special type of a piezo-scanner for fine positioning and by three sets of piezo-actuators (Microslide, Omicron) for coarse motion. The goniometer stage enables parallel shifts in three directions and tilt rotation with respect to the

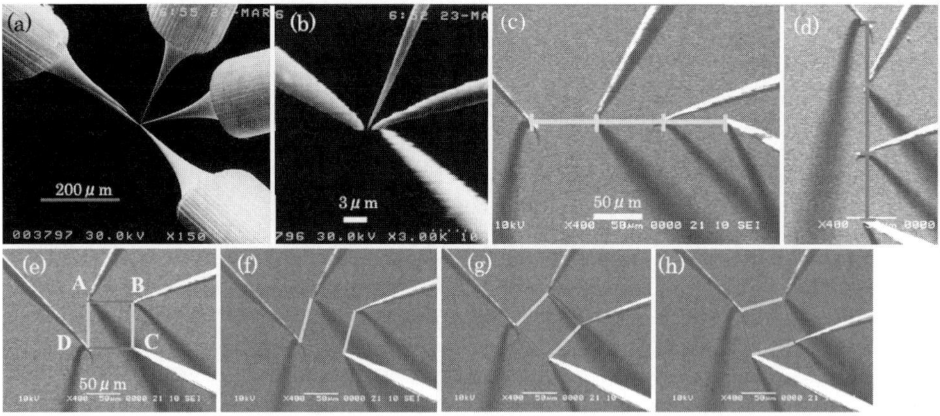

FIGURE 3. SEM images of the four W tips in various arrangements in the four-tip STM [17,19].

SEM electron beam. The sample can be rotated azimuthally by 360° with respect to the stage. These positioning mechanisms enable fine adjustments with respect to the SEM electron beam, required to perform reflection-high-energy electron diffraction (RHEED) and scanning reflection electron microscopy (SREM) observations of the sample surface simultaneously. These supplementary electron microscopy/diffraction techniques are indispensable not only for positioning the four tips properly, but also for confirming the surface structures of sample. The STM tips and sample can be exchanged and installed by transfer rods from load-lock chambers without breaking vacuum.

Figure 3 shows a series of scanning electron micrographs (SEM) of the four tips arranged in various configurations [17–19]. The tips are chemically etched W wires. The probe spacing can be changed from 1 mm to ca. 200 nm in Figure 3(a–c). They can be arranged on a line equidistantly (linear μ-4PP method) (Figure 3(c,d)) in arbitrary directions, or in a square arrangement (square-μ-4PP method) (Figure 3 (e–h)). The square can be rotated with respect to the sample surface (rotational square μ-4PP method) by repositioning each tip [19]. This is useful for measuring anisotropic conductivity as described in the following sections.

This apparatus enables usual STM operation by each tip independently, and also four-point probe conductivity measurements with various probe arrangements and spacing as shown in Figure 3. The STM images obtained by each tip do not yet have atomic resolutions, but monatomic steps are resolved. The four tips approach the sample surface simultaneously with feedback control by tunnel current detection. After that, the tips are brought into direct contact with the sample surface, and then the 4PP conductivity measurement is performed. The pre-amplifier is switched from the tunnel-current mode to the 4PP-conductivity measurement mode. The control system for four-tip STM is still in its infancy. Each tips is independently controlled, but not in an integrated way. If the positions of all tips are controlled

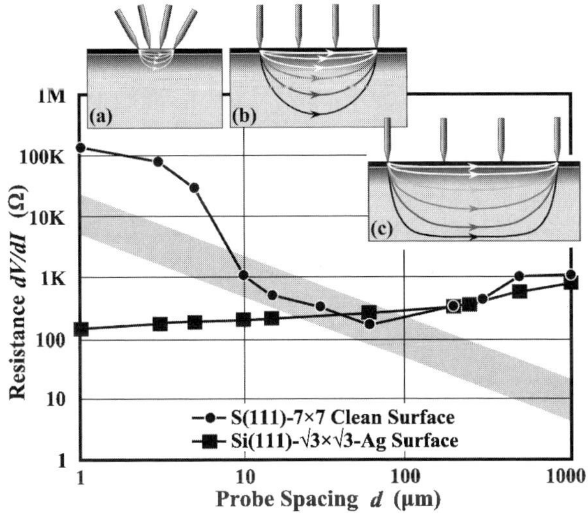

FIGURE 4. Probe-spacing d dependence of the resistance of a Si crystal, measured at room temperature by the independently driven four-tip STM, for the Si(111)-7 × 7 clean surface (filled circles) and Si(111)-$\sqrt{3} \times \sqrt{3}$-Ag surface (filled squares). The insets are schematic illustrations of the current distribution in the sample for the 7 × 7 surface [13,17].

by their xyz coordinates at nanometer-scale precision by a single controller, we do not need SEM for tip positioning anymore. A method for navigating two STM tip has been developed by using a special sample [20].

3 Surface Sensitivity versus Probe Spacing

Here we illustrate the effect of probe spacing on surface conductivity measurements, by comparing the results of two typical surface superstructures on a Si(111) crystal. One is Si(111)-7 × 7 clean surface, obtained by flash heating up to 1,250° in UHV, and another is Si(111)-$\sqrt{3} \times \sqrt{3}$-Ag superstructure, obtained by depositing one atomic layer of Ag on the 7 × 7 surface kept at 450°C. Their atomic arrangements and surface electronic states are already well understood; the details can be found in, e.g., [9,21]. The latter surface has a two-dimensional free-electron-like metallic surface-state band, while the former surface has a localized metallic surface state (dangling-bond state). Since the characters of their electronic structures are so different, they can be considered as good model systems for the surface conductivity measurement.

We did the systematic 4PP measurements for these two surfaces on the same crystal by changing the probe spacing from 1 mm to 1 μm in a linear 4PP arrangement; the four tips can be independently driven in our four-tip STM, so that we can arrange the probes arbitrarily on the sample surface to change the probe spacing continuously. The result is shown in Figure 4 [13,14,17]. The horizontal axis is the probe spacing d (the spacing between neighboring two probes), and the vertical axis is the differential resistance derived from the I-V curves around zero-current (the values are without geometrical-factor corrections). Data points of filled circles are for the 7×7 clean surface, and the filled squares are for the $\sqrt{3} \times \sqrt{3}$-Ag surface. The result clearly shows that the probe-spacing dependence of resistance is quite different between the two surfaces. For the 7×7 surface, the resistance changes significantly in a characteristic way depending on the probe spacing; it increases drastically at $d < 10$ μm. For the $\sqrt{3} \times \sqrt{3}$-Ag surface, on the other hand, the change in resistance with d is much more moderate; the resistance gradually decreases with decreasing d, which is opposite to the tendency for the 7×7 surface. These differences mean that the nature of electron conduction is quite different between the two surfaces. While the difference in resistance between the two surfaces is negligibly small at $d \sim 1$ mm, the difference becomes as large as two or three orders of magnitudes at $d < 10$ μm. It is now evident that the miniaturization of four-point probes makes the resistance measurements very sensitive to the surface structures, in agreement with Figure 1. The results with smaller probe spacings should be more intrinsic to the surface structures because of negligible contribution from the bulk region.

The results are qualitatively described by Ohm's law in classical electromagnetism. If we assume the sample to be a homogeneous and isotropic semi-infinite three-dimensional resistive material, the resistance R measured by the linear 4PP method should be

$$R = \frac{\rho}{2\pi d}, \tag{1}$$

where ρ is its bulk resistivity [Ωcm]. The measured resistance should be inversely proportional to the probe spacing d. This relation is shown in Figure 4 by a shaded band (our Si crystal had the resistivity of $\rho = 5 \sim 15$ Ωcm). The experimental data for the 7×7 surface are consistent with this theoretical prediction only at 10 μm $< d <$ 100 μm, while the data points deviate upward from the theory at larger d and smaller d. This feature with comparison to Eq. (1) was confirmed also for silicon crystals of different bulk resistivities. Therefore, we can say that the sample crystal (0.4 mm thick) is regarded as a homogeneous semi-infinite bulk when the probe spacing d is between 10 and 100 μm, because the current distribution does not practically reach to the bottom of the crystal as schematically shown by inset (b) in Figure 4. Furthermore, the influence of the surface layer is negligible, because the majority of the measuring current passes through the bulk region because of comparatively large probe spacings.

When d is much larger ($d >$ 100 μm), on the other hand, the current penetrates deeper in the crystal and reaches its bottom as shown in an inset (c) of Figure 4. The current distribution may be compressed due to the finite thickness of the sample. This effectively raises the measured resistance compared with the case of a semi-infinite bulk, corresponding to the data points at $d >$ 100 μm, deviating upward from the prediction of Eq. (1). The results in the large d regime are roughly described by a standard form of $R = \rho L/S$, where L is the length of the measured area and S is the cross section of the sample. Since $L = 3d$ (the spacing between the outer pair of probes), and the measuring current spreads laterally on the order of $3d$, $S \sim 3d \cdot t$, where t is the thickness of the sample (0.4 mm). Therefore, $R \sim \rho/t$, then R is calculated from the ρ to be $120 \sim 400\ \Omega$, which is roughly consistent with the measured data at $d > 100\ \mu\text{m}$ in Figure 4.

At a smaller d regime ($d < 10\ \mu\text{m}$), which is comparable to the thickness of the surface space-charge layer ($\sim 1\ \mu\text{m}$), the current mainly flows near the surface as shown in inset (a); the penetration depth of the current distribution in the sample is similar to the probe spacing in usual cases. Therefore, the data points at $d < 10\ \mu\text{m}$ mean that the resistance at the surface region is larger than that of the bulk, because the data at $d < 10\ \mu\text{m}$ deviate upward from the prediction Eq. (1). This conclusion is reasonable when one recalls the fact that the surface space-charge layer beneath the clean 7×7 surface is always a depletion layer, irrespective of the bulk doping concentration and type. This is because the Fermi level at surface is strongly pinned by the dangling-bond state located at the middle of the band gap [22,23]. Thus the surface region has a higher resistance for the 7×7 surface compared with the bulk region. However, from this data only, the dominant contribution to the measured resistance, the surface state or space-charge layer, cannot be determined unambiguously.

In contrast, the d dependence of resistance for the $\sqrt{3} \times \sqrt{3}$-Ag surface does not fit Eq. (1) at all. According to Ohm's law, when the resistance of an infinite two-dimensional sheet is measured by the linear 4PP with probe spacing d, the measured resistance R is

$$R = \frac{\ln 2}{\pi} R_S, \tag{2}$$

where R_S is the sheet resistance [Ω] [24]. This means that the measured resistance should be constant, independent of the probe spacing d. The experimental data points for the $\sqrt{3} \times \sqrt{3}$-Ag surface in Figure 4 roughly follow this tendency, rather than Eq. (1). The $\sqrt{3} \times \sqrt{3}$-Ag surface is known to have a two-dimensional free-electron-like surface-state band which is metallic and conductive; in this case, furthermore, the surface space-charge layer is always a weakly hole-accumulation layer. This is why the surface region has a much higher conductivity than in the bulk [9,25]. The contribution from the surface-state band dominates the measured conductance in this case [26]. Therefore, the conduction is two-dimensional, rather than three-dimensional.

In this way, by changing the probe spacing from macroscopic distances to microscopic ones, one can switch the conductivity measurement from the bulk-sensitive mode to surface-sensitive one, so that one can clearly distinguish between 2D conduction and 3D conduction.

4 Anisotropy in Surface Conductivity

Since the probe spacing is on the order of μm, which is much smaller than the sample size (~mm), the sample can be regarded as a (semi-)infinitely large object.

This allows us to solve the Poisson equation analytically to derive the potential distribution under the 4PP measurement. Equations (1) and (2) in the previous section are the results for isotropic samples measured by the linear μ-4PP probe method. When the sample is anisotropic, the Poisson equation can be similarly solved, providing non-trivial results, as summarized below.

First, we consider a semi-infinitely large three-dimensional (3D) sample having conductivity σ_x ($\Omega^{-1}m^{-1}$), σ_y ($\Omega^{-1}m^{-1}$), and σ_z ($\Omega^{-1}m^{-1}$) along x-, y-, and z-directions, respectively, with the surface on (xy) plane. The Poisson equation tells us that the resistance measured by the linear 4PP method is given by

$$V/I = \frac{1}{\sqrt{\sigma_y \sigma_z}} \cdot \frac{1}{2\pi d}, \tag{3}$$

where the current probes are aligned along x-axis. The resistance is inversely proportional to the probe spacing d, as with the isotropic samples (Eq. (1)). Interestingly, σ_x does not appear in Eq. (3), which is counterintuitive. The conductivity along the direction of probe alignment does not affect the measured result.

In the case of anisotropic and infinitely large two-dimensional (2D) samples having sheet conductivity σ_x (Ω^{-1}/Square) and σ_y (Ω^{-1}/Square) along x and y directions, respectively, the measured resistance by the linear 4PP method is

$$V/I = \frac{ln2}{\pi \sqrt{\sigma_x \sigma_y}}, \tag{4}$$

if the current probes are aligned along x-axis. The resistance does not depend on the probe spacing d, similar to the isotropic sample Eq. (2). Again, we notice that the measured resistance does not change even when the probes are aligned along y-axis (see Figure 3(c,d)); the σ_x and σ_y are just exchanged in Eq. (4), giving the same value. The linear 4PP method just gives a geometrical mean of σ_x and σ_y. This means that *the linear 4PP method is not effective to detect the anisotropy in conductivity*. We cannot obtain the σ_x and σ_y separately. By solving Poisson equation, we have found that the potential distribution along the line defined by the current-source probes is analytically the same irrespective of the anisotropy in conductivity for the 2D sample and of the aligned direction of the probes.

Similar calculations solving the Poisson equation give a voltage drop measured in "square four-point probe method (square 4PP)" as in Figure 3(e–h), in which the four probes are arranged in square with a side d. The current is injected by an adjacent pair of probes and the voltage is measured by another pair of probes. In this configuration, just by changing the pairs of current and voltage probes, we can perform two measurements; the current probes are A and B (A and D), and the voltage probes are C and D (B and C) in Figure 3(e). In contrast to the linear 4PP method, this square 4PP method is found to be useful for samples having anisotropic conductivity.

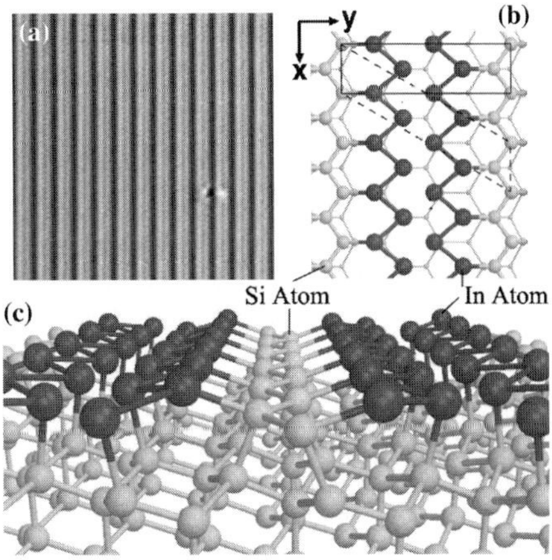

FIGURE 5. Si(111)-4 × 1-In surface superstructure. (a) Its STM image at room temperature. (b) Atomic structure model in plan view and (c) in ball-and-stick model. Reproduced with permission from [32].

For 3D and 2D samples as defined above, the resistances measured by the square 4PP method are, respectively, given by

$$V/I = \frac{1}{\sqrt{\sigma_y \sigma_z}} \left\{ 1 - \left(1 + \frac{\sigma_y}{\sigma_x} \right)^{-\frac{1}{2}} \right\} \frac{1}{\pi s}, \tag{5}$$

$$V/I = \frac{1}{2\pi \sqrt{\sigma_x \sigma_y}} ln \left(1 + \frac{\sigma_y}{\sigma_x} \right), \tag{6}$$

where the current probes are aligned along x-axis in both cases. As seen from these equations, the square 4PP method enables detecting the anisotropy. As shown by Eq. (6), by changing the probe pairs of current source and voltage detection to make the current probes aligned along y-axis, the measured resistance is different from that in Eq. (6); the ratio σ_y/σ_x is inverted, which gives different values of resistance. This should be contrasted with Eqs. (3) and (4). Then we can obtain two equations with the two measured values. By solving the simultaneous equations, we can obtain the values of σ_x and σ_y *separately*. Thus, by solving the Poisson equation, it is shown that the potential distribution in the region except on the line connecting the current-source probes is different depending on the probe alignment direction with respect to the sample surface. This is the reason the square 4PP method works for anisotropic conductors.

Now we illustrate an example of an anisotropic surface, Si(111)-4 × 1-In surface superstructure (Figure 5). The surface-state band structure of this surface shows a quasi-one-dimensional (1D) metallic character at room temperature, and its Fermi surface (Fermi line) is determined by angle-resolved photoemission spectroscopy (ARPES) [27–29]. Its atomic arrangement is a massive array of metallic quantum wires composed of four lines of In atom chains running along $[10\bar{1}]$ direction, each wire being separated by a Si-atom chain in-between the metallic wires [30–32] (see Figure 5). We can, therefore, expect a higher conductivity along the In wires than across the wires. This surface superstructure is contrasted to the isotropic

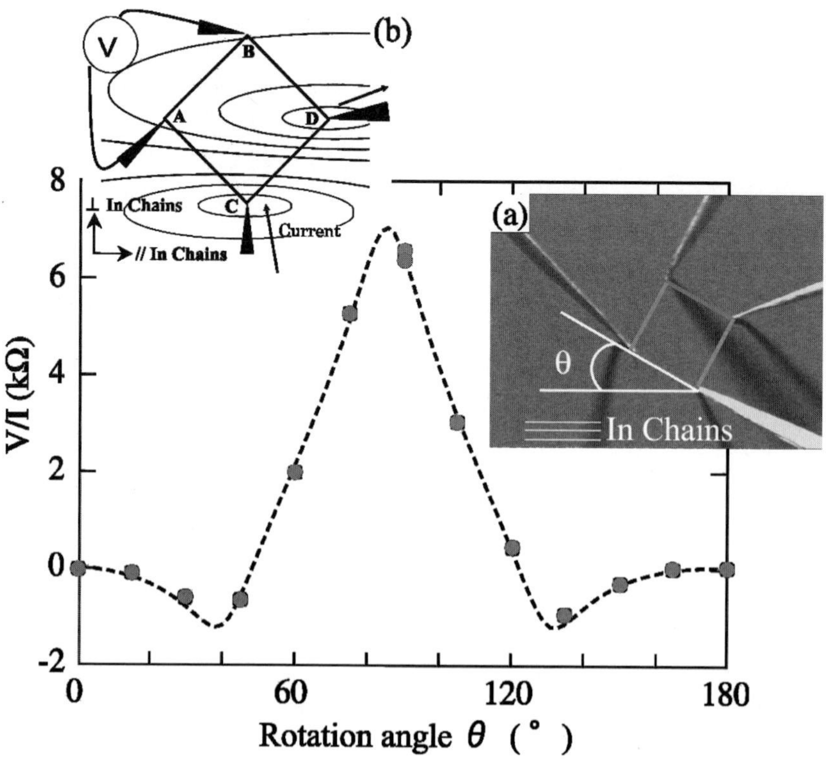

FIGURE 6. Angle dependence of the measured resistance *V/I* of the single-domain Si(111)-4 × 1-In surface. Experimental data are fitted by Eq. (7). Inset (a) is a SEM image during the rotational square μ-4PP measurements with 60-μm probe spacing. Inset (b) is a map of equi-potential lines calculated by assuming $\sigma_{//}/\sigma_{\perp} = 20$, showing an apparent 'negative' resistance at $\theta = 135°$ [19].

Si(111)-$\sqrt{3} \times \sqrt{3}$-Ag surface mentioned in the previous section. The $\sqrt{3} \times \sqrt{3}$-Ag surface has a completely circular Fermi surface (isotropic 2D metal) [33–35], while the 4 × 1-In surface has almost a straight Fermi surface (quasi-1D metal) [27].

The 4PP in-square arrangement was rotated in a step of 15° with respect to the metallic chains as shown in Figure 3(e–h). The results are shown in Figure 6 as a function of the rotation angle θ. The rotation angle θ is defined by an angle between a line connecting two current-source probes and the In-chain direction. At $\theta = 0°(90°)$, the current/voltage probes are along (perpendicular to) the In atomic wires. As expected, the apparent resistance is much higher when the current/voltage probes are perpendicular to the wires. Negative values of V/I at some θ's come from deformed electro-static potential contours due to the high anisotropy in conductance (see inset (b) in Figure 6).

In contrast, the resistance measured by the linear μ-4PP method was essentially the same, irrespective of the probe orientation with respect to the In chains. Other orientations of the probes gave the same results. Thus, the distinction between the linear and square 4PP methods is now clear [19].

Although, in principle, two values of resistance obtained by the square 4PP (at $\theta = 0°$ and $90°$, for example) in Figure 6, combined with Eq. (6), are enough for deriving the conductivity along the atomic wires ($\sigma_{//}$) and that across the wires ($\sigma_{\perp}$) separately, all data points in Figure 6 are used to improve the measurement precision. Poisson equation gives an analytical form for resistance as a function of θ measured by this "rotational square μ-4PP method":

$$\frac{V}{I} = \frac{1}{2\pi\sqrt{\sigma_{//}\sigma_{\perp}}} \cdot \ln\sqrt{\frac{(\sigma_{//}/\sigma_{\perp}+1)^2 - 4\cos^2\theta\sin^2\theta(\sigma_{//}/\sigma_{\perp}-1)^2}{(\sin^2\theta + \sigma_{//}/\sigma_{\perp}\cos^2\theta)^2}}. \quad (7)$$

Equation (6) is derived by inserting $\theta = 90°$ in Eq. (7). By fitting Eq. (7) with the experimental data in Figure 6, $\sigma_{//}$ and $\sigma_{\perp}$ were determined to be $7.2(\pm 0.6) \times 10^{-4}$ and $1.2(\pm 0.1)\times 10^{-5}$ S/square, respectively, its anistropy $\sigma_{//}/\sigma_{\perp}$ being ~ 60 [19].

We now discuss quantitatively the origin of the anisotropic conductivity of the 4×1-In surface. As mentioned before, in general, the measured conductance of a semiconductor crystal is a sum of contributions from three current channels, $\sigma = \sigma_{ss} + \sigma_{sc} + \sigma_b$; (1) surface-state bands on the topmost atomic layers (σ_{ss}), (2) bulk-state bands in a surface space-charge layer beneath the surface (σ_{sc}), and (3) bulk-state bands in the inner crystal (σ_b). Since the σ_{sc} and σ_b are isotropic, the measured anisotropy comes from the σ_{ss} only. Since we know the band bending beneath the surface, we can calculate the σ_{sc} by solving the Poisson equation, and it turns out that the σ_{sc} is comparable to the measured $\sigma_{\perp}$ and much smaller than the measured $\sigma_{//}$ [19]. This analysis also indicates that the measured anisotropic conductivity comes from the Indium atomic wires. From a quantitative analysis using the Boltzmann equation, the measured value of $\sigma_{//}$ is reasonably explained by the anisotropic surface-state bands of the 4×1-In surface superstructure [19].

The same square 4PP method was applied to metallic Au atomic chain arrays on a vicinal Si(557) surface [36], too. In this case, the conductivity along the metal atomic chains was much smaller than that expected from the surface-state bands and non-metallic, meaning failure of the Boltzmann picture of transport. This is due to atomic protrusions randomly adsorbed on the chains, which divide the chains into metallic segments and make the conducting along the chains a thermal-activation type.

Thus, even though we cannot yet measure the conductivity through individual atomic chains directly, we can estimate it through the anisotropy in sheet conductivity of the massive array of atomic chains.

5 Resistance Across an Atomic Step

As described in the previous sections, the μ-4PP methods allow us to detect and measure the electrical conductivity through the topmost atomic layers on crystal surfaces. However, real surfaces always have atomic steps like ones shown in

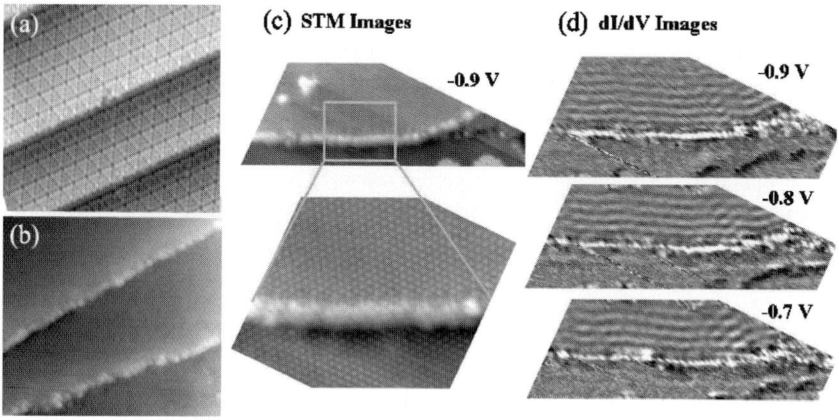

FIGURE 7. STM images of Si(111)-7 × 7 clean (a) and Si(111)-$\sqrt{3} \times \sqrt{3}$-Ag (b) surfaces with some monatomic steps. (c) Magnified STM images near an atomic step on the $\sqrt{3} \times \sqrt{3}$-Ag surface. (d) Differential conductance (*dI/dV*) images at the same area as in (c), showing electron standing waves [37]. Three images taken with different bias voltages (the empty-state images) are shown.

Figure 7(a,b), where the topmost layer is broken off, and the electrical conduction through the layer would be interrupted. But the current flow would not be blocked completely at the steps; the conduction electrons would pass through the steps with some probability by some means such as tunneling effect or thermal effect. Therefore, a finite electrical resistance would be caused by an atomic step. In this section, by using our four-tip STM and monolithic micro-4PP, direct measurements of this resistance across the atomic steps are performed [37].

We have adopted three different methods—the first one is a "conventional" method of analyzing Friedel oscillations near step edges in the spectroscopic STM (*dI/dV*) images, which enables us to obtain the transmission probability of electron wavefunction and the resulting conductivity across an atomic step from Landauer formula. The second method is direct measurements of resistance across step-bunched regions where hundreds of monatomic steps are accumulated [14].The third one is measurements of anisotropy in surface conductivity of a vicinal crystal surface on which atomic steps are regularly aligned with almost equal spacing in one direction. All of these methods yield consistent values of the conductivity across a monatomic step of 1-m-long, ca. $5 \times 10^3\ \Omega^{-1}\mathrm{m}^{-1}$. Electron transport across an atomic step is modeled as a tunneling process through an energy barrier whose height approximately equals to the work function.

The sample used was the Si(111)-$\sqrt{3} \times \sqrt{3}$-Ag surface superstructure, which is produced by one atomic-layer Ag adsorption on a Si(111) crystal surface, as mentioned before. This surface is known to have a nearly free-electron-like metallic surface-state band [9,11,21]. The band has a parabolic dispersion with an effective mass $m^* = 0.13 \cdot m_e$ and crosses the Fermi level [33]. Since the Fermi surface is completely circlar [34], the surface state composes an isotropic two-dimensional metallic electron system of monatomic-layer thick. Since the surface state is located within the bulk-band gap, it is electronically isolated from the substrate bulk crystal, so that the carriers in the surface state are not easily scattered into the substrate.

5.1 Method 1: Analysis of Electron Standing Waves

Figure 7 shows the STM (c) and spectroscopic dI/dV (d) images of the $\sqrt{3} \times \sqrt{3}$-Ag surface near a monatomic step at 65 K. It can be seen that the $\sqrt{3} \times \sqrt{3}$-Ag surface superstructure almost extends up to the step edge. In the dI/dV images, the modulation of the local density of states (LDOS), a so-called electron standing wave (Friedel oscillation), is caused by interference between the incoming and reflected electron waves near the step. Its physical properties have been well studied on noble metal surfaces [38–42] as well as $\sqrt{3} \times \sqrt{3}$-Ag surfaces [43].

Assuming, for simplicity, that the electron wave, traveling along x direction on the surface toward the step, is a plane wave

$$\Psi_i \sim \exp(ikx), \tag{8}$$

the wave reflected by the step is written by

$$\Psi_r \sim \exp(-ikx - i\eta), \tag{9}$$

where η is a phase shift at the reflection. Then the LDOS (dI/dV) image is

$$\mathrm{LDOS}(x) = |\Psi_i + \Psi_r|^2 \sim |\exp(ikx) + \exp(-ikx - i\eta)|^2 \sim \cos(2kx - \eta). \tag{10}$$

Since the wavenumber k changes with energy E due to the band dispersion, the wavelength of standing wave $\lambda(= \pi/k)$ varies with the bias voltage at STM imaging as shown in Figure 7(d). The energy versus wave number dispersion obtained through the bias-dependent standing waves is consistent with that obtained by ARPES [33–44]. The actual LDOS shows a damped oscillation due to some decoherence processes, which can be fitted using a function,

$$\mathrm{LDOS}(x) \sim \exp\left(-\frac{x}{L}\right) \cdot \cos(2kx - \eta), \tag{11}$$

where L is the decay length [41]. The important parameter for determining the resistance across the step is the reflection phase shift η, which is derived at each bias voltage by fitting the standing wave image with Eq. (11). Our results [37] show that η is almost constant $\eta = -(0.8 \pm 0.05)\pi$ in the measured energy range including the Fermi level within the experimental error.

Once the reflection phase shift at the step is determined, the transmission probability T of electron wavefunction can be obtained through the δ-potential model [42], and then the conductance through the step can be derived from the Landauer formula [45]. The atomic step makes a potential barrier for the surface-state electrons, which is reasonably represented by a δ-potential because the step-edge potential barrier is atomically thin. The transmission coefficient T and the reflection phase shift η in this model are

$$T = 1 - \frac{1}{1 + \tan^2(\eta + \pi)}. \tag{12}$$

Inserting the experimentally determined value of η into this equation, T was determined to be 0.3 ± 0.15. *The conduction electrons in the $\sqrt{3} \times \sqrt{3}$-Ag layer (surface state) pass through a monatomic step with a probability of about 30%.* It should be noted here that the transmission probability is the same for the step-up and step-down directions. This is supported by experimentally observed linear *I-V* curves in the present 4PP resistance measurements across the steps as described in the following subsections and theoretically by the WKB tunneling regime typically used in STM theories.

Once the transmission probability T is obtained, the conductivity σ across the step can be calculated by Landauer formula. In general, the formula varies with the dimensions of specimens. The conductance of 1D, 2D, and 3D conductors through 0D, 1D, and 2D thin potential barriers is, respectively,

$$1D: \qquad \sigma_{1D} = \frac{2e^2}{h} T \quad [\Omega^{-1}], \tag{13}$$

$$2D: \qquad \sigma_{2D} = \frac{2e^2}{h} \frac{k_F}{\pi} T \quad [\Omega^{-1} m^{-1}], \tag{14}$$

$$3D: \qquad \sigma_{3D} = \frac{2e^2}{h} \frac{k_F^2}{4\pi} T \quad [\Omega^{-1} m^{-2}], \tag{15}$$

where e, h, and k_F are the elementary charge, Planck's constant, and Fermi wavenumber, respectively. Eq. (13) is a well-known formula for conductance quantization at nanowires and point contacts. Eq. (15) is also used for the analysis of tunneling devices of GaAs/GaAlAs superlattices (sometimes referred to as Tsu-Esaki formula). On the other hand, Eq. (14) has not been utilized much previously due to the lack of appropriate real physical systems. The present case of 2D metallic monolayers (2D conductor) separated by a monatomic step (a 1D potential wall) is a model case for the σ_{2D} formula Eq. (14). For the 1D case, the conductance is written by universal constants only, while for the 2D and 3D cases, a material-dependent parameter k_F is needed. The σ_{2D} and σ_{3D} are the conductance per unit length of barrier line and unit area of barrier plane, respectively.

Inserting T determined from Eq. (12) and k_F value (~ 0.1 Å^{-1}) determined in the previous ARPES measurements for this surface [33,35,44] into Eq. (14), the σ_{2D} across a monatomic step on this particular surface is determined to be

$$\sigma_{step} = 9(\pm 4) \times 10^3 \quad [\Omega^{-1} \mathrm{m}^{-1}]. \tag{16}$$

This is a conductance across a 1-m-long step. It should be noted that in the Landauer regime, the conductance is independent of temperature when the transmission coefficient (the reflection phase shift) is independent of energy [46]. The experimental result supports this picture [37].

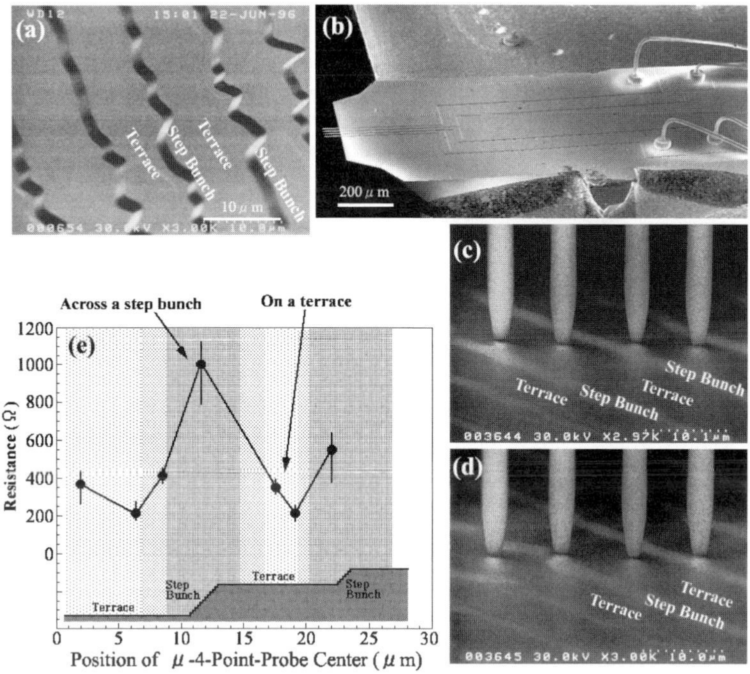

FIGURE 8. Resistance measurements across a step bunch on the Si(111)-$\sqrt{3} \times \sqrt{3}$-Ag surface with a monolithic μ-4PP. (a) SEM image of the step-bunched surface. (b) SEM image of the μ-4PP chip. (c,d) SEM images of the μ4PP contacting the sample surface in UHV. The resistance is measured across a step bunch in (c) while the resistance on a terrace is measured in (d). (e) Line profile of resistance measured by making the μ4PP move along a line across step bunches on the surface [14].

5.2 Method 2: Direct Measurements Across a Step Bunch

For the second method for measuring the resistance across atomic steps, we have conducted linear μ-4PP measurements across a step-bunching region on the $\sqrt{3} \times \sqrt{3}$-Ag surface at room temperature by using monolithic micro-four-point probe [14]. Figure 8(a) shows a SEM image of the step-bunched surface of Si(111), on which atomically flat terrace regions of 10~20 μm wide and step bunches of ca. 2 μm wide are alternately arranged. Each step bunch contains about 300 monatomic steps, so that the average terrace width in the step bunch region is around 5~10 nm. The morphology is schematically shown in Figure 9(b). Preparation of such surfaces is described in detail in Ref. [47]. Figure 8(b) is an SEM image of the monolithic μ-4PP chip, which is now commercially available [48]. While the probe spacing is fixed in this device, one can choose the probe spacing ranging from 1 μm to 100 μm. The probes consist of four sharpened silicon oxide cantilevers extending from a silicon support chip. On the substrate surface a metal layer is deposited to make conducting paths. The metal layer covers the very end of four cantilevers

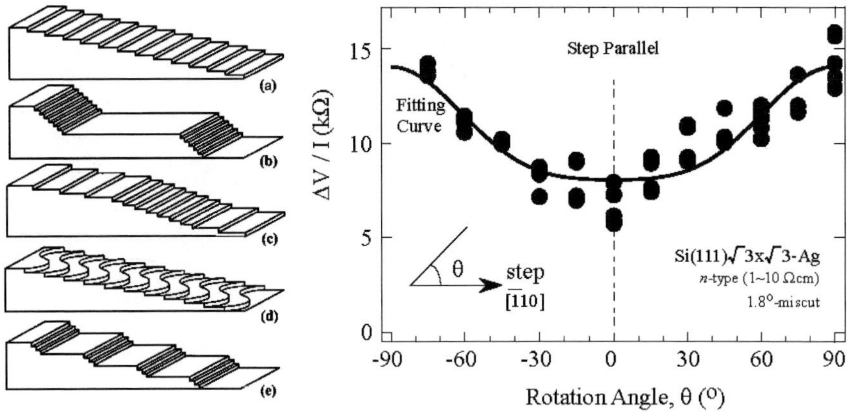

FIGURE 9. **(Left)** Various surface morphology, determined by monatomic-steps configurations: (a) regular-step surface, (b) step-bunched surface, (c) hill-and-valley surface, (d) step-wandering surface, and (e) multi-step surface. (Right) Resistance of Si(111)-$\sqrt{3} \times \sqrt{3}$-Ag surface measured as a function of the rotation angle of square in the square micro-four-point probe method with the four-tip STM. The substrate is a vicinal Si(111) wafer (n-type, $1 \sim 10\ \Omega$cm) with a miscut angle of 1.8° at room temperature. The probe spacing (a side of the square) was 60 μm [37].

so that they can make direct contact with the sample surface. The probes are very flexible, which makes the contact with sample surfaces straightforward and very soft even when the surface plane is not aligned with the probes.

The μ-4PP chip was mounted on three sets of piezo-actuators for fine movements in XYZ directions, which enabled the probe positioning as precisely as ~10 nm accuracy with a 5 mm travel distance for each direction. The probes were made to approach the sample surface with an angle of about 30° from the surface with the aid of *in-situ* SEM observation, so that gentle contact with the selected area on the surface can be established. Figure 8(c,d) shows SEM images of the μ-4PP contacting the sample surface in UHV. In Figure 8(d), there is no step bunch running between the inner probes, measuring a resistance on an almost step-free terrace, while in Figure 8(c), the inner probes measure a voltage drop across a step bunch. In this way, we can select the surface area under measurement by shifting the probe position, so that we can detect the influence of steps on the conductivity.

Figure 8(e) shows a "line profile" of resistance measured in this way with a μ-4PP of probe spacing $d = 8$ μm shifting along a line across step bunches on the Si(111)-$\sqrt{3} \times \sqrt{3}$-Ag surface at room temperature. The horizontal axis shows the position of the probe on the line; the data points are plotted at the position of the center of the μ-4PP. The surface morphology observed by SEM is schematically shown at the bottom in the graph; the vertical scale is exaggerated. These results show that the resistance changes drastically from place to place. As expected, when the inner pair of probes cross over a step bunch as shown in Figure 8(c) (such a situation occurs at dark shaded areas in Figure 8(e)), the measured resistance is higher by about 800 Ω than that on the almost step-free terrace region (Figure 8(d)). Through theoretical simulations [49], the resistance of the step-bunching region R_{bunch} was determined to be ~1, 700 Ω.

Since any facets are formed at the step-bunch region, we can simply say that the measured resistance across a step-bunching region R_{bunch} is a sum of the resistance across individual monatomic steps R_{single}; $R_{bunch} = N \cdot R_{single}$, where $N(\sim 300)$ is the number of steps in the bunch. Since $R_{single} = 1/(L \cdot \sigma_{step})$, where σ_{step} is the conductance across a single atomic step of unit length [$\Omega^{-1}\text{m}^{-1}$] and L is the step length [m] through which the measurement current passes. Because, according to our simple simulation, the measurement current spreads laterally on the surface as wide as twice the current probe spacing (8μm × 3 × 2 ~ 50μm), $L \sim 50$ μm. Therefore, we obtain

$$\sigma_{step} \sim 4 \times 10^3 \quad [\Omega^{-1}\text{m}^{-1}]. \tag{17}$$

This value shows fairly good agreement with the conductivity obtained by the standing-wave analysis (Eq. (16)) in the previous subsection.

5.3 Method 3: Conductivity Anisotropy on a Vicinal Surface

For the third method of step-resistance measurement, we have used a vicinal Si surface that was miscut by 0.9° or 1.8° from the (111) surface toward $[\bar{1}\bar{1}2]$ direction. By proper heating procedures, we can obtain a so-called "regular-step" surface, as illustrated in Figure 9(a), where monatomic steps are aligned equidistantly and parallel to each other along $[\bar{1}10]$ direction, as opposed to the step-bunched surface used in the previous section (Figure 9(b)). Since the step distance (10 ~ 20 μm) is much smaller than the probe spacing (5 ~ 100 μm) in our μ-4PP measurements, the sample surface is regarded as a homogeneous, but anisotropic sheet conductor. That is, due to the parallel step array, the conductivity normal to the step array ($\sigma_{\perp}$) is expected to be lower than the conductivity parallel to the step array ($\sigma_{//}$). Therefore, any anisotropy in conductivity detected by the square μ-4PP method can be attributed to the steps, and we can determine the resistance caused by individual atomic steps from the anisotropy. This is possible since all channels for electrical conduction near the surface (surface-state bands on the topmost atomic layers, bulk-state bands in the space-charge layer beneath the surface, and bulk-state bands in the crystal bulk) are isotropic. If $\sigma_{\perp}$ and $\sigma_{//}$ are obtained separately, we can deduce the resistance across a single step.

Owing to the square μ-4PP method by the four-tip STM described in section 4, we can measure the $\sigma_{\perp}$ and $\sigma_{//}$ separately.

The results for the 1.8° miscut wafer are shown in Figure 9 (right panel) as a function of the rotation angle of the square relative to the step array direction. By fitting the results with Eq. (7), we have obtained sheet conductivities $\sigma_{//} = 23 \times 10^{-6}$ Ω^{-1}/square and $\sigma_{\perp} = 14 \times 10^{-6}$ Ω^{-1}/square separately. The anisotropy $\sigma_{//}/\sigma_{\perp} \sim 1.6$, which is much smaller than the quasi-1D metallic surface of Si(111)-4 × 1-In mentioned in section 4.

As described above, the difference of sheet resistance between step-perpendicular ($1/\sigma_\perp$) and step-parallel ($1/\sigma_{//}$) directions is ascribed to the resistance associated with monatomic steps. The conductivity across a step of unit length σ_{step} [$\Omega^{-1}\mathrm{m}^{-1}$] can be related to the measured anisotropic sheet conductivities by

$$\frac{1}{\sigma_\perp} - \frac{1}{\sigma_{//}} = \frac{n_{step}}{\sigma_{step}}, \tag{18}$$

where n_{step} [m^{-1}] is the step density on the sample surface. Our vicinal Si(111) wafer with a miscut angle of 1.8° has $n_{step} \sim 10^8\mathrm{m}^{-1}$. Therefore, we can obtain

$$\sigma_{step} \sim 3 \times 10^3 \quad [\Omega^{-1}\mathrm{m}^{-1}]. \tag{19}$$

5.4 Conduction Mechanism Across an Atomic Step

Now we can compare the results of three different measurements of step resistance—Eqs. (16), (17), and (19)—giving consistent values of about 5×10^3 $\Omega^{-1}\mathrm{m}^{-1}$ for the conductance across a one-meter-long monatomic step on the Si(111)-$\sqrt{3} \times \sqrt{3}$-Ag. Since these three measurement methods are independent, we can say that the similar values of result are quite reliable. Especially, it should be noticed that because the indirect method of standing-wave analysis and the direct methods with step-bunched and vicinal surfaces gives consistent results, we can give a clear picture of carrier conduction across a step: the atomic step acts as a thin potential barrier for the conduction electrons in the surface state, and the electrons pass through the step by tunneling effect and flow out on the next terraces.

According to the δ-potential model used in section 5.1, we can obtain a product of the potential barrier height (V_0) and barrier width (a) from the reflection phase shift η:

$$V_0 \cdot a \sim 10 \qquad [\mathrm{eV} \cdot \text{Å}]. \tag{20}$$

Assuming $a \sim 2.35$ Å, the atomic bond length in Si crystal, we obtain $V_0 \sim 4.5$ eV. This is similar to the workfunction of Si. In other words, the potential barrier that the conduction electrons at the Fermi level feel is the energy barrier to the vacuum level. This gives a reasonable picture that the electrons tunnel through the vacuum at the step edges. This can be a general picture for the conduction across an atomic step on other surface structures.

However, an alternative model for electronic transport across steps is tunneling via an evanescent state of the substrate crystal [50]. Since the surface state of the $\sqrt{3} \times \sqrt{3}$-Ag is located within the band gap of the substrate Si crystal, the carriers cannot be easily scattered into the substrate bulk state. However, there is a virtual gap state in the band gap, through which the surface-state electrons can pass across a step. An electronic state localized at the step exists, which also form an evanescent state on the terraces only near the step. The finite resistance across a step can be regarded as a result of mismatch in Bloch waves on the terraces at both sides of the step, and its value is determined by atomic-scale wavefunction near the step edge.

From our experimental results Eqs. (16–19), we can make an important conclusion. When electrical current flows through a monatomic step 1 cm, 1 mm, 1 μm, or 1 nm in length, the resistance across the step should be 0.02, 0.2, 200, and $2 \times 10^5\ \Omega$, respectively. *The resistance is negligible on a macroscopic scale, but quite large at nanometer-length scales.* This means that the step control is crucial in nanometer-scale electronic devices.

6 Silicide Nanowires and Carbon Nanotubes

The four-tip STM is applicable to the conductivity measurements not only for the surfaces, but also for individual microscopic objects such as silicide nanowires [51,52] and carbon nanotubes [53]. Figure 10(a) is an SEM image of the four W tips in contact with a cobalt-slicide nanowire on a Si(110) substrate. The $CoSi_2$ nanowires can be formed and self-assembled simply by depositing Co onto the heated Si(110) surface [54]. The typical size is approximately 60 nm wide and 5 μm long. The wires are epitaxial and single crystalline, and embedded in the substrate as shown in a cross-sectional TEM image of Figure 10(b) [54]. These kinds of nanowires have received much attention recently as promising elements for future nanoscale devices such as high-performance transistors and interconnects [55].

By putting two or four tips on one of the nanowires, the conductivity of the individual wire was measured with the two- and four-point probe method with varied tip spacing. When one of the current-source probes is on the nanowire and the other is on the substrate, the current-voltage curve shows non-linearity due to a Schottky barrier between the nanowire and substrate. This means that the nanowire is electrically isolated from the substrate so that the current flowing through it does not leak into the substrate. This satisfies an important requisite for the device application.

Individual carbon nanotubes were also measured by the four-tip STM [53]. Figure 11(a) shows a SEM image of a multiwalled carbon nanotube bridging between Ta electrode pads. The two W-tips contact the pad for the two-terminal

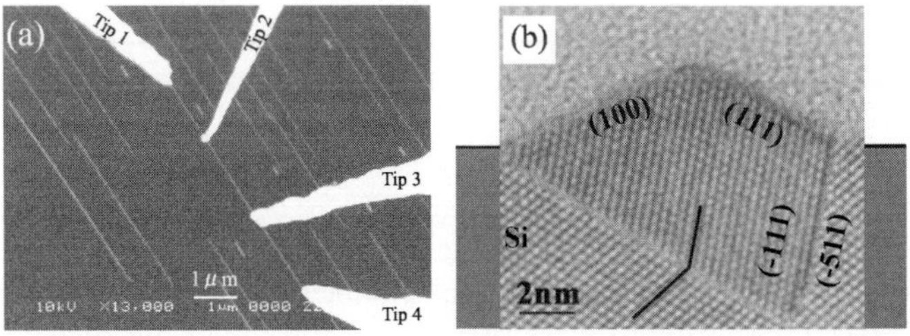

FIGURE 10. (a) SEM image of the four W tips in contact with a Co-slicide ($CoSi_2$) nano-wire on Si(110) substrate [52]. (b) A cross-sectional TEM image. Reproduced with permission from [54].

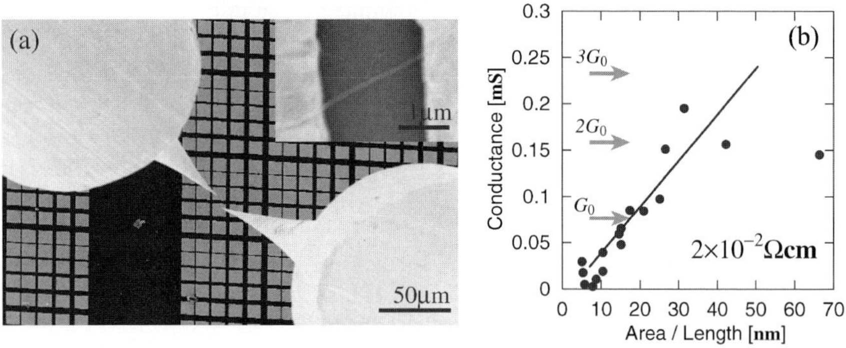

FIGURE 11. (a) SEM image of a carbon nanotube bridging between Ta electrode pads (a magnified image of the inset) with the two W tips contacting to the pads. (b) The measured conductance of several carbon nanotubes as a function of the cross-sectional area divided by length [53].

conductance measurements. Commercial carbon nanotubes having various diameters (0.05~0.2 μm) were randomly dispersed on the patterned square Ta pads formed on a SiO_2/Si wafer susbtrate. The spacing between the pads was 0.8~5 μm so that carbon nanotubes of different lengths were measured. From the length dependence of resistance, the contact resistance between the carbon nanotubes and metal electrodes was shown to be sufficiently small. Figure 11(b) shows the measured conductance versus A/L (cross section/length of carbon nanotubes). The data points show a linear dependence up to $A/L \sim 30$ nm, following Ohm's law $G = \sigma \cdot A/L$, where σ is the conductivity of carbon nanotubes. Arrows indicate the quantum conductance ($G_0 = 2e^2/h$); the measured conductance is not quantized. This indicates the occurrence of diffusive transport through the multi-walled carbon nanotubes (longer than 0.8 μm) at room temperature. A nonlinear current-voltage characteristic was found; the conductance increased steeply with current. This can be interpreted as being due to a change of the current distribution in the cross section of the nanotubes; the outermost shell mainly contributes to the conduction at lower current, while the inner shells gradually begin to contribute with increasing current, so that the total conductance increases with current. A single nanotube could carry a current of more than 0.3 mA.

7 Concluding Remarks

The μ-4PP method by four-tip STM described here is a unique and powerful tool in surface science, especially for studying surface transport, and is expected to be increasingly important, because electrical conduction through one or two atomic layers on crystal surfaces may play an essential role in nanometer-scale science and technology. Of course, the probes can be applied not only for the study of surface transport, but also for transport properties of microscopic and nanometer-scale structures and devices.

An important issue for future development of the four-tip STM is the probe spacing. At the present the minimum probe spacing in the multi-tip STM is approximately 100 nm [2]. This is due to the radius of tip apex of electrochemically etched metal wires. This probe spacing is not small enough for observing ballistic transport and quantum interference effects because the coherence length of conduction carriers is shorter than 100 nm in many cases. For this reason, continuous efforts are being made to shorten the probe spacing down to ca. 10 nm and develop a low-temperature multi-tip STM in which both the sample and tips are cooled down to liquid He temperature. To make the probe spacing shorter, carbon nanotube tips have been developed in which a carbon nanotube is glued at the end of W-tip [56,57]. Since the radius of the (multiwalled) carbon nanotubes is usually ca. 10 nm and the aspect ratio is much higher than usual W-tips, two carbon nanotube tips can be brought together into an approximately 10-nm spacing. Another feature of the carbon nanotube tip is its mechanical flexibility, which can reduce damage to delicate organic and biological molecules, and make the tips withstand numerous direct contacts with the sample. These properties are quite convenient for transport measurements by multi-tip STM at nanometer scales. However, there have been problems in the carbon nanotube tips; high contact resistance between the supporting metal tip and the attached carbon nanotube strongly disturbs electron transport at the STM/STS measurements. Moreover, adsorbates contained in the carbon nanotube degrade the surface cleanness of the specimen under STM operation.

A novel technique for overcoming these difficulties has been developed—the carbon nanotube together with the supporting metal tip is fully coated with a thin metal layer [58]. Figure 12(a,b) shows TEM images of a W-coated carbon nanotube tip. The W-layer of ca. 3-nm thickness was deposited by pulsed laser deposition (PLD) method. The W-layer fully covers the tip even at the end. Figure

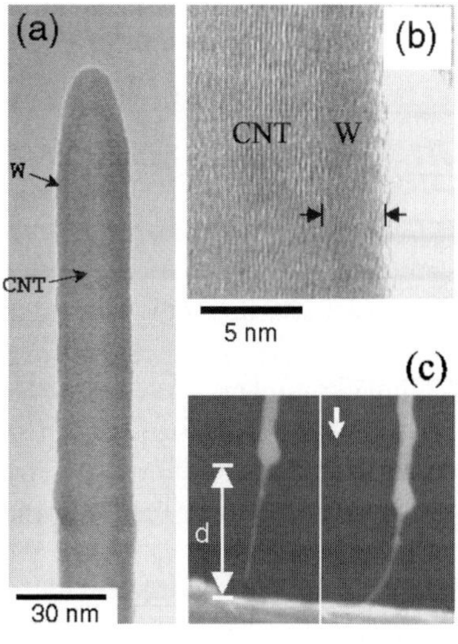

FIGURE 12. (a) TEM image and (b) high-resolution TEM image of a W-coated carbon nanotube tip synthesized by pulsed laser deposition method. (c) SEM images of the W-coated carbon nanotube tip, directly contacting to a sample surface. Reproduced with permission from [58].

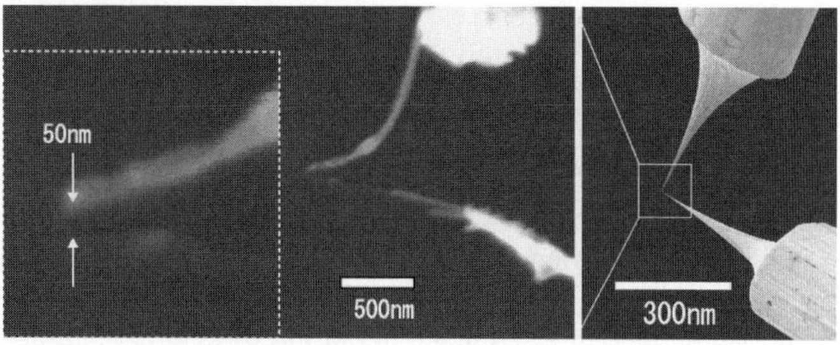

FIGURE 13. SEM images of the two W-tips with W-coated carbon nanotubes at the ends, approaching each other to reach ca. 50-nm probe spacing [60].

12(c) shows flexibility and robustness of the W-coated carbon nanotube tip upon the direct contact to a sample surface. We have also confirmed that the electrical resistance at the glued point between the carbon nanotube and supporting W-tip is stably reduced by the metal coating; especially PtIr coating is the most efficient for this purpose [59]. Atomic-resolution STM imaging and STS spectra were acquired with the W-coated carbon nanotube tip on the first attempt [58].

With metal-coated carbon nanotube tips, we have succeeded in bringing the two tips together to within >ca. 50 nm, as shown in Figure 13 [60]. Two W-coated carbon nanotube tips were installed in our four-tip STM, and the tips were observed by SEM during the approach. Since the resolution of SEM is not enough for observing smaller probe spacing, we believe that the minimum spacing can be reduced to ca. 10 nm.

The reduced probe spacing down to the ca. 10-nm range will make it possible to measure not only quantum transport phenomena at nanometer scales, but also a novel physical property, Green's function [61,62]. To illustrate this concept, imagine a situation in which two tips, tip 1 and tip 2, are simultaneously kept in tunnel contact with the sample surface at a tip spacing less than the carrier coherence length. Each tip carries tunneling current I_i with bias voltage V_i $(i = 1, 2)$. The tunnel conductance σ_i of each tip is defined by the derivative of I_i with respect to V_i,

$$\sigma_i = \frac{\partial I_i}{\partial V_i} \propto \mid \Psi(r_i) \mid^2 = \rho(r_i), \tag{21}$$

which is proportional to the local density of states (or the square of absolute value of wavefunction) at the tip positions r_i on the sample. This is a usual single-tip STS measurement.

When the bias voltage V_1 of tip 1 is changed, the tunneling current I_2 of tip 2 can change because of nonlocal nature of electron systems. The *transconductance* σ_{21} is then defined by the derivative of I_2 with respect to V_1;

$$\sigma_{21} = \frac{\partial I_2}{\partial V_1} \propto \mid G(r_1, r_2; \epsilon = eV_1) \mid^2, \tag{22}$$

which is proportional to the square of absolute value of Green's function at an energy of eV_1, $G(r_1, r_2; \epsilon = eV_1)$. Since Green's function generally contains information about nonlocal properties as well as local ones, we can deduce various response functions of the sample electronic systems; dynamical properties as well as transport properties of the samples can be derived. If we can measure the transconductance with four tips, we will be able to obtain the multi-particle Green's function so that information about electron correlation can also be obtained. Since this measurement is possible only when the probe spacing is less than the coherence length of carriers, the results inevitably reflect the nanometer-scale properties. Therefore, four-tip STM with nanometer probe spacing can be an ultimate tool for measuring transport and dynamics properties at nanometer scales.

The PLD method, furthermore, enables us to coat the carbon nanotube tips with various kinds of materials, such as magnetic, dielectric, and superconducting coatings. These tips can be used not only for the 4PP conductivity measurements, but also for other measurements e.g., field-effect transport, spin-dependent conduction, and so on. The "Green's function STM" with such functionalized tips may be the final goal of the multi-probe STM.

Acknowledgments. I gratefully acknowledge the present and former members of my research group at the University of Tokyo; many of them have greatly contributed to the construction of the four-tip STM apparatus and its application to various samples. I also acknowledge the collaboration with Prof. Mitsuhiro Katayama's group on the metal-coated carbon nanotube tips. This work has been supported in part by Grants-In-Aid from the Japanese Society for the Promotion of Science. I have been supported also by SENTAN Project (Development of System and Technology for Advanced Measurement and Analysis) of the Japan Science and Technology Corporation Agency (JST).

References

1. S. Tsukamoto, B. Siu, N. Nakagiri, Rev. Sci. Instrum. **62**, 1767 (1991).
2. M. Aono, C.-S. Jiang, T. Nakayama, T. Okuda, S. Qiao, M. Sakurai, C. Thirstrup, Z.-H. Wu, J. Surf. Sci. Soc. Jpn. **19**, 698 (1998), in Japanese.
3. H. Okamoto and D. M. Chen, Rev. Sci. Instr. **72**, 4398 (2001).
4. H. Grube, B. C. Harrison, J. Jia, and J. J. Boland, Rev. Sci. Instr. **72**, 4388 (2001); ibid. **73**, 1343 (2002) [ERRATA].
5. H. Watanabe , C . Manabe , T . Shigematsu and M. Shimizu: Appl. Phys. Lett. **78**, 2928 (2001); **79**, 2462 (2001).
6. K. Takami, M. Akai-Kasaya, A. Saito, M. Aono, and Y. Kuwahara, Jpn. J. Appl. Phys. **44**, L120 (2005).
7. M. Ishikawa, M. Yoshimura, and K. Ueda, Jpn. J. Appl. Phys. **44**, 1502 (2005).
8. Omicron NanoTechnology GmbH (http://www.omicron.de/), MultiProbe, Inc (http://www.multiprobe.com/), Zyvex Co. (http://www.zyvex.com/).
9. S. Hasegawa, X. Tong, S. Takeda, N. Sato, and T. Nagao, Prog. Surf. Sci. **60**, 89(1999).

10. S. Hasegawa and S. Ino, Int. J. Mod. Phys **7**, 3817 (1993).
11. S. Hasegawa, J. Phys.: Cond. Matters **12**, R463 (2000).
12. S. Hasegawa and F. Grey, Surf. Sci. **500**, 84 (2002).
13. S. Hasegawa, I. Shiraki, F. Tanabe, R. Hobara, T. Kanagawa, T. Tanikawa, I. Matsuda, C. L. Petersen, T. M. Hansen, P. Boggild, F. Grey, Surf. Rev. Lett. **10**, 963 (2003).
14. S. Hasegawa, I. Shiraki, T. Tanikawa, C. L. Petersen, T. M. Hansen, P. Goggild, and F. Grey, J. Phys.: Cond. Matters **14**, 8379 (2002).
15. M. Henzler, in: J. M. Blakely (Ed.), *Surface Physics of Materials I*, Academic Press, New York, 1975, p. 241.
16. P. Boggild, T. M. Hansen, O. Kuhn, F. Grey, T. Junno, and L. Montelius, Rev. Sci. Instrum. **71**, 2781 (2000).
17. I. Shiraki, F. Tanabe, R. Hobara, T. Nagao, S. Hasegawa, Surf. Sci. **493**, 643 (2001).
18. S. Hasegawa, I. Shiraki, F. Tanabe, and R. Hobara, Current Appl. Phys. **2**, 465 (2002).
19. T. Kanagawa, R. Hobara, I. Matsuda, T. Tanikawa, A. Natori, and S. Hasegawa, Phys. Rev. Lett. **91**, 036805 (2003).
20. H. Okamoto and D. M. Chen, J. Vac. Sci. Technol. A **19**, 1822 (2001).
21. For review, S. Hasegawa, N. Sato, I. Shiraki, C. L. Petersen, P. Boggild, T. M. Hansen, T. Nagao, F. Grey, Jpn. J. Appl. Phys. **39**, 3815 (2000).
22. F. J. Himpsel, G. Hollinger, R. A. Pollak, Phys. Rev. B **28**, 7014 (1983).
23. J.Viernow, M. Henzler, W. L. O'Brien, F. K. Men, F. M. Leibsle, D. Y. Petrovykh, J. L. Lin, and F. J. Himpsel, Phys. Rev B **57**, 2321 (1998).
24. F. M. Smits, Bell System Technical Journal **37**, 711 (1958).
25. S. Hasegawa, X. Tong, C.-S. Jiang, Y. Nakajima, and T. Nagao, Surf. Sci. **386**, 322 (1997).
26. C. L. Petersen, F. Grey, I. Shiraki, S. Hasegawa, Appl. Phys. Lett. **77**, 3782 (2000).
27. H. W. Yeom, S. Takeda, E. Rotenberg, I. Matsuda, K. Horikoshi, J. Schaefer, C. M. Lee, S. D. Kevan, T. Ohta, T. Nagao, and S. Hasegawa, Phys. Rev. Lett. **82**, 4898 (1999).
28. T. Abukawa, M. Sasaki, F. Hisamatsu, T. Goto, T. Kinoshita, A. Kakizaki, and S. Kono, Surf. Sci. **325**, 33 (1995).
29. O. Gallus, Th. Pillo, M. Hengsberger, P. Segovia, Y. Baer, Eur. Phys. J. B, **20**, 313 (2001).
30. O. Bunk, G. Falkenberg, J. H. Zeysing, L. Lottermoser, R. L. Johnson, M. Nielsen, F. Berg-Rasmussen, J. Baker, and R. Feidenhans'l, Phys. Rev. B **59**, 12228 (1999).
31. J. Nakamura, S. Watanabe, and M. Aono, Phys. Rev. B **63** 193307 (2001).
32. J.-H. Cho, D.-H. Oh, K. S. Kim, and L. Kleinman, Phys. Rev. B **64** 235302 (2001).
33. T. Hirahara, I. Matsuda, M. Ueno, and S. Hasegawa, Surf. Sci. **563**, 191 (2004).
34. T. Hirahara, I. Matsuda, and S. Hasegawa, e-J. Surf. Sci. Nanotech. **2**, 141 (2004).
35. J. N. Crain, K. N. Altmann, C. Bromberger, and F. J. Himpsel, Phys. Rev. B **66**, 205302 (2002)
36. H. Okino, R. Hobara, I. Matsuda, T. Kanagawa, S. Hasegawa, J. Okabayashi, S. Toyoda, M. Oshima, and K. Ono, Phys. Rev. **70**, 113404 (2004).
37. I. Matsuda, M. Ueno, T. Hirahara, R. Hobara, H. Morikawa, and S. Hasegawa, Phys. Rev. Lett. **93**, 236801 (2004).
38. M. F. Crommie, C. P. Lutz, and D. M. Eigler, Nature (London) **363**, 524 (1993).
39. Y. Hasegawa and P. Avouris, Phys. Rev. Lett. **71**, 1071 (1993).
40. M. F. Crommie, C. P. Lutz, D.M. Eigler, and E. J. Heller, Surf. Rev. Lett. **2**, 127 (1995).
41. P. Avouris, I.-W. Lyo, R. E.Walkup, and Y. Hasegawa, J. Vac. Sci. Technol. B **12**, 1447 (1994).

42. L. Bürgi, O. Jeandupeux, A. Hirstein, H. Brune, and K. Kern, Phys. Rev. Lett. **81**, 5370 (1998).
43. N. Sato, S. Takeda, T. Nagao, and S. Hasegawa, Phys. Rev. B **59** (1999) 2035.
44. X. Tong, S. Ohuchi, N. Sato, T. Tanikawa, T. Nagao, I. Matsuda, Y. Aoyagı, and S. Hasegawa, Phys. Rev. B **64**, 205316 (2001).
45. J. H. Davies, *The Physics of Low-Dimensional Semiconductors* (Cambridge University Press, Cambridge, England, 1998).
46. S. Datta, *Electronic Transport in Mesoscopic Systems* (Cambridge University Press, Cambridge, England, 2002).
47. T. Ogino, Surf. Sci. **386**, 137 (1997).
48. http://www.capres.com.
49. T. M. Hansen, K. Stokbro, O. Hansen, T. Hassenkam, I. Shiraki, S. Hasegawa, P. Boggild, Rev. Sci. Inst. **74**, 3701 (2003).
50. K. Kobayashi, Surf. Sci. **583**, 16 (2005).
51. M. Aono, *et al.*, unpublished (http://www.nims.go.jp/nanomat_lab/).
52. H. Okino, I. Matsuda, R. Hobara, Y. Hosomura, and S. Hasegawa, P. A. Bennett, Appl. Phys. Lett., **86**, 233108 (2005).
53. R. Hobara, S. Yoshimoto, T. Ikuno, M. Katayama, N. Yamauchi, W. Wongwiriyapan, S. Honda, I. Matsuda, S. Hasegawa, and K. Oura, Jpn. J. Appl. Phys. **43**, L1081 (2004).
54. Z. He, D.J. Smith, and P.A. Bennett, Phys. Rev. Lett. **93**, 256102 (2004).
55. Z. Zhong, D. Wang, Y. Cui, M. W. Bockrath, and C. M. Lieber, Science **302**, 1377 (2003).
56. Y. Shingaya, et al., Physica B **323**, 153 (2002).
57. M. Ishikawa, et al., Japan. J. Appl. Phys. **41**, 4908 (2002).
58. T. Ikuno, M. Katayama, M. Kishida, K. Kamada, Y. Murata, T. Yasuda, S. Honda, J. Lee, H. Mori, and K. Oura, Jpn. J. Appl. Phys. **43**, L644 (2004).
59. S. Yoshimoto, Y. Murata, M. Kishida, R. Hobara, D. Maeda, T. Yasuda, T. Ikuno, S. Honda, H. Okado, I. Matsuda, M. Katayama, and S. Hasegawa, Jpn. J. Appl. Phys. **44**, L1563 (2005).
60. Y. Murata, S. Yoshimoto, M. Kishida, D. Maeda, T. Yasuda, T. Ikuno, S. Honda, H. Okado, R. Hobara, I. Matsuda, S. Hasegawa, K. Oura, and M. Katayama, Jpn. J. Appl. Phys. **44**, No. 7B 5336 (2005).
61. Q. Niu, M. C. Chang, and C. K. Shih, Phys. Rev. B **51**, 5502 (1995).
62. J. M. Byers and M. E. Flatte: Phys. Rev. Lett. **74**, 306 (1995).

II.8
Dynamic Force Microscopy and Spectroscopy in Vacuum

Udo D. Schwarz and Hendrik Hölscher

Dynamic force microscopy (DFM) operated in ultrahigh vacuum, which is often also called *noncontact atomic force microscopy* (NC-AFM), is able to image the atomic structure of surfaces, including observation of point defects independent from the sample's conductivity. Within the last ten years, a variety of materials, including conductors, semiconductors, and insulators, have been imaged down to the atomic scale. Atomic arrangements and individual defects such as vacancies or impurity atoms have been observed, and chemically different species have been distinguished due to their different energy dissipation properties in DFM. In these atomic-scale images, the contrast is mainly due to short-range interatomic bonding forces, repulsive atom–atom forces, or van der Waals forces. However, in many experiments, electrostatic forces play a significant role in contrast formation, and electrostatic charge distributions can be imaged with high resolution. For example, the charge distribution around charged monoatomic vacancies can be mapped, as well as the charge distribution around individual doping atoms in semiconductors. At higher doping levels, the charge clouds around individual doping atoms overlap, and DFM images the resulting modulation in the surface potential of the sample. Finally, using dynamic force spectroscopy, the relative contributions of long-ranged electrostatic and van der Waals forces as opposed to short-ranged chemical bonding forces can be distinguished and analyzed site-specifically.

1 Introduction

Since the development of the first *atomic force microscope* (AFM) by Binnig, Quate, and Gerber in 1986 [1], this microscopy method has been established as a standard tool in surface physics. Its lateral resolution, however, was restricted for a long time by the finite contact area between tip and sample. Fortunately, this limitation was overcome in 1995, when atomic resolution was achieved independently by Giessibl [2] and Sugawara et al. [3] using the so-called *noncontact mode* of the atomic force microscope. This approach, which is often referred to as noncontact atomic force microscopy (NC-AFM), is based on a specific detection scheme, where the cantilever is oscillated close to the sample surface in ultrahigh vacuum (UHV) [4]. Due to the oscillating behavior of the cantilever, NC-AFM is alternatively denoted as *dynamic force microscopy* (DFM). We will follow this second notation throughout the rest of the chapter.

Since the pioneering work was published in 1995, true atomic resolution—the resolution of point defects—has been obtained on many different types of surfaces, including conductors, semiconductors, and insulators. These remarkable results initiated intense research, and the field of DFM/NC-AFM has been growing rapidly during the last few years (see, e.g., [5–8] for reviews). The aim of this chapter is first to give an overview of the basic principles of DFM/NC-AFM, followed by a discussion on what properties can be detected exactly, which includes an analysis of contrast mechanisms with a special emphasise on the local detection of charges. Thereby, we restrict ourselves to the "classical" high-resolution mode of DFM/NC-AFM, where measurements are applied in ultrahigh vacuum and with constant bias voltage.

2 Basic Principles of Dynamic Force Microscopy

Obtaining high-resolution images with an atomic force microscope requires clean sample surfaces without unwanted adsorbates. Therefore, such experiments have to be performed in ultrahigh vacuum. Most DFM experiments performed in UHV use the so-called *frequency modulation* (FM) *detection scheme* introduced by Albrecht et al. [4]. In this mode, the cantilever is "self-oscillated", i.e., an electronic feedback system is used to drive the cantilever at its actual resonance frequency, which depends on the forces acting on the tip. This is in contrast to amplitude modulation (AM) modes [9–11], which are in widespread use mainly at ambient conditions and in liquids,[1] where the cantilever is externally driven with a fixed driving frequency.

As mentioned earlier, the FM technique enables the imaging of single-point defects, such as individual atomic vacancies on clean sample surfaces in vacuum. Its resolution is comparable to that obtained in scanning tunneling microscopy (STM), but without STM's restriction to conducting surfaces. The above-introduced terms *noncontact mode* and *noncontact atomic force microscopy* have been established due to the common belief that a destructive repulsive contact between tip and sample is effectively prevented by this technique. This section presents the basic principles of the experimental setup, the origin of the detected frequency shift, as well as methods to calculate it.

2.1 Experimental Setup

Figure 1 shows schematically the basic setup of an atomic force microscope operated in frequency-modulation mode. The movement of the cantilever is measured by detecting the position of a laser beam reflected from the cantilever backside

[1] AM modes in air and liquid include the popular *tapping mode* of AFM, where the tip gets into contact with the sample surface during an individual oscillation cycle.

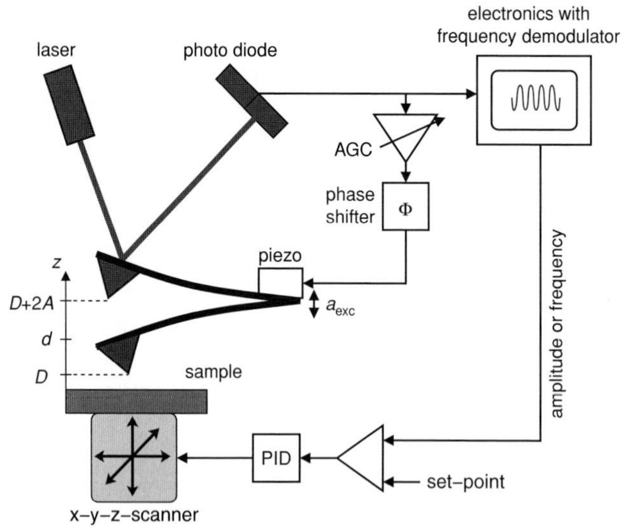

FIGURE 1. Schematic setup of an atomic force microscope operated in frequency modulation mode, as it is often realized for UHV applications. The key feature is the positive feedback used to "self-oscillate" the cantilever. A laser beam is reflected at the backside of a cantilever that has an integrated sharp tip at its free end. Bending of the cantilever causes a deflection of the reflected beam, which is detected by a photo diode. The resulting signal is subsequently amplified and phase-shifted before it is used to drive the dither piezo oscillating the cantilever. The measured quantity is the frequency shift due to the tip-sample interaction, which serves as the control signal for the cantilever-sample distance.

on a photo diode. This signal is then fed into an amplifier featuring the possibility of *automatic gain control* (AGC) and subsequently used to excite a piezo that drives the cantilever. The time delay between the excitation signal and cantilever deflection is adjusted by a time ("phase") shifter to a value $t_0 = 1/(4f_0)$, where f_0 represents the eigenfrequency of the free cantilever. This choice corresponds to a phase shift of $\approx 90^\circ$, leading to an oscillation near resonance. Two different modes have been established: The *constant-amplitude* mode [4], where the oscillation amplitude A is kept at a constant value by the AGC, and the *constant-excitation mode* [12], where the excitation amplitude a_{exc} of the dither piezo is kept constant. In the following, we will focus mainly on the constant-amplitude mode, which is the dominant mode of operation in most studies published to date.

The key feature of the described FM setup is the positive-feedback loop. It ensures that the cantilever always oscillates at its actual resonance frequency f, in contrast to the AM modes where the cantilever is driven externally by a frequency generator at a fixed frequency f_d. Constant-frequency driving naturally fails at maintaining resonance conditions during DFM measurements since the tip-sample interaction alters the cantilever's resonance frequency. The actual resonance frequency f of a cantilever oscillating close to the sample surface is therefore different from the eigenfrequency f_0 of the free cantilever and depends on the nearest tip-sample distance as well as on the oscillation amplitude. The positive-feedback loop sketched in Figure 1 elegantly solves the problem of continuously readjusting the cantilever driving frequency during tip-sample approach by using the cantilever itself as the frequency-determining element. The measured quantity is then the *frequency shift* Δf representing the difference between actual resonance frequency and eigenfrequency of the free cantilever (i.e., $\Delta f = f - f_0$).

The FM detection method received its name from the frequency demodulator (FM-detector) used to detect the frequency shift. The cantilever driving, however, is independent from this part of the setup. Other setups are based on *phase-locked loops* (PLLs), which output the actual oscillating frequency of the cantilever. This signal may then be used as input for a numerically controlled oscillator, which drives the cantilever exactly at the frequency measured by the PLL [13].

For imaging, the frequency shift Δf is used to control the cantilever sample distance. Thus, the frequency shift is constant and the acquired data represents planes of constant Δf, which can be related to the surface topography in most cases (see, e.g., section 3.3 for exceptions). Alternatively, the recording of the frequency shift as a function of the tip-sample distance or the oscillation amplitude can be used to determine the tip-sample force with high resolution (see section 4).

2.2 Origin of the Frequency Shift

Since the frequency shift is the measured quantity in FM-DFM, we will start our discussion with an analysis of its origin. Valuable insight into cantilever dynamics can be gained by looking at the tip potential displayed in Figure 2. If the

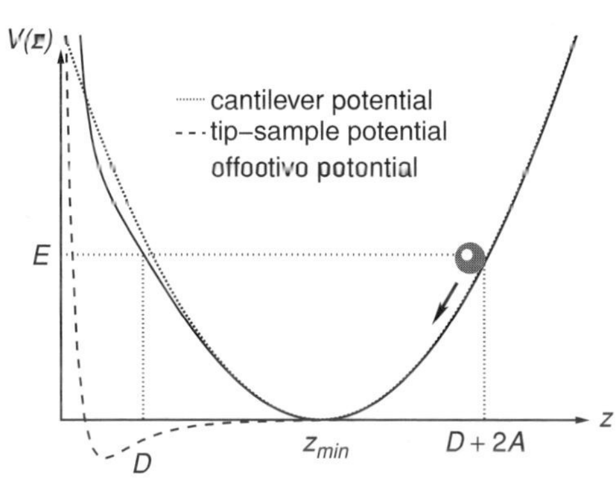

FIGURE 2. Plot of the various potentials involved in the cantilever oscillation as a function of the tip-sample distance z. A is the oscillation amplitude, while D represents the distance where the tip approaches the surface the closest. The frequency shift experienced in dynamic force microscopy is caused by the tip-sample interaction potential (dashed line), which alters the harmonic cantilever potential (dotted line). Therefore, the tip moves in an asymmetric effective potential (solid line), what leads to an anharmonic oscillation behavior. For better visualization, the mechanical analog of the situation is indicated by a ball rolling frictionless on an equivalently shaped surface. z_{min} is the minimum position of the effective potential.

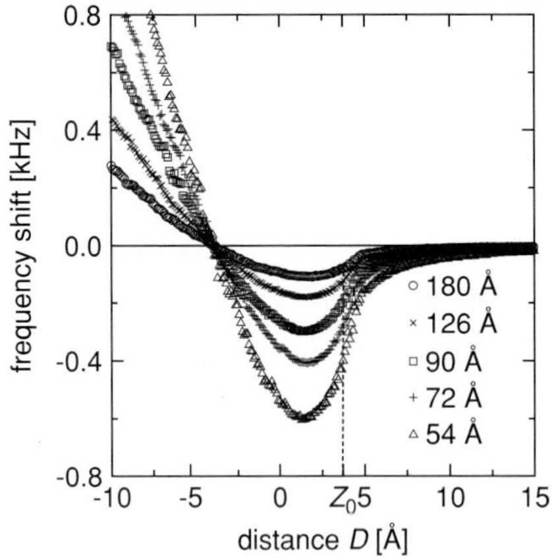

FIGURE 3. Experimental frequency shift versus distance curves acquired with a silicon tip and a graphite sample for different amplitudes (54–180 Å) in UHV. The experiments were carried out with an atomic force microscope designed for operation at low temperatures [14]. The rectangular-shaped silicon cantilever used for this experiment featured a spring constant of 38 N/m and an eigenfrequency of 171 kHz. The sample was cooled to $T = 80$ K by liquid nitrogen cooling. Tip and sample were grounded during the recording of the data set [15]. The curves are shifted along the z-axes to make them comparable. The zero line and the contact point z_0 are defined by the force laws described in section 4.

cantilever is far away from the sample surface, the tip moves in a symmetric parabolic potential (dotted line), and its oscillation is harmonic. In such a case, the tip motion is sinusoidal and the resonance frequency is given by the eigenfrequency f_0 of the cantilever. If, however, the cantilever approaches the sample surface, the original symmetric potential is modified to an asymmetric effective potential V_{eff} (solid line) given by the sum of the parabolic potential and the tip-sample interaction potential V_{ts} (dashed line). As a result of this modification, the oscillation becomes anharmonic, and the resonance frequency of the cantilever depends now on the oscillation amplitude A. Since the effective potential experienced by the tip changes also with the nearest distance D, the frequency shift is a functional of both parameters (i.e., $\Delta f = \Delta f(D, A)$).

Figure 3 displays experimental frequency shift versus distance curves for different oscillation amplitudes. The observed behavior is expected from the simple model introduced above. All curves exhibit a similar overall shape, but differ in magnitude depending on the oscillation amplitude and the nearest tip-sample distance. During the approach of the cantilever toward the sample surface, the frequency shift decreases and reaches a minimum. With a further reduction of the nearest tip-sample distance, the frequency shift increases again and becomes positive.

For smaller oscillation amplitudes, the minimum of the $\Delta f(z)$-curves is deeper and the slope after the minimum is steeper than for larger amplitudes, i.e., the overall effect is larger for smaller amplitudes. This can be explained by the model as well: a decrease of the amplitude A for a fixed nearest distance D moves the minimum of the effective potential closer to the sample surface. Therefore, the relative perturbation of the harmonic cantilever potential increases, which causes the observed increase of the absolute value of the frequency shift.

2.3 Calculation of the Frequency Shift

As described earlier in section 2.1, the prominent feature of the FM technique is that the cantilever is "self-driven" by a positive-feedback loop. To account for this feature, the corresponding equation of motion differs from the well-known case of an externally driven oscillator [16]. It is a differential equation with time delay [17–19],

$$m^*\ddot{z}(t) + \frac{2\pi f_0 m^*}{Q}\dot{z}(t) + c_z z(t) = F_{\text{ts}}[z(t), \dot{z}(t)] + \underbrace{g c_z z(t - t_0)}_{\text{driving}}, \tag{1}$$

where $z(t)$ represents the position of the tip at the time t and F_{ts} the tip-sample interaction force. c_z, m^*, and Q denote the spring constant, the effective mass, and the quality factor of the cantilever, respectively. The term on the right reflects the positive feedback of the system, which includes the amplification of the displacement signal by the *gain factor* g measured at the retarded time $t - t_0$.

The frequency shift can be calculated from the above equation of motion with the assumption that the tip oscillations are nearly sinusoidal. In this case, we can use the the ansatz $z(t) = A\cos(2\pi f t)$ describing the stationary solutions of Eq. (1). This approach is justified if the tip-sample interaction force is much smaller than the retracting cantilever force. Additionally, we assume that the microscope is driven in the *constant amplitude* mode, where the AGC keeps the oscillation amplitude A constant by continuously adjusting the gain factor g. The solutions for the *constant excitation* mode look slightly different. A more detailed discussion can be found in [20,21]

The equation obtained after introducing this ansatz into Eq. (1) can be simplified by multiplying with $\cos(2\pi f t)$, followed by integration over one period of oscillation $t = [0, 1/f]$. Alternatively, integration is performed after multiplication with $\sin(2\pi f t)$.[2] The result is a set of two coupled trigonometric equations [19],

$$g\cos(2\pi f t_0) = \frac{f_0^2 - f^2}{f_0^2} - \frac{2f}{Ac_z}\int_0^{1/f} F_{\text{ts}}[z(t), \dot{z}(t)]\cos(2\pi f t)\,dt, \tag{2}$$

$$g\sin(2\pi f t_0) = -\frac{1}{Q}\frac{f}{f_0} - \frac{2f}{Ac_z}\int_0^{1/f} F_{\text{ts}}[z(t), \dot{z}(t)]\sin(2\pi f t)\,dt, \tag{3}$$

which might be solved numerically.

[2] This treatment of the equation is equivalent to a Fourier transformation of the tip-sample force.

Fortunately, detailed analysis shows that the results of an FM-DFM experiment are mainly determined by the tip-sample force and depend only very little on the time delay if t_0 has been set by the experimentalist to an optimal value before approaching the tip toward the sample surface. These values of the time delay are specific resonance values corresponding to 90° phase shift (i.e., $t_0 = 1/4 f_0$), which can easily be found by a minimization of the gain factor as a function of the time delay. As a consequence, it can be assumed that $\cos(2\pi f t_0) \approx 0$ and $\sin(2\pi f t_0) \approx 1$, allowing the decoupling of Eqs. (2) and (3):

$$\Delta f \cong -\frac{f_0^2}{Ac_z} \int_0^{1/f_0} F_{\text{ts}}[z(t), \dot{z}(t)] \cos(2\pi f_0 t)\, dt, \tag{4}$$

$$|g| \cong \frac{1}{Q} + \frac{2 f_0}{Ac_z} \int_0^{1/f_0} F_{\text{ts}}[z(t), \dot{z}(t)] \sin(2\pi f_0 t)\, dt. \tag{5}$$

Since we did not make any assumptions about the specific force law describing the tip-sample interaction F_{ts}, these equations are valid for every type of interaction as long as the resulting cantilever oscillations are nearly sinusoidal. Assuming reasonable tip-sample forces, it can be shown that the frequency shift is determined by the average of the tip-sample force between forward and backward movement, whereas the gain factor is directly related to dissipative processes like hysteresis or viscous damping [19]. This behavior is caused by the odd- and even-weighting functions (sine and cosine) occurring inside the integrals in Eqs. (4) and (5) [22].

2.4 Frequency Shift for Specific Force Laws

For the further analysis of the FM-DFM mode, handy formulas for the calculation of the frequency shift for given force laws are helpful. For example, tip-sample forces are often described by inverse power laws of the type

$$F_n(z) = -\frac{C_n}{z^n}, \tag{6}$$

where C_n represents a constant that depends on factors such as the nature of the force, tip and sample materials, and the exact geometry of the contact. Introducing this force law into Eq. (4), comparatively simple analytic solutions of the integral can be found for small values of n:

$$\Delta f(n=1) = \frac{f_0 C_1}{c_z A^2}\left(1 - \frac{d}{\sqrt{d^2 - A^2}}\right), \tag{7}$$

$$\Delta f(n=2) = -\frac{f_0 C_2}{c_z} \frac{1}{(d^2 - A^2)^{3/2}}, \tag{8}$$

$$\Delta f(n=3) = -\frac{3}{2} \frac{f_0 C_3}{c_z} \frac{d}{(d^2 - A^2)^{5/2}}, \tag{9}$$

$$\Delta f(n=4) = -\frac{1}{2} \frac{f_0 C_4}{c_z} \frac{A^2 + 4d^2}{(d^2 - A^2)^{7/2}}, \tag{10}$$

$$\Delta f(n=5) = -\frac{5}{8}\frac{f_0 C_5}{c_z}\frac{d(3A^2+4d^2)}{(d^2-A^2)^{9/2}}, \tag{11}$$

$$\Delta f(n=6) = -\frac{3}{8}\frac{f_0 C_6}{c_z}\frac{A^4+12A^2d^2+8d^4}{(d^2-A^2)^{11/2}}. \tag{12}$$

These equations are especially valuable for the measurement of long-range forces such as van der Waals forces and magnetic and electrostatic forces, which are frequently well described by a force law following Eq. (6).

More generally, the integral in Eq. (4) can be further simplified with the assumption that the decay length of the tip-sample interaction λ is much smaller than the oscillation amplitude ($\lambda \ll A$). This so-called *large amplitude* limit was first analyzed by Giessibl [23]. This condition is fulfilled for typical FM-DFM experiments, where oscillation amplitudes are larger than 5 nm. In this case, the "right" part of the effective potential with $z_{min} \leq z \leq D + 2A$ remains almost perfectly parabolic; only its "left" part, where $D \leq z \leq z_{min}$, is deformed (cf. Figure 2). We first transform the integral Eq. (4) with $u = \cos(2\pi f_0 t)$ to

$$\Delta f = -\frac{f_0}{\pi c_z A}\int_{-1}^{1} F_{ts}[D + A(1+u)]\frac{u}{\sqrt{1-u^2}}\,du. \tag{13}$$

Because of the assumed large amplitude, the weighting term in this integral Eq. (13) can be expanded at $u = -1$ to $u/\sqrt{1-u^2} \approx 1/(\sqrt{2}\sqrt{1-x})$. Applying the transformation $z = D + A(1+u)$ and extending the upper limit of the integral to infinity, the frequency shift can be simplified to [24]

$$\Delta f = \frac{1}{\sqrt{2\pi}}\frac{f_0}{c_z A^{3/2}}\int_{D}^{\infty}\frac{F_{ts}(z)}{\sqrt{z-D}}\,dz. \tag{14}$$

It is interesting to note that the integral in this equation is independent of the amplitude, as it depends solely on the tip-sample force and the tip-sample distance D. The experimental parameters (c_z, f_0, and A) appear only as pre-factors. We can therefore define a new quantity γ, which is referred to as *normalized frequency shift* [23]:

$$\gamma(z) = \frac{c_z A^{3/2}}{f_0}\Delta f(z). \tag{15}$$

This is a very useful quantity if experiments obtained with different oscillation amplitudes and/or cantilevers are to be compared. Additionally, it allows us elegantly to link theoretical calculations with experiments. The validity of Eq. (15) is nicely demonstrated by its application to the frequency shift curves presented earlier in Figure 3. As shown in Figure 4, all curves obtained for different amplitudes collapse into one universal γ-curve, which depends only on the actual tip-sample distance D.

Another aspect of Eq. (14) is that it can be used to calculate the frequency shift for inverse power force laws in the large amplitude limit. If we introduce the force

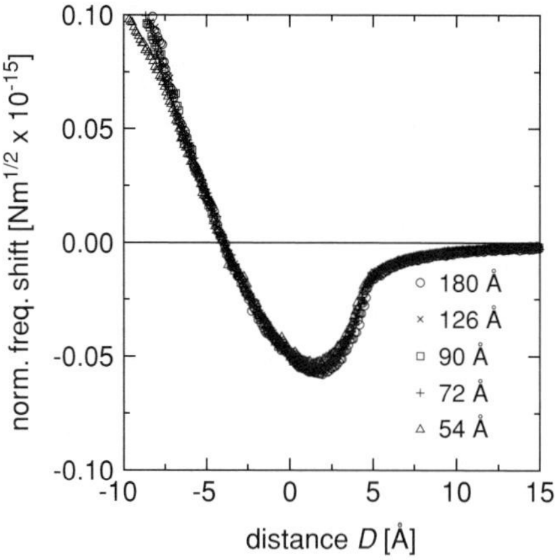

FIGURE 4. Application of Eq. (15) to the frequency shift curves shown in Figure 3. The normalized frequency shift γ is almost identical for all amplitudes ranging from 54 to 180 Å.

TABLE 1. Some Often Used Values for the Integral $I_1(n)$.

n	1	2	3	4	5	6	7	8	13
$I_1(n)$	π	$\frac{1}{2}\pi$	$\frac{3}{8}\pi$	$\frac{5}{16}\pi$	0.273π	0.246π	0.226π	0.209π	0.161π

law Eq. (6) into this equation, the frequency shift can be calculated from [23],

$$\Delta f \approx -\frac{f_0}{\sqrt{2\pi}\, c_z A^{3/2}} \frac{C_n}{D^{n-1/2}} I_1(n), \tag{16}$$

where $I_1(n)$ is given by

$$I_1(n) := \int_{-\infty}^{\infty} \frac{1}{(1+y^2)^n}\, dy = \begin{cases} \pi & n = 1 \\ \pi \frac{1.3.....(2n-3)}{2^{n-1}\cdot(n-1)!} & n > 1 \end{cases}. \tag{17}$$

Selected values of the integral function $I_1(n)$ are listed in Table 1. It can be shown that the accuracy of this approximation increases not only with larger oscillation amplitudes A, but also with larger values of n. A discussion on the accuracy of the different formulas can be found in [25].

3 Application of Dynamic Force Microscopy

3.1 Mechanisms of DFM Contrast Formation on Semiconductors

Semiconductors were the first class of material to be resolved with atomic resolution using DFM [2,3]. Even though semiconductors can quite easily be imaged down to the atomic scale by alternative methods such as scanning tunneling microscopy (STM), they are still frequently investigated by DFM due to the complementary information obtained. Moreover, semiconductors are likely to be

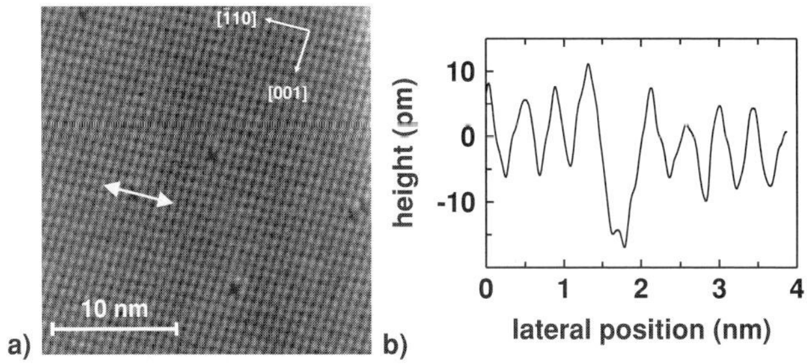

FIGURE 5. (a) Atomic-scale FM-DFM image of InAs(110) with missing protrusions, which can be identified as As vacancies [26]. The size of an individual unit cell is 0.606 nm x 0.427 nm. (b) Section along the white line with the double arrow. It was taken across one of the vacancies, which manifests as a depression of about 15pm depth.

the class of materials where the DFM contrast mechanism is best understood due to numerous theoretical studies, including computer simulations that target semiconducting surfaces. Therefore, we will first discuss the DFM contrast formation on semiconductors before we generalize for other surfaces. To illustrate the various aspects of contrast formation, we will concentrate on III-V semiconductors as a model system in general and indium arsenide in particular.

3.1.1 Basic principles: Imaging the Total Charge Density with Non-Reactive Tips

The measurement presented in Figure 5 demonstrates the level of resolution that can be achieved with DFM applied in ultrahigh vacuum. The data has been obtained on the (110) surface of an n-doped InAs single crystal (sulfur $\approx 3 \times 10^{18}$ cm^{-3}), which has been cleaved in ultrahigh vacuum [26]. Figure 5(a) represents a topographic image, where each of the protrusions reflects an arsenic atom. Individual defects are readily visible. The cross section presented in Figure 5(b) shows that the noise level is below 2 pm rms.

InAs crystallizes in the ZnS structure, i.e., the (110) surface is made up of zigzag chains of alternating In and As atoms along the <1$\bar{1}$0>-direction. Therefore, one might expect that both atomic species, In and As, should be visible. According to the bond rotation model, however, which is valid for most III-V semiconductors, the As atoms relax outward and the In atoms inward, lifting the As sublattice by 80 pm above the In sublattice (cf. Figure 6(a)) [26]. As a result, the In atoms are not visible, and the rows of bright protrusions can be identified as the As atoms where the total valence charge density has its maximum (Figures 5(a) and 6(b)) [26,27]. A similar observation was made by Ohta et al. on InP [28]. More detailed analysis reveals that the observed contrast is essentially caused by short-range (3–6 Å) interaction forces between tip and sample, which are mainly governed by the total electron density at the surface. Thus, if the probe tip only negligibly influences the electron density of the sample surface, as it does in Figs. 5 and 6(b), the resulting

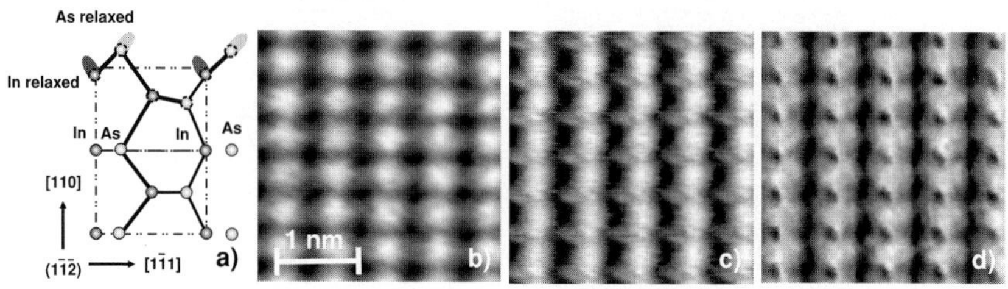

FIGURE 6. (a) Relaxed (110) surface of InAs according to the bond rotation model for III-V semiconductors (side view). The In atoms, having empty dangling bonds, are located about 80 pm lower than the As atoms featuring filled dangling bonds. (b) DFM image of the InAs(110) surface acquired with a non-reactive tip. Only one feature per unit cell is visible. The bright protrusions can be identified as As atoms [27]. (c, d) Two examples of dynamic mode SFM images where two features per surface unit cell are resolved on InAs(110)-(1 $\times$ 1). The second feature, an additional protrusion in (c) and a depression in (d), is located between the bright protrusions and can be attributed to the presence of the lower lying In atoms.

DFM image can be interpreted as a *map of the total electron density at the sample surface*, which is usually regarded as representing the "topography" of the sample.

3.1.2 Imaging with Reactive tips: Orbital Hybridization and the Onset of Bond Formation

In many cases, however, the atom at the tip apex exhibits reactive bonds, which are likely to interact with electron orbitals of surface atoms. In this case, the electron density at the sample surface will be significantly disturbed by the presence of the tip at close tip-sample distances. The images presented in Figure 6(c) and 6(d) can be explained by such an interaction. In both images, a second feature emerges at exactly the position where the In atom is located [26,29]. The nature of this second feature, however, is quite different: It manifests as an additional protrusion or "shoulder" in (c), while depressions are observed in (d). This at the first sight puzzling contradiction can be solved by assuming different types of tips (e.g., tips that have either an In atom, a Si atom, or an As atom at the tip end), which exhibit very different ways of interacting with the surface atoms of the sample. More information on this subject can be found in [26, 27]; computer simulations for the very similar case of GaAs are presented in [30, 31]. Later, the effect was confirmed for GaAs by Uehara et al. [32,33]. Using distance-dependent DFM imaging, they also showed that effects arising from In atoms are only visible at small tip-sample distances. As a result, we can conclude that atomically resolved DFM images represent a mixture between the *sample topography* (i.e., the total electron density at the sample surface) and the *reactivity* of the tip with specific sites at the sample surface, which is usually very short-range. This reactivity is essentially due to *orbital hybridization* [34,35], which is sometimes also regarded as reflecting the *onset of chemical bonding* [36]. In the present case, this feature allows to visualize positions of atoms which are hidden under normal circumstances.

To summarize, the presence of the tip near the sample surface is a factor that can significantly influence properties such as the electron distribution at the sample surface. The varying interaction strength between tip and sample during approach can then lead to a significant distance-dependence of the image contrast: at "larger" distances, where the onset of atomic resolution during tip-sample approach takes place, comparatively long-range electrostatic and van der Waals forces dominate the image contrast. Upon further approach, increasing levels of orbital hybridization involving both tip- and sample states might give rise to different contrasts on samples at intermediate and close tip-sample distances. Experiments performed by Morita et al. on the Si(111)$\sqrt{3} \times \sqrt{3}$-Ag surface, for example, illustrate this behavior elegantly [34,35]. This general mechanism gets even more complex if the atom at the tip end is replaced by another chemical species that exhibits different interaction strengths with the atoms at the sample surface; examples are given in [35,37,38].

3.1.3 Tip-Induced Surface Relaxation and Adhesion Hysteresis

Another important effect influencing atomic-scale contrast formation is the tip-sample interaction that not only can alter the electron distribution at the surface, but also the actual location of the surface atoms. Attractive and repulsive forces acting between tip and sample can lead to a tip-induced relaxation of the position of the surface atoms relative to the bulk atoms, which increases with decreasing tip-sample distance. Even though this effect is significant, it is difficult to quantify from experimental data since all surface atoms visible in an image are likely to relax in the same manner. As a consequence, effects due to surface atom relaxation cannot easily be separated from the "regular" increase in resolution caused by the reduced tip-surface separation. However, such effects might get evident near defects, where some atoms are more weakerly bonded to the bulk then others.

This is the case in Figure 7, where a point defect on InAs is shown that has been imaged at two different tip-sample distances [26]. As discussed above, the protrusions reflect the positions of the As atoms. The two neighboring As lattice sites marked with X exhibit a different corrugation amplitude compared to their surrounding. We also note that the contrast inverts from (a) to (b) relative to the surrounding As atoms after the tip-sample distance has been decreased (resulting in an increase of the magnitude of the tip-sample interaction). From the symmetry with respect to the As atoms, it can be concluded that the point defect is located on the In sub-lattice. The observed distance dependence of the contrast indicates the presence of an In vacancy in the surface layer [26]. A missing In atom reduces the number of bonds of two neighboring surface As atoms from three to two. According to calculations carried out for GaAs(110), this results in an inward relaxation for surface cation vacancies [39]. With the tip approaching the surface, the additional attractive tip-sample interaction provides enough energy to pull the two weakly bonded As atoms toward the tip, e.g., into the vacuum region. Although

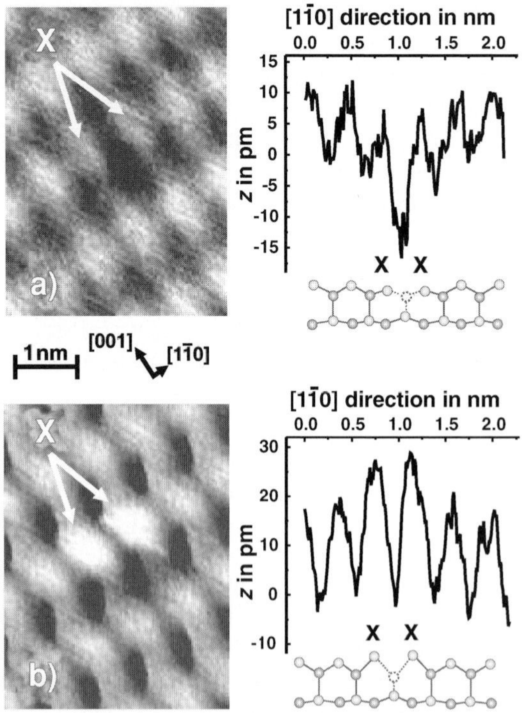

FIGURE 7. DFM images of the same point defect on InAs(110) acquired at two different tip-sample distances. The position of the two As lattice sites, which are influenced by the defect, are marked with X (X-sites). The symmetry suggests that the defect is located on an In lattice site, and the distance dependence indicates a surface In vacancy. The effective relaxation of the X-site As atoms with respect to the undisturbed As sublattice, as explained in more detail in the text, is influenced by the magnitude of the tip-sample interaction (see the line sections and the corresponding sketches below).

only the relaxation of sample atoms around a defect relative to the plane defined by the other surface As atoms can be directly observed in DFM, all other surface atoms will also relax due to the tip-sample interaction. In particular, relaxation of the foremost atom at the tip apex has to be considered, because of its exposed position. This has been confirmed by ab initio calculations for GaAs(110), where the tip atoms as well as the atoms of the defect free surface were allowed to relax [30].

Additional complication of the described effect occurs if the tip and the surface deform in such a way during an individual oscillation cycle that the force felt by the tip during approach and retraction is different. This does not only have an effect on the observed contrast that is difficult to quantify, it also can give rise to a significant energy dissipation mechanism [27]. Denoted as *adhesion hysteresis*, it has been numerously discussed in theoretical work about DFM (see also Section 4.3 or [40,41] for reviews), but experimental proof has been difficult so far [42–44].

3.1.4 Charge Imaging

Like all forces acting between tip and sample, electrostatic forces cause a shift of the resonance frequency if the oscillating tip approaches the surface.

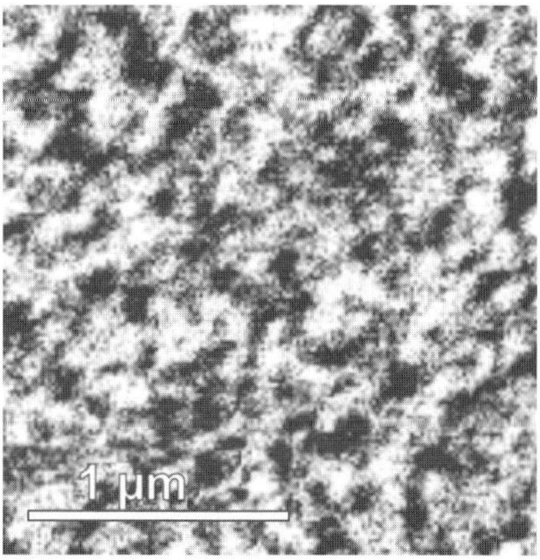

FIGURE 8. Large-scale DFM image acquired on an atomically flat terrace of n-InAs(110) [46]. The image has been taken directly after cleavage under ultrahigh vacuum conditions. No bias voltage has been applied during data acquisition. The observed corrugation is on the order of 0.1–0.2 nm and can be attributed to the presence of near-surface doping atoms, which lead to a locally varying electrostatic tip-sample interaction.

As a consequence, they are readily detectable by DFM. In most cases, electrostatic forces exist even if both tip and sample are grounded or on the exact same potential. This is because tip and sample can be regarded as two electrodes of a capacitor. If they are electrically connected via their back sides and have different work functions, electrons will flow between tip and sample until their Fermi levels are equalized. As a result, an electric field, and consequently, an attractive electrostatic force, exists between them at zero bias. This *contact potential difference* can be balanced by applying an appropriate bias voltage [45]. But even then, local variations in the surface potential can cause a local variation of the resonance frequency that has a visible effect on DFM images. Again, we will explore this aspect with the example of InAs.

As mentioned earlier, the InAs sample shown in Figs. 5 to 7 has been doped with sulfur in a concentration of approximately $3 \times 10^{18} \mathrm{cm}^{-3}$. Assuming an equal distribution of sulfur everywhere in the crystal, this is equivalent to a separation of 7 nm between individual sulfur atoms. In reality, however, sulfur atoms are statistically distributed, leading to a variation of the dopant density on the nanometer scale. Since sulfur acts as electron donor, this causes a modulation of the electron density on the same scale, which modifies the surface potential and thus shows up in DFM images. An example for this effect is given in Figure 8 [46]. The 2×2 - µm image shows an atomically flat terrace on n-InAs, which appears to have a corrugation of around 0.1–0.2 nm. We can interpret this "apparent corrugation" according to the above discussion as reflecting the *modulation of the electron density* (charge imaging). In the present case, this is equivalent with the distribution of the doping atoms. The "brightness" of the contrast is influenced by how close the doping atoms are to the surface. The cloudy structure arises from the fact that the screening length, i.e., the distance on which an individual doping atom significantly disturbs the charge density, is 8 nm in sulfur-doped n-InAs. This distance is larger than the mean distance between individual sulfur atoms of 7 nm. Therefore, the charge

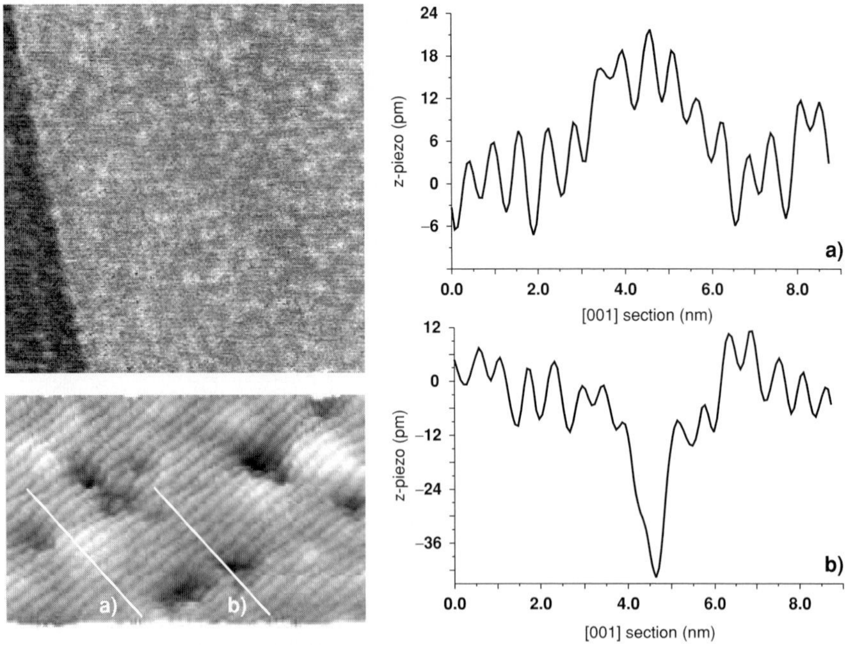

FIGURE 9. **Upper-left**: Large-scale DFM image (120 × 120 nm) of p-InAs(110) with a monoatomic step on the left hand side. The most prominent features are the approximately 150 bright spots, which can be identified as being caused by the Coulomb potentials of individual near surface doping atoms. **Lower-left:** Atomically resolved DFM image of the same sample displayed in a quasi-3D-representation. Image size is 16 × 10 nm. Bright areas with an intact atomic structure (see the cross section along (a)) correspond again to ionized (negatively charged) near surface doping atoms (acceptors). Dark areas, in contrast, have at least one missing protrusion in their center (cf. (b)) and can be attributed to positively charged As vacancies.

clouds around the doping atoms overlap, which leads to the observed "cloudy" contrast.

Even though the charge distribution in Figure 8 is on the nanometer scale, the high resolution of DFM operated in ultrahigh vacuum allows charge imaging down to the atomic scale, if an appropriate sample is investigated. On p-InAs, screening lengths are shorter than on n-InAs (2 nm), so that charge clouds around doping atoms should not overlap on p-InAs if the sample crystal exhibits a similar doping concentration than the crystal imaged in Figure 8. This can be seen in Figure 9. The picture in the upper-left corner represents an 120 × 120-nm area of the sample. The ∼150 different white spots belong to individual zinc doping atoms (concentration: $\approx 4\text{–}6 \times 10^{18}\,\text{cm}^{-3}$) located at different depths below the surface (up to a maximum distance of about 2 nm, which is determined by the screening length).

The lower-left image in Figure 9 represents a zoom to the atomic scale. The regular atomic corrugation is modulated by bright and dark areas. Bright areas

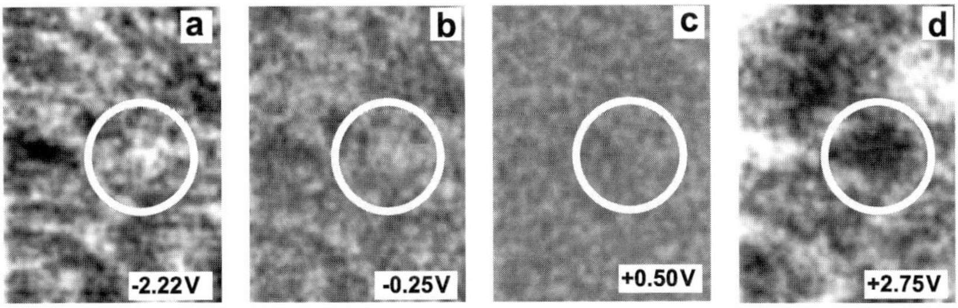

FIGURE 10. Series of images demonstrating the bias dependence of the frequency shift at identical sample locations, which can be interpreted by the presence of an accumulation, depletion, or inversion zone underneath the probe tip. Bias voltages are applied with respect to the grounded cantilever. A detailed description of the experiment can be found in [46].

feature an intact atomic structure and correspond to ionized (negatively charged) near surface doping atoms (acceptors). As they attract holes, which represent the majority carriers on p-InAs, the surface appears elevated (see the cross section along line a). Dark areas have at least one missing protrusion in their center (see Figure (b)), therefore representing As vacancies. Since As vacancies on p-InAs are known to be positively charged (in contrast to As vacancies on n-InAs as the ones shown in Figure 5, which are neutral), they repel holes and thus appear dark. A more detailed discussion of the contrast formation can be found in [46].

An apparent problem in charge imaging is that effects due to topography and charge are difficult to distinguish in the images presented above. In contrast to all other forces acting between tip and sample, however, the electrostatic interaction is easily modified by applying appropriate bias voltages. This has significant influence on the electrostatic surface potential, but not on the topography. Therefore, information acquired at different bias voltages can be used to separate electrostatic interactions from other interactions.

An example illustrating the bias dependence is shown in Figure 10. The sample is the same n-doped InAs crystal that has already been presented in Figs. 5 to 8. The contrast variation within the encircled area is used to discuss how an externally applied bias voltage U_{bias} influences the image contrast. This bias dependence of the contrast can be interpreted by the presence of an accumulation, depletion, or inversion zone underneath the probe tip. (a) At $U_{\text{bias}} = -2.22$ V, electrons (majority carriers) are accumulated at the surface. Bright areas correspond to a high density of doping atoms. (b) The contrast becomes weaker at smaller absolute values of U_{bias}. (c) At $U_{\text{bias}} = +0.50$ V, the contrast vanishes. The surface is depopulated of mobile charge carriers. (d) At $U_{\text{bias}} = +2.75$ V, holes (minority carriers) are generated near the surface. Inversion occurs and the contrast inverts. By subtracting images acquired with vanishing electrostatic contrast (compensated contact potential) from images acquired with substantial applied bias voltage, charge accumulations or depletions can be visualized, as it has been shown by Sugawara et al. [47] and Morita et al. [48] for Si-doped n-GaAs. A useful further discussion of the bias dependence of DFM images is given by Arai and Tomitori in [49].

Finally, we would like to close this section with mentioning that a very convenient way of visualizing directly charge distributions at surfaces using DFM techniques has been developed with the so-called *Kelvin probe microscopy*, which involves the oscillation of the applied bias voltage. Using lock-in methods, the changing contribution caused by surface charges can be separated from the constant topography part. A detailed discussion of Kelvin probe microscopy is provided in contribution by Glatzel.

3.2 Insulators

While the general mechanism of contrast formation in FM-DFM is very similar for all materials, there are differences due to the chemical nature of the specimen investigated. Most insulators do not feature dangling bonds in the way that semiconductors do, and the details of atomic-scale contrast mechanisms vary from material to material. Ionic crystals, e.g., are bonded by Coulomb interactions. Therefore, direct Coulomb interaction replaces orbital hybridization as the main interaction responsible for atomic scale contrast on ionic crystals. Other sources of interaction that have to be considered are image forces (the charge of ionized surface atoms induces a polarization in the conducting tip), capacitance forces, and forces due to tip and surface charging. While tip charging is usually not a major problem for conducting tips, charging of the sample surface might be significant on highly insulating ionic crystals due to charge separation during the cleavage process.

As an example, DFM images acquired on NiO(110) are presented. In Figure 11(a), atomic resolution at a step edge is shown [50]. The step height is 0.22 nm, which is equivalent to a monoatomic step or, with other words, half of a unit cell. Atomic resolution is visible both on the upper and on the lower terrace. In addition, a defect that comprises at least three atoms is located in the upper right corner. Figure 11(b) displays a high resolution image of the atomic structure

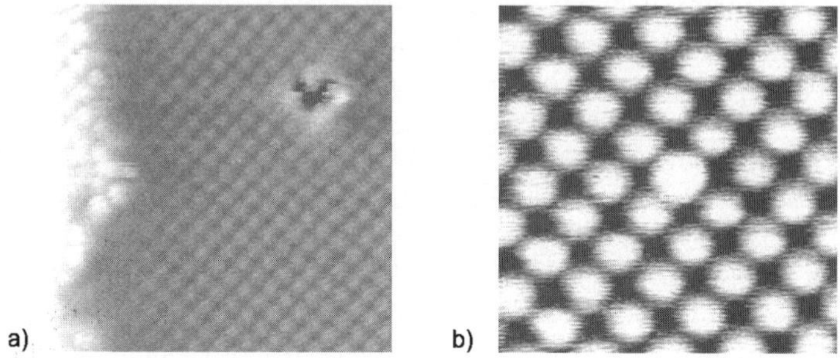

FIGURE 11. (a) Atomically resolved step edge of NiO(001). In the upper-right corner, a defect comprising at least three missing atoms is visible. (b) High-resolution image of NiO(001). In the center of the image, an impurity atom that is 10 pm higher that the surrounding atoms is visible.

on NiO. The atom right in the center is about 10 pm higher that all surrounding atoms, i.e., it reflects an impurity atom of a different chemical species. Its exact nature, however, cannot be further specified. Moreover, we note that in both images Figures. 11(a) and 11(b), only one atom per unit cell is visible. Without knowledge of the exact tip structure, it is not possible to determine which one of the ions—nickel or oxygen—is imaged [51]. The unambiguous interpretation of atomically resolved DFM images on insulators has so far only been achieved in very few cases, such as, e.g., for some experiments performed on the $CaF_2(111)$ surface [52].

3.3 Van der Waals Materials

On most materials, atomic resolution is enabled by strong short-range interactions such as Coulomb forces or chemical forces arising from orbital hybridization. As a last example for contrast mechanisms in DFM, we will explore in this section what resolution can be achieved on weakly interacting surfaces such as van der Waals materials. Due to the weak interaction strength, atomic resolution has not yet been observed on van der Waals materials at room temperature due to the insufficient signal-to-noise ratio of FM-DFM at these temperatures, which is most prominently caused by thermal drift and thermally induced cantilever oscillations. If the microscope is cooled to low temperatures, however, these noise sources can be suppressed, so that the sensitivity becomes sufficient to image the atomic structure.

A first example is given in Figure 12(a), which shows an atomic-scale image of graphite. In graphite, a layered material, the carbon atoms are covalently bonded and arranged in a honeycomb structure within the (0001) plane. In the high-resolution DFM images presented in Figure 12(a), however, a large maximum and two different minima have been resolved instead of the expected hexagonally arranged protrusions [53]. A simulation using the Lennard–Jones potential, in which the attractive part is given by the short-range interatomic van der Waals force,

FIGURE 12. (a) Atomic-scale image of graphite(0001). Note that the protrusions reflect the locations of the minima of the surface potential (hollow sites), while the atoms are situated in the apparent depressions (see text for details). (b) Atomically resolved image of xenon. Here, protrusions actually mark the locations of the individual xenon atoms. Both images are 5.3 nm x 5.3 nm in size and have been acquired at a temperature of $T = 22K$.

explained this finding: the large *maximum* could be assigned to the hollow site, while the *minima represented the locations of the atoms* [54,55]. The two different types of minima reflect whether the carbon atom has a neighbor in the layer below or not. In other words, the contrast on graphite is inverted: atoms manifest as depressions, while the highest protrusion indicates the location of the potential minima. This surprising finding arises from the fact that the nearest-neighbor distance in graphite is only 142 pm and thus very small. As a result, the attractive force on the foremost tip atom is maximal if it is positioned directly in the center of the hexagon, where all six carbon atoms forming the hexagon attract the probing apex atom simultaneously. On the other hand, the attraction is minimal if the tip is positioned directly above a surface atom. This mechanism should be again distance-dependent: if the tip is approached further and further, forces above the locations of the atoms should get larger than the forces at the hollow sites, and hexagonally arranged protrusions that actually reflect the atomic positions should become visible. This assumption agrees with the results of Hembacher et al. [56], who observed the complete hexagonal rings of carbon atoms as maxima at close distances.

While experiments on graphite basically take advantage of the increased stability and signal-to-noise ratio at low temperatures, solid xenon, a pure van der Waals crystal, can only be observed at sufficiently low temperatures due to its melting temperature of 161 K [57]. Since xenon is additionally a good insulator, DFM applied in ultrahigh vacuum is the only real space method available today that allows the study of solid xenon on the atomic scale. Figure 12(b) presents an atomically resolved image of a well-ordered xenon film adsorbed on graphite(0001). The sixfold symmetry and the distance between the protrusions corresponds well with the nearest neighbor distance in the closed packed (111) plane of bulk xenon, which crystallizes in the face centered cubic structure. A comparison between experiment and simulation confirmed that the protrusions correspond to the position of the xenon atoms [58].

4 Dynamic Force Spectroscopy

Ever since it has been invented in 1986, atomic force microscope has been widely used to study tip-sample interactions for various material combinations. If performed as a function of the tip-sample separation, such measurements are referred to as *force spectroscopy*. Unfortunately, force spectroscopy is often strongly complicated by an instantaneous jump of the tip to the sample surface. The origin of this "*jump to contact*" is an instability in the effective tip-sample potential at the position where the actual force gradient of the tip-sample interaction is larger than the spring constant of the cantilever [59].

This problem is avoided when operating the AFM in FM mode. If the oscillation amplitude is large enough, the "jump to contact" of the tip is prevented by the restoring force of the cantilever [23]. In section 2.3, we showed how the frequency shift can be calculated for a given tip-sample interaction law. In general, however, the inverse problem will be of more interest: *How can the tip-sample interaction*

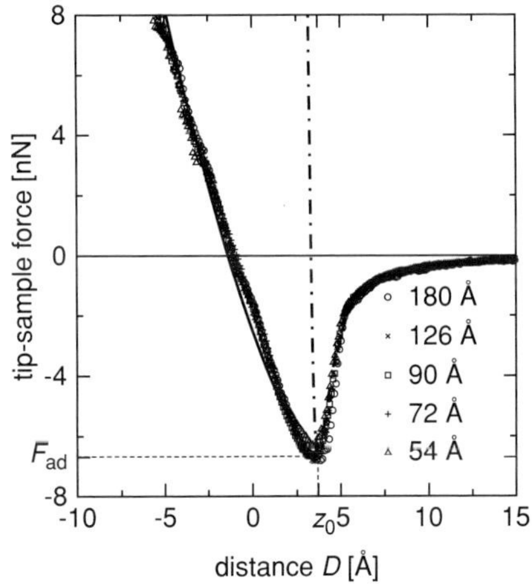

FIGURE 13. Example for the application of dynamic force spectroscopy. The gray symbols reflect the tip-sample force as calculated from the experimental data shown in Figure 3 using Eq. (18). The force F_{ts} Eq. (19) is plotted by a dashed-dotted line; the solid line represents the best fit based on the contact force F_c Eq. (20). The border between "contact" and "noncontact" force is marked by the position z_0; F_{ad} indicates the adhesion force (cf. Sect. 4.2).

be determined from frequency shift data? The answer to this question is given in this section and will lead us to the *dynamic force spectroscopy* (DFS) technique, which is a direct extension of FM-DFM.

4.1 Determining Forces from Frequencies

For typical DFM setups, it is straightforward to measure the frequency shift with a fixed oscillation amplitude A as a function of the actual tip-sample distance D (i.e., the distance of closest approach of the tip during an individual oscillation cycle). An example of such behavior is illustrated in Figure 13 (see section 2.3). The calculation of the tip-sample interactions from such data sets is possible by reversing the integral in Eq. (14). As shown by Dürig [24], this results in

$$F_{ts}(D) = \sqrt{2}\frac{c_z A^{3/2}}{f_0}\frac{\partial}{\partial D}\int_D^\infty \frac{\Delta f(z)}{\sqrt{z-D}}\,dz, \tag{18}$$

allowing a direct calculation of the tip-sample interaction force from frequency shift versus distance curves.

Application of this formula to the experimental frequency shift versus distance curves presented in section 2.2 is shown in Figure 13. The obtained force curves are almost identical although obtained with different oscillation amplitudes. The fact that tip-sample potentials and forces can be determined with this accuracy without "jump-to-contact" demonstrates the advantage of dynamic force spectroscopy compared to force-distance curves acquired without oscillating the cantilever, as DFS can probe the complete attractive and repulsive regime of the force curves without discontinuity.

Since Eq. (14) was derived with the condition that the oscillation amplitude is considerably larger than the decay length of the tip-sample interaction, the same restriction applies to Eq. (18). However, using more advanced algorithms it is also possible to determine forces from dynamic force spectroscopy experiments without restriction to large amplitudes. The numerical approach of Gotsmann et al. [18,60], e.g., works in every regime, as do the semi-analytical methods introduced by Dürig [61], Giessibl [62], and Sader and Jarvis [63].

Not only is the z-resolution of DFS remarkable—if highly stable DFM setups such as microscopes operated at low temperatures are used, the lateral resolution of dynamic force spectroscopy can be driven down to the atomic scale. Lantz et al. [64] measured frequency shift versus distance curves at different lattice sites of the Si(111)-(7 × 7) surface (see Figure 14). In this way, they were able to distinguish differences in the bonding forces between inequivalent adatoms of the 7 × 7 surface reconstruction of silicon.

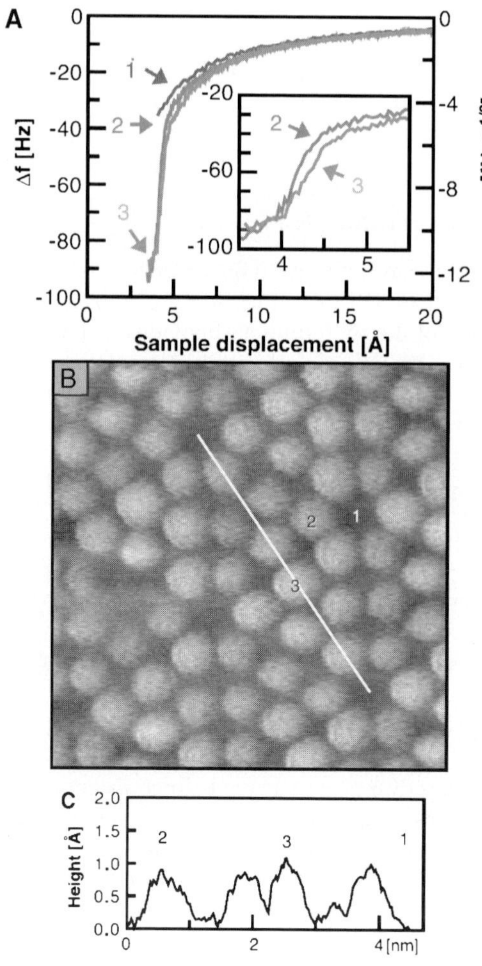

FIGURE 14. Dynamic force spectroscopy performed with atomic resolution on Si(111)-(7 × 7). The frequency shift versus distance curves plotted in (A) were measured at the positions marked in the topographical image shown in (B). The atoms labeled "2" and "3" represent inequivalent adatoms (cf. also the cross section displayed in C); DFS is able to distinguish differences in their bonding strength (see insert in A). Experiments have been performed at a temperature of $T = 7.2$ K. (Reprinted with permission from [64]. Copyright 2001, AAAS.) (See also Plate 3 in the Color Plate Section.)

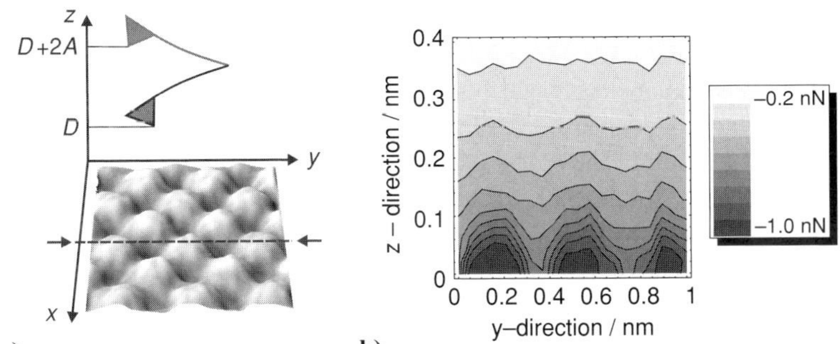

a) b)

FIGURE 15. (a) Principle of 3D force spectroscopy with the example of NiO(110) (cf. Sect. 3.2). The cantilever oscillates near the sample surface and measures the frequency shift at all locations defined by a dense raster grid inside an xyz-box. The surface topography of the NiO sample, as obtained immediately before recording the spectroscopy field, is displayed in a perspective representation (image size: 10 Å × 10 Å). (b) The reconstructed force field of NiO(001) taken along the line indicated in a). The positions of the atoms are clearly visible.

This concept can be extended to achieve true three-dimensional force spectroscopy by mapping the complete force field above the sample surface [16]. Figure 15(a) illustrates the measurement principle. Frequency shift vs. distance curves are recorded on a matrix of points perpendicular to the sample surface. Using Eq. (18), the complete three-dimensional force field between tip and sample can be recovered with atomic resolution. Figure 15(b) shows a cut through the force field as a two-dimensional map, from which the decay of the force field into the vacuum can be recovered.

Recently, it became possible to use dynamic force microscopy for nanomanipulation. Atomic-scale manipulation has been demonstrated by Oyabu et al. [44], who removed individual atoms from a Si(111)-7 × 7 surface with the AFM tip. Subsequently, they were able to re-deposit the same atoms from the tip back onto the surface. Loppacher et al. [65] achieved pushing on different parts of an isolated Cu-TBBP molecule, which is known to possess four rotatable legs. They measured force-distance curves while one of the legs was pushed by the AFM tip and turned by 90°. By this means, they were able to quantify the energy that was dissipated during the "switching" of this molecule between different conformational states. The possibility simultaneously to exert and measure forces during single-atom or -molecule manipulation represents an exciting future application for high-resolution DFM experiments.

4.2 Analysis of Tip-Sample Interaction Forces

Since the above-presented DFS methods can be used to measure tip-sample interactions with high resolution, this approach opens a direct way to compare experiments with theoretical models and predictions. As an example, we will analyze the force curves presented in Figure 13.

Following a suggestion by Giessibl [23], we describe the force between tip and sample by a combination of a long-range (van der Waals) and a short-range (Lennard–Jones) term. The long-range part reflects the van der Waals interaction of the tip with the sample, modeled as a sphere with radius R above an atomically flat, infinitely extended plane. The short-range Lennard–Jones part represents the interaction of the last tip apex atom with the surface. With these assumptions, we obtain

$$F_{\mathrm{ts}}(z) = \underbrace{-\frac{A_{\mathrm{H}}R}{6z^2}}_{\text{long-range}} + \underbrace{\frac{12E_0}{r_0}\left(\left(\frac{r_0}{z}\right)^{13} - \left(\frac{r_0}{z}\right)^{7}\right)}_{\text{short-range}}. \tag{19}$$

A_{H} is the Hamaker constant, and E_0 and r_0 are the binding energy and the equilibrium distance of the Lennard–Jones force, respectively. Note that this approach does not consider elastic contact forces between tip and sample. Therefore, we will call F_{ts} the "noncontact force" in the following.

A fit of this equation to the experimental tip-sample force data is indicated in Figure 13 by a dashed-dotted line; the parameters used are $A_{\mathrm{H}}R = 2.4 \times 10^{-27}$ Jm, $r_0 = 3.4$ Å, and $E_0 = 3$ eV [15]. The regime right from the minimum fits well to the experimental results, but the deep and wide minimum of the experimental curves cannot be described accurately with the noncontact force. This is caused by the steep increase of the Lennard-Jones force in the repulsive regime ($F_{\mathrm{ts}} \propto 1/r^{13}$ for $z < r_0$), which does not consider elastic deformation.

Left of the minimum, however, the tip-sample force is mainly determined by elastic contact forces, as the tip starts to tap on the surface. This elastic contact behavior can be described with the assumption that the overall shape of tip and sample changes only slightly until point contact is reached and that, after the formation of this point contact, the tip-sample forces are given by the well-known Hertz theory (see, e.g., [66,67]). The Hertz theory considers only the repulsive elastic force between tip and sample, but no additional deformation due to attractive forces (see, e.g., [68] for a discussion about this topic). The easiest way to consider attractive forces is by a simple offset (Hertz-plus-offset model). This approach coincides with Maugis' approximation [69] to a model introduced earlier by Derjaguin, Muller, and Toporov (DMT model) [70]. It results in a force law of the type

$$F_{\mathrm{c}} = G_0(z_0 - z)^{3/2} + F_{\mathrm{ad}} \quad \text{for} \quad z \leq z_0. \tag{20}$$

The first term in this equation describes the elastic behavior of a Hertzian contact, where z_0 is the point of contact and G_0 a constant that depends on the elasticity of tip and sample and on the shape of the tip [67]. The offset F_{ad} is the adhesion force between tip and sample surface. Since the experimentally measured tip-sample force shows a reasonable agreement with the noncontact force F_{ts} until it reaches its minimum, we define the contact point by this minimum, i.e., $z_0 = \min\{F_{\mathrm{ts}}(z)\} = 3.7$Å, and therefore $F_{\mathrm{ad}} = F_{\mathrm{ts}}(z_0) = -6.7$nN. With this choice, the connection between noncontact and contact force and their corresponding force gradients is continuous.

A fit of Eq. (20) to the experimental data is given in Figure 13 by a solid line ($G_0 = 5.8 \times 10^5$ nN/m$^{3/2}$). The experimental force curves agree well with the contact force law for distances $D < z_0$. Therefore, we can summarize that the overall behavior of the experimentally obtained force curves can be adequately explained by a combination of long-range (van der Waals), short-range (Lennard–Jones), and contact (Hertz/DMT) forces. Furthermore, this result demonstrates that not only noncontact, but also elastic contact forces can be quantitatively measured by dynamic force spectroscopy.

4.3 Measurement of Energy Dissipation

Besides the measurement of tip-sample forces, another physical mechanism exists that can be analyzed with dynamic mode microscopy. As explained in section 2.3, the gain factor g (and thus the excitation amplitude a_{exc}, which is directly proportional to g) represents a measure for the energy dissipation due to non-conservative tip-sample interactions. It can be shown that in self-excitation mode, conservative and dissipative interactions can be strictly separated if the time delay t_0 has been adjusted to match an excitation-oscillation phase difference of 90° [19,22]. Energy dissipation is partially caused by the cantilever itself (intrinsic damping). More interestingly, however, energy is "lost" due to the tip-sample interaction.

Conservative forces acting at the tip-sample junction can, at least in vacuum, be understood in terms combinations of van der Waals, electrostatic and chemical interactions. In contrast, the dissipative processes taking place when the tip comes close to the sample surface are comparatively poorly understood. Stowe et al. [72] modified the experimental setup of an dynamic force microscope and oscillated the cantilever parallel to the sample surface. They observed that if a voltage potential is applied between tip and sample, charges are induced in the sample surface, which will follow the tip motion. Due to the finite resistance of the sample material, energy will be dissipated during the charge movement. This effect has been used to image the doping level of semiconductors. But even in absence of external electromagnetic fields, energy dissipation was observed if tip and sample approached each other within less than about one nanometer. Clearly, mechanical surface relaxations must give rise to energy losses. One could think of the AFM tip as a small hammer, hitting the surface and inducing phonon excitations. From a continuum mechanics point of view, we assume that the mechanical relaxation of the surface is not only governed by elastic responses. Viscoelastic effects of soft surfaces are also likely to render a significant contribution to energy dissipation.

In the atomistic view, the last tip atom can be envisaged changing position while yielding to the tip-sample force field. A fully reversible change of position would not result in a loss of energy. Still, it has been pointed out by Sasaki and Tsukada [73] that a change in atom position would result in a change in the force interaction itself. Therefore, it is possible that the tip atom changes position at different tip-surface distances during approach and retraction, effectively causing an

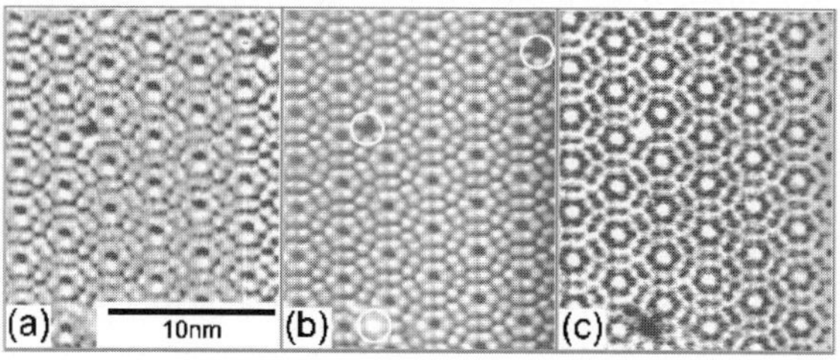

FIGURE 16. FM-DFM image of the Si(111)-(7 × 7) surface. The atomic structure can be seen simultaneously in the topography (a), the tunneling current (b), and the dissipation channel (c). Defects are marked by circles and visible in all three channels. (Reprinted with permission from [71]. Copyright 2000, Elsevier.)

atomic-scale hysteresis to develop. This model has been extended by Kantorovich and Trevethan [74], who included thermal activation processes.

The dissipation channel (i.e., the recorded values of either the gain factor g or the excitation amplitude a_{exc}, respectively, as a function of the lateral position) can be also used to image surfaces with atomic resolution as shown by Lüthi et al. [75] and Guggisberg et al. [45]. Instead of feedbacking the distance on the frequency shift, the excitation amplitude in FM mode can even be used as the control signal. The DFM image of the Si(111)-(7 × 7) reconstruction shown in Figure 16 was successfully recorded this mode. The step edges of monoatomic NaCl islands on single crystalline copper have also rendered atomic resolution contrast in the dissipation channel [76].

Acknowledgments. The authors would like to thank Wolf Allers, Alexander Schwarz, and Shenja Langkat for their experimental results presented throughout this chapter. This contribution was prepared during the stay of U. D. Schwarz at the Center for Nanotechnology (CeNTech) of the University of Münster. H. Hölscher acknowledges financial support from the German Federal Ministry of Education and Research (BMBF) (grant No. 03N8704) and the Deutsche Forschungsgemeinschaft (DFG) (grant No. HO 2237/2-1), U. D. Schwarz from the National Science Foundation (grant No. DMR-0414944) and the Petroleum Research Fund (grant No. 42259-AC5).

References

1. G. Binnig, C. F. Quate, and C. Gerber, Phys. Rev. Lett. **56**, 930 (1986).
2. F.-J. Giessibl, Science **267**, 68 (1995).
3. Y. Sugawara, M. Otha, H. Ueyama, and S. Morita, Science **270**, 1646 (1995).
4. T. Albrecht, P. Grütter, D. Horne, and D. Rugar, J. Appl. Phys. **69**, 668 (1991).
5. R. Garcia and R. Pérez, Surf. Sci. Rep. **47**, 197 (2002).
6. *Noncontact Atomic Force Microscopy*, edited by S. Morita, R. Wiesendanger, and E. Meyer (Springer-Verlag, Heidelberg, 2002).

7. F.-J. Giessibl, Rev. Mod. Phys. **75**, 949 (2003).
8. H. Hölscher and A. Schirmeisen, in *Advances in Imaging and Electron Physics*, edited by P. W. Hawkes (Academic Press Ltd., London, 2005).
9. Q. D. Zhong, D. Inniss, K. Kjoller, and V. B. Elings, Surf. Sci. Lett. **290**, L688 (1993).
10. P. K. Hansma, J. P. Cleveland, M. Radmacher, D. A. Walters, P. E. Hillner, M. Bezanilla, M. Fritz, D. Vie, H. G. Hansma, C. B. Prater, J. Massie, L. Fukunaga, L. Gurley, and V. B. Elings, Appl. Phys. Lett. **64**, 1738 (1994).
11. C. A. J. Putman, K. O. Vanderwerf, B. G. Degrooth, N. F. Vanhulst, and J. Greve, Appl. Phys. Lett. **64**, 2454 (1994).
12. H. Ueyama, Y. Sugawara, and S. Morita, Appl. Phys. A **66**, S295 (1998).
13. C. Loppacher, M. Bammerlin, F. Battiston, M. Guggisberg, D. Müller, R. L. H. R. Hidber, E. Meyer, and H.-J. Güntherodt, Appl. Phys. A. **66**, S215 (1998).
14. W. Allers, A. Schwarz, U. D. Schwarz, and R. Wiesendanger, Rev. Sci. Instrum. **69**, 221 (1998).
15. H. Hölscher, A. Schwarz, W. Allers, U. D. Schwarz, and R. Wiesendanger, Phys. Rev. B **61**, 12678 (2000).
16. H. Hölscher, B. Gotsmann, W. Allers, U. D. Schwarz, H. Fuchs, and R. Wiesendanger, Phys. Rev. Lett. **88**, 019601 (2002).
17. B. Gotsmann, D. Krüger, and H. Fuchs, Europhys. Lett. **39**, 153 (1997); ibid **41**, 581 (1998).
18. B. Gotsmann, B. Ancykowski, C. Seidel, and H. Fuchs, Appl. Surf. Sci. **140**, 314 (1999).
19. H. Hölscher, B. Gotsmann, W. Allers, U. D. Schwarz, H. Fuchs, and R. Wiesendanger, Phys. Rev. B **64**, 75402 (2001).
20. B. Gotsmann and H. Fuchs, Appl. Surf. Sci. **188**, 355 (2002).
21. H. Hölscher, B. Gotsmann, and A. Schirmeisen, Phys. Rev. B **68**, 153401 (2003).
22. U. Dürig, N. J. of Phys. **2**, 5.1 (2000).
23. F.-J. Giessibl, Phys. Rev. B **56**, 16010 (1997).
24. U. Dürig, Appl. Phys. Lett. **75**, 433 (1999).
25. H. Hölscher, U. D. Schwarz, and R. Wiesendanger, Appl. Surf. Sci. **140**, 344 (1999).
26. A. Schwarz, W. Allers, U. D. Schwarz, and R. Wiesendanger, Phys. Rev. B **61**, 2837 (2000).
27. V. Caciuc, H. Hölscher, S. Blügel, and H. Fuchs, Nanotechnology **16**, S59 (2005).
28. M. Ohta, H. Ueyama, Y. Sugawara, and S. Morita, Jpn. J. Appl. Phys. **34**, L1692 (1995).
29. A. Schwarz, W. Allers, U. D. Schwarz, and R. Wiesendanger, Appl. Surf. Sci. **140**, 293 (1999).
30. S. H. Ke, T. Uda, R. Pérez, I. Štich, and K. Terakura, Phys. Rev. B **60**, 11631 (1999).
31. S. H. Ke, T. Uda, I. Štich, and K. Terakura, Phys. Rev. B **63**, 245323 (2001).
32. N. Uehara, H. Hosoi, K. Sueoka, and K. Mukasa, Nanotechnology **15**, S97 (2004).
33. N. Uehara, H. Hosoi, K. Sueoka, and K. Mukasa, Jap. J. Appl. Phys. **43**, 4676 (2004).
34. T. Minobe, T. Uchihashi, T. Tsukamoto, S. Orisaka, Y. Sugawara, and S. Morita, Appl. Surf. Sci. **140**, 298 (1999).
35. S. Morita and Y. Sugawara, Ultramicroscopy **91**, 89 (2002).
36. R. Pérez, I. Štich, M. C. Payne, and K. Terakura, Phys. Rev. B **58**, 10835 (1998).
37. K. Yokoyama, T. Ochi, Y. Sugawara, and S. Morita, Phys. Rev. Lett. **83**, 5023 (1999).
38. S. Morita and Y. Sugawara, in *Noncontact Atomic Force Microscopy*, edited by S. Morita, R. Wiesendanger, and E. Meyer (Springer-Verlag, Heidelberg, Germany, 2002), pp. 47–77.
39. G. Schwarz, A. Kley, J. Neugebauer, and M. Scheffler, Phys. Rev. B **58**, 1392 (1998).

40. M. Gauthier, L. Kantorovich, and M. Tsukada, in *Noncontact Atomic Force Microscopy*, edited by S. Morita, R. Wiesendanger, and E. Meyer (Springer-Verlag, Heidelberg, Germany, 2002), pp. 371–394.
41. H.-J. Hug and A. Baratoff, in *Noncontact Atomic Force Microscopy*, edited by S. Morita, R. Wiesendanger, and E. Meyer (Springer-Verlag, Heidelberg, Germany, 2002), pp. 395–431.
42. T. Uchihashi, Y. Sugawara, T. Tsukamoto, M. Otha, S. Morita, and M. Suzuki, Phys. Rev. B **56**, 9834 (1997).
43. S. Morita, Y. Sugawara, K. Yokoyama, and T. Uchihashi, Nanotechnology **11**, 120 (2000).
44. N. Oyabu, O. Custance, I. Yi, Y. Sugawara, and S. Morita, Phys. Rev. Lett. **90**, 176102 (2003).
45. M. Guggisberg, M. Bammerlin, C. Loppacher, O. Pfeiffer, A. Abdurixit, V. Barwich, R. Bennewitz, A. Baratoff, E. Meyer, and H.-J. Güntherodt, Phys. Rev. B **61**, 11151 (2000).
46. A. Schwarz, W. Allers, U. D. Schwarz, and R. Wiesendanger, Phys. Rev. B **62**, 13617 (2000).
47. Y. Sugawara, T. Uchihashi, M. Abe, and S. Morita, Appl. Surf. Sci. **140**, 371 (1999).
48. S. Morita, M. Abe, K. Yokoyama, and Y. Sugawara, J. Cryst. Growth **210**, 408 (2000).
49. T. Arai and M. Tomitori, in *Noncontact Atomic Force Microscopy*, edited by S. Morita, R. Wiesendanger, and E. Meyer (Springer-Verlag, Heidelberg, Germany, 2002), pp. 79–92.
50. W. Allers, S. Langkat, and R. Wiesendanger, Appl. Phys. A **72**, S27 (2001).
51. A. S. Foster and A. L. Shluger, Surf. Sci. **490**, 211 (2001).
52. A. S. Foster, C. Barth, A. L. Shluger, and M. Reichling, Phys. Rev. Lett. **86**, 2373 (2001).
53. W. Allers, A. Schwarz, U. D. Schwarz, and R. Wiesendanger, Appl. Surf. Sci. **140**, 247 (1999).
54. H. Hölscher, W. Allers, U. D. Schwarz, A. Schwarz, and R. Wiesendanger, Phys. Rev. B. **62**, 6967 (2000).
55. M. Ashino, A. Schwarz, H. Hölscher, U. D. Schwarz, and R. Wiesendanger, Nanotechnology **16**, S134 (2005).
56. S. Hembacher, F. J. Giessibl, J. Mannhart, and C. F. Quate, Proc. Natl. Acad. Sci. **100**, 12539 (2003).
57. W. Allers, A. Schwarz, U. D. Schwarz, and R. Wiesendanger, Europhys. Lett. **49**, 276 (1999).
58. Hölscher, W. Allers, U. D. Schwarz, A. Schwarz, and R. Wiesendanger, Appl. Phys. A **72**, S35 (2001).
59. N. A. Burnham and R. J. Colton, J. Vac. Sci. Technol. A.
60. B. Gotsmann and H. Fuchs, Phys. Rev. Lett. **86**, 2597 (2001).
61. U. Dürig, Appl. Phys. Lett. **76**, 1203 (2000).
62. F.-J. Giessibl, Appl. Phys. Lett. **78**, 123 (2001).
63. J. E. Sader and S. P. Jarvis, Appl. Phys. Lett. **84**, 1801 (2004).
64. M. A. Lantz, H. Hug, R. Hoffmann, P. J. A. van Schendel, P. Kappenberger, S. Martin, A. Baratoff, and H.-J. Güntherodt, Science **291**, 2580 (2001).
65. C. Loppacher, M. Guggisberg, O. Pfeiffer, E. Meyer, M. Bammerlin, R. Lüthi, R. Schlittler, J. K. Gimzewski, H. Tang, and C. Joachim, Phys. Rev. Lett. **90**, 066107 (2003).

66. K. L. Johnson, *Contact Mechanics* (Cambridge University Press, Cambridge, UK, 1985).
67. L. D. Landau and E. M. Lifschitz, *Lehrbuch der theoretischen Physik VII: Elastizitätstheorie* (Akademie-Verlag, Berlin, 1991).
68. U. D. Schwarz, J. Colloid Interface Sci. **261**, 99 (2003).
69. D. Maugis, J. Colloid Interface Sci. **150**, 243 (1992).
70. B. V. Derjaguin, V. M. Muller, and Y. P. Toporov, J. Colloid Interface Sci. **53**, 314 (1975).
71. M. Guggisberg, M. Bammerlin, A. Baratoff, R. Lüthi, C. Loppacher, F. Battiston, J. Lü, R. Bennewitz, E. Meyer, and H.-J. Güntherodt, Surf. Sci. **461**, 255 (2000).
72. T. Stowe, T. W. Kenny, D. J. Thomson, and D. Rugar, Appl. Phys. Lett. **75**, 2785 (1999).
73. N. Sasaki and M. Tsukada, Jpn. J. Appl. Phys **39**, L1334 (2000).
74. L. N. Kantorovich and T. Trevethan, Phys. Rev. Lett. **93**, 236102 (2004).
75. R. Lüthi, E. Meyer, M. Bammerlin, A. Baratoff, L. Howald, C. Gerber, and H.-J. Güntherodt, Surf. Rev. Lett. **4**, 1025 (1997).
76. R. Bennewitz, A. S. Foster, L. N. Kantorovich, M. Bammerlin, C. Loppacher, S. Schär, M. Guggisberg, E. Meyer, and A. L. Shluger, Phys. Rev. B **62**, 2074 (2000).

II.9
Scanning Tunneling Microscopy and Spectroscopy of Manganites

CHRISTOPH RENNER AND HENRIK M. RØNNOW

The quest to unravel the properties of manganites has fueled several important concepts in contemporary solid-state physics, involving intricately coupled degrees of freedom (lattice, charge, spin and orbital) and electronic inhomogeneity or phase separation between states of different electronic character. With simultaneous spatial and spectral resolution, scanning tunneling microscopy (STM) can provide direct information on each of these states and on their spatial extent. Here we highlight some recent advances and outline the potential for further STM studies of manganites.

1 Introduction

Manganese oxides (manganites) have been the focus of revived interest [1–3] following the rediscovery of a very large (colossal) negative magnetoresistance (CMR), i.e., an orders of magnitude reduction in resistance upon applying a magnetic field, in some manganites in the early 1990s [4–7]. Beyond their technological prospects [8], manganites are challenging the established understanding of the electronic behavior of materials. Along with the high-temperature superconducting copper oxides, they have spurred the development of novel concepts of confinement, phase separation, and electronic inhomogeneity as central topics for transition metal oxides (TMOs) in which strong electron correlation leads to novel electronic properties.

Manganites belong to the Ruddlesden–Popper series [9] with the general formula $T_{\mathrm{n}-\mathrm{n}x}D_{1+\mathrm{n}x}\mathrm{Mn_nO_{3n+1}}$, where T is a trivalent metal or lanthanide cation (e.g., Bi, La, Pr, Nd) and D is a divalent metal or alkaline-earth cation (e.g., Pb, Ca, Sr, Ba). They can be viewed as a stacking of n layers of corner sharing MnO_6 octahedra cut from the perovskite structure ($\mathrm{n} = \infty$) and separated by cation oxide rock-salt layers (Figure 1).

The remarkable properties of the manganites derive from the outer d-shell electrons. With increasing divalent cation doping, x, the Mn valency is shifted from Mn^{3+} toward Mn^{4+}. While both ions hold three electrons in the t_{2g} orbitals, Mn^{3+} has one extra electron in the e_g orbital that is unstable toward a Jahn–Teller

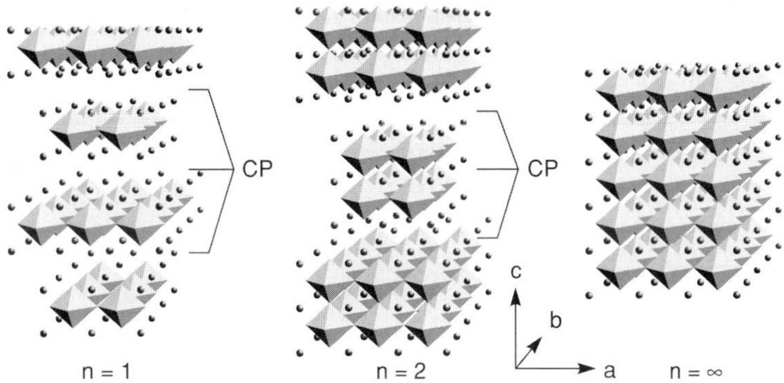

FIGURE 1. Generic structure of the n = 1, n = 2, and n = ∞ members of the Ruddlesden–Popper series $T_{n-nx}D_{1+nx}Mn_nO_{3n+1}$ of manganese oxides (manganites). The oxygen octahedra are each centered on a Mn ion, and the black spheres represent the trivalent lanthanide (T) and divalent alkaline-earth (D) cation lattice sites. The n = ∞ compounds with perovskite structure do not cleave, making them very difficult to study by STM, whereas n = 1 and n = 2 compounds do have cleaving planes (CP) within the rock-salt oxide layers, making them ideal for STM investigations.

distortion. Due to strong Hund's coupling, both ions are in high-spin states, $S = 3/2$ in Mn^{4+} and $S = 2$ in Mn^{3+}. It immediately becomes apparent how systems of mixed-valence between these two types of ions are likely to host a series of intertwined degrees of freedom: lattice, charge, spin, and orbital. The competition between the typical energy scales associated with each of these degrees of freedom leads to a rich phase diagram[1] dominated by different-ordered phases with distinct effects on the electronic transport properties, and several metal-insulator transitions (MITs), both as a function of doping and temperature. CMR essentially results as a proximity effect to these transitions, in the vicinity of which even moderate magnetic fields are sufficient to influence the phase lines and domain structure on microscopic or macroscopic length scales.

The proposal of electronic phase separation (EPS) [10] challenges most experimental techniques which average over macroscopic sample sizes. In contrast, local probes like STM have the potential to search directly for such inhomogeneities and to investigate the intrinsic properties of the different phases involved. As such, the manganites provide an ideal arena for developing scanning probe techniques as an important tool in unraveling the physics of complex correlated electron materials.

This article aims at providing a snapshot of the key issues related to the understanding of the remarkable properties and possible electronic inhomogeneities of manganites, and how they can be addressed by local probe techniques. It is not intended as a comprehensive review of scanning tunneling probe studies of

[1] Several examples of phase diagrams—with references—for manganites with n = 1, n = 2 and n = ∞ can be found in Dagotto et al [10].

manganites and a colossal magnetoresistance. We shall mainly focus on STM, describing its particular capabilities as well as some of the potential pitfalls one should keep in mind when evaluating these very sensitive experiments. Spatial resolution is the main thrust of this article. We start the discussion at the lowest resolution in section 2, with temperature-dependent tunneling spectroscopy of the various gap features appearing around the Fermi energy. In section 3, we introduce scanning tunneling spectroscopy and scanning tunneling potentiometry, two spatially resolved spectroscopic techniques especially suitable for tackling the question of electronic phase separation. Section 4 is devoted to atomic-scale imaging of manganites, the ultimate resolution needed to study the Jahn–Teller–driven lattice distortions, the various phases of ordered electronic degrees of freedom (charge, spin and orbital), and possible electronic (in-)homogeneities in real space. Finally, in section 5 we conclude with a summary of the current experimental situation and a brief outlook on where STM investigations of manganites may lead in the future.

2 Temperature-Dependent Tunneling Spectroscopy

Tunneling spectroscopy is undoubtedly one of the most powerful techniques for probing the electronic density of states (DOS) in the vicinity of the Fermi energy, E_F. While photoemission [11] can only measure the occupied DOS below E_F with limited energy resolution and no direct spatial resolution,[2] local probe tunneling spectroscopy can resolve the occupied ($eV \leq E_F$) *and* unoccupied ($eV > E_F$) DOS[3] with very high spatial resolution ($\sim$0.1 nm) and optimal energy resolution ($\sim k_B T$). We refer the reader to other publications [14, 15] for detailed descriptions of basic STM imaging and spectroscopy applications.

The complex phase diagram of manganites contains a range of electronic phases and phase transitions, many of them separating a metallic from an insulating phase. The most fiercely debated metal-insulator transitions are those associated with charge ordering in Mn^{4+} rich manganites ($x \geq 0.5$, section 2.1), and with CMR in manganites with dilute Mn^{4+} concentrations ($0.2 < x < 0.5$, section 2.2). It may seem ill-advised to study such phase transitions by STM spectroscopy, since in principle the STM cannot operate on insulating samples. However, it turns out that a number of MITs in manganites can be readily probed by STM, in a temperature range where the insulating phase is not too resistive ($\rho < 1\ \Omega$cm) and the gap at E_F in the local DOS is not too large.

---

[2] The unoccupied DOS above E_F can in principle be accessed by inverse photoemission [12], though with much coarser energy resolution.

[3] That STM spectroscopy is measuring the local DOS is justified, e.g., by Tersoff and Hamann [13], though their simple model neglects electronic correlations and therefore does not a priori hold for strongly correlated systems. This question definitely deserves future attention. However, for the purpose of the present discussion, we shall assume that a similar relationship to the one-particle spectral function holds for correlated systems.

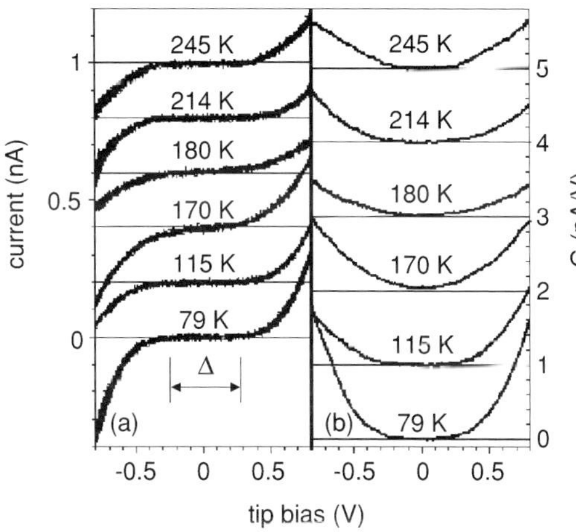

FIGURE 2. Temperature-dependent tunneling spectroscopy of polycrystalline $Nd_{0.5}Sr_{0.5}MnO_3$. (a) $I(V)$ characteristics measured by STM. (b) Numerical derivatives $dI/dV(V)$ of the measured $I(V)$ spectra. Spectra at $T > 79$ K are offset by 0.2 nA and 1 nA/V, respectively. Adapted with permission from [16]. Copyright 1997, Institute of Physics.

The central goal of temperature dependent tunneling experiments is to detect the gap, Δ, around the Fermi energy associated with the insulating phase in the current-voltage, $I(V)$, tunneling characteristics, and/or in the differential tunneling conductance, *dI/dV(V)*. In this article, we define the gap Δ as the energy range around E_F where the tunneling conductance does not depend on bias voltage and is equal to zero. Incidentally, this is the gap probed by optical spectroscopy involving electron-hole excitations across the gap. In the tunneling spectra, E_F corresponds to the bias voltage $V = 0$.

2.1 Charge-Ordering Metal-Insulator Phase Transition

A number of STM spectroscopy studies of manganites are concerned with the charge-ordering MIT in compounds with $x \geq 0.5$. Biswas et al. [16] measured $I(V)$ characteristics on polycrystalline $Nd_{0.5}Sr_{0.5}MnO_3$ (NSMO) as a function of temperature (Figure 2(a)). Below the charge ordering transition temperature, $T_{CO} \simeq 130$ K, the tunneling spectra reveal a gap $\Delta \simeq 0.5$ eV in the local DOS around E_F. As the temperature is increased above 79 K, Biswas et al. observe a gradual reduction of the gap. By the time the temperature exceeds T_{CO}, the $I(V)$ characteristics acquire a finite slope at $V = 0$, which Biswas et al. contend marks the closing of the gap. Because this happens near T_{CO} and since Δ compares well with the nearest-neighbor Coulomb repulsion energy [17], they ascribe it to the charge-ordered state. However, upon further increasing the temperature, a new gap appears to be opening in the 214 K and 245 K spectra. This is below the Curie temperature $T_c = 250$ K, where the sample should be metallic, and the origin of this gap remains to be explained.

Yao et al. [18] measured a very similar temperature dependence of the $I(V)$ characteristics on $Pr_{0.5}Sr_{0.5}MnO_3$ epitaxial thin films, another manganite with a charge-ordering MIT, deposited on $LaAlO_3$ (100) substrates. Similarly to the above results on NSMO, they measured a gap $\Delta \simeq 0.4$ eV, which closes at $T_N \simeq 160$ K and another unexplained gap which opens below $T_c = 260$ K. Yao et al. question if a gap in the tunneling spectrum below T_{CO} is sufficient to infer the existence of a charge-ordered state. Indeed, in the absence of any complementary measurements, such as atomic-scale topographic STM imaging discussed in section 4, the observation of a gap around E_F is not indisputable proof of charge ordering by itself. In an effort to gather independent and more robust evidence of the origin of the gap around E_F at $T < T_{CO}$, Biswas et al. [19] studied the magnetic field dependence of Δ in NSMO. The charge-ordered state was shown to melt in response to an applied magnetic field [20], and they indeed observed the collapse of the gap, giving them confidence that Δ is really associated with charge ordering.

The above results are based on a graphical analysis of the differential tunneling conductance curves obtained by computing the numerical derivative of the measured $I(V)$ spectra (Figure 2(b)). The analysis is simple and straightforward: with increasing temperature, the conductance at zero bias eventually becomes finite, which is interpreted as the closing of a gap in the local DOS around E_F. However, such an interpretation neglects the impact of thermal smearing, possible tip-induced effects at large tunneling bias, and collective effects due to electron correlations. It may be justified in a band-gap insulator, where the single-electron gap remains essentially unaltered by temperature, and thermal smearing only becomes significant when the temperature reaches values comparable to the gap amplitude. This would be the case, for example, in a semiconductor, where states thermally excited from the valence into the conduction band account for the increasing conductance with increasing temperature. In this case, a finite slope in the $I(V)$ curves at $V = 0$ does not imply that states have been transferred into the gap nor that the material has become metallic.

In Mott–Hubbard insulators with strong electron–electron and electron–phonon interactions, like manganites, the situation is even more complex. In this case, the gap is a collective result of strongly correlated electrons, and is itself damped by temperature. This is what has been dubbed the "bad insulator" crossover by Kotliar and Vollhardt [22]. The upshot is a remarkably stronger temperature dependence. This is illustrated in Figure 3, where the zero-bias slope of the $I(V)$ curve, σ_0, is nonzero at elevated temperature although the gap remains constant over the entire temperature range shown. Therefore, $\sigma_0 > 0$ is not a reliable measure of the local metallicity of the sample surface. Note that the pseudogap in high-temperature superconductors shows a similar enhanced temperature dependence [23,24], leading to the appealing suggestion they both are a consequence of strong correlation and not a band structure effect.

2.2 Colossal Magnetoresistance Metal-Insulator Phase Transition

Colossal magnetoresistance is one of the main reasons behind the surge of interest in manganites. Several STM experiments focus on the MIT associated with this phenomenon in dilute Mn^{4+} compounds with $x < 0.5$. Spectroscopy of

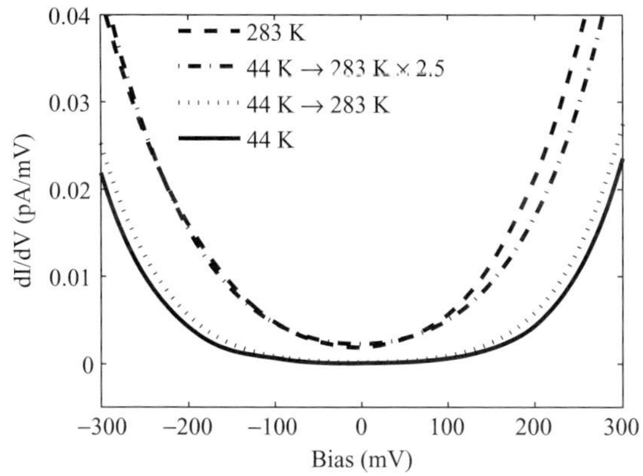

FIGURE 3. Tunneling spectra of a gapped manganite measured at 44K (solid line) and at 283 K (dashed line) [21]. Transforming the 44 K spectrum by the regular Fermi–Dirac function for 283 K (dotted line) has little effect, but allowing 2.5 times more smearing (i.e., taking the Fermi–Dirac function with an effective temperature of 700 K) for this Mott–Hubbard-type system produces remarkable agreement (dot-dashed line) with the measured 283K spectrum, which should therefore not be interpreted as metallic.

$La_{1-x}Ca_xMnO_3$ (LCMO) [25] and $La_{1-x}Sr_xMnO_3$ (LSMO) [26] thin films with $x = 0.3$ reveal a gap around E_F at high temperature which vanishes upon cooling the sample below the ferromagnetic transition temperature T_c. A similar gap was measured in LCMO with $x = 0.3$ and LSMO with $x = 0.3$ and $x = 0.33$ by Wei et al. [27]. However, in contrast with these former two studies, in the experiments by Wei et al. the gap persists at all temperatures, including in the low-temperature ferromagnetic metallic phase. Wei et al. contend this gap is a measure of the Jahn–Teller distortions, corroborating their argument with the fact that the gap amplitude compares well with the calculated Jahn–Teller coupling energy. Further investigations are needed for a full understanding of this gap feature and the apparent discrepancy between the above tunneling experiments.

Double exchange, in conjunction with strong electron–phonon interactions, has been proposed as a possible mechanism for CMR and the concomitant ferromagnetic state. A direct consequence of double exchange and the Jahn–Teller lattice distortions is a large spin splitting of the conduction band. The ensuing well separated minority and majority spin subbands result in a half-metallic conduction band, with nearly perfectly spin polarized electrons in the ground state. This idea is supported by different theoretical approaches, and the half-metallic DOS has been calculated [28] for LCMO with $x = 0.25$. Peaks in the local DOS (near ± 1.75 V in Figure 4(a)) compatible with this theoretical description have been reported by Wei et al. [27] on LCMO ($x = 0.3$) thin films in the ferromagnetic state at 77 K. Similar peaks were measured in LSMO ($x = 0.33$) thin films [29], which also exhibit CMR. As expected within the double exchange scenario, these peaks are absent from tunneling spectra of $LaMnO_3$ thin films, which do not exhibit CMR

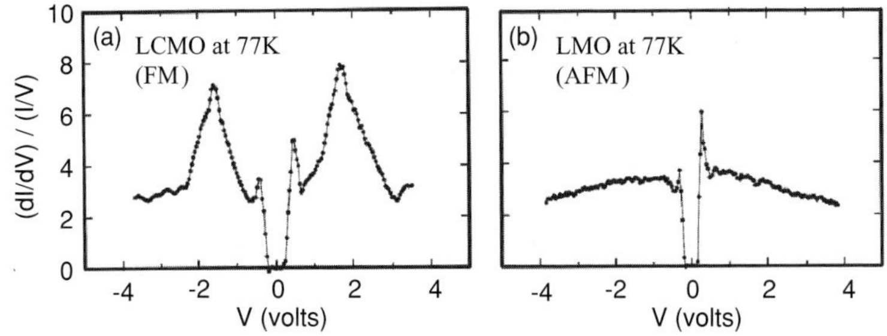

FIGURE 4. STM tunneling spectra at 77 K of a (a) ferromagnetic (FM) $La_{0.7}Ca_{0.3}MnO_3$ thin film and (b) antiferromagnetic (AFM) $LaMnO_3$ thin film. Reprinted with permission from [27]. Copyright 1997, American Physical Society.

(Figure 4(b)). Likewise, the half-metallic peaks in LCMO consistently disappear in the paramagnetic metallic state. However, in LSMO they have already vanished unexpectedly at room temperature [29], which is well below the Curie temperature of this compound, $T_c = 360$ K. This inconsistency as well as the absence of such peaks in other STM spectra measured on similar compounds [25,26] calls for further scrutiny to determine the origin of the high energy peaks in the tunneling DOS.

3 Spatially Resolved Tunneling Spectroscopy

Electronic phase separation (EPS) is the subject of a vivid debate prompted by theoretical models which ascribe to such inhomogeneities a key role in the occurrence of colossal magnetoresistance in manganites [10] and high-temperature superconductivity in cuprates [30], two of the most challenging problems in contemporary solid-state physics. Both in the manganites and the cuprates, the debated EPS involves spatial segregation of electronic charges, where the local carrier density is enhanced in some regions and depleted in others. The charge segregation can be over nanometer distances as well as microns. Owing to its combined topographic and spectroscopic resolution, STM is optimally suited to study EPS—as long as each phase remains sufficiently conducting ($\rho < 1\,\Omega\,\text{cm}$).

STM offers two ways of probing EPS in real space: Firstly, scanning tunneling spectroscopy (STS, section 3.1), which consists of measuring the local tunneling characteristics and plotting the gap at E_F [31] or the differential tunneling conductance at a given energy eV [25] as a function of position. Secondly, scanning tunneling potentiometry (STP, section 3.2), whose output is a map of the local surface potential shaped by the path of an electric current traversing the sample.

3.1 Scanning Tunneling Spectroscopy

Electronic phase separation (EPS) can be visualized in real space with very high spatial resolution using scanning tunneling spectroscopy (STS). However, because

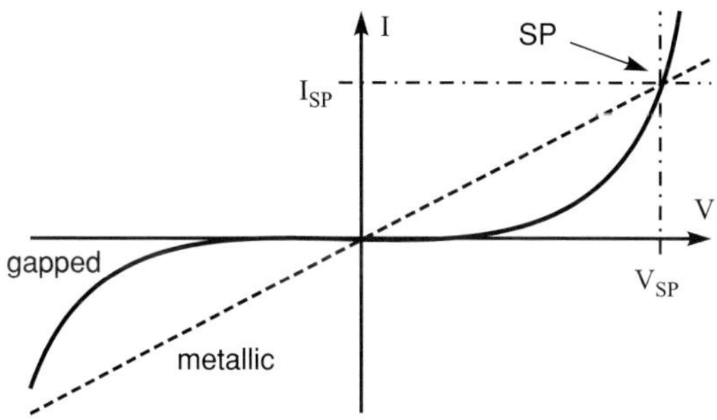

FIGURE 5. Linear $I(V)$ tunneling characteristics (dashed line) expected over metallic regions and nonlinear $I(V)$ characteristics (solid line) expected over regions with a gap in the local DOS when using an ideal free electron metal STM tip for voltages small compared to the work function. All spectra pass through the origin and the set point (SP) of the STM tunnel junction. Note that the corresponding *dI/dV(V)* curves do cross at $|V| < |V\mathrm{sp}|$ (see Figure 10).

of the operating principle of the STM, metallic regions only distinguished by their local DOS cannot be delineated in this way. When both the sample and the tip are metallic and at small voltages compared with the local work function, the differential tunneling conductance, $dI/dV(V)$, is independent of the bias voltage and equal to the junction conductance $\sigma_t \equiv I_{SP}/V_{SP}$, where I_{SP} and V_{SP} are the setpoint current and voltage, respectively (Figure 5). During normal operation, the STM feedback loop is regulating the tip-to-sample distance in order to maintain a constant current at the setpoint voltage. Hence, σ_t and *dI/dV(V)* are independent of position along the sample surface, and any EPS will only generate contrast in topographic micrographs.[4] For nonlinear tunneling characteristics on the other hand, the local differential tunneling conductance is mostly different from σ_t. In this case, EPS will produce contrast both in STS maps and topographic micrographs at some appropriately selected bias voltage (e.g., $V = V_{SP}$ or $V = 0$—suitable for STS maps only—in Figure 5).

Different types of EPS have been reported and are being considered theoretically: EPS may either take the form of a granular patchwork of sizable regions with distinct electronic properties [10], implying charge segregation over large distances, or charge stripes [32], where charge segregation is limited to a few lattice sites. The Coulomb energy cost, which is a major argument against the formation of large domains, is greatly reduced in the stripe configuration. However, the observation of stripe-like EPS by STM has so far only been reported in a few isolated instances in manganites [33] and HTS cuprates [34].

The majority of STM experiments claiming to find evidence of EPS, both in manganites [25,26] and cuprates [31], conclude that granular phase separation oc-

[4] Besides topography, an alternative method to probe EPS in this case is scanning tunneling potentiometry discussed in section 3.2.

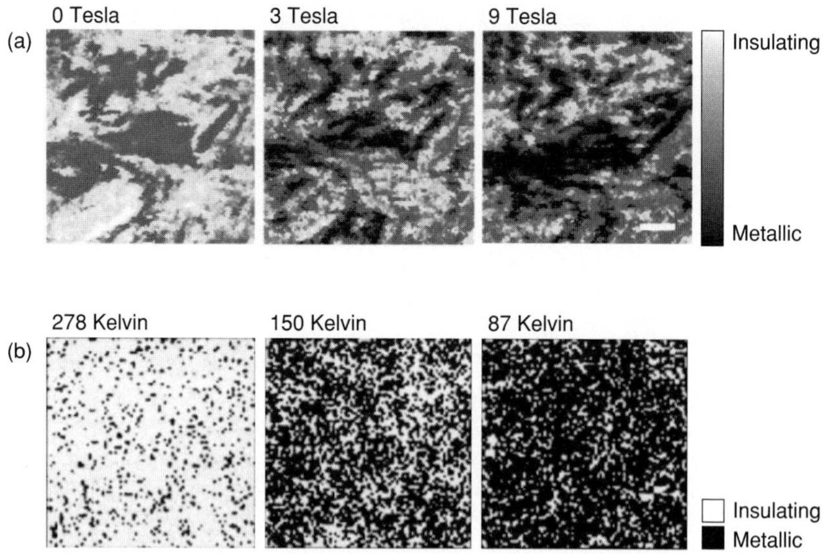

FIGURE 6. STS micrographs suggesting granular electronic phase separation. (a) 610 × 610 nm^2 STS images of an epitaxial $La_{0.73}Ca_{0.27}MnO_3$ thin film measured as a function of magnetic field just below the Curie temperature (scale bar: 100 nm). The gray scale represents the slope of the tunneling $I(V)$ characteristics at $V = 3$V. Adapted with permission from [25]. Copyright 1999, AAAS. (b) 500 × 500 nm^2 STS images of an epitaxial $La_{0.7}Sr_{0.3}MnO_3$ thin film measured as a function of temperature in zero magnetic field. The contrast is derived from the slope σ_0 of the tunneling $I(V)$ characteristics at $V = 0$, black corresponding to $\sigma_0 > 5.9 \times 10^{-3}$ nA/V and white to $\sigma_0 < 5.9 \times 10^{-3}$ nA/V. Adapted with permission from [26]. Copyright 2002, American Physical Society.

curs, with coexisting nanometer- to micrometer-size regions exhibiting different electronic characters as illustrated in Figure 6. Fäth et al. [25] measured $dI/dV(V)$ maps at $V = 3$V as a function of magnetic field slightly below the Curie temperature on $La_{.73}Ca_{0.27}MnO_3$ thin films grown on $SrTiO_3$ substrates with and without a $YBa_2Cu_3O_y$ buffer layer. They interpret the inhomogeneous conductance maps they measure (Figure 6(a)) in terms of a percolative phase transition [35,36]. They argue that, because the spatially averaged conductance maps and bulk CMR values are in excellent agreement, the STM results are representative of the bulk EPS. However, in a percolative MIT transition, the net conductance is determined by the least resistive percolation path, and is not expected to scale with the spatial average over all local conductances. Furthermore, if the conductance maps were to show percolation linked to the bulk transport, their granular structure should evolve with magnetic field, unlike the nearly field-independent texture revealed in Figure 6(a).

A step toward a more quantitative analysis was taken by Becker et al. [26], who mapped the zero-bias conductivity of $La_{0.7}Sr_{0.3}MnO_3$ thin films grown on MgO substrates as a function of temperature (Figure 6(b)). By introducing a zero-bias conductance threshold criterion to distinguish metallic and insulating regions, the authors argued that they had observed the formation of percolating networks, which they could fit to bulk resistivity data through resistor-network simulations. However, the use of a threshold criteria to partition the scanned area into two types of regions can be misleading. Any statistical conductance distribution with a finite width, whose average value is shifting with temperature, will seem to show percolation at a temperature that can be chosen arbitrarily by tuning the threshold value. Undisputable evidence of the coexistence of two phases with distinct electronic properties would be provided by a temperature dependent distribution, rather than a threshold, of a selected tunneling feature, such as the conductance at zero bias or the gap, which becomes bimodal near the phase transition [37].

Scanning tunneling spectroscopy can be extremely sensitive to the surface morphology [38,39]. Therefore, one should pay particular attention to possible correlations between spatial inhomogeneities in the local tunneling conductance and the surface texture, especially the local surface gradients, to avoid mistaking topography-induced fluctuations for intrinsic phase separation or percolation. Both STS studies discussed above were performed on thin films exhibiting granular surfaces, with a substantial (on the length scale typical for STM) peak-to-peak roughness of about 20 nm. The importance of a very flat surface to differentiate spectroscopic inhomogeneities induced by topography and those intrinsic to the metal-insulator transition is illustrated in a recent study by Akiyama et al. [38]. They measured STS conductance maps at $V = 2$ V on the very smooth surface of 20–30 nm thick epitaxial LSMO ($x = 0.3$) films deposited by laser-MBE on Nb doped $SrTiO_3$ (100) substrates (Figure 7). Within the terraces, they find extremely homogeneous STS maps (Figure 7(b)). The few fluctuations they observe are all

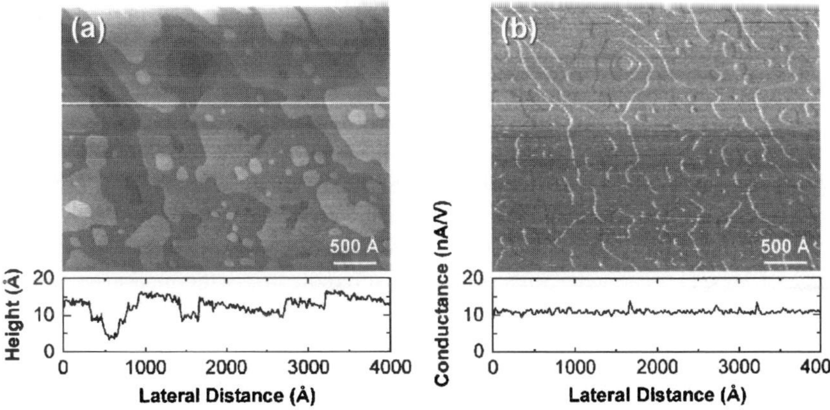

FIGURE 7. (a) STM micrograph of a 20–30 nm thick epitaxial $La_{0.7}Sr_{0.3}MnO_3$ film deposited by laser-MBE on a Nb-$SrTiO_3$ (100) substrate measured at 300 K. (b) Simultaneously recorded STS *dI/dV(V)* map measured at $V = 2$ V. Note the remarkable homogeneity of the STS map, showing absolutely no structure except at the unit-cell high ($\simeq$0.4 nm) steps. The amplitude profiles shown in the lower panels are measured along the white line in each corresponding image. Reprinted with permission from [38]. Copyright 2001, American Institute of Physics.

correlated with the unit-cell high ($\simeq$0.4 nm) steps (Figure 7(a)) delimiting the terraces. Interestingly, the largest effect is found at steps running perpendicular to the scanning direction, with the sign depending on whether it is an up or a down step. Almost no spectroscopic contrast is seen along steps running parallel to the scanning (horizontal) direction. This result is a striking illustration of the real possibility that spectroscopic inhomogeneities are entirely attributable to the surface morphology. It clearly excludes static mesoscopic electronic phase separation as a generic feature of CMR manganites.

3.2 Scanning Tunneling Potentiometry

In a homogeneous sample, the local potential associated with the transport current will drop uniformly throughout the sample, with a constant gradient parallel to the applied electric field. On the other hand, if the sample is phase separated into low- and high-resistance regions, the current will channel along some path of lowest resistance. In that case, the surface potential will not drop homogeneously through the sample, and its gradient will vary both in amplitude and direction as a function of position.

Scanning tunneling potentiometry (STP), first introduced by Muralt and Pohl [40] in 1986, enables the mapping of the local surface potential, V_{loc}, with high resolution. The principle of STP is depicted schematically in Figure 8(a). A floating potential V_s is applied to two opposite sides of the sample, and the resulting electric field drives the current through the sample. The tunneling bias voltage, V_b, is applied to the sample via a bridge circuit while the STM tip is grounded. Experimentally, V_{loc} is defined as the tunneling bias voltage V_b one has to apply to the tunnel junction in order for the tunneling current, I_t, to be zero.

V_{loc} can be measured in two ways (Figure 8(b)): (i) An entire $I(V_b)$ spectrum is acquired at each sampling point along the specimen surface, and V_{loc} is calculated from the intersection with the $I_t = 0$ axis (according to the definition $I_t(V_{\text{loc}}) = 0$). (ii) A single point $I_t(V_b = 0)$ is acquired at each sampling point along the sample surface, and V_{loc} is calculated by a linear extrapolation $V_{\text{loc}} = R_t \cdot I_t(0)$. Method (i) relies on the acquisition of a full spectrum for each data point in the STP map, making it rather slow. But it is extremely robust and does not rely on any particular assumptions, neither on the nature of the local tunnel junction, nor on the line shape of the $I(V)$ characteristics. Method (ii) requires only the measurement of

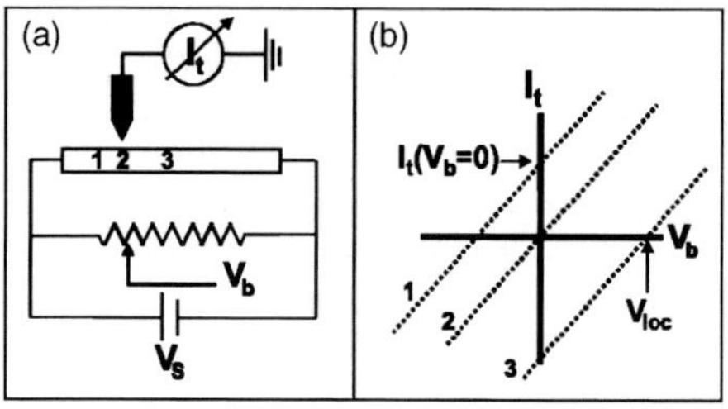

FIGURE 8. (a) Principle of scanning tunneling potentiometry (see text for details). (b) The tunneling $I(V)$ characteristics shift as the surface potential, V_{loc}, changes depending on position (1–3) along the sample surface. Reprinted with permission from [41]. Copyright 2002, American Institute of Physics.

a single point for each data point in the STP map, and is therefore much faster. However, it is susceptible to noise and spatial variations in the junction resistance R_t, and only works in the linear regime of the $I(V)$ characteristics. In both cases, the spatial resolution of the surface potential maps is limited by the accuracy with which the voltage can be measured, not by the spatial resolution of the STM.

Nonuniform STP maps measured on polycrystalline [42] and textured epitaxial [43] $La_{0.7}Sr_{0.3}MnO_3$ thin films grown on MgO substrates were ascribed to electronic phase separation. STP images of the room temperature ferromagnetic metallic phase of those specimens revealed extended areas where V_{loc} was essentially constant, separated by sharp steps where V_{loc} changes abruptly. These steps were found to correlate with topographic features, predominantly the grain boundaries, indicating the nonuniform drop of the surface potential is not due to EPS but to the granular structure of the sample.

Inhomogeneous current flow in thin films was also found to be caused by strain, as studied by Paranjape et al. [44] in an experiment where they tuned the amount of strain in $La_{0.7}Ca_{0.3}MnO_3$ (LCMO) by selecting different substrates and varying the film thickness. They measured 50 and 200 nm thin films grown epitaxially on $SrTiO_3$ (STO) and $NdGaO_3$ (NGO) substrates by pulsed laser deposition. The 50 nm thick LCMO film grown on STO was subject to the largest strain due to the lattice mismatch between the film and the substrate. Relaxed, nearly strain-free LCMO films were obtained either by growing thicker films (200 nm) on STO or growing the thin films (50 nm) on the lattice-matched NGO substrate. Paranjape et al. observed a direct correlation between the amount of strain and the degree of inhomogeneity in the surface potential. An important result of their work is that, even within flat terraces, local strain fluctuations can cause inhomogeneities which therefore do not necessarily reflect intrinsic EPS. Strain induced inhomogeneities of the local transport properties were also observed in $La_{0.67}Ca_{0.33}MnO_3$ thin films at low temperature using magnetic force microscopy [45].

In contrast to the inhomogeneous surface potential distribution found in polycrystalline and epitaxial textured or strained thin films, very uniform surface potential maps were measured on epitaxial $La_{0.7}Sr_{0.3}MnO_3$ (LSMO) thin films [41,42]. Grévin et al. [41] investigated single crystalline LSMO thin films (26 nm thick) grown in situ on $SrTiO_3$ (110) substrates by magnetron sputtering. The key findings of their measurements, done in ultrahigh vacuum at room temperature, are presented in Figure 9. STP maps of the metallic ferromagnetic state (Figure 9(b)) show essentially no correlation to topographic features such as steps (Figure 9(a)). Except for a small voltage drop near $x = 170$ nm due to a large structural defect, V_{loc} decreases linearly from left to right (Figure 9(c)), with a slope (13mV over 500 nm), in good agreement with the gradient expected from the applied voltage (263 V/cm).

The above STP experiments on epitaxial LSMO ($x = 0.3$) thin films [41] are in perfect agreement with STS experiments in similar epitaxial thin films [38], but are entirely at odds with the picture emerging from the STS experiments on granular LCMO ($x = 0.3$) [25] and granular LSMO ($x = 0.3$) [26] thin films discussed in section 3.1. In particular, the uniform voltage drop observed over 500 nm on several occasions in LSMO epitaxial thin films [41,42] precludes the existence of

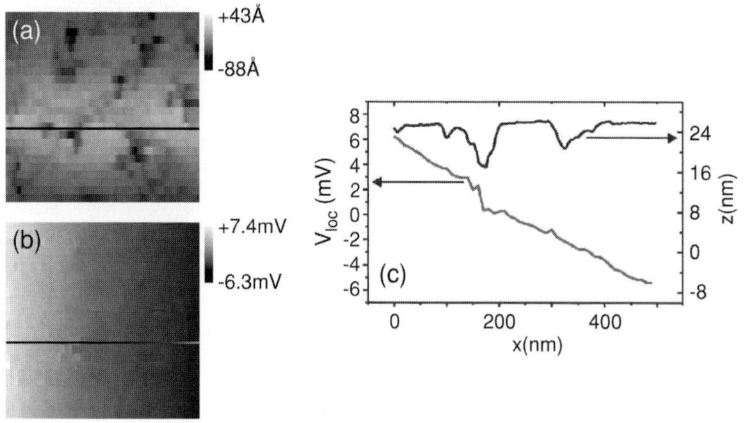

FIGURE 9. Room temperature STP of an epitaxial $La_{0.7}Sr_{0.3}MnO_3$ thin film grown in situ. (a) 500×500 nm^2 area topography and (b) simultaneously measured map of the surface potential V_{loc}(x,y). (c) Topographic (top) and surface potential (bottom) traces extracted along the line shown in the corresponding images. Adapted with permission from [41]. Copyright 2002, American Institute of Physics.

an intricate network of metallic and more insulating clusters on this length scale. However, they do not exclude the existence of nanometer-scale (cluster size below the spatial resolution of STP) or dynamic (cluster fluctuating on a time scale too fast to be resolved by STP) phase separation in these systems. In any case, STP experiments on LSMO epitaxial thin films (like the STS experiments discussed in section 3.1) clearly exclude static mesoscopic electronic phase separation as a generic feature of CMR manganites.

4 Atomic-Scale Topographic Imaging

Under optimal conditions, scanning tunneling microscopy offers the unique ability to uncover structural, spectroscopic, and magnetic features with subnanometer length scale resolution in real space. This capability is of particular interest in manganites, which host a wide variety of crystalline, electronic, and magnetic phases that often exhibit some degree of atomic-scale order. Atomic-scale STM imaging has been achieved on a number of transition metal oxides (TMOs) [46–48], but despite extensive efforts, it remains elusive on most CMR and HTS transition metal oxides.[5] In manganites, this finest resolution has only been achieved on

[5] Colossal magnetoresistance and high-temperature superconductor transition metal oxides which have been imaged with atomic-scale resolution by STM include $Bi_{0.24}Ca_{0.76}MnO_3$ [49], $La_{1.4}Sr_{1.6}Mn_2O_7$ [21], $Bi_2Sr_2Cu_2O_6$ [50,51], $Bi_2Sr_2CaCu_2O_8$ [52,53], $Bi_2Sr_2Ca_2Cu_3O_{10}$ [54] and $YBa_2Cu_3O_7$ [55,56], mostly on single crystals.

single crystals of the pseudocubic perovskite [49,57] $T_{1-x}D_x MnO_3$ (section 4.1) and its bilayer $T_{2-2x}D_{1+2x}Mn_2O_7$ polymorph [21] (section 4.2). No high-resolution imaging has been published on manganite thin films to date, including epitaxial thin films.

4.1 Pseudocubic Perovskite Manganites ($n = \infty$)

The perovskite structure does not have a natural cleaving plane, making it very challenging to prepare a flat and clean surface suitable for STM experiments. However, against all odds, atomic resolution micrographs were first achieved on the pseudocubic member ($n = \infty$) of the manganese oxide Ruddlesden–Popper series. Biswas et al. [57] obtained some atomic-scale contrast on a $La_{0.6}Pb_{0.4}MnO_3$ single crystal at 320 K. Very systematic and reproducible atomic-scale resolution was reported by Renner et al. [49] on $Bi_{1-x}Ca_xMnO_3$ (BCMO) single crystals grown by the self flux method at a nominal doping of $x = 0.76$.

Mn^{4+}-rich BCMO ($x = 0.76$) exhibits a charge-ordering metal-insulator phase transition at $T_{CO} = 250$ K, which was characterized with unprecedented high resolution by Renner et al. [49] using scanning tunneling microscopy and spectroscopy. They achieved atomic resolution imaging in ultra-high vacuum between 88 K and 300 K on as grown surfaces, gently cleaned in air without any additional in situ conditioning. Depending on temperature and tip position along the surface, the STM micrographs revealed three distinct atomic-scale patterns: a square lattice with $a_0 \simeq 3.8$ Å, a $\sqrt{2}a_0 \times \sqrt{2}a_0$ checkerboard lattice, and stripes with a periodicity of about $2\sqrt{2}a_0$. The square lattice was found to be characteristic of the paramagnetic metallic phase at $T > T_{CO}$, whereas the other two patterns were associated with the charge-ordered insulating phase at $T < T_{CO}$ [33,49].

4.1.1 Checkerboard Charge-Ordered Phase

The topographic micrographs and tunneling spectra of $Bi_{0.24}Ca_{0.76}MnO_3$ both change remarkably as a function of temperature in the vicinity of T_{CO}. STM micrographs of the paramagnetic (PM) metallic phase above T_{CO} revealed a square lattice ($a_0 = 3.8 \pm 0.1$ Å) consistent with the pseudocubic coordination of the Mn ions (Figure 10(a)). The tunneling $I(V)$ characteristics (Figure 10(d)) measured in those regions exhibit a finite differential conductance at $V = 0$, consistent with the metallic nature of BCMO at $T = 299$ K [58]. When BCMO is cooled below T_{CO}, a $\sqrt{2}a_0 \times \sqrt{2}a_0$ checkerboard lattice develops in the topography, with a doubling of the unit cell along the diagonal (Figure 10(b)), and the $I(V)$ characteristics turn insulating, with a gap on the order of 700 meV becoming clearly visible at E_F (Figure 10(d,e)). Upon closer examination, the low-temperature micrographs (Figure 10(b)) revealed substantial distortions, both in lateral position and amplitude (Figure 11). The lattice sites in Figure 11(a) are shifted away from the vertices of a regular square lattice into a bimodal distribution of short (3.0 ± 0.1 Å) and long (4.5 ± 0.1 Å) nearest-neighbor distances (Figure 11(c)). Remarkably, these short and long interatomic distances are not randomly distributed, but alternate regularly

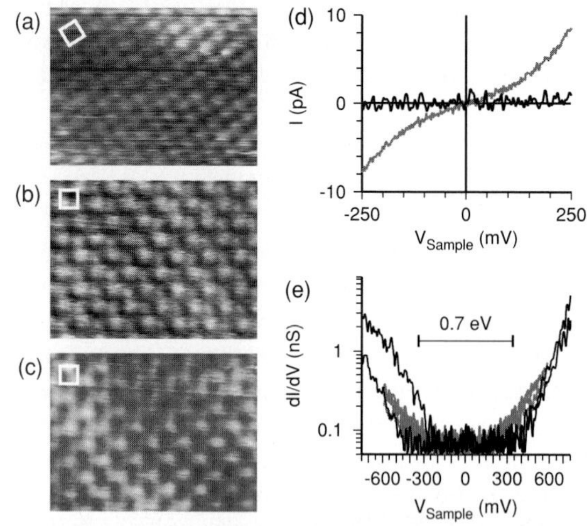

FIGURE 10. Nanometer-scale STM characterization of the charge-ordering phase transition in $Bi_{0.24}Ca_{0.76}MnO_3$ single crystals ($T_{CO} = 250$ K). 4.5×3.5 nm^2 constant, current micrographs measured in (a) the paramagnetic metallic phase at 299 K, and the charge-ordered insulating phase (b) at 146 K and (c) at 299 K. The white squares depict the $a_0 = 3.8$ Å unit cell. (d) $I(V)$ characteristics measured in the paramagnetic metallic phase at 299 K (gray line) and in the insulating charge-ordered state at 146 K (black line). (e) Logarithmic plot of the differential tunneling conductance (numerical derivatives of measured $I(V)$ curves normalized to the 299 K junction conductance $R = V/I$) demonstrating that the gap is essentially the same ($\simeq 0.7$ eV) in the charge-ordered phases at 299 K (black lines) and 146 K (gray lines). Adapted with permission from [49]. Copyright 2002, Nature Publishing.

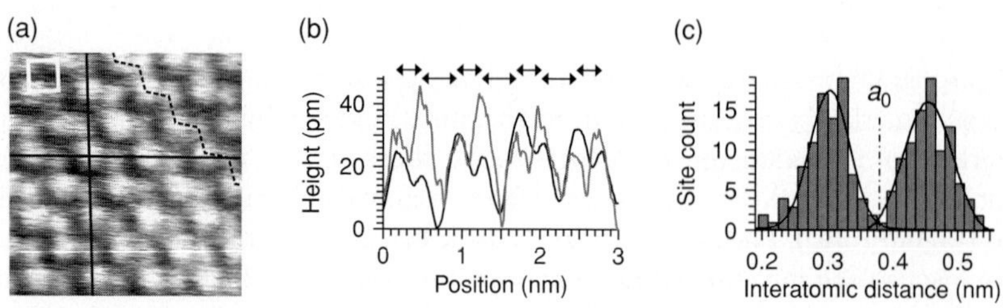

FIGURE 11. Lattice distortions (small polarons) observed by STM in the charge-ordered phase of $Bi_{0.24}Ca_{0.76}MnO_3$ single crystals at 146 K. (a) 3×3 nm^2 constant current STM micrograph. The dashed line highlights one of the long range-order zig-zag structures obtained when connecting short interatomic distances, and the white square shows the a_0 unit cell. (b) Amplitude profiles measured along the vertical (gray line) and horizontal (black line) solid lines in (a). Arrows emphasize alternating short and long interatomic distances. (c) Bimodal distribution of the lattice distortions into short (3.0 ± 0.1Å) and long (4.5 ± 0.1Å) interatomic distances ($a_0 = 3.77$Å). Adapted with permission from [49]. Copyright 2002, Nature Publishing.

across the entire imaged area, as shown in Figure 11(b), to form long-range ordered zig-zag chains running along the diagonal of the square basal plane (Figure 11(a)). Note that this charge and structural zig-zag is different from the spin and orbital zig-zag proposed by Goodenough [59] for the CE-type antiferromagnetic phase found in $x = 0.5$ compounds [60,61]. Its amplitude and periodicity are half those of the spin and orbital zig-zag. The height profiles reveal a concomitant change in the relative amplitude of neighbouring atomic sites (Figure 11(b)).

Renner et al. [49] ascribe the checkerboard pattern to alternating Mn^{3+} and Mn^{4+}, as posited in the simplest (integer charge) description of mixed-valence manganites.[6] The $\sqrt{2}a_0 \times \sqrt{2}a_0$ modulation resolved in the STM micrographs below T_{CO} (Figures. 10 and 11) is the first and only real space atomic-scale observation of the charge-ordered phase in a manganite. It graphically illustrates the existence of a checkerboard charge-ordered phase in a doped Mott insulator. However, the lateral distortions, in excess of 0.5 Å, are too large to ascribe the STM contrast to the Mn ions alone. X-ray structural analysis of charge-ordered $La_{0.5}Ca_{0.5}MnO_3$ single crystals [63], which are isostructural to BCMO, showed that the Jahn–Teller distorted MnO_6 octa hedra were tilted by an angle $\theta \simeq 10°$, thus shifting the apical oxygen away from the vertices of the square lattice by a distance $d_{\|} \equiv a_{MO} \sin\theta = 0.34$ Å, where $a_{MO} = 1.94$ Å is the Mn-O bond length. Assuming that tunneling occurs through the apical oxygen, Renner et al. explained the bimodal distribution shown in Figure 11(c) in terms of the non-uniform tilting of the octahedra along the zig-zag chains seen by X-ray diffraction [63], where neighboring apical oxygens are either brought closer or moved apart by roughly $2d_{\|}$. Given that the undistorted lattice parameter is equal to 3.77 Å, the displaced oxygens should generate a lattice with alternating interatomic distances of 3.09 Å and 4.45 Å, in excellent agreement with the STM findings [49].

The contrast in constant-current STM micrographs is a convolution of atomic structure and density of states. Hence, two contributions may account for the nearest-neighbor amplitude difference which develops in the charge-ordered phase (Figure 11(b)): (i) the minute difference in the distance separating the apical oxygen from the Mn ion between a Jahn–Teller distorted MnO_6 octahedron centered on Mn^{3+} and an undistorted one centered on Mn^{4+}, and (ii) the local charge difference between a MnO_6 octahedron centered on a Mn^{3+} ion and a Mn^{4+} ion. The Jahn–Teller distortion can at most account for a modulation of 0.1 Å, the effective contribution to the topography being further reduced due to the 10° tilting of the octahedra. Hence, because the actual average amplitude modulation exceeds 0.1 Å, it must entail some contribution from the distinctive spatial charge modulation of the charge-ordered phase.

---

[6] The exact distribution of charge around the Mn ions and the possibility of fractional valence is under discussion (see, e.g., Herrero-Martin et al. [62]). We simply name the two distinct ionization states observed by STM using the non-hybridized labels Mn^{3+} and Mn^{4+}.

4.1.2 Stripe Charge-Ordered Phase

The high-resolution STM micrographs of BCMO discussed in the previous paragraph are not consistent with the nominal doping ($x = 0.76$) of the single crystals investigated by Renner et al. [49]. The checkerboard shown in Figure 10(b,c) suggests that Mn^{4+} and Mn^{3+} are present in a ratio of 1:1, corresponding to $x = 0.5$. Diffraction data [64] for crystals with the same nominal doping prepared following the same procedure show double stripes with a periodicity of $\sim 4\sqrt{2}a_0$ instead of the $\sqrt{2}a_0$ periodicity seen by STM. This apparent discrepancy was pointed out by Renner et al. [49], who suggested two possible explanations. One is that the surface is stabilizing the checker-board phase. Theory indeed shows the checkerboard to be energetically more favorable, even in systems away from half doping [65]. Another possibility is that, even though the samples studied by Renner et al. have a bulk charge-ordering transition as sharp as any reported, the doping may not be $x = 0.76$ in all regions at the surface. The observation of stripe-like charge ordering in BCMO (Figure 12) by Renner et al. [33] suggests the doping may indeed be different in some locations.

Commensurate single stripes built from a 3:1 ratio of Mn^{4+} to Mn^{3+}, running parallel or diagonal to the square Mn lattice in the *ab*-plane, would have a periodicity of $4a_0$ or $2\sqrt{2}a_0$, respectively. The stripes seen by STM have a periodicity of the order of 1.1 nm (Figure 12(b)), which is consistent with diagonal single stripes. However, a detailed assessment of the stripe structure can only be done based on atomically resolved micrographs. So far, STM has been unable to resolve atoms in the striped regions, and the meandering shown in Figure 12(a) indicates the latter may have a more complex registration to the lattice. However, Figure 12 is definitely not consistent with the double stripes $4\sqrt{2}a_0$ inferred from high-resolution X-ray scattering [64] and low-energy electron diffraction [66]. The modulation periodicity of $2\sqrt{2}a_0$ observed by STM clearly suggests single stripes. STM tunneling spectroscopy of the stripes reveals a gap around E_F that is smaller than the

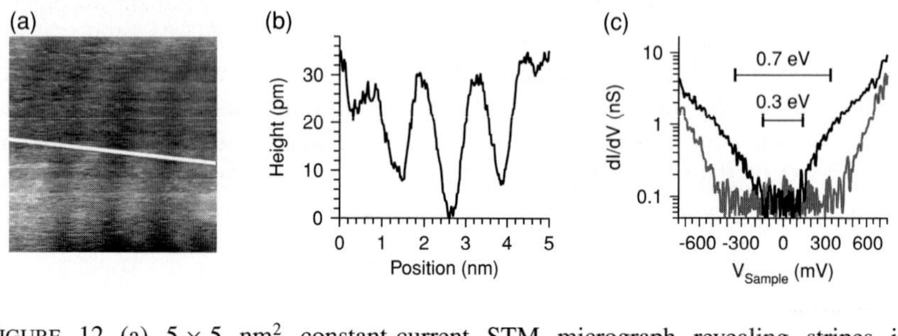

FIGURE 12. (a) 5×5 nm^2 constant-current STM micrograph revealing stripes in $Bi_{0.24}Ca_{0.76}MnO_3$ at 145 K. (b) Amplitude profile extracted along the white line in showing the stripe periodicity to be 1.1 nm $\simeq 2\sqrt{2}a_0$. (c) Logarithmic plot of the differential tunneling conductance to highlight the $\sim$0.3 eV gap measured in regions with the stripe phase. For comparison, the $\sim$0.7 eV gap measured in the $\sqrt{2}a_0 \times \sqrt{2}a_0$ checkerboard regions is also shown. Reprinted with permission from [33]. Copyright 2005, Elsevier B.V.

gap measured in the checkerboard phase [33] (Figure 12(c)), which is consistent with the checkerboard ground state being energetically more favorable than the stripe configuration.

4.1.3 Electronic Phase Separation

STM micrographs acquired below T_{CO} predominantly show the $\sqrt{2}a_0 \times \sqrt{2}a_0$ checkerboard, whereas the more homogeneous a_0 square lattice is primarily seen above T_{CO}. However, occasionally a checkerboard pattern is observed at room temperature [49] (Figure 10(c)). STM data (Figure 10) indisputably shows that the checkerboards seen above and below T_{CO} are the signature of one and the same phase. They are identical from a structural (Figure 10(b,c)) and from a spectroscopic (Figure 10(e)) point of view, exhibiting the same topographic contrast and ~700 meV gap at the Fermi energy. It is not surprising to find the high and low temperature phases coexisting. The first-order nature of the transition means that even far above T_{CO}, the charge-ordered phase can be locally nucleated by, for example, surface geometry and defects. Note that the interface separating the two phases (Figure 13(a), dashed line) is atomically sharp—the transition from one spatial charge distribution to the other occurs over one lattice distance a_0—and stable in time.

The simultaneous observation of the PM metallic and charge-ordered insulating phases in a single STM micrograph above T_{CO} (Figure 13) provides further insight into the charge-ordering MIT and the nature of the checkerboard phase [49]. On an atomically flat and clean surface, constant-current STM is primarily sensitive to the local DOS as a function of position. The amplitude profile shown in Figure 13(b) graphically illustrates how the homogeneous charge density of the PM metallic phase is redistributed (charge-ordered) over atomic lengths scales in the $\sqrt{2}a_0 \times \sqrt{2}a_0$ checkerboard region. Indeed, the amplitude is almost equal on each lattice site in the PM metallic region (position >2 nm in Figure 13(b)), and alternates between high and low on nearest-neighboring lattice sites in the charge-ordered

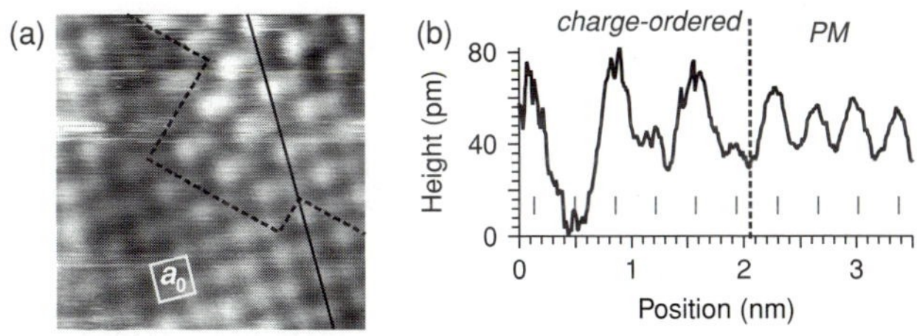

FIGURE 13. (a) 3.5 × 3.5 nm^2 constant current STM micrograph of $Bi_{0.24}Ca_{0.76}MnO_3$ at 299 K. It shows the atomically sharp boundary separating coexisting paramagnetic metallic and charge-ordered insulating regions. (b) Amplitude profile extracted along the black line in graphically illustrating the charge redistribution on the nearest-neighboring lattice sites in the charge-ordered phase. The dashed line depicts the atomically sharp domain boundary. Adapted with permission from [49]. Copyright 2002, Nature Publishing.

region (position <2 nm in Figure 13 (b)). This observation is strong evidence that the checkerboard does indeed reflect charge ordering, with a characteristic gap $\Delta_{CO} \simeq 700$ meV.

The electronic phase separation shown in Figure 13 is between phases of the same doping that are stable at different temperatures. It does not stem from charge segregation, where the insulating and metallic region correspond to different doping levels—as mentioned above, the total DOS is the same in both regions. The insulating nature of the charge-ordered phase is due to localization, where the charge is trapped on a frozen lattice distortion. Detailed analysis shows that the average interatomic distance can be determined to better than ± 0.1Å, but both phases exhibit significant distributions with standard deviations of ± 0.6Å in the PM metallic phase and ± 0.3Å in the charge-ordered insulating phase. This may reflect polaronic fluctuations in the PM metallic state, which are frozen below the MIT, very similar to what has been seen by neutron powder diffraction [67,68]. The STM micrographs shown in Figure 10(b,c) and Figure 11 are the first real-space observations of small Jahn–Teller polarons below T_{CO}. Hence, STM provides a unique microscopic insight, graphically emphasizing the importance of considering polaronic lattice distortions and concomitant charge localization in describing and understanding the charge-ordered insulating phase [69].

4.2 Bilayer Manganites (n = 2)

Compared to the pseudocubic perovskite compounds ($n = \infty$), the planar structure of the bilayer polymorph has two important implications. Firstly, fluctuations—both magnetic and electronic—generally increase when reducing the dimensionality. It also loosens the topological constraints for domain structure formation. Both of these effects probably contribute to the higher CMR ratios reported for bilayer manganites [10,70]. Secondly, and this is of particular importance for STM and other surface-based techniques, the bilayer manganites offer a natural crystal truncation plane in the rock-salt oxide layer, as illustrated in Figure 1. This allows mechanical cleaving of single crystals, which enables the preparation of extremely flat surfaces over micron-sized areas [21].

The vast majority of experimental efforts on bilayer materials have been concentrating on $La_{2-2x}Sr_{1+2x}Mn_2O_7$ (LSM_2O) (see, e.g., Mitchell et al. [71]). Somewhat surprisingly, considering the ease with which flat surfaces can be prepared, few STM data have been reported on LSM_2O to date. Tunneling spectra have been obtained using non-spatially resolved techniques, such as point-contact junctions [72] and photolithographically produced mesoscopic stacks of intrinsic tunnel junctions (MESAs) [73]. Using MESAs of about 20 double layers of LSM_2O with $x = 0.3$, Nachtrab et al. [73] obtained spectra with a flat zero-bias slope and two maxima in $dI/dV(V)$ at $V = \pm 1.4$ V. They interpreted these spectra as corresponding to a gap of about 75 meV per bilayer. To facilitate comparison with other gap measurements, one should not define the gap from the peak positions but from the range where $dI/dV(V) \sim 0$, which would yield an even smaller gap—much smaller than the gap reported from point-contact [72] and vacuum [21] tunneling.

One way to reconcile these experiments would be to hypothesize that the voltage drop is not divided between the internal layers of the MESA, but dominated by the contact to the top-electrode, in which case the voltage scale of the spectra should not be divided by the number of layers. The raw spectra obtained for the 20 layer stacks do indeed look very similar to gold point-contact measurements reported up to $V = 1$ V on an $x = 0.36$ sample [72].

Freeland et al. [72] interpret point-contact tunneling spectra and resonant X-ray scattering in terms of a single insulating antiferromagnetic surface bilayer. However, for slightly lower doping ($x = 0.30$), spin-polarized, scanning electron microscope data clearly demonstrate that even the topmost layer is ferromagnetic [74], and optical conductivity shows that the insulating c-axis behavior goes at least a couple of layers deep [75]. This apparent disagreement may be related to the different doping levels of the bilayer specimen, and it would be interesting to see these investigations extended to coinciding doping levels.

Recently, variable-temperature STM was performed on in situ cleaved LSM_2O single crystals with $x = 0.32$. The cleaved surfaces were found to be extremely flat, with a peak-to-peak roughness below 0.05 nm as illustrated in Figure 14(a). The topography further revealed steps whose heights amount to multiples of half the unit cell c-axis length (Figure 14(a)). Occasionally, the micrographs revealed a square atomic-scale pattern, whose dimensions were consistent with the ab-plane lattice parameters of the orthorhombic LSM_2O crystal (Figure 14(b)). These very smooth and clean surfaces enabled very reproducible tunneling spectroscopy over a wide temperature range. The analysis of these spectra and the topography is presented in Ref. [21].

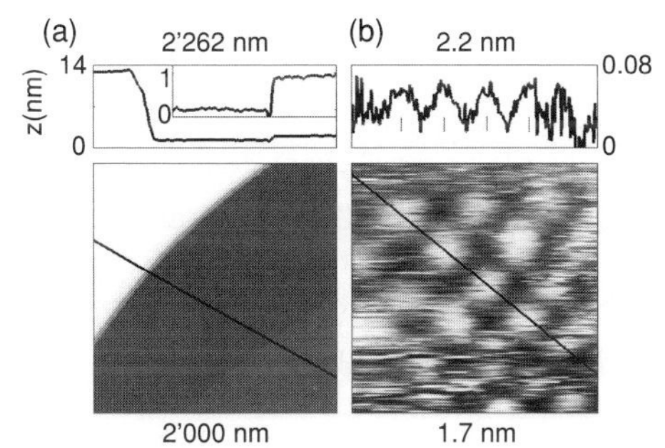

FIGURE 14. STM micrographs of $La_{1.4}Sr_{1.6}Mn_2O_7$. (a) Micron-sized atomically flat surfaces obtained by in situ cleaving in ultrahigh vacuum ($5 \cdot 10^{-10}$Torr). From the amplitude trace, the large step is seen to correspond to seven unit cells, and the small step separating the two terraces in the bottom right corresponds to half a unit cell (see enlarged z-scale in the inset). (b) 1.7×1.7 nm^2 constant-current STM micrograph showing atomic resolution. The dimensions and square symmetry are in accordance with the crystal ab-plane lattice parameters $a_0 = 3.86$Å. In both micrographs, the amplitude traces are measured along the corresponding black line. See Rønnow et al. [21].

5 Summary and Outlook

As with any experimental technique applied to a given problem, there are both technical and fundamental challenges to studying manganites using STM. For instance, electronic inhomogeneity and phase separation can be strongly linked to defects, disorder, and grain boundaries in the samples as well as steps and islands on the surface. It may therefore be asked, what can be learned about intrinsic phase separation from local studies of nonperfect samples? If the coexistence of two or more phases is observed on the same surface, this is an indicator that the involved phases are sufficiently close in energy that weak boundary-forces can enforce selection between the two. If, on the other hand, a surface is seen to be homogeneous over a certain length scale, then this must mean that this system does not display intrinsic phase separation. While inhomogeneity can be caused or pinned by defects and disorder, homogeneity cannot occur as a coincidence.

A related potential objection to surface techniques in general is the fact that in some cases surface phenomena dominate and completely alter the physics, so that it is no longer representative of bulk behavior. Indeed, surface specific effects may occur, as e.g. reported for $La_{0.65}Sr_{0.35}MnO_3$ by Dulli et al. [76] who, based on photoemission, deduce that one or more $n = 1$ layered phases are formed on the surface.[7] It is therefore important that surface-based observations be cross-checked against bulk probes. For instance, Sr_2RuO_4 and BCMO surfaces show similar checkerboard patterns in STM micrographs. Whereas it was explained in terms of a surface reconstruction in Sr_2RuO_4 [48], consistency with X-ray diffraction [63], neutron scattering [67] and optical conductivity [58], techniques which are all much less surface sensitive, demonstrated it to be due to bulk charge ordering in BCMO.

STM is a powerful tool providing a unique insight into the phenomenon of colossal magnetoresistance observed in manganese perovskite oxides. It has already revealed some of the structural and electronic phase transitions associated with this intriguing macroscopic effect, including direct real-space imaging of charge ordering. In reviewing existing STM work on the manganites, we have pointed out unresolved questions and discussed various pitfalls, mainly associated with the need for resolution and very high quality surfaces, and thereby provided guidelines for future studies. We would like to stress that these pioneering works have proved the potential of STM and should be systematically extended to cover the relevant doping ranges, thereby solidifying the interpretation of these early studies and strengthening their impact on the understanding of the manganite physics.

[7] Such a surface reconstruction may be related to the high-temperature annealing that was applied. In STM, atomic resolution was achieved on as grown surfaces [49] ($n = \infty$) and surfaces cleaved in situ at room temperature [21] ($n = 2$) without any further in situ processing. So far, despite extensive attempts, atomic resolution has not been achieved on surfaces which were etched and/or annealed in UHV prior to the STM, possibly due to oxygen loss.

In particular, we conclude that atomic resolution provides an extremely robust and detailed insight into the different electronic phases and associated lattice distortions [49]. Hence, while obtaining atomic resolution is at times cumbersome, it is worth the extra effort. Beyond currently reported STM observations exploiting only the charge component of the tunneling current, two of the key variables, namely, the spin and orbital electronic degrees of freedom, have not yet been resolved with atomic resolution in real space on manganites. In particular, further developments in spin-polarized STM [77] applied to the question of colossal magnetoresistance [38] would be of the utmost importance in unravelling the many distinct paramagnetic, ferromagnetic, and antiferromagnetic phases at the core of this intriguing phenomenon.

Acknowledgments. We gratefully acknowledge G. Aeppli, S.-W. Cheong, and T. Kimura for enjoyable and sustained collaborations probing the manganites with scanning tunneling microscopy. We also benefited from numerous enlightening discussions, advice, and exchange of ideas with C. Berthod, S. Billinge, A. Biswas, E. Dagotto, A. Fisher, G. Kotliar, D. Khomskii, E. Kotomin, P. Littlewood, N. Mathur, J. Mesot, A. Millis, V. Moshnyaga, T. Perring, K. Samwer, and M. Stoneham. Special thanks go to A. Fisher and C. Bird for their careful reading and pertinent comments on the manuscript.

References

1. J. M. D. Coey, M. Viret, and S. von Molnar, Adv. Phys. **48**, 167 (1999).
2. A. P. Ramirez, J. Phys.: Condens. Matter **9**, 8171 (1997).
3. M. B. Salamon and M. Jaime, Rev. Mod. Phys. **73**, 583 (2001).
4. K. Chahara, T. Ohno, M. Kasai, and Y. Kozono, Appl. Phys. Lett. **63**, 1990 (1993).
5. R. von Helmolt, J. Wecker, B. Holzapfel, L. Schultz, and K. Samwer, Phys. Rev. Lett. **71**, 2331 (1993).
6. S. Jin, T. H. Tiefel, M. McCormack, R. A. Fastnacht, R. Ramesh, and L. H. Chen, Science **264**, 413 (1994).
7. R. M. Kusters, J. Singleton, D. A. Keen, R. McGreevy, and W. Hayes, Physica B **155**, 362 (1989).
8. T. Venkatesan, M. Rajeswari, Z. W. Dong, S. B. Ogale, and R. Ramesh, Philos. Trans. R. Soc. Lond. A **356**, 1661 (1998).
9. S. N. Ruddlesden and R. Popper, Acta Cryst. **11**, 54 (1958).
10. E. Dagotto, *Nanoscale Phase Separation and Colossal Magnetoresistance*, volume 136 of *Springer Series in Solid-State Sciences. The Physics of Manganites and Related Compounds*, Springer, 2003.
11. D. S. Dessau, Y. D. Chuang, A. Gromko, T. Saitoh, T. Kimura, and Y. Tokura, J. Electron Spectrosc. Relat. Phenom. **117**, 265 (2001).
12. Y. Kuwata, S. Suga, S. Imada, A. Sekiyama, S. Ueda, T. Iwasaki, H. Harada, T. Muro, T. Fukawa, K. Ashida, H. Yoshioka, T. Terauchi, J. Sameshima, H. Kuwahara, Y. Moritomo, and Y. Tokura, J. Electron Spectrosc. Relat. Phenom. **88**, 281 (1998).
13. J. Tersoff and D. R. Hamann, Phys. Rev. B **31**, 805 (1985).

14. Scanning tunneling spectroscopy, in *J. Electron Spectrosc. Relat. Phenom*, edited by W.-D. Schneider, pages 1–224, 2000, (special edition).
15. R. Wiesendanger, *Scanning Probe Microscopy and Spectroscopy: Methods and Applications*, Cambridge University Press, Cambridge, 1994.
16. A. Biswas, A. K. Raychaudhuriyk, R. Mahendirany, A. Guhay, R. Maheshz, and C. N. R. Raoz, J. Phys.: Condens. Matter **9**, L355 (1997).
17. S. K. Mishra, R. Pandit, and S. Satpathy, J. Phys.: Condens. Matter **11**, 8561 (1999).
18. B. C. Yao, J. B. Lee, C. C. Chi, M. K. Wu, and H. C. Yang, J. Magn. Magn. Mater. **209**, 103 (2000).
19. A. Biswas, A. Arulraj, A. K. Raychaudhuri, and C. N. R. Rao, J. Phys.: Condens. Matter **12**, L101 (2000).
20. Y. Tomioka, H. Kuwahara, A. Asamitsu, M. Kasai, and Y. Tokura, Appl. Phys. Lett. **70**, 3609 (1997).
21. H. M. Rønnow, Ch. Renner, G. Aeppli, T. Kimura, and Y. Tokura, Nature **440**, 1025 (2006).
22. G. Kotliar and D. Vollhardt, Physics Today **March**, 53 (2004).
23. Ø. Fischer, M. Kugler, I. Maggio-Aprile, C. Berthod, and Ch. Renner, Scanning Tunneling Microscopy of HTS. Rev. Mod. Phys. (2006).
24. Ch. Renner, B. Revaz, J. Y. Genoud, K. Kadowaki, and Ø. Fischer, Phys. Rev. Lett. **80**, 149 (1998).
25. M. Fäth, S. Freisem, A. A. Menovsky, Y. Tomioka, J. Aarts, and J. A. Mydosh, Science **285**, 1540 (1999).
26. T. Becker, C. Streng, Y. Luo, V. Moshnyaga, B. Damaschke, N. Shannon, and K. Samwer, Phys. Rev. Lett. **89**, 237203 (2002).
27. J. Y. T. Wei, N. C. Yeh, and R. P. Vasquez, Phys. Rev. Lett. **79**, 5150 (1997).
28. W. E. Pickett and D. J. Singh, Phys. Rev. B **53**, 1146 (1996).
29. J. Y. T. Wei, N. C. Yeh, R. P. Vasquez, and A. Gupta, J. Appl. Phys. **83**, 7366 (1998).
30. J. C. Phillips, A. Saxena, and A. R. Bishop, Rep. Prog. Phys. **66**, 2111 (2003).
31. S. H. Pan, J. P. O'Neal, R. L. Badzey, C. Chamon, H. Ding, J. R. Engelbrecht, Z. Wang, H. Eisaki, S. Uchida, A. K. Guptak, K. W. Ng, E. W. Hudson, K. M. Lang, and J. C. Davis, Nature **413**, 282 (2001).
32. V. J. Emery, S. A. Kivelson, and J. M. Tranquada, Proc. Natl. Acad. Sci. USA **96**, 8814 (1999).
33. Ch. Renner, G. Aeppli, and H. M. Rønnow, Mater. Sci. Eng. C **25**, 775 (2005).
34. C. Howald, H. Eisaki, N. Kaneko, M. Greven, and A. Kapitulnik, Phys. Rev. B **67**, 014533 (2003).
35. J. Burgy, E. Dagotto, and M. Mayr, Phys. Rev. B **67**, 014410 (2003).
36. A. Weisse, J. Loos, and H. Fehske, Phys. Rev. B **68**, 024402 (2003).
37. Further analysis of the conductance maps does reveal some degree of bimodal splitting of the zero-bias conductance distribution. K. Samwer and B. Damaschke, private communication.
38. R. Akiyama, H. Tanaka, T. Matsumoto, and T. Kawai, Appl. Phys. Lett. **79**, 4378 (2001).
39. S. F. Chen, P. I. Lin, J. Y. Juang, T. M. Uen, K. H. Wu, Y. S. Gou, and J. Y. Lin, Appl. Phys. Lett. **82**, 1242 (2003).
40. P. Muralt and D. W. Pohl, Appl. Phys. Lett. **48**, 514 (1986).
41. B. Grévin, I. Maggio-Aprile, A. Bentzen, O. Kuffer, I. Joumard, and Ø. Fischer, Appl. Phys. Lett. **80**, 3979 (2002).
42. J. J. Versluijs, F. Ott, and J. M. D. Coey, Appl. Phys. Lett. **75**, 1152 (1999).

43. B. Grévin, I. Maggio-Aprile, A. Bentzen, L. Ranno, A. Llobet, and Ø. Fischer, Phys. Rev. B **62**, 8596 (2000).
44. M. Paranjape, A. K. Raychaudhuri, N. D. Mathur, and M. G. Blamire, Phys. Rev. B **67**, 214415 (2003).
45. A. Biswas, M. Rajeswari, R. C. Srivastava, T. Venkatesan, R. L. Greene, Q. Lu, A. L. de Lozanne, and A. J. Millis, Phys. Rev. B **63**, 184424 (2001).
46. D. A. Bonnell, Prog. Surf. Sci. **57**, 187 (1998).
47. T. Hanaguri, C. Lupien, Y. Kohsaka, D. H. Lee, M. Azuma, M. Takano, H. Takagi, and J. C. Davis, Nature **430**, 1001 (2004).
48. R. Matzdorf, Z. Fang, Ismail, J. D. Zhang, T. Kimura, Y. Tokura, K. Terakura, and E. W. Plummer, Science **289**, 746 (2000).
49. Ch. Renner, G. Aeppli, B. G. Kim, Y. A. Soh, and S.-W. Cheong, Nature **416**, 518 (2002).
50. M. Kugler, Ø. Fischer, Ch. Renner, S. Ono, and Y. Ando, Phys. Rev. Lett. **86**, 4911 (2001).
51. L. Shan, A. Ejov, A. Volodin, V. V. Moshchalkov, H. H. Wen, and C. T. Lin, Europhys. Lett. **61**, 681 (2003).
52. S. H. Pan, E. W. Hudson, J. Ma, and J. C. Davis, Appl. Phys. Lett. **73**, 58 (1998).
53. Ch. Renner and Ø. Fischer, Phys. Rev. B **51**, 9208 (1995).
54. M. Kugler, G. Levy de Castro, E. Giannini, A. Piriou, A.A. Manuel, C. Hess, and Ø. Fischer, J. Phys. Chem. Solids **67**, 353 (2006).
55. D. J. Derro, E. W. Hudson, K. M. Lang, S. H. Pan, J. C. Davis, J. T. Markert, and A. L. de Lozanne, Phys. Rev. Lett. **88**, 097002 (2002).
56. S. H. Pan, E. W. Hudson, and J. C. Davis, Rev. Sci. Instrum. **70**, 1459 (1999).
57. A. Biswas, S. Elizabeth, A. K. Raychaudhuri, and H. L. Bhat, Phys. Rev. B **59**, 5368 (1999).
58. H. L. Liu, S. L. Cooper, and S.-W. Cheong, Phys. Rev. Lett. **81**, 4684 (1998).
59. J. B. Goodenough, Phys. Rev. **100**, 564 (1955).
60. B. J. Sternlieb, J. P. Hill, U. C. Wildgruber, G. M. Luke, B. Nachumi, Y. Moritomo, and Y. Tokura, Phys. Rev. Lett. **76**, 2169 (1996).
61. Y. Tokura and N. Nagaosa, Science **288**, 462 (2000).
62. J. Herrero-Martin, J. Garcia, G. Subias, J. Blasco, and M. C. Sanchez, Phys. Rev. B **70**, 024408 (2004).
63. P. G. Radaelli, D. E. Cox, M. Marezio, and S.-W. Cheong, Phys. Rev. B **55**, 3015 (1997).
64. Y. Su, C. H. Du, P. D. Hatton, S. P. Collins, and S.-W. Cheong, Phys. Rev. B **59**, 11687 (1999).
65. M. Y. Kagan, K. I. Kugel, and D. I. Khomskii, J. Exp. Theor. Phys. **93**, 415 (2001).
66. S. Mori, C. H. Chen, and S.-W. Cheong, Nature **392**, 473 (1998).
67. S. J. L. Billinge, R. G. DiFrancesco, G. H. Kwei, J. J. Neumeier, and J. D. Thompson, Phys. Rev. Lett. **77**, 715 (1996).
68. P. G. Radaelli, M. Marezio, H. Y. Hwang, S. W. Cheong, and B. Batlogg, Phys. Rev. B **54**, 8992 (1996).
69. A. J. Millis, P. B. Littlewood, , and B. I. Shraiman, Phys. Rev. Lett. **74**, 5144 (1995).
70. Y. Moritomo, A. Asamitsu, H. Kuwahara, and Y. Tokura, Nature **380**, 141 (1996).
71. J. F. Mitchell, D. N. Argyriou, A. Berger, K. E. Gray, R. Osborn, and U. Welp, J. Phys. Chem. B **105**, 10731 (2001).
72. J. W. Freeland, K. E. Gray, L. Ozyuzer, P. Berghuis, B. Elvira, J. Kavich, H. Zheng, and J. F. Mitchell, Nature Materials **4**, 62 (2005).

73. T. Nachtrab, S. Heim, M. Mössle, R. Kleiner, O. Waldmann, R. Koch, P. Müller, T. Kimura, and Y. Tokura, Phys. Rev. B **65**, 012410 (2001).
74. M. Konoto, T. Kohashi, K. Koike, T. Arima, Y. Kaneko, T. Kimura, and Y. Tokura, Phys. Rev. Lett. **93**, 107201 (2004).
75. T. Ishikawa, K. Tobe, T. Kimura, T. Katsufuji, and Y. Tokura, Phys. Rev. B **62**, 12354 (2000).
76. H. Dulli, P. A. Dowben, S. H. Liou, and E. W. Plummer, Phys. Rev. B **62**, 14629 (2000).
77. R. Wiesendanger and M. Bode, Solid State Comm. **119**, 341 (2001).

Index

Printed in the United States of America.